Annals of the International Society of Dynamic Games
Volume 7

Series Editor
Tamer Başar

Editorial Board

Annals of the International Society of Dynamical Games

Advances in Dynamic Games

Applications to Economics, Finance, Optimization, and Stochastic Control

Andrzej S. Nowak
Krzysztof Szajowski
Editors

Birkhäuser
Boston • Basel • Berlin

Andrzej S. Nowak
Wrocław University of Technology
Institute of Mathematics
Wybrzeże Wypiańskiego 27
50-370 Wrocław
Poland
and
Faculty of Mathematics, Computer Science,
and Econometrics
University of Zielona Góra
Podgórna 50
65-246 Zielona Góra
Poland

Krzysztof Szajowski
Wrocław University of Technology
Institute of Mathematics
Wybrzeże Wypiańskiego 27
50-370 Wrocław
Poland

AMS Subject Classifications: 91A-xx, 91A05, 91A06, 91A10, 91A12, 91A13, 91A15, 91A18, 91A20, 91A22, 91A23, 91A25, 91A28, 91A30, 91A35, 91A40, 91A43, 91A46, 91A50, 91A60, 91A65, 91A70, 91A80, 91A99

Library of Congress Cataloging-in-Publication Data
International Symposium of Dynamic Games and Applications (9th : 2000 : Adelaide, S. Aust.)
Advances in dynamic games : applications to economics, finance, optimization, and stochastic control / Andrzej S. Nowak, Krzysztof Szajowski, editors.
p. cm. – (Annals of the International Society of Dynamic Games ; [v. 7])
Papers based on presentations at the 9th International Symposium on Dynamic Games and Applications held in Adelaide, South Australia in Dec. 2000.
ISBN 0-8176-4362-1 (alk. paper)
1. Game theory–Congresses. I. Nowak, Andrzej S. II. Szajowski, Krzysztof. III. Title. IV. Series.

HB144.I583 2000
330'.01'5193–dc22 2004048826

ISBN 0-8176-4362-1 Printed on acid-free paper.

Birkhäuser

Printed in the United States of America. (KeS/SB)

9 8 7 6 5 4 3 2 1 SPIN 10988183

www.birkhauser.com

Contents

Part III Stopping Games

Part IV Applications of Dynamic Games to Economics, Finance and Queuing Theory

Preface

Modern game theory has evolved enormously since its inception in the 1920s in the works of Borel and von Neumann. The branch of game theory known as dynamic games descended from the pioneering work on differential games by R. Isaacs, L. S. Pontryagin and his school, and from seminal papers on extensive form games by Kuhn and on stochastic games by Shapley. Since those early developmental decades, dynamic game theory has had a significant impact on such diverse disciplines as applied mathematics, economics, systems theory, engineering, operations research, biology, ecology, and the environmental sciences. On the other hand, a large variety of mathematical methods from differential equations to stochastic processes has been applied to formulate and solve many different problems.

This new edited book focuses on various aspects of dynamic game theory, providing authoritative, state-of-the-art information and serving as a guide to the vitality of the field and its applications. Most of the selected, peer-reviewed papers are based on presentations at the 9th International Symposium on Dynamic Games and Applications held in Adelaide, South Australia in December 2000. This conference took place under the auspices of the International Society of Dynamic Games (ISDG), established in 1990. The conference has been cosponsored by Centre for Industrial and Applicable Mathematics (CIAM), University of South Australia, IEEE Control Systems Society, Institute of Mathematics, Wrocław University of Technology (Poland), Faculty of Mathematics, Computer Science and Econometrics, University of Zielona Góra (Poland), ISDG Organizing Society, and the University of South Australia. Every paper that appears in this volume has passed through a stringent reviewing process, as is the case with publications for archival journals.

A variety of topics of current interest are presented. They are divided in to six parts: the first (five papers) treat repeated games and stochastic games, and the second (three papers) covers differential dynamic games. The third part of the volume (nine papers) is devoted to the various extensions of stopping games, which are also known as Dynkin's games. In the fourth part there are seven papers on applications of dynamic games to economics, finance, and queuing theory. The final two parts contain five papers which are devoted to algorithms and numerical solution approaches for dynamic games, and the section on Parrondo's games (five papers).

We wish to thank all the associate editors and the referees for their valuable contributions that made this volume possible.

Wrocław and Zielona Góra *Andrzej S. Nowak*
Wrocław *Krzysztof Szajowski*

Contributors

Mohammed Abbad, Département de Mathematiques et Informatique, Faculté des Sciences B.P. 1014, Université Mohammed V, 10000 Rabat, Morocco

Derek Abbott, Centre for Biomedical Engineering (CBME) and Department of Electrical and Electronic Engineering, The University of Adelaide, Adelaide, SA 5005, Australia

Andrew Allison, Centre for Biomedical Engineering (CBME) and Department of Electrical and Electronic Engineering, University of Adelaide, Adelaide, SA 5005, Australia

Tansu Alpcan, Coordinated Science Laboratory, University of Illinois, 1308 West Main Street, Urbana, IL 61801, USA

Eitan Altman, INRIA, B.P. 93, 2004 route des Lucioles, 06902 Sophia-Antipolis Cedex, France

Bruno Amable, Faculté des Sciences Economies, Université Paris X-Nanterre, 200 av. de la République, 92000 Nanterre, France

Jean-Pierre Aubin, Centre de Recherche Viabilité, Jeux, Contrôle, Université Paris-Dauphine, 75775 Paris cx (16), France

Rachid El Azouzi, University of Avignon, LIA, 339, chemin des Meinajaries, Agroparc B.P. 1228, 84911 Avignon Cedex 9, France

Tamer Başar, Coordinated Science Laboratory, University of Illinois, 1308 West Main Street, Urbana, IL 61801, USA

Pierre Bernhard, Laboratoire I3S, UNSA and CNRS, 2000 route des Lucioles, Les Algorithmes – bât. Euclide 8, BP.121, 106903 Sophia Antipolis-Cedex, France

Dean A. Carlson, Mathematical Reviews 416 Fourth Street, P.O.Box 8604, Ann Arbor, MI 48107-8604, USA

Shu-Heng Chen, AI-ECON Research Center, Department of Economics, National Chengchi University, 64 Chi-Nan Rd., Sec.2, Taipei 11623, Taiwan

Andre Costa, School of Applied Mathematics, University of Adelaide, Adelaide, SA 5005, Australia

Victor Domansky, St. Petersburg Institute for Economics and Mathematics, Russian Academy of Science, Tchaikovskogo 1, 191187, St. Petersburg, Russia

John Duffy, Department of Economics, University of Pittsburgh, 4S01 Posvar Hall, 230 S. Bouquet Street, Pittsburgh, PA 15260, USA

Ekkehard C. Ernst, Directorate General Economics, European Central Bank, Kaiserstrasse 29, 60311 Frankfurt, Germany

Mark Fackrell, Department of Mathematics and Statistics, University of Melbourne, Victoria, 3010, Australia

Maurizio Falcone, Dipartimento di Matematica, Università di Roma "La Sapienza", P. Aldo Moro 2, 00185 Roma, Italy

Elżbieta Z. Ferenstein, Faculty of Mathematics and Information Science, Warsaw University of Technology, Plac Politechniki 1, 00-661 Warsaw, Poland; and Polish-Japanese Institute of Information Technology, Koszykowa 86, 02-008 Warsaw, Poland

Thomas S. Ferguson, Department of Mathematics, University of California at Los Angeles, 405 Hilgard Ave., Los Angeles, CA 90095-1361, USA

Adrian P. Flitney, Centre for Biomedical Engineering (CBME) and Department of Electrical and Electronic Engineering, The University of Adelaide, Adelaide, SA 5005, Australia

Gregory P. Harmer, Centre for Biomedical Engineering (CBME) and Department of Electrical and Electronic Engineering, University of Adelaide, Adelaide, SA 5005, Australia

Alain Haurie, HEC-Management Studies, Faculty of Economics and Social Science, 40 Blvd. du Pont-d'Arve, CH-1211 Geneva 4, Switzerland

Hisao Kameda, Institute of Information Science and Electronics, University of Tsukuba, Tsukuba Science City, Ibaraki 305-8573, Japan

Ioannis Karatzas, Department of Mathematics and Statistics, Columbia University, New York, NY 10027, USA

Masami Kurano, Department of Mathematics, Chiba University, Inage-ku, Chiba 263-8522, Japan

Heinz-Uwe Küenle, Brandenburgische Technische Universität Cottbus, PF 10 13 44, D-03013 Cottbus, Germany

Vladimir V. Mazalov, Institute of Applied Mathematical Research, Karelian Research Center of Russian Acad. Sci., Pushkinakaya st. 11, Petrozavodsk, 185610, Russia

Koichi Mizukami, Graduate School of Engineering, Department of Computer Science, Hiroshima Kokusai Gakuin University, 20-1 Nakano 6, Aki-ku, Hiroshima, 739-0321, Japan

Hiroaki Mukaidani, Graduate School of Education, Hiroshima University, 1–1–1, Kagamiyama, Higashi–Hiroshima, 739–8524, Japan

Jun-ichi Nakagami, Department of Mathematics and Informatics, Chiba University, Inage-ku, Chiba 263-8522, Japan

Joseph Ng, Centre for Biomedical Engineering (CBME) and Department of Electrical and Electronic Engineering, University of Adelaide, Adelaide, SA 5005, Australia

Andrzej S. Nowak, Wrocław University of Technology, Institute of Mathematics Wybrzeże Wypianskiego 27, PL-50-370 Wrocław Poland; and Faculty of Mathematics, Computer Science and Econometrics, University of Zielona Góra, Podgorna 50, 65-246 Zielona Góra, Poland

Koji Okuguchi, Department of Economics and Information, Gifu Shotoku Gakuen University, Gifu-shi, Gifu-ken 500-8288, Japan

Stefano Palombarini, Faculté des Sciences Economies, Université Paris VIII, 2 rue de la Liberté, 93526 Saint-Denis Cedex 02, France

Juan M.R. Parrondo, Departamento de Física Atómica, Molecular y Nuclear, Universidad Complutense de Madrid, 28040 Madrid, Spain

Charles Pearce, Department of Applied Mathematics, The University of Adelaide, Adelaide, SA 5005, Australia

Leon A. Petrosjan, Faculty of Applied Mathematics, St. Petersburg State University, Bibliotechnaya pl. 2, Petrodvorets 199504, St. Petersburg, Russia

Zdzisław Porosiński, Institute of Mathematics, Wrocław University of Technology, Wybrzeże Wyspiańskiego 27, 50-370 Wrocław, Poland

David Ramsey, Institute of Mathematics, Wrocław University of Technology, Wybrzeże Wyspiańskiego 27, 50-370 Wrocław, Poland

Minoru Sakaguchi, 3-26-4 Midorigaoka, Toyonaka, Osaka 560-0002, Japan

Piercesare Secchi, Dipartimento di Matematica, Politecnico di Milano, Piazza Leonardo da Vinci 32, I-20133 Milano, Italia

Martin Shubik, Cowles Foundation for Research in Economics, Yale University, New Haven, CT 06520, USA

Eilon Solan, Department of Managerial Economics and Decision Sciences, Kellogg School of Management, Northwestern University; and School of Mathematical Sciences, Tel Aviv University, Tel Aviv 69978, Israel

Sylvain Sorin, Equipe Combinatoire et Optimisation, UFR 921, Université Pierre et Marie Curie-Paris 6, 4 place fussieu, 75230 Paris, France; and Laboratoire d'Econometrie, Ecole Polytechnique, 1 rue Descartes, 75005 Paris, France

Paolo Stefani, CASPUR, P. Aldo Moro 2, 00185 Roma, Italy

William D. Sudderth, School of Statistics, University of Minnesota, Church Street SE 224, Minneapolis, MN 55455, USA

Krzysztof Szajowski, Institute of Mathematics, Wrocław University of Technology, Wybrzeże Wyspiańskiego 27, 50-370 Wrocław, Poland

Peter G. Taylor, Department of Mathematics and Statistics, University of Melbourne, Victoria, 3010, Australia

Raúl Toral, Departamento de Física, Universitat de les Illes Balears; and Instituto Mediterráneo de Estudios Avanzados, IMEDEA (CSIC-UIB), 07071 Palma de Mallorca, Spain

Nicolas Vieille, Département Finance et Economie, HEC School of Management (HEC), 78 Jouy-en-Josas, France

Piotr Więcek, Institute of Mathematics, Wrocław University of Technology, Wybrzeże Wyspiańskiego 27, 50-370 Wrocław, Poland

Agnieszka Wiszniewska-Matyszkiel, Institute of Applied Mathematics and Mechanics, Warsaw University, Banacha 2, 02-097 Warsaw, Poland

Hua Xu, Graduate School of Business Sciences, The University of Tsukuba, 3–29–1, Otsuka, Bunkyou–ku, Tokyo, 112–0012, Japan

Masami Yasuda, Department of Mathematics and Informatics, Chiba University, Inage-ku, Chiba 263-8522, Japan

Chia-Hsuan Yeh, Department of Information Management, Yuan Ze University, Chungli, Taoyuan 320, Taiwan

Yuji Yoshida, Faculty of Economics and Business Administration, The University of Kitakyushu, Kitakyushu 802-8577, Japan

Anatoliy A. Zabelin, Chita State Pedagogical University, Babushkin st. 121, Chita 672090, Russia

Georges Zaccour, GERAD and Ecole des H.E.C. Montréal, 300 Cote S. Catherine, H3T 2A7, Montreal, Canada

PART I
Repeated and Stochastic Games

Information and the Existence of Stationary Markovian Equilibrium

Ioannis Karatzas
Department of Mathematics and Statistics
Columbia University
New York, NY 10027
ik@math.columbia.edu

Martin Shubik
Cowles Foundation for Research in Economics
Yale University
New Haven, CT 06520
martin.shubik@yale.edu

William D. Sudderth
School of Statistics
University of Minnesota
Minneapolis, MN 55455
bill@stat.umn.edu

Abstract

We describe conditions for the existence of a stationary Markovian equilibrium when total production or total endowment is a random variable. Apart from regularity assumptions, there are two crucial conditions: (i) *low information*—agents are ignorant of both total endowment and their own endowments when they make decisions in a given period, and (ii) *proportional endowments*—the endowment of each agent is in proportion, possibly random, to the total endowment. When these conditions hold, there is a stationary equilibrium. When they do not hold, such an equilibrium need not exist.

1 Introduction

This paper is part of an effort to investigate a mass-market economy with stochastic elements, in which the optimization problems faced by each of a continuum of agents are modeled as parallel dynamic programming problems. The model used is a strategic market game at the highest level of aggregation, in order to concentrate on the monetary aspects of a stochastic environment. Although there are several previous papers which provide economic motivation and modeling details [2]–[4], we have attempted to make this paper as self-contained as possible. However, we shall make use of several results established in these earlier works.

We consider an economy with a stochastic supply of goods, where: (i) the endowment of each agent is in proportion (possibly random) to the total amount of goods available; and (ii) the agents must bid for goods in each period without knowing either the total supply of goods available, or the realization of their own random endowments.

For such an economy, we shall show the existence of a stationary equilibrium, where the optimal amount bid by an agent in each period depends only on the agent's current wealth. In equilibrium, there will be a stationary distribution of wealth among agents, although prices and wealth-levels of individual agents will fluctuate randomly with time. This will be true whether or not the opportunity is available for agents to borrow from, or deposit in, an outside (government) bank.

When either the individual endowments are not proportional to the total available supply of goods, or the agents have additional information (in the form of advance knowledge of the total supply of goods), there need not exist such an equilibrium. This will be illustrated by two examples. One interpretation of these results is that *better short-term forecasting can be destabilizing*. We plan further investigation of these "high information" phenomena in a subsequent paper.

The next section has some preliminary discussion of our model. Sections 3 and 4 treat the model without lending, sections 5 and 6 are on the model with lending and possible bankruptcy, whereas the final section 7 treats five simple examples that illustrate the existence and non-existence of stationary equilibrium.

2 Preliminaries

For simplicity we omit production from consideration. Instead, we consider an economy where all consumption goods are bought for cash (*fiat money*) in a competitive market. Each individual agent begins with an initial endowment of money and a claim to the proceeds from consumption goods that are sold in the market. The goods enter the economy in each period as if they were "manna" from an undescribed production process, and are owned by the individual agents. However, the agents are required to offer the goods in the market, and do not receive the proceeds until the start of the subsequent period. The assumption that all goods go through the market is probably a better approximation of the realities in a modern economy than the reverse, where each agent can consume everything directly, without the interface of markets or prices.

Our model has a continuum of agents indexed by the unit interval $I = [0, 1]$, and distributed according to a non-atomic probability measure φ on the σ-algebra $\mathcal{B}(I)$ of Borel subsets of I. Time runs in discrete time-periods $n = 0, 1, \cdots$. At the beginning of each time-period n, every agent $\alpha \in I$ receives an endowment $Y_n^\alpha(\omega)$ in units of a nondurable commodity. The random variables $\{Y_n^\alpha; \alpha \in I, n \in \mathbb{N}\}$, and all other random variables encountered in this paper, are defined on a given probability space $(\Omega, \mathcal{F}, \mathbb{P})$.

We shall consider the *no-lending model* of [3], and also the *lending with possible bankruptcy model* of [2]. Unlike these earlier papers, it will no longer be assumed that total production Q is constant from period to period, but instead that production

$$Q_n(\omega) = \int Y_n^\alpha(\omega)\varphi(d\alpha)$$

in period n is a random variable, for all $n = 1, 2, \cdots$.

The following assumption will be in force throughout sections 2–6.

Assumption 2.1. (a) The total-production variables $Q_1, Q_2, \cdots$ are I.I.D. (independent and identically distributed) with common distribution ζ. It will also be assumed that the Q_n's are strictly positive with finite mean.

(b) The individual endowment variables $Y_n^\alpha(\omega)$ are proportional to the $Q_n(\omega)$, in the sense that

$$Y_n^\alpha(\omega) = Z_n^\alpha(\omega)Q_n(\omega) \qquad \text{for all} \qquad \alpha \in I, n \in \mathbb{N}, \omega \in \Omega. \tag{1}$$

Here the sequences $\{Z_1^\alpha, Z_2^\alpha, \ldots\}$ and $\{Q_1, Q_2, \ldots\}$ are independent; $Z_n^\alpha \geq 0$, $E(Z_n^\alpha) = 1$; and $Z_1^\alpha, Z_2^\alpha, \cdots$ are I.I.D. with common distribution λ^α, for each $\alpha \in I$.

This is the simplest set of assumptions that permit *both* the total-production random variables to fluctuate with time, *and* a stationary equilibrium to exist; their negation precludes the existence of such an equilibrium, as Example 7.4 below demonstrates. A consequence of these assumptions is that

$$E(Y_n^\alpha) = E(Z_n^\alpha) \cdot E(Q_n) = E(Q_n). \tag{2}$$

3 The Model without Lending

For $\alpha \in I$ and $n \in \mathbb{N}$, let $S_{n-1}^\alpha(\omega)$ and $\mathcal{F}_{n-1}^\alpha$ denote respectively the wealth and information σ-algebra available to agent α at the beginning of period n. As in [3], agent α bids an $\mathcal{F}_{n-1}^\alpha$-measurable amount $b_n^\alpha(\omega) \in [0, S_{n-1}^\alpha(\omega)]$ of money for the consumption good *before* knowing the value of $Q_n(\omega)$ or $Y_n^\alpha(\omega)$. We call this the *low-information* condition. (In other words, the information σ-algebra $\mathcal{F}_{n-1}^\alpha$ available to the agent at the beginning of period n, measures the values of past quantities including $S_0^\alpha, S_k^\alpha, Q_k, Z_k^\alpha, b_k^\alpha$ for $k = 1, \cdots, n-1$, but *not* of Q_n, Y_n^α.)

Once all agents have placed their bids, the total amount of fiat money bid for the consumption good is given by

$$B_n(\omega) = \int b_n^\alpha(\omega)\varphi(d\alpha),$$

and a new price is formed as

$$p_n(\omega) = \frac{B_n(\omega)}{Q_n(\omega)}$$

for period $t = n$. Each agent α receives an amount

$$x_n^\alpha(\omega) = \frac{b_n^\alpha(\omega)}{p_n(\omega)} = \frac{b_n^\alpha(\omega)}{B_n(\omega)} \cdot Q_n(\omega)$$

of goods, equal to his bid's worth in the price of the goods for period $t = n$, as well as

$$p_n(\omega)Y_n^\alpha(\omega) = \frac{B_n(\omega)}{Q_n(\omega)} \cdot Z_n^\alpha(\omega)Q_n(\omega) = B_n(\omega)Z_n^\alpha(\omega) \tag{3}$$

in cash income, and then enters the next period with wealth in fiat money

$$S_n^\alpha(\omega) = S_{n-1}^\alpha(\omega) - b_n^\alpha(\omega) + B_n(\omega)Z_n^\alpha(\omega). \tag{4}$$

Each agent $\alpha \in I$ is assumed to have a *utility function* $u^\alpha : [0, \infty) \to [0, \infty)$ for consumption of goods; this function is continuous and continuously differentiable, strictly concave, strictly increasing, and satisfies $u^\alpha(0) = 0$, $(u^\alpha)'_+(0) \in (0, \infty)$. The utility earned by agent α in period n is $u^\alpha(x_n^\alpha(\omega))$, and the agent seeks to maximize the expected value of his total discounted utility

$$\sum_{n=0}^{\infty} \beta^n u^\alpha \left(x_{n+1}^\alpha(\omega)\right).$$

A *strategy* π^α for agent α specifies the sequence of bids $\{b_n^\alpha\}_{n=1}^\infty$. The strategy π^α is called *stationary*, if it specifies the bids in terms of a single function $c^\alpha : [0, \infty) \to [0, \infty)$ of wealth, in the form

$$b_n^\alpha(\omega) = c^\alpha(S_{n-1}^\alpha(\omega)), \quad n \geq 1, \tag{5}$$

where $c^\alpha(s) \in [0, s]$, $\forall s \geq 0$. We call $c^\alpha(\cdot)$ the *consumption function for* the strategy π^α.

The *wealth distribution in period n* is the random measure $\nu_n(\cdot, \omega)$ given by

$$\nu_n(A, \omega) = \varphi\left(\{\alpha \in I : S_n^\alpha(\omega) \in A\}\right), \qquad A \in \mathcal{B}([0, \infty)). \tag{6}$$

We are now ready to define the type of equilibrium that we want to study in this note.

Definition 3.1. A collection of stationary strategies $\{\pi^\alpha, \alpha \in I\}$ and a probability distribution μ on $\mathcal{B}((0, \infty))$ form a *stationary equilibrium*, if
(a) given that $\nu_0 = \mu$ and that every agent α plays strategy π^α, we have $\nu_n = \mu$ for all $n \geq 1$, and

(b) given that $\nu_0 = \mu$, the strategy π^α is optimal for agent α, when every other agent β plays π^β ($\beta \in I$, $\beta \neq \alpha$), for each $\alpha \in I$.

Unlike [3], there is no mention of price in Definition 3.1. This is, in part, because the sequence of prices $\{p_n\}$ will not be constant – even in stationary equilibrium – for the model studied here. Indeed, if the consumption function for π^α is the same across all agents $\alpha \in I$, and equal to $c^\alpha(\cdot) \equiv c(\cdot)$, then

$$p_n(\omega) = \frac{\int b_n^\alpha(\omega)\varphi(d\alpha)}{Q_n(\omega)} = \frac{\int c(s)\mu(ds)}{Q_n(\omega)},$$

where the sequence of total bids

$$B_n(\omega) \equiv B := \int c(s)\mu(ds)$$

is constant in equilibrium; see Theorem 4.1 below. Thus, the prices $\{p_n\}$ form then a sequence of I.I.D. random variables, because the $\{Q_n\}$ do so by assumption. The constant B will play the same mathematical role that was played by the price p in the earlier works [3] and [2], but of course the interpretation here will be different.

4 Existence of Stationary Equilibrium for the Model without Lending

The methods of the paper [3] can be adapted, to construct a stationary equilibrium for the present model. As in [3], we consider first the one-person game faced by an agent α, assuming that the economy is in stationary equilibrium. For ease of notation we suppress the superscript α while discussing the one-person game. Furthermore, we also assume that the agents are *homogeneous*, in the sense that they all have the same utility function $u(\cdot)$ and the same distribution λ for their income variables. This assumption makes the existence proof more transparent, but is not necessary; the proof in [3] works for many types of agents, and can be adapted to the present context as well.

We introduce a new utility function defined by

$$\tilde{u}(b) := E\left[u(bQ(\omega))\right] = \int u(bq)\zeta(dq), \quad b \geq 0. \tag{7}$$

Observe that the expected utility earned by an agent who bids b when faced by a random price $p(\omega) = B/Q(\omega)$, can be written

$$E\left[u\left(\frac{b}{p(\omega)}\right)\right] = E\left[u\left(\frac{b}{B}Q(\omega)\right)\right] = \tilde{u}\left(\frac{b}{B}\right). \tag{8}$$

It is straightforward to verify that $\tilde{u}(\cdot)$ has all the properties, such as strict concavity, that were assumed for $u(\cdot)$.

Let $V(\cdot)$ be the value function for an agent playing in equilibrium. In essence, the agent faces a discounted dynamic programming problem and, just as in [3], the value function $V(\cdot)$ satisfies the Bellman equation

$$V(s) = \sup_{0 \le b \le s} \left[\tilde{u}\left(\frac{b}{B}\right) + \beta \cdot E[V(s - b + BZ)] \right]. \tag{9}$$

This dynamic programming problem is of the type studied in [3], and Theorem 4.1 of that paper has information about it. In particular, there is a unique optimal stationary plan $\pi = \pi(B)$ corresponding to a consumption function $c : [0, \infty) \to [0, \infty)$. We sometimes write this function as $c(s) = c(s; B)$, to make explicit its dependence on the quantity B.

Consider now the Markov chain $\{S_n\}$ of successive fortunes for an agent who plays the optimal strategy π given by $c(\cdot)$. Then we have

$$S_{n+1} = S_n - c(S_n; B) + BZ_{n+1}, \quad n \in \mathbb{N}_0 \tag{10}$$

where $Z_1, Z_2, \ldots$ are I.I.D. with common distribution λ. By Theorem 5.1 of [3], this chain has a unique stationary distribution $\mu(\cdot) = \mu(\cdot\,; B)$ defined on $\mathcal{B}([0, \infty))$. Now assume that Z has a finite second moment: $E(Z^2) < \infty$. Then, by Theorem 5.7 of [3], the stationary distribution μ has a finite mean, namely $\int_{(0,\infty)} s\,\mu(ds) < \infty$. The following lemma expresses the fact that the total amount bid by all agents is equal to B, when the wealth distribution is $\mu(\cdot\,; B)$.

Lemma 4.1. $\int c(s; B)\mu(ds; B) = B$.

Proof. Assume that S_0 has the stationary distribution μ. Then take expectations in (10) to obtain

$$E(S_{n+1}) = E(S_n) - \int c(s; B)\mu(ds; B) + B \cdot E(Z), \quad n \in \mathbb{N}_0$$

and the desired formula follows, since $E(S_{n+1}) = E(S_n)$ by stationarity and $E(Z) = 1$ by assumption of our model. □

Theorem 4.1. *For each $B > 0$, there is a stationary equilibrium for the no-lending model, with wealth distribution $\mu(\cdot) = \mu(\cdot; B)$, and with stationary strategies $\pi^\alpha \equiv \pi(B)$ for all agents $\alpha \in I$.*

Proof. Construct the variables $Z_n^\alpha(\omega) = Z_n(\alpha, \omega)$ using the technique of Feldman and Gilles [1], so that

$Z_1(\alpha, \cdot), Z_2(\alpha, \cdot), \ldots$ are I.I.D. with distribution λ, for every $\alpha \in I$, and

$Z_1(\cdot, \omega), Z_2(\cdot, \omega), \ldots$ are I.I.D. with distribution λ, for every $\omega \in \Omega$.

Then the chain $\{S_n(\alpha, \omega)\}$ has the same dynamics for each fixed $\omega \in \Omega$ as it does for each fixed $\alpha \in I$. The distribution μ is stationary for the chain when α is fixed, and will therefore be a stationary wealth distribution for the many-person game if the total bids $B_1(\omega), B_2(\omega), \ldots$ remain equal to B. Now, if $S_0(\cdot,\omega)$ has distribution μ, then

$$B_1(\omega) = \int c(S_0(\alpha,\omega))\varphi(d\alpha) = \int c(s)\mu(ds) = B,$$

by Lemma 4.1. By induction, $B_n(\omega) = B$ for all $n \geq 1$ and $\omega \in \Omega$. Hence, the wealth-distributions ν_n are all equal to μ.

The optimality of $\pi^\alpha = \pi(B)$ follows from its optimality in the one-person game together with the fact that a single player cannot affect the value of the total bid. □

5 The Model with Lending and Possible Bankruptcy

We now assume that there is a Central Bank which gives loans and accepts deposits. The bank sets two interest rates in each time-period n, namely $r_{1n}(\omega) = 1+\rho_{1n}(\omega)$ to be paid by borrowers and $r_{2n}(\omega) = 1 + \rho_{2n}(\omega)$ to be paid to depositors. These rates are assumed to satisfy

$$1 \leq r_{2n}(\omega) \leq r_{1n}(\omega), \quad r_{2n}(\omega) \leq 1/\beta, \tag{11}$$

for all $n \in \mathbb{N}$, $\omega \in \Omega$.

Agents are required to pay their debts back at the beginning of the next period, when they have sufficient funds to do so. However, it can happen that they are unable to pay back their debts in full, and are thus forced to pay a *bankruptcy penalty* in units of utility, before they are allowed to continue in the game. For this reason, we assume now that each agent α has a utility function $u^\alpha : \mathbb{R} \to \mathbb{R}$ defined on the entire real line, and satisfies all the other assumptions made above. For $x < 0$, the quantity $u^\alpha(x)$ is negative and measures the "disutility" for agent α of going bankrupt by an amount x; for $x > 0$, the quantity $u^\alpha(x)$ is positive and measures the utility derived by α from consuming x units of the commodity, just as before.

Suppose that an agent $\alpha \in I$ begins in period n with wealth $S^\alpha_{n-1}(\omega)$. If $S^\alpha_{n-1}(\omega) < 0$, then agent α has an unpaid debt from the previous period and is assessed a penalty of $u(S^\alpha_{n-1}(\omega)/p_{n-1}(\omega))$. The debt is then forgiven, and the agent continues play from wealth-position 0. If $S^\alpha_{n-1}(\omega) \geq 0$, then agent α is not in debt and plays from position $S^\alpha_{n-1}(\omega)$. In both cases, an agent α, possibly after being punished, plays from the wealth-position $(S^\alpha_{n-1}(\omega))^+ = \max\{S^\alpha_{n-1}(\omega), 0\}$.

Based on knowledge of past quantities $S^\alpha_0, S^\alpha_k, Z^\alpha_k, Q_k, r_{1k}, r_{2k}$ for $k = 1, \cdots, n-1$, agent α chooses a bid

$$b^\alpha_n(\omega) \in [0, (S^\alpha_{n-1}(\omega))^+ + k^\alpha],$$

where $k^\alpha \geq 0$ is an upper bound on loans to agent α. As before, agent α must bid in ignorance of both the total endowment $Q_n(\omega)$ and his personal endowment $Y_n^\alpha(\omega)$ for period n.

The total bid B_n, the price p_n, and agent α's quantities of goods x_n^α and fiat money $p_n Y_n^\alpha = B_n Z_n^\alpha$ are formed exactly as in the no-lending model of Section 3. Formula (4) for the dynamics now takes the form

$$S_n^\alpha = \begin{cases} -r_{1n}(b_n^\alpha - (S_{n-1}^\alpha)^+) + B_n Z_n^\alpha, & \text{if} \quad (S_{n-1}^\alpha)^+ \leq b_n^\alpha, \\ r_{2n}((S_{n-1}^\alpha)^+ - b_n^\alpha) + B_n Z_n^\alpha, & \text{if} \quad (S_{n-1}^\alpha)^+ > b_n^\alpha. \end{cases} \tag{12}$$

The wealth-distribution ν_n on day n is defined by formula (6) as before, but with the understanding that the set A now ranges over Borel subsets of the whole real line, since some agents may have negative wealth. An agent α's utility in period n is now given by

$$\xi_n^\alpha(\omega) = \begin{cases} u^\alpha(x_n^\alpha(\omega)), & \text{if} \quad S_{n-1}^\alpha(\omega) \geq 0, \\ u^\alpha(x_n^\alpha(\omega)) + u^\alpha(S_{n-1}^\alpha(\omega)/p_{n-1}(\omega)), & \text{if} \quad S_{n-1}^\alpha(\omega) < 0. \end{cases}$$

As before, agent α seeks to maximize the expected value of his total discounted utility

$$\sum_{n=0}^{\infty} \beta^n \xi_n^\alpha(\omega).$$

We extend now the definition of stationary equilibrium to the model with lending.

Definition 5.1. A *stationary equilibrium* for the model with lending, consists of a wealth distribution μ (i.e. a probability distribution) on the Borel subsets of the real line, of interest rates r_1, r_2 with $1 \leq r_2 \leq r_1, r_2 \leq 1/\beta$, and of a collection of stationary strategies $\{\pi^\alpha, \alpha \in I\}$ such that, if the bank sets interest rates r_1 and r_2 in every period, and if the initial wealth distribution is $\nu_0 = \mu$, then

(a) $\nu_n = \mu$ for all $n \geq 1$ when every agent α plays strategy π^α, and

(b) the strategy π^α is optimal for agent α when every other agent β plays $\pi^\beta (\beta \in I, \beta \neq \alpha)$, for each $\alpha \in I$.

Suppose that the model is in stationary equilibrium, and that each stationary strategy π^α specifies its bids b_n^α by the same consumption function $c^\alpha \equiv c(\cdot)$ with $0 \leq c^\alpha(s) \leq s + k$ for all $s \geq 0$, and the same upper-bound $k^\alpha \equiv k$ on loans, for all $\alpha \in I$. Then the total bid,

$$B = B_n(\omega) = \int c(s^+)\mu(ds),$$

remains constant from period to period, while the prices $\{p_n\}$ form an I.I.D. sequence, just as in the no-lending model of Section 3.

6 Existence of Stationary Equilibrium for the Model with Lending and Possible Bankruptcy

The methods and results of [2] can be used here, as those of [3] were used in Section 4. We consider the one-person game faced by an agent when the economy is in stationary equilibrium. We suppress the superscript α and assume that agents are homogeneous, with common utility function $u(\cdot)$, income distribution λ, and loan limit k. We define the utility function $\tilde{u}(\cdot)$ as in (7) and observe that (8) remains valid. Formula (12) for the dynamics can be written in the simpler form

$$S_n = g\left((S_{n-1})^+ - b_n\right) + BZ_n, \quad n \in \mathbb{N} \tag{13}$$

where

$$g(x) = \begin{cases} r_1 x, & x < 0, \\ r_2 x, & x \geq 0. \end{cases}$$

The Bellman equation becomes

$$V(s) = \begin{cases} \sup_{0 \leq b \leq s+k} [\tilde{u}(b/B) + \beta \cdot E[V(g(s-b) + BZ]], & s \geq 0 \\ \tilde{u}(s/B) + V(0), \quad s < 0. \end{cases} \tag{14}$$

This equation is of the type studied in [2], and all the major results of that paper have counterparts here. For example, Theorem 4.2 of [2] applies, to tell us that there is a unique stationary optimal strategy $\pi = \pi(B)$ corresponding to a consumption function $c(\cdot) = c(\cdot\,; B)$. The Markov chain $\{S_n\}$ for the fortunes of an agent who plays $\pi(B)$ evolves according to the dynamics

$$S_{n+1} = g((S_n)^+ - c((S_n)^+; B)) + BZ_{n+1}, \quad n \in \mathbb{N}_0\,. \tag{15}$$

Conditions for this chain to have a stationary distribution μ with finite mean are available in Theorem 4.3 of [2]. For μ to be the wealth-distribution of a stationary equilibrium, we must also assume that the bank balances its books under μ.

Assumption 6.1. (i) The Markov chain $\{S_n\}$ of (15) has an invariant distribution μ with finite mean.

(ii) Under the wealth-distribution μ, the total amount of money paid back to the bank by borrowers in a given period, is equal to the sum of the total amount borrowed, plus the amount of interest paid by the bank to lenders. This condition can be written as

$$\int\int [Bz \wedge r_1 d(s^+)]\, \mu(ds)\lambda(dz) = \int d(s^+)\, \mu(ds) + \rho_2 \int \ell(s^+)\, \mu(ds),$$

where $d(s) = (c(s) - s)^+$ and $\ell(s) = (s - c(s))^+$ are the amounts borrowed and deposited, respectively, under the stationary strategy $c(\cdot)$, by an agent with wealth $s \geq 0$.

Theorem 6.1. *If Assumption* 6.1 *holds, then there is a stationary equilibrium with wealth distribution* μ, *and interest rates* r_1, r_2 *in which every agent plays the plan* π.

The proof of this result is the same as that of Theorem 4.1, once the following lemma is established. Its proof is similar to that of Lemma 5.1 in [2].

Lemma 6.1. $\int c(s^+, B)\,\mu(ds) = B$.

Theorem 6.1 is intuitively appealing, and useful for verifying examples of stationary equilibria. However, it is inadequate as an existence result, because condition (ii) of Assumption 2.1 is delicate and difficult to check. There are two existence results in [2], Theorems 7.1 and 7.2, that do not rely on such an assumption. Here we present the analogue of the second of them.

Theorem 6.2. *Suppose that the variables* $\{Z_n^\alpha\}$ *are uniformly bounded, and that the derivative of the utility function* $u(\cdot)$ *is bounded away from zero. Then a stationary equilibrium exists.*

The proof is similar to that of Theorem 7.2 in [2], with the constant B again playing the mathematical role played by the price p in [2]. The utility function $\tilde{u}(\cdot)$ replaces $u(\cdot)$ in the argument, and the hypothesis that $\inf u'(\cdot) > 0$ implies that the same is true for $\tilde{u}(\cdot)$.

7 Examples

Here we present five examples. The first two illustrate the existence Theorem 4.1 for the model without lending.

Example 7.1. Suppose that the utility function is linear, namely $u(x) = x$. The endowment variables $\{Y_n^\alpha\}$ satisfy the proportionality assumption (1) but are otherwise arbitrary. The function $\tilde{u}(\cdot)$ of (7) is also linear, since

$$\tilde{u}(b) = \int bq\,\zeta(dq) = b \cdot E(Q).$$

We shall show that the optimal policy π of Theorem 4.1 is given by the "spend all" consumption function $c(s) = s$. To see this, let $I(\cdot)$ be the return function for π. Then $I(\cdot)$ satisfies

$$I(s) = \tilde{u}(s/B) + \beta \cdot E[I(s - s + BZ)] = \frac{s}{B}E(Q) + I^*, \tag{16}$$

where $I^* := \beta \cdot E[I(BZ)]$. To prove optimality, we have to check that $I(\cdot)$ also satisfies the Bellman equation

$$I(s) = \max_{0 \le b \le s} \left[\tilde{u}(b/B) + \beta \cdot E[I(s - b + BZ)]\right].$$

Now

$$E[I(s-b+BZ)] = E\left[\frac{s-b+BZ}{B}\cdot E(Q)+I^*\right] = \frac{s-b+B}{B}\cdot EQ+I^*$$

so that the function

$$b \mapsto \tilde{u}(b/B)+\beta\cdot E[I(s-b+BZ)] = b(1-\beta)\frac{E(Q)}{B} + \beta\left[\left(\frac{s}{B}+1\right)E(Q)+I^*\right]$$

attains its maximum $((s/B)+\beta)\,E(Q)+\beta\,I^*$ on $[0,s]$ at $b=c(s)=s$. In order for this maximum to agree with the expression of (16), we need $I^* = [\beta/(1-\beta)]\,E(Q)$; this, in turn, yields $I(s)=[(s/B)+(\beta/(1-\beta)]\,E(Q)$, in agreement with $I^*=I(0)$. Hence, the Bellman equation holds and π is optimal. Notice that under π, we have

$$S_{n+1} = S_n - S_n + BZ_{n+1} = BZ_{n+1}, \quad n \in \mathbb{N}_0,$$

and the stationary distribution μ is that of BZ_1.

Example 7.2. Assume that the utility function is

$$u(b) = \begin{cases} b, & 0 \le b \le 1, \\ 1, & b > 1, \end{cases}$$

that the distribution ζ of the I.I.D. endowment variables $\{Q_n\}$ is the two-point distribution

$$\zeta(\{1/2\}) = \zeta(\{1\}) = 1/2,$$

and that the distribution λ of the I.I.D. proportions $\{Z_n\}$ of the total endowment is the two-point distribution

$$\lambda(\{0\}) = 3/4, \lambda(\{4\}) = 1/4.$$

Suppose also that the total bid B is 1. Then the price $p=B/Q$ fluctuates between $p_1=1$ (when $Q=1$) and $p_2=2$ (when $Q=1/2$). The modified utility function of (7) is given by

$$\tilde{u}(b) = \tfrac{1}{2}u(b)+\tfrac{1}{2}u(b/2) = \begin{cases} (3b)/4, & 0 \le b \le 1, \\ (2+b)/4, & 1 \le b \le 2, \\ 1, & b \ge 2. \end{cases} \tag{17}$$

Clearly, an agent with this utility function should never bid more than 2. However, for small values of β, it is optimal to bid all up to a maximum of 2. In fact we shall show that, for $0 < \beta < 3/7$, the policy π with consumption function of the form

$$c(s) = \begin{cases} s, & 0 \le s \le 2, \\ 2, & s \ge 2. \end{cases}$$

is optimal. To establish the optimality of this π, it suffices to show that the return function $I(\cdot)$ satisfies the Bellman equation (9). First observe that $I(\cdot)$ satisfies the functional equation

$$I(s) = \begin{cases} ((3s)/4) + \beta \cdot EI(Z), & 0 \le s \le 1, \\ ((2+s)/4) + \beta \cdot EI(Z), & 1 \le s \le 2, \\ 1 + (\beta/4)I(s+2) + (3\beta/4)I(s-2), & s \ge 2. \end{cases} \tag{18}$$

In particular,

$$I'(s) = \begin{cases} 3/4, & 0 < s < 1, \\ 1/4, & 1 < s < 2, \\ (\beta/4)\, I'(s+2) + (3\beta/4)\, I'(s-2), & \text{for non-integers } s > 2. \end{cases} \tag{19}$$

As a step toward the verification of the Bellman equation, we shall see that the function $I(\cdot)$ satisfies the concavity condition:

$$1 \ge \beta I'_+(4) + 3\beta I'_+(0). \tag{20}$$

Note that, if we can compute $I(0) = \beta \cdot EI(Z)$, then, by (18), we know the function $I(\cdot)$ on the interval [0, 2].

Now write $a_k := I(2k)$, $k = 0, 1, \dots$ and, by (18), we have the recursion

$$a_k = 1 + \frac{\beta}{4} a_{k+1} + \frac{3\beta}{4} a_{k-1}, \qquad k \ge 1. \tag{21}$$

A particular solution of (21) is $a_k \equiv 1/(1-\beta)$, so the general *bounded* solution is given by

$$I(2k) = a_k = [1/(1-\beta)] + A\theta^k, \qquad k = 1, 2, \dots \tag{22}$$

for a suitable real constant A. Here $\theta = (2 - \sqrt{4 - 3\beta^2})/\beta$ is the root of the equation

$$f(\xi) := \xi^2 - (4/\beta)\xi + 3 = 0$$

in the interval (0, 1), and we have $\theta < \beta$.

Using (18) and (22), we see that $I(2) = 1 + I(0)$ and hence $I(0) = \beta/(1-\beta) + A\theta$. Also $I(0) = (\beta/4)I(4) + (3\beta/4)I(0)$. Thus $(1 - (3\beta/4))\,I(0) = (\beta/4)I(4)$, or

$$\left(1 - \frac{3\beta}{4}\right)\left(\frac{\beta}{1-\beta} + A\theta\right) = \frac{\beta}{4}\left(A\theta^2 + \frac{1}{1-\beta}\right).$$

Hence, $A = -1/(1-\theta)$, $I(0) = [\beta/(1-\beta)] - [\theta/(1-\theta)] > 0$, and $I(1) = (3/4) + I(0) = [1/(1-\beta)] - [\theta/(1-\theta)] - (1/4)$.

More generally, with $d_k := I(2k+1)$, $k = 0, 1, \ldots$, we have the recursion

$$d_k = 1 + (\beta/4)d_{k+1} + (3\beta/4)d_{k-1}, \qquad k = 1, 2, \ldots$$

with general solution

$$d_k = [1/(1-\beta)] + D\theta^k, \qquad k = 1, 2, \ldots.$$

Plugging this last expression into the equality $I(3) = 1 + (\beta/4)I(5) + (3\beta/4)I(1)$, and substituting the value of $I(1)$ from above, we obtain $D = -[\theta/(1-\theta)] - (1/4)$.

With these computations in place, we are now in a position to check the concavity condition (20). Indeed, $I'_+(4) = I(5) - I(4) = d_2 - a_2 = (D - A)\theta^2 = [A(\theta - 1) - (1/4)]\,\theta^2 = (3/4)\theta^2 = (3/4)\,((4/\beta)\theta - 3) = (3\theta/\beta) - (9/4)$. Thus

$$\beta I'_+(4) + 3\beta I'_+(0) = 3\theta - (9/4)\beta + 3\beta \cdot (3/4)3\theta < 1$$

amounts to $f(1/3) < 0$, or $(1/9) - (4/3\beta) + 3 < 0$, and this last condition is equivalent to our assumption that $\beta < 3/7$.

We are now prepared to complete the proof that π is optimal by showing that its return function $I(\cdot)$ satisfies the Bellman equation (9). Equivalently, we have to check that the function

$$\psi_s(b) := \tilde{u}(b) + \beta E I(s - b + Z) = \tilde{u}(b) + \frac{\beta}{4} I(s - b + 4) + \frac{3\beta}{4} I(s - b)$$

attains its maximum over $b \in [0, s]$ at $b^* = c(s)$. We consider three cases.

Case I: $0 \le s \le 1$. In this case, for $0 < b < s$:

$$\psi_s(b) = (3/4)b + (\beta/4)I(s - b + 4) + (3\beta/4)I(s - b)$$

and

$$\psi'_s(b) = (3/4) - \left[(\beta/4)I'_+(4) + (3\beta/4)I'_+(0)\right] > 0.$$

Thus $\psi'_s(s-) > 0$, and $b^* = s \equiv c(s)$ is the location of the maximum.

Case II: $1 < s \leq 2$. Here we use (16) and (18) to obtain

$$\psi_s'(b) = \begin{cases} \frac{3}{4} - \left[\frac{\beta}{4} I_+'(5) + \frac{3\beta}{4} I_+'(1)\right], & 0 < b < s-1, \\ \frac{3}{4} - \left[\frac{\beta}{4} I_+'(4) + \frac{3\beta}{4} I_+'(0)\right], & s-1 < b < 1, \\ \frac{1}{4} - \left[\frac{\beta}{4} I_+'(4) + \frac{3\beta}{4} I_+'(0)\right], & 1 < b < s. \end{cases}$$

In particular, $\psi_s'(\cdot) > 0$ on $[0, s]$, thus $b^* = s \equiv c(s)$, as follows from Lemma 7.1 below.

Case III: $s > 2$. In this case, we have

$$\psi_s'(b) = \begin{cases} \frac{3}{4} - \left[\frac{\beta}{4} I_+'(s+3) + \frac{3\beta}{4} I_+'(s-1)\right], & 0 < b < 1, \\ \frac{1}{4} - \left[\frac{\beta}{4} I_+'(s+2) + \frac{3\beta}{4} I_+'(s-2)\right], & 1 < b < 2, \\ -\frac{\beta}{4} I_+'(\lfloor s-b \rfloor + 4) - \frac{3\beta}{4} I_+'(\lfloor s-b \rfloor), & b > 2. \end{cases}$$

The function $\psi(\cdot)$ now attains its maximum at $b* = 2 \equiv c(s)$, since $\psi'(\cdot) > 0$ on $(0, 2)$ and $\psi'(\cdot) < 0$ on $(2, s)$ as follows from Lemma 7.1 below.

Lemma 7.1. *The function $I(\cdot)$ satisfies:*

$$I_+'(0) > I_+'(1) > I_+'(2) > \cdots > 0.$$

Proof. The first three of these inequalities amount to

$$3/4 > 1/4 > (\beta/4)I_+'(4) + (3\beta/4)I_+'(0) = I_+'(0),$$

and have been established already. So we have to prove

$$I_+'(2k) > I_+'(2k+1) > I_+'(2(k+1)) > 0, \quad \text{for} \quad k \geq 2. \tag{23}$$

Now

$$I_+'(2k) = I(2k+1) - I(2k) = d_k - a_k = (D - A)\theta^k,$$

and

$$I_+'(2k+1) = I(2(k+1)) - I(2k+1) = a_{k+1} - d_k = (A\theta - D)\theta^k.$$

So the inequalities of (23) amount to

$$D - A > A\theta - D > \theta(D - A) > 0.$$

But $D - A = A(\theta - 1) - 1/4 = 3/4$, $A\theta - D = 1/4$, and these inequalities reduce to $0 < \theta < 1/3$, which has already been proved. □

The optimality of the strategy π for an agent playing in equilibrium with $B = 1$ has now been established. The stationary distribution μ for the corresponding Markov chain as in (10) is supported by the set of even integers $\{0, 2, \cdots\}$ and is given by

$$\mu(\{0\}) = 1/2, \quad \mu(\{2\}) = 1/6, \quad \mu(\{2k\}) = (2/3)(1/3)^{k-1} \quad \text{for} \quad k \geq 2.$$

(The calculation of μ is explained in some detail in Example 2.5 of [3].) Note that, with this distribution, the total amount bid in equilibrium is $B \equiv \int_0^\infty s\,\mu(ds) = 1$, as postulated. The family of stationary strategies $\pi^\alpha \equiv \pi$ and the wealth distribution μ form a stationary equilibrium as in Theorem 4.1 for $0 < \beta < 3/7$. There will also exist equilibria for other values of β, but we shall not calculate them here.

The next example provides a simple illustration of Theorem 6.1.

Example 7.3. Let the utility function be

$$u(b) = \begin{cases} b, & b \geq 0, \\ 2b, & b < 0. \end{cases}$$

Suppose that the common distribution ζ of the random variables $\{Q_n\}$ is $\zeta(\{1\}) = \zeta(\{3\}) = 1/2$, and that the distribution λ of the variables $\{Z_n\}$ is $\lambda(\{0\}) = \lambda(\{2\}) = 1/2$. The modified utility function $\tilde{u}(\cdot)$ is then

$$\tilde{u}(b) = \frac{1}{2}u(b) + \frac{1}{2}u(3b) = \begin{cases} 2b, & b \geq 0, \\ 4b, & b < 0. \end{cases}$$

Take the interest rates to be $r_1 = r_2 = 2$ and the bound on lending to be $k = 1$. Finally assume that the total bid B is 1.

Although the penalty for default is heavy, as reflected by the larger value of $u'(b)$ for $b < 0$, it is to be expected that *an agent will choose to make large bids for β sufficiently small.* Indeed, we shall show that the optimal strategy π for $0 < \beta < 1/3$ is to borrow up to the limit and spend everything, corresponding to $c(s) = s + 1$ for all $s \geq 0$, as he is then not very concerned about the penalty for default. (Recall that an agent with wealth $s < 0$ is punished in amount $u(s)$ and then plays from position 0. Thus, a strategy need only specify bids for nonnegative values of s.)

Let $I(\cdot)$ be the return function for π. Then this function must satisfy

$$\begin{aligned} I(s) &= \tilde{u}(s + 1) + \beta\, E[I(2(s - (s + 1)) + Z)] \\ &= 2s + 2 + (\beta/2)\,[I(-2) + I(0)] \end{aligned}$$

for $s \geq 0$, and

$$I(s) = \tilde{u}(s) + I(0) = 4s + I(0)$$

for $s < 0$. Thus

$$I'(s) = \begin{cases} 2, & s > 0, \\ 4, & s < 0. \end{cases}$$

To verify that $I(\cdot)$ satisfies the Bellman equation (14), consider the function

$$\begin{aligned} \psi_s(b) &= \tilde{u}(b) + \beta\, E[I(2(s-b)+Z)] \\ &= 2b + (\beta/2)\,[I(2(s-b)) + I(2(s-b+1))] \\ &= \begin{cases} 2b - 4\beta b + c_1, & 0 \le b < s, \\ 2b - 6\beta b + c_2, & s < b < s+1, \end{cases} \end{aligned}$$

where $c_1 = c_1(s)$ and $c_2 = c_2(s)$ are constants. Thus

$$\psi_s'(b) = \begin{cases} 1 - 2\beta, & 0 < b < s, \\ 1 - 3\beta, & s < b < s+1, \end{cases}$$

and we see that $\psi_s(\cdot)$ attains its maximum on $[0, s+1]$ at $s+1$, thanks to our assumption that $0 < \beta < 1/3$. It follows that $I(\cdot)$ satisfies the Bellman equation, and that π is optimal. The Markov chain $\{S_n\}$ of (15) becomes

$$S_{n+1} = 2[(S_n)^+ - ((S_n)^+ + 1)] + Z = Z - 2.$$

The stationary wealth-distribution, namely, the distribution of $Z-2$, assigns mass $\mu(\{-2\}) = 1/2$ at -2 and mass $\mu(\{0\}) = 1/2$ at 0. Obviously, clause (i) of Assumption 6.1 is satisfied. Clause (ii) is also satisfied, because every agent borrows one unit of money and spends it; one-half of the agents receive no income and pay back back nothing, whereas the other half receive an income of 2 units of money, all of which they pay back to the bank since the interest rate is $r_1 = 2$. As there are no lenders, the books balance. Theorem 6.1 now says that we have a stationary equilibrium, in which half of the agents are in debt for 2 units of money, and the other half hold no money at the beginning of each period. All the money is held by the bank.

Suppose now that the discount factor is larger, so that agents will be more concerned about the penalties for default. In particular, assume that $1/3 < \beta < 1/2$. Then an argument similar to that above shows that an optimal strategy is for an agent to borrow nothing and spend what he has; that is, the optimal strategy π corresponds to $c(s) = s$ for every $s \ge 0$. This induces the Markov chain,

$$S_{n+1} = 2[(S_n)^+ - (S_n)^+] + Z = Z,$$

with stationary distribution equal to the distribution λ of Z, which assigns mass $\lambda(\{0\}) = \lambda(\{2\}) = 1/2$ each to 0 and 2. This time the books obviously balance, since no one borrows and no one pays back. In fact, the bank has no role to play.

For the next example we drop the assumption that individual endowments are proportional to total production (Assumption 2.1, part (b)) and show that a stationary equilibrium need not exist.

Example 7.4. For simplicity, we return to the no-lending model of Section 3 for this example. Assume that the utility function is $u(b) = b$, and let the distribution ζ of the variables $\{Q_n\}$ be the two-point distribution $\zeta(\{1\}) = \zeta(\{3\}) = 1/2$. Suppose that when $Q_n = 1$, the variables $\{Z_n^\alpha, \alpha \in I\}$ are equal to 0 or 2 with probability 1/2 each, but that when $Q_n = 3$, each of the Z_n^α is equal to 1. Thus the $\{Q_n\}$ and the $\{Z_n^\alpha\}$ are *not* independent, as we had postulated in Assumption 2.1. We claim that no stationary equilibrium can exist in this case.

Now suppose, by way of contradiction, that a stationary equilibrium exists, with wealth distribution μ and optimal stationary strategies $\{\pi^\alpha,\ \alpha \in I\}$ corresponding to consumption functions $c^\alpha(\cdot), \alpha \in I$. The total bid in each period is then $B = \int c^\alpha(s)\,\mu(ds)$ and the prices $p_n = B/Q_n$ are independent, and equal to B and $B/3$ with probability 1/2 each.

Consider next the spend-all strategy π' with consumption function $c(s) = s$. We will sketch the proof that π' is the unique optimal strategy. First we calculate the return function $I(\cdot)$ for π': this function satisfies

$$I(s) = E\left[(s/B)\cdot Q + \beta \cdot I(BZ)\right] = (2s/B) + \beta \cdot E[I(BZ)].$$

It is easy to check that $I(\cdot)$ also satisfies the Bellman equation

$$I(s) = \sup_{0 \le b \le s} E\left[(b/B)\cdot Q + \beta \cdot I(s - b + BZ)\right],$$

and that the supremum above is uniquely attained at $b = s$. It follows that π' is the unique optimal strategy. Thus, we must have $\pi^\alpha = \pi'$ for all $\alpha \in I$, which means that every agent α spends his entire wealth at every time-period n and enters the next period with wealth

$$S_{n+1}^\alpha(\omega) = BZ_{n+1}^\alpha(\omega).$$

But the distribution of Z_{n+1}^α depends on the value of Q_n. Thus, the distribution of wealth varies with the value of Q_n and cannot be identically equal to the equilibrium distribution μ, as we had assumed.

In our final example, we assume that agents know the value of the production variable for each time-period, before placing their bids. It is not surprising then, that agents playing optimally will take advantage of this additional information, and therefore that a stationary equilibrium need not exist. What sort of equilibrium is appropriate for this "high information" model is a question that we plan to investigate in future work.

Example 7.5. As in the previous example, we consider a no-lending model with the linear utility $u(x) = x$ and with the distribution ζ of the variables $\{Q_n\}$ given by $\zeta(\{1\}) = \zeta(\{3\}) = 1/2$. We assume that the individual endowments are proportional (so that, as in Assumption 2.1, the variables $\{Z_n^\alpha\}$ are independent of the $\{Q_n\}$'s), and that agents know the value of the 'production variable' Q_n for the time-period $t = n$, before making their bids for that period. Again, we claim that no stationary equilibrium can exist in this case.

Suppose, by way of contradiction, that a stationary equilibrium does exist, with wealth distribution μ and optimal stationary strategies $\{\pi^\alpha,\ \alpha \in I\}$ corresponding to consumption functions $\{c^\alpha(\cdot),\ \alpha \in I\}$. Let $B = \int c^\alpha(s)\,\mu(ds)$ be the total bid in each period, so that the price $p_n = B/Q_n$ in period n is $B/3$ if $Q_n = 3$ and is B if $Q_n = 1$. It is not difficult to show that, in a period when the price is low (i.e., when $Q_n = 3$), the optimal bid for an agent is $c(s) = s$. Thus, we must have $c^\alpha(s) = s$ for all α and s. However, in a period when the price is high (i.e. when $Q_n = 1$), an agent who spends one unit of money receives in utility $(1/B)$, whereas an agent who saves the money and spends it in the next period expects to receive $\beta\,[(1/2B) + (3/2B)] = (2\beta/B)$. Thus, for $\beta \in ((1/2), 1)$, it is optimal for an agent to spend nothing in a period when the price is high. But then $c^\alpha(s) = 0$ for all $\alpha \in I$ and $s \geq 0$, a contradiction.

Acknowledgements

Our research was supported by National Science Foundation Grants DMS-00-99690 (Karatzas) and DMS-97-03285 (Sudderth), by the Cowles Foundation at Yale University, and by the Santa Fe Institute.

REFERENCES

[1] Feldman, M. and Gilles Ch., An Expository Note on Individual Risk Without Aggregate Uncertainty, *Journal of Economic Theory* **35** (1985) 26–32.

[2] Geanakoplos, J., Karatzas I., Shubik M. and Sudderth W.D., A Strategic Market Game with Active Bankruptcy, *Journal of Mathematical Economics*, **34** (2000) 359–396.

[3] Karatzas, I., Shubik M. and Sudderth W.D., Construction of Stationary Markov Equilibria in a Strategic Market Game, *Mathematics of Operations Research*, **19** (1994) 975–1006.

[4] Karatzas, I., Shubik M. and Sudderth W.D., A Strategic Market Game with Secured Lending. *Journal of Mathematical Economics*, **28** (1997) 207–247.

Markov Games under a Geometric Drift Condition

Heinz-Uwe Küenle
Brandenburgische Technische Universität Cottbus
PF 10 13 44, D-03013 Cottbus, Germany
kueenle@math.tu-cottbus.de

Abstract

Zero-sum stochastic games with the expected average cost criterion and unbounded stage cost are studied. It is assumed that the transition probabilities of the Markov chains induced by stationary strategies satisfy a certain geometric drift condition. Under additional assumptions concerning especially the existence of ε-optimal strategies in corresponding one-stage games it is shown that the average optimality equation has a solution and that both players have ε-optimal stationary strategies.

Key words. Markov games, Borel state space, average cost criterion, geometric drift condition, unbounded costs.

1 Introduction

In this paper two-person stochastic games with standard Borel state space, standard Borel action spaces, and the expected average cost criterion are considered. Such a zero-sum stochastic game can be described in the following way: The state x_n of a dynamic system is periodically observed at times $n = 1, 2, \ldots$. After an observation at time n the first player chooses an action a_n from the action set $\boldsymbol{A}(x_n)$ and afterwards the second player chooses an action b_n from the action set $\boldsymbol{B}(x_n)$ dependent on the complete history of the system at this time. The first player must pay cost $k(x_n, a_n, b_n)$ to the second player, and the system moves to a new state x_{n+1} from the state space $\mathbf{X}$ according to the transition probability $p(\cdot|x_n, a_n, b_n)$.

Stochastic games with Borel state space and average cost criterion are considered by several authors. Related results are given by Maitra and Sudderth [10], [11], [12], Nowak [14], Rieder [16] and Küenle [8] in the case of bounded costs (payoffs). The case of unbounded payoffs is treated by Nowak [15], Jaśkiewicz and Nowak [4], Hernández-Lerma and Lasserre [3], Küenle [6] and Küenle and Schurath [9]. The assumptions in these papers are compared in [9]. The assumptions in our paper concerning the transition probabilities are related to Nowak's assumptions in [15], [4]: Nowak assumes that there is a Borel set $C \in \mathbf{X}$ and for every stationary strategy pair $(\pi^\infty, \rho^\infty)$ a measure μ such that C is μ-small with respect to the Markov chain induced by this strategy pair. We assume that C is only a μ-petite set with respect to a resolvent of this Markov chain; on the other hand, we demand that μ is independent of the corresponding strategy pair. (For the definition of "small sets" and "petite sets" see [13], for example.) Since in our paper the assumptions concern the resolvents of the corresponding Markov chains instead of the one-step transition probabilities as in the above mentioned papers, it is possible that the Markov chains are periodic, for instance. Furthermore, in [15] and [3] the existence of a density of the transition probability is assumed while in [4], [9] and in this paper such a density is not used.

The paper is organized as follows: in Section 2 the mathematical model of Markov games with arbitrary state and action spaces is presented. Section 3 contains the assumptions on the transition probabilities and the stage costs and also some preliminary results. In Section 4 we study the expected average cost of a fixed stationary strategy pair. We show that the so-called Poisson equation has a solution. Under additional assumptions (which are satisfied if the action spaces are finite or if certain semi-continuity and compactness conditions are fulfilled, for instance) we prove in Section 5 that the average cost optimality equation has a solution and both players have ε-optimal stationary strategies for every $\varepsilon > 0$.

2 The Mathematical Model

Stochastic games considered in this paper are defined by nine objects:

Definition 2.1. $\mathcal{M} = ((\mathbf{X}, \sigma_{\mathbf{X}}), (\mathbf{A}, \sigma_{\mathbf{A}}), \boldsymbol{A}, (\mathbf{B}, \sigma_{\mathbf{B}}), \boldsymbol{B}, p, k, \mathbf{E}, \mathbf{F})$ is called a *Markov game* if the elements of this tuple have the following meaning:

- $(\mathbf{X}, \sigma_{\mathbf{X}})$ is a standard Borel space, called the *state space*.
- $(\mathbf{A}, \sigma_{\mathbf{A}})$ is a standard Borel space and $\boldsymbol{A} : \mathbf{X} \to \sigma_{\mathbf{A}}$ is a set-valued map which has a $\sigma_{\mathbf{X}}$ - $\sigma_{\mathbf{A}}$-measurable selector. $\mathbf{A}$ is called the *action space of the first player* and $\boldsymbol{A}(x)$ is called the *admissible action set of the first player at state* $x \in \mathbf{X}$. We assume $\{(x, a) : x \in \mathbf{X}, a \in \boldsymbol{A}(x)\} \subseteq \sigma_{\mathbf{X}\times\mathbf{A}}$.
- $(\mathbf{B}, \sigma_{\mathbf{B}})$ is a standard Borel space and $\boldsymbol{B} : \mathbf{X} \times \mathbf{A} \to \sigma_{\mathbf{B}}$ is a set-valued map which has a $\sigma_{\mathbf{X}}$ - $\sigma_{\mathbf{B}}$-measurable selector. $\mathbf{B}$ is called the *action space of the second player* and $\boldsymbol{B}(x)$ is called the *admissible action set of the second player at state* $x \in \mathbf{X}$. We assume $\{(x, b) : x \in \mathbf{X}, a \in \boldsymbol{B}(x)\} \subseteq \sigma_{\mathbf{X}\times\mathbf{B}}$.
- p is a transition probability from $\sigma_{\mathbf{X}\times\mathbf{A}\times\mathbf{B}}$ to $\sigma_{\mathbf{X}}$, the *transition law*.
- k is a $\sigma_{\mathbf{X}\times\mathbf{A}\times\mathbf{B}}$-measurable function, called *stage cost function of the first player*.
- Assume that $(\mathbf{Y}, \sigma_{\mathbf{Y}})$ is a standard Borel space. Then we denote by $\overline{\sigma}_{\mathbf{Y}}$ the σ-algebra of the $\sigma_{\mathbf{Y}}$-universally measurable sets. Let $\mathbf{H}_n = (\mathbf{X} \times \mathbf{A} \times \mathbf{B})^n \times \mathbf{X}$ for $n \geq 1$, $\mathbf{H}_0 = \mathbf{X}$. $h \in \mathbf{H}_n$ is called the *history at time* n.

  A transition probability π_n from $\overline{\sigma}_{\mathbf{H}_n}$ to $\overline{\sigma}_{\mathbf{A}}$ with

$$\pi_n(\boldsymbol{A}(x_n)|x_0, a_0, b_0, \ldots, x_n) = 1$$

for all $(x_0, a_0, b_0, \ldots, x_n) \in \mathbf{H}_n$ is called a *decision rule of the first player at time n*.

A transition probability ρ_n from $\overline{\sigma}_{\mathbf{H}_n \times \mathbf{A}}$ to $\overline{\sigma}_{\mathbf{B}}$ with

$$\rho_n(\boldsymbol{B}(x_n)|x_0, a_0, b_0, \ldots, x_n, a_n) = 1$$

for all $(x_0, a_0, b_0, \ldots, x_n, a_n) \in \mathbf{H}_n \times \mathbf{A}$ is called a *decision rule of the second player at time n*.

A decision rule of the first [second] player is called *Markov* iff a transition probability $\tilde{\pi}_n$ from $\overline{\sigma}_{\mathbf{H}_n}$ to $\overline{\sigma}_{\mathbf{A}}$ [$\tilde{\rho}_n$ from $\overline{\sigma}_{\mathbf{H}_n \times \mathbf{A}}$ to $\overline{\sigma}_{\mathbf{B}}$] exists with

$$\pi_n(\cdot|x_0, a_0, b_0, \ldots, x_n) = \tilde{\pi}_n(\cdot|x_n)$$

$$[\rho_n(\cdot|x_0, a_0, b_0, \ldots, x_n, a_n) = \tilde{\rho}_n(\cdot|x_n, a_n)]$$

for all $(x_0, a_0, b_0, \ldots, x_n, a_n) \in \mathbf{H}_n \times \mathbf{A}$. (Notation: We identify π_n as $\tilde{\pi}_n$ and ρ_n as $\tilde{\rho}_n$.)

E and **F** denote nonempty sets of Markov decision rules.

A decision rule of the first [second] player is called *deterministic* if a function $e_n : \mathbf{H}_n \to \mathbf{A}$ [$f_n : \mathbf{H}_n \times \mathbf{A} \to \mathbf{B}$] exists with $\pi_n(e_n(h_n)|h_n) = 1$ for all $h_n \in \mathbf{H}_n$ [$\rho_n(f_n(h_n, a_n)|h_n, a_n) = 1$ for all $(h_n, a_n) \in \mathbf{H}_n \times \mathbf{A}$].

A sequence $\Pi = (\pi_n)$ or $P = (\rho_n)$ of decision rules of the first or second player is called a *strategy* of that player.

Strategies are called *deterministic*, or *Markov* iff all their decision rules have the corresponding property.

A Markov strategy $\Pi = (\pi_n)$ or $P = (\rho_n)$ is called *stationary* iff $\pi_0 = \pi_1 = \pi_2 = \ldots$ or $\rho_0 = \rho_1 = \rho_2 = \ldots$. (Notation: $\Pi = \pi^\infty$ or $P = \rho^\infty$.) We assume in this paper that the sets of all admissible strategies are $\mathbf{E}^\infty$ and $\mathbf{F}^\infty$. Hence, only Markov strategies are allowed. But by means of dynamic programming methods it is possible to get corresponding results also for Markov games with larger sets of admissible strategies. If **E** and **F** are the sets of all Markov decision rules (in the above sense) then we have a Markov game with perfect (or complete) information. In this case the action set of the second player may depend also on the present action of the first player. If **E** is the set of all Markov decision rules but **F** is the set of all Markov decision rules which do not depend on the present action of the first player, then we have a usual Markov game with independent action choice. Let $\Omega := \mathbf{X} \times \mathbf{A} \times \mathbf{B} \times \mathbf{X} \times \mathbf{A} \times \mathbf{B} \times \ldots$ and $K^N(\omega) := \sum_{j=0}^N k(x_j, a_j, b_j)$ for $\omega = (x_0, a_0, b_0, x_1, \ldots) \in \Omega$, $N \in \mathbb{N}$. By means of a modification of the Ionescu–Tulcea Theorem (see [17]), it follows that there exists a suitable σ-algebra $\mathcal{F}$ in Ω and for every initial state $x \in \mathbf{X}$ and strategy pair (Π, P), $\Pi = (\pi_n)$, $P = (\rho_n)$, a unique probability measure $\mathsf{P}_{x,\Pi,P}$ on $\mathcal{F}$ according to the transition probabilities π_n, ρ_n and p. Furthermore, K^N is $\mathcal{F}$-measurable for all $N \in \mathbb{N}$. We set

$$V_{\Pi P}^N(x) = \int_\Omega K^N(\omega)\mathsf{P}_{x,\Pi,P}(d\omega) \tag{1}$$

and

$$\Phi_{\Pi P}(x) = \liminf_{N\to\infty} \frac{1}{N+1} V_{\Pi P}^N(x) \tag{2}$$

if the corresponding integrals exist.

Definition 2.2. Let $\varepsilon \geq 0$. A strategy pair (Π^*, P^*) is called *ε-optimal* iff

$$\Phi_{\Pi^* P} - \varepsilon \leq \Phi_{\Pi^* P^*} \leq \Phi_{\Pi P^*} + \varepsilon$$

for all strategy pairs (Π, P).

A 0-optimal strategy pair is called *optimal.*

3 Assumptions and Preliminary Results

In this paper we use the same notation for a sub-stochastic kernel and for the "expectation operator" with respect to this kernel, that means:
If $(\mathbf{Y}, \sigma_{\mathbf{Y}})$ and $(\mathbf{Z}, \sigma_{\mathbf{Z}})$ are standard Borel spaces, $v : \mathbf{Y} \times \mathbf{Z} \to \mathbb{R}$ a $\overline{\sigma}_{\mathbf{Y}\times\mathbf{Z}}$-measurable function, and q a sub-stochastic kernel from $(\mathbf{Y}, \overline{\sigma}_{\mathbf{Y}})$ to $(\mathbf{Z}, \overline{\sigma}_{\mathbf{Z}})$, then we put

$$qv(y) := \int_{\mathbf{Z}} q(dz|y)v(y, z),$$

for all $y \in \mathbf{Y}$, if this integral is well-defined.

We assume in the following that u and v are universally measurable functions for which the corresponding integrals are well-defined. If $v : \mathbf{X}\times\mathbf{A}\times\mathbf{B}\times\mathbf{X} \to \mathbb{R}$, then we have for example

$$pv(x, a, b) = \int_{\mathbf{X}} p(d\xi|x, a, b)v(x, a, b, \xi),$$

for all $(x, a, b) \in \mathbf{X} \times \mathbf{A} \times \mathbf{B}$. If $u : \mathbf{X} \to \mathbb{R}$ then pu means

$$pu(x, a, b) = \int_{\mathbf{X}} p(d\xi|x, a, b)u(\xi),$$

for all $(x, a, b) \in \mathbf{X} \times \mathbf{A} \times \mathbf{B}$. For $\pi \in \mathbf{E}$, $\rho \in \mathbf{F}$ we get

$$\pi\rho pv(x) = \int_{\mathbf{A}} \pi(da|x) \int_{\mathbf{B}} \rho(db|x, a) \int_{\mathbf{X}} p(d\xi|x, a, b)v(x, a, b, \xi),$$

for all $x \in \mathbf{X}$. Furthermore, we define the operator T by

$$Tu = k + pu, \tag{3}$$

for $u : \mathbf{X} \to \mathbb{R}$, that means

$$Tu(x, a, b) = k(x, a, b) + \int_{\mathbf{X}} p(d\xi|x, a, b)u(\xi),$$

for all $x \in \mathbf{X}$, $a \in \mathbf{A}$, $b \in \mathbf{B}$. $\pi\rho T$ is then the operator with

$$\begin{aligned}\pi\rho Tu(x) &= \pi\rho k(x) + \pi\rho p u(x) \\ &= \int_{\mathbf{A}} \pi(da|x) \int_{\mathbf{B}} \rho(db|x, a) \left(k(x, a, b) + \int_{\mathbf{X}} p(d\xi|x, a, b)u(\xi) \right),\end{aligned}$$

for all $x \in \mathbf{X}$. This operator is well-known in stochastic dynamic programming and Markov games. It is often denoted by $T_{\pi\rho}$.

Let $\Pi = (\pi_n) \in \mathbf{E}^\infty$, $P = (\rho_n) \in \mathbf{F}^\infty$. If $V^N_{\Pi P}$ exists then we get

$$V^N_{\Pi P} = \pi_0\rho_0 k + \sum_{j=1}^{N} \pi_0\rho_0 p \cdots \pi_{j-1}\rho_{j-1} p \pi_j \rho_j k.$$

For $\pi \in \mathbf{E}$, $\rho \in \mathbf{F}$ we put $(\pi\rho p)^n := \pi\rho p(\pi\rho p)^{n-1}$ where $(\pi\rho p)^0$ denotes the identity. Let $\vartheta \in (0, 1)$. We set for every $\pi \in \mathbf{E}$, $\rho \in \mathbf{F}$, $x \in \mathbf{X}$, and $Y \in \overline{\sigma}_{\mathbf{X}}$

$$Q_{\vartheta,\pi,\rho}(Y|x) := (1 - \vartheta) \sum_{n=0}^{\infty} \vartheta^n (\pi\rho p)^n \mathbf{I}_Y(x),$$

where $\mathbf{I}_Y$ is the characteristic function of the set Y.

We remark that for a stationary strategy pair $(\pi^\infty, \rho^\infty)$ the transition probability $Q_{\vartheta,\pi,\rho}$ is a resolvent of the corresponding Markov chain.

Assumption 3.1. There are a nontrivial measure μ on $\sigma_{\mathbf{X}}$, a set $C \in \sigma_{\mathbf{X}}$, a $\sigma_{\mathbf{X}}$-measurable function $W \geq 1$, and constants $\vartheta \in (0, 1)$, $\alpha \in (0, 1)$, and $\beta \in \mathbb{R}$ with the following properties:

(a)
$$Q_{\vartheta,\pi,\rho} \geq \mathbf{I}_C \cdot \mu,$$

for all $\pi \in \mathbf{E}$ and $\rho \in \mathbf{F}$,

(b)
$$pW \leq \alpha W + \mathbf{I}_C \beta,$$

(c)
$$\sup_{x \in \mathbf{X}, a \in \mathbf{A}(x), b \in \mathbf{B}(x,a)} \frac{|k(x, a, b)|}{W(x)} < \infty.$$

Assumption 3.1 (a) means that C is a "petite set", (b) is called "geometric drift towards C" (see Meyn and Tweedie [13]). We assume in this paper that Assumption 3.1 is satisfied.

For a measurable function $u : \mathbf{X} \to \mathbb{R}$ we denote by μu the integral $\mu u := \int_{\mathbf{X}} \mu(d\xi)u(\xi)$ (if it exists).

Lemma 3.1. *There are a $\sigma_{\mathbf{X}}$-measurable function V with $1 \le W \le V \le W +$ const and a constant $\lambda \in (0, 1)$ with*

$$Q_{\vartheta,\pi,\rho} V \le \lambda V + \mathbf{I}_C \cdot \mu V \tag{4}$$

and

$$\vartheta p V \le \lambda V. \tag{5}$$

Proof. Without loss of generality we assume $\beta > 0$.

Let $\beta' := [\vartheta/(1-\vartheta)]\beta$, $W' := W + \beta'$, and $\alpha' := (\beta' + \alpha)/(\beta' + 1)$. Then it holds $\alpha' \in (\alpha, 1)$ and

$$\begin{aligned} pW' &= pW + \beta' \\ &\le \alpha W + \beta' + \beta \mathbf{I}_C \\ &\le \alpha' W - (\alpha' - \alpha) W + \alpha'\beta' + (1-\alpha')\beta' + \beta \mathbf{I}_C \\ &\le \alpha' W' - (\alpha' - \alpha) + (1-\alpha')\beta' + \beta \mathbf{I}_C \\ &= \alpha' W' + \beta' + \alpha - \alpha'(\beta' + 1) + \beta \mathbf{I}_C \\ &= \alpha' W' + \beta \mathbf{I}_C. \end{aligned} \tag{6}$$

Let now $W'' := W' - \beta' \mathbf{I}_C = W + \beta'(1 - \mathbf{I}_C)$. Then we get from (6)

$$\begin{aligned} p(W'' + \beta' \mathbf{I}_C) &= pW' \\ &\le \alpha' W' + \beta \mathbf{I}_C \\ &= \alpha' W'' + \alpha'\beta' \mathbf{I}_C + \beta \mathbf{I}_C \\ &= \alpha' W'' + \alpha'\beta' \mathbf{I}_C + \frac{1-\vartheta}{\vartheta}\beta' \mathbf{I}_C \\ &= \alpha' W'' + \frac{\alpha'\vartheta + 1 - \vartheta}{\vartheta}\beta' \mathbf{I}_C \\ &\le \alpha' W'' + \frac{\beta'}{\vartheta}\mathbf{I}_C. \end{aligned} \tag{7}$$

We put $\alpha'' := (1-\vartheta)/(1-\alpha'\vartheta)$. Then it holds $\alpha' = (\alpha'' + \vartheta - 1)/(\alpha''\vartheta)$. For $\beta'' := \alpha''\beta'$ it follows

$$pW'' \le \frac{\alpha'' + \vartheta - 1}{\alpha''\vartheta} W'' - \frac{\beta''}{\alpha''} p\mathbf{I}_C + \frac{\beta''}{\alpha''\vartheta}\mathbf{I}_C.$$

Hence,

$$\alpha''\vartheta p W'' \le (\alpha'' + \vartheta - 1) W'' - \vartheta\beta'' p\mathbf{I}_C + \beta''\mathbf{I}_C.$$

Then

$$(1-\vartheta)W'' \le \alpha''W'' + \beta''\mathbf{I}_C - \vartheta p(\alpha''W'' + \beta''\mathbf{I}_C).$$

This implies

$$(1-\vartheta)W'' \le \alpha''W'' + \beta''\mathbf{I}_C - \vartheta\pi\rho p(\alpha''W'' + \beta''\mathbf{I}_C),$$

for every $\pi \in \mathbf{E}$, $\rho \in \mathbf{F}$. Hence,

$$\begin{aligned} Q_{\vartheta,\pi,\rho}W'' &= \sum_{n=0}^{\infty}(1-\vartheta)\vartheta^n(\pi\rho p)^n W'' \\ &\le \sum_{n=0}^{\infty}\vartheta^n(\pi\rho p)^n(\alpha''W'' + \beta''\mathbf{I}_C) - \sum_{n=1}^{\infty}\vartheta^n(\pi\rho p)^n(\alpha''W'' + \beta''\mathbf{I}_C) \\ &= \alpha''W'' + \beta''\mathbf{I}_C. \end{aligned} \tag{8}$$

We choose $\vartheta' \in (\vartheta, 1)$ and set

$$\eta := \max\left\{\frac{\beta''}{\mu(\mathbf{X})}, \frac{\beta'}{\vartheta' - \vartheta}\right\}, \quad \lambda' := \frac{\alpha'' + \eta}{1+\eta}, \quad \lambda := \max\{\lambda', \vartheta'\}.$$

It follows $\alpha'' < \lambda' \le \lambda < 1$ and $\lambda' - \alpha'' = (1-\lambda')\eta$. Hence,

$$(\lambda - \alpha'')W'' \ge \lambda' - \alpha'' \ge (1-\lambda')\eta \ge (1-\lambda)\eta. \tag{9}$$

We put $V := W'' + \eta$. Obviously, $V \ge W'' \ge 1$ and $V \ge \eta$. Then it follows

$$\begin{aligned} Q_{\vartheta,\pi,\rho}V &= Q_{\vartheta,\pi,\rho}W'' + \eta \\ &\le \alpha''W'' + \mathbf{I}_C \cdot \beta'' + \eta \\ &\le \alpha''W'' + \mathbf{I}_C \cdot \eta\mu(\mathbf{X}) + \eta \\ &\le \alpha''W'' + \mathbf{I}_C \cdot \mu V + \eta \\ &\le \alpha''W'' + \mathbf{I}_C \cdot \mu V + (\lambda - \alpha'')W'' + \lambda\eta \text{ (see (9))} \\ &= \lambda(W'' + \eta) + \mathbf{I}_C \cdot \mu V \\ &= \lambda V + \mathbf{I}_C \cdot \mu V. \end{aligned}$$

Hence, (4) is proven.

From $\eta \ge \beta'/(\vartheta' - \vartheta)$ it follows

$$\vartheta'\eta \ge \vartheta\eta + \beta'. \tag{10}$$

Then

$$\begin{aligned} \vartheta pV &= \vartheta pW'' + \vartheta\eta \\ &\le \alpha'\vartheta W'' + \beta' + \vartheta\eta \text{ (see (7))} \end{aligned}$$

$$
\begin{aligned}
&\leq \alpha' \vartheta W'' + \vartheta' \eta \text{ (see (10))} \\
&\leq \vartheta'(W'' + \eta) \\
&= \vartheta' V \\
&\leq \lambda V.
\end{aligned}
$$

Hence, (5) is also proven. □

4 Properties of Stationary Strategy Pairs

For a function $u : \mathbf{X} \to \mathbb{R}$ we put $\|u\|_V := \sup_{x \in \mathbf{X}}(|u(x)|/V(x))$. Furthermore, we denote by $\mathfrak{V}$ the set of all $\sigma_{\mathbf{X}}$-universally measurable functions u with $\|u\|_V < \infty$. In the following we will assume that on $\mathfrak{V}$ the metric is given which is induced by the weighted supremum norm $\|\cdot\|_V$. Then $\mathfrak{V}$ is complete.

Lemma 4.1. *Let $\Pi = (\pi_n)$, $P = (\rho_n)$. Then*

$$
\left\| \sup_{n \in \mathbb{N}, \Pi \in \mathbf{E}^\infty, P \in \mathbf{F}^\infty} \pi_0 \rho_0 p \cdots \pi_n \rho_n p V \right\|_V < \infty.
$$

Proof. From Assumption 3.1(b) it follows

$$
\pi_0 \rho_0 p \cdots \pi_n \rho_n p W \leq \alpha^{n+1} W + \frac{1}{1-\alpha} \beta.
$$

By Lemma 3.1 there is a constant d with

$$
\begin{aligned}
\pi_0 \rho_0 p \cdots \pi_n \rho_n p V &\leq \pi_0 \rho_0 p \cdots \pi_n \rho_n p W + d \\
&\leq \alpha^{n+1} W + d + \frac{1}{1-\alpha} \beta \leq \alpha^{n+1} V + d + \frac{1}{1-\alpha} \beta
\end{aligned}
$$

for all $n \in \mathbb{N}$. This implies the statement. □

Let T_w be the operator given by

$$
T_w u(x, a, b) := (1 - \vartheta)(\vartheta k(x, a, b) + w(x)) + \vartheta p u(x, a, b)
$$

for all $u \in \mathfrak{V}$, $x \in \mathbf{X}$, $a \in \mathbf{A}$, $b \in \mathbf{B}$. We note that T_w has essentially the same structure as the cost operator T used in stochastic dynamic programming and stochastic game theory (see (3)). This implies that some of our proofs are very similar to known proofs. Therefore, we restrict ourselves to only a few remarks in these cases. (A very good exposition of basic ideas and recent developments in stochastic dynamic programming can be found in the books of Hernández-Lerma and Lasserre [1], [2].)

Obviously,

$$
T_w u = (1 - \vartheta) \vartheta T \left(\frac{u}{1 - \vartheta} \right) + (1 - \vartheta) w. \tag{11}
$$

Lemma 4.2. *Let $w \in \mathfrak{V}$ arbitrary, $\pi \in \mathbf{E}$, $\rho \in \mathbf{F}$. Then the functional equation*

$$u = \pi\rho T_w u \tag{12}$$

has a unique solution $u_w \in \mathfrak{V}$ and it holds for $u_w := S_{\pi\rho}w$,

$$S_{\pi\rho}w = \lim_{n\to\infty}(\pi\rho T_w)^n u = (1-\vartheta)\sum_{n=0}^{\infty}\vartheta^n(\pi\rho p)^n(\vartheta\pi\rho k + w), \tag{13}$$

for every $u \in \mathfrak{V}$.

Proof. We note that $\pi\rho T_w \mathfrak{V} \subseteq \mathfrak{V}$. From (5) it follows that $\pi\rho T_w$ is contracting on $\mathfrak{V}$ with modulus λ. The rest of the proof follows by Banach's Fixed Point Theorem. □

We can consider $S_{\pi\rho}$ as an operator $S_{\pi\rho}\colon \mathfrak{V} \to \mathfrak{V}$. Let $S_{\gamma,\pi,\rho}$ be the operator defined by

$$S_{\gamma,\pi,\rho}w := -(1-\mathbf{I}_C)\gamma + S_{\pi\rho}w - \mathbf{I}_C\mu w \tag{14}$$

for $\pi \in \mathbf{E}$, $\rho \in \mathbf{F}$, $w \in \mathfrak{V}$. The following lemma gives some properties of this operator.

Lemma 4.3.
(a) $S_{\gamma,\pi,\rho}\mathfrak{V} \subseteq \mathfrak{V}$.
(b) $S_{\gamma,\pi,\rho}$ is isotonic.
(c) $S_{\gamma,\pi,\rho}$ is contracting.

Proof.
(a) is obvious.
(b) Using (13) we get

$$\begin{aligned} S_{\gamma,\pi,\rho}w &= -(1-\mathbf{I}_C)\gamma + (1-\vartheta)\sum_{n=0}^{\infty}\vartheta^n(\pi\rho p)^n(\vartheta\pi\rho k + w) - \mathbf{I}_C\mu w \\ &= -(1-\mathbf{I}_C)\gamma + (1-\vartheta)\sum_{n=0}^{\infty}\vartheta^{n+1}(\pi\rho p)^n\pi\rho k + (Q_{\vartheta,\pi,\rho} - \mathbf{I}_C\mu)w. \end{aligned} \tag{15}$$

From Assumption 3.1 (a) it follows the statement.
(c) By Lemma 3.1 and (15) we get for $u, v \in \mathfrak{V}$

$$\begin{aligned} |S_{\gamma,\pi,\rho}u - S_{\gamma,\pi,\rho}v| &= |(Q_{\vartheta,\pi,\rho} - \mathbf{I}_C\mu)(u-v)| \\ &\le (Q_{\vartheta,\pi,\rho} - \mathbf{I}_C\mu)V\|u-v\|_V \\ &\le \lambda V\|u-v\|_V. \end{aligned} \tag{16}$$

□

Lemma 4.4. *The operator $S_{\gamma,\pi,\rho}$ has in $\mathfrak{V}$ a unique fixed point $u_{\gamma,\pi,\rho}$. $\mu u_{\gamma,\pi,\rho}$ is continuous and non-increasing in γ.*

Proof. The existence and uniqueness of the fixed point follows from Lemma 4.3 by Banach's Fixed Point Theorem.

From $S_{\gamma,\pi,\rho}v \geq S_{\gamma',\pi,\rho}v$ for $\gamma \leq \gamma'$, and the isotonicity of $S_{\gamma,\pi,\rho}$ it follows $u_{\gamma,\pi,\rho} \geq u_{\gamma',\pi,\rho}$. Hence, $\mu u_{\gamma,\pi,\rho} \geq \mu u_{\gamma',\pi,\rho}$.

Furthermore, for arbitrary γ, γ'

$$\begin{aligned}|u_{\gamma,\pi,\rho} - u_{\gamma',\pi,\rho}| &= |(1 - \mathbf{I}_C)(\gamma' - \gamma) + (Q_{\vartheta,\pi,\rho} - \mathbf{I}_C\mu)(u_{\gamma,\pi,\rho} - u_{\gamma',\pi,\rho})| \\ &\leq |\gamma - \gamma'|V + \lambda\|u_{\gamma,\pi,\rho} - u_{\gamma',\pi,\rho}\|_V V.\end{aligned}$$

Hence,

$$\|u_{\gamma,\pi,\rho} - u_{\gamma',\pi,\rho}\|_V \leq |\gamma - \gamma'| + \lambda\|u_{\gamma,\pi,\rho} - u_{\gamma',\pi,\rho}\|_V$$

and

$$|\mu u_{\gamma,\pi,\rho} - \mu u_{\gamma',\pi,\rho}| \leq \|u_{\gamma,\pi,\rho} - u_{\gamma',\pi,\rho}\|_V \mu V \leq \frac{|\gamma - \gamma'|}{1-\lambda}\mu V.$$

□

Theorem 4.1. *There are $g = const$ and $v \in \mathfrak{V}$ with*

$$g + v = \pi\rho T v. \tag{17}$$

It holds

$$g = \Phi_{\pi^\infty\rho^\infty}.$$

Proof. From Lemma 4.4 it follows that there is a γ^* with $\gamma^* = \mu u_{\gamma^*,\pi,\rho}$. Hence,

$$\begin{aligned}u_{\gamma^*,\pi,\rho} &= S_{\gamma^*,\pi,\rho}u_{\gamma^*,\pi,\rho} \\ &= -(1 - \mathbf{I}_C)\gamma^* + S_{\pi\rho}u_{\gamma^*,\pi,\rho} - \mathbf{I}_C\mu u_{\gamma^*,\pi,\rho} \\ &= S_{\pi\rho}u_{\gamma^*,\pi,\rho} - \gamma^*.\end{aligned} \tag{18}$$

Let $w^* := u_{\gamma^*,\pi,\rho}$. If we put $w = w^*$ in (12) then we get

$$S_{\pi\rho}w^* = (1 - \vartheta)(\vartheta\pi\rho k + w^*) + \vartheta\pi\rho p S_{\pi\rho}w^*.$$

It follows by (18)

$$w^* + \gamma^* = (1 - \vartheta)(\vartheta\pi\rho k + w^*) + \vartheta\pi\rho p(w^* + \gamma^*).$$

Therefore,

$$\vartheta w^* + (1 - \vartheta)\gamma^* = (1 - \vartheta)\vartheta\pi\rho k + \vartheta\pi\rho p w^*.$$

For $g = \frac{\gamma^*}{\vartheta}$, $v = \frac{w^*}{1-\vartheta}$ we get (17). From (17) it follows

$$Ng = \sum_{n=0}^{N-1} (\pi\rho p)^n \pi\rho k + (\pi\rho p)^N v - v.$$

If we consider Lemma 4.1 we get

$$g = \lim_{N\to\infty} \frac{1}{N} \sum_{n=0}^{N-1} (\pi\rho p)^n \pi\rho k = \Phi_{\pi^\infty\rho^\infty}.$$

□

5 The Optimality Equation

In this section we use a further operator. Let $w : \mathbf{X} \times \mathbf{A} \times \mathbf{B} \to \mathbb{R}$ a $\overline{\sigma}_{\mathbf{X}\times\mathbf{A}\times\mathbf{B}}$-measurable function, such that $\pi\rho w$ exists for all $\pi \in \mathbf{E}$, $\rho \in \mathbf{F}$. Then we set

$$Lw(x) := \inf_{\pi\in\mathbf{E}} \sup_{\rho\in\mathbf{F}} \pi\rho w(x)$$

for all $x \in \mathbf{X}$.

Assumption 5.1. Let $\mathfrak{K} \subseteq \mathfrak{V}$ a complete subspace of the metric space $\mathfrak{V}$ with the following properties:

(a) $LT\mathfrak{K} \subseteq \mathfrak{K}$.
(b) $\nu_1 u_1 + \nu_2 u_2 \in \mathfrak{K}$ for $u_1, u_2 \in \mathfrak{K}$, $\nu_1, \nu_2 \in [0, \infty)$.
(c) $\nu \mathbf{I}_C \in \mathfrak{K}$ and $\nu \in \mathfrak{K}$ for all $\nu \in \mathbb{R}$.
(d) For every $u \in \mathfrak{K}$, $\varepsilon > 0$ there are decision rules $\pi_\varepsilon \in \mathbf{E}$, $\rho_\varepsilon \in \mathbf{F}$ with

$$\pi_\varepsilon \rho T u - \varepsilon \le \pi_\varepsilon \rho_\varepsilon T u \le \pi \rho_\varepsilon T u + \varepsilon$$

for all $\pi \in \mathbf{E}$, $\rho \in \mathbf{F}$.

Assumption 5.1 is satisfied if $\mathcal{M}$ is a Markov game with perfect information and **A** and **B** are denumerable, or if $\mathcal{M}$ is a Markov game with independent action choice and **A** and **B** are finite. We give a further assumption which implies Assumption 5.1. The proof of this implication is similarly to corresponding proofs in [5] or [16].

Assumption 5.2.

(a) $\mathcal{M}$ is a Markov game with independent action choice or perfect information.
(b) $\boldsymbol{A}$ and $\boldsymbol{B}$ are compact-valued, and $\boldsymbol{B}$ has a Castaing representation (that means, there is a sequence (g_n) of $\sigma_{\mathbf{X}}$-$\sigma_{\mathbf{B}}$-measurable maps such that $\{g_1, g_2, \dots\}$ is dense in $\boldsymbol{B}$ for every $x \in \mathbf{X}$).
(c) $k(x, \cdot)$ is lower semi-continuous for every $x \in \mathbf{X}$.
(d) $pu(x, \cdot)$ is lower semi-continuous for every $x \in \mathbf{X}$ and every Borel measurable $u \in \mathfrak{V}$.

In these cases $\mathfrak{K}$ is the set of all Borel measurable functions from $\mathfrak{V}$. Further conditions under which Assumption 5.1 is satisfied can be derived from the results in [7], for example.

We assume in this section that Assumption 5.1 is fulfilled. Then we get

$$\begin{aligned}\inf_{\pi\in\mathbf{E}}\sup_{\rho\in\mathbf{F}}\pi\rho Tu-\varepsilon &\leq \sup_{\rho\in\mathbf{F}}\pi_\varepsilon\rho Tu-\varepsilon\\ &\leq \pi_\varepsilon\rho_\varepsilon Tu\\ &\leq \inf_{\pi\in\mathbf{E}}\pi\rho_\varepsilon Tu+\varepsilon\\ &\leq \sup_{\rho\in\mathbf{F}}\inf_{\pi\in\mathbf{E}}\pi\rho Tu\\ &\leq \inf_{\pi\in\mathbf{E}}\sup_{\rho\in\mathbf{F}}\pi\rho Tu.\end{aligned}$$

It follows for $\varepsilon\to\infty$

$$\sup_{\rho\in\mathbf{F}}\inf_{\pi\in\mathbf{E}}\pi\rho Tu=\inf_{\pi\in\mathbf{E}}\sup_{\rho\in\mathbf{F}}\pi\rho Tu=LTu. \tag{19}$$

Hence,

$$|\pi_\varepsilon\rho_\varepsilon Tu-LTu|\leq\varepsilon.$$

It follows

$$\pi_\varepsilon\rho Tu-2\varepsilon\leq LTu\leq\pi\rho_\varepsilon Tu+2\varepsilon. \tag{20}$$

Lemma 5.1. *The functional equation*

$$\begin{aligned}u&=\inf_{\pi\in\mathbf{E}}\sup_{\rho\in\mathbf{F}}\{(1-\vartheta)(\vartheta\pi\rho k+w)+\vartheta\pi\rho pu\}\\ &=LT_wu=(1-\vartheta)\vartheta LT\left(\frac{u}{1-\vartheta}\right)+(1-\vartheta)w\end{aligned} \tag{21}$$

has for every $w\in\mathfrak{K}$ in $\mathfrak{K}$ a unique solution $u^*=:Sw$.

Proof. Let $w\in\mathfrak{K}$. Then it follows from Assumption 5.1 $LT_w\mathfrak{K}\subseteq\mathfrak{K}$. Because $\pi\rho T_w$ is contracting on $\mathfrak{V}$, it holds for $u,v\in\mathfrak{K}$

$$\pi\rho T_wu\leq\pi\rho T_wv+\lambda\|u-v\|_V V.$$

Since L is isotonic, it follows

$$LT_wu\leq LT_wv+\lambda\|u-v\|_V V.$$

Because u and v can be interchanged, we get that LT_w is also contracting. By Banach's Fixed Point Theorem it follows the statement. □

We consider S as an operator $S : \mathfrak{K} \to \mathfrak{K}$ according to Lemma 5.1.

Lemma 5.2. *For every $w \in \mathfrak{K}$ and $\varepsilon > 0$ there are $\pi_\varepsilon \in \mathbf{E}$, $\rho_\varepsilon \in \mathbf{F}$ with*

$$S_{\pi_\varepsilon,\rho} w - \varepsilon V \le Sw \le S_{\pi,\rho_\varepsilon} w + \varepsilon V \tag{22}$$

for all $\pi \in \mathbf{E}$, $\rho \in \mathbf{F}$. Furthermore, it holds

$$Sw = \inf_{\pi \in \mathbf{E}} \sup_{\rho \in \mathbf{F}} S_{\pi\rho} w. \tag{23}$$

Proof. Let $\varepsilon' := (1-\lambda)\varepsilon$. It follows from Assumption 5.1(d) that there are $\pi_\varepsilon \in \mathbf{E}$, $\rho_\varepsilon \in \mathbf{F}$ with

$$\begin{aligned}\pi_\varepsilon \rho T\left(\frac{u_w}{1-\vartheta}\right) - \frac{\varepsilon'}{\vartheta(1-\vartheta)} V &\le LT\left(\frac{u_w}{1-\vartheta}\right)\\ &\le \pi\rho_\varepsilon T\left(\frac{u_w}{1-\vartheta}\right) + \frac{\varepsilon'}{\vartheta(1-\vartheta)} V,\end{aligned} \tag{24}$$

where $u_w = Sw$ (see (20). Hence,

$$\pi_\varepsilon \rho T_w u_w - \varepsilon' V \le LT_w u_w = u_w \le \pi\rho_\varepsilon T_w u_w + \varepsilon' V, \tag{25}$$

for all $\pi \in \mathbf{E}$, $\rho \in \mathbf{F}$. Assume that

$$(\pi_\varepsilon \rho T_w)^n u_w - \frac{\varepsilon'}{1-\lambda} V \le u_w \le (\pi\rho_\varepsilon T_w)^n u_w + \frac{\varepsilon'}{1-\lambda} V \tag{26}$$

for $n \in \mathbb{N}$. Then it follows from (25)

$$\begin{aligned}u_w &\le \pi\rho_\varepsilon T_w \left((\pi\rho_\varepsilon T_w)^n u_w + \frac{\varepsilon'}{1-\lambda} V\right) + \varepsilon' V\\ &\le (\pi\rho_\varepsilon T_w)^{n+1} u_w + \frac{\lambda\varepsilon'}{1-\lambda} V + \varepsilon' V\\ &= (\pi\rho_\varepsilon T_w)^{n+1} u_w + \frac{\varepsilon'}{1-\lambda} V.\end{aligned} \tag{27}$$

Analogously,

$$u_w \ge (\pi_\varepsilon \rho T_w)^{n+1} u_w - \frac{\varepsilon'}{1-\lambda} V. \tag{28}$$

From (27) and (28) it follows by mathematical induction that (26) holds for all $n \in \mathbb{N}$. For $n \to \infty$ we get (22), and (23) follows analogously to (19). □

We define a new operator S_γ by

$$S_\gamma w := -(1 - \mathbf{I}_C)\gamma + Sw - \mathbf{I}_C \mu w$$

for $w \in \mathfrak{K}$, $\gamma \in \mathbb{R}$. The following lemma gives some properties of this operator.

Lemma 5.3.

(a) $S_\gamma \mathfrak{K} \subseteq \mathfrak{K}$.

(b) S_γ *is isotonic.*

(c) S_γ *is contracting with modulus* λ.

(d) S_γ *has in* $\mathfrak{K}$ *a unique fixed point* v_γ. *It holds* $\lim_{n\to\infty}(S_\gamma)^n u = v_\gamma$ *for every* $u \in \mathfrak{K}$. v_γ *is isotonic and continuous in* γ.

Proof.

(a) is obvious.

(b) From (14) and (23) it follows

$$S_\gamma w = \inf_{\pi\in\mathbf{E}} \sup_{\rho\in\mathbf{F}} S_{\gamma,\pi,\rho} w.$$

By Lemma 4.3 we get the statement.

(c) Let $w', w'' \in \mathfrak{K}$. By Lemma 5.2 it follows that for every $\varepsilon > 0$ there are $\pi''_\varepsilon \in \mathbf{E}$, $\rho'_\varepsilon \in \mathbf{F}$, with

$$Sw' \le S_{\pi,\rho'_\varepsilon} w' + \varepsilon V$$
$$Sw'' \ge S_{\pi''_\varepsilon,\rho} w'' - \varepsilon V$$

for all $\pi \in \mathbf{E}$, $\rho \in \mathbf{F}$. Hence,

$$\begin{aligned} S_\gamma w' - S_\gamma w'' &= -(1-\mathbf{I}_C)\gamma + Sw' - \mathbf{I}_C \mu w' \\ &\quad - (-(1-\mathbf{I}_C)\gamma + Sw'' - \mathbf{I}_C \mu w'') \\ &\le -(1-\mathbf{I}_C)\gamma + S_{\pi''_\varepsilon,\rho'_\varepsilon} w' - \mathbf{I}_C \mu w' \\ &\quad - (-(1-\mathbf{I}_C)\gamma + S_{\pi''_\varepsilon,\rho'_\varepsilon} w'' - \mathbf{I}_C \mu w'') + 2\varepsilon V \\ &= S_{\gamma,\pi''_\varepsilon,\rho'_\varepsilon} w' - S_{\gamma,\pi''_\varepsilon,\rho'_\varepsilon} w'' + 2\varepsilon V \\ &\le \lambda V \|w' - w''\|_V + 2\varepsilon V, \end{aligned}$$

since $S_{\gamma,\pi''_\varepsilon,\rho'_\varepsilon}$ is contracting (see Lemma 4.3). For $\varepsilon \to 0$ we get

$$S_\gamma w' - S_\gamma w'' \le \lambda V \|w' - w''\|_V. \tag{29}$$

Because w' and w'' can be interchanged, we get the statement.

(d) The existence of a unique fixed point $v_\gamma \in \mathfrak{K}$ and $\lim_{n\to\infty}(S_\gamma)^n u = v_\gamma$ for every $u \in \mathfrak{K}$ follows from Banach's Fixed Point Theorem. For $\gamma' \le \gamma$ it holds

$$S_\gamma w \le S_{\gamma'} w = S_\gamma w + (1-\mathbf{I}_C)(\gamma - \gamma') \le S_\gamma w + (\gamma - \gamma')V.$$

Assume that for $n > 1$

$$S_\gamma^{n-1} v_{\gamma'} \le v_{\gamma'} \le S_\gamma^{n-1} v_{\gamma'} + \frac{\gamma - \gamma'}{1-\lambda} V.$$

Then it follows

$$
\begin{aligned}
S_\gamma^n v_{\gamma'} &\le S_{\gamma'} S_\gamma^{n-1} v_{\gamma'} \le S_{\gamma'} v_{\gamma'} = v_{\gamma'} \le S_{\gamma'} \left(S_\gamma^{n-1} v_{\gamma'} + \frac{\gamma - \gamma'}{1-\lambda} V \right) \\
&\le S_\gamma \left(S_\gamma^{n-1} v_{\gamma'} + \frac{\gamma - \gamma'}{1-\lambda} V \right) + (\gamma - \gamma')V \\
&\le S_\gamma^n v_{\gamma'} + \frac{\lambda(\gamma - \gamma')}{1-\lambda} V + (\gamma - \gamma')V \text{ (see (29))} \\
&= S_\gamma^n v_{\gamma'} + \frac{\gamma - \gamma'}{1-\lambda} V.
\end{aligned}
$$

Hence, by mathematical induction we get that this inequality holds for all $n \in \mathbb{N}$. For $n \to \infty$ it follows

$$v_\gamma \le v_{\gamma'} \le v_\gamma + \frac{\gamma - \gamma'}{1-\lambda} V.$$

This implies the rest of the statement. □

Theorem 5.1.
(a) The optimality equation

$$g + v = LTv \tag{30}$$

has a solution (g^*, v^*) with $g^* = const$ and $v^* \in \mathfrak{K}$.
(b) Let $(\tilde{g}, \tilde{v})$ be a solution of (30) with $\tilde{g} = const$ and $\tilde{v} \in \mathfrak{K}$. Then it holds

$$\tilde{g} = g^* = \inf_{\Pi \in \mathbf{E}^\infty} \sup_{P \in \mathbf{F}^\infty} \Phi_{\Pi P}$$

and

$$\tilde{v} = v^* + const.$$

(c) For every $\varepsilon > 0$ there is an ε-optimal stationary strategy pair.

Proof. From Lemma 5.3 it follows that μv_γ is non-increasing in γ. Therefore, there is a unique γ^* with $\gamma^* = \mu v_{\gamma^*}$.

$$
\begin{aligned}
v_{\gamma^*} &= S_{\gamma^*} v_{\gamma^*} \\
&= -(1 - \mathbf{I}_C)\gamma^* + S v_{\gamma^*} - \mathbf{I}_C \mu v_{\gamma^*} \\
&= S v_{\gamma^*} - \gamma^*.
\end{aligned} \tag{31}
$$

Let $w^* := v_{\gamma^*}$. If we put $w = w^*$ in (21) then we get

$$Sw^* = L((1 - \vartheta)(\vartheta k + w^*) + \vartheta p S w^*).$$

It follows by (31)

$$w^* + \gamma^* = L((1-\vartheta)(\vartheta k + w^*) + \vartheta p(w^* + \gamma^*)).$$

Therefore,

$$\vartheta w^* + (1-\vartheta)\gamma^* = L((1-\vartheta)\vartheta k + \vartheta p w^*).$$

For $g^* = \frac{\gamma^*}{\vartheta}$, $v^* = \frac{w^*}{1-\vartheta}$ we get (a).

Let $(\tilde{g}, \tilde{v})$ be a solution of (30) with $\tilde{g} = const$ and $\tilde{v} \in \mathfrak{K}$. From Assumption 5.1 it follows that for every $\varepsilon > 0$ there are $\pi_\varepsilon \in \mathbf{E}$, $\rho_\varepsilon \in \mathbf{F}$, with

$$\pi_\varepsilon \rho_n T\tilde{v} - \varepsilon - \tilde{g} \le \tilde{v} \le \pi_n \rho_\varepsilon T\tilde{v} + \varepsilon - \tilde{g},$$

for all $\Pi = (\pi_n) \in \mathbf{E}^\infty$, $P = (\rho_n) \in \mathbf{F}^\infty$. It follows

$$\begin{aligned} \pi_\varepsilon \rho_0 T \pi_\varepsilon \rho_1 T \cdots \pi_\varepsilon \rho_N T\tilde{v} - (N+1)(\tilde{g} + \varepsilon) &\le \tilde{v} \\ &\le \pi_0 \rho_\varepsilon T \pi_1 \rho_\varepsilon T \cdots \pi_N \rho_\varepsilon T\tilde{v} + (N+1)(-\tilde{g} + \varepsilon). \end{aligned}$$

Using Lemma 4.1, for $N \to \infty$, we get in a usual way

$$\Phi_{\Pi\rho_\varepsilon^\infty} - \varepsilon \le \tilde{g} \le \Phi_{\pi_\varepsilon^\infty P} + \varepsilon,$$

for all $\Pi \in \mathbf{E}^\infty$, $P \in \mathbf{F}^\infty$. This implies

$$g^* = \tilde{g} = \inf_{\Pi \in \mathbf{E}^\infty} \sup_{P \in \mathbf{F}^\infty} \Phi_{\Pi P},$$

and the ε-optimality of $(\pi_\varepsilon^\infty, \rho_\varepsilon^\infty)$. Hence, (c) and the first part of (b) are proven.

Let $\tilde{w} := (1-\vartheta)\tilde{v} + c$ with $c := (1/\mu(\mathbf{X}))(\vartheta g^* - (1-\vartheta)\mu\tilde{v})$. Then it holds

$$\mu\tilde{w} = (1-\vartheta)\tilde{v} + \mu(\mathbf{X}) = \vartheta g^* = \gamma^*.$$

Furthermore, by (30) we get

$$\frac{\gamma^*}{\vartheta} + \frac{\tilde{w}}{1-\vartheta} = LT\left(\frac{\tilde{w}}{1-\vartheta}\right).$$

It follows

$$(1-\vartheta)\gamma^* + \vartheta\tilde{w} = L(\vartheta(1-\vartheta)k + \vartheta p\tilde{w}).$$

For $u := \gamma^* + \tilde{w}$ we get

$$u = (1-\vartheta)\vartheta LT\left(\frac{u}{1-\vartheta}\right) + (1-\vartheta)\tilde{w}.$$

Since $S\tilde{w}$ is the unique solution of this functional equation (see Lemma 5.1), it follows

$$\tilde{w} = S\tilde{w} - \gamma^* = S_{\gamma^*}\tilde{w}.$$

The solution of this functional equation is also unique (see Lemma 5.3). Hence, $\tilde{w} = v_{\gamma^*}$ and

$$\tilde{v} = v^* + \frac{c}{1-\vartheta}. \qquad \square$$

Acknowledgement

The author thanks A. S. Nowak and the anonymous referee for many helpful comments.

REFERENCES

[1] Hernández-Lerma, O.; Lasserre, J. B.: *Discrete-Time Markov Control Processes: Basic Optimality Criteria.* Springer-Verlag, New York, 1996

[2] Hernández-Lerma, O.; Lasserre, J. B.: *Further Topics on Discrete-Time Markov Control Processes.* Springer-Verlag, New York, 1999

[3] Hernández-Lerma, O.; Lasserre, J. B.: Zero-sum stochastic games in Borel spaces: average payoff criteria. *SIAM J. Control Optim.*, 39, 1520–1539, (2000)

[4] Jaśkiewicz, A.; Nowak, A. S.: On the optimality equation for zero-sum ergodic stochastic games. *Math. Methods Oper. Res.*, 54, 291–301, (2001)

[5] Küenle, H.-U.: *Stochastische Spiele und Entscheidungsmodelle.* Teubner-Texte zur Mathematik 89. Teubner-Verlag, Leipzig, 1986

[6] Küenle, H.-U.: Stochastic games with complete information and average cost criterion. *Advances in Dynamic Games and Applications.* Vol 5, pp 325–338, Birkhäuser, Boston, 2000

[7] Küenle, H.-U.: Equilibrium strategies in stochastic games with additive cost and transition structure. Internat. *Game Theory Rev.*, 1, 131–147, (1999)

[8] Küenle, H.-U.: On multichain Markov games. Advances in dynamic games and applications. *Ann. Int. Soc. Dynam. Games*, Vol. 6, pp 147–163, Birkhäuser, Boston, 2001

[9] Küenle, H.-U., Schurath, R.: The optimality equation and ε-optimal strategies in Markov games with average reward criterion. *Math. Methods Oper. Res.*, 56, 451–471, (2002)

[10] Maitra, A.; Sudderth, W.: Borel stochastic games with limsup payoff. *Ann. Probab.*, 21, 861–885, (1993)

[11] Maitra, A.; Sudderth, W.: Finitely additive and measurable stochastic games. *Int. J. Game Theory*, 22, 201–223, (1993)

[12] Maitra, A.; Sudderth, W.: Finitely additive stochastic games with Borel measurable payoffs. *Int. J. Game Theory*, 27, 257–267, (1998)

[13] Meyn, S. P.; Tweedie, R. L.: *Markov Chains and Stochastic Stability.* Communication and Control Engineering Series. Springer-Verlag, London, 1993

[14] Nowak, A. S.: Zero-sum average payoff stochastic games with general state space. *Games and Econ. Behavior*, 7, 221–232, (1994)

[15] Nowak, A. S.: Optimal strategies in a class of zero-sum ergodic stochastic games. *Math. Methods Oper. Res.*, 50, 399–420, (1999)

[16] Rieder, U.: Average optimality in Markov games with general state space. *Proc. 3rd International Conf. on Approximation and Optimization*, Puebla, 1995

[17] Thiemann, J. G. F.: *Analytic spaces and dynamic programming.* CWI Tract, Vol. 14. Centrum voor Wiskunde en Informatica, Amsterdam, 1984

A Simple Two-Person Stochastic Game with Money

Piercesare Secchi
Dipartimento di Matematica
Politecnico di Milano
Piazza Leonardo da Vinci 32
I-20133 Milano, Italia
secchi@mate.polimi.it

William D. Sudderth
School of Statistics
University of Minnesota
Minneapolis, MN 55455
bill@stat.umn.edu

Abstract

Two players hold money and bid each day for a nondurable consumer good whose worth to each player is measured by a concave utility function. The money recirculates to the players according to a rule that treats them symmetrically. When the total reward is discounted and the discount factor is small, there is a Nash equilibrium in which the players make large bids. For a discount factor close to one and also for a game with long run average reward, there is a Nash equilibrium with small bids.

1 Introduction

Each player holds an integral amount of money with the total amount of money equal to a fixed quantity M. Every day, one unit of a nondurable commodity is brought to a market and each player bids some integral part of his or her money for the good. Portions of the good are awarded to the players based on their bids and they consume their portion on the same day.

A player's bid may or may not be accepted in payment for a portion of the good. If the bid is accepted, the player pays the bid and the sum of these payments on a given day is the value or the price of the good on that day. The value is returned to the players at the end of the day with each player receiving a random share having the same distribution. One interpretation is that the players own shares of the market that are stochastically equivalent; another is that they own stochastically equivalent shares of the good which they bring to the market.

The utility of the good to the players is measured by a concave utility function $u : [0, 1] \to \Re$ such that $u(0) = 0$. If a player receives y_n units of the good on day n for $n = 1, 2, \ldots$, then the player's total discounted reward is

$$\sum_{n=1}^{\infty} \beta^{n-1} u(y_n),$$

where $0 < \beta < 1$ is a discount factor, and the player's long run average reward is

$$\limsup_{n\to\infty} \frac{1}{n} \sum_{k=1}^{n} u(y_k).$$

We will consider below both discounted and average reward games.

We also consider two schemes for apportioning the good to the players based on their bids. In the first version, called "winner-takes-all", the player who makes the larger bid receives the entire good or if their bids are the same positive number, they each receive an equal share. The second version is called "proportional rewards" because in it each player receives a fraction of the good in proportion to his or her bid.

The strategic dilemma of a player of these games was aptly described by Shubik and Whitt [10], namely, whether "to spend more now to get more real goods now or to spend less now to get more real goods in the future." It is natural to expect that a wise player would save less when future values are discounted by a small β and save more when β is near 1 or when the long run average reward payoff is used. We will construct pure strategy Nash equilibria that seem to justify this intuition. We focus here on two person games, but several, and perhaps all, of our results generalize to n person games.

Our work stems from that of Shubik and Whitt [10], who, with the aim of providing satisfactory connections between theories in macroeconomics and microeconomics, introduced a non-stochastic proportional reward game in which each player receives every day a fixed, nonrandom, fraction of the current value of the good. More recently Karatzas, Shubik and Sudderth [4,5] and Geanakoplos, Karatzas, Shubik and Sudderth [3] have studied proportional reward games with a continuum of players and with borrowing and lending. Constant sum versions of the games studied here were considered in two earlier papers [8,9]. In [8], we introduced a constant sum, winner-takes-all game with discounted rewards where money is a finite but continuous quantity; we proved the existence of the value for the game and found good strategies for the players when the discount factor is less than or equal to 2/3. In [9], we assumed money to be finite and discrete, and we focused on constant sum versions of the winner-takes-all game and the proportional rewards game both with discounted and long run average rewards; we found the value of our games and optimal strategies for small values of the discount parameter β as well as for β near 1 and long run average reward.

In the next section, the games to be studied are defined and notation and terminology are introduced. In Section 3 we show that "bold" strategies for each player form a Nash equilibrium when β is small, and in Section 4 we show that "timid" strategies are Nash for β near 1. In Section 5 we introduce a slight change in the rules for apportioning the good – namely, each player receives half of the good when both bid nothing. This leads to asymmetric equilibria when β is near 1. In the final section we study the average reward game.

2 Formulation

Let M be a positive integer corresponding to the total amount of money held by the two players who are denoted I and II. The *state space* is $S = \{0, 1, \ldots, M\}$ and a state $x \in S$ will represent the cash held by player I at some stage of the game whereas $M - x$ corresponds to that held by player II. Actions represent bids and, for each $x \in S$, the action sets for players I and II, respectively, are $A_x = \{0, 1, \ldots, x\}$ and $B_x = \{0, 1, \ldots, M - x\}$.

Let $Z = \{(x, a, b) : x \in S, a \in A_x, b \in B_x\}$. To complete the definition of the stochastic game, we must specify a *daily reward function* $r : Z \to \Re$, a *law of motion* q that assigns to each triple $(x, a, b) \in Z$ a probability distribution defined on S, and a discount factor $\beta \in (0, 1)$ if the game is discounted. We will consider two versions called *winner-takes-all* and *proportional rewards*. In both versions, the players have the same utility function $u : [0, 1] \to [0, \infty)$ for the good. The function u is assumed to be concave, and strictly increasing with $u(0) = 0$.

In the winner-takes-all game, the daily reward function for player I is

$$r(x, a, b) = \begin{cases} u(1) & \text{if} \quad a > b, \\ u(1/2) & \text{if} \quad a = b > 0, \\ 0 & \text{if} \quad a < b \text{ or } a = b = 0, \end{cases} \tag{1}$$

whereas the daily reward function for player II is defined symmetrically as

$$\tilde{r}(x, a, b) = r(M - x, b, a),$$

for every $(x, a, b) \in Z$. Thus, if $a > b$, player I receives all the good which has price $p = a$; if $a < b$, player II receives all the good at price $p = b$; and, if $a = b > 0$, each player receives half the good and the price is $p = a + b$.

In the proportional-rewards game, the reward function for player I is

$$r(x, a, b) = \begin{cases} u(a/(a + b)) & \text{if} \quad a + b > 0, \\ 0 & \text{if} \quad a = b = 0, \end{cases} \tag{2}$$

and the reward function for player II is defined symmetrically as before. The price of the good in this game is the total bid $p = a + b$.

The law of motion can be viewed as a means of redistributing the money spent for the good, that is, the price p, to the players. Suppose that, for each possible

$p = 0, 1, \ldots, M$, there is a probability distribution π_p on $\{0, 1, \ldots, p\}$ for the amount that will go to each of the players. Let $Y^{(p)}$ be a random variable with distribution π_p. If $Y^{(p)}$ represents the amount for player I, then $p - Y^{(p)}$ represents that for player II. We assume that $Y^{(p)}$ and $p - Y^{(p)}$ have the same distribution. For example, $Y^{(p)}$ could be the uniform distribution on $\{0, 1, \ldots, p\}$, or binomial $(p, 1/2)$, or equal to p or 0 with probability 1/2 each. We will also assume that $Y^{(p)}$ is stochastically increasing in p. That is, $Y^{(p)}$ is stochastically smaller than $Y^{(p+1)}$ for $p = 0, 1, \ldots, M - 1$. This assumption, which holds for the three special cases above, implies that larger bids are riskier.

Consider first the law of motion for the winner-takes-all game. Let X_1 be a random variable with distribution $q(x, a, b)$ and we can specify q by setting

$$X_1 = \begin{cases} x - a + Y^{(a)} & \text{if} \quad a > b, \\ x + Y^{(b)} & \text{if} \quad a < b, \\ x - a + Y^{(a+b)} & \text{if} \quad a = b. \end{cases} \tag{3}$$

Thus player I's new fortune equals his old fortune, less what he spends, plus his share of the proceeds from selling the good.

The rule takes a simpler form in the proportional-rewards game where we let

$$X_1 = x - a + Y^{(a+b)}, \tag{4}$$

for a random variable with distribution $q(x, a, b)$.

In either game play begins at a state $X_0 = x$ and proceeds in stages that we often call "days". If on day $n \geq 1$, the state of the game is $X_{n-1} = x_{n-1}$ and player I chooses action $a_n \in A_{x_{n-1}}$ while simultaneously player II chooses $b_n \in B_{x_{n-1}}$, then I receives the daily reward $r_n = r(x_{n-1}, a_n, b_n)$ and II receives $\tilde{r}_n = \tilde{r}(x_{n-1}, a_n, b_n)$. The game then moves to the new state X_n with probability distribution $q(x_{n-1}, a_n, b_n)$.

A strategy μ for player I (respectively, ν for player II) specifies the distribution of each action a_{n+1} (respectively, b_{n+1}) as a function of the partial history $(x, a_1, b_1, \ldots, a_n, b_n, x_n)$. Strategies for the two players together with the law of motion determine the distribution $P_{x,\mu,\nu}$ of the stochastic process $x, a_1, b_1, X_1, a_2, b_2, \ldots$ and the expected discounted rewards

$$v_\beta(x, \mu, \nu) = E_{x,\mu,\nu}\left(\sum_{n=1}^{\infty} \beta^{n-1} r_n\right), \tag{5}$$

$$\tilde{v}_\beta(x, \mu, \nu) = E_{x,\mu,\nu}\left(\sum_{n=1}^{\infty} \beta^{n-1} \tilde{r}_n\right),$$

for players I and II respectively. Likewise the expected average rewards are determined. (For an introduction to stochastic games, see [2] or [6]. A more advanced treatment is in [7].)

Let σ and τ be functions defined on S such that, for every $x \in S$, $\sigma(x)$ is a probability distribution on A_x and $\tau(x)$ is a probability distribution on B_x. The functions σ and τ are called *selectors* for players I and II respectively. The selectors σ and τ determine stationary strategies $\sigma^\infty(x)$ and $\tau^\infty(x)$ at each x; namely $\sigma^\infty(x)$ (respectively $\tau^\infty(x)$) uses the mixed action $\sigma(y)$ (respectively, $\tau(y)$) whenever y is the current state.

Associated with each stationary strategy $\sigma^\infty(x)$ for player I is a symmetric stationary strategy $\tilde{\sigma}^\infty(x)$ for player II where $\tilde{\sigma}(x) = \sigma(M - x)$. (Recall that when the state is x, player I has x units of money and player II has $M - x$.) It is well-known that there exists a Nash equilibrium in stationary strategies for a discounted stochastic game with finite state space, finite action sets, and finitely many players. For our games the strategies can be taken to be symmetric.

Theorem 2.1. *For either a discounted winner-takes-all game or a discounted proportional rewards game, there exists a selector σ for player I such that, for every $x \in S$, $(\sigma^\infty(x), \tilde{\sigma}^\infty(x))$ is a Nash equilibrium.*

Proof. The proof follows that of Theorem 1 in [11] where the Brower fixed point theorem is applied to a certain mapping $\phi : \Pi \to \Pi$ where Π is the compact, convex set of all pairs of selectors $\{(\sigma(\cdot), \tau(\cdot))\}$ considered as vectors in a Euclidean space. It is easy to see that the subset of all symmetric pairs $\{(\sigma(\cdot), \tilde{\sigma}(\cdot))\}$ is also compact, convex and is mapped into itself by ϕ. The rest of the proof is the same as in [11]. □

Most of the particular strategies we consider will be pure stationary strategies $a^\infty(x)$ where $a(x) \in A_x$ for all $x \in S$. We write $\tilde{a}^\infty(x)$ for the symmetric pure stationary strategy where $\tilde{a}(x) = a(M - x)$.

Nash equilibria need not exist for average reward games. However, there is a deep result of Vieille [12,13] showing the existence of equilibrium payoff vectors in a strong sense for two person games. We will give simple direct arguments in Section 6 to identify Nash equilibria for the games studied here.

3 Bold Play for Small Beta

If the discount factor is small, then there is an urgency for the players to spend now rather than save for the future. Consider first the winner-takes-all game and call a player "rich" if his or her fortune exceeds $M/2$ and "poor" otherwise. On any given day, a rich player can obtain the entire good by bidding one more unit of money than the poor player has available. Now the larger the bid of the rich player, the more money he or she risks losing to the poor player. This suggests that the poor player should bid as much as possible in order to maximize the monetary losses of the rich player.

Define the *bold* selector b by

$$b(x) = \begin{cases} M - x + 1 & \text{if} \quad x > M/2, \\ x & \text{if} \quad x \leq M/2, \end{cases}$$

and call the pure stationary strategy $b^{\infty}(x)$ the *bold strategy* for player I at x. The symmetric strategy $\tilde{b}^{\infty}(x)$ is the *bold strategy* for player II at x.

Theorem 3.1. *For*

$$\beta \leq \frac{u(1) - u(1/2)}{2u(1) - u(1/2)} \tag{6}$$

and every $x \in S$, the bold strategies $(b^{\infty}(x), \tilde{b}^{\infty}(x))$ form a Nash equilibrium in the winner-takes-all game.

It suffices to show that $b^{\infty}(x)$ is an optimal strategy for I when II plays $\tilde{b}^{\infty}(x)$. (The optimality of $\tilde{b}^{\infty}(x)$ against $b^{\infty}(x)$ will then follow by symmetry.) So assume that player II plays the action $\tilde{b}(x)$ at each x. Thus player I faces a discounted dynamic programming problem with state space S, action sets $A_x, x \in S$, daily reward function $r(x, \cdot, \tilde{b}(x))$ and law of motion $q(x, \cdot, \tilde{b}(x))$. For $x \in S$ and $a \in A_x$, let $X(x, a)$ be a random variable with distribution $q(x, a, \tilde{b}(x))$: that is $X(x, a)$ is distributed like X_1 in (3) with $b = \tilde{b}(x)$. For functions $Q : S \to \Re$, define the operator L by

$$(LQ)(x) = \sup_{a \in A_x} [r(x, a, \tilde{b}(x)) + \beta E Q(X(x, a))].$$

Let V be the optimal reward function for the dynamic programming problem. Then, for $x \in S$, V satisfies the Bellman equation

$$V(x) = (LV)(x) = \sup_{a \in A_x} [r(x, a, \tilde{b}(x)) + \beta E V(X(x, a))],$$

and a strategy is optimal if and only if it uses actions that achieve the supremum in this equation. In particular, $b^{\infty}(x)$ is optimal for all x if and only if

$$V(x) = r(x, b(x), \tilde{b}(x)) + \beta E V(X(x, b(x))) \tag{7}$$

for all $x \in S$. Equality (7) will be established with the help of a lemma.

Lemma 3.1. *Let $Q : S \to [0, \infty)$ be non-decreasing and bounded above by $u(1)/(1 - \beta)$. Then*

(i) $(LQ)(x) = r(x, b(x), \tilde{b}(x)) + \beta E Q(X(x, b(x)))$, for $x \in S$,
(ii) LQ is non-decreasing and bounded above by $u(1)/(1 - \beta)$.

Proof. To prove (i), define

$$\psi(x,a) = r(x,a,\tilde{b}(x)) + \beta E Q(X(x,a))$$

$$= \begin{cases} u(1) + \beta E Q(x - a + Y^{(a)}) & \text{if } \ a > \tilde{b}(x), \\ u(1/2) + \beta E Q(x - \tilde{b}(x) + Y^{(2\tilde{b}(x))}) & \text{if } \ a = \tilde{b}(x), \\ u(0) + \beta E Q(x + Y^{(\tilde{b}(x))}) & \text{if } \ a < \tilde{b}(x), \end{cases} \tag{8}$$

for $x \in S$, $a \in A_x$. With this notation, (i) can be written as $(LQ)(x) = \psi(x, b(x))$.

We consider four cases.

Case 1: $x = M$.

For $a = 1, 2, \ldots, M$,

$$\psi(M,a) = u(1) + \beta E Q(M - a + Y^{(a)}) = u(1) + \beta E Q(M - Y^{(a)})$$

because $Y^{(a)}$ and $a - Y^{(a)}$ have, by assumption, the same distribution. Also $Y^{(1)}$ is stochastically smaller than each $Y^{(a)}$, $a \geq 2$, and Q is non-decreasing. Hence,

$$\psi(M,1) = \psi(M, b(M)) = \sup_{a \geq 1} \psi(M,a).$$

Also

$$\psi(M,0) = u(0) + \beta Q(M) \leq 0 + \frac{\beta}{1-\beta} u(1) \leq u(1) \leq \psi(M,1),$$

where the second inequality holds because

$$\beta \leq \frac{u(1) - u(1/2)}{2u(1) - u(1/2)} \leq \frac{1}{2}.$$

Thus $(LQ)(M) = \psi(M,1) = \psi(M, b(M))$.

Case 2: $M/2 < x < M$.

First we rewrite (8) as

$$\psi(x,a) = \begin{cases} u(1) + \beta E Q(x - a + Y^{(a)}) & \text{if } \ a \geq M - x + 1, \\ u(1/2) + \beta E Q(x - a + Y^{(2(M-x))}) & \text{if } \ a = M - x, \\ u(0) + \beta E Q(x + Y^{(M-x)}) & \text{if } \ a < M - x. \end{cases}$$

Because $Y^{(M-x+1)}$ is stochastically smaller than $Y^{(a)}$ for $a > M - x + 1$, we have, as in Case 1,

$$\psi(x, M - x + 1) = \sup_{a \geq M-x+1} \psi(x,a).$$

Moreover

$$\psi(x, M-x) \leq u\left(\frac{1}{2}\right) + \frac{\beta}{1-\beta}u(1) \leq u(1) \leq \psi(x, M-x+1),$$

where the second inequality is equivalent to our assumption (6). Finally, for $a < M - x$,

$$\psi(x, a) \leq u(0) + \frac{\beta}{1-\beta}u(1) \leq u(1) \leq \psi(x, M-x+1),$$

where the second inequality has already been established above in Case 1. So again we have $(LQ)(x) = \psi(x, M-x+1) = \psi(x, b(x))$.

Case 3: $x = M/2$.

Here (8) becomes

$$\psi\left(\frac{M}{2}, a\right) = \begin{cases} u(1/2) + \beta EQ(Y^{(M)}) & \text{if} \quad a = M/2, \\ u(0) + \beta EQ(M/2 + Y^{(a)}) & \text{if} \quad a < M/2. \end{cases}$$

For $a < M/2$,

$$\psi\left(\frac{M}{2}, a\right) \leq u(0) + \frac{\beta}{1-\beta}u(1) \leq u\left(\frac{1}{2}\right) \leq \psi\left(\frac{M}{2}, \frac{M}{2}\right)$$

where the second inequality holds because

$$\beta \leq \frac{u(1) - u(1/2)}{2u(1) - u(1/2)} = \frac{u(1) - u(1/2)}{u(1) + (u(1) - u(1/2))} \leq \frac{u(1/2) - u(0)}{u(1) + (u(1/2) - u(0))}$$

and $u(1) - u(1/2) \leq u(1/2) - u(0)$ by the concavity of u. Thus

$$(LQ)\left(\frac{M}{2}\right) = \psi\left(\frac{M}{2}, \frac{M}{2}\right) = \psi\left(\frac{M}{2}, b\left(\frac{M}{2}\right)\right).$$

Case 4: $0 \leq x < M/2$.

In this case

$$\psi(x, a) = u(0) + \beta Q(x + Y^{(M-x+1)})$$

is the same for all $a \in A_x$. So trivially, $(LQ)(x) = \psi(x, b(x))$.

This completes the proof of part (i).

To see that LQ is non-decreasing we look at four cases again.

Case 1_a: $x + 1 < M/2$.
By part (i),

$$\begin{aligned}(LQ)(x+1) &= r(x+1, b(x+1), \tilde{b}(x+1)) + \beta EQ(X(x+1, b(x+1))) \\ &= u(0) + \beta EQ(x+1+Y^{(\tilde{b}(x+1))}) \\ &= u(0) + \beta EQ(x+1+Y^{(x+2)}) \\ &\geq u(0) + \beta EQ(x+Y^{(x+1)}) \\ &= (LQ)(x).\end{aligned}$$

The inequality holds because $x + Y^{(x+1)}$ is stochastically smaller than $x + 1 + Y^{(x+2)}$.
Case 2_a: $x + 1 = M/2$.
In this case,

$$\begin{aligned}(LQ)(x+1) = (LQ)(M/2) &= u(1/2) + \beta EQ(Y^{(M)}) \\ &\geq u(1/2) \\ &\geq u(0) + \frac{\beta}{1-\beta}u(1) \\ &\geq u(0) + \beta EQ(x+Y^{(x+1)}) \\ &= (LQ)(x).\end{aligned}$$

The second inequality was established in Case 3 above.

Case 3_a: $x = M/2$.
Here

$$\begin{aligned}(LQ)\left(\frac{M}{2}+1\right) &= u(1) + \beta EQ(1+Y^{(M/2)}) \\ &\geq u(1) \\ &\geq u\left(\frac{1}{2}\right) + \frac{\beta}{1-\beta}u(1) \\ &\geq u(1/2) + \beta EQ(Y^{(M)}) \\ &= (LQ)(M/2).\end{aligned}$$

Case 4_a: $M > x > M/2$.
By definition of the operator L,

$$\begin{aligned}(LQ)(x+1) &\geq r(x+1, b(x), \tilde{b}(x+1)) + \beta EQ(X(x+1, b(x))) \\ &= u(1) + \beta EQ(x+1-b(x)+Y^{(b(x))})\end{aligned}$$

$$\geq u(1) + \beta E Q(x - b(x) + Y^{(b(x))})$$
$$= (LQ)(x),$$

where $r(x+1, b(x), \tilde{b}(x+1)) = u(1)$ since $b(x) = M - x + 1 = b(x+1) + 1 > \tilde{b}(x+1)$.

This completes the proof that LQ is non-decreasing.

To see that LQ is bounded above by $u(1)/(1-\beta)$, note that

$$(LQ)(x) = r(x, b(x), \tilde{b}(x)) + \beta E Q(X(x, b(x)))$$
$$\leq u(1) + \frac{\beta}{1-\beta} u(1) = \frac{u(1)}{1-\beta}.$$

□

Proof. (*of Theorem 3.1*). Apply Lemma 3.1 first with $Q \equiv 0$ and then use induction to conclude that $L^n 0$ satisfies (i) and (ii) for all $n \geq 1$. But $L^n 0$ is the optimal n-day reward for the dynamic programming problem and converges pointwise to V. Hence, V satisfies the hypothesis of Lemma 3.1 because $L^n 0$ does for all n. Equality (7) now follows from part (i) of the lemma and the Bellman equation. □

Consider next the proportional-rewards game. Because larger bids result in larger portions of the good, a player can guarantee the largest portion only by making the largest possible bid – except when he or she has all the money. This suggests a new form of bold play based on the selector

$$b_1(x) = \begin{cases} x & \text{if} \quad x < M, \\ 1 & \text{if} \quad x = M, \end{cases}$$

for player I.

Theorem 3.2. *If*

$$\beta \leq \frac{u\left(\frac{M-1}{M}\right) - u\left(\frac{M-2}{M-1}\right)}{u(1) + u\left(\frac{M-1}{M}\right) - u\left(\frac{M-2}{M-1}\right)},$$

then, for every $x \in S$, $(b_1^\infty(x), \tilde{b}_1^\infty(x))$ is a Nash equilibrium in the proportional rewards game.

Proof. Suppose $Q : S \to [0, \infty)$ is non-decreasing and bounded above by $u(1)/(1-\beta)$, and

$$(L_1 Q)(x) = \sup_{a \in A_x} [r(x, a, \tilde{b}_1(x)) + \beta E Q(X(x, a))]$$

where r is given by (2) and $X(x, a)$ is the X_1 of (4) with $b = \tilde{b}_1(x)$. We can prove the analogues of Lemma 3.1 (i) and (ii) by similar, but easier arguments. The rest of the proof is the same as that of Theorem 3.1. □

It is clear from the proofs that Theorems 3.1 and 3.2 can be extended to the situation where players I and II have different utility functions u_1, u_2 and discount factors β_1, β_2. For example, the conclusion of Theorem 3.1 is still true if

$$\beta_i \le \frac{u_i(1) - u_i(1/2)}{2u_i(1) - u_i(1/2)}$$

for $i = 1, 2$.

It would be interesting to know whether the Nash equilibria in Theorems 3.1 and 3.2 are unique.

4 Timid Play for Beta Near 1

If the discount factor is large, then consumption tomorrow is worth almost as much as consumption today. This suggests a conservative strategy based on small bids. Since a bid of zero brings a player no part of the good, the most conservative strategy of interest, which we call *timid*, is for a player to bid 1 whenever he or she has at least one unit of money. The corresponding selector for player I is

$$t(x) = \begin{cases} 1 & \text{if} \quad x \ge 1, \\ 0 & \text{if} \quad x = 0. \end{cases} \tag{9}$$

Indeed the symmetric stationary strategies $(t^\infty(x), \tilde{t}^\infty(x))$ form a Nash equilibrium in both versions of our game for β close to 1, if we rule out one extreme possibility.

Recall that when the total price of the good is 2, as in the case when both players bid 1, the cash income to each player is a random variable $Y^{(2)}$ with possible values $\{0, 1, 2\}$ and satisfying the symmetry condition that $Y^{(2)}$ and $2 - Y^{(2)}$ have the same distribution. Let $\pi_2 = (\pi_{2,0}, \pi_{2,1}, \pi_{2,2}) = (\pi_{2,0}, \pi_{2,1}, \pi_{2,0})$ be the distribution of $Y^{(2)}$. For the rest of this section, we assume that $Y^{(2)}$ is not concentrated at 1 or, equivalently, $\pi_{2,0} > 0$.

Theorem 4.1. *There exists a $\beta^* \in (0, 1)$ such that, for all $\beta \in [\beta^*, 1)$ and all $x \in S$, the timid strategies $(t^\infty(x), \tilde{t}^\infty(x))$ form a Nash equilibrium in the winner-takes-all game and in the proportional-rewards game.*

The proofs for the two games are similar and the proof for the proportional-rewards game is slightly more difficult. So we will give the details only for that case.

Let $T(x) = T_\beta(x)$ be the expected discounted reward to player I in the proportional-rewards game with initial state x when I plays $t^\infty(x)$ and II plays $\tilde{t}^\infty(x)$. The key to the proof is an estimate on the increments of T given in the following lemma.

Lemma 4.1. *Let* $0 < \alpha < 1$. *There exists* $\beta^* = \beta^*(\alpha)$ *in* $(0, 1)$ *such that, for all* $\beta \in [\beta^*, 1)$ *and all* $x \in \{1, \ldots, M\}$,

$$\frac{2\alpha\,[u(1) - u\,(1/2)]}{\beta} \leq T_\beta(x) - T_\beta(x-1) \leq \frac{2u(1/2)}{\beta}.$$

Proof. Suppose players I and II play timidly starting from x and that, in another realization of the same game, players I′ and II′ play timidly starting from $x - 1$. Assume that the games are coupled so that the position of player I is always one unit larger than that of player I′ until the day after the first time, t_1, that either player I reaches M or I′ reaches 0. Players I and I′ each receive 1/2 unit of the good on every day up to and including day t_1. On the next day after t_1, player I receives 1/2 unit more of the good than does I′. Indeed, when I reaches M, I receives in utility $u(1) - u(1/2)$ more than I′, and when I′ reaches 0, I receives $u(1/2)(\geq u(1) - u(1/2))$ more than I′. Furthermore, the games can be coupled so that on the day after t_1 the probability is $1/2$ that players I and I′ have the same fortune and the probability is also 1/2 that I remains one unit ahead. (For example, if I is at M and I′ at $M - 1$, the motion in both games can be modeled by an urn with $\pi_{2,0}$ white balls, $\pi_{2,1}/2$ pink balls, $\pi_{2,1}/2$ red balls, and $\pi_{2,0}$ black balls. The draw of a white ball moves I′ to $M - 2$ and I to $M - 1$; a pink leaves I′ at $M - 1$ and moves I to $M - 1$; a red leaves I′ at $M - 1$ and I at M; and a black moves I′ to M and leaves I at M.) If I remains one unit ahead of I′, then, just as before, they each receive 1/2 of the good each day until the next time, t_2, that one of the players reaches the boundary. And so on.

Thus, if t_n is the n-th time that one of the players reaches the boundary, then

$$\begin{aligned} T_\beta(x) - T_\beta(x-1) &\leq u\left(\frac{1}{2}\right) E\left[\beta^{t_1} + \frac{1}{2}\left[\beta^{t_2} + \frac{1}{2}\left[\beta^{t_3} + \ldots\right]\right]\right] \\ &\leq u\left(\frac{1}{2}\right) \sum_{n=0}^{\infty} \left(\frac{\beta}{2}\right)^n \\ &= \frac{2u\,(1/2)}{2-\beta} \\ &\leq \frac{2u\,(1/2)}{\beta}, \end{aligned}$$

which establishes the right hand inequality. To obtain the other inequality, first choose a positive integer n such that $1 - 2^{-n} > \alpha$. Then calculate

$$\begin{aligned} T_\beta(x) - T_\beta(x-1) &\geq \left[u(1) - u\left(\frac{1}{2}\right)\right] E\left[\beta^{t_1} + \frac{1}{2}\left[\beta^{t_2} + \frac{1}{2}\left[\beta^{t_3} + \ldots\right]\right]\right] \\ &\geq \left[u(1) - u\left(\frac{1}{2}\right)\right] E\left[\beta^{t_1} + \frac{1}{2}\beta^{t_2} + \ldots + \frac{1}{2^{n-1}}\beta^{t_n}\right]. \end{aligned}$$

As β grows to 1, the last quantity converges to

$$\left[u(1) - u\left(\frac{1}{2}\right)\right](2(1-2^{-n})) > \left[u(1) - u\left(\frac{1}{2}\right)\right](2\alpha).$$

Hence there is a $\beta^* = \beta^*(\alpha)$ such that the left hand inequality holds for all $\beta \in [\beta^*, 1)$. □

Proof. (*of Theorem 4.1 for the proportional rewards game*). First choose

$$\alpha = \frac{u\,(M/(M+1)) - u\,(1/2)}{u(1) - u\,(1/2)}$$

and let $\beta^* = \beta^*(\alpha)$ be as in the statement of Lemma 4.1. Then fix the discount factor $\beta \in [\beta^*, 1)$.

To prove that $(t^\infty(x), \tilde{t}^\infty(x))$ is a Nash equilibrium, it suffices to show that $t^\infty(x)$ is optimal for player I when player II plays $\tilde{t}^\infty(x)$. So fix the strategy of II to be $\tilde{t}^\infty(x)$ at every $x \in S$ and consider the discounted dynamic programming problem faced by player I. The function $T(x) = T_\beta(x)$ is the reward function for player I from the strategy $t^\infty(x)$. A necessary and sufficient condition for the optimality of $t^\infty(x)$ at every x is that T satisfy the Bellman equation; that is

$$T(x) = \sup_{a \in A_x} [r(x, a, \tilde{t}(x)) + \beta E T(X(x, a))] \tag{10}$$

where r is the reward function of (2) and $X(x, a)$ is distributed like the X_1 of (4) when $b = \tilde{t}(x)$. In other words we need to show that the supremum on the right side of (10) is attained when $a = t(x)$ for each $x \in S$.

To simplify notation, let

$$\begin{aligned}
\psi(x, a) &= r(x, a, \tilde{t}(x)) + \beta E T(X(x, a)) \\
&= \begin{cases} u(a/(a+1)) + \beta E T(x - a + Y^{(a+1)}) & \text{if} \quad x < M, \\ u(1) + \beta E T(M - a + Y^{(a)}) & \text{if} \quad x = M \text{ and } a > 0, \\ u(0) + \beta T(M) & \text{if} \quad x = M \text{ and } a = 0. \end{cases}
\end{aligned}$$

We consider four cases.

Case 1: $x = M$.

For $a \geq 1$, the random variable $-a + Y^{(a)}$ has the same distribution as $-Y^{(a)}$ and is stochastically smaller that $-Y^{(1)}$. Also the function T is increasing by Lemma 4.1. Hence,

$$\sup_{a \in \{1, \ldots, x\}} \psi(M, a) = \psi(M, 1).$$

Also,

$$\psi(M,0) = \beta T(M) < T(M) = \psi(M,1).$$

Case 2: $x = 0$.

This case is trivial since only the timid action $t(0) = 0$ is available at 0.

Case 3: $2 \le x \le M - 1$.

We have

$$\psi(x,a) = \begin{cases} u(a/(a+1)) + \beta ET(x - a + Y^{(a+1)}) & \text{if} \quad 1 \le a \le x, \\ \beta ET(x + Y^{(1)}) & \text{if} \quad a = 0. \end{cases}$$

We will first show that

$$\sup_{a \in \{1,2,\dots,x\}} \psi(x,a) = \psi(x,1). \tag{11}$$

Observe that, for $a = 2, \dots, x$,

$$\begin{aligned} &ET(x - 1 + Y^{(2)}) - ET(x - a + Y^{(a+1)}) \\ &\quad = ET(x + 1 - Y^{(2)}) - ET(x + 1 - Y^{(a+1)}) \\ &\quad \ge ET(x + 1 - Y^{(2)}) - ET(x + 1 - Y^{(3)}) \end{aligned}$$

because $-a + Y^{(a)}$ has the same distribution as $-Y^{(a)}$ and $-Y^{(a)}$ is stochastically decreasing in a. Also

$$u\left(\frac{a}{a+1}\right) - u\left(\frac{1}{2}\right) \le u\left(\frac{M}{M+1}\right) - u\left(\frac{1}{2}\right)$$

because u is increasing. Therefore, for $a = 2, \dots, x$,

$$u\left(\frac{a}{a+1}\right) + \beta ET(x - a + Y^{(a+1)}) \le u\left(\frac{1}{2}\right) + \beta ET(x - 1 + Y^{(2)})$$

if

$$\frac{1}{\beta}\left[u\left(\frac{M}{M+1}\right) - u\left(\frac{1}{2}\right)\right] \le ET(x + 1 - Y^{(2)}) - ET(x + 1 - Y^{(3)}). \tag{12}$$

So (11) will be established once we have proved (12).

For the proof of (12), let $\pi_2 = (\pi_{2,0}, \pi_{2,1}, \pi_{2,2}) = (\pi_{2,0}, \pi_{2,1}, \pi_{2,0})$ be the distribution of $Y^{(2)}$, and let $\pi_3 = (\pi_{3,0}, \pi_{3,1}, \pi_{3,2}, \pi_{3,3}) = (\pi_{3,0}, \pi_{3,1}, \pi_{3,1}, \pi_{3,0})$ be the distribution of $Y^{(3)}$. Notice that

$$\pi_{2,1} = 1 - 2\pi_{2,0} \text{ and } \pi_{3,1} = (1/2) - \pi_{3,0}$$

and, because $Y^{(2)}$ is stochastically smaller than $Y^{(3)}$, we also have the condition $\pi_{3,0} \leq \pi_{2,0}$. Now the right-hand-side of (12) is a linear function of $(\pi_{2,0}, \pi_{3,0})$ so that it suffices to check the inequality on the extreme points of the set

$$\Delta = \left\{(\pi_{2,0}, \pi_{3,0}) : 0 \leq \pi_{3,0} \leq \pi_{2,0} \leq (1/2)\right\}.$$

We will check each of the three extreme points.

For $(\pi_{2,0}, \pi_{3,0}) = (0, 0)$, we have $\pi_{2,1} = 1, \pi_{3,1} = 1/2$, and the right side of (12) equals

$$\begin{aligned} T(x) - \tfrac{1}{2}[T(x-1) + T(x)] &= \tfrac{1}{2}[T(x) - T(x-1)] \\ &\geq \frac{\alpha\,[u(1) - u\,(1/2)]}{\beta} \\ &= \frac{1}{\beta}\left[u\left(\frac{M}{M+1}\right) - u\left(\frac{1}{2}\right)\right] \end{aligned}$$

where the inequality holds by Lemma 4.1. For $(\pi_{2,0}, \pi_{3,0}) = (1/2, 0)$, we have $\pi_{2,1} = 0, \pi_{3,1} = 1/2$, and the right side of (12) is

$$\tfrac{1}{2}[T(x+1) + T(x-1)] - \tfrac{1}{2}[T(x) + T(x-1)] = \tfrac{1}{2}[T(x+1) - T(x)],$$

which is greater than or equal to the left side of (12) just as before. Finally for $(\pi_{2,0}, \pi_{3,0}) = (1/2, 1/2)$, we have $\pi_{2,1} = \pi_{3,1} = 0$ and the right side of (12) is

$$\begin{aligned} &\tfrac{1}{2}[T(x+1) + T(x-1)] - \tfrac{1}{2}[T(x+1) + T(x-2)] \\ &\qquad = \tfrac{1}{2}[T(x-1) - T(x-2)] \end{aligned}$$

which is greater than or equal to the right side of (12) as above.

The proof of (11) is now complete. To finish Case 3, we need to show that $\psi(x, 0) \leq \psi(x, 1)$; that is, we must show

$$\beta ET(x + Y^{(1)}) \leq u\,(1/2) + \beta ET(x - 1 + Y^{(2)})$$

or

$$ET(x + Y^{(1)}) - ET(x - 1 + Y^{(2)}) \leq \frac{u\,(1/2)}{\beta}. \tag{13}$$

By our symmetry assumption $Y^{(1)}$ must equal 0 and 1 with probability $1/2$ each; we use $\pi_2 = (\pi_{2,0}, \pi_{2,1}, \pi_{2,0})$ for the distribution of $Y^{(2)}$ as before. The left-hand-side of (13) is linear in $\pi_{2,0}$. So we need only check the extreme points $\pi_{2,0} = 0$ and $\pi_{2,0} = 1/2$. For $\pi_{2,0} = 0$, the left side of (13) equals

$$\frac{1}{2}[T(x) + T(x+1)] - T(x) = \frac{1}{2}[T(x+1) - T(x)] \leq \frac{u\,(1/2)}{\beta}$$

by Lemma 4.1. For $\pi_{2,0} = 1/2$, we have

$$\frac{1}{2}[T(x) + T(x+1)] - \frac{1}{2}[T(x-1) + T(x+1)] = \frac{1}{2}[T(x) - T(x-1)] \le \frac{u\,(1/2)}{\beta}.$$

Case 4: $x = 1$.

We need to show $\psi(x, 0) \le \psi(x, 1)$, which is equivalent to (13) above. The proof is the same as that already given for (13) in Case 3.

This completes the proof of Theorem 4.1 for the proportional-rewards game. □

As with Theorems 3.1 and 3.2 it is easy to extend Theorem 4.1 to a situation where the players have different utility functions and discount factors. Also we suspect that timid play provides the unique Nash equilibrium for sufficiently large β.

Remark 4.1. When $\pi_{2,0} = 0$ and both players use the timid strategy, their initial fortunes donot change during the course of the game except when one player has all the money and the other has none. Therefore, for $1 \le x \le M - 1$, $T_\beta(x) = u(1/2)/(1-\beta)$, a constant, and it is easy to see that, for β near 1, T_β need not satisfy the Bellman equation (10) at every x; for instance, for $M \ge 3$ and $x = M - 1$. Thus the timid strategies need not form a Nash equilibrium in this case.

5 A More Generous Rule for Distribution of the Good

In this section we consider a change in the rule for distributing the good to the players. In previous sections, the rule has been that a player who bids zero for the good always receives a zero portion. In particular, $r(x, 0, 0) = u(0) = 0$. Now we assume instead that when both players bid zero, they each receive half of the good. That is,

$$r(x, 0, 0) = \tilde{r}(x, 0, 0) = u\,(1/2)$$

for all $x \in S$. We call this the *generous* rule and make no other changes in our games.

The change to the generous rule has relatively little effect when the discount factor is small. Recall from Section 3 that large bets are good for small β. Indeed Theorems 3.1 and 3.2 on the symmetric bold strategies remain true and the proofs are essentially the same.

It is amusing, however, that a form of the Prisoner's Dilemma is now embedded in our game. To see that this is so, we introduce the zero selector

$$z(x) = 0, \quad \text{for } x \in S,$$

and observe that, because u is concave, the total of the discounted rewards to the two players is maximized by the zero stationary strategies $(z^{\infty}(x), z^{\infty}(x))$ at every x. Under these strategies the return to each player is $u(1/2)/(1-\beta)$ and, if u is strictly concave, the sum of the returns will be smaller under strategies for which the players do not have equal portions of the good every day. Nevertheless the zero strategies need not form a Nash equilibrium.

Example 5.1. Suppose $M = 2$, u is strictly concave, and β is small. As explained above, the strategies $(z^{\infty}(1), z^{\infty}(1))$ yield $u(1/2)/(1-\beta)$. A switch to bold play by one of the players gives him an immediate return of $u(1)$ which clearly exceeds $u(1/2)/(1-\beta)$ for sufficiently small β. By the theorems of Section 3, the bold strategies $(b^{\infty}(1), \tilde{b}^{\infty}(1))$ form a Nash equilibrium, but have a smaller return for both players than the zero strategies.

For β near 1, a new strategy is of interest based on the selector

$$m(x) = \begin{cases} 1 & \text{if} \quad x = 1, 2, \ldots, M-1, \\ 0 & \text{if} \quad x = 0 \text{ and } x = M. \end{cases}$$

We call m the *meek* selector and the corresponding stationary strategies are called the *meek* strategies. Notice that a meek strategy $m^{\infty}(x)$ differs from the timid strategy $t^{\infty}(x)$ only at state M where the meek bid is zero and the timid bid is one.

When both players play meekly, they each receive half of the good every day and a total discounted return of $u(1/2)/(1-\beta)$, the same as when both play zero strategies. However, for β near 1, meek play is not optimal for one player when the other plays meekly. In fact, timid play is optimal against meek play and meek against timid; the result is a surprising asymmetric equilibrium.

We continue to assume in this section that $\pi_{2,0} > 0$ where $\pi_2 = (\pi_{2,0}, \pi_{2,1}, \pi_{2,0})$ is the distribution of $Y^{(2)}$.

Theorem 5.1. *Assume that the utility function u is strictly concave. Then for the winner-takes-all game and the proportional rewards game with the generous rule for distribution, there exists a $\beta^* \in (0, 1)$ such that, for all $\beta \in [\beta^*, 1)$ and all $x \in S$, the meek-timid strategies $(m^{\infty}(x), \tilde{t}^{\infty}(x))$ and the timid-meek strategies $(t^{\infty}(x), \tilde{m}^{\infty}(x))$ form Nash equilibria.*

The equilibria in Sections 3 and 4, unlike the meek-timid equilibria of Theorem 5.1, used symmetric strategies for the two players. Also, by Theorem 2.1 we know that a symmetric equilibrium exists. However, there may not exist a symmetric equilibrium based on *pure* stationary strategies as the following example illustrates.

Example 5.2. Suppose $M = 1$ and u is strictly concave. There are only two pure stationary strategies available to each player. Player I has the meek strategy $m^{\infty}(\cdot)$, which is now the same as the zero strategy, and the bold strategy $b^\infty(\cdot)$, which is the same as timid. Player II has the symmetric strategies $\tilde{m}^\infty(\cdot)$ and $\tilde{b}^\infty(\cdot)$. Let

$$\beta^* = \frac{2\,[u(1) - u\,(1/2)]}{u(1)}$$

It can be shown by a straightforward argument that, for $0 < \beta \le \beta^*$, the bold strategies $(b^\infty(x), \tilde{b}^\infty(x))$ form a Nash equilibrium, but for $\beta^* < \beta < 1$, neither $(b^\infty(x), \tilde{b}^\infty(x)) \equiv (t^\infty(x), \tilde{t}^\infty(x))$ nor $(m^\infty(x), \tilde{m}^\infty(x)) \equiv (z^\infty(x), z^\infty(x))$ is a Nash equilibrium. Indeed, for $\beta^* < \beta < 1$, there are three stationary Nash equilibria: namely, the meek–timid $(m^\infty(x), \tilde{t}^\infty(x))$ and the timid–meek $(t^\infty(x), \tilde{m}^\infty(x))$ and a symmetric equilibrium $(s^\infty(x), \tilde{s}^\infty(x))$ where $s(0) = 0$ and $s(1)$ is the randomized action that equals 0 with probability

$$\gamma = \frac{2u\,(1/2) - 2u(1) + \beta u(1)}{\beta\,[2u\,(1/2) - u(1)]}$$

and equals 1 with probability $1 - \gamma$.

Here we prove Theorem 5.1 for the winner-takes-all game; with slight modifications, along the lines of the proof of Theorem 4.1, it is not difficult to prove the proportional rewards case.

Proof. (*of Theorem 5.1 for the winner-takes-all game*). Suppose player I uses the meek strategy and player II uses the timid strategy. For a given initial state x, let $X_0 = x, X_1, X_2, \ldots$ be the Markov chain of successive states corresponding to the fortunes of player I. The chain $\{X_n\}$ is absorbed at state M because both players bid zero there.

Assume $x \ge 1$ and let $\{Y_n\}$ be another Markov chain starting from $Y_0 = x - 1$ and having the same transition probabilities as $\{X_n\}$. Assume that the two chains are coupled so that Y_n stays one unit behind X_n until either X_n reaches M or Y_n reaches 0. If X_n reaches M first, then Y_n continues until it too is absorbed at M. If Y_n reaches 0 first, then the processes are coupled so that on the next day $X_{n+1} = Y_{n+1}$ with probability 1/2, $Y_{n+1} = 0$ and $X_{n+1} = 1$ with probability $\pi_{2,1}/2$, and $Y_{n+1} = 1$ and $X_{n+1} = 2$ with probability $\pi_{2,0}$. (This motion can be modeled by a simple urn as in the proof of Lemma 4.1.) Once X_n and Y_n coincide, they remain equal.

This coupling will be used to obtain information about the functions

$$\begin{aligned} v(x) &= v_\beta(x) = v_\beta(x, m^\infty(x), \tilde{t}^\infty(x)), \\ w(x) &= w_\beta(x) = \tilde{v}_\beta(x, m^\infty(x), \tilde{t}^\infty(x)), \end{aligned} \tag{14}$$

corresponding to the expected discounted rewards (as in (5)) to player I and II, respectively, when I plays meekly and II plays timidly. Notice that each player receives $u(1/2)$ on every day when the state x is not zero. If $x = 0$, then I bids

$m(0) = 0$ and II bids $\tilde{t}(0) = t(M) = 1$ so that I receives $u(0) = 0$ and II receives $u(1)$. So, to compare the functions v and w at adjacent states x and $x - 1$, we introduce the random variable

$$N = \text{number of visits } Y_n \text{ makes to 0 prior to the random time } \xi,$$

where

$$\xi = \inf\{n : X_n = Y_n\}.$$

It turns out that the expected value of N does not depend on the initial state $X_0 = x$.

For the rest of the proof of Theorem 5.1, we write E_x and P_x for $E_{x,m^\infty(x),\tilde{t}^\infty(x)}$ and $P_{x,m^\infty(x),\tilde{t}^\infty(x)}$ respectively.

Lemma 5.1. *For all* $x = 1, 2, \ldots, M$, $E_x N = 2$.

Proof. As a preliminary step, we consider the random variable N_1 that equals the number of visits paid to state 0 by the process $\{Y_n\}$ before it is absorbed at M. Let γ be the expected value of N_1 when $Y_0 = 0$. Then

$$\gamma = 1 + \tfrac{1}{2}\gamma + \tfrac{1}{2}E[N_1 | Y_0 = 1]$$

and

$$E[N_1 | Y_0 = 1] = pE[N_1 | Y_0 = 0] = p\gamma,$$

where,

$$p = P[\{Y_n\} \text{ reaches 0 before } M | Y_0 = 1] = (M-1)/M,$$

is a familiar gambler's ruin probability. Thus

$$\gamma = 1 + \frac{1}{2}\gamma + \frac{1}{2}\frac{M-1}{M}\gamma$$

and so $\gamma = 2M$.

Now let $\rho_x = E_x N$, $x = 1, 2, \ldots, M$. Then

$$\rho_M = E[N_1 | Y_0 = M-1] = \frac{1}{M}E[N_1 | Y_0 = 0] = \frac{1}{M}\gamma = 2.$$

Next consider $\rho_1 = E_1 N$. Recall how the process behaves when $X_0 = 1$ and $Y_0 = 0$ and also that, starting from state 2, the process $\{X_n\}$ will reach 1 before M with probability $(M-2)/(M-1)$. Hence,

$$\rho_1 = 1 + \frac{1}{2}\cdot 0 + \frac{\pi_{2,1}}{2}\rho_1 + \pi_{2,0}\left[\frac{M-2}{M-1}\rho_1 + \frac{1}{M-1}\rho_M\right].$$

Substitute $\rho_M = 2$ and $\pi_{2,1} = 1 - 2\pi_{2,0}$, and it easily follows that $\rho_1 = 2$.

For $x = 2, 3, \ldots, M-1$, we also have

$$\rho_x = \frac{M-x}{M-1}\rho_1 + \frac{x-1}{M-1}\rho_M = 2.$$

□

Define

$$t_0 = \inf\{n \geq 0 : Y_n = 0\},$$

and, for $n \geq 1$,

$$t_n = \inf\{n > t_{n-1} : Y_n = 0\}$$

so that t_n is the time of the n-th visit (if there is one) by the process $\{Y_n\}$ to 0. Then the random variable N can be written as

$$N = \sum_{n=0}^{\infty} I_{[t_n < \xi]}$$

where $I_{[t_n<\xi]}$ equals 1 on the event $[t_n < \xi]$ and equals 0 on the complement.

Lemma 5.2. *Given $\epsilon > 0$, there exists $\beta_1 = \beta_1(\epsilon) \in (0, 1)$ such that, for all $\beta \in [\beta_1, 1)$ and all $x = 1, 2, \ldots, M$,*

(i) $\frac{1}{\beta}\left[2u\left(\frac{1}{2}\right) - \epsilon\right] \leq v(x) - v(x-1) \leq \frac{2}{\beta}u\left(\frac{1}{2}\right)$,

(ii) $\frac{1}{\beta}\left[2\left(u(1) - u\left(\frac{1}{2}\right)\right) - \epsilon\right] \leq w(x) - w(x-1) \leq \frac{2}{\beta}\left[u(1) - u\left(\frac{1}{2}\right)\right]$.

Proof. It follows from the discussion after the definition of v and w in (14) that

$$v(x) - v(x-1) = u\left(\frac{1}{2}\right) E_x\left[\sum_{n=0}^{\infty} \beta^{t_n} I_{[t_n<\xi]}\right], \tag{15}$$

$$w(x) - w(x-1) = \left[u(1) - u\left(\frac{1}{2}\right)\right] E_x\left[\sum_{n=0}^{\infty} \beta^{t_n} I_{[t_n<\xi]}\right].$$

By the first equality, we have for all $\beta \in (0, 1)$,

$$\begin{aligned} v(x) - v(x-1) &\leq u\left(\frac{1}{2}\right) E_x\left[\sum_{n=0}^{\infty} I_{[t_n<\xi]}\right] \\ &= u\left(\frac{1}{2}\right) E_x N = 2u\left(\frac{1}{2}\right) < \frac{2}{\beta}u\left(\frac{1}{2}\right), \end{aligned}$$

which proves the second inequality of *(i)*. To see that the first inequality of *(i)* holds for β sufficiently close to 1, first apply the monotone convergence theorem to get

$$\lim_{\beta\uparrow 1} E_x\left[\sum_{n=0}^{\infty} \beta^{t_n} I_{[t_n<\xi]}\right] = E_x N = 2,$$

and then observe that

$$\lim_{\beta\uparrow 1}\frac{1}{\beta}\left[2u\left(\frac{1}{2}\right)-\epsilon\right]=2u\left(\frac{1}{2}\right)-\epsilon.$$

The proof of *(ii)* is similar. □

We will now argue that the meek strategy is optimal for player I when player II uses the timid strategy and β is sufficiently close to 1. We assume that II plays $\tilde{t}^{\infty}(x)$ at each state x so that I faces a discounted dynamic programming problem. As in the proof of Theorem 4.1, it sufficies to show that player I's return function v from the meek strategy satisfies the Bellman equation

$$v(x)=\sup_{a\in A_x}[r(x,a,\tilde{t}(x))+\beta Ev(X(x,a))], \tag{16}$$

for $x\in S$, where

$$r(x,a,b)=\begin{cases}u(1) & \text{if} \quad a>b,\\ u(1/2) & \text{if} \quad a=b,\\ 0 & \text{if} \quad a<b,\end{cases}$$

and

$$X(x,a)=\begin{cases}x-a+Y^{(a)} & \text{if} \quad a>\tilde{t}(x),\\ x-1+Y^{(2)} & \text{if} \quad a=\tilde{t}(x)=1,\\ x & \text{if} \quad a=\tilde{t}(x)=0,\\ x+Y^{(1)} & \text{if} \quad a=0<\tilde{t}(x).\end{cases}$$

Recall that $Y^{(2)}$ has distribution $\pi_2=(\pi_{2,0},\pi_{2,1},\pi_{2,0})$ and, because $Y^{(1)}$ is distributed as $1-Y^{(1)}$, $Y^{(1)}$ equals 0 and 1 with probability 1/2 each.

Let

$$\psi(x,a)=r(x,a,\tilde{t}(x))+\beta Ev(X(x,a))$$

so that (16) can be written as

$$\psi(x,m(x))=\sup_{a\in A_x}\psi(x,a) \tag{17}$$

for $x\in S$. To prove (17), we consider three cases.

Case 1: $x=0$.

This case is trivial because $A_0=\{0\}$.

Case 2: $x=M$.

Since $m(M)=0$, we have to show

$$\psi(M,0)\geq\psi(M,a),$$

for $a = 1, 2, \ldots, M$, where

$$\psi(M, 0) = u(1/2) + \beta v(M)$$

and

$$\psi(M, a) = u(1) + \beta E v(M - a + Y^{(a)})$$

for $a \geq 1$. By Lemma 5.2, v is increasing for β sufficiently near 1; also $-a + Y^{(a)}$ is distributed the same as $-Y^{(a)}$ and $Y^{(a)}$ is stochastically larger than $Y^{(1)}$ for $a \geq 1$. Hence, for $a \geq 1$, $\psi(M, a) \leq \psi(M, 1)$ and we need only show that $\psi(M, 1) \leq \psi(M, 0)$. That is, it suffices to show

$$\begin{aligned} u(1/2) + \beta v(M) &\geq u(1) + \beta E v(M - 1 + Y^{(1)}) \\ &= u(1) + (\beta/2)[v(M - 1) + v(M)] \end{aligned}$$

or, equivalently

$$v(M) - v(M - 1) \geq (2/\beta)[u(1) - u(1/2)].$$

By the strict concavity of u,

$$2[u(1) - u(1/2)] < 2u(1/2) - \epsilon$$

for a sufficiently small $\epsilon > 0$. Now use Lemma 5.2 again.

Case 3: $1 \leq x \leq M - 1$.
For these x, $m(x) = \tilde{t}(x) = 1$ and

$$\psi(x, a) = \begin{cases} u(1) + \beta E v(x - a + Y^{(a)}) & \text{if } \ x \geq 2 \text{ and } a \geq 2, \\ u(1/2) + \beta E v(x - 1 + Y^{(2)}) & \text{if } \ x \geq 1 \text{ and } a = 1, \\ \beta E v(x + Y^{(1)}) & \text{if } \ x \geq 1 \text{ and } a = 0. \end{cases}$$

Since v is increasing and $-a + Y^{(a)}$ is stochastically smaller than $-2 + Y^{(2)}$ for $a > 2$, we have $\psi(x, 2) = \sup_{2 \leq a \leq x} \psi(x, a)$ for $x \geq 2$. It remains to check that $\psi(x, 2) \leq \psi(x, 1)$ and $\psi(x, 0) \leq \psi(x, 1)$. Rewrite the first inequality as

$$u(1/2) + \beta E v(x - 1 + Y^{(2)}) \geq u(1) + \beta E v(x - 2 + Y^{(2)})$$

or

$$E[v(x - 1 + Y^{(2)}) - v(x - 2 + Y^{(2)})] \geq (1/\beta)[u(1) - u(1/2)].$$

This holds for β near 1 since, by Lemma 5.2,

$$v(y) - v(y - 1) \geq 1/\beta[2u(1/2) - \epsilon] \geq 2/\beta[u(1) - u(1/2)]$$

for ϵ sufficiently small and $y = 1, 2, \dots, M$. The remaining inequality is

$$u(1/2) + \beta E v(x - 1 + Y^{(2)}) \geq \beta E v(x + Y^{(1)})$$

or

$$\begin{aligned}(1/\beta)u(1/2) &\geq Ev(x + Y^{(1)}) - Ev(x - 1 + Y^{(2)}) \\ &= \tfrac{1}{2}[v(x) + v(x+1)] - [\pi_{2,0}v(x-1) + \pi_{2,1}v(x) + \pi_{2,0}v(x+1)] \\ &= \pi_{2,0}[v(x) + v(x+1)] + (\pi_{2,1}/2)[v(x) + v(x+1)] \\ &\quad - \pi_{2,0}[v(x-1) + v(x+1)] - \pi_{2,1}v(x) \\ &= \pi_{2,0}[v(x) - v(x-1)] + (\pi_{2,1}/2)[v(x+1) - v(x)].\end{aligned}$$

By Lemma 5.2, the increments $v(x) - v(x-1)$ and $v(x+1) - v(x)$ are each less than or equal to $(2/\beta)u(1/2)$ and $\pi_{2,0} + \pi_{2,1}/2 = 1/2$.

The proof that meek is optimal against timid is now complete. The proof that timid is optimal against meek is quite similar and we omit it. □

Once again, as with the theorems of Sections 3 and 4, one may extend Theorem 5.1 to a situation where players have different utilities and discount factors.

6 Long Run Average Reward

Consider one of our games beginning at state $x \in S$. Suppose player I uses strategy μ and player II uses strategy ν. Their respective long run average rewards are defined to be

$$v(x, \mu, \nu) = E_{x,\mu,\nu}\left[\limsup_{n\to\infty} \frac{1}{n}\sum_{k=1}^{n} r_k\right],$$

$$\tilde{v}(x, \mu, \nu) = E_{x,\mu,\nu}\left[\limsup_{n\to\infty} \frac{1}{n}\sum_{k=1}^{n} \tilde{r}_k\right],$$

where r_n and $\tilde{r}_n$ are their rewards on day n for each n.

As is well-known, the long run average rewards from stationary strategies can be calculated as the limit of their discounted rewards as β increases to 1.

Lemma 6.1. *Let $x \in S$ and let $\sigma^\infty(x)$ and $\tau^\infty(x)$ be stationary strategies for players I and II, respectively. Then*

$$v(x, \sigma^\infty(x), \tau^\infty(x)) = \lim_{\beta\uparrow 1}(1-\beta)v_\beta(x, \sigma^\infty(x), \tau^\infty(x)),$$

$$\tilde{v}(x, \sigma^\infty(x), \tau^\infty(x)) = \lim_{\beta\uparrow 1}(1-\beta)\tilde{v}_\beta(x, \sigma^\infty(x), \tau^\infty(x)),$$

where the discounted rewards v_β, $\tilde{v}_\beta$ are defined in (5).

Proof. The process of states $\{X_n\}$ is a Markov chain under $P = P_{x,\sigma^\infty(x),\tau^\infty(x)}$. So, by the ergodic theorem for Markov chains, the sum

$$\frac{1}{n}\sum_{k=1}^{n} r_k = \frac{1}{n}\sum_{k=0}^{n-1} r(X_k, \sigma(X_k), \tau(X_k))$$

converges almost surely to a limit variable, say L, for n going to infinity. By a result of Hardy and Littlewood (cf. Appendix H in [2]), the sum

$$(1-\beta)\sum_{n=1}^{\infty} \beta^{n-1} r_n$$

converges almost surely to the same variable L. The first equality now follows from the dominated convergence theorem; the second follows by symmetry from the first. □

We can use the lemma together with our theorems for discounted games to find equilibria for the corresponding average reward games.

Theorem 6.1. *For each $x \in S$, the timid strategies $(t^\infty(x), \tilde{t}^\infty(x))$ form a Nash equilibrium in the proportional-rewards and winner-takes-all games with long run average reward.*

Proof. Let β^* be as in Theorem 4.1. Then, for $x \in S$ and $\beta \in [\beta^*, 1)$, the timid strategy is optimal against $\tilde{t}(x)$ in the β-discounted game. So, for any strategy μ for player I,

$$v_\beta(x, \mu, \tilde{t}^\infty(x)) \le v_\beta(x, t^\infty(x), \tilde{t}^\infty(x)).$$

If μ is stationary, then, by Lemma 6.1,

$$v(x, \mu, \tilde{t}^\infty(x)) \le v(x, t^\infty(x), \tilde{t}^\infty(x)).$$

This sufficies to show $t^\infty(x)$ is optimal against $\tilde{t}^\infty(x)$ in the average reward game, because there exists an optimal stationary strategy for player I against $\tilde{t}^\infty(x)$ by a result of Blackwell [1]. By symmetry $\tilde{t}^\infty(x)$ is optimal against $t^\infty(x)$. □

We suspect that the timid strategies are the unique stationary equilibria for the games of Theorem 6.1.

Theorem 6.2. *Assume that the utility function u is strictly concave. For the winner-takes-all game and the proportional rewards game with the generous rule for distribution and long run average rewards, both the meek-timid strategies $((m^\infty(x), \tilde{t}^\infty(x)))$ and the timid-meek strategies $(t^\infty(x), \tilde{m}^\infty(x))$ form Nash equilibria.*

Proof. Use Theorem 5.1 and then the same proof as for Theorem 6.1. □

The meek–timid and timid–meek equilibria are not unique in Theorem 6.2. Another stationary equilibrium is formed by the zero strategies $(z^\infty(x), z^\infty(x))$. All three of these equilibria have the property that, with probability 1, a state is eventually reached where both players bid zero. Indeed, this property is shared by all stationary equilibria for the game of Theorem 6.2 when we rule out the possibility for the law of motion to be concentrated at a point, except when both players bid zero. In fact, we assume for the next theorem that, for $1 \le k \le M/2$, the random variable $Y^{(2k)}$, which controls the distribution of money among the players when they make the same bid k, is not concentrated at k.

Theorem 6.3. *Assume that the utility function u is strictly concave. Let $x \in S$ and $(\sigma^\infty(x), \tau^\infty(x))$ form a stationary Nash equilibrium for a winner-takes-all game or a proportional reward game with the generous rule for distribution, long run average reward and initial state x. Then*

$$P_{x,\sigma^\infty(x),\tau^\infty(x)}[\sigma(X_n) = \tau(X_n) = 0 \textit{ eventually}] = 1.$$

Before proving the theorem, we show that both players receive a long run average reward of $u(1/2)$ from any Nash equilibrium with stationary strategies. The proof of this lemma is based on the argument that the zero strategy guarantees to a player a long run average reward of $u(1/2)$ regardless of the opponent's strategy; incidentally, this fact implies that the zero strategies $(z^\infty(x), z^\infty(x))$ form a Nash equilibrium.

Lemma 6.2. *Let $(\sigma^\infty(x), \tau^\infty(x))$ form a Nash equilibrium with stationary strategies for the games of Theorem 6.3 with initial state $x \in S$. Then*

$$v(x, \sigma^\infty(x), \tau^\infty(x)) = \tilde{v}(x, \sigma^\infty(x), \tau^\infty(x)) = u(1/2).$$

Proof. By Lemma 6.1

$$v(x, \sigma^\infty(x), \tau^\infty(x)) = \lim_{\beta \uparrow 1}(1-\beta) v_\beta(x, \sigma^\infty(x), \tau^\infty(x)),$$
$$\tilde{v}(x, \sigma^\infty(x), \tau^\infty(x)) = \lim_{\beta \uparrow 1}(1-\beta) \tilde{v}_\beta(x, \sigma^\infty(x), \tau^\infty(x)).$$

Since u is concave, for $\beta < 1$,

$$v_\beta(x, \sigma^\infty(x), \tau^\infty(x)) + \tilde{v}_\beta(x, \sigma^\infty(x), \tau^\infty(x)) \le \frac{2u(1/2)}{1-\beta}$$

and thus

$$v(x, \sigma^\infty(x), \tau^\infty(x)) + \tilde{v}(x, \sigma^\infty(x), \tau^\infty(x)) \le 2u(1/2).$$

Suppose that $\tilde{v}(x, \sigma^{\infty}(x), \tau^{\infty}(x)) > u(1/2)$. Then $v(x, \sigma^{\infty}(x), \tau^{\infty}(x)) < u(1/2)$. However, if player I uses the zero strategy $z^{\infty}(x)$ against $\tau^{\infty}(x)$, the process of states $\{X_n\}$ becomes a sub-martingale under $P_{x,z^{\infty}(x),\tau^{\infty}(x)}$ and thus converges almost surely to a random variable X with values in S. Since S is discrete, this amounts to say that

$$P_{x,z^{\infty}(x),\tau^{\infty}(x)}[X_n = X \text{ eventually}] = 1.$$

Therefore player II's bid is eventually equal to 0 with probability one; thus the daily reward to player I is eventually equal to $u(1/2)$ and

$$v(x, z^{\infty}(x), \tau^{\infty}(x)) = u(1/2) > v(x, \sigma^{\infty}(x), \tau^{\infty}(x)).$$

The last inequality shows that $z^{\infty}(x)$ is a better reply for player I to $\tau^{\infty}(x)$ than $\sigma^{\infty}(x)$ and contradicts the assumption that $(\sigma^{\infty}(x), \tau^{\infty}(x))$ form a Nash equilibrium; hence

$$\tilde{v}(x, \sigma^{\infty}(x), \tau^{\infty}(x)) \leq u(1/2).$$

However it cannot be that $\tilde{v}(x, \sigma^{\infty}(x), \tau^{\infty}(x)) < u(1/2)$ because the zero strategy is available also to player II and, inverting the role of the players in the argument above, it's easy to see that $z^{\infty}(x)$ would be a better reply against $\sigma^{\infty}(x)$ than $\tau^{\infty}(x)$ by guaranteeing to player II a long run average reward of $u(1/2)$. Hence $\tilde{v}(x, \sigma^{\infty}(x), \tau^{\infty}(x)) = u(1/2)$. Analogously, $v(x, \sigma^{\infty}(x), \tau^{\infty}(x)) = u(1/2)$.

□

Proof. (*of Theorem 6.3*). The process $\{X_n\}$ is a Markov chain under $P_{x,\sigma^{\infty}(x),\tau^{\infty}(x)}$. Since the state space of $\{X_n\}$ is finite, the chain will eventually be absorbed in a class of positive recurrent states; without loss of generality we may assume that the initial state x of the game belongs to a class C of positive recurrent states. The theorem will then be proved if we show that $\sigma(x) = \tau(x) = 0$.

Let λ denote the stationary distribution of $\{X_n\}$ over C and let X be a random variable with values in C and distribution λ. The ergodic theorem for Markov chains implies that the long run average rewards of player I and player II, respectively, can be represented as

$$v(x, \sigma^{\infty}, \tau^{\infty}(x)) = E[r(X, \sigma(X), \tau(X))],$$
$$\tilde{v}(x, \sigma^{\infty}, \tau^{\infty}(x)) = E[\tilde{r}(X, \sigma(X), \tau(X))]$$

where r is defined in (1) or (2), depending on whether the game is a winner-takes-all or a proportional reward one, and $\tilde{r}$ is defined by symmetry. Therefore, by Lemma 6.2,

$$2u(1/2) = E[r(X, \sigma(X), \tau(X)) + \tilde{r}(X, \sigma(X), \tau(X))]. \tag{18}$$

Whenever player I acquires the right for a fraction f of good, player II gets $1 - f$; this fact depends on the generous rule to the effect that the entire good is always divided among the players regardless of their bids. Hence, for all $y \in S$, $r(y, \sigma(y), \tau(y)) = u(f)$ if and only if $\tilde{r}(y, \sigma(y), \tau(y)) = u(1 - f)$; moreover, strict concavity of the utility u implies that

$$r(y, \sigma(y), \tau(y)) + \tilde{r}(y, \sigma(y), \tau(y)) < 2u(1/2)$$

unless $r(y, \sigma(y), \tau(y)) = u(1/2)$. Finally note that the random variables $r(X, \sigma(X), \tau(X))$ and $\tilde{r}(X, \sigma(X), \tau(X))$ are discrete. These facts and (18) imply that

$$P[r(X, \sigma(X), \tau(X)) = u(1/2)] = 1. \tag{19}$$

We now show that the class C of recurrent states must be a singleton. Let x_* be the smallest element of C. It follows from (19) that $\sigma(x_*) = \tau(x_*) = k$(say), where $k \leq M/2$. If $k > 0$, then the chain moves from x_* to a random state distributed like the random variable $x_* - k + Y^{(2k)}$, which is less than x_* with positive probability because of the assumption that $Y^{(2k)}$ is not concentrated at k. But this contradicts the minimality of x_* in C. Hence, $k = 0$, and x_* is an absorbing state for the Markov chain. Therefore, $C = \{x_*\}$.

To finish the proof of the theorem, it is enough to observe that under the assumption that $Y^{(2k)}$ is not concentrated at k, for $1 \leq k \leq M/2$, the process $\{X_n\}$ stays at x forever, if and only if $\sigma(x) = \tau(x) = 0$. □

Acknowledgement

We are grateful to Martin Shubik who introduced us to strategic market games and to Ashok Maitra for helpful conversations about the games of this paper. We also want to thank Laura Pontiggia who discovered an error in our statement of Theorem 3.2 in an earlier draft of this paper.

REFERENCES

[1] Blackwell, D., Discrete Dynamic Programming, *The Annals of Mathematical Statistics*, **33** (1962) 719–726.

[2] Filar, J. and Vrieze, K., *Competitive Markov Decision Processes*, Springer Verlag, New York, (1997).

[3] Geanakoplos, J., Karatzas, I., Shubik, M. and Sudderth, W.D., A strategic market game with active bankruptcy, *Cowles Foundation Discussion Paper*, (1998).

[4] Karatzas, I., Shubik, M. and Sudderth, W.D., Construction of stationary Markov equilibria in a strategic market game, *Mathematics of Operation Research* **19**(4) (1992) 975–1006.

[5] Karatzas, I., Shubik, M. and Sudderth, W.D., A strategic market with secured lending, *Journal of Mathematical Economics* **28** (1997) 207–247.

[6] Maitra, A. and Sudderth, W.D., *Discrete Gambling and Stochastic Games*, Springer Verlag, New York, (1996).

[7] Mertens, J.F., Sorin, S. and Zamir, S., Repeated games, *CORE Discussion Paper* #9420-9422, Université Catholique de Louvain, (1994).

[8] Secchi, P. and Sudderth, W.D., How to bid for a pizza, *Quaderno di Dipartimento* #42(5-96), Dipartimento di Economia Politica e Metodi Quantitativi, Università di Pavia, (1996).

[9] Secchi, P. and Sudderth, W.D., A two-person strategic market game, *Quaderno di Dipartimento* #83(6-98), Dipartimento di Economia Politica e Metodi Quantitativi, Università di Pavia, (1998).

[10] Shubik, M. and Whitt, W., Fiat money in an economy with one nondurable good and no credit. A non-cooperative sequential game, in *Topics in Differential Games*, edited by A. Blaquiere, 401–449, North Holland, Amsterdam, (1973).

[11] Sobel, M., Non-cooperative stochastic games. *The Annals of Mathematical Statistics* **42** (1971) 1930–1935.

[12] Vieille, N., Two-player stochastic game I: a reduction. *Israel Journal of Mathematics* **119** (2000) 55–91.

[13] Vieille, N., Two-player stochastic game II: the case of recursive games. *Israel Journal of Mathematics* **119** (2000) 93–126.

New Approaches and Recent Advances in Two-Person Zero-Sum Repeated Games

Sylvain Sorin*
Laboratoire d'Econométrie
Ecole Polytechnique
1 rue Descartes
75005 Paris, France

and

Equipe Combinatoire et Optimisation
UFR 921, Université Pierre et Marie Curie - Paris 6
4 place Jussieu 75230 Paris, France

1 Preliminaries

In repeated games where the payoff is accumulated along the play, the players face a problem since they have to take into account the impact of their choices both on the current payoff and on the future of the game.

When considering long games this leads to two alternative cases. Whenever the previous problem can be solved in a "robust" way the game possesses a **uniform value**. In the other situation optimal strategies are very sensitive to the exact specification of the duration of the process. The **asymptotic approach** consists in studying the values of games with finite expected length along a sequence with length going to infinity and the questions are then the existence of a limit and its dependence w.r.t. the sequence.

A typical example is the famous Big Match (Blackwell and Ferguson, 1968) described by the following matrix:

	α	β
a	1^*	0^*
b	0	1

This corresponds to a stochastic game where, as soon as Player 1 plays a, the game reaches an absorbing state with a constant payoff corresponding to the entry played at that stage. Both the n-stage value v_n and the λ-discounted value v_λ are equal to 1/2 and are also independent of the additional information transmitted along the play to the players. Moreover, under standard signaling (meaning that the past play is public knowledge), or with only known past payoffs, the uniform value exists.

*Prepared for a plenary lecture at the International Symposium on Dynamic Games and Applications, Adelaide, Australia, December 18-21, 2000.

This is no longer the case with general signals: for example, when Player 1 has no information on Player 2's moves the max min is 0 (Kohlberg, 1974). It follows that the existence of a uniform value for stochastic games depends on the signalling structure on moves (Mertens and Neyman, 1981; Coulomb, 1992, 1999, 2001). On the other hand, the asymptotic behavior does not (Shapley, 1953).

We now describe more precisely the model of two-person zero-sum repeated game Γ that we consider. We are given a parameter space M and a function g from $I \times J \times M$ to $\mathbb{R}$: for each $m \in M$ this defines a two-person zero-sum game with action spaces I and J for players 1 and 2 respectively and payoff function g. To simplify the presentation we will first consider the case where all sets are finite, avoiding measurability issues. The initial parameter m_1 is chosen at random and the players receive some initial information about it, say a_1 (resp. b_1) for Player 1 (resp. Player 2). This choice is performed according to some probability π on $M \times A \times B$, where A and B are the signal sets of each player. In addition, after each stage the players obtain some further information about the previous choice of actions and both the previous and the current values of the parameter. This is represented by a map Q from $M \times I \times J$ to probabilities on $M \times A \times B$. At stage t given the state m_t and the moves (i_t, j_t), a triple $(m_{t+1}, a_{t+1}, b_{t+1})$ is chosen at random according to the distribution $Q(m_t, i_t, j_t)$. The new parameter is m_{t+1}, and the signal a_{t+1} (resp. b_{t+1}) is transmitted to Player 1 (resp. Player 2). A play of the game is thus a sequence $m_1, a_1, b_1, i_1, j_1, m_2, a_2, b_2, i_2, j_2, \ldots$ while the information of Player 1 before his play at stage t is a private history of the form $(a_1, i_1, a_2, i_2, \ldots, a_t)$ and similarly for Player 2. The corresponding sequence of payoffs is $g_1, g_2, \ldots$ with $g_t = g(i_t, j_t, m_t)$. (Note that it is not known to the players except if included in the signals.)

A strategy σ for Player 1 is a map from private histories to $\Delta(I)$, the space of probabilities on the set I of actions and τ is defined similarly for Player 2. Such a couple (σ, τ) induces, together with the components of the game, π and Q, a distribution on plays, hence on the sequence of payoffs.

There are basically two ways of handling the game repeated a large number of times that are described as follows (see Aumann and Maschler, 1995, Mertens, Sorin and Zamir, 1994):

1) The first one corresponds to the "compact case". One considers a sequence of evaluations of the stream of payoffs converging to the "uniform distribution on the set of integers, $\mathbb{N}$". For each specific evaluation, under natural assumptions on the action spaces and on the reward and transition mappings, the strategy spaces will be compact for a topology for which the payoff function will be continuous, hence the value will exist.

Two typical examples correspond to:

1.1) the **finite n-stage** game Γ_n with payoff given by the average of the first n rewards:

$$\gamma_n(\sigma, \tau) = E_{\sigma,\tau}\left(\frac{1}{n}\sum\nolimits_{t=1}^{n} g_t\right).$$

In the finite case (all sets considered being finite), this reduces to a game with finitely many pure strategies.

1.2) the λ-**discounted game** Γ_λ with payoff equal to the discounted sum of the rewards:

$$\gamma_\lambda(\sigma, \tau) = E_{\sigma,\tau}\left(\sum\nolimits_{t=1}^{\infty} \lambda(1-\lambda)^{t-1} g_t\right).$$

The values of these games are denoted by v_n and v_λ respectively. The study of their asymptotic behavior, as n goes to ∞ or λ goes to 0 is the study of the **asymptotic game**.

Extensions consider games with random duration or random duration process (Neyman, 2003, Neyman and Sorin, 2001).

2) An alternative analysis considers the whole family of "long games". It does not specify payoffs in some infinite game like $\liminf \frac{1}{n}\sum_{t=1}^{n} g_t$ or a measurable function defined on plays (see Maitra and Sudderth, 1998), but requires uniformity properties of the strategies.

Explicitly, $\underline{v}$ is the **maxmin** if the two following conditions are satisfied:

- Player 1 can **guarantee** $\underline{v}$: for any $\varepsilon > 0$, there exists a strategy σ of Player 1 and an integer N such that for any $n \geq N$ and any strategy τ of Player 2:

$$\gamma_n(\sigma, \tau) \geq \underline{v} - \varepsilon.$$

  (It follows from the uniformity in τ that if Player 1 can guarantee f both $\liminf_{n\to\infty} v_n$ and $\liminf_{\lambda\to 0} v_\lambda$ will be greater than f.)
- Player 2 can **defend** $\underline{v}$: for any $\varepsilon > 0$ and any strategy σ of Player 1, there exist an integer N and a strategy τ of Player 2 such that for all $n \geq N$:

$$\gamma_n(\sigma, \tau) \leq \underline{v} + \varepsilon.$$

  (Note that to satisfy this requirement is stronger than to contradict the previous condition; hence the existence of $\underline{v}$ is an issue.)

  A dual definition holds for the **minmax** $\overline{v}$. Whenever $\underline{v} = \overline{v}$, the game has a **uniform value**, denoted by v_∞. Remark that the existence of v_∞ implies:

$$v_\infty = \lim_{n\to\infty} v_n = \lim_{\lambda\to 0} v_\lambda.$$

We will be concerned here mainly with the asymptotic approach that relies more on the recursive structure and the related value operator and less on the construction of strategies. We will describe several recent ideas and results: the extension and the study of the Shapley operator, the variational approach to the asymptotic game, the use of the dual game, the limit game and the relation with differential games of fixed duration.

2 The Operator Approach

In this section the games are analyzed through the recursive relations that basically extend the dynamic programming principle.

2.1 Operators and Games

2.1.1 Stochastic Games and Shapley Operator

Consider a stochastic game with parameter space Ω, action spaces I and J and real bounded payoff function g from $\Omega \times I \times J$. This corresponds to the model of Section 1 with $M = \Omega$ and where the initial probability π is the law of a parameter ω_1 announced to both. In addition at each stage $t+1$, the transition $Q(\cdot|\omega_t, i_t, j_t)$ determines the new parameter ω_{t+1} and the signal for each player a_{t+1} or b_{t+1} contains at least the information ω_{t+1}. It follows that Ω will be the natural state space on which v_n and v_λ are defined.

Explicitly, let $X = \Delta(I)$ and $Y = \Delta(J)$ and extend by bilinearity g and Q to $X \times Y$. The Shapley (1953) operator Ψ acts on the set $\mathcal{F}$ of real bounded measurable functions f on Ω as follows:

$$\Psi(f)(\omega) = \mathtt{val}_{X\times Y}\left\{g(\omega, x, y) + \int_\Omega f(\omega')Q(d\omega'|\omega, x, y)\right\} \tag{1}$$

where $\mathtt{val}_{X\times Y}$ stands for the value operator:

$$\mathtt{val}_{X\times Y} = \max_X \min_Y = \min_Y \max_X .$$

A basic property is that Ψ is non-expansive on $\mathcal{F}$ endowed with the uniform norm:

$$\|\Psi(f) - \Psi(g)\| \leq \|f - g\| = \sup_{\omega\in\Omega} |f(\omega) - g(\omega)|.$$

Ψ determines the family of values through:

$$nv_n = \Psi^n(0), \qquad \frac{v_\lambda}{\lambda} = \Psi\left((1-\lambda)\frac{v_\lambda}{\lambda}\right). \tag{2}$$

The same relations hold for general state and action spaces when dealing with a complete subspace of $\mathcal{F}$ on which Ψ is well defined and which is stable under Ψ.

2.1.2 Non-expansive Mappings

The asymptotic approach of the game is thus related to the following problems: given T a non-expansive mapping on a linear normed space Z, study the iterates $T^n(0)/n = v_n$ as n goes to ∞ and the behavior of $\lambda z_\lambda = v_\lambda$, where z_λ is the fixed point of the mapping $z \mapsto T((1-\lambda)z)$, as λ goes to 0.

Kohlberg and Neyman (1981) proved the existence of a linear functional, f, of norm 1 on Z such that:

$$\lim_{n\to\infty} f(v_n) = \lim_{n\to\infty} \|v_n\| = \lim_{\lambda\to 0} f(v_\lambda) = \lim_{\lambda\to 0} \|v_\lambda\| = \inf_{z\in Z} \|T(z) - z\| \qquad (3)$$

Then they deduce that if Z is reflexive and strictly convex, there is weak convergence to one point, and if the dual space Z^* has a Frechet differentiable norm, the convergence is strong.

In our framework the norm is the uniform norm on a space of real bounded functions and is not strictly convex, see however section 5.3.

Neyman (2003) proved that if v_λ is of bounded variation in the sense that for any sequence λ_i decreasing to 0,

$$\sum_i \|v_{\lambda_{i+1}} - v_{\lambda_i}\| < \infty, \qquad (4)$$

then $\lim_{n\to\infty} v_n = \lim_{\lambda\to 0} v_\lambda$.

2.1.3 ε-Weighted and Projective Operators

Back to the framework of Section 2.1.1, it is also natural to introduce the ε-weighted operator:

$$\Phi(\varepsilon, f)(\omega) = \mathtt{val}_{X\times Y}\left\{\varepsilon g(\omega, x, y) + (1-\varepsilon)\int_\Omega f(\omega')Q(d\omega'|\omega, x, y)\right\}, \qquad (5)$$

related to the initial Shapley operator by:

$$\Phi(\varepsilon, f) = \varepsilon\Psi\left(\frac{(1-\varepsilon)f}{\varepsilon}\right). \qquad (6)$$

Then one has:

$$v_n = \Phi\left((1/n), v_{n-1}\right), \qquad v_\lambda = \Phi(\lambda, v_\lambda) \qquad (7)$$

which are the basic recursive equations for the values. The asymptotic study relies thus on the behavior of $\Phi(\varepsilon, \cdot)$, as ε goes to 0. Obviously, if v_n or v_λ converges uniformly, the limit w will satisfy:

$$w = \Phi(0, w) \qquad (8)$$

hence $\Phi(\varepsilon, \cdot)$ also appears as a perturbation of the "projective" operator $\mathcal{P}$ which gives the evaluation today of the payoff f tomorrow:

$$\mathcal{P}(f)(\omega) = \Phi(0, f)(\omega) = \mathtt{val}_{X\times Y}\int_\Omega f(\omega')Q(d\omega'|\omega, x, y). \qquad (9)$$

Explicitly, one has

$$\Phi(\varepsilon, f)(\omega) = \mathtt{val}_{X\times Y}\left\{\int_\Omega f(\omega')Q(d\omega'|\omega, x, y) + \varepsilon h(\omega, x, y)\right\}, \tag{10}$$

with $h(\omega, x, y) = g(\omega, x, y) - \int_\Omega f(\omega')Q(d\omega'|\omega, x, y)$.

One can also consider $\Phi(0, f)$ as a function of Ψ defined by:

$$\Phi(0, f)(\omega) = \lim_{\varepsilon\to 0} \varepsilon\Psi\left(\frac{(1-\varepsilon)}{\varepsilon} f\right)(\omega) = \lim_{\varepsilon\to 0} \varepsilon\Psi\left(\frac{f}{\varepsilon}\right)(\omega) \tag{11}$$

thus $\Phi(0, .)$ is the recession operator associated to Ψ. Note that this operator is independent of g and relates to the stochastic game only through the transition Q.

2.1.4 Games with Incomplete Information

A similar representation is available in the framework of repeated games with incomplete information, Aumann and Maschler (1995).

We will describe here the simple case of independent information and standard signaling. In the setup of Section 1, the parameter space M is a product $K\times L$ endowed with a product probability $\pi = p \otimes q \in \Delta(K)\times\Delta(L)$ and the initial signals are $a_1 = k_1$, $b_1 = \ell_1$. Hence the players have partial private information on the parameter (k_1, ℓ_1). This one is fixed for the duration of the play ($(k_t, \ell_t) = (k_1, \ell_1)$) and the signals to the players reveal the previous moves $a_{t+1} = b_{t+1} = (i_t, j_t)$. A one-stage strategy of Player 1 is an element x in $\mathbf{X} = \Delta(I)^K$ (resp. y in $\mathbf{Y} = \Delta(J)^L$ for Player 2).

We represent now this game as a stochastic game. The basic state space is $\chi = \Delta(K)\times\Delta(L)$ and corresponds to the beliefs of the players on the parameter along the play. The transition is given by a map Π from $\chi\times\mathbf{X}\times\mathbf{Y}$ to probabilities on χ with $\Pi((p(i), q(j))|(p, q), x, y) = x(i)y(j)$, where $p(i)$ is the conditional probability on K given the move i and $x(i)$ the probability of this move (and similarly for the other variable). Explicitly: $x(i) = \sum_k p^k x_i^k$ and $p^k(i) = (p^k x_i^k)/(x(i))$. Ψ is now an operator on the set of real bounded saddle (concave/convex) functions on χ, Rosenberg and Sorin (2001):

$$\Psi(f)(p, q) = \mathtt{val}_{\mathbf{X}\times\mathbf{Y}}\left\{g(p, q, x, y) + \int_\chi f(p', q')\Pi(d(p', q')|(p, q), x, y)\right\} \tag{12}$$

with $g(p, q, x, y) = \sum_{k,\ell} p^k q^\ell g\big(k, \ell, x^k, y^\ell\big)$. Then one establishes recursive formula for v_n and v_λ, Mertens, Sorin and Zamir (1994), similar to the ones described in section 2.1.1.

Note that by the definition of Π, the state variable is updated as a function of the one-stage strategies of the players, which are not public information during the play. The argument is thus first to prove the existence of a value (v_n or v_λ) and then using the minmax theorem to construct an equivalent game, in the sense of having the same sequence of values, in which one-stage strategies are announced. This last game is now reducible to a stochastic game.

2.1.5 General Recursive Structure

More generally a recursive structure holds for games described in Section 1 and we follow the construction in Mertens, Sorin and Zamir (1994), Sections III.1, III.2, IV.3.

Consider for example a game with lack of information on one side (described as in Section 2.1.4 with L of cardinal 1) and with signals so that the conditional probabilities of Player 2 on the parameter space are unknown to Player 1, but Player 1 has probabilities on them. In addition Player 2 has probabilities on those beliefs of Player 1 and so on.

The recursive structure thus relies on the construction of the universal belief space, Mertens and Zamir (1985), that represents this infinite hierarchy of beliefs: $\Xi = M \times \Theta^1 \times \Theta^2$, where Θ^i, homeomorphic to $\Delta(M \times \Theta^{-i})$, is the type set of Player i, The signaling structure in the game, just before the moves at stage t, describes an information scheme that induces a consistent distribution on Ξ. This is referred to as the entrance law $\mathcal{P}_t \in \Delta(\Xi)$. The entrance law $\mathcal{P}_t$ and the (behavioral) strategies at stage t (say α_t and β_t) from type set to mixed move set determine the current payoff and the new entrance law $\mathcal{P}_{t+1} = H(\mathcal{P}_t, \alpha_t, \beta_t)$. This updating rule is the basis of the recursive structure. The stationary aspect of the repeated game is expressed by the fact that H does not depend on t. The Shapley operator is defined on the set of real bounded functions on $\Delta(\Xi)$ by:

$$\Psi(f)(\mathcal{P}) = \sup_{\alpha} \inf_{\beta} \{g(\mathcal{P}, \alpha, \beta) + f(H(\mathcal{P}, \alpha, \beta))\},$$

(there is no indication at this level that sup inf commutes for all f) and the usual relations hold, see Mertens, Sorin and Zamir (1994) Section IV.3:

$$(n+1)v_{n+1}(\mathcal{P}) = \mathtt{val}_{\alpha \times \beta} \{g(\mathcal{P}, \alpha, \beta) + n v_n(H(\mathcal{P}, \alpha, \beta))\},$$

$$v_\lambda(\mathcal{P}) = \mathtt{val}_{\alpha \times \beta} \{\lambda g(\mathcal{P}, \alpha, \beta) + (1-\lambda) v_\lambda(H(\mathcal{P}, \alpha, \beta))\},$$

where $\mathtt{val}_{\alpha \times \beta} = \sup_\alpha \inf_\beta = \inf_\beta \sup_\alpha$ is the value operator for the "one stage game on $\mathcal{P}$".

We have here a "deterministic" stochastic game: in the framework of a regular stochastic game, it would correspond to working at the level of distributions on the state space, $\Delta(\Omega)$.

2.2 Variational Inequalities

We use here the previous formulations to obtain properties on the asymptotic behavior of the values, following Rosenberg and Sorin (2001).

2.2.1 A Basic Inequality

We first introduce sets of functions that will correspond to upper and lower bounds on the sequences of values. This allows us, for certain classes of games, to identify the asymptotic value through variational inequalities. The starting point is the next inequality.

Given $\delta > 0$, assume that the function f from Ω to $\mathbb{R}$ satisfies, for all R large enough

$$\Psi(Rf) \leq (R+1)f + \delta.$$

This gives $\Psi(R(f+\delta)) \leq (R+1)(f+\delta)$ and implies:

$$\limsup_{n\to\infty} v_n \leq f + \delta,$$

as well as

$$\limsup_{\lambda\to 0} v_\lambda \leq f + \delta.$$

In particular if f belongs to the set $\mathcal{C}^+$ of functions satisfying the stronger condition: for all $\delta > 0$ there exists R_δ such that $R \geq R_\delta$ implies

$$\Psi(Rf) \leq (R+1)f + \delta \tag{13}$$

and one obtains that both $\limsup_{n\to\infty} v_n$ and $\limsup_{\lambda\to 0} v_\lambda$ are less than f.

2.2.2 Finite State Space

We first apply the above results to **absorbing games**: these are stochastic games where all states except one are absorbing, hence the state can change at most once. It follows that the study on Ω can be reduced to that at one point. In this case, one has easily:

i) $\Psi(f) \leq \|g\|_\infty + f$,

ii) $\Psi(Rf) - (R+1)f$ is strictly decreasing in f:

In fact, let $g - f = d > 0$, then

$$\Psi(Rg) - \Psi(Rf) \leq \Psi(R(f+d)) - \Psi(Rf) \leq Rd = (R+1)(g-f) - d,$$

so that

$$\Psi(Rf) - (R+1)f - (\Psi(Rg) - (R+1)g) \geq d,$$

iii) $\Psi(Rf) - (R+1)f$ is decreasing in R, for $f \geq 0$:

$$\Psi((R+R')f) - (R+R'+1)f \leq \Psi(Rf) - (R+1)f.$$

Define $\mathcal{C}^-$ in a way symmetric to (13). From $i), ii)$ and $iii)$, there exists an element $f \in \mathcal{C}^+ \cap \mathcal{C}^-$ and it is thus equal to both $\lim_{n\to\infty} v_n$ and $\lim_{\lambda\to 0} v_\lambda$. This extends the initial proof of Kohlberg (1974).

In the framework of **recursive games** where the payoff in all non absorbing states (say Ω_0) is 0, the Shapley operator is defined on real functions f on Ω_0 (with an obvious extension $\overline{f}$ to Ω) by:

$$\Psi(f)(\omega) = \mathtt{val}_{X\times Y} \int_\Omega \overline{f}(\omega') Q(d\omega'|\omega, x, y).$$

It follows that condition (13) reduces to

$$\Psi(f) \leq f \quad \text{and moreover} \quad f(\omega) < 0 \quad \text{implies} \quad \Psi(f)(\omega) < f(\omega), \tag{14}$$

which defines a set $\mathcal{E}^+$. Everett (1957) has shown that the closure of the set $\mathcal{E}^+$ and of its symmetric $\mathcal{E}^-$ have a non-empty intersection from which one deduces that $\lim_{n\to\infty} v_n = \lim_{\lambda\to 0} v_\lambda$ exists and is the only element of this intersection.

A recent result of Rosenberg and Vieille (2000) drastically extends this property to recursive games with lack of information on both sides. The proof relies on an explicit construction of strategies. Let w an accumulation point of the family v_λ as λ goes to 0. Player 1 will play optimally in the discounted game with a small discount factor if w is larger than $\varepsilon > 0$ at the current value of the parameter and optimally in the projective game $\Psi(0, w)$ otherwise. The sub-martingale property of the value function and a bound on the upcrossings of $[0, \varepsilon]$ are used to prove that $\liminf_{n\to\infty} v_n \geq w$, hence the result.

2.2.3 Simple Convergence

More generally, when Ω is not finite, one can introduce the larger class of functions $\mathcal{S}^+$ where in condition (13) only simple convergence is required:
for all $\delta > 0$ and all ω, there exists $R_{\delta,\omega}$ such that $R \geq R_{\delta,\omega}$ implies

$$\Psi(Rf)(\omega) \leq (R+1)f(\omega) + \delta \tag{15}$$

or

$$\theta^+(f)(\omega) = \limsup_{R\to\infty} \{\Psi(Rf)(\omega) - (R+1)f(\omega)\} \leq 0. \tag{16}$$

In the case of continuous functions on a compact set Ω, an argument similar to point $ii)$ above implies that $f^+ \geq f^-$ for any functions $f^+ \in \mathcal{S}^+$ and $f^- \in \mathcal{S}^-$ (defined similarly with $\theta^-(f^-) \leq 0$). Hence the intersection of the closures of $\mathcal{S}^+$ and $\mathcal{S}^-$ contains at most one point.

This argument suffices for the class of games with incomplete information on both sides: any accumulation point w of the family v_λ as $\lambda \to 0$ belongs to the closure of $\mathcal{S}^+$, hence by symmetry the existence of a limit follows. A similar argument holds for $\limsup_{n\to\infty} v_n$.

In the framework of (finite) absorbing games with incomplete information on one side, where the parameter is both changing and unknown, Rosenberg (2000) used similar tools in a very sophisticated way to obtain the first general results of existence of an asymptotic value concerning this class of games. First she shows that any w as above belongs to the closure of $\mathcal{S}^+$. Then that at any point (p, q), $\limsup_{\lambda \to 0} v_\lambda(p, q) \leq w(p, q)$ which again implies convergence. A similar analysis is done for $\lim_{n \to \infty} v_n$.

Remarks

1) Many of the results above only used the following two properties of the operator Ψ, Sorin (2004):
   Ψ is monotonic
   Ψ reduces the constants: for all $\delta > 0$, $\Psi(f + \delta) \leq \Psi(f) + \delta$.
2) The initial and basic proof of convergence of $\lim_{\lambda \to 0} v_\lambda$ for stochastic games relies on the finiteness of the sets involved (Ω, I and J). Bewley and Kohlberg (1976a) used an algebraic approach and proved that v_λ is an algebraic function of λ, from which existence of $\lim_{\lambda \to 0} v_\lambda$ and equality with $\lim_{n \to \infty} v_n$ follows.
3) The results sketched above correspond to three levels of proofs:
   a) The non emptiness of the intersection of the closure of $\mathcal{C}^+$ and $\mathcal{C}^-$. This set contains then one point, namely $\lim_{n \to \infty} v_n = \lim_{\lambda \to 0} v_\lambda$.
   b) For continuous functions on a compact set Ω: any accumulation point of the family of values (as the length goes to ∞) belongs to the intersection of the closures of $\mathcal{S}^+$ and $\mathcal{S}^-$, which contains at most one element.
   c) A property of some accumulation point (related to $\mathcal{S}^+$ or $\mathcal{S}^-$) and a contradiction if two accumulation points differ.

2.3 The Derived Game

We follow here Rosenberg and Sorin (2001). Still dealing with the Shapley operator, condition (15) can be written in a simpler form. This relies, using the expression (10), on the existence of a limit:

$$\varphi(f)(\omega) = \lim_{\varepsilon \to 0^+} \frac{\Phi(\varepsilon, f)(\omega) - \Phi(0, f)(\omega)}{\varepsilon}$$

extending a result of Mills (1956), see also Mertens, Sorin and Zamir (1994), Section 1.1, Ex. 6. More precisely $\varphi(f)(\omega)$ is the value of the "derived game" with payoff $h(\omega, x, y)$, see (10), played on the product of the subsets of optimal strategies in $\Phi(0, f)$. The relation with (16) is given by:

$$\theta^+(f) = \theta^-(f) = \begin{cases} \varphi(f) & \text{if } \Phi(0, f) = f \\ +\infty & \text{if } \Phi(0, f) > f \\ -\infty & \text{if } \Phi(0, f) < f \end{cases}$$

In the setup of games with incomplete information, the family $v_\lambda(p, q)$ is uniformly Lipschitz and any accumulation point as $\lambda \to 0$ is a saddle function $w(p, q)$ satisfying: $\Phi(0, w) = w$. Thus one wants to identify one point in the set of solutions of equation (8) which contains in fact all saddle functions.

For this purpose, one considers the set $\mathcal{A}^+$ of continuous saddle functions f on $\Delta(K)\times\Delta(L)$ such that for any positive strictly concave perturbation η on $\Delta(K)$: $\varphi(f + \eta) \leq 0$. The proof that w belongs to $\mathcal{A}^+$, which is included in the closure of $\mathcal{S}^+$, shows then the convergence of the family v_λ. A similar argument holds for v_n, which in addition implies equality of the limits. Note that the proof relies on the explicit description of $\varphi(f)$ as the value of the derived game.

In addition one obtains the following geometric property. Given f on $\Delta(K)$, say that p is an extreme point of f, $p \in \mathcal{E}f$, if $(p, f(p))$ cannot be expressed as a convex combination of a finite family $\{(p_i, f(p_i))\}$. Then one shows that for any $f \in \mathcal{A}^-$, $f(p, q) \leq u(p, q)$ holds at any extreme point p of $f(\cdot, q)$, where u is the value of the non-revealing game or equivalently:

$$u(p, q) = \mathtt{val}_{\Delta(I)\times\Delta(J)} \sum\nolimits_{k,\ell} p^k q^\ell g(k, \ell, x, y).$$

Hence $v = \lim_{n\to\infty} v_n = \lim_{\lambda\to 0} v_\lambda$ is a saddle continuous function satisfying both inequalities:

$$\begin{aligned} &p \in \mathcal{E}v(\cdot, q) \Rightarrow v(p, q) \leq u(p, q) \\ &q \in \mathcal{E}v(p, \cdot) \Rightarrow v(p, q) \geq u(p, q) \end{aligned} \tag{17}$$

and it is easy to see that it is the only one, Laraki (2001a).

Given a function f on a compact convex set C, let us denote by $\mathtt{Cav}_C f$ the smallest function concave and greater than f on C. By noticing that, for a function f continuous and concave on a compact convex set C, the property $p \in \mathcal{E}f \Rightarrow f(p) \leq g(p)$ is equivalent to $f = \mathtt{Cav}_C \min(g, f)$ one recovers the famous characterization of v due to Mertens and Zamir (1971):

$$v = \mathtt{Cav}_{\Delta(K)} \min(u, v) = \mathtt{Vex}_{\Delta(L)} \max(u, v) \tag{18}$$

where $\mathtt{Cav}_{\Delta(K)} f(p, q)$ stands for $\mathtt{Cav}_{\Delta(K)} f(., q)(p)$.

2.4 The Splitting Game

This section deals again with games with incomplete information on both sides as defined in Section 2.1.4 and follows Laraki (2001a). The operator approach is then extended to more general games. Recall that the recursive formula for the discounted value is:

$$v_\lambda(p, q) = \Phi(\lambda, v_\lambda)(p, q) = \mathtt{val}_{\mathbf{X}\times\mathbf{Y}}\{\lambda g(p, q, x, y) + (1 - \lambda)E(v_\lambda(p', q'))\}$$

where (p', q') is the new posterior distribution and E stands for the expectation induced by p, q, x, y. Denoting by $\mathbf{X}_\lambda(p, q)$ the set of optimal strategies of Player 1 in $\Phi(\lambda, v_\lambda)(p, q)$ and using the concavity of v_λ one deduces:

$$\max_{\mathbf{X}_\lambda(p,q)} \min_Y \{g(p, q, x, y) - E(v_\lambda(p', q))\} \geq 0.$$

Let w be an accumulation point of the family $\{v_\lambda\}$ as λ goes to 0 and let $p \in \mathcal{E}w(., q)$. Then the set of optimal strategies for Player 1 in $\Phi(0, w)(p, q)$ is included in the set $NR^1(p)$ of non-revealing strategies (namely with $x(i)\|p(i) - p\| = 0$, recall (8)), hence by uppersemicontinuity one gets from above:

$$\max_{NR^1(p)} \min_Y \{g(p, q, x, y) - E(w(p', q))\}$$
$$= \max_X \min_Y \{g(p, q, x, y) - w(p, q)\} \geq 0$$

hence $u(p, q) \geq w(p, q)$ which is condition (17).

We now generalize this approach.

Since the payoff $g(p, q, x, y) = \sum p^k q^\ell x_i^k y_j^\ell g(k, \ell, i, j)$ is linear it can be written as $\sum_{i,j} x(i) y(j) g(p(i), q(j), i, j)$ so that the Shapley operator is:

$$\Psi(f)(p, q) = \mathtt{val}_{X \times Y} \left\{ \sum\nolimits_{i,j} [g(p(i), q(j), i, j) + f(p(i), q(j))] x(i) y(j) \right\}$$

and one can consider that Player 1's strategy set is the family of random variables from I to $\Delta(K)$ with expectation p. In short, rather than deducing the state variables from the strategies, the state variables are now taken as strategies. The second step is to change the payoffs (introducing a perturbation of the order of $n^{-1/2}$ in v_n) and to replace $g(p(i), q(j), i, j)$ by the value of the "local game" at this state $u(p(i), q(j))$. There is no reason now to keep the range of the martingale finite so that the operator becomes:

$$\mathcal{S}(f)(p, q) = \mathtt{val}_{\Delta_p^2(K) \times \Delta_q^2(L)} E\{u(\widetilde{p}, \widetilde{q}) + f(\widetilde{p}, \widetilde{q})\},$$

where $\Delta_p^2(K)$ is the set of random variables $\widetilde{p}$ with values in $\Delta(K)$ and expectation p. The corresponding game is called the "splitting game". The recursive formula is now different:

$$\frac{w_\lambda}{\lambda} = \mathcal{S}\left(\frac{(1-\lambda)}{\lambda} w_\lambda\right)$$

but with the same proof as above, it leads to the existence of $w = \lim_{\lambda \to 0} w_\lambda$ satisfying the same functional equations:

$$w = \mathtt{Cav}_{\Delta(K)} \min(u, w) = \mathtt{Vex}_{\Delta(L)} \max(u, w).$$

These tools then allow us to extend the operator $u \mapsto M(u) = w$ (existence and uniqueness of a solution) to more general products of compact convex sets $P \times Q$ and functions u, see Laraki (2001b).

3 Duality Properties

3.1 Incomplete Information, Convexity and Duality

Consider a two-person zero-sum game with incomplete information on one side defined by sets of actions S and T, a finite parameter space K, a probability p on K and for each k a real payoff function G^k on $S \times T$. Assume S and T convex and for each k, G^k bounded and bilinear on $S \times T$. The game is played as follows: $k \in K$ is selected according to p and told to Player 1 (the maximizer) while Player 2 only knows p.

In normal form, Player 1 chooses $\mathbf{s} = \{s^k\}$ in S^K, Player 2 chooses t in T and the payoff is

$$G^p(\mathbf{s}, t) = \sum_k p^k G^k\big(s^k, t\big).$$

Assume that this game has a value

$$v(p) = \sup_{S^K} \inf_T G^p(\mathbf{s}, t) = \inf_T \sup_{S^K} G^p(\mathbf{s}, t),$$

then v is concave and continuous on the set $\Delta(K)$ of probabilities on K.

Following De Meyer (1996a) one introduces, given $z \in \mathbb{R}^k$, the "dual game" $G^*(z)$ where Player 1 chooses k, then Player 1 plays s in S (resp. Player 2 plays t in T) and the payoff is

$$h[z](k, s; t) = G^k(s, t) - z^k.$$

Translating in normal form, Player 1 chooses $(p, \mathbf{s})$ in $\Delta(K) \times S^K$, Player 2 chooses t in T and the payoff is $\sum_k p^k h[z]\big(k, s^k; t\big) = G^p(\mathbf{s}, t) - \langle p, z \rangle$.

Then the game $G^*(z)$ has a value $w(z)$, which is convex and continuous on $\mathbb{R}^K$ and the following duality relations holds:

$$w(z) = \max_{p \in \Delta(K)} \{v(p) - \langle p, z \rangle\} = \Lambda_s(v)(z), \tag{19}$$

$$v(p) = \inf_{z \in \mathbb{R}^K} \{w(z) + \langle p, z \rangle\} = \Lambda_i(w)(p), \tag{20}$$

Two consequences are:

Property 3.1. Given z, let p achieve the maximum in (19) and $\mathbf{s}$ be ε-optimal in G^p: then $(p, \mathbf{s})$ is ε-optimal in $G^*(z)$.

Given p, let z achieve the infimum up to ε in (20) and t be ε-optimal in $G^*(z)$: then t is also 2ε-optimal in G^p.

Property 3.2. Let G' be another game on $K \times S \times T$ with corresponding primal and dual values v' and w'. Since Fenchel's transform is an isometry one has

$$\|v - v'\|_{\Delta(K)} = \|w - w'\|_{\mathbb{R}^K}.$$

3.2 The Dual of a Repeated Game with Incomplete Information

We consider now repeated games with incomplete information on one side as introduced in 2.1.4. (with L reduced to one point), and study their duals, following De Meyer (1996b). Obviously the previous analysis applies when working with mixed strategies in the normalized form.

3.2.1 Dual Recursive Formula

The use of the dual game will be of interest for two purposes: construction of optimal strategies for the uninformed player and asymptotic analysis. In both cases the starting point is the recursive formula in the original game.

$$\begin{aligned} F(p) &= \Phi(\varepsilon, f)(p) \\ &= \mathtt{val}_{x\in\mathbf{X}, y\in\mathbf{Y}} \left\{ \varepsilon \sum_k p^k x^k G^k y + (1-\varepsilon) \sum_i x(i) f(p(i)) \right\}, \end{aligned} \tag{21}$$

where we write G^k_{ij} for $g(k, i, j)$. Then one obtains:

$$\begin{aligned} F^*(z) &= \max_{p\in\Delta(K)} \{F(p) - \langle p, z\rangle\} \\ &= \max_{p,x} \min_{y} \left\{ \varepsilon \sum_k p^k x^k G^k y + (1-\varepsilon) \sum_i x(i) f(p(i)) - \langle p, z\rangle \right\}. \end{aligned}$$

We represent now the couple (p, x) in $\Delta(K)\times\Delta(I)^K$ as an element π in $\Delta(K\times I)$: p is the marginal on K and x^k the conditional probability on I given k:

$$F^*(z) = \max_{\pi} \min_{y} \left\{ \varepsilon \sum_{i,k} \pi(i,k) G^k_i y + (1-\varepsilon) \sum_i \pi(i) f(\pi(.|i)) - \langle p, z\rangle \right\}.$$

The concavity of f w.r.t. p implies the concavity of $\sum_i \pi(i) f(\pi(.|i))$ w.r.t. π. This allows us to use the minmax theorem leading to:

$$F^* = \min_{y} \max_{\pi} \left\{ \varepsilon \sum_{i,k} \pi(i,k) G^k_i y + (1-\varepsilon) \sum_i \pi(i) f(\pi(.|i)) - \langle p, z\rangle \right\},$$

hence, since $p^k = \sum_i \pi(i)\pi(k|i)$:

$$\begin{aligned} F^*(z) = \min_{y} \max_{\pi(i)} \Bigg\{ \sum_i \pi(i)(1-\varepsilon) \max_{\pi(.|i)} \Bigg[& f(\pi(.|i)) \\ & - \left\langle \pi(.|i), \frac{1}{1-\varepsilon} z - \frac{\varepsilon}{1-\varepsilon} G_i y \right\rangle \Bigg] \Bigg\} \end{aligned}$$

where $G_i y$ is the vector $\{G_i^k y\}$. This finally leads to the dual recursive formula:

$$F^*(z) = \min_y \max_i (1-\varepsilon) f^* \left(\frac{1}{1-\varepsilon} z - \frac{\varepsilon}{1-\varepsilon} G_i y \right). \tag{22}$$

The main advantage of dealing with (22) rather than with (21) is that the state variable is known by Player 2 (who controls y and observes i) and evolves smoothly from z to $z + (\varepsilon/(1-\varepsilon))(z - G_i y)$.

3.2.2 Properties of Optimal Strategies

Rosenberg (1998) extended the previous duality to games having at the same time incomplete information and stochastic transition on the parameters. There are then two duality operators (D^1 and D^2) corresponding to the private information of each player. D^1 associates to each function on $\Omega \times \Delta(K) \times \Delta(L)$ a function on $\Omega \times \Delta(K) \times \mathbb{R}^L$. The duality is taken with respect to the unknown parameter of Player 1 replacing q by a vector in $\mathbb{R}^L$. The extension of formula (22) to each dual game allows us to deduce properties of optimal strategies in this dual game for each player. In the discounted case, Player 1 has stationary optimal strategies on a private state space of the form $\Omega \times \Delta(K) \times \mathbb{R}^L$. The component on Ω is the publicly known stochastic parameter; the second component p is the posterior distribution on $\Delta(K)$ that is induced by the use of x: it corresponds to the transmission of information to Player 2; the last one is a vector payoff indexed by the unknown parameter $\ell \in L$ that summarizes the past sequence of payoffs. Similarly, in the finitely repeated game, Player 1 has an optimal strategy which is Markovian on $\Omega \times \Delta(K) \times \mathbb{R}^L$. Obviously dual properties hold for Player 2.

Recall that as soon as lack of information on both sides is present the recursive formula does not allow us to construct inductively optimal strategies (except in specific classes, like games with almost perfect information where a construction similar to the one above could be done, Ponssard and Sorin (1982)). It simply expresses a property satisfied by an alternative game having the same sequence of values, but not the same signals along the play, hence not the same strategy sets. However the use of the dual game allows us, through Property 3.1 (Section 3.1), to deduce optimal strategies in the primal game from optimal strategies in the dual game, and hence to recover an inductive procedure for constructing optimal strategies.

Further properties of the duality operators have been obtained in Laraki (2000). First one can apply the (partial 2) duality operator D^2 to the (partial 1) dual game $D^1(\Gamma)$, then the duality transformations commute and other representations of the global dual game $D^1 \circ D^2(\Gamma) = D^2 \circ D^1(\Gamma)$ are established.

3.2.3 Asymptotic Analysis and Approximate Fixed Points

This section follows De Meyer and Rosenberg (1999). Going back to the class of games with incomplete information on one side, Aumann and Maschler's theorem on the convergence of the families v_n or v_λ to $\mathtt{Cav}_{\Delta(K)}u$ will appear as a consequence of the convergence of the conjugate functions w_n (value of the dual game Γ_n^*) or $w_\lambda(z)$ (for Γ_λ^*) to the Fenchel conjugate of u.

Explicitly let

$$\Lambda_s(u)(z) = \max_{\Delta(K)}\{u(p) - \langle p, z\rangle\},$$

then the bi-conjugate

$$\Lambda_i \circ \Lambda_s(u)(z) = \min_{z \in \mathbb{R}^K}\{\Lambda_s(u)(z) - \langle p, z\rangle\}$$

equals $\mathtt{Cav}_{\Delta(K)}u$. Using property 3.2 in Section 3.1 it is enough to prove the convergence of w_n or w_λ to $\Lambda_s(u)$. Heuristically one deduces from (22) that the limit w should satisfy:

$$w(z) = (1-\varepsilon)\min_Y \max_X \left\{ w(z) + \frac{\varepsilon}{(1-\varepsilon)}\langle \nabla w(z), z - xGy\rangle \right\},$$

which leads to the partial differential equation:

$$-w(z) + \langle \nabla w(z), z\rangle + u(-\nabla w(z)\} = 0, \tag{23}$$

where we recall that $u(q) = \min_Y \max_X \left\{\sum_k q^k x G^k y\right\}$.

Fenchel duality gives:

$$\Lambda_s u(z) - u(-q) = \langle q, z\rangle$$

for $-q \in \partial \Lambda_s u(z)$, which shows that $\Lambda_s u$ is a solution (in a weak sense) of (23). The actual proof uses a general property of approximate operators and fixed points that we described now. Consider a family of operators Ψ_n on a Banach space Z with the following contracting property:

$$\|\Psi_{n+1}(f) - \Psi_{n+1}(g)\| \le \left(\frac{n}{n+1}\right)^a \|f - g\|,$$

for some positive constant a, and n large enough. For example, Ψ is non-expansive and $\Psi_{n+1}(.) = \Phi\left(\frac{1}{n+1}, .\right)$. Define a sequence in Z by $f_0 = 0$ and $f_{n+1} = \Psi_{n+1}(f_n)$. Then if a sequence g_n satisfies an approximate induction in the sense that, for some positive b and n large enough:

$$\|\Psi_{n+1}(g_n) - g_{n+1}\| \le \frac{1}{(n+1)^{1+b}},$$

and g_n converges to g, then f_n converges to g also.

The result on the convergence of w_n follows by choosing g_n as a smooth perturbation of $\Lambda_s u$, like $g_n(z) = E\left[\Lambda_s u\left(z + (X/\sqrt{n})\right)\right]$, X being a cantered reduced normal random variable.

A similar property holds for the sequence w_λ.

3.2.4 Speed of Convergence

The recursive formula and its dual also play a crucial role in the recent deep and astonishing result of Mertens (1998). Given a game with incomplete information on one side with finite state and action spaces but allowing for measurable signal spaces the speed of convergence of v_n to its limit is bounded by $C\big((\ln n)/n\big)^{1/3}$ and this is the best bound. (Recall that the corresponding order of magnitude is $n^{-1/2}$ for standard signaling and $n^{-1/3}$ for state independent signals – even allowing for lack of information on both sides.)

3.3 The Differential Dual Game

This section follows Laraki (2002) and starts again from equation (22). The recursive formula for the value w_n of the dual of the n stage game can be written, since $w_n(z)$ is convex, as:

$$w_n(z) = \min_y \max_x \left(1 - \frac{1}{n}\right) w_{n-1}\left(\frac{1}{(1-(1/n))}\left(z - \frac{1}{n}xGy\right)\right). \qquad (24)$$

This leads us to consider w_n as the nth discretization of the upper value of a differential game.

Explicitly consider the differential game (of fixed duration) on $[0, 1]$ with dynamic $\zeta(t) \in \mathbb{R}^K$ given by:

$$d\zeta/dt = x_t G y_t, \quad \zeta(0) = -z$$

$x_t \in X$, $y_t \in Y$ and terminal payoff $\max_k \zeta^k(1)$.

Given a partition $\Pi = \{t_0 = 0, ..., t_k, ...\}$ with $\theta_k = t_k - t_{k-1}$ and $\sum_{k=1}^{\infty} \theta_k = 1$ we consider the discretization of the game adapted to Π. Let $W^+_\Pi(t_k, \zeta)$ denote the upper value (correspondingly to the case where Player 2 plays first) of the game starting at time t_k from state ζ. It satisfies:

$$W^+_\Pi(t_k, \zeta) = \min_y \max_x W^+_\Pi(t_{k+1}, \zeta + \theta_{k+1} xGy).$$

In particular if Π_n is the uniform discretization with mesh $(1/n)$ one obtains:

$$W^+_{\Pi_n}(0, \zeta) = \min_y \max_x W^+_{\Pi_n}\left(\frac{1}{n}, \zeta + \frac{1}{n}xGy\right)$$

and by time homogeneity, $W^+_{\Pi_n}\left(\frac{1}{n}, \zeta\right) = \left(1 - \frac{1}{n}\right) W^+_{\Pi_{n-1}}\left(0, \frac{\zeta}{1-(1/n)}\right)$, so that:

$$W^+_{\Pi_n}(0, \zeta) = \min_y \max_x \left(1 - \frac{1}{n}\right) W^+_{\Pi_{n-1}}\left(0, \frac{\zeta + (1/n)xGy}{1-(1/n)}\right). \qquad (25)$$

Hence (24) and (25) prove that $w_n(z)$ and $W^+_{\Pi_n}(0, -z)$ satisfy the same recursive equation. They have the same initial value for $n = 1$ hence they coincide.

Basic results of the theory of differential games (see e.g. Souganidis (1999)) show that the game starting at time t from state ζ has a value $\varphi(t, \zeta)$, which is the only viscosity solution, uniformly continuous in ζ uniformly in t, of the following partial differential equation with boundary condition:

$$\frac{\partial \varphi}{\partial t} + u(\nabla \varphi) = 0, \quad \varphi(1, \zeta) = \max_k \zeta^k. \tag{26}$$

One thus obtains $\varphi(0, -z) = \lim_{n\to\infty} W^+_{\Pi_n}(0, -z) = \lim_{n\to\infty} w_n(z) = w(z)$. The time homogeneity property gives $\varphi(t, \zeta) = (1-t)\varphi\left(0, \zeta/(1-t)\right)$, so that w is a solution of

$$f(x) - \langle x, \nabla f(x)\rangle - u(-\nabla f(x)) = 0, \quad \lim_{\alpha\to 0} \alpha f(x/\alpha) = \max_k \left\{ -x^k \right\}$$

which is the previous equation (23) but with a limit (recession) condition.

One can identify the solution of (26), written with $\psi(t, \zeta) = \varphi(1-t, \zeta)$ as satisfying:

$$\frac{\partial \psi}{\partial t} + L(\nabla \psi) = 0 \quad \psi(0, \zeta) = b(\zeta)$$

with L continuous, b uniformly Lipschitz and convex. Hence, using Hopf's representation formula, one obtains:

$$\psi(t, \zeta) = \sup_{p\in\mathbb{R}^K} \inf_{q\in\mathbb{R}^K} \{b(q) + \langle p, \zeta - q\rangle - tL(p)\}$$

which gives here:

$$\psi(t, \zeta) = \sup_{p\in\mathbb{R}^K} \inf_{q\in\mathbb{R}^K} \left\{ \max_k q^k + \langle p, \zeta - q\rangle + tu(p) \right\}$$

and finally $w(z) = \psi(1, -z) = \sup_{p\in\Delta(K)}\{u(p) - \langle p, z\rangle\} = \Lambda_s u(z)$, as in section 3.2.3.

In addition the results in Souganidis (1985) concerning the approximation schemes give a speed of convergence of $(\delta(\Pi))^{1/2}$ of W_Π to φ (where $\delta(\Pi)$ is the mesh of the subdivision Π), hence by duality one obtains Aumann and Maschler's (1995) bound:

$$\|v_n - \mathrm{Cav}_{\Delta(K)} u\| \le \frac{C}{\sqrt{n}}, \qquad \|v_\lambda - \mathrm{Cav}_{\Delta(K)} u\| \le C\sqrt{\lambda}$$

for some constant C.

A last result is a direct identification of the limit. Since w is the conjugate of a concave continuous function v on $\Delta(K)$ and $\varphi(t,\zeta) = (1-t)w\,(-\zeta/(1-t))$ the conditions on φ can be translated as conditions on v. More precisely the first order conditions in terms of local sub- and super-differentials imply that φ is a viscosity subsolution (resp. supersolution) of (26) if and only if v satisfies the first (resp. second) inequality in the variational system (17). In our framework this gives

$$p\in\mathcal{E}v \Rightarrow v(p) \leq u(p) \;\text{ and }\; v(p) \geq u(p),\ \forall p,$$

so that $v = \mathtt{Cav}_{\Delta(K)}u$.

4 The Game in Continuous Time

4.1 Repeated Games and Discretization

The main idea here is to consider a repeated game (in the compact case, i.e. with finite expected length) as a game played between time 0 and 1, the length of stage n being simply its relative weight in the evaluation. Non-negative numbers θ_n with $\sum_{n=1}^{\infty}\theta_n = 1$ define a partition Π of $[0,1]$ with $t_0 = 0$ and $t_n = \sum_{m\leq n}\theta_m$. The repeated game with payoff $\sum_n g_n\theta_n$ corresponds to the game in continuous time where changes in the moves can occur only at times t_m. The finite n-stage game is represented by the uniform partition Π_n with mesh $(1/n)$ while the λ-discounted game is associated to the partition Π_λ with $t_m = 1-(1-\lambda)^m$. In the framework of section 2.1.5. one obtains a recursive formula for the value $W_\Pi(t,\mathcal{P})$ of the game starting at time t with state variable $\mathcal{P}$:

$$W_\Pi(t_n,\mathcal{P}) = \mathtt{val}_{\alpha\times\beta}(\theta_{n+1}g(\mathcal{P},\alpha,\beta) + W_\Pi(t_{n+1}, H(\mathcal{P},\alpha,\beta))).$$

The fact that the payoff is time-independent is expressed by the relation:

$$W_\Pi(t_n,.) = (1-t_n)W_{\Pi[t_n]}(0,.)$$

where $\Pi[t_m]$ is the renormalization to the whole interval $[0,1]$ of Π restricted to $[t_m,1]$. By enlarging the state space and incorporating the payoff as new parameter, say ζ, we obtain new functions $L_\Pi(t,\zeta,\mathcal{P})$ with

$$L_\Pi(t_n,\zeta,\mathcal{P}) = \mathtt{val}_{\alpha\times\beta}L_\Pi(t_{n+1},\zeta+\theta_{n+1}g(\mathcal{P},\alpha,\beta), H(\mathcal{P},\alpha,\beta))$$

and

$$L_\Pi(t_n,\zeta,\mathcal{P}) = (1-t_n)L_{\Pi[t_n]}\left(0,\frac{\zeta}{1-t_n},\mathcal{P}\right).$$

This time normalization explains why the PDE obtained as a limit is homogeneous.

A first heuristic approach in this spirit is in Mertens and Zamir (1976a) where they study, for a specific example of repeated game with lack of information on one side, the limit of the "normalized error term" $\eta_n(p) = \sqrt{n}\,(v_n(p) - v_\infty(p))$, on $[0, 1]$. From the recursive formula for v_n, they deduce another one for the sequence η_n and obtain the following equation for the limit: $\varphi\varphi'' + 1 = 0$. It follows then that $\varphi(p)$ is the normal density evaluated at its p-quantile.

Consider now a simple variation of the Big Match game where Player 1 knows the true game while Player 2 does not and the payoffs are as follows:

	α	β
a	1*	0*
b	0	0

Game 1: Probability p

	α	β
a	0*	0*
b	0	1

Game 2: Probability $1 - p$

Sorin (1984) derives from the recursive formula the following equation for the limit of v_n: $(2 - p)\varphi(p) = (1 - p) - (1 - p)^2\varphi'(p)$ which leads to $\varphi(p) = (1 - p)\,\{1 - \exp(-p/(1 - p))\}$. Note that this function is not algebraic, which could not be the case for stochastic games nor for games with incomplete information on one side. (Moreover it is also equal to the max min $\underline{v}$.)

4.2 The Limit Game

The recursive formula may also, by exhibiting properties of optimal strategies, allow us to define an auxiliary game in continuous time, considered as a representation of the "limit game" on $[0, 1]$. Two examples are as follows.

A first class, Sorin (1984), corresponds to specific absorbing games with incomplete information on one side of the form:

a	a_1^{k*}	$\dots^{*}$	a_J^{k*}
b	b_1^k	$\dots$	b_J^k

Game k: Probability p^k

From the recursive formula one deduces that both players can be restricted to strategies independent of the past. One constructs then a game on $[0, 1]$ where Player 1's strategies are stopping times ρ^k corresponding to the first occurrence of a in game k, while Player 2's strategies are measurable functions f from $[0, 1]$ to $\Delta(J)$. The payoff is the integral from 0 to 1 of the instantaneous payoff at time t, $\sum_k p^k g_t^k(\rho^k, f)$ with

$$g_t^k(\rho, f) = \int_0^t a^k(f(s))\rho(ds) + (1 - \rho([0, t]))b^k(f(t)),$$

where $a^k(f) = \sum_j a_j^k f_j$ and similarly for b^k. This game has a value v and it is easy to show that $v = \lim_{n\to\infty} v_n = \lim_{\lambda\to 0} v_\lambda$. In fact discretizations of ε-optimal strategies in the limit game define strategies in G_n or G_λ and the payoff is continuous.

A much more elaborate construction is in De Meyer (1999). The starting point is again the asymptotic expansion of v_n for games with incomplete information on one side. More precisely the dual recursive formula for $\eta_n = \sqrt{n}\,(v_n - v_\infty)$ leads on one hand to an heuristic second-order PDE (E) and on the other to properties of optimal strategies for both players. One shows that any regular solution of (E) would be the limit of η_n. De Meyer constructs a family of games $\chi(z,t)$ on $[0,1]$, endowed with a Brownian filtration where strategies for each player are adapted stochastic processes and the payoff is defined through a stochastic integral on $[0,1]$. The existence of a value $W(z,t)$ and optimal strategies in $\chi(z,t)$ are then established. One deduces that, under optimal strategies of the players, the state variable $Z_s(z,t)$ in $\chi(z,t)$, $t \le s \le 1$, follows a stochastic differential equation. The value being constant on such trajectories one obtains that $W(z,0)$ is a solution to (E) – where the regularity remains to be proved.

Note that this approach is somehow a dual of the one used in differential games where the initial model is in continuous time and is analyzed through discretization. Here the game on $[0,1]$ is an idealization of discrete time model with a large number of stages.

4.3 Repeated Games and Differential Games

The first example of resolution of a repeated game trough a differential game is due to Vieille (1992). Consider a repeated game with vector payoffs described by a function g from $I \times J$ to $\mathbb{R}^K$. Given a compact set C in $\mathbb{R}^K$ let $f(z) = -d(z,C)$ where d is the euclidean distance and defines the n stage repeated game with standard information G_n. The sequence of payoffs is $g_1 = g(i_1 j_1), \cdots, g_n$ with average $\overline{g}_n$ and the reward is $f(\overline{g}_n)$.

The game was introduced by Blackwell (1956) who proved the existence of a uniform value (in the sense of Section 1) when C is convex or $K = 1$. He gave also an example of a game in $\mathbb{R}^2$ with no uniform value.

We consider here the asymptotic approach. The value of the G_n is $v_n = V_{\Pi_n}(0,0)$ where V_{Π_n} satisfies $V_{\Pi_n}(1,z) = f(z)$ and:

$$V_{\Pi_n}(t_k, z) = \mathtt{val}_{X\times Y} E_{x,y}\{V_{\Pi_n}(t_{k+1}, z + \theta_{k+1} G_{ij})\}$$

with $X = \Delta(I)$, $Y = \Delta(J)$. The idea is to replace the above equation by the two equations:

$$W^-_{\Pi_n}(t_k, z) = \max_X \min_Y W^-_{\Pi_n}(t_{k+1}, z + \theta_{k+1} G_{ij}),$$

$$W^+_{\Pi_n}(t_k, z) = \min_Y \max_X W^+_{\Pi_n}(t_{k+1}, z + \theta_{k+1} G_{ij}),$$

hence to approximate v_n by the lower and upper values of the discretization of a differential game Γ played on $X \times Y$ between time 0 and 1, with terminal payoff $f\big(\int_0^1 g_u du\big)$ and deterministic differential dynamic given by:

$$\frac{dz}{dt} = x_t G y_t.$$

The main results used are, see e.g. Souganidis (1999):

1) $W^-_{\Pi_n}$ and $W^+_{\Pi_n}$ converge to some functions W^- and W^+ as n goes to ∞,

2) W^- is a viscosity solution on [0, 1] of the equation:

$$\frac{\partial U}{\partial t} + \max_X \min_Y \langle \nabla U, xGy \rangle = 0, \qquad U(1, z) = g(z)$$

which is condition (26) with a new limit condition,

3) this solution is unique.

A similar result for W^+ and the property: $\max_X \min_Y xGy = \min_Y \max_X xGy$ finally imply: $W^- = W^+$ and we denote this value by W.

Hence if $W(0, 0) = 0$, for any $\varepsilon > 0$ there exists N such that if $n \geq N$ Player 1 can force an outcome within ε of C in the lower n^{th} discretization Γ^-_n. The fact that Player 1 can do the same in the original game where the payoff is random relies on a uniform law of large numbers. For L large enough, playing i.i.d. the mixed move x in the mth block between stages mL (included) and $(m+1)L$ (excluded) will generate in G_{nL} an average path near the one generated by x at stage m of Γ^-_n.

Otherwise, $W(0, 0) \leq 2\delta < 0$, in this case Player 2 can avoid a δ-neighborhood of C and a symmetric argument applies.

Altogether the above construction shows that any set is either weakly approachable ($\overline{g}_n$ will be near C with high probability) or weakly excludable ($\overline{g}_n$ will be near the complement of a neighborhood C with high probability)

A second example, Laraki (2002), was described in the earlier section 3.3.

Note that in both cases the random aspect due to the use of mixed moves was eliminated, either by taking expectation or by working with the dual game.

5 Alternative Methods and Further Results

5.1 Dynamic Programming Setup

In the framework of dynamic programming (one person stochastic game), Lehrer and Sorin (1992) gave an example where $\lim_{n\to\infty} v_n$ and $\lim_{\lambda\to 0} v_\lambda$ both exist and differ.

They also proved that uniform convergence (on Ω) of v_n is equivalent to uniform convergence of v_λ and then the limits are the same.

However this condition alone does not imply existence of the uniform value, v_∞, see Lehrer and Monderer (1994), Monderer and Sorin (1993).

5.2 A Limit Game with Double Scale

Another example of a game where the play in Γ_n between stages $t_1 n$ and $t_2 n$ is approximated by the play in the limit game between time t_1 and t_2 is in Sorin (1989). The framework is simple since there are no signals. However one cannot work directly in continuous time because of the presence of two properties: some moves are exceptional in the sense that they induce some change in the state and the number of times they occur has to be taken into account; as for the other moves only the frequency matters. The analysis is done through a "semi normalization" of Γ_n by a game G_L. Each stage ℓ in L corresponds to a large block of stages in Γ_n and the strategies used in G_L at stage ℓ are the summary of the ones used on the block ℓ in Γ_n according to the above classification. One then shows that both $\liminf_{n\to\infty} v_n$ and $\liminf_{\lambda\to 0} v_\lambda$ are greater than $\limsup_{L\to\infty} \mathtt{val} G_L$ and the result follows.

One should add that these sets of reduced strategies were introduced by Mertens and Zamir (1976b) for the uniform approach: they proved the existence of the min max $\overline{v}$ and of the max min $\underline{v}$ and showed that they may differ. See also Waternaux (1983).

5.3 Non-expansive Mappings and Convexity

A proof of the convergence of v_n in the framework of one-sided incomplete information repeated games; using Kohlberg and Neyman's Theorem (result 2.1.2), was achieved by Mertens; see Mertens, Sorin and Zamir (1994), Chapter V, Exercise 5. Convergence of the sequence of norms $\|v_n\|$ implies convergence of the dual values hence of the primal values via Fenchel 's transform. Let v be the limit. Then the linear functional f appearing in Kohlberg and Neyman's result is identified at each extreme point of v and leads to $v = \mathtt{Cav}u$.

5.4 Asymptotic and Uniform Approaches

There are several deep connections between the two approaches (recall Section 1), in addition to the fact that the existence of a uniform value implies convergence of the limiting values under very general conditions (even with private information upon the duration) (Neyman (2003), Neyman and Sorin (2001)).

a) Under standard signaling ($a_t = b_t = (i_t, j_t)$) a bounded variation condition on the discounted values, see (4), is a sufficient condition for the existence of a uniform value in stochastic games, Mertens and Neyman (1981). In addition an optimal strategy is constructed stage after stage by computing at stage t a discounted factor λ_t as a function of the past history of payoffs and then playing once optimally in Γ_{λ_t}.
b) A general conjecture states that in (finitely generated) games where Player 1's information includes Player 2's information the equality: max min $=$

$\lim_{n\to\infty} v_n = \lim_{\lambda\to 0} v_\lambda$ holds, Sorin (1984, 1985), Mertens (1987). Somehow Player 1 could, using an optimal strategy of Player 2 in the limit game, define a map from histories to [0, 1]. Given the behavior of Player 2 at stage n, this map induces a time t and Player 1 plays an optimal strategy in the limit game at this time.

Finally recent results along the uniform approach include:

- proof of existence and characterization of max min and min max in absorbing games with signals, Coulomb (1999, 2001),
- proof of existence of max min and equality with $\lim_{n\to\infty} v_n$ and $\lim_{\lambda\to 0} v_\lambda$ in recursive games with lack of information on one side, Rosenberg and Vieille (2000), see point b) above.

REFERENCES

[1] Aumann R.J. and Maschler M. (1995), *Repeated Games with Incomplete Information*, M.I.T. Press (with the collaboration of R. Stearns).

[2] Bewley T. and Kohlberg E. (1976a), The asymptotic theory of stochastic games, *Mathematics of Operations Research*, **1**, 197–208.

[3] Bewley T. and Kohlberg E. (1976b), The asymptotic solution of a recursion equation occurring in stochastic games, *Mathematics of Operations Research*, **1**, 321–336.

[4] Blackwell D. (1956), An analog of the minmax theorem for vector payoffs, *Pacific Journal of Mathematics*, **6**, 1–8.

[5] Blackwell D. and Ferguson T. (1968), The Big Match, *Annals of Mathematical Statistics*, **39**, 159–163.

[6] Coulomb J.-M. (1992), Repeated games with absorbing states and no signals, *International Journal of Game Theory*, **21**, 161–174.

[7] Coulomb J.-M. (1996), A note on 'Big Match', *ESAIM: Probability and Statistics*, **1**, 89–93, http://www.edpsciences.com/ps/.

[8] Coulomb, J.-M. (1999), Generalized Big Match, *Mathematics of Operations Research*, **24**, 795–816.

[9] Coulomb, J.-M. (2001), Repeated games with absorbing states and signaling structure, *Mathematics of Operations Research*, **26**, 286–303.

[10] De Meyer B. (1996a), Repeated games and partial differential equations, *Mathematics of Operations Research*, **21**, 209–236.

[11] De Meyer B. (1996b), Repeated games, duality and the Central Limit theorem, *Mathematics of Operations Research*, **21**, 237–251.

[12] De Meyer B. (1999), From repeated games to Brownian games, *Annales de l'Institut Henri Poincaré, Probabilités et Statistiques*, **35**, 1–48.

[13] De Meyer B. and Rosenberg D. (1999), "Cav u" and the dual game, *Mathematics of Operations Research*, **24**, 619–626.

[14] Everett H. (1957), Recursive games, in *Contributions to the Theory of Games, III*, M. Dresher, A.W. Tucker and P. Wolfe (eds.), Annals of Mathematical Studies, 39, Princeton University Press, 47–78.

[15] Kohlberg E. (1974), Repeated games with absorbing states, *Annals of Statistics*, **2**, 724–738.

[16] Kohlberg E. and Neyman A. (1981), Asymptotic behavior of non expansive mappings in normed linear spaces, *Israel Journal of Mathematics*, **38**, 269–275.

[17] Laraki R. (2000), Duality and games with incomplete information, preprint.

[18] Laraki R. (2001a), Variational inequalities, systems of functional equations and incomplete information repeated games, *SIAM Journal of Control and Optimization*, **40**, 516–524.

[19] Laraki R. (2001b), The splitting game and applications, *International Journal of Game Theory*, **30**, 359–376.

[20] Laraki R. (2002), Repeated games with lack of information on one side: the dual differential approach, *Mathematics of Operations Research*, **27**, 419–440.

[21] Lehrer E. and Monderer D. (1994), Discounting versus averaging in dynamic programming, *Games and Economic Behavior*, **6**, 97–113.

[22] Lehrer E. and Sorin S. (1992), A uniform Tauberian theorem in dynamic programming, *Mathematics of Operations Research*, **17**, 303–307.

[23] Maitra A. and Sudderth W. (1998), Finitely additive stochastic games with Borel measurable payoffs, *International Journal of Game Theory*, **27**, 257–267.

[24] Mertens J.-F. (1972), The value of two-person zero-sum repeated games: the extensive case, *International Journal of Game Theory*, **1**, 217–227.

[25] Mertens J.-F. (1987), Repeated games, in *Proceedings of the International Congress of Mathematicians, Berkeley, 1986*, A. M. Gleason (ed.), American Mathematical Society, 1528–1577.

[26] Mertens J.-F. (1998), The speed of convergence in repeated games with incomplete information on one side, *International Journal of Game Theory*, **27**, 343–359.

[27] Mertens J.-F. (2002), Stochastic games, in *Handbook of Game Theory, 3*, R. J. Aumann and S. Hart (eds.), North-Holland, 1809–1832.

[28] Mertens J.-F. and Neyman A. (1981), Stochastic games, *International Journal of Game Theory*, **10**, 53–66.

[29] Mertens J.-F., S. Sorin and Zamir S. (1994), *Repeated Games*, CORE D.P. 9420-21–22.

[30] Mertens J.-F. and Zamir S. (1971), The value of two-person zero-sum repeated games with lack of information on both sides, *International Journal of Game Theory*, **1**, 39–64.

[31] Mertens J.-F. and Zamir S. (1976a), The normal distribution and repeated games, *International Journal of Game Theory*, **5**, 187–197.

[32] Mertens J.-F. and Zamir S. (1976b), On a repeated game without a recursive structure, *International Journal of Game Theory*, **5**, 173–182.

[33] Mertens J.-F. and Zamir S. (1985), Formulation of Bayesian analysis for games with incomplete information, *International Journal of Game Theory*, **14**, 1–29.

[34] Mills H. D. (1956), Marginal values of matrix games and linear programs, in *Linear Inequalities and Related Systems*, H. W. Kuhn and A. W. Tucker (eds.), Annals of Mathematical Studies, 38, Princeton University Press, 183–193.

[35] Monderer D. and Sorin S. (1993), Asymptotic properties in dynamic programming, *International Journal of Game Theory*, **22**, 1–11.

[36] Neyman A. (2003), Stochastic games and non-expansive maps, Chapter 26 in *Stochastic Games and Applications*, A. Neyman and S. Sorin (eds.), NATO Science Series C 570, Kluwer Academic Publishers.

[37] Neyman A. and Sorin S. (2001), Zero-sum two-person games with public uncertain duration process, Cahier du Laboratoire d'Econometrie, Ecole Polytechnique, 2001–013.

[38] Ponssard J.-P. and Sorin S. (1982), Optimal behavioral strategies in zero-sum games with almost perfect information, *Mathematics of Operations Research*, **7**, 14–31.

[39] Rosenberg D. (1998), Duality and Markovian strategies, *International Journal of Game Theory*, **27**, 577–597.

[40] Rosenberg D. (2000) Zero-sum absorbing games with incomplete information on one side: asymptotic analysis, *SIAM Journal on Control and Optimization*, **39**, 208–225.

[41] Rosenberg D. and Sorin S. (2001), An operator approach to zero-sum repeated games, *Israel Journal of Mathematics*, **121**, 221–246.

[42] Rosenberg D. and Vieille N. (2000), The maxmin of recursive games with lack of information on one side, *Mathematics of Operations Research*, **25**, 23–35.

[43] Shapley L. S. (1953), Stochastic games, *Proceedings of the National Academy of Sciences of the U.S.A*, **39**, 1095–1100.

[44] Sorin S. (1984), Big Match with lack of information on one side (Part I), *International Journal of Game Theory*, **13**, 201–255.

[45] Sorin S. (1985), Big Match with lack of information on one side (Part II), *International Journal of Game Theory*, **14**, 173–204.

[46] Sorin S. (1989), On repeated games without a recursive structure: existence of lim v_n, *International Journal of Game Theory*, **18**, 45–55.

[47] Sorin S. (2002), *A First Course on Zero-Sum Repeated Games*, Springer.

[48] Sorin S. (2003), The operator approach to zero-sum stochastic games, Chapter 27 in *Stochastic Games and Applications*, A. Neyman and S. Sorin (eds.), NATO Science Series C 570, Kluwer Academic Publishers.

[49] Sorin S. (2004), Asymptotic properties of monotonic non-expansive mappings, *Discrete Events Dynamic Systems*, **14**, 109–122.

[50] Souganidis P.E. (1985), Approximation schemes for viscosity solutions of Hamilton–Jacobi equations, *Journal of Differential Equations*, **17**, 781–791.

[51] Souganidis P.E. (1999), Two player zero sum differential games and viscosity solutions, in *Stochastic and Differential Games*, M. Bardi, T.E.S. Raghavan and T. Parthasarathy (eds.), Birkhauser, 70–104.

[52] Vieille N. (1992), Weak approachability, *Mathematics of Operations Research*, **17**, 781–791.

[53] Waternaux C.(1983), Solution for a class of repeated games without recursive structure, *International Journal of Game Theory*, **12**, 129–160.

[54] Zamir S. (1973), On the notion of value for games with infinitely many stages, *Annals of Statistics*, **1**, 791–796.

Notes on Risk-Sensitive Nash Equilibria

Andrzej S. Nowak
Faculty of Mathematics, Computer Science, and Econometrics
University of Zielona Góra
65–246 Zielona Góra, Poland
a.nowak@wmie.uz.zgora.pl

Abstract

We discuss the risk-sensitive Nash equilibrium concept in static noncooperative games and two-stage stochastic games of resource extraction. Two equilibrium theorems are established for the latter class of games. Provided examples explain the meaning of risk-sensitive equilibria in games with random moves.

1 Introduction

Mixed strategies have been widely accepted since the beginning of game theory. When at least one player chooses a mixed strategy to play a game, then a probability distribution of the random variables being the payoffs of the players is determined. The classical approach resulting from the von Neumann and Morgenstern utility theory suggests using the mathematical expectation for evaluating different strategy profiles of the players. The *expected payoff*, also called in the sequel the *mean* or the *statistical payoff* of the player, has a natural interpretation based on the *law of large numbers*. If the game is repeated infinitely many times and the players always implement a fixed mixed strategy profile, then the average payoff really received by any player converges with probability one to his/her mean payoff. Using the standard approach one can say that there is no difference between getting nothing for sure and winning or losing 1000 with probability 0.5. Many people would say that in the latter case a player faces a big *risk* which can easily be reflected by the variance of the outcome in this very simple game. The so-called *risk-neutral players* ignore the risk. They do not calculate it in any form, but everybody agrees that there always is a risk when pure strategies are selected at random. A modification of the expected utilities by introducing the variance as it is done, for example, in portfolio theory [10] is not a good idea for studying games. An equilibrium may not exist (see Example 3.5). It seems that the most fruitful approach is based on a more delicate criterion involving all the moments of the random payoffs. We first describe the idea in the one-person game context [5,16]. Let $\lambda \neq 0$, hereafter referred to as the *risk-sensitivity factor* of the decision maker, be fixed. Define the utility function

$$U_\lambda(x) := \operatorname{sgn}(\lambda) e^{\lambda x}. \tag{1}$$

Assume that the decision maker grades a bounded random reward X via the expectation of $U_\lambda(X)$. A *certain equivalent* of X is a number $E(\lambda, X)$ such that $U_\lambda(E(\lambda, X)) = E(U_\lambda(X))$. Therefore, for a person with risk-sensitive factor λ getting the random reward X is equivalent to obtain $E(\lambda, X)$ for sure. From (1), it follows that

$$E(\lambda, X) = (1/\lambda)\log(E(e^{\lambda X})). \tag{2}$$

If $\lambda < 0$, then by Jensen's inequality $E(\lambda, X) \leq E(X)$ and a decision maker having negative risk factor λ and grading a random reward X according to the certain equivalent $E(\lambda, X)$ is *risk-averse*. If λ is close enough to zero, then from the Taylor series expansion for $\log(1+t)$ and $e^{\lambda t}$ and (2), we obtain

$$E(\lambda, X) \approx E(X) + (\lambda/2)V(X), \tag{3}$$

where $V(X)$ is the variance of X. The right-side of (3) is the standard utility used in portfolio management [10], but is not good for studying games, see Example 3.5. Clearly, $E(\lambda, X)$ depends on all moments of X. The utility functions of the form (2) were first applied to control theory in [7,8]. Nowadays they have a lot of further applications to various dynamic optimization models, portfolio management, economic theory and stochastic control, see for example [2–4,17] and their references. There are few publications dealing with risk-sensitive criteria in dynamic games, see [2,9,11]. Our aim in this note is two-fold. We give a detailed description of a static game with risk-sensitive players and observe that the existence of a Nash equilibrium follows easily from Glicksberg's extension [6] of Nash's theorem [12]. To get some insight into the nature of the concept, we solve some examples. In this way we can observe some interesting relations between the statistical payoffs and the variances in both the risk-sensitive and risk-neutral cases. The players accepting von Neumann's approach do not care about the variances, but we wish to know the quantities which they ignore. To avoid any misunderstanding, we emphasize that we do not compare the incomparable utility functions. In both cases one can talk about the mean payoffs and the variances and may wish to see some changes in the situation of, for example, risk-neutral players when they become risk-averse. It seems that some results on the dependence of equilibria (even in matrix games) on the risk-sensitivity factors of the players would be desirable. Our second and more important aim is to study risk-sensitive equilibria in a class of two-stage stochastic games of resource extraction where the players restrict themselves to pure strategies, which is desirable in such games. However, choosing pure strategies they determine a probability distribution of a random state for the second stage of the game. Risk-sensitivity is a natural problem in such games. We also solve some examples to show the meaning of the risk-sensitive Nash equilibrium in such models.

2 Risk-Sensitive Mixed Nash Equilibria in Static Games

Let G denote an n-person game in strategic form where the strategy space X_i of every player i is compact metric and his/her payoff function p_i is *continuous* on the product space $Y := X_1 \times X_2 \times \cdots \times X_n$. By M_i, we denote the set of all mixed strategies of player i. Put $M := M_1 \times M_2 \times \cdots \times M_n$. Let $\mu \in M$ and $\nu_i \in M_i$. Then (μ_{-i}, ν_i) denotes the strategy profile μ with μ_i replaced by ν_i. For any strategy profile $\mu = (\mu_1, \mu_2, \ldots, \mu_n) \in M$ and any bounded continuous function $u : Y \mapsto R$, which can be considered as a random variable, we let $E^{\mu}(u)$ denote the *expected value* of u, namely:

$$E^{\mu}[u] := \int_{X_1}\int_{X_2}\cdots\int_{X_n} u(x_1, x_2, \ldots, x_n)\mu_1(dx_1)\mu_2(dx_2)\cdots\mu_n(dx_n).$$

Clearly, the risk-neutral von Neumann–Morgenstern utility (or payoff) of player i is $p_i(\mu) := E^{\mu}(p_i)$. Assuming that all the players are maximizers we recall that a strategy profile $\mu^* = (\mu_1^*, \mu_2^*, \ldots, \mu_n^*)$ is a *Nash equilibrium* for the game G if and only if

$$p_i(\mu^*) \geq p_i(\mu^*_{-i}, \nu_i)$$

for every player i and $\nu_i \in M_i$. If λ_i is the risk sensitivity factor of i, then his/her utility is of the form

$$u_i(\mu) := (1/\lambda_i)\log(E^{\mu}[\exp(\lambda_i p_i)]) \tag{4}$$

where $\mu = (\mu_1, \mu_2, \ldots, \mu_n) \in M$. As already mentioned, the factors $\lambda_i < 0$ for risk-sensitive maximizing players. The Nash equilibrium concept can be applied to this new game, denoted in the sequel by G^*. Since every λ_i is negative, it is easy to see that a strategy profile $\mu^* = (\mu_1^*, \mu_2^*, \ldots, \mu_n^*)$ is a risk-sensitive Nash equilibrium (*RSNE*) if and only if

$$p_i^*(\mu^*) := E^{\mu^*}[\exp(\lambda_i p_i))] \leq p_i^*(\mu^*_{-i}, \nu_i) := E^{(\mu^*_{-i}, \nu_i)}[\exp(\lambda_i p_i))] \tag{5}$$

for every player i and $\nu_i \in M_i$.

Remark 2.1. At this point we wish to emphasize that any game G in strategic form can be transformed into the form G^* with the risk-sensitive utilities (4) and from (5), we infer that an *RSNE* is simply a Nash equilibrium in the game with payoff functions

$$p_i^*(\mu) := E^{\mu}[\exp(\lambda_i p_i))]$$

where $\mu = (\mu_1, \mu_2, \ldots, \mu_n) \in M$ and *all the players try to minimize* their payoffs.

From Glicksberg's theorem [6], we conclude the following result.

Theorem 2.1. *If the strategy spaces X_i are compact metric, the payoff functions $p_i : Y \mapsto R$ are continuous, and all the players are risk-sensitive, i.e., every $\lambda_i < 0$, then there exists a risk-sensitive Nash equilibrium.*

We close this section by fixing the following notation. The *variance* of player i's payoff with respect to $\mu \in M$ is denoted by $V_i(\mu)$. Clearly,

$$V_i(\mu) = E^{\mu}[p_i^2] - (E^{\mu}[p_i])^2.$$

3 Risk-Sensitive Nash Equilibria in Matrix Games: Examples

We consider two-person games labeled $G_1, \dots, G_4$. The corresponding games where the players are the minimizers (recall Remark 2.1) are denoted by $G_1^*, \dots, G_4^*$. In all examples below, the risk-neutral Nash equilibria (or saddle points) are denoted by (μ_i^o, ν_j^o) $(i = 1, \dots, k_1,\ j = 1, \dots, m_1)$ and the risk-sensitive Nash equilibria are denoted by (μ_i^*, ν_j^*) $(i = 1, \dots, k_2,\ j = 1, \dots, m_2)$. We restrict attention to extreme equilibrium points.

Remark 3.1. In the following examples some interpretations of *RSNE* are given in terms of mean (statistical) payoffs and variances. The *RSNE* are not *NE* in the games G_i. (Also, it should be noted that G^* may not be a zero-sum game when G is.) The strategic stability is given in terms of the risk-sensitive utility functions. However, looking at the means and variances we get some insight into the nature of the *RSNE* concept.

Example 3.1. In G_1 the payoff matrices for players 1 and 2 are:

$$A = \log 2 \begin{bmatrix} -1 & 1 \\ 3 & -3 \end{bmatrix}, \qquad B = \log 2 \begin{bmatrix} -1 & 3 \\ 1 & -3 \end{bmatrix}.$$

We have the only risk neutral Nash equilibrium (*NE* for short) (μ_1^o, ν_1^o) with $\mu_1^o = \nu_1^o = (0.5, 0.5)$. The mean (or statistical) payoffs $p_i(\mu_1^o, \nu_1^o) = 0$ for $i = 1, 2$. The variance is the same for both players. Namely $V_i(\mu_1^o, \nu_1^o) = 5\log^2 2$. Assume that the risk factor $\lambda_i = -1$ for each player. Then the payoff matrices in the game G_1^* are:

$$A^* = \begin{bmatrix} 2 & 0.5 \\ 0.125 & 8 \end{bmatrix}, \qquad B^* = \begin{bmatrix} 2 & 0.125 \\ 0.5 & 8 \end{bmatrix}.$$

In the only *RSNE* (μ_1^*, ν_1^*), we have $\mu_1^* = \nu_1^* = (0.8, 0.2)$. The mean payoffs $p_i(\mu_1^*, \nu_1^*) = -0.12\log 2 < p_i(\mu_1^o, \nu_1^o) = 0$ for $i = 1, 2$, but the variance is lower than before. Namely, $V_i(\mu_1^*, \nu_1^*) = 2.5856\log^2 2$. Using (μ_1^*, ν_1^*), the players have little lower statistical payoffs but the risk significantly goes down for both players.

Example 3.2. The situation is somewhat different in the following game G_2. The payoff matrices are:

$$A = \log 2 \begin{bmatrix} -1 & 1 \\ 3 & -3 \end{bmatrix}, \qquad B = \log 2 \begin{bmatrix} 3 & -1 \\ -3 & 1 \end{bmatrix}.$$

The only *NE* (μ_1^o, ν_1^o) is $\mu_1^o = \nu_1^o = (0.5, 0.5)$, the mean payoffs are $p_i(\mu_1^o, \nu_1^o) = 0$, and $V_i(\mu_1^o, \nu_1^o) = 5\log^2 2$, for $i = 1, 2$. Let $\lambda_i = -1$ for each player. The payoff matrices in the game G_1^* are:

$$A^* = \begin{bmatrix} 2 & 0.5 \\ 0.125 & 8 \end{bmatrix}, \qquad B^* = \begin{bmatrix} 0.125 & 2 \\ 8 & 0.5 \end{bmatrix}.$$

The *RSNE* (μ_1^*, ν_1^*) is as in Example 3.1. For the mean payoffs, we have

$$p_1(\mu_1^*, \nu_1^*) = -0.12\log 2 < p_1(\mu_1^o, \nu_1^o) = 0,$$
$$p_2(\mu_1^*, \nu_1^*) = 1.32\log 2 > p_2(\mu_1^o, \nu_1^o) = 0.$$

Next, we have $V_1(\mu_1^*, \nu_1^*) = 2.5856\log^2 2$ as before, but $V_2(\mu_1^*, \nu_1^*) = 5.6576 > V_2(\mu_1^o, \nu_1^o)$. This inequality is consistent with risk-sensitivity of player 2. He/she accepts higher risk (variance) for higher mean payoff. The situation of player 1 is as before.

Example 3.3. Consider a zero-sum game G_3 where the payoff matrix for player 1 is:

$$A = \log 2 \begin{bmatrix} 16 & -8 & 9 & -3 \\ -20 & 4 & 9 & -3 \\ 25 & 1 & 18 & -6 \\ -11 & 13 & -18 & 6 \end{bmatrix}.$$

The payoff matrix of player 2 is $-A$. The extreme *NE* are: (μ_i^o, ν_1^o), $i = 1, \dots, 5$, where:

$$\nu_1^o = (0, 0, 0.25, 0.75), \quad \mu_1^o = \frac{1}{144}(0, 28, 51, 65), \quad \mu_2^o = \frac{1}{12}(7, 1, 0, 4),$$
$$\mu_3^o = \frac{1}{36}(17, 7, 0, 12), \quad \mu_4^o = \frac{1}{48}(28, 0, 3, 17), \quad \mu_5^o = (0, 0, 0.5, 0.5).$$

The value of this game is zero. Hence, $p_1(\mu_i^o, \nu_1^o) = p_2(\mu_i^o, \nu_1^o) = 0$ for $i = 1, \dots, 5$. Assume that $\lambda_i = -0.1$ for $i = 1, 2$. (Note that in the game G_3 with payoff matrices A and $-A$ both players are maximizers.) The corresponding nonzero-sum game G_3^* has the payoff matrices:

$$A^* = \begin{bmatrix} 0.3299 & 1.7411 & 0.5359 & 1.2311 \\ 4 & 0.7579 & 0.5359 & 1.2311 \\ 0.1768 & 0.933 & 0.2872 & 1.5157 \\ 2.1435 & 0.4061 & 3.4822 & 0.6598 \end{bmatrix},$$

$$B^* = \begin{bmatrix} 3.0314 & 0.5743 & 1.8661 & 0.8123 \\ 0.25 & 1.3195 & 1.8661 & 0.8123 \\ 5.6569 & 1.0718 & 3.4822 & 0.6598 \\ 0.4665 & 2.4623 & 0.2872 & 1.5157 \end{bmatrix}.$$

There are 2 extreme *RSNE* (μ_1^*, ν_1^*) and (μ_2^*, ν_1^*) with

$$\mu_1^* = (0.5383, 0, 0, 0.4617), \quad \mu_2^* = (0.283, 0.2553, 0, 0.4617),$$
$$\nu_1^* = (0, 0, 0.1624, 0.8376).$$

The mean payoffs are the same, that is, $p_1(\mu_i^*, \nu_1^*) = 0.4048 \log 2$ for $i = 1, 2$. (Such a situation happens in many examples.) Also $V_1(\mu_i^*, \nu_1^*) = 49.1905 \log^2 2$ for $i = 1, 2$. This common variance is smaller than the variance of any risk-neutral equilibrium (μ_i^o, ν_1^o), $i = 1, \ldots, 5$. Namely, we have: $V_1(\mu_1^o, \nu_1^o) = 92.25 \log^2 2$, $V_1(\mu_2^o, \nu_1^o) = V_1(\mu_3^o, \nu_1^o) = 54 \log^2 2$, $V_1(\mu_4^o, \nu_1^o) = 60.75 \log^2 2$, and $V_1(\mu_5^o, \nu_1^o) = 108 \log^2 2$. The variances for player 2 are same. The mean payoff of player 1 corresponding to *RSNE* is very close to zero but negative. It seems that player 1 statistically loses compared with his risk-neutral counterpart but the variance is much lower than in the situation where any of the risk-neutral equilibrium is played.

The situation when G is zero-sum may be more complex than above.

Example 3.4. Let the payoff matrix for player 1 in the game G_4 be:

$$A = \log 2 \begin{bmatrix} 1 & 2 & 3 & -2 & -1 \\ -1 & -1 & -3 & 0 & 1 \\ 3 & 0 & 1 & -2 & -3 \\ -3 & -1 & -1 & 0 & 3 \\ -1 & -2 & -3 & 2 & 1 \end{bmatrix}.$$

Clearly, $-A$ is the payoff matrix for player 2. The value of this game is zero and there are four extreme *NE*: (μ_1^o, ν_j^o), $j = 1, \ldots, 4$, where $\mu_1 = (0.5, 0, 0, 0, 0.5)$ and $\nu_1 = (0.5, 0, 0, 0, 0.5)$, $\nu_2 = (0, 0.5, 0, 0.5, 0)$, $\nu_3 = (0, 0, 0.4, 0.6, 0)$, $\nu_4 = (2/7, 2/7, 0, 3/7, 0)$. The variances are: $V_1(\mu_1^o, \nu_1^o) = \log^2 2$, $V_1(\mu_1^o, \nu_2^o) = 4 \log^2 2$, $V_1(\mu_1^o, \nu_3^o) = 6 \log^2 2$, $V_1(\mu_1^o, \nu_4^o) = (22/7) \log^2 2$. Now assume that

$\lambda_1 = \lambda_2 = -1$. The corresponding nonzero-sum game G_4^* is:

$$A^* = \begin{bmatrix} 0.5 & 0.25 & 0.125 & 4 & 2 \\ 2 & 2 & 8 & 1 & 0.5 \\ 0.125 & 1 & 0.5 & 4 & 8 \\ 8 & 2 & 2 & 1 & 0.125 \\ 2 & 4 & 8 & 0.25 & 0.5 \end{bmatrix},$$

$$B^* = \begin{bmatrix} 2 & 4 & 8 & 0.25 & 0.5 \\ 0.5 & 0.5 & 0.125 & 1 & 2 \\ 8 & 1 & 2 & 0.25 & 0.125 \\ 0.125 & 0.5 & 0.5 & 1 & 8 \\ 0.5 & 0.25 & 0.125 & 4 & 2 \end{bmatrix}.$$

This game has 2 extreme equilibria: $(\mu_1^*, \nu_1^*) = (\mu_1^o, \nu_1^o)$, (μ_2^*, ν_2^*) with $\nu_2^* = \nu_1^o$ and $\mu_2^* = (1/24)(12, 7, 0, 0, 5)$, which make the mean payoffs zero and have the variance $\log^2 2$. Therefore they are indifferent. Moreover, there are 2 additional *RSNE* (μ_i^*, ν_i^*), $i = 3, 4$, where:

$$\mu_3^* = \frac{1}{129}(14, 111, 4, 0, 0), \qquad \nu_3^* = \frac{1}{55}(24, 12, 0, 19, 0),$$

$$\mu_4^* = \frac{1}{15381}(4278, 6363, 1092, 0, 3648),$$

$$\nu_4^* = \frac{1}{12427}(6596, 1530, 0, 4080, 221).$$

We have

$$p_1(\mu_3^*, \nu_3^*) = -0.5243 \log 2, \qquad V_1(\mu_3^*, \nu_3^*) = 0.745 \log^2 2,$$

$$p_1(\mu_4^*, \nu_4^*) = -0.1963 \log 2, \qquad V_1(\mu_4^*, \nu_4^*) = 1.8963 \log^2 2.$$

The variances for player 2 are same. It is difficult to define a relation between these equilibria. In some sense (μ_3^*, ν_3^*) may be regarded as the best. If we define the variability coefficient c_i as:

$$c_i := \frac{\sqrt{V_1(\mu_i^*, \nu_i^*)}}{|p_i(\mu_i^*, \nu_i^*)|},$$

then $c_3 = 1.646259$, while $c_4 = 7.01509$.

We close this section with a remark concerning utilities of the form $w_i(\mu, \nu) = E^{(\mu,\nu)}(p_i) - V_i(\mu, \nu)$ for player i where $\mu \in M_1$ and $\nu \in M_2$. Such utilities also represent the risk-aversion of the players and are inspired by optimization techniques in, for example, portfolio analysis or minimum variance control problems. The next example shows that an equilibrium need not exist even in a simple matrix game case.

Example 3.5. Consider the matrix:

$$A = \begin{bmatrix} 1 & -1 \\ -3 & 3 \end{bmatrix}.$$

This matrix represents the values of the function p_1. Assume that $p_2 = -p_1$. Let $\mu = (a, 1-a)$, $\nu = (b, 1-b)$, where $a \in [0, 1]$, $b \in [0, 1]$. Then

$$w_1(\mu, \nu) = (4a-3)(2b-1) + 8a - 9 + (4a-3)^2(2b-1)^2$$

and

$$w_2(\mu, \nu) = -(4a-3)(2b-1) + 8a - 9 + (4a-3)^2(2b-1)^2.$$

It is easy to see that this game has no Nash equilibrium point.

4 Risk-Sensitive Nash Equilibria in Games of Resource Extraction

In this section we restrict attention to pure strategies for the players in a simple two-stage stochastic game where randomness is connected with a transition probability function.

An n-person nonzero-sum *game of resource extraction* is defined by the objects:

(i) $S = [0, 1]$ is the *state space* or the set of all levels of some resource.

(ii) $A_i(s) = [0, a_i(s)]$ is the *set of actions* available to player i in state s.

We assume that $a_i : S \mapsto S$ is an increasing continuous function such that $a_i(0) = 0$. The quantity $a_i(s)$ is the *consumption capacity* of player i in state $s \in S$. For each $s \in S$, let

$$A(s) = A_1(s) \times A_2(s) \times \cdots \times A_n(s).$$

Define

$$D := \{(s, x) : s \in S, x \in A(s)\}.$$

Let F_i denote the set of all Borel measurable functions $f_i : S \mapsto S$ such that $f_i(s) \in A_i(s)$ for all $s \in S$.

(iii) $u_i : S \mapsto [0, \infty)$ is a bounded continuous *utility* or *payoff function* for player i such that $u_i(0) = 0$.

(iv) q is a transition probability from D to S, called the *law of motion* among states. If s is a state at the first stage of the game and the players select an $x = (x_1, x_2, \ldots, x_n) \in A(s)$, then $q(\cdot|s, x)$ is the probability distribution of the second state and player i's utility is $u_i(x_i)$ (depends on his consumption only).

Our further assumptions are:
C1: For every $s \in S$, $\sum_{i=1}^{n} a_i(s) \leq s$.
C2: u_i is a concave twice differentiable and increasing function.
C3: The transition probability is such that $q(\{0\}|0, 0, \ldots, 0) = 1$, and for $s > 0$ is of the form

$$q(\cdot|s, x) = (1 - g(s(x)))\delta_0(\cdot) + g(s(x))\eta(\cdot),$$

where:

(a) η is a probability measure on $(0, 1]$ and δ_0 is the Dirac delta at the state $s = 0$,
(b) $g(0) = 0$ and $g : S \mapsto [0, 1]$ is a twice differentiable and increasing function,
(c) For each $x = (x_1, x_2, \ldots, x_n) \in X(s)$,

$$s(x) := s - \sum_{i=1}^{n} x_i$$

and is called the *joint investment of the players for the second period* when the game starts in state s.

Clearly, condition **C3** says that if the game starts in $s \in S$ and the players select an $x \in X(s)$, then the distribution $q(\cdot|s, x)$ over the next stock depends only on $s(x)$. **C1** is a feasibility assumption.

Remark 4.1. *(a)* Under **C3**, we have

$$1 - g(s(x)) = q(\{0\}|s, x),$$

that is, $1 - g(s(x))$ is the probability that under the joint investment $s(x)$ the next state will be zero.

(b) A typical example of the functions g is:

$$g(s(x)) = c(1 - e^{-\alpha s(x)}),$$

where $\alpha > 0$ and $c > 0$. One can also consider g of the form $g(s(x)) = s(x)/(\alpha + s(x))$ or $g(s(x)) = \beta(s(x))^c$ with $\alpha > 0$, $\beta > 0$ and $c > 0$.

Fix an initial state $s > 0$. Let $\pi_i = (x_i, f_i)$, $x_i \in A_i(s)$, $f_i \in F_i$. Put $\pi = (\pi_1, \pi_2, \ldots, \pi_n)$. In the risk-neutral case the expected utility of player i is:

$$p_i(\pi) = u_i(x_i) + \int_S u_i(f_i(y))q(dy|s, x),$$

where $x = (x_1, x_2, \ldots, x_n)$. Since $u_i(0) = 0$, from **C3**, it follows that

$$p_i(\pi) = u_i(x_i) + g(s(x)) \int_0^1 u_i(f_i(y))\eta(dy).$$

Since u_i is increasing for every i, it is easily seen that in every Nash equilibrium $\pi^o = (\pi_1^o, \pi_2^o, \ldots, \pi_n^o) = ((x_1^o, f_1^o), (x_2^o, f_2^o), \ldots, (x_n^o, f_n^o))$, we must have $f_i^o(y) = a_i(y)$ for every i and $y \in S$, that is, at the second stage every player consumes as much as possible. Therefore, to find a Nash equilibrium in the risk-neutral case, we put $f^o(y) = a_i(y)$ for each i and y and consider the game on $A(s)$ with the utility w_i for player i defined as:

$$w_i(x) = u_i(x_i) + g(s(x)) \int_0^1 u_i(a_i(y))\eta(dy),$$

where $x = (x_1, x_2, \ldots, x_n) \in A(s)$. We denote this game by G. When all the players are risk-averse then we fix some negative risk factors λ_i and consider the utility functions of the form:

$$p_i^*(\pi) := \frac{1}{\lambda_i} \log \int_S \exp[\lambda_i(u_i(x_i) + u_i(f_i(y)))]q(dy|s, x)$$

where $x = (x_1, x_2, \ldots, x_n) \in A(s)$ and $\pi = ((x_1, f_1), (x_2, f_2), \ldots, (x_n, f_n))$. Since $u_i(0) = 0$, under our assumptions on q, we get

$$\begin{aligned} p_i^*(\pi) : &= \frac{1}{\lambda_i} \log \left[\exp[\lambda_i u_i(x_i)] \int_S \exp[\lambda_i u_i(f_i(y))]\right] q(dy|s, x) \\ &= \frac{1}{\lambda_i} \log \left[\exp[\lambda_i u_i(x_i)] \left(1 - g(s(x)) J_i(f_i)\right)\right]. \end{aligned}$$

where

$$J_i(f_i) = 1 - \int_S \exp[\lambda_i u_i(f_i(y))]\eta(dy).$$

It can also be easily seen that in any Nash equilibrium (if it exists)

$$\pi^* = ((x_1^*, f_1^*), (x_2^*, f_2^*), \ldots, (x_n^*, f_n^*)),$$

we have $f_i^*(y) = a_i(y)$ for every i and $y \in S$. Define $J_i^* := J_i(f_i^*) = J_i(a_i)$ and

$$w_i^*(x) := \exp[\lambda_i u_i(x_i)](1 - g(s(x)) J_i^*) \tag{6}$$

where $x = (x_1, x_2, \ldots, x_n) \in A(s)$. Since all factors λ_i are negative a risk-sensitive Nash equilibrium (*RSNE*) exists if and only if the noncooperative game with payoff functions (6) has a Nash equilibrium x^* assuming that all players in this auxiliary game are *minimizers*.

Observe that $0 < J_i^* < 1$ for every i. Assume that g is exponential, that is, $g(s(x)) = 1 - \exp[-\alpha s(x)]$, $\alpha > 0$. From (6), it follows that

$$w_i^*(x) := \exp[\lambda_i u_i(x_i)](1 - J_i^*) + J_i^* \exp[\lambda_i u_i(x_i) - \alpha s(x)].$$

Then w_i^* is convex with respect to x_i and continuous on $A(s)$, $s > 0$. Since the players are minimizers in the game G^* with the payoff functions w_i^*, we conclude from Nash's theorem [12] that G^* has an equilibrium point, say $x^* = (x_1^*, x_2^*, \ldots, x_n^*) \in A(s)$. From the above discussion, we infer now that $\pi^* = ((x_1^*, f_1^*), (x_2^*, f_2^*), \ldots, (x_n^*, f_n^*))$ with $f_i^*(y) = a_i(y)$ for every i and $y \in S$, is a *RSNE*. We have proved the following result:

Theorem 4.1. *Every two-stage stochastic game of resource extraction satisfying conditions* **C1-C3** *with the exponential function g has a RSNE point.*

In general the functions w_i^* do not satisfy the assumptions of Nash's theorem (w_i^* need not be convex in x_i), but since u_i and g are functions of one variable using (6), one can give some necessary and sufficient conditions for the existence of a *RSNE* in terms of the derivatives of u_i and g. Actually, we use such an approach to solve two examples illustrating the theorem given below.

We now formulate two conditions concerning the functions u_i and g:

$$\lambda_i u_i'(x_i^*)[1 - J_i^* g(s(x^*))] + g'(s(x^*)) J_i^* = 0 \tag{7}$$

and

$$\begin{aligned} &2\lambda_i J_i^* g'(s(x^*)) u_i'(x_i^*) - J_i^* g''(s(x^*)) \\ &\quad + \lambda_i [1 - J_i^* g(s(x^*))](u_i''(x_i^*) + \lambda_i [u_i'(x_i^*)]^2) > 0, \end{aligned} \tag{8}$$

where $s(x^*) = s - x_1^* - \ldots - x_n^*$, $i = 1, 2, \ldots, n$.

Theorem 4.2. *Consider a two-stage stochastic game of resource extraction satisfying conditions* **C1–C3**. *Let $s > 0$ be fixed initial state and*

$$\pi^* = (x^*, f^*) := ((x_1^*, f_1^*), (x_2^*, f_2^*), \ldots, (x_n^*, f_n^*)).$$

If $\pi^* = (x^*, f^*)$ is a RSNE with $x^* \in \mathrm{Int}A(s)$, then x^* is a solution to the system of equations (7). If $x^* \in \mathrm{Int}A(s)$ satisfies both (7) and (8), then π^* is a RSNE.

The proof of this theorem follows from the above transformation of the problem to solving for Nash equilibria in the game G^* and the form (6) of the payoff functions in this auxiliary game. First two examples in the next section are solved using the necessary condition (7).

5 Games of Resource Extraction: Examples

We shall solve some two-person symmetric games in which $A_i(s) = [0, s/2]$, $u_i(a) = 4a - 4a^2 =: u(a)$ for $s \in S, a \in A_i(s)$. We assume that the transition probability q is of the form

$$q(\cdot|s, x_1, x_2) = (s - x_1 - x_2)\eta(\cdot) + (x_1 + x_2 + 1 - s)\delta_0(\cdot),$$

where $x = (x_1, x_2) \in A_1(s) \times A_2(s)$. To find an equilibrium in both the risk-neutral and risk-sensitive cases we can assume that on the second stage the players implement a symmetric strategy $\varphi^*(y) = f_i^*(y) = y/2,\ y \in S,\ i = 1, 2$. Then

$$u^*(y) := u(y/2) = 2y - y^2, \quad y \in S,$$

is the random payoff for each player at the second stage. This payoff and the probability measure η will determine the utility in both the risk-neutral and risk-sensitive cases. Assuming that at the second stage the players have to use φ^* at any equilibrium, we conclude that in the risk-neutral case it is sufficient to solve for pure Nash equilibria in the game G with the utility of player i of the form:

$$w_i(x_1, x_2) := 4x_i - 4x_i^2 + (s - x_1 - x_2) \int_0^1 (2y - y^2)\eta(dy), \tag{9}$$

where $x_1, x_2 \in [0, s/2], s > 0$. The value $w_i(x_1, x_2)$ will be called the *mean or statistical payoff* of player i with the interpretation in mind based on the law of large numbers. We also need the variance of a strategy pair (x_1, x_2). Note that

$$\begin{aligned} V_i(x_1, x_2) :&= V_i[(x_1, \varphi^*), (x_2, \varphi^*)] \\ &= (s - x_1 - x_2)^2 \left[\int_0^1 (2y - y^2)^2 \eta(dy) - \left(\int_0^1 (2y - y^2)\eta(dy) \right)^2 \right]. \end{aligned}$$

Assume now that the players are risk-sensitive and $\lambda_i = -1$ for $i = 1, 2$. To find an equilibrium in the risk-sensitive case we can solve for Nash equilibria in the game G^* on $[0, s/2] \times [0, s/2]$ with the payoff function for player i as follows:

$$\begin{aligned} w_i^*(x_1, x_2) &:= \exp[\lambda_i(4x_i - 4x_i^2)] \int_0^1 \exp[\lambda_i(2y - y^2)] q(dy|s, x_1, x_2) \qquad (10) \\ &= \exp[4x_i^2 - 4x_i] \left[1 - (s - x_1 - x_2) \left(1 - \int_0^1 \exp[y^2 - 2y]\eta(dy) \right) \right] \end{aligned}$$

and under the assumption that the players are minimizers. Denote by (x_1^o, x_2^o) $[(x_1^*, x_2^*)]$ a Nash equilibrium in the game G $[G^*]$. Since both games G and G^* are symmetric we shall obtain equilibria in the form (x^o, x^o) and (x^*, x^*) respectively. For such equilibrium points the mean payoffs and variances are the same for both players. Therefore, we accept the following notation:

$$p^o := p_i(x^o, x^o), \quad p^* := p_i(x^*, x^*) \quad \text{and} \quad V^o := V_i(x^o, x^o),$$
$$V^* := V_i(x^*, x^*).$$

Now we are ready to present some results of our calculations. We assume below that the initial state $s = 1$.

Example 5.1. Assume that $\lambda_i = -1$, for $i = 1, 2$, and η has the density function $\rho(y) := 2y$, $y \in (0, 1]$. Then

$$\int_0^1 u^*(y)\eta(dy) = \int_0^1 (2y - y^2)2y dy = 5/6 \tag{11}$$

$$\int_0^1 \exp[\lambda_i u^*(y)]\eta(dy) = \int_0^1 2y \exp(y^2 - 2y)dy \approx 0.444038454. \tag{12}$$

Using (9), (10), (11) and (12), we obtain the risk-neutral symmetric Nash equilibrium (*NE* in short) (x^o, x^o) where $x^o = 19/48 \approx 0.3958333$ and the risk-sensitive symmetric Nash equilibrium (*RSNE*) (x^*, x^*) with $x^* \approx 0.424099072$. The corresponding mean payoffs and variances are $p^o \approx 1.130208333$, $V^o \approx 1.687885802 \cdot 10^{-3}$ and $p^* \approx 1.103457741$, $V^* \approx 0.896147685 \cdot 10^{-3}$. We see that in the risk-sensitive case the variance is smaller and the mean payoffs are a little lower. This situation is consistent with common sense of risk-sensitivity.

Example 5.2. We now change the measure η by assuming that this is the uniform distribution on $(0, 1]$. Then

$$\int_0^1 u^*(y)\eta(dy) = \int_0^1 (2y - y^2)dy = 2/3,$$

$$\int_0^1 \exp[\lambda_i u^*(y)]\eta(dy) = \int_0^1 \exp[y^2 - 2y]dy \approx 0.538079507.$$

By the same method as above, we obtain the symmetric *NE* point with $x^o = 5/12$, $p^o \approx 1.08333$, and $V^o = 1/405 \approx 2.4691358 \cdot 10^{-3}$. Next we get the symmetric

RSNE with $x^* \approx 0.438799714$, $p^* \approx 1.06661848$, and $V^* \approx 1.3317244 \cdot 10^{-3}$. The uniform distribution on $(0, 1]$ has bigger variance than η with the density $\rho(y) = 2y$, $y \in S$. This explains why the players using symmetric risk-sensitive equilibrium point consume more in the first stage of the game in this example compared with the previous one. In both examples the players consume more in the first stage in the risk-sensitive cases compared with their risk-neutral counterparts. Such behavior is not surprising because of the chance move before the second consumption.

It is interesting to note that in Examples 5.1 and 5.2 there is also a *RSNE* with no risk. The players consume everything in the first stage ($x^* = 1/2$). However, the mean payoffs associated with this equilibrium are low, i.e., $p^* = 1$.

Example 5.3. We now assume that player 1 is risk-averse ($\lambda_1 = -1$) while player 2 is risk-neutral. Then (x_1^*, x_2^*) is an equilibrium in the reduced one-stage game (we assume that on the second stage both players apply φ^*), if and only if

$$w_1^*(x_1^*, x_2^*) \leq w_1^*(x_1, x_2^*) \quad \text{and} \quad w_2(x_1^*, x_2^*) \geq w_2(x_1^*, x_2),$$

for every $x_1, x_2 \in [0, 1/2]$. Assume as in Example 5.1 that η has the density $\rho(y) = 2y$, $y \in (0, 1]$. Then we obtain $x_2^* = 19/48$ and $x_1^* = 0.422707156$ ($x_2^* = x_2^o$, $x_1^* > x^o = 19/48$ from Example 5.1). Next $p_1(x_1^*, x_2^*) \approx 1.127319524 < p_1(x_1^o, x_2^o) = p^o$ (see Example 5.1) but the variance also decreases, that is, $V_1(x_1^*, x_2^*) \approx 1.28051598 \cdot 10^{-3} < V^o$. For player 2 we have $V_2(x_1^*, x_2^*) = V_1(x_1^*, x_2^*)$ and $p_2(x_1^*, x_2^*) \approx 1.107813491 < p_2(x_1^o, x_2^o) = p^o$.

Remark 5.1. The result on risk-sensitive Nash equilibria in stochastic games of resource extraction can be extended to n-stage games by the backward induction method (as in the risk-neutral case [15]) when the function g is exponential. Research on an infinite horizon discounted risk-sensitive games of this type is in progress [14]. A broad discussion (with some additional references) of risk-neutral equilibria in discounted stochastic games of capital accumulation/resource extraction can be found in [1,13,15].

REFERENCES

[1] Balbus, L. and Nowak, A.S. Construction of Nash equilibria in symmetric stochastic games of capital accumulation. *Math. Methods of Oper. Res.* **60** No.1 (2004).

[2] Başar, T. Nash equilibria of risk-sensitive nonlinear stochastic differential games. *J. Optim. Theory Appl.* **100** (1999), 479–498.

[3] Bielecki, T.R. Risk-sensitive dynamic asset management. *Appl. Math. Optim.* **39** (1999), 337–360.

[4] Cavazos-Cadena, R. Solution to the risk-sensitive average cost optimality equation in a class of Markov decision processes with finite state space. *Math. Methods of Oper. Res.* **57** (2003), 263–285.

[5] Fishburn, P.C. *Utility Theory for Decision Making.* Wiley, New York, 1970.

[6] Glicksberg, I.E. A further generalization of the Kakutani fixed point theorem with application to Nash equilibrium points. *Proc. Amer. Math. Soc.* **3** (1952), 170–174.

[7] Howard, R.A. and Matheson, J.E. Risk-sensitive Markov decision processes. *Management Sci.* **18** (1972), 356–369.

[8] Jacobson, D.H. Optimal stochastic linear systems with exponential performance criteria and their relation to deterministic differential games. *IEEE Trans. Automatic Control* **18** (1973), 124–131.

[9] Klompstra, M.B. Nash equilibria in risk-sensitive dynamic games. *IEEE Trans. Automatic Control* **45** (2000), 1397–1401.

[10] Markowitz, H. Portfolio selection. *J. of Finance* **7** (1952), 77–91.

[11] Monahan, G.E. and Sobel, M.J. Risk-sensitive dynamic market share attraction games. *Games and Econ. Behavior* **20** (1997), 149–160.

[12] Nash, J.F. Equilibrium points in n-person games. *Proc. Natl. Acad. Sci. U.S.A.* **36** (1950), 48–49.

[13] Nowak, A.S. On a new class of nonzero-sum discounted stochastic games having stationary Nash equilibrium points. *Int. J. of Game Theory* **32** (2003), 121–132.

[14] Nowak, A.S. Risk-sensitive equilibria in stochastic games of resource extraction. (2004), preprint.

[15] Nowak, A.S. and Szajowski, P. On Nash equilibria in stochastic games of capital accumulation. *Game Theory and Appl.* **9** (2003), 118–129.

[16] Pratt, J.W. Risk aversion in the small and in the large. *Econometrica* **32** (1964), 122–136.

[17] Whittle, P. *Risk-Sensitive Optimal Control.* Wiley, New York, 1990.

Continuous Convex Stochastic Games of Capital Accumulation

Piotr Więcek
Institute of Mathematics
Wrocław University of Technology
Wybrzeże Wyspiańskiego 27
50-370 Wrocław, Poland
wiecek@im.pwr.wroc.pl

Abstract

We present a generalization of Amir's continuous model of nonsymmetric infinite-horizon discounted stochastic game of capital accumulation. We show that the game has a pure-strategy equilibrium in strategies that are nondecreasing and have Lipschitz property. To prove that, we use a technique based on an approximation of continuous model by the analogous discrete one.

1 Introduction

We present a fully continuous model of stochastic game of capital accumulation. This kind of economic game, also known as resource extraction game, introduced by Levhari and Mirman [5] was extensively studied in literature in recent years. The outline of results includes stationary-equilibrium-existence theorems for a deterministic version of this game in Sundaram [10] and two variants of stochastic game with symmetric players proposed by Majumdar and Sundaram [6] and Dutta and Sundaram [4].

The actual progenitor of this paper is however the one by Amir [1]. He extended the works of Dutta, Majumdar and Sundaram to the nonsymmetric case. This generalization was achieved at the expense of some additional structural assumptions (continuity and convexity of law of motion between states, bounded spaces of players' actions). However, this enabled the author to show some important features of stationary equilibrium strategies, such as continuity, monotonicity and Lipschitz property.

Recently some extensions of Amir's model were presented in papers of Nowak [7] and Nowak and Szajowski [9]. In their models, some of the assumptions of Amir such as bounds on players' action spaces were relaxed. Instead of them, some special structure of law of motion was assumed (transition probability function being a combination of finitely many stochastic kernels depending on the state variable only). All of these papers treated the games with continuum of states. (A review of other results for games with continuum of states can be found in [8].)

In our previous paper [11], we presented a model of a stochastic game of capital accumulation similar to that of Amir's, but with countable state space. Under assumptions similar to his, we have shown that optimal strategies for a discrete model are nondecreasing and have Lipschitz property with respect to their expected value. Moreover, in every state they are concentrated in at most two neighboring points of the players' action spaces. As will be shortly seen, the result obtained there will be essential for our further considerations.

In this paper we slightly generalize Amir's result on the continuous model, removing "strict" from the assumptions on the model required in his paper. In the proof we use a non-standard method of approximation of the continuous model by the discrete one, together with adapting the result found in our previous paper [11]. This way of solving the continuous model is certainly not easier than the standard approach, which did not need an intermediate step in form of the discrete game-model. In exchange, it does not require from a reader the knowledge of such an advanced solution technique. Finally, the method used here gives a link between real-life discrete economies and continuous models widely used in economic literature, showing that at least in this model of capital-accumulation-game, the existence of equilibrium and general features of optimal strategies for the two models coincide.

The organization of this paper is as follows. In section 2 we present the assumptions of the model along with the main theorem, in section 3 we recall the result from [11] in a version, which will be used in the proof of the main theorem, while section 4 contains some lemmata used there. Finally, in section 5 we present the proof itself.

2 The Model and the Main Result

The model presented in this section is a continuous counterpart of the discrete one discussed in our previous paper [11]. It is also a slight generalization of the model of Amir [1]. We shall show, that our earlier results obtained for the discrete model can be directly used to prove the existence of a pure stationary Nash equilibrium in its continuous version.

The situation we consider is the following: Two players jointly own a productive asset characterized by some stochastic input–output technology. At each of infinitely many periods of the game, they decide independently and simultaneously, what part of the available stock should be utilized for consumption and what part for investment. The objective of each player is to maximize discounted sum of utilities from his own consumption over an infinite horizon. The players have different utility functions, discount factors and one-period consumption capacities. The model, given in the form of a nonzero-sum two-person stochastic game (in the sequel denoted by G), is described with the help of four items as follows:

1. The state space for the game is the interval $[0, \infty)$.
2. The sets of actions available to players 1 and 2 in state $x \in [0, \infty)$ are intervals $[0, K_1(x)]$ and $[0, K_2(x)]$ respectively, where $K_i(x)$ is player i's one-period consumption capacity, as a function of available stock x.
3. Player i's payoff is given by
$$E \sum_{t=0}^{\infty} \beta_i^t u_i(c_t^i),$$
where c_t^i is his action in period t, u_i his utility function and $\beta_i \in [0, 1)$ his discount factor. The expectation here is taken over the induced probability measure on all histories.
4. The transition law is described by
$$x_{t+1} \sim q\left(\cdot | x_t - c_t^1 - c_t^2\right).$$

A general *strategy* for player 1 in game G is a sequence $\pi = (\pi_1, \pi_2, \dots)$, where π_n is a conditional probability $\pi_n(\cdot|h_n)$ on the set $A^1 = \bigcup_{x \in [0,\infty)} [0, K_1(x)]$ of his possible actions, depending on all the histories of the game up to its n-th stage $h_n = (x_1, c_1^1, c_1^2, \dots, x_{n-1}, c_{n-1}^1, c_{n-1}^2, x_n)$, such that $\pi_n([0, K_1(x_n)]|h_n) = 1$. The class of all strategies for player 1 is denoted by Π^1.

Let F^1 be the set of all transition probabilities $f : [0, \infty) \to P(A^1)$ such that $f(x)(\cdot) \in P([0, K_1(x)])$ for each $x \in [0, \infty)$. (Here and in the sequel $P(S)$ denotes the set of all probability measures on S.) Then a strategy of the form $\pi = (f, f, \dots)$, where $f \in F^1$ will be called *stationary* and identified with f. We will interpret f as a strategy for player 1 that prescribes him to take, at any moment t, action c_t^1 being a realization of $f(x)$, provided x is a state at that moment. Similarly, we define the set Π^2 (F^2) of all strategies (stationary strategies) for player 2. A strategy $\pi = (\pi_1, \pi_2, \dots)$ is called *pure* if each conditional probability $\pi_n(\cdot|h_n)$ is concentrated at exactly one point. So, in any pure strategy $\pi = (\pi_1, \pi_2, \dots)$ for player i, each π_n is in fact a function that transforms any history of form $h_n = (x_1, c_1^1, c_1^2, \dots, x_n)$ to the set $[0, K_i(x_n)]$.

Let $H = [0, \infty) \times A^1 \times A^2 \times [0, \infty) \times \cdots$ be the space of all infinite histories of the game. For every initial state $x_0 = x \in [0, \infty)$ and all strategies $\pi \in \Pi^1$ and $\gamma \in \Pi^2$ we define (with the help of Ionescu-Tulcea's theorem) the unique probability measure $P_x^{\pi\gamma}$ defined on subsets of H consisting of histories starting at x. Then, for each initial state $x \in [0, \infty)$, any strategies $\pi \in \Pi^1$ and $\gamma \in \Pi^2$ and the discount factor $\beta_i \in (0, 1)$ the *expected discounted reward* for the player i is

$$J^i(x, \pi, \gamma) = E_x^{\pi\gamma}\left[\sum_{t=0}^{\infty} \beta^t u_i(c_t^i)\right].$$

A pair of (stationary) strategies (f^1, f^2) is called the *(stationary) Nash equilibrium* for the discounted stochastic game, iff for every $\pi \in \Pi^1$, $\gamma \in \Pi^2$ and $x \in [0, \infty)$ we have:

$$J^1(x, f^1, f^2) \geq J^1(x, \pi, f^2) \quad \text{and} \quad J^2(x, f^1, f^2) \geq J^2(x, f^1, \gamma).$$

The functions $V^1_{f^2}$ and $V^2_{f^1}$ are called the players' *value functions* for optimally responding to f_2 and f_1 respectively (sometimes we will call them simply "value functions corresponding to (f_1, f_2)".)

The assumptions we require about the model are slightly weaker in comparison with the ones given in Amir's paper [1]. Namely, we assume that utility functions u_i, transition probability q and consumption-capacity functions K_i satisfy the following:

(B1) $u_i : [0, \infty) \to [0, \infty)$ is a nondecreasing concave function.

(B2) q is a transition probability from $[0, \infty)$ to itself. Let $F(\cdot|y)$ denote the cumulative distribution function associated with $q(\cdot|y)$ (i.e. for $x \geq 0$ $F(x|y) = q([0, x]|y))$. We assume that:

(a) For each $x \geq 0$, $F(x|\cdot)$ is a nonincreasing function (F is first-order stochastically increasing in y.)
(b) For each $x \geq 0$, $F(x|\cdot)$ is a continuous convex function on $[0, \infty)$.
(c) $F(0|0) = 1$.

(B3) For $i = 1, 2$, the function $K_i(\cdot)$ is nondecreasing, uniformly bounded above by some constant $C_i \in [0, \infty)$, and satisfies $K_i(0) = 0$,

$$\frac{K_i(x_1) - K_i(x_2)}{x_1 - x_2} \leq 1, \quad \forall x_1, x_2 \in [0, \infty), x_1 \neq x_2$$

and $K_1(x) + K_2(x) \leq x$ for all $x \geq 0$.

Before the formulation of the main result, we need to define the effective pure strategy and value-function spaces.

$$LCM_i = \Big\{ f : [0, \infty) \to [0, C_i] : f(x) \leq K_i(x) \text{ for all } x \geq 0 \text{ and}$$
$$0 \leq \frac{f(x_1) - f(x_2)}{x_1 - x_2} \leq 1 \text{ for all distinct } x_1, x_2 \in [0, \infty) \Big\}.$$

$$CM_i = \Big\{ v : [0, \infty) \to [0, \infty) \text{ such that } v \leq \frac{u_i(C_i)}{1 - \beta_i},$$
$$v \text{ is continuous on } (0, \infty) \text{ and nondecreasing on } [0, \infty) \Big\}.$$

Theorem 2.1. *The game G has a pure stationary Nash equilibrium which is an element of $LCM_1 \times LCM_2$. Moreover, corresponding value functions $(V_1, V_2) \in CM_1 \times CM_2$.*

Remark 2.1. Theorem 2.1 expresses the result obtained by Amir [1], but under weaker assumptions. He needed to assume additionally strict concavity of the functions u_i in (B1) and strict monotonicity of the functions u_i, $F(\cdot|y)$ and K_i in (B1)–(B3). It has not been necessary as our result shows.

3 Discrete Counterparts of Game G

Our considerations of the game G are fundamentally related to a sequence $\{G_n^0, n = 1, 2, \ldots\}$ of discrete counterparts of G defined in the following way:

- The state space for the game G_n^0 is the set
$$S_n = \left\{0, \frac{1}{2^n}, \frac{2}{2^n}, \ldots\right\}.$$
- The action space for player i, $i = 1, 2$ in state $x \in S_n$ is the set
$$A_{in}(x) = \left\{0, \frac{1}{2^n}, \ldots, K_{in}\right\},$$
where one-period consumption capacities K_{in} are defined by the formula:
$$K_{in}(x) = \frac{\lfloor 2^n K_i(x) \rfloor}{2^n}.$$
- The utility function for player i, u_{in}^0 is defined as follows: $u_{in}^0(x) = u_i(\lfloor 2^n x \rfloor / 2^n)$ for all $x \in S_n$.
- The law of motion between states is the same as in G, but with different cumulative distribution function, described by the formula: $F_n^0(x|y) = F(\lfloor 2^n x \rfloor / 2^n | \lfloor 2^n y \rfloor / 2^n)$ for all $x \in [0, \infty)$, $y \in S_n$.

The result about games G_n^0 that we shall apply in our proof of theorem 2.1, will make use of the following definitions. The effective strategy space for player i is the space of strategies whose support in every state is concentrated in two adjoining points of his action space, satisfying Lipschitz property and nondecreasing in their expected value:

$$LTM_{in} = \left\{ f : S_n \to P\left(\left\{0, \frac{1}{2^n}, \ldots, C_{in}\right\}\right) : \text{ for all } x \in S_n \right.$$
$$f(x) = \alpha_x \delta[a_x] + (1 - \alpha_x)\delta\left[a_x + \frac{1}{2^n}\right] \text{ for some } 0 \le \alpha_x \le 1$$
$$\text{and } a_x \in S_n,\ 0 \le a_x < K_{in}(x),$$
$$\left. \text{and } 0 \le \frac{E(\widetilde{f}(x_1)) - E(\widetilde{f}(x_2))}{x_1 - x_2} \le 1 \text{ for all distinct } x_1, x_2 \in S_n \right\},$$

where $C_{in} = \lfloor 2^n C_i \rfloor / 2^2$, $\delta[a]$ denotes the probability measure with total mass concentrated in point a, while for $x \in \mathrm{N}$ and for all f, $\widetilde{f}(x)$ means a random variable with distribution described by $f(x)$.

The corresponding space of value functions in the game G_n^0 for player i using strategies in LTM_{in} will be:

$$M_{in} = \Big\{ v : S_n \to [0, \infty) \text{ such that } 0 \le v \le \frac{u_i(C_{in})}{1 - \beta_i} \\ \text{and } v \text{ is nondecreasing} \Big\}.$$

One can easily see that the model of each game G_n^0 coincides with the game G studied in [11] and that they satisfy the same assumptions, regarding the inessential difference that the state space in G_n^0 is S_n instead of $\{0, 1, 2, \dots\}$ considered for G in [11] (and consequently players' action spaces are rescaled in the same way). Therefore, we may rewrite Theorem 2.1 from that paper in the following form.

Theorem 3.1. *For every n the game G_n^0 has a stationary equilibrium which is an element of $LTM_{1n} \times LTM_{2n}$. Furthermore, the corresponding value functions $(V_1^0, V_2^0) \in M_{1n} \times M_{2n}$.*

4 Helpful Lemmata

The proof of Theorem 2.1 will be based on approximation of the game G by games G_n that can be solved with the help of the equilibrium–existence theorem for the discrete model presented in the previous section. It will be convenient to proceed to it through four lemmata. We start, however with the following definition of the auxiliary games G_n, $n = 1, 2, \dots$:

- The state and action spaces in the game G_n are the same as in G.
- The utility function u_i^n for player i is defined by the formula: $u_i^n(x) = u_i(\lfloor 2^n x \rfloor / 2^n)$ for all $x \in [0, \infty)$.
- The law of motion between states is the same as in G, but with different cumulative distribution function: $F_n(x|y) = F(\lfloor 2^n x \rfloor / 2^n | \lfloor 2^n y \rfloor / 2^n)$ for all $x, y \in [0, \infty)$.

Lemma 4.1. *For each $n \in N$ the game G_n has a stationary equilibrium (f_{1n}, f_{2n}) such that for $i = 1, 2$ and $x \ge 0$, f_{in} is totally concentrated in at most two adjoining points of the set $[0, K_i(x)] \cap S_n$ and satisfies the inequality*

$$0 \le E\widetilde{f_{in}}\left(x + \frac{1}{2^n}\right) - E\widetilde{f_{in}}(x) \le \frac{1}{2^n} \tag{1}$$

for $x \ge 0$. The corresponding value functions (V_{1n}, V_{2n}) for (f_{1n}, f_{2n}) are continuous from the right, uniformly bounded (with respect to n) by $u_1(C_1)/(1 - \beta_1)$ and $u_2(C_2)/(1 - \beta_2)$ respectively, and nondecreasing.

Proof. Fix $n \in \mathrm{N}$. Theorem 3.1 guarantees the existence of a stationary equilibrium, say (f_{1n}^0, f_{2n}^0) in $LTM_{1n} \times LTM_{2n}$ in game G_n^0 with corresponding value functions $V_{1n}^0 \in M_{1n}$, $V_{2n}^0 \in M_{2n}$. It follows from definition of LTM_{in} that for $k \geq 0$:

$$0 \leq E\widetilde{f_{in}^0}\left(\frac{k+1}{2^n}\right) - E\widetilde{f_{in}^0}\left(\frac{k}{2^n}\right) \leq \frac{1}{2^n}, \quad i = 1, 2.$$

On the other hand the definition of M_{in} implies that functions V_{in}^0, $i = 1, 2$ are nondecreasing and bounded by $u_i(C_i)/(1-\beta_i)$ (notice that $C_{in} \leq C_i$).

Hence, a simple analysis of the definition of game G_n shows that the pair (f_{1n}, f_{2n}) defined on $[0, \infty)$ by

$$f_{in}(x) = f_{in}^0\left(\frac{\lfloor 2^n x \rfloor}{2^n}\right), \quad i = 1, 2$$

is a stationary equilibrium in G_n satisfying the hypothesis of the lemma. One can easily see that the corresponding value functions (V_{1n}, V_{2n}) for (f_{1n}, f_{2n}) are of the form

$$V_{in}(x) = V_{in}^0\left(\frac{\lfloor 2^n x \rfloor}{2^n}\right), \quad i = 1, 2,$$

and they are uniformly bounded by $u_i(C_i)/(1-\beta_i)$ with respect to n, nondecreasing and obviously continuous from the right. □

For the rest of the paper let f_{1n}, f_{2n}, V_{1n} and V_{2n} denote any strategies and value functions satisfying Lemma 4.1. In Lemma 4.3, we give another important property of value functions V_{1n}, V_{2n}.

Lemma 4.2. *If u_1 is discontinuous in 0, then for every $x > 0$, $\widetilde{f_{1n}}(x) \neq 0$ with probability 1 for sufficiently large n.*

Proof. Suppose by contradiction, that for some $x > 0$ and every n $f_{1n}(x) = \alpha_n(x)\delta[0] + (1-\alpha_n(x))\delta\left[1/2^n\right]$ with $\alpha_n > 0$. Then the payoff in game G_n for player 1 using strategy $\pi_0 = (\delta[0], f_{1n}, f_{1n}, \ldots)$ in state x cannot be smaller than his payoff, when he uses strategy $\pi_{1n} = \left(\delta\left[1/2^n\right], f_{1n}, f_{1n}, \ldots\right)$. In the latter part of the proof we will show that this is not the case for large enough n.

The payoff for player 1 using strategy π_0 against f_{2n} of player 2 is given by the formula

$$E\left[u_{1n}(0) + \beta_1 \int V_{1n}(x')\, dF_n(x'|x - \widetilde{f_{2n}}(x))\right],$$

while the payoff for player 1 using strategy π_{1n} is given by

$$E\left[u_{1n}\left(\frac{1}{2^n}\right) + \beta_1 \int V_{1n}(x')\, dF_n\left(x'|x - \frac{1}{2^n} - \widetilde{f_{2n}}(x)\right)\right].$$

Using integration by parts we can deduce as follows:

$$\left|\beta_1 E\left[\int V_{1n}(x')\,dF_n\left(x'|x-\frac{1}{2^n}-\widetilde{f_{2n}}(x)\right)-\int V_{1n}(x')\,dF_n(x'|x-\widetilde{f_{2n}}(x))\right]\right|$$

$$=\left|\beta_1 E\left[V_{1n}(0)\left(F_n(0|x-\widetilde{f_{2n}}(x))-F_n\left(0|x-\frac{1}{2^n}-\widetilde{f_{2n}}(x)\right)\right)\right.\right.$$

$$\left.\left.-\int\left(F_n\left(x'|x-\frac{1}{2^n}-\widetilde{f_{2n}}(x)\right)-F_n(x'|x-\widetilde{f_{2n}}(x))\right)dV_{1n}\right]\right|$$

$$\leq 2\beta_1 E\left[\sup_{y\in[0,\infty)}V_{1n}(y)\sup_{y\in[0,\infty)}\left|F_n\left(y|x-\frac{1}{2^n}-\widetilde{f_{2n}}(x)\right)\right.\right.$$

$$\left.\left.-F_n(y|x-\widetilde{f_{2n}}(x))\right|\right]\leq\frac{2\beta_1 u_1(C_1)}{1-\beta_1}\sup_{y\in[0,\infty)}\sup_{t\in[0,\infty)}\left|F_n(y|t)\right.$$

$$\left.-F_n\left(y|t+\frac{1}{2^n}\right)\right|\leq\frac{2\beta_1 u_1(C_1)}{1-\beta_1}\sup_{y\in[0,\infty)}\sup_{t\in[0,\infty)}\left|F(y|t)\right.$$

$$\left.-F\left(y|t+\frac{2}{2^n}\right)\right|.$$

This is, by convexity of F, no greater than

$$\frac{2\beta_1 u_1(C_1)}{1-\beta_1}\sup_{y\in[0,\infty)}\left|F(y|0)F\left(y\middle|\frac{2}{2^n}\right)\right|=\frac{2\beta_1 u_1(C_1)}{1-\beta_1}\left(1-F\left(0\middle|\frac{2}{2^n}\right)\right),$$

which converges to zero as n goes to infinity.

On the other hand, by the discontinuity assumption, there exists an $\varepsilon>0$ such that $u_{1n}(1/2^n)-u_{1n}(0)>\varepsilon$ for all n and thereby for n large enough

$$E\left[\left(u_{1n}\left(\frac{1}{2^n}\right)+\beta_1\int V_{1n}(x')\,dF_n\left(x'|x-\frac{1}{2^n}-\widetilde{f_{2n}}(x)\right)\right)\right.$$

$$\left.-\left(u_{1n}(0)+\beta_1\int V_{1n}(x')\,dF_n(x'|x-\widetilde{f_{2n}}(x))\right)\right]>\frac{\varepsilon}{2},$$

which yields a contradiction. □

Lemma 4.3. *For $i=1,2$,*

$$\sup_{x\in(0,\infty)}\left[V_{in}\left(x+\frac{1}{2^n}\right)-V_{in}(x)\right]\longrightarrow_{n\to\infty}0.$$

Proof. Fix $x > 0$ and let $i = 1$. V_{1n} satisfies the equation:

$$V_{1n}(x) = E\left[u_{1n}(\widetilde{f_{1n}}(x)) + \beta_1 \int V_{1n}(x')\, dF_n(x'|x - \widetilde{f_{1n}}(x) - \widetilde{f_{2n}}(x))\right]$$

and therefore

$$V_{1n}\left(x + \frac{1}{2^n}\right) - V_{1n}(x) = E\left[u_{1n}\left(\widetilde{f_{1n}}\left(x + \frac{1}{2^n}\right)\right) - u_{1n}(\widetilde{f_{1n}}(x))\right]$$
$$+ \beta_1 E\left[\int V_{1n}(x')\, dF_n\left(x'|x + \frac{1}{2^n} - \widetilde{f_{1n}}\left(x + \frac{1}{2^n}\right) - \widetilde{f_{2n}}\left(x + \frac{1}{2^n}\right)\right)\right.$$
$$\left. - \int V_{1n}(x')\, dF_n(x'|x - \widetilde{f_{1n}}(x) - \widetilde{f_{2n}}(x))\right].$$

By Lemma 4.1, for $j = 1, 2$, $f_{jn}\left(x + \frac{1}{2^n}\right)$ and $f_{jn}(x)$ are probability measures totally concentrated in at most two points of the set S_n. Hence, because of (1), the inequality

$$\widetilde{f_{jn}}\left(x + \frac{1}{2^n}\right) - \widetilde{f_{jn}}(x) \le \frac{2}{2^n}, \quad j = 1, 2, \tag{2}$$

holds with probability 1.

Now, by Lemma 4.2 whenever u_1 is discontinuous in 0, $\widetilde{f_{1n}}(x) \ge 1/2^n$ with probability 1 for large enough n. Thereby in this case, we can conclude that for such an n

$$\left|E\left[u_{1n}\left(\widetilde{f_{1n}}\left(x + \frac{1}{2^n}\right)\right) - u_{1n}(\widetilde{f_{1n}}(x))\right]\right| \le \sup_{y \in \left[\frac{1}{2^n}, \infty\right)} \left[u_{1n}\left(y + \frac{2}{2^n}\right)\right.$$
$$\left. - u_{1n}(y)\right] \le \sup_{y \in (0,\infty)} \left[u_1\left(y + \frac{3}{2^n}\right) - u_1(y)\right],$$

by the construction of game G_n. On the other hand, if u_1 is continuous in 0, then similarly

$$\left|E\left[u_{1n}\left(\widetilde{f_{1n}}\left(x + \frac{1}{2^n}\right)\right) - u_{1n}(\widetilde{f_{1n}}(x))\right]\right|$$
$$\le \sup_{y \in (0,\infty)} \left[u_{1n}\left(y + \frac{2}{2^n}\right) - u_{1n}(y)\right] \le \sup_{y \in (0,\infty)} \left[u_1\left(y + \frac{3}{2^n}\right) - u_1(y)\right].$$

But, the concavity of u_1 implies that

$$\sup_{y \in (0,\infty)} \left[u_1\left(y + \frac{3}{2^n}\right) - u_1(y)\right] = \left[u_1\left(\frac{3}{2^n}\right) - u_1(0^+)\right],$$

which converges to zero as n increases to infinity.

On the other hand, using integration by parts, we can deduce

$$\Big|\beta_1 E\Big[\int V_{1n}(x')\,dF_n\Big(x'|x+\frac{1}{2^n}-\widetilde{f_{1n}}\Big(x+\frac{1}{2^n}\Big)-\widetilde{f_{2n}}\Big(x+\frac{1}{2^n}\Big)\Big)$$

$$-\int V_{1n}(x')\,dF_n(x'|x-\widetilde{f_{1n}}(x)-\widetilde{f_{2n}}(x))\Big]\Big| = \Big|\beta_1 E\Big[V_{1n}(0)$$

$$\Big(F_n(0|x-\widetilde{f_{1n}}(x)-\widetilde{f_{2n}}(x))-F_n\Big(0|x+\frac{1}{2^n}-\widetilde{f_{1n}}\Big(x+\frac{1}{2^n}\Big)$$

$$-\widetilde{f_{2n}}\Big(x+\frac{1}{2^n}\Big)\Big)\Big)-\int\Big(F_n\Big(x'|x+\frac{1}{2^n}-\widetilde{f_{1n}}\Big(x+\frac{1}{2^n}\Big)$$

$$-\widetilde{f_{2n}}\Big(x+\frac{1}{2^n}\Big)\Big)-F_n(x'|x-\widetilde{f_{1n}}(x)-\widetilde{f_{2n}}(x))\Big)\,dV_{1n}\Big]\Big|$$

$$\le 2\beta_1 E\Big[\sup_{y\in[0,\infty)} V_{1n}(y)\sup_{y\in[0,\infty)}\Big|F_n\Big(y|x+\frac{1}{2^n}-\widetilde{f_{1n}}\Big(x+\frac{1}{2^n}\Big)$$

$$-\widetilde{f_{2n}}\Big(x+\frac{1}{2^n}\Big)\Big)-F_n(y|x-\widetilde{f_{1n}}(x)-\widetilde{f_{2n}}(x))\Big|\Big].$$

Using (2) we easily get that

$$\Big|\Big(x+\frac{1}{2^n}-\widetilde{f_{1n}}\Big(x+\frac{1}{2^n}\Big)-\widetilde{f_{2n}}\Big(x+\frac{1}{2^n}\Big)\Big)-(x-\widetilde{f_{1n}}(x)-\widetilde{f_{2n}}(x))\Big|\le\frac{5}{2^n}$$

with probability 1. Hence, with the help of Lemma 4.1,

$$E\Big[\sup_{y\in[0,\infty)} V_{1n}(y)\sup_{y\in[0,\infty)}\Big|F_n\Big(y|x+\frac{1}{2^n}-\widetilde{f_{1n}}\Big(x+\frac{1}{2^n}\Big)-\widetilde{f_{2n}}\Big(x+\frac{1}{2^n}\Big)\Big)$$

$$-F_n(y|x-\widetilde{f_{1n}}(x)-\widetilde{f_{2n}}(x))\Big|\Big]$$

$$\le\frac{u_1(C_1)}{1-\beta_1}\sup_{y\in[0,\infty)}\sup_{t\in[0,\infty)}\Big|F_n\Big(y|t+\frac{5}{2^n}\Big)-F_n(y|t)\Big|$$

$$\le\frac{u_1(C_1)}{1-\beta_1}\sup_{y\in[0,\infty)}\sup_{t\in[0,\infty)}\Big|F\Big(y|t+\frac{6}{2^n}\Big)-F(y|t)\Big|,$$

which, by convexity of F, is no greater than

$$\frac{u_1(C_1)}{1-\beta_1}\sup_{y\in[0,\infty)}\Big|F\Big(y\Big|\frac{6}{2^n}\Big)-F(y|0)\Big|=\frac{u_1(C_1)}{1-\beta_1}\Big(1-F\Big(0\Big|\frac{6}{2^n}\Big)\Big).$$

But this converges to zero as n increases to infinity, ending the proof. □

Lemma 4.4. *Let $x \in R$. Let $\{d_n\}$ and $\{e_n\}$ be two sequences of real numbers satisfying for all n, $0 \leq d_n + e_n \leq x$, such that $d_n \to d_0$ and $e_n \to e_0$. Then the sequence of distributions $\overline{F_n}(x') = F_n(x'|x - d_n - e_n)$ is weakly convergent to $\overline{F}(x') = F(x'|x - d_0 - e_0)$.*

Proof. First, notice that the functions F_n are defined in such a way, that

$$F\left(x' - \frac{1}{2^n}|x - d_n - e_n\right) \leq F_n(x'|x - d_n - e_n) \leq F\left(x'|x - d_n - e_n - \frac{1}{2^n}\right)$$

for all nonnegative x and x' and natural n (here we put $x - d_n - e_n - (1/2^n) \equiv 0$, if this value is negative). Therefore, to show that $\overline{F_n} \Rightarrow \overline{F}$ it suffices to prove the statement: for any $y_0 \in$ R, if $F(x|y_0)$ is continuous in $x = x_0$ then $F(x|y)$ is continuous in $(x, y) = (x_0, y_0)$.

We will consider two cases.

Case 1. $y_0 > 0$: Suppose by contradiction that $F(x|y)$ is discontinuous in (x_0, y_0) although $F(x|y_0)$ is continuous in x_0. In other words, there exist sequences $\{x_n\}_{n=1}^{\infty}$ and $\{y_n\}_{n=1}^{\infty}$ such that $(x_n, y_n) \to (x_0, y_0)$ and $|F(x_n|y_n) - F(x_0|y_0)| > \varepsilon$ for $n = 1, 2, \ldots$, and some $\varepsilon > 0$.

Since $F(x|y_0)$ is continuous in x_0, there exists an $N \in$ N such that

$$|F(x_n|y_0) - F(x_0|y_0)| < \varepsilon/2, \quad \text{for all} \quad n \geq N.$$

However, we assume that

$$\varepsilon < |F(x_n|y_n) - F(x_0|y_0)| \leq |F(x_n|y_0) - F(x_0|y_0)| + |F(x_n|y_n) - F(x_n|y_0)|,$$

whence $|F(x_n|y_n) - F(x_n|y_0)| > \frac{\varepsilon}{2}$ for $n > N$.

On the other hand, for every $\delta > 0$ there exists an $M_\delta \geq N$ such that $\max\{|y_n - y_0|, |x_n - x_0|\} < \delta$ for all $n \geq M_\delta$. Hence,

$$\frac{|F(x_n|y_n) - F(x_n|y_0)|}{|y_n - y_0|} > \frac{\varepsilon}{2\delta}, \quad \text{for all} \quad n \geq M_\delta.$$

However, by (a) and (b) of assumption (B2), $F(x|y)$ is convex nonincreasing in y and thereby for any fixed $n \geq M_\delta$

$$\frac{|F(x_n|y) - F(x_n|y_0)|}{|y - y_0|} > \frac{\varepsilon}{2\delta}$$

for any $y < \min\{y_0, y_n\}$. This means that for small enough δ

$$|F(x_n|y) - F(x_n|y_0)| > (\varepsilon/2\delta)|y - y_0| > 1,$$

which is a contradiction.

Case 2. $y_0 = 0$: As in Case 1, suppose by contradiction that there exist sequences $\{x_n\}_{n=1}^{\infty}$ and $\{y_n\}_{n=1}^{\infty}$ such that $(x_n, y_n) \to (x_0, y_0)$ and

$$|F(x_n|y_n) - F(x_0|y_0)| > \varepsilon, \tag{3}$$

for $n = 1, 2, \ldots$, and some $\varepsilon > 0$.

Since

$$F(x_0|y_0) = F(x_0|0) = 1 \quad \text{for all} \quad x_0 > 0, \tag{4}$$

(3) can be rewritten as

$$F(x_n|y_n) \leq 1 - \varepsilon. \tag{5}$$

Obviously (4) implies also that (5) can be satisfied only if $y_n > 0$, for $n = 1, 2, \ldots$.

Now, note that, since $F(x|y)$ is nondecreasing in x, whenever there exist some $x_n > x_0$ satisfying (5) for some y_n, then also every $x_n' < x_0$ satisfies (5) for y_n. Therefore, we may assume without loss of generality that $x_1 < x_2 < x_3 < \ldots < x_0$.

Fix $n_0 \in \mathrm{N}$. Since $y_n \to y_0 = 0$ and $y_n > 0$ for $n = 1, 2, \ldots$, the continuity of $F(x_{n_0}|\cdot)$ implies that there exists an $n_1 > n_0$ such that

$$F(x_{n_0}|y_{n_1}) > 1 - (\varepsilon/2).$$

On the other hand (5) implies that

$$F(x_{n_1}|y_{n_1}) \leq 1 - \varepsilon < 1 - (\varepsilon/2) < F(x_{n_0}|y_{n_1})$$

which is a contradiction, because $F(\cdot|y_{n_1})$ is nondecreasing. □

5 Proof of Theorem 2.1

Standard dynamic programming arguments imply that for all $n \in \mathrm{N}$ and all $x \in [0, \infty)$ the following equation holds:

$$\begin{aligned} V_{1n}(x) &= E\left[u_{1n}(\widetilde{f_{1n}}(x)) + \beta_1 \int V_{1n}(x')\, dF_n(x'|x - \widetilde{f_{1n}}(x) - \widetilde{f_{2n}}(x))\right] \\ &= \max_{c \in [0, K_1(x)]} E\left[u_{1n}(c) + \beta_1 \int V_{1n}(x')\, dF_n(x'|x - c - \widetilde{f_{2n}}(x))\right]. \end{aligned} \tag{6}$$

The functions V_{1n} $(n = 1, 2, \ldots)$ are uniformly bounded so we can use the diagonal method to show that there exist a function V_1 on $S_\infty = \cup_{n=1}^{\infty} S_n$ and a subsequence V_{1n_k} convergent to V_1 on S_∞. By Lemma 4.1, V_1 is nondecreasing and bounded by $u_1(C_1)/(1 - \beta_1)$. Additionally by Lemma 4.3, V_1 is continuous on S_∞. Therefore, since S_∞ is dense in $[0, \infty)$ and all of the functions V_{1n_k} are nondecreasing, we can obtain convergence in all of the points of $[0, \infty)$ by putting $V_1(x) = \lim_{x' \to x^+, x' \in S_\infty} V_1(x')$. Of course, V_1 of this form is continuous on $[0, \infty)$, and thereby $V_1 \in CM_1$. Without loss of generality we may assume that $V_{1n} \to V_1$. In an analogous way, $V_2 \in CM_2$ is defined, and pointwise convergence $V_{2n} \to V_2$ on $[0, \infty)$ may be assumed.

Now, recall from Lemma 4.1 that for all $x \geq 0$ and for $i = 1, 2$, the strategy $f_{in}(x)$ is concentrated in at most two adjoining points of the bounded set $[0, K_i(x)] \cap \widetilde{S_n}$. Hence, one may use similiar arguments to show that there exist subsequences $\widetilde{f_{1n_m}}$ and $\widetilde{f_{2n_{m_r}}}$ convergent pointwise to $f_1 \in LCM_1$ and $f_2 \in LCM_2$ with probability 1. In this case we can also assume that $f_{1n} \to f_1$ and $f_{2n} \to f_2$.

The theorem will be proved if we show that for $x \in [0, \infty)$

$$V_1(x) = u_1(f_1(x)) + \beta_1 \int V_1(x')\, dF(x'|x - f_1(x) - f_2(x)) \tag{7}$$

$$= \max_{c \in [0, K_1(x)]} \left[u_1(c) + \beta_1 \int V_1(x')\, dF(x'|x - c - f_2(x)) \right],$$

and the analogous optimality equation for player 2:

$$V_2(x) = u_2(f_2(x)) + \beta_1 \int V_2(x')\, dF(x'|x - f_1(x) - f_2(x))$$

$$= \max_{c \in [0, K_2(x)]} \left[u_2(c) + \beta_1 \int V_2(x')\, dF(x'|x - f_1(x) - c) \right].$$

In the following, we shall restrict our attention to the first of these equations. (The other one is proved in the same way.)

We already know that the LHS of (6) converges to the LHS of (7). Obviously also $E[u_{1n}(\widetilde{f_{1n}}(x))] \to u_1(f_1(x))$ for $x \geq 0$ as n goes to infinity, since $\widetilde{f_{1n}} \to f_1$ with probability 1 and (by construction) $u_{1n} \to u_1$ on $[0, \infty)$.

As we have already noticed, V_1 is continuous on $(0, \infty)$. Hence, since all the functions V_{1n} are nondecreasing, by Polya's theorem (see e.g. [2], p. 173) the convergence of V_n is uniform on every closed interval of $(0, \infty)$. Therefore, for every $\varepsilon > 0$ there exist $M_1, M_2 \in (0, \infty)$, such that

$$\Bigg| \beta_1 E\left[\int V_{1n}(x')\, dF_n(x'|x - \widetilde{f_{1n}}(x) - \widetilde{f_{2n}}(x)) \right]$$

$$- \beta_1 \int V_1(x')\, dF(x'|x - f_1(x) - f_2(x)) \Bigg|$$

$$\leq \Bigg| \beta_1 E\left[\int_{M_1}^{M_2} V_{1n}(x')\, dF_n(x'|x - \widetilde{f_{1n}}(x) - \widetilde{f_{2n}}(x)) \right]$$

$$- \beta_1 \int_{M_1}^{M_2} V_1(x')\, dF(x'|x - f_1(x) - f_2(x)) \Bigg| + \frac{4\varepsilon u_1(C_1)}{1 - \beta_1}$$

$$+\beta_1 \left| E\left[V_1(0)F(0|x - f_1(x) - f_2(x)) - V_{1n}(0)F(0|x - \widetilde{f_{1n}}(x)\right.\right.$$
$$\left.\left. - \widetilde{f_{2n}}(x))\right]\right| \le \beta_1 \sup_{x\in[M_1,M_2]} |V_{1n}(x) - V_1(x)|$$
$$+ \left|\beta_1 E\left[\int_{M_1}^{M_2} V_1(x')\, dF_n(x'|x - \widetilde{f_{1n}}(x) - \widetilde{f_{2n}}(x))\right.\right.$$
$$\left.\left. - \int_{M_1}^{M_2} V_1(x')\, dF(x'|x - f_1(x) - f_2(x))\right]\right| + \frac{4\varepsilon u_1(C_1)}{1-\beta_1}$$
$$+ \beta_1 \frac{u_1(0)}{1-\beta_1} \left| F(0|x - f_1(x) - f_2(x)) - F(0|x - \widetilde{f_{1n}}(x) - \widetilde{f_{2n}}(x))\right|.$$

(In the last inequality we use the fact that for $n \ge 1$, $V_{1n}(0) = V_1(0) = u_1(0) + \beta_1 u_1(0) + \beta_1^2 u_1(0) + \ldots = u_1(0)/(1-\beta_1)$, as a consequence of assumption (c) of (B2).) But using Lemma 4.4 one may easily see that the RHS of the above inequality converges to zero as $n \to \infty$.

We are left with showing that the maximand in (6) converges to the one in (7). Suppose the contrary, i.e., that for some $x \in [0, \infty)$ there exists a $c \in [0, K_1(x)]$ such that

$$V_1(x) < u_1(c) + \beta_1 \int V_1(x')\, dF(x'|x - c - f_2(x)).$$

However, since the convergence of V_{1n} is uniform on every closed subset of $(0, \infty)$, and thereby $V_{1n}(x_n) \to V_1(x)$ for every sequence $\{x_n\}_{n=1}^{\infty}$ of points from $(0, \infty)$ convergent to some $x \in (0, \infty)$ (note that the weak convergence assumption is satisfied due to Lemma 4.4), we may use Theorem 5.5 in [3] to get

$$u_1(c) + \beta_1 \int V_1(x')\, dF(x'|x - c - f_2(x))$$
$$= \lim_{n\to\infty} E\left[u_{1n}(c) + \beta_1 \int V_{1n}(x')\, dF_n(x'|x - c - \widetilde{f_{2n}}(x))\right]$$
$$\le \lim_{n\to\infty} E\left[u_{1n}(\widetilde{f_{1n}}(x)) + \beta_1 \int V_{1n}(x')\, dF_n(x'|x - \widetilde{f_{1n}}(x) - \widetilde{f_{2n}}(x))\right]$$
$$= \lim_{n\to\infty} V_{1n}(x) = V_1(x),$$

which is a contradiction.

Acknowledgement

The author would like to thank Professor Andrzej Nowak for proposing the problem and some helpful comments and Professor Tadeusz Radzik for his help and advice.

REFERENCES

[1] Amir R. (1996) Continuous Stochastic Games of Capital Accumulation with Convex Transitions. *Games Econ. Behavior* 15, 111–131.

[2] Bartle R. (1976) *The Elements of Real Analysis.* Wiley, New York.

[3] Billingsley P. (1968) *Convergence of Probability Measures.* Wiley, New York.

[4] Dutta P.K., Sundaram R.K. (1992) Markovian Equilibrium in a Class of Stochastic Games: Existence Theorems for Discounted and Undiscounted Models. *Econ. Theory* 2, 197–214.

[5] Levhari D., Mirman L. (1980) The Great Fish War: An Example Using a Dynamic Cournot–Nash Solution. *Bell J. Econ* 11, 322–334.

[6] Majumdar M., Sundaram R.K. (1991) Symmetric Stochastic Games of Resource Extraction: Existence of Nonrandomized Stationary Equilibrium. in *Stochastic Games and Related Topics* (Raghavan *et. al.* Eds.). Kluwer, Dordrecht.

[7] Nowak A.S. (2003) On a New Class of Nonzero-sum Discounted Stochastic Games Having Stationary Nash Equilibrium Points. *Internat. J. Game Theory* 32: 121–132.

[8] Nowak A.S. (2003) N-person Stochastic Games: Extensions of the Finite State Space Case and Correlation. In: *Stochastic Games and Applications* (Neyman A., Sorin S. Eds.). Kluwer, Dordrecht, pp. 93–106.

[9] Nowak A.S., Szajowski P. (2003) On Nash Equilibria in Stochastic Games of Capital Accumulation. *Game Theory Appl.* 9: 118–129.

[10] Sundaram R.K. (1989) Perfect Equilibrium in Non-Randomized Strategies in a Class of Symmetric Dynamic Games. *J. Econ. Theory* 47, 153–177.

[11] Więcek P. (2003) Convex Stochastic Games of Capital Accumulation with Nondivisible Money Unit. *Scientiae Mathematicae Japonicae* 57, 397–411.

PART II
Differential Dynamic Games

Dynamic Core of Fuzzy Dynamical Cooperative Games

Jean-Pierre Aubin
Université Paris Dauphine
Place du Maréchal, De Lattre de Tessigny
75775 Paris Cedex 16, France

Abstract

We use in this paper the viability/capturability approach for studying the problem of characterizing the dynamic core of a dynamic cooperative game defined in a characteristic function form. In order to allow coalitions to evolve, we embed them in the set of fuzzy coalitions. Hence, we define the dynamic core as a set-valued map associating with each fuzzy coalition and each time the set of allotments is such that their payoffs at that time to the fuzzy coalition are larger than or equal to the one assigned by the characteristic function of the game. We shall characterize this core through the (generalized) derivatives of a valuation function associated with the game. We shall provide its explicit formula, characterize its epigraph as a viable-capture basin of the epigraph of the characteristic function of the fuzzy dynamical cooperative game, use the tangential properties of such basins for proving that the valuation function is a solution to a Hamilton–Jacobi–Isaacs partial differential equation and use this function and its derivatives for characterizing the dynamic core.

1 Introduction

This paper takes up on a recent line of research, *dynamic cooperative game theory*, opened by Leon Petrosjan (see for instance [57, Petrosjan] and [58, Petrosjan & Zenkevitch]), Alain Haurie ([50, Haurie]), Jerzy Filar and others. We quote the first lines of [42, Filar & Petrosjan]: *"Bulk of the literature dealing with cooperative games (in characteristic function form) do not address issues related to the evolution of a solution concept over time. However, most conflict situations are not "one shot" games but continue over some time horizon which may be limited* a priori *by the game rules, or terminate when some specified conditions are attained."* This paper, however, deals also with the evolution of coalitions.

1.1 Fuzzy Coalitions and their Evolution

For that purpose, since cooperative games deal with the behavior of coalitions of players, the first definition of a coalition which comes to mind being that of a subset of players $A \subset N$ is not adequate for tackling dynamical models of evolution of coalitions since the 2^n coalitions range over a finite set, preventing us from using analytical techniques. One way to overcome this difficulty is to embed the family of subsets of a (discrete) set N of n players to the space $\mathbf{R}^n$ through the map χ associating with any coalition $A \in \mathcal{P}(N)$ its characteristic function[1] $\chi_A \in \{0, 1\}^n \subset \mathbf{R}^n$.

By definition, the family of fuzzy sets[2] is the convex hull $[0, 1]^n$ of the power set $\{0, 1\}^n$ in $\mathbf{R}^n$. Therefore, we can write any fuzzy set in the form

$$x = \sum_{A \in \mathcal{P}(N)} m_A \chi_A \text{ where } m_A \geq 0 \ \& \sum_{A \in \mathcal{P}(N)} m_A = 1.$$

The memberships are then equal to

$$\forall i \in N, x_i = \sum_{A \ni i} m_A.$$

Consequently, if m_A is regarded as the probability for the set A to be formed, the membership of the player i to the fuzzy set x is the sum of the probabilities of the coalitions to which player i belongs. Player i participates fully in x if $x_i = 1$, does not participate at all if $x_i = 0$ and participates in a fuzzy way if $x_i \in]0, 1[$.

Actually, this idea of using fuzzy coalitions has already been used in the framework of cooperative games with and without side-payments described by a characteristic function[3] $\mathbf{u}$ assigning to each fuzzy coalition $x \in [0, 1]^n$ a lower bound $\mathbf{u}(x)$ to gains or payoffs $y := \langle p, x \rangle$ associated with an allotment $p \in \mathbf{R}^n_+$ (see [3,4, Aubin], [2, Aubin, Chapter 12], [7, Aubin, Chapter 13], the books [54, Mares] and [55, Mishizaki & Sokawa], and [27–29, Basile], [26, Basile, De Simone & Graziano], [1, Allouch & Florenzano], [43, Florenzano]). Fuzzy coalitions have been used in dynamical models of cooperative games in [16, Aubin & Cellina, Chapter 4], economic theory in [11, Aubin, Chapter 5] and in [14, Aubin].

For instance, it has been shown that in the framework of static cooperative games with side payments involving fuzzy coalitions, *the concepts of Shapley*

---

[1]This canonical embedding is more adapted to the nature of the power set $\mathcal{P}(N)$ than the universal embedding of a discrete set M of m elements to $\mathbf{R}^m$ by the Dirac measure associating with any $j \in M$ the jth element of the canonical basis of $\mathbf{R}^m$. The convex hull of the image of M by this embedding is the probability simplex of $\mathbf{R}^m$. Hence fuzzy sets offer a "dedicated convexification" procedure of the discrete power set $M := \mathcal{P}(N)$ instead of the universal convexification procedure of frequencies, probabilities, mixed strategies derived from its embedding in $\mathbf{R}^m = \mathbf{R}^{2^n}$.

[2]This concept of fuzzy set was introduced in 1965 by L. A. Zadeh. Since then, it has been wildly successful, even in many areas outside mathematics!. Lately, we found in *"La lutte finale"*, Michel Lafon (1994), p. 69 by A. Bercoff the following quotation of the late François Mitterand, president of the French Republic (1981–1995): *"Aujourd'hui, nous nageons dans la poésie pure des sous ensembles flous"* … (Today, we swim in the pure poetry of fuzzy subsets)!

[3]Not to be confused with characteristic functions of sets!

value and core coincide with the (generalized) gradient[4] $\partial\mathbf{u}(x_N)$ *of the "characteristic function"* $\mathbf{u} : [0,1]^n \mapsto \mathbf{R}_+$ *at the "grand coalition"* $x_N := (1, \ldots, 1)$, *the characteristic function of* $N := \{1, 2, \ldots, n\}$.

For instance, when the characteristic function of the static cooperative game $\mathbf{u}$ is concave, positively homogeneous and continuous on the interior of $\mathbf{R}^n_+$, one checks[5] that the generalized gradient $\partial\mathbf{u}(x_N)$ is not empty and coincides with the subset of allotments $p := (p_1, \ldots, p_n) \in \mathbf{R}^n_+$ accepted by all fuzzy coalitions in the sense that

$$\forall x \in [0,1]^n, \ \langle p, x\rangle = \sum_{i=1}^{n} p_i x_i \geq \mathbf{u}(x), \tag{1}$$

and that, for the grand coalition $x_N := (1, \ldots, 1)$,

$$\langle p, x_N\rangle = \sum_{i=1}^{n} p_i = \mathbf{u}(x_N).$$

1.2 Fuzzy Dynamic Cooperative Games

In a dynamical context, (fuzzy) coalitions evolve, so that static conditions (1) should be replaced by conditions[6] stating that for any evolution $t \mapsto x(t)$ of fuzzy coalitions, the payoff $y(t) := \langle p(t), x(t)\rangle$ should be larger than or equal to $\mathbf{u}(x(t))$ according (at least) to one of the three following rules:

(a) at a prescribed final time T of the end of the game:

$$y(T) := \sum_{i=1}^{n} p_i(T) x_i(T) \geq \mathbf{u}(x(T));$$

(b) during the whole time span of the game:

$$\forall t \in [0,T], \ y(t) := \sum_{i=1}^{n} p_i(t) x_i(t) \geq \mathbf{u}(x(t));$$

(c) at the first winning time $t^* \in [0,T]$ when

$$y(t^*) := \sum_{i=1}^{n} p_i(t^*) x_i(t^*) \geq \mathbf{u}(x(t^*));$$

at which time the game stops.

---

[4]The differences between these concepts for usual games is explained by the different ways one "fuzzyfies" a characteristic function defined on the set of usual coalitions. See [3,4, Aubin], [2, Aubin, Chapter 12] and [7, Aubin, Chapter 13].

[5]See [3,4, Aubin], [2, Aubin, Chapter 12] and [7, Aubin, Chapter 13].

[6]Naturally, the privileged role played by the grand coalition in the static case must be abandoned, since the coalitions evolve, so that the grand coalition eventually loses its status.

Summarizing, the above conditions require to find – for each of the above three rules of the game – an evolution of an allotment $p(t) \in \mathbf{R}^n$ such that, for all evolutions of fuzzy coalitions $x(t) \in [0,1]^n$ starting at x, the corresponding rule of the game:

$$\begin{cases} i) & \sum_{i=1}^n p_i(T)x_i(T) \geq \mathbf{u}(x(T)), \\ ii) & \forall t \in [0,T], \sum_{i=1}^n p_i(t)x_i(t) \geq \mathbf{u}(x(t)), \\ iii) & \exists t^* \in [0,T] \text{ such that } \sum_{i=1}^n p_i(t^*)x_i(t^*) \geq \mathbf{u}(x(t^*)), \end{cases} \tag{2}$$

must be satisfied.

Therefore, for each one of the above three rules of the game (2), a concept of dynamical core should provide a set-valued map $\Gamma : \mathbf{R}_+ \times [0,1]^n \rightsquigarrow \mathbf{R}^n$ associating with each time t and any fuzzy coalition x a set $\Gamma(t,x)$ of allotments $p \in \mathbf{R}^n_+$ such that, taking $p(t) \in \Gamma(T-t, x(t))$, and in particular, $p(0) \in \Gamma(T, x(0))$, the condition chosen above is satisfied. This is the purpose of this paper.

Naturally, this makes sense only once the dynamics of the coalitions and of the allotments are given. We shall assume that

(a) the evolution of coalitions $x(t) \in \mathbf{R}^n$ is governed by differential inclusions,

$$x'(t) := f(x(t), v(t)), \text{ where } v(t) \in Q(x(t)),$$

where $v(t)$ are perturbations,

(b) static constraints

$$\forall x \in [0,1]^n, \; p \in P(x) \subset \mathbf{R}^n_+,$$

and dynamic constraints on the velocities of the allotments $p(t) \in \mathbf{R}^n_+$ of the form,

$$\langle p'(t), x(t)\rangle = -\mathbf{m}(x(t), p(t), v(t))\langle p(t), x(t)\rangle,$$

stating that the cost $\langle p', x\rangle$ of the instantaneous change of allotment of a coalition is proportional to it by a discount factor $\mathbf{m}(x,p)$,

(c) from which we deduce the velocity $y'(t) = \langle p(t), f(x(t), v(t))\rangle - \mathbf{m}(x(t), p(t))y(t)$ of the payoff $y(t) := \langle p(t), x(t)\rangle$ of the fuzzy coalition $x(t)$.

The evolution of the fuzzy coalitions is thus parameterized by allotments and perturbations, i.e., is governed by a dynamic game:

$$\begin{cases} i) & x'(t) = f(x(t), v(t)), \\ ii) & y'(t) = \langle p(t), f(x(t), v(t))\rangle - \mathbf{m}(x(t), p(t))y(t), \\ iii) & \text{where } p(t) \in P(x(t)) \;\; \& \;\; v(t) \in Q(x(t)). \end{cases} \tag{3}$$

A feedback $\widetilde{p}$ is a selection of the set-valued map P in the sense that for any $x \in [0,1]^n$, $\widetilde{p}(x) \in P(x)$. We thus associate with any feedback $\widetilde{p}$ the set $\mathcal{C}_{\widetilde{p}}(x)$ of triples $(x(\cdot), y(\cdot), v(\cdot))$ solutions to

$$\begin{cases} i) & x'(t) = f(x(t), v(t)), \\ ii) & y'(t) = \langle \widetilde{p}(x(t)), f(x(t), v(t)) \rangle - y(t)\mathbf{m}(x(t), \widetilde{p}(x(t)), v(t)), \\ & \text{where } v(t) \in Q(x(t)), \end{cases} \tag{4}$$

1.3 Characterization of the Dynamical Core

We shall characterize the dynamical core of the fuzzy dynamical cooperative game in terms of the derivatives of a valuation function that we now define.

For each rule of the game (2), the set $\mathcal{V}^\sharp$ of initial conditions (T, x, y) such that there exists a feedback $x \mapsto \widetilde{p}(x) \in P(x)$ such that, for all perturbations $t \in [0, T] \mapsto v(t) \in Q(x(t))$, for all solutions to system (4) of differential equations satisfying $x(0) = x$, $y(0) = y$, the corresponding condition (2) is satisfied, is called the guaranteed valuation set[7].

Knowing it, we deduce the valuation function

$$V^\sharp(T, x) := \inf\{y | (T, x, y) \in \mathcal{V}\}$$

providing the cheapest initial payoff allowing to satisfy the viability/capturability conditions (2). It satisfies the initial condition:

$$V^\sharp(0, x) := \mathbf{u}(x).$$

In each of the three cases, we shall compute explicitly the valuation functions as infsup of underlying criteria we shall uncover. For that purpose, we associate with the characteristic function $\mathbf{u} : [0,1]^n \mapsto \mathbf{R} \cup \{+\infty\}$ of the dynamical cooperative game the functional

$$\begin{cases} J_{\mathbf{u}}(t; (x(\cdot), v(\cdot)); \widetilde{p})(x) := e^{\int_0^t \mathbf{m}(x(s), \widetilde{p}(x(s)), v(s))ds} \mathbf{u}(x(t)) \\ - \int_0^t e^{\int_0^\tau \mathbf{m}(x(s), \widetilde{p}(x(s)), v(s))ds} \langle \widetilde{p}(x(\tau)), f(x(\tau), v(\tau)) \rangle d\tau. \end{cases}$$

We shall associate with it and with each of the three rules of the game the three corresponding valuation functions[8]:

[7]One can also define the conditional valuation set $\mathcal{V}^\flat$ of initial conditions (T, x, y) such that for all perturbations v, there exists an evolution of the allotment $p(\cdot)$ such that viability/capturability conditions (2) are satisfied. We omit this study for the sake of brevity, since it is parallel to the one of guaranteed valuation sets.

[8]The notations $(\mathbf{0}, \mathbf{u}_\infty)$, $(\mathbf{u}, \mathbf{u}_\infty)$, $(\mathbf{0}, \mathbf{u})$ will be explained later.

(a) **Prescribed end rule**: We obtain

$$V^{\sharp}_{(\mathbf{0},\mathbf{u}_\infty)}(T,x) := \inf_{\widetilde{p}(x)\in P(x)} \sup_{(x(\cdot),v(\cdot))\in\mathcal{C}_{\widetilde{p}}(x)} J_{\mathbf{u}}(T;(x(\cdot),v(\cdot));\widetilde{p})(x); \tag{5}$$

(b) **Time span rule**: We obtain

$$V^{\sharp}_{(\mathbf{u},\mathbf{u}_\infty)}(T,x) := \inf_{\widetilde{p}(x)\in P(x)} \sup_{(x(\cdot),v(\cdot))\in\mathcal{C}_{\widetilde{p}}(x)} \sup_{t\in[0,T]} J_{\mathbf{u}}(t;(x(\cdot),v(\cdot));\widetilde{p})(x); \tag{6}$$

(c) **First winning time rule**: We obtain

$$V^{\sharp}_{(\mathbf{0},\mathbf{u})}(T,x) := \inf_{\widetilde{p}(x)\in P(x)} \sup_{(x(\cdot),v(\cdot))\in\mathcal{C}_{\widetilde{p}}(x)} \inf_{t\in[0,T]} J_{\mathbf{u}}(t;(x(\cdot),v(\cdot));\widetilde{p})(x). \tag{7}$$

Although these functions are only lower semicontinuous, one can define epi-derivatives of lower semicontinuous functions (or generalized gradients) in adequate ways and compute the core Γ: for instance, when the valuation function is differentiable, we shall prove that Γ associates with any $(t,x)\in\mathbf{R}_+\times\mathbf{R}^n$ the subset $\Gamma(t,x)$ of allotments $p\in P(x)$ satisfying

$$\sup_{v\in Q(x)}\left(\sum_{i=1}^{n}\left(\frac{\partial V^{\sharp}(t,x)}{\partial x_i}-p_i\right)f_i(x,v)+\mathbf{m}(x,p,v)V^{\sharp}(t,x)\right)\le\frac{\partial V^{\sharp}(t,x)}{\partial t}.$$

The valuation function $V^{\sharp}$ is actually a solution to the nonlinear Hamilton–Jacobi–Isaacs partial differential equation

$$-\frac{\partial\mathbf{v}(t,x)}{\partial t}+\inf_{p\in P(x)}\sup_{v\in Q(x)}\left(\sum_{i=1}^{n}\left(\frac{\partial\mathbf{v}(t,x)}{\partial x_i}-p_i\right)f_i(x,v)+\mathbf{m}(x,p,v)\mathbf{v}(t,x)\right)=0$$

satisfying the initial condition

$$\mathbf{v}(0,x)=\mathbf{u}(x),$$

on each of the subsets.

(a) **Prescribed end rule**:

$$\Omega_{(\mathbf{0},\mathbf{u}_\infty)}(\mathbf{v}) := \{(t,x)|t>0\,\&\,\mathbf{v}(t,x)\ge 0\};$$

(b) **Time span rule:**

$$\Omega_{(\mathbf{u},\mathbf{u}_\infty)}(\mathbf{v}) := \{(t,x)|t>0\ \&\ \mathbf{v}(t,x)\ge\mathbf{u}(x)\};$$

(c) **First winning time rule:**

$$\Omega_{(\mathbf{0},\mathbf{u})}(\mathbf{v}) := \{(t,x) | t > 0 \;\&\; \mathbf{u}(x) > \mathbf{v}(t,x) \geq 0\}.$$

Remark: The Static Case as a Limiting Case – Let us consider the case when $\mathbf{m}(x,p,v) = 0$ (self-financing of fuzzy coalitions) and when the evolution of coalitions is governed by $f(x,v) = v$ and $Q(x) = rB$. Then the dynamical core is the subset $\Gamma(t,x)$ of allotments $p \in P(x)$ satisfying on $\Omega(V^\sharp)$ the equation[9]

$$r \left\| \frac{\partial V^\sharp(t,x)}{\partial x} - p \right\| = \frac{\partial V^\sharp(t,x)}{\partial t}.$$

Now, assuming that the data and the solution are smooth we deduce formally that letting the radius $r \to \infty$, we obtain as a limiting case that $p = (\partial V^\sharp(t,x))/\partial x$ and that $(\partial V^\sharp(t,x))/\partial t = 0$. Since $V^\sharp(0,x) = \mathbf{u}(x)$, we infer that in this case $\Gamma(t,x) = (\partial \mathbf{u}(x))/\partial x$, i.e., the Shapley value of the fuzzy static cooperative game when the characteristic function $\mathbf{u}$ is differentiable and positively homogenous, and the core of the fuzzy static cooperative game when the characteristic function $\mathbf{u}$ is concave, continuous and positively homogenous. □

Actually, the solution of the above partial differential equation is taken in the "contingent sense", where the directional derivatives are the contingent epi-derivatives $D_\uparrow \mathbf{v}(t,x)$ of $\mathbf{v}$ at (t,x). They are defined by

$$D_\uparrow \mathbf{v}(t,x)(\lambda, v) := \liminf_{h \to 0+, u \to v} \frac{\mathbf{v}(t+h\lambda, x+hu)}{h},$$

(see for instance [21, Aubin & Frankowska] and [64, Rockafellar & Wets]). In this case, for each rule of the game, the dynamical core Γ of the corresponding fuzzy dynamical cooperative game is equal to

$$\begin{cases} \Gamma(t,x) := \{p \in P(x), \text{ such that} \\ \sup_{v \in Q(x)} (D_\uparrow V^\sharp(t,x)(-1, f(x,v)) - \langle p, f(x,v) \rangle + \mathbf{m}(x,p,v) \\ V^\sharp(t,x)) \leq 0\}, \end{cases}$$

where $V^\sharp$ is the corresponding value function. We shall prove that for each feedback $\widetilde{p}(t,x) \in \Gamma(t,x)$, selection of the dynamical core Γ, all evolutions $(x(\cdot), v(\cdot))$ of the system

$$\begin{cases} i) & x'(t) = f(x(t), v(t)), \\ ii) & y'(t) = \langle \widetilde{p}(T-t, x(t)), x(t) \rangle - \mathbf{m}(x(t), \widetilde{p}(T-t, x(t)) y(t), \\ iii) & v(t) \in Q(x(t)), \end{cases} \tag{8}$$

satisfy the corresponding condition (2).

[9] when $p = 0$, we find the eikonal equation.

1.4 Outline

The paper is organized as follows:

(a) We shall present fuzzy dynamical cooperative games (allotments and pay-off, dynamics, dynamical constraints on allotments, dynamics of the pay-off, characteristic functions), raise the questions and provide some answers: underlying criteria, Hamilton–Jacobi–Isaacs variational inequalities and the derivation of the dynamical core,
(b) outline the viability/capturability strategy,
(c) study and characterize guaranteed viable-capture basins of targets under dynamical games and use these results for studying intertemporal dynamical games problems,
(d) characterize guaranteed viable-capture basins in terms of tangential conditions, deduce that the valuation function is the solution to Hamilton–Jacobi–Isaacs variational inequalities and derive the regulation map and the adjustment law.

2 Fuzzy Dynamical Cooperative Game

Let us consider the set N of n players $i = 1, \dots, n$ and the set $[0, 1]^n$ of fuzzy coalitions[10]. The components of the state variable $x := (x_1, \dots, x_n) \in [0, 1]^n$ are the rates of participation in the fuzzy coalition x of player $i = 1, \dots, n$.

[10]The choice of "cooperative" fuzzy coalitions $x \in [0, 1]^n$ is arbitrary.
We could, for instance, introduce negative memberships when players enter a coalition with aggressive intents. This is mandatory if one wants to be realistic! A positive membership is interpreted as a cooperative participation of the player i in the coalition, while a negative membership is interpreted as a non-cooperative participation of the ith player in the generalized coalition. In what follows, one can replace the cube $[0, 1]^n$ by any product $\prod_{i=1}^{n} [\lambda_i, \mu_i]$ for describing the cooperative or noncooperative behavior of the consumers.
We can still enrich the description of the players by representing each player i by what psychologists call her 'behavior profile' as in [23, Aubin, Louis-Guerin & Zavalloni]. We consider q 'behavioral qualities' $k = 1, \dots, q$, each with a unit of measurement. We also suppose that a behavioral quantity can be measured (evaluated) in terms of a real number (positive or negative) of units. A behavior profile is a vector $a = (a_1, \dots, a_q) \in \mathbf{R}^q$ which specifies the quantities a_k of the q qualities k attributed to the player. Thus, instead of representing each player by a letter of the alphabet, she is described as an element of the vector space $\mathbf{R}^q$. We then suppose that each player may implement all, none, or only some of her behavioral qualities when she participates in a social coalition. Consider n players represented by their behavior profiles in $\mathbf{R}^q$. Any matrix $x = (x_i^k)$ describing the levels of participation $x_i^k \in [-1, +1]$ of the behavioral qualities k for the n players i is called a **social coalition**. Extension of the following results to social coalitions is straightforward.
Technically, the choice of the scaling $[0, 1]$ inherited from the tradition built on integration and measure theory is not adequate for describing convex sets. When dealing with convex sets, we have to replace the characteristic functions by indicators taking their values in

An **allotment** is an element,

$$p := (p_1, \dots, p_n) \in \mathbf{R}^n_+,$$

describing the payoff of player $i = 1, \dots, n$ in the game. The associated **payoff of the coalition** (or the coalition-payoff) y is defined by,

$$y := \langle p, x \rangle = \sum_{i=1}^{n} p_i x_i.$$

We now introduce constraints on the allotments gathered by a fuzzy coalition x. They range over a subset $P(x) = P(x_1, \dots, x_n) \subset \mathbf{R}^n_+$ that can be a constant set $P \subset \mathbf{R}^n$ or can depend on the fuzzy coalitions.

We assume that the velocity of the fuzzy coalitions of players are uncertain, in a contingent (i.e., nonstochastic) or "tychastic" way: they depend upon a parameter—usually known under the name or a **perturbation**, or **disturbance**, or *tyche*—$v(t) \in Q(x(t)) \subset \mathcal{V}$ and evolve according to the perturbed—or tychastic—system of differential equations,

$$\forall i = 1, \dots, n, x_i'(t) = f_i(x(t), v(t)),$$

or, in a more compact form,

$$x'(t) = f(x(t), v(t)), v(t) \in Q(x(t)).$$

In agreement with the tradition, the standard example of dynamics should be defined by,

$$f(x, v) := v \& Q(x) := rB, \text{ the ball of radius } r,$$

allowing the coalitions to evolve in all directions without any constraint. It seems to us more reasonable to take into account dedicated dynamics governing the evolution of coalitions.

$[0, +\infty]$ and take their convex combinations to provide an alternative allowing us to speak of "fuzzy" convex sets. Therefore, "toll-sets" are nonnegative cost functions assigning to each element its cost of belonging, $+\infty$, if it does not belong to the toll set. The set of elements with finite positive cost do form the "fuzzy boundary" of the toll set, the set of elements with zero cost its "core". This has been done to adapt viability theory to "fuzzy viability theory". See Chapter 10 of [10, Aubin] and [20, Aubin & Dordan] for more details. Actually, the Cramer transform,

$$C_\mu(p) := \sup_{x \in \mathbf{R}^n} \left(\langle p, x \rangle - \log \left(\int_{\mathbf{R}^n} e^{\langle x, y \rangle} d\mu(y) \right) \right),$$

maps probability measures to toll sets. In particular, it transforms convolution products of density functions to inf-convolutions of extended functions, Gaussian functions to squares of norms etc. See [20, Aubin & Dordan] for more information on this topic.

Remark: Tychastic Differential Equations and Control – The set-valued map $Q : X \rightsquigarrow \mathcal{V}$ translates mathematically the concept of uncertainty in a contingent (i.e., nonstochastic) or tychastic way to adopt Charles Peirce's terminology[11]: Contingent uncertainty depends upon a parameter – usually known under the name or a perturbation, or disturbance, that could also be called a *tyche*, ranging over a given subset (that could be a fuzzy subset, as it is advocated in [20, Aubin & Dordan]). The size of this subset captures mathematically the concept of "contingency" – instead of "volatility", that by now has a specific meaning in mathematical finance. The larger the subsets $Q(x)$, the more contingent or "tychastic" the system.

Indeed, we are investigating properties (such as the viability/capturability properties) that hold true for every *tyche* – instead of "random", terminology already confiscated by probability theory – or every perturbation or disturbance, and, "robust", in the sense of robust control in control theory.

Controlling a system for solving a problem (such as viability, capturability, intertemporal optimality) whatever the perturbation is the branch of dynamical games known among control specialists as "robust control", that we propose to call "tychastic control" in contrast to "stochastic control".

In tychastic control problems, we have two kinds of uncertainties, one described by the set-valued map Q, describing the unknown contingent uncertainty, and the one described by the set-valued map P, providing a set of available regulation parameters (regulons), here, the allotments, describing what biologists call "pleiotropism". The larger the set-valued map P, the more able is the system to find a regulation parameter or a control to satisfy a given property whatever the perturbation in Q. In some sense, the set-valued map P is an antidote to cure the negative effects of unknown perturbations.

"Tychastic equations" seem to us reasonable candidates for encapsulating the idea underlying robust control and "games against Nature", on which there is an abundant literature. They provide an alternative way of representing uncertainty to usual "stochastic differential equations":

$$\forall i = 1, \ldots, n, dx_i(t) = f_i(x_i(t))dt + \sigma_i(x_i(t))dW(t),$$

in a stochastic environment.

However, these two choices can be reconciled in the framework of stochastic differential inclusions (see [17,18, Aubin & Da Prato], [19, Aubin, Da Prato & Frankowska], [40, Da Prato & Frankowska], [33, Buckdahn, Cardaliaguet & Quincampoix], [30,31, Buckdahn, Quincampoix & Rascanu], [32, Buckdahn, Peng, Quincampoix & Rainer], [49, Gautier & Thibault] etc.).

[11] See [56, Peirce] among other references of this prolific and profound philosopher. He associates with the Greek concept of necessity, *ananke*, the concept of *anancastic evolution*, anticipating the "chance and necessity" framework that has motivated viability theory in the first place.

Invariance of a set under a tychastic differential equation requires that **for all** tyche $v(\cdot)$, the associated solution is viable in the set whereas under a stochastic differential equation, it requires that the stochastic process is viable **for almost all** ω. Thanks to the equivalence formulas between Itô and Stratonovitch stochastic integrals and to the Strook & Varadhan "Support Theorem"(see for instance [41, Doss], [67,68, Zabczyk]), and under convenient assumptions, stochastic viability problems are equivalent to invariance problems for tychastic systems and thus, viability problems for stochastic control systems are equivalent to guaranteed viability problems for dynamical games. □

In the context of dynamical cooperative games, a fuzzy coalition $x(t)$ at time t is allowed to change the allotment $p(t)$ at time t in such a way that

$$\sum_{i=1}^{n} p_i'(t)x_i(t) = -\mathbf{m}(x(t), p(t), v(t))\langle p(t), x(t)\rangle,$$

imposing instantaneous exchange of allotments of payments among players is allowed only in the extent that the variation $\langle p'(t), x(t)\rangle$ of the payoff of the fuzzy coalition is equal to the payoff discounted by a given factor depending upon the fuzzy coalition, the allotment and, possibly, the perturbation.

Therefore, under this behavioral rule of the decision maker, the velocity of the evolution of the payoff is equal to $y'(t) = \langle p(t), f(x(t), v(t))\rangle - y(t)\mathbf{m}(x(t), p(t), v(t))$.

An important instance is the case when $\mathbf{m}(x, p, v) = 0$, i.e., when allotments are **self-financed** by the fuzzy coalition x: This means that along the evolutions of the coalitions and the allotments, the velocity of the payoff satisfies $\langle p'(t), x(t)\rangle = 0$.

Hence, the evolution of the coalitions $x(t)$ of the shares, of the allotment and of the payoff $y(t)$ is governed by the two-person **dynamical game** (3), with which we associate with any feedback $\widetilde{p}$ the set $\mathcal{C}_{\widetilde{p}}(x)$ of triple $(x(\cdot), y(\cdot), v(\cdot))$ solutions to the tychastic equations (4):

$$\begin{cases} i) & x'(t) = f(x(t), v(t)), \\ ii) & y'(t) = \langle \widetilde{p}(x(t)), f(x(t), v(t))\rangle - y(t)\mathbf{m}(x(t), \widetilde{p}(x(t)), v(t)), \\ & \text{where } v(t) \in Q(x(t)). \end{cases}$$

By construction, since $y' = \langle p', x\rangle + \langle p, x'\rangle$ whenever $y = \langle p, x\rangle$ in all cases, the constraints,

$$y(t) = \langle p(t), x(t)\rangle,$$

are automatically satisfied whenever the initial conditions satisfy $y = \langle p, x\rangle$.

Remark: Path-Dependent Problems – Actually, our study (up to Hamilton–Jacobi–Isaacs partial differential equations) does not depend upon the fact that the evolution of the coalitions is governed by differential equations. It holds true for discretized systems, for path-dependent dynamics, as well as for other kinds of dynamics. We shall use only the properties of the set-valued map $(x, \widetilde{p}) \rightsquigarrow C_{\widetilde{p}}(x)$, that are shared by solution maps of other dynamical systems. □

2.1 Characteristic Functions of the Cooperative Game

Let us recall that a function $\mathbf{u} : X \mapsto \mathbf{R} \cup \{+\infty\}$ is called an *extended (real-valued) function*. Its *domain* is the set of points at which $\mathbf{u}$ is finite:

$$\text{Dom}\,(\mathbf{u}) := \{x \in X | \mathbf{u}(x) < +\infty\},$$

that embodies underlying state constraints: in particular, we shall assume that $\mathbf{u}(x) := +\infty$ whenever $x \notin [0, 1]^n$ to take into account that x is a fuzzy coalition[12].

Actually, in order to treat the three rules of the game (2) as particular cases of a more general framework, we introduce two nonnegative extended functions $\mathbf{b}$ and $\mathbf{c}$ (characteristic functions of the cooperative games) satisfying

$$\forall (t, x) \in \mathbf{R}_+ \times \mathbf{R}^n_+ \times \mathbf{R}^n, 0 \leq \mathbf{b}(t, x) \leq \mathbf{c}(t, x) \leq +\infty.$$

By associating with the initial characteristic function $\mathbf{u}$ of the game adequate pairs $(\mathbf{b}, \mathbf{c})$ of extended functions, we shall replace the requirements (2) by the requirement:

$$\begin{cases} i) & \forall t \in [0, t^*], y(t) \geq \mathbf{b}(T - t, x(t)) \text{ (dynamical constraints)}, \\ ii) & y(t^*) \geq \mathbf{c}(T - t^*, x(t^*)) \text{ (objective)}. \end{cases} \tag{9}$$

We extend the functions $\mathbf{b}$ and $\mathbf{c}$ as functions from $\mathbf{R} \times \mathbf{R}^n \times \mathbf{R}^n$ to $\mathbf{R}_+ \cup \{+\infty\}$ by setting,

$$\forall t < 0, \mathbf{b}(t, x) = \mathbf{c}(t, x) = +\infty,$$

so that nonnegativity constraints on time are automatically taken into account.

For instance, *problems with prescribed final time are obtained with objective functions satisfying the condition*

$$\forall t > 0, \mathbf{c}(t, x) := +\infty.$$

In this case, $t^* = T$ and condition (9) boils down to

$$\begin{cases} i) & \forall t \in [0, T], y(t) \geq b(T - t, x(t)), \\ ii) & y(T) \geq c(0, x(T)). \end{cases}$$

Indeed, since $y(t^*)$ is finite and since $\mathbf{c}(T - t^*, x(t^*))$ is infinite whenever $T - t^* > 0$, we infer from inequality (9)ii) that $T - t^*$ must be equal to 0. □

[12] If we assume that $\mathbf{u}$ is positively homogenous, it is enough to assume that $\mathbf{u}(x) := +\infty$ whenever $x \notin \mathbf{R}^n_+$.

Allowing the characteristic functions to take infinite values (i.e., to be extended), allows us to acclimate many examples.

For example, *the three rules (2) associated with a same characteristic function* $\mathbf{u} : [0,1]^n \mapsto \mathbf{R} \cup \{+\infty\}$ *can be written in the form (9) by adequate choices of pairs* $(\mathbf{b}, \mathbf{c})$ *of functions associated with* $\mathbf{u}$. Indeed, denoting by u_∞ the function defined by

$$\mathbf{u}_\infty(t,x) := \begin{cases} \mathbf{u}(x) & \text{if} \quad t = 0, \\ +\infty & \text{if} \quad t > 0, \end{cases}$$

and by $\mathbf{0}$ the function defined by

$$\mathbf{0}(t,x) = \begin{cases} 0 & \text{if} \quad t \geq 0, \\ +\infty & \text{if not,} \end{cases}$$

we can recover the three rules of the game

1. We take $\mathbf{b}(t,x) := \mathbf{0}(t,x)$ and $\mathbf{c}(t,x) = \mathbf{u}_\infty(t,x)$, we obtain the prescribed final time rule (2)i).
2. We take $\mathbf{b}(t,x) := \mathbf{u}(x)$ and $\mathbf{c}(t,x) := \mathbf{u}_\infty(t,x)$, we obtain the span time rule (2)ii).
3. We take $\mathbf{b}(t,x) := \mathbf{0}(t,x)$ and $\mathbf{c}(t,x) = \mathbf{u}(x)$, we obtain the first winning time rule (2)iii).

Using a pair $(\mathbf{b}, \mathbf{c})$ of time-dependent extended characteristic functions for describing rules allows to consider more general fuzzy dynamical cooperative games than the ones using time-independent characteristic functions. The problem of cooperative dynamical games is now that at each instant $t \in [0, t^*]$, the payoff $y(t) := \langle p(t), x(t) \rangle$ of the fuzzy coalition at time t is larger than or equal to the characteristic function of the dynamical game associating with any time t and any coalition $x(t)$ a lower bound $\mathbf{b}(T - t, x(t))$ on the payoff that the fuzzy coalition $x(t)$ may accept. Furthermore, one can impose a final constraint at the end of the game another lower bound $\mathbf{c}(T - t^*, x(t^*))$ on the payoff that the fuzzy coalition $x(t^*)$ when the game can stop.

2.2 The Valuation Function

By now, we have all the elements for setting the problem we shall study.

Definition 2.1. Let us consider the dynamical game (3) governing the evolution of the coalitions, the allotment and the payoff.

1. the first problem is to find the guaranteed valuation subset $\mathcal{V}^\sharp_{(,\mathbf{c})} \subset \mathbf{R}_+ \times \mathbf{R}^n \times \mathbf{R}_+$ of triples (T, x, y) made of the final time T, the initial fuzzy coalition x and the initial payoff y such that there exists a feedback $x \mapsto \widetilde{p}(x) \in P(x)$

such that, for all perturbations $t \in [0, T] \mapsto v(t) \in Q(x(t))$, for all solutions to system (4) of differential equations satisfying $x(0) = x$, $y(0) = y$, there exists a time $t^* \in [0, T]$ such that conditions (9):

$$\begin{cases} i) & \forall t \in [0, t^*],\ y(t) \geq \mathbf{b}(T - t, x(t)), \\ ii) & y(t^*) \geq \mathbf{c}(T - t^*, x(t^*)), \end{cases}$$

are satisfied.

2. Associate with any final time T and initial coalition x the smallest payoff $V^\sharp(T, x)$:

$$V^\sharp_{(\mathbf{b},\mathbf{c})}(T, x) := \inf_{(T,x,y) \in \mathcal{V}^\sharp_{(,\mathbf{c})}} y. \tag{10}$$

The function $(T, x) \mapsto V^\sharp_{(\mathbf{b},\mathbf{c})}(T, x)$ is called the guaranteed valuation function of the allotment, i.e., the minimal initial payoff y satisfying the two constraints (9).

Formulas (5), (6) and (7) for the valuation functions for each of the three rules of the game (2) that we mentioned in the preceding section are particular cases of the valuation function $V_{(\mathbf{b},\mathbf{c})}$ that we instanced above.

2.3 Formula for the Valuation Function

We shall associate with the objective function $\mathbf{c}$ the functional

$$\begin{cases} J_{\mathbf{c}}(t; (x(\cdot), v(\cdot)); \widetilde{p})(T, x) := e^{\int_0^t \mathbf{m}(x(s),\widetilde{p}(x(s)),v(s))ds}\mathbf{c}(T - t, x(t)) \\ -\displaystyle\int_0^t e^{\int_0^\tau \mathbf{m}(x(s),\widetilde{p}(x(s)),v(s))ds}\langle \widetilde{p}(x(\tau)), f(x(\tau), v(\tau))\rangle d\tau, \end{cases}$$

(where t ranges over $[0, T]$),

$$I_{\mathbf{b}}(t; (x(\cdot), v(\cdot)); \widetilde{p})(T, x) := \sup_{s \in [0,t]} J_{\mathbf{b}}(s; (x(\cdot), v(\cdot)); \widetilde{p})(T, x),$$

and

$$\begin{cases} L_{(\mathbf{b},\mathbf{c})}(t; (x(\cdot), v(\cdot)); \widetilde{p})(T, x) \\ := \max\left(J_{\mathbf{c}}(t; (x(\cdot), v(\cdot)); \widetilde{p})(T, x), I_{\mathbf{b}}(t; (x(\cdot), v(\cdot)); \widetilde{p})(T, x)\right). \end{cases}$$

We shall prove the

Theorem 2.1. *The guaranteed valuation function $(T, x, p) \mapsto V^\sharp_{(\mathbf{b},\mathbf{c})}(T, x)$ is equal to*

$$V^\sharp_{(\mathbf{b},\mathbf{c})}(T, x) = \inf_{\widetilde{p}(x) \in P(x)} \sup_{(x(\cdot),v(\cdot)) \in \mathcal{C}_{\widetilde{p}}(x)} \inf_{t \in [0,T]} L_{(\mathbf{b},\mathbf{c})}(t; (x(\cdot), v(\cdot)); \widetilde{p})(T, x),$$

satisfies the initial condition,

$$V^{\sharp}_{(\mathbf{b},\mathbf{c})}(0,x)=\mathbf{c}(0,x),$$

and inequalities,

$$\forall (T,x)\in \mathbf{R}_{+}\times \mathbf{R}^{n}\times \mathbf{R}^{n}, 0\leq \mathbf{b}(T,x)\leq V^{\sharp}_{(,\mathbf{c})}(T,x)\leq \mathbf{c}(T,x).$$

2.4 Examples of Valuation Functions

Let us consider a given time-independent function $\mathbf{u}:\mathbf{R}^{n}\mapsto \mathbf{R}\cup\{+\infty\}$, with which we associate the functional

$$\begin{cases} J_{\mathbf{u}}(t;(x(\cdot),v(\cdot));\widetilde{p})(x):=e^{\int_0^t \mathbf{m}(x(s),\widetilde{p}(x(s)),v(s))ds}\mathbf{u}(x(t)) \\ -\int_0^t e^{\int_0^{\tau} \mathbf{m}(x(s),\widetilde{p}(x(s)),v(s))ds}\langle \widetilde{p}(x(\tau)), f(x(\tau),v(\tau))\rangle d\tau. \end{cases}$$

We shall associate with it three pairs of time-dependent functions $(\mathbf{b},\mathbf{c})$ and obtain the valuation functions for the three rules of the game:

1. We assume that $f(x,v)\leq 0$ and we take $\mathbf{b}(t,x):=0$ and $\mathbf{c}(t,x):=\mathbf{u}_{\infty}(t,x)$. In this case, we obtain

$$V^{\sharp}_{(\mathbf{0},\mathbf{u}_{\infty})}(T,x):=\inf_{\widetilde{p}(x)\in P(x)}\ \sup_{(x(\cdot),v(\cdot))\in \mathcal{C}_{\widetilde{p}}(x)} J_{\mathbf{u}}(T;(x(\cdot),v(\cdot));\widetilde{p})(x).$$

2. We take $\mathbf{b}(t,x):=\mathbf{u}(x)$ and $\mathbf{c}(t,x)=\mathbf{u}_{\infty}(t,x)$. In this case, we obtain

$$V^{\sharp}_{(\mathbf{u},\mathbf{u}_{\infty})}(T,x):=\inf_{\widetilde{p}(x)\in P(x)}\ \sup_{(x(\cdot),v(\cdot))\in \mathcal{C}_{\widetilde{p}}(x)}\ \sup_{t\in[0,T]} J_{\mathbf{u}}(t;(x(\cdot),v(\cdot));\widetilde{p})(x).$$

3. We assume that $f(x,v)\leq 0$ and we take $\mathbf{b}(t,x):=0$ and $\mathbf{c}(t,x)=\mathbf{u}(x)$ In this case, we obtain

$$V^{\sharp}_{(\mathbf{0},\mathbf{u})}(T,x):=\inf_{\widetilde{p}(x)\in P(x)}\ \sup_{(x(\cdot),v(\cdot))\in \mathcal{C}_{\widetilde{p}}(x)}\ \inf_{t\in[0,T]} J_{\mathbf{u}}(t;(x(\cdot),v(\cdot));\widetilde{p})(x).$$

Indeed, when $\mathbf{b}=0$ and $f(x,v)\leq 0$, we observe that

$$\begin{aligned} & J_{\mathbf{0}}(t;(x(\cdot),v(\cdot));\widetilde{p})(T,x) \\ & = -\int_0^t e^{\int_0^{\tau} \mathbf{m}(x(s),\widetilde{p}(x(s)),v(s))ds}\langle \widetilde{p}(x(\tau)), f(x(\tau),v(\tau))\rangle d\tau \end{aligned}$$

so that,

$$\begin{cases} I_{\mathbf{0}}(t;\,(x(\cdot),\,v(\cdot));\,\widetilde{p})(T,x) \\ = -\sup_{s\in[0,t]} \int_0^s e^{\int_0^\tau \mathbf{m}(x(s),\widetilde{p}(x(s)),v(s))ds} \langle \widetilde{p}(x(\tau)),\, f(x(\tau),\, v(\tau))\rangle d\tau. \end{cases}$$

If $f(x,v) \le 0$, we infer that

$$\begin{aligned} & I_{\mathbf{0}}(t;\,(x(\cdot),\,v(\cdot));\,\widetilde{p})(T,x) f \\ & = -\int_0^t e^{\int_0^\tau \mathbf{m}(x(s),\widetilde{p}(x(s)),v(s))ds} \langle \widetilde{p}(x(\tau)),\, f(x(\tau),\, v(\tau))\rangle d\tau \end{aligned}$$

Therefore, for any nonnegative cost function $\mathbf{c}$, we have $L_{(\mathbf{0},\mathbf{c})} = J_{\mathbf{c}}$ and thus,

$$V^{\sharp}_{(\mathbf{0},\mathbf{c})}(T,x) = \inf_{\widetilde{p}(x)\in P(x)} \sup_{(x(\cdot),v(\cdot))\in\mathcal{C}_{\widetilde{p}}(x)} \inf_{t\in[0,T]} J_{\mathbf{c}}(t;\,(x(\cdot),\,v(\cdot));\,\widetilde{p})(T,x). \quad (11)$$

When $\mathbf{c}(t,x) := \mathbf{u}(x)$, we find the example of the first winning time problem.

When $\mathbf{b}(t,x) := \mathbf{0}(t,x)$ and $\mathbf{c}(t,x) := \mathbf{u}_\infty(t,x)$, the two above remarks imply that

$$V^{\sharp}_{(\mathbf{0},\mathbf{u}_\infty)}(T,x) = \inf_{\widetilde{u}(x)\in P(x)} \sup_{(x(\cdot),v(\cdot))\in\mathcal{C}_{\widetilde{u}}(x)} J_{\mathbf{u}_\infty}(T;\,(x(\cdot),\,v(\cdot));\,\widetilde{u})(x).$$

When we take $\mathbf{c}(t,x) := \mathbf{u}_\infty(t,x)$ that takes infinite values for $t > 0$, we have seen that

$$J_{\mathbf{u}_\infty}(t;\,(x(\cdot),\,v(\cdot));\,\widetilde{p})(T,x) := \begin{cases} J_{\mathbf{u}}(T;\,(x(\cdot),\,v(\cdot));\,\widetilde{p})(T,x) & \text{if} \quad t = T, \\ +\infty & \text{if} \quad t \in [0,T[, \end{cases}$$

so that

$$\inf_{t\in[0,T]} J_{\mathbf{u}_\infty}(t;\,(x(\cdot),\,v(\cdot));\,\widetilde{p})(T,x) := J_{\mathbf{u}}(T;\,(x(\cdot),\,v(\cdot));\,\widetilde{p})(x).$$

When $\mathbf{b}(t,x) := \mathbf{u}(x)$ and $\mathbf{c}(t,x) := \mathbf{u}_\infty(t,x)$, we infer that

$$L_{(\mathbf{u},\mathbf{u}_\infty)}(t;\,(x(\cdot),\,v(\cdot));\,\widetilde{p})(T,x) := \begin{cases} I_{\mathbf{u}}(T;\,(x(\cdot),\,v(\cdot));\,\widetilde{p})(x) & \text{if} \quad t = T, \\ +\infty & \text{if} \quad t < T, \end{cases}$$

so that

$$\inf_{t\in[0,T]} L_{(\mathbf{u},\mathbf{u}_\infty)}(t;\,(x(\cdot),\,v(\cdot));\,\widetilde{p})(T,x) = I_{\mathbf{u}}(T;\,(x(\cdot),\,v(\cdot));\,\widetilde{p})(x).$$

Consequently, we deduce that

$$V^{\sharp}_{(\mathbf{u},\mathbf{u}_\infty)}(T,x) := \inf_{\widetilde{p}(x)\in P(x)} \sup_{(x(\cdot),v(\cdot))\in\mathcal{C}_{\widetilde{p}}(x)} I_{\mathbf{u}}(T;\,(x(\cdot),\,v(\cdot));\,\widetilde{p})(x).$$

2.5 Hamilton–Jacobi–Isaacs Variational Inequalities

Let us associate with a nonnegative extended function $\mathbf{v}$ the subset

$$\Omega_{(\mathbf{b},\mathbf{c})}(\mathbf{v}) := \{(t,x) \in \mathbf{R}_+ \times \mathbf{R}^n \times \mathbf{R}^n, \text{ such that } \mathbf{b}(t,x) \leq \mathbf{v}(t,x) < \mathbf{c}(t,x)\},$$

which depends of the function $\mathbf{v}$.

Example. When for all $t > 0$, $\mathbf{c}(t,x) := +\infty$, and when $\mathbf{b}(0,x) := \mathbf{c}(0,x)$, we observe that

$$\Omega_{(\mathbf{b},\mathbf{c})}(\mathbf{v}) := \{(t,x) \in \mathbf{R}_+ \times \mathbf{R}^n \times \mathbf{R}^n \text{ such that } t > 0 \;\&\; \mathbf{b}(t,x) \leq \mathbf{v}(t,x)\}. \quad \square$$

Then the guaranteed value-function $V^{\sharp}_{(\mathbf{b},\mathbf{c})}$ is a "generalized" solution $\mathbf{v}$ to the Hamilton–Jacobi–Isaacs variational inequality: for every $(t,x) \in \Omega_{(\mathbf{b},\mathbf{c})}(\mathbf{v})$,

$$-\frac{\partial \mathbf{v}(t,x)}{\partial t} + \inf_{p \in P(x)} \sup_{v \in Q(x)} \left(\sum_{i=1}^{n} \left(\frac{\partial \mathbf{v}(t,x)}{\partial x_i} - p_i \right) f_i(x,v) + \mathbf{m}(x,p,v)\mathbf{v}(t,x) \right) = 0,$$

satisfying the initial condition,

$$\mathbf{v}(0,x) = \mathbf{c}(0,x).$$

This is a free boundary problem, well studied in mechanics and physics: the domain $\Omega_{(\mathbf{b},\mathbf{c})}(\mathbf{v})$ on which we look for a solution $\mathbf{v}$ to the Hamilton–Jacobi partial differential equation depends upon the unknown solution $\mathbf{v}$.

Observe that the Hamilton–Jacobi partial differential equation itself depends only upon the dynamic of the system (f, P, Q) and the map $\mathbf{m}$, whereas the domain $\Omega_{(\mathbf{b},\mathbf{c})}(\mathbf{v})$ depends only upon the pair $(\mathbf{b}, \mathbf{c})$ describing the characteristic functions of the fuzzy dynamical cooperative game. Changing them, the valuation function is a solution of the same Hamilton–Jacobi partial differential equation, but defined on a different "free set" $\Omega_{(\mathbf{b},\mathbf{c})}(\mathbf{v})$ depending on $\mathbf{v}$.

The usefulness and relevance of the Hamilton–Jacobi–Isaacs variational inequality is that it provides the dynamical core of the game – through dynamical feedbacks – that we are looking for. Indeed, we introduce the dynamical core map Γ associating with any $(t,x) \in \mathbf{R}_+ \times \mathbf{R}^n$ the subset $\Gamma(t,x)$ of allotments $p \in P(x)$ satisfying

$$\sup_{v \in Q(x)} \left(\left\langle \frac{\partial V^{\sharp}_{(\mathbf{b},\mathbf{c})}(t,x)}{\partial x} - p, f(x,v) \right\rangle + \mathbf{m}(x,p,v) V^{\sharp}_{(\mathbf{b},\mathbf{c})}(t,x) \right) \leq \frac{\partial V^{\sharp}_{(\mathbf{b},\mathbf{c})}(t,x)}{\partial t}.$$

Namely, knowing the guaranteed valuation function and its derivatives, a guaranteed evolution is obtained in the following way: Starting from an initial fuzzy coalition x_0 such that $\mathbf{b}(T, x_0) \leq V^{\sharp}_{(\mathbf{b},\mathbf{c})}T, x_0) < c(T, x_0, p_0)$, solutions to the new system:

$$\begin{cases} i) & \forall i = 0, \ldots, n, x_i'(t) = f_i(x(t), v(t)), \\ ii) & y'(t) = -y(t)\mathbf{m}(x(t), p(t), v(t)) + \langle p(t), f(x(t), v(t))\rangle, \\ iii) & p(t) \in \Gamma(T - t, x(t)), \end{cases}$$

regulate the guaranteed solutions of the cooperative dynamical game until the first time $t^* \in [0, T]$ when

$$V^{\sharp}_{(\mathbf{b},\mathbf{c})}(T - t^*, x(t^*)) = \mathbf{c}(T - t^*, x(t^*)).$$

Actually, the guaranteed valuation function is seldom differentiable, but, generally only lower semicontinuous: This happens whenever state constraints are involved, i.e., whenever the cost function takes infinite values. However, one can define generalized directional derivatives – contingent epiderivatives – of any lower semicontinuous function. Replacing the classical derivatives by contingent epiderivatives in the Hamilton–Jacobi–Isaacs variational inequalities above and in the definition of the regulation map, the same conclusions hold true for Frankowska's episolutions[13]. In particular, we can still build the dynamical core of the fuzzy dynamic cooperative game. By duality, one can formulate and prove equivalent statements involving subdifferentials and superdifferentials in the "viscosity solution format".

We shall derive from Theorem 5.2 below the following

Theorem 2.2. *Let us assume that the maps f, & $\mathbf{m}$ are Lipschitz, that the set-valued maps P and Q are Lipschitz and bounded and that the functions* $\mathbf{b}$ *and* $\mathbf{c}$ *are lower semicontinuous.*

Then

1. *the dynamic core $\Gamma_{(\mathbf{b},\mathbf{c})}$ of the fuzzy dynamical cooperative with rules defined by $(\mathbf{b}, \mathbf{c})$ is equal to*

$$\begin{cases} \Gamma_{(\mathbf{b},\mathbf{c})}(t, x) := \{p \in P(x), \text{ such that} \\ \sup_{v \in Q(x)} \big(D_\uparrow V^{\sharp}_{(\mathbf{b},\mathbf{c})}(t, x)(-1, f(x, v)) - \langle p, f(x, v)\rangle \\ \qquad +\mathbf{m}(x, p, v)V^{\sharp}_{(\mathbf{b},\mathbf{c})}(t, x)\big) \leq 0\}, \end{cases}$$

defined on $\Omega_{(\mathbf{b},\mathbf{c})}(V_{(\mathbf{b},\mathbf{c})})$,

[13] Hélène Frankowska proved that the epigraph of the value function of an optimal control problem – assumed to be only lower semicontinuous – is invariant and backward viable under a (natural) auxiliary system. Furthermore, when it is continuous, she proved that its epigraph is viable and its hypograph invariant ([46–48, Frankowska]). By duality, she proved that the latter property is equivalent to the fact that the value function is a viscosity solution of the associated Hamilton–Jacobi equation in the sense of M. Crandall and P.-L. Lions.

2. *the guaranteed valuation function* $V^{\sharp}_{(\mathbf{b},\mathbf{c})}$ *is the smallest of the lower semicontinuous "episolutions"* $\mathbf{v}$ *to the Hamilton–Jacobi–Isaacs contingent inequalities*

$$\begin{cases} i) & \mathbf{b}(t,x) \leq \mathbf{v}(t,x) \leq \mathbf{c}(t,x), \\ ii) & \text{if } \mathbf{v}(t,x) < \mathbf{c}(t,x), \\ & \inf_{p\in P(x)} \sup_{v\in Q(x)} (D_{\uparrow}\mathbf{v}(t,x)(-1, f(x,v)) - \langle p, f(x,v)\rangle \\ & \qquad\qquad +\mathbf{m}(x,p,v)\mathbf{v}(t,x)) \leq 0, \end{cases} \tag{12}$$

satisfying the initial condition $\mathbf{v}(0,x) = \mathbf{c}(0,x)$ *and such that there exists a Lipschitz selection* $\widetilde{p}$ *of the set-valued map* Γ *defined by*

$$\begin{cases} \Gamma(t,x) := \{p \in P(x), \text{ such that} \\ \sup_{v\in Q(x)} (D_{\uparrow}\mathbf{v}(t,x)(-1, f(x,v)) - \langle p, f(x,v)\rangle \\ \qquad\qquad +\mathbf{m}(x,p,v)\mathbf{v}(t,x)) \leq 0\}. \end{cases}$$

Remark. We describe only the equivalent dual version of episolutions to the above Hamilton–Jacobi–Isaacs partial differential equation. We introduce the Hamiltonian H defined by

$$H(t,x,p_t,p_x,y) := -p_t + \inf_{u\in P(x)} \sup_{v\in Q(x)} \left(\langle p_x - u, f(x,v)\rangle + \mathbf{m}(x,u,v)y\right).$$

We recall that the subdifferential $\partial_{-}\mathbf{v}(t,x)$ of the extended function $\mathbf{v}$ at (t,x) is the set of pairs (p_t, p_x), such that

$$\forall (\lambda, v) \in \mathbf{R} \times X,\ p_t\lambda + \langle p_x, v\rangle \leq D_{\uparrow}\mathbf{v}(t,x)(\lambda, v).$$

Hence, the function $\mathbf{v}$ is an episolution of (12) if and only if $\mathbf{v}$ satisfies

$$\forall (t,x) \in \Omega_{(\mathbf{b},\mathbf{c})}(\mathbf{v}), \forall (p_t,p_x) \in \partial_{-}\mathbf{v}(t,x),\ H(t,x,p_t,p_x,\mathbf{v}(t,x)) \leq 0. \quad \square$$

The proofs of the above results require a more abstract geometric approach that we shall now describe.

3 The Viability/Capturability Strategy

3.1 Epigraphs of Extended Functions

The *epigraph* of an extended function $\mathbf{v} : X \mapsto \mathbf{R} \cup \{+\infty$ is defined by

$$\mathcal{E}p(\mathbf{v}) := \{(x,\lambda) \in X \times \mathbf{R} | \mathbf{v}(x) \leq \lambda\}.$$

We recall that *an extended function* $\mathbf{v}$ *is convex (resp. positively homogeneous) if and only if its epigraph is convex (resp. a cone)* and that the epigraph of $\mathbf{v}$ is closed if and only if $\mathbf{v}$ is lower semicontinuous:

$$\forall x \in X, \mathbf{v}(x) = \liminf_{y \to x} \mathbf{v}(y).$$

The definition of the guaranteed valuation function $V^{\sharp}_{(\mathbf{b},\mathbf{c})}$ from the guaranteed valuation subset $\mathcal{V}^{\sharp}_{(,\mathbf{c})}$ fits the following definition:

Definition 3.1. We associate with a subset $\mathcal{V} \subset X \times \mathbf{R}_+$ the function $\mathbf{v}_{\mathcal{V}} : X \mapsto \mathbf{R}_+ \cup \{+\infty\}$ defined by

$$\mathbf{v}_{\mathcal{V}}(x) := \inf_{(x,w)\in\mathcal{V}} w \in \overline{\mathbf{R}},$$

that we shall call its **southern border.**

We shall say that $[\mathcal{V}]^{\uparrow} := \mathcal{E}p(\mathbf{v}_{\mathcal{V}})$ is the **southern closure**[14] of $\mathcal{V}$.

We recall the convention $\inf(\emptyset) := +\infty$.

We observe that

$$\mathcal{V} + \{0\} \times \mathbf{R}_+ \subset \mathcal{E}p(\mathbf{v}_{\mathcal{V}}) \subset \overline{\mathcal{V} + \{0\} \times \mathbf{R}_+},$$

and that if $\mathcal{V} \subset X \times \mathbf{R}_+$ is a closed subset, then its southern closure $\mathbf{v}_{\mathcal{V}}$ is lower semicontinuous and the three above sets are closed and equal.

We shall need the following.

Lemma 3.1. *Let $\mathcal{V}_{i\in I}$ be a family of subsets $\mathcal{V}_i \subset X \times \mathbf{R}$. Then the southern border of the union of the $\mathcal{V}_i$ is the infimum of the southern borders of the set $\mathcal{V}_i$:*

$$\mathbf{v}_{\bigcup_{i\in I}\mathcal{V}_i} = \inf_{i\in I} \mathbf{v}_{\mathcal{V}_i},$$

or, equivalently,

$$\mathcal{E}p(\inf_{i\in I} \mathbf{v}_{\mathcal{V}_i}) = \left[\bigcup_{i\in I} \mathcal{V}_i\right]^{\uparrow}.$$

In particular, the epigraph of the pointwise infimum $\inf_{i\in I} \mathbf{v}_i$ of a family of functions $\mathbf{v}_i$ is the southern closure of the union of their epigraphs:

$$\mathcal{E}p(\inf_{i\in I} \mathbf{v}_i) = \left[\bigcup_{i\in I} \mathcal{E}p(\mathbf{v}_i)\right]^{\uparrow}.$$

Proof. Indeed,

$$\mathbf{v}_{\bigcup_{i\in I}\mathcal{V}_i}(x) = \inf_{(x,y)\in\bigcup_{i\in I}\mathcal{V}_i} y = \inf_{i\in I} \inf_{(x,y)\in\mathcal{V}_i} y = \inf_{i\in I} \mathbf{v}_{\mathcal{V}_i}(x).$$

□

[14]When $\mathcal{V} = \mathcal{V} + \{0\} \times \mathbf{R}_+$, this is the **vertical closure** introduced in [64, Rockafellar & Wets].

3.2 The Epigraphical Approach

With these definitions, we can translate the viability/capturability conditions (9) in the following geometric form:

$$\begin{cases} i) & \forall t \in [0, t^*], (T - t, x(t), y(t)) \in \mathcal{E}p(\mathbf{b}) \\ & \text{(viability constraint),} \\ ii) & (T - t^*, x(t^*), y(t^*)) \in \mathcal{E}p(\mathbf{c}) \\ & \text{(capturability of a target).} \end{cases} \tag{13}$$

This "epigraphical approach" proposed by J.-J. Moreau and R.T. Rockafellar in convex analysis in the early 60's[15], has been used in optimal control by H. Frankowska in a series of papers [46–48, Frankowska] and [22, Aubin & Frankowska] for studying the value function of optimal control problems and characterizing it as a generalized solution (episolutions and/or viscosity solutions) of (first-order) Hamilton–Jacobi–Bellman equations, in [6,16,8,10, Aubin] for characterizing and constructing Lyapunov functions, in [35–38, Cardaliaguet] for characterizing the minimal time function, in [59, Pujal] and [24, Aubin, Pujal & Saint-Pierre] in finance and other authors since. This is this approach that we adopt and adapt here, since the concepts of "capturability of a target" and of "viability" of a constrained set allows us to study this problem in a new light (see for instance [10, Aubin] and [11, Aubin] for economic applications) for studying the evolution of the state of a tychastic control system subjected to viability constraints in control theory and in dynamical games against nature or robust control (see [60, Quincampoix], [35–38, Cardaliaguet], [39, Cardaliaguet, Quincampoix & Saint-Pierre]. Numerical algorithms for finding viability kernels have been designed in [66, Saint-Pierre] and adapted to our type of problems in [59, Pujal].

The properties and characterizations of the valuation function are thus derived from the ones of guaranteed viable-capture basins, that are easier to study – and that have been studied – in the framework of plain constrained sets K and targets $C \subset K$ (see [12,13, Aubin] and [15, Aubin & Catté] for recent results on that topic).

3.3 Introducing Auxiliary Dynamical Games

We observe that the evolution of $(T - t, x(t), y(t))$ made up of the backward time $\tau(t) := T - t$, of fuzzy coalitions $x(t)$ of the players, of allotments and of the payoff $y(t)$ is governed by the dynamical game

[15] see for instance [21, Aubin & Frankowska] and [64, Rockafellar & Wets] among many other references.

$$\begin{cases} i) & \tau'(t) = -1, \\ ii) & \forall i = 0, \dots, n, x_i'(t) = f_i(x(t), v(t)), \\ iii) & y'(t) = -y(t)\mathbf{m}(x(t), p(t), v(t)) + \langle p(t), f(x(t), v(t))\rangle, \\ & \text{where } p(t) \in P(x(t)) \ \& \ v(t) \in Q(x(t)), \end{cases} \tag{14}$$

starting at (T, x, y). We summarize it in the form of the dynamical game

$$\begin{cases} i) & z'(t) \in g(z(t), u(t), v(t)), \\ ii) & u(t) \in P(z(t)) \ \& \ v(t) \in Q(z(t)), \end{cases}$$

where $z := (\tau, x, y) \in \mathbf{R} \times \mathbf{R}^n \times \mathbf{R}$, where the controls $u := p$ are the allotments, where the map $g : \mathbf{R} \times \mathbf{R}^n \times \mathbf{R} \rightsquigarrow \mathbf{R} \times \mathbf{R}^n \times \mathbf{R}^n \times \mathbf{R}$ is defined by $g(z, v)$

$$= (-1, f(x, v), u, -\mathbf{m}(x, u, v)y + \langle u, f(x, v)\rangle),$$

where u ranges over $P(z) := P(x)$ and v over $Q(z) := Q(x)$.

We say that a selection $z \mapsto \widetilde{p}(z) \in P(z)$ is a feedback, regarded as a strategy. One associates with such a feedback chosen by the decision maker or the player the evolutions governed by the tychastic differential equation,

$$z'(t) = g(z(t), \widetilde{p}(z(t)), v(t)),$$

starting at time 0 at z.

3.4 Introducing Guaranteed Capture Basins

We now define the guaranteed viable-capture basin that are involved in the definition of guaranteed valuation subsets.

Definition 3.2. Let K and $C \subset K$ be two subsets of Z.

The guaranteed viable-capture basin of the target C viable in K is the set of elements $z \in K$ such that there exists a continuous feedback $\widetilde{p}(z) \in P(z)$ such that for every $v(\cdot) \in Q(z(\cdot))$, for every solutions $z(\cdot)$ to $z' = g(z, \widetilde{p}(z), v)$, there exists $t^* \in \mathbf{R}_+$ such that the viability/capturability conditions,

$$\begin{cases} i) & \forall t \in [0, t^*], & z(t) \in K, \\ ii) & & z(t^*) \in C, \end{cases}$$

are satisfied.

We thus observe that

Proposition 3.1. *The guaranteed valuation subset $\mathcal{V}^{\sharp}_{(,\mathbf{c})}$ defined in Definition 2.1 is the southern border of the guaranteed viable-capture basin under the dynamical game (14) of the epigraph of the function* $\mathbf{c}$ *viable in the epigraph of the function* $\mathbf{b}$.

The characterization of this subset and the study of its properties is one of the major topics of the viability approach to dynamical games theory that we summarize in the two next sections.

3.5 The Strategy

Since we have related the guaranteed valuation problem to the much simpler – although more abstract – study of guaranteed viable-capture basin of a target and other guaranteed viability/capturability issues for dynamical games,

1. we first "solve" these "viability/capturability problems" for dynamical games at this general level, and in particular, study the tangential conditions enjoyed by the guaranteed viable-capture basins (see Theorem 5.1 below),
2. and use set-valued analysis and nonsmooth analysis for translating the general results of viability theory to the corresponding results of the auxiliary dynamical game, in particular translating tangential conditions to give a meaning to the concept of a generalized solution (Frankowska's episolutions or, by duality, viscosity solutions) to Hamilton–Jacobi–Isaacs variational inequalities (see theorems 4.1 and 5.2 below).

4 Guaranteed Viability/Capturability under Dynamical Games

4.1 Guaranteed Viable-Capture Basins

We summarize the main results on guaranteed viability/capturability of a target under dynamical games that we need to prove the results stated in the preceding section.

We denote by $X, \mathcal{U}$ and $\mathcal{V}$ three finite dimensional vector spaces, and we introduce a set-valued map $F : X \times \mathcal{U} \times \mathcal{V} \rightsquigarrow X$, a set-valued map $P : X \rightsquigarrow \mathcal{U}$ and a set-valued map $Q : X \rightsquigarrow \mathcal{V}$.

We consider a dynamical game described by

$$\begin{cases} i) & x'(t) \in F(x(t), u(t), v(t)), \\ ii) & u(t) \in P(x(t)), \\ iii) & v(t) \in Q(x(t)), \end{cases} \tag{15}$$

which is, so to speak, a control system regulated by two parameters, $u(t)$ and $v(t)$, the first one regarded as a regulating parameter, controlled by a player, the second one regarded as a perturbation, or a disturbance, or a tyche, chosen in a unknown way by "Nature".

We introduce a class $\widetilde{\mathcal{P}}$ of continuous selections $x \mapsto \widetilde{u}(x) \in P(x)$, that are used as feedbacks or strategies by the player controlling the parameters u.

We associate with such a feedback $\widetilde{u}(x) \in P(x)$ the set $\mathcal{C}_{\widetilde{u}}(x)$ of solutions $(x(\cdot), v(\cdot)) \in \mathcal{C}(0, \infty; X) \times L^1(0, \infty; \mathcal{U})$ to the parameterized system,

$$\begin{cases} i) & x'(t) \in F(x(t), \widetilde{u}(x(t)), v(t)), \\ ii) & v(t) \in Q(x(t)), \end{cases} \tag{16}$$

starting at x.

We may identify the above dynamical game with the set-valued map $(x, \widetilde{u}) \rightsquigarrow \mathcal{C}_{\widetilde{u}}(x)$, that we regard as an evolutionary game.

Definition 4.1. Let $C \subset K \subset X$ be two subsets, C being regarded as a target, K as a constrained set.

The subset $\text{Abs}_{\widetilde{u}}(K, C)$ of initial states $x_0 \in K$ such that C is reached in finite time before possibly leaving K by all solutions to (16) starting at x_0 is called the invariance-absorption basin of C in K.

The subset,

$$[\text{Capt}_P \, \text{Abs}_Q](K, C) := \bigcup_{\widetilde{u} \in \widetilde{\mathcal{P}}} \text{Abs}_{\widetilde{u}}(K, C),$$

of elements $x \in K$, such that there exists a feedback $\widetilde{u} \in \widetilde{\mathcal{P}}$, such that for every solution $(x(\cdot), v(\cdot)) \in \mathcal{C}_{\widetilde{u}}(x)$, there exists $t^* \in \mathbf{R}_+$ satisfying the viability/capturability conditions,

$$\begin{cases} i) & \forall t \in [0, t^*], \quad x(t) \in K, \\ ii) & \qquad\qquad\qquad\; x(t^*) \in C, \end{cases}$$

is called the guaranteed viable-capture basin of a target under the evolutionary game $(x, \widetilde{u}) \rightsquigarrow \mathcal{C}_{\widetilde{u}}(x)$ defined on $X \times \widetilde{\mathcal{P}}$ (that, naturally, depends upon the choice of the family $\widetilde{\mathcal{P}}$ of feedbacks).

4.2 Intertemporal Games

We introduce the following four features:

1. a discount factor,

$$\mathbf{m} : (x, u, v) \in X \times \mathcal{U} \times \mathcal{V} \mapsto \mathbf{m}(x, u, v) \in \mathbf{R};$$

2. a "Lagrangian",

$$\mathbf{l} : (x, u, v) \in X \times \mathcal{U} \times \mathcal{V} \mapsto \mathbf{l}(x, u, v) \in \mathbf{R}_+;$$

3. two nonnegative extended cost functions $\mathbf{b}$ and $\mathbf{c}$ from $\mathbf{R}_+ \times X$ to $\mathbf{R}_+ \cup \{+\infty\}$ satisfying,

$$\forall (t, x) \in \mathbf{R}_+ \times X, 0 \leq \mathbf{b}(t, x) \leq \mathbf{c}(t, x) \leq +\infty,$$

that we shall extend to cost functions (denoted by) $\mathbf{b}$ (constrained function) and $\mathbf{c}$ (objective function) from $\mathbf{R} \times X$ to $\mathbf{R}_+ \cup \{+\infty\}$ by setting,

$$\mathbf{b}(t, x) = \mathbf{c}(t, x) := +\infty, \text{ whenever } t < 0.$$

We next fix a horizon or an final time T and associate with it the cost functionals,

$$\begin{cases} J_{\mathbf{c}}(t;(x(\cdot),v(\cdot));\widetilde{u})(T,x) := e^{\int_0^t \mathbf{m}(x(s),\widetilde{u}(x(s)),v(s))ds}\mathbf{c}(T-t,x(t)) \\ + \displaystyle\int_0^t e^{\int_0^\tau \mathbf{m}(x(s),\widetilde{u}(x(s)),v(s))ds}\mathbf{l}(x(\tau),\widetilde{u}(x(\tau)),v(\tau))d\tau, \end{cases}$$

(where t ranges over $[0,T]$),

$$I_{\mathbf{b}}(t;(x(\cdot),v(\cdot));\widetilde{u})(T,x) := \sup_{s\in[0,t]} J_{\mathbf{b}}(s;(x(\cdot),v(\cdot));\widetilde{u})(T,x),$$

and

$$L_{(\mathbf{b},\mathbf{c})}(t;(x(\cdot),v(\cdot));\widetilde{u})(T,x) := \max(J_{\mathbf{c}}(t;(x(\cdot),v(\cdot));\widetilde{u})(T,x), I_{\mathbf{b}}(t;(x(\cdot),v(\cdot));\widetilde{u})(T,x))$$

We associate with it the guaranteed valuation function

$$V^{\sharp}_{(\mathbf{b},\mathbf{c})}(T,x) := \inf_{\widetilde{u}\in\widetilde{\mathcal{P}}} \sup_{(x(\cdot),v(\cdot))\in\mathcal{C}_{\widetilde{u}}(x)} \inf_{t\in[0,T]} L_{(\mathbf{b},\mathbf{c})}(t;(x(\cdot),v(\cdot));\widetilde{u})(T,x). \quad (17)$$

The function $V^{\sharp}_{(\mathbf{b},\mathbf{c})}$ is called the guaranteed valuation function associated with $\mathbf{l}$, $\mathbf{m}$ and the cost functions $\mathbf{b}$ and $\mathbf{c}$.

Let us consider the extended dynamical game of the form:

$$\begin{cases} i) & \tau'(t) = -1, \\ ii) & x'(t) \in F(x(t),u(t),v(t)), \\ iii) & y'(t) = -y(t)\mathbf{m}(x(t),u(t),v(t)) - \mathbf{l}(x(t),u(t),v(t)), \\ & \text{where } v(t) \in Q(x(t)). \end{cases} \quad (18)$$

We associate with such a feedback $\widetilde{u}(x) \in P(x)$ the set $\mathcal{B}_{\widetilde{u}}(T,x,y)$ of solutions $(T-\cdot, x(\cdot), v(\cdot), y(\cdot))$ to the auxiliary system

$$\begin{cases} i) & \tau'(t) = -1, \\ ii) & x'(t) \in F(x(t),\widetilde{u}(x(t)),v(t)), \\ iii) & y'(t) = -y(t)\mathbf{m}(x(t),\widetilde{u}(x(t)),v(t)) - \mathbf{l}(x(t),\widetilde{u}(x(t)),v(t)), \\ & \text{where } v(t) \in Q(x(t)). \end{cases}$$

Theorem 4.1. *Let us assume that the extended functions* $\mathbf{b}$ *and* $\mathbf{c}$ *are nontrivial and non-negative.*

The guaranteed valuation function $V^{\sharp}_{(\mathbf{b},\mathbf{c})}$ *defined by (17) is the southern border of the guaranteed viable-capture basin* $[Capt_P\, Abs_Q](\mathcal{E}p(\mathbf{b}), \mathcal{E}p(\mathbf{c})$ *of the epigraph* $\mathcal{E}p(\mathbf{c})$ *of* $\mathbf{c}$ *under the dynamical game (18) viable in the epigraph* $\mathcal{E}p(\mathbf{b})$ *of* $\mathbf{b}$*:*

$$V^{\sharp}_{(\mathbf{b},\mathbf{c})}(T,x) := \inf_{(x,T,y)\in[Capt_P\, Abs_Q](\mathcal{E}p(\mathbf{b}),\mathcal{E}p(\mathbf{c})} y$$

In other words, Theorem 4.1 states that

$$\mathcal{E}p(V^{\sharp}_{(\mathbf{b},\mathbf{c})}) = [[\mathrm{Capt}_P \,\mathrm{Abs}_Q](\mathcal{E}p(\mathbf{b}), \mathcal{E}p(\mathbf{c})]^{\uparrow}.$$

Since the guaranteed viable-capture basin,

$$[\mathrm{Capt}_P \,\mathrm{Abs}_Q](\mathcal{E}p(\mathbf{b}), \mathcal{E}p(\mathbf{c})) := \bigcup_{\widetilde{u}\in\widetilde{\mathcal{P}}} \mathrm{Abs}_{\widetilde{u}}(\mathcal{E}p(\mathbf{b}), \mathcal{E}p(\mathbf{c})).$$

Lemma 3.1 implies that the southern border,

$$V^{\sharp}_{(\mathbf{b},\mathbf{c})}(T, x) := \inf_{(x,T,y)\in[\mathrm{Capt}_P \,\mathrm{Abs}_Q](\mathcal{E}p(\mathbf{b}),\mathcal{E}p(\mathbf{c}))} y,$$

of $[\mathrm{Capt}_P \,\mathrm{Abs}_Q](\mathcal{E}p(\mathbf{b}), \mathcal{E}p(\mathbf{c})$ is the pointwise infimum,

$$V^{\sharp}(T, x) = \inf_{\widetilde{u}\in\widetilde{\mathcal{P}}} U_{((\mathbf{b},\mathbf{c});\widetilde{u})}(T, x),$$

of the southern borders,

$$U_{(\mathbf{b},\mathbf{c};\widetilde{u})}(T, x) := \inf_{(x,T,y)\in\mathrm{Abs}_{\widetilde{u}}(\mathcal{E}p(\mathbf{b}),\mathcal{E}p(\mathbf{c}))} y.$$

It remains to prove that

$$U_{(\mathbf{b},\mathbf{c};\widetilde{u})}(T, x) = \sup_{(x(\cdot),v(\cdot))\in\mathcal{C}_{\widetilde{u}}(x)} \inf_{t\in[0,T]} L_{(\mathbf{b},\mathbf{c})}(t; (x(\cdot), v(\cdot)); \widetilde{u})(T, x), \tag{19}$$

to derive the Theorem 4.1.

This is purpose of

Theorem 4.2. *Let us assume that the extended functions* $\mathbf{b}$ *and* $\mathbf{c}$ *are nontrivial and non-negative.*

The valuation function $U_{(\mathbf{b},\mathbf{c};\widetilde{u})}$ *is equal to the southern border of the invariant-absorption basin Abs* $(\mathcal{E}p(\mathbf{b}), \mathcal{E}p(\mathbf{c}))$ *of* $\mathcal{E}p(\mathbf{c})$*:*

$$U_{(\mathbf{b},\mathbf{c};\widetilde{u})}(T, x) := \inf_{(x,T,y)\in Abs(\mathcal{E}p(\mathbf{b}),\mathcal{E}p(\mathbf{c}))} y$$

We refer to [24, Aubin, Pujal & Saint-Pierre] for the proof of this theorem.

5 Hamilton–Jacobi–Isaacs Equations

5.1 Lipschitz Dynamical Games

We shall assume that the dynamical game (15) is Lipschitz in the sense that the set-valued maps P and Q are Lipschitz with compact values and that the set-valued map F is Lipschitz with closed values.

Let $\widetilde{\mathcal{P}}_\lambda$ be the set of Lipschitz selections with constant λ of the set-valued map P: for every $x, \widetilde{u}(x) \in P(x)$.

The subset,

$$[\mathrm{Capt}_{P_\lambda} \mathrm{Abs}_Q](K, C) := \bigcup_{\widetilde{u} \in \widetilde{\mathcal{P}}_\lambda} \mathrm{Abs}_{\widetilde{u}}(K, C),$$

is called the λ-guaranteed viable-capture basin of a target under the evolutionary game $(x, \widetilde{u}) \rightsquigarrow \mathcal{C}_{\widetilde{u}}(x)$.

One can prove that when the game is Lipschitz, the set-valued map $(x, \widetilde{u}) \in X \times \widetilde{\mathcal{P}}_\lambda \rightsquigarrow \mathcal{C}_{\widetilde{u}}(x) \subset \mathcal{C}(0, \infty; X)$ is lower semicontinuous and consequently, that the λ-guaranteed viable-capture basin is closed.

We recall that the contingent cone to a subset K at a point $x \in K$, introduced in the early thirties independently by Bouligand and Severi, adapts to any subset the concept of tangent space to manifolds: A direction $v \in X$ belongs to $T_K(x)$ if there exist sequences $h_n > 0$ and $v_n \in X$ converging to 0 and v respectively such that

$$\forall n \geq 0, x + h_n v_n \in K.$$

Using the Viability and the Invariance Theorems, one can prove the following tangential properties of guaranteed viability kernels with targets:

Theorem 5.1. *Let us assume that the dynamical game (P, Q, F) is Lipschitz, that $C \subset K$ and K are closed subsets of X and that $K \backslash C$ is a repeller under all the maps $(x, \widetilde{u}) \rightsquigarrow \mathcal{C}_{\widetilde{u}}(x)$.*

Then the λ-guaranteed viable-capture basin $[\mathrm{Capt}_{P_\lambda} \mathrm{Abs}_Q](K, C)$ of target C viable in K is the largest of the closed subsets D satisfying $C \subset D \subset K$ and

1. *the tangential property*[16]

$$\forall x \in D\backslash C, \exists u \in P(x) \text{ such that } \forall v \in Q(x), F(x, u, v) \subset T_D(x); \qquad (20)$$

2. *there exists a λ-Lipschitz selection of the guaranteed regulation map Γ_D defined by*

$$\forall x \in D\backslash C, \Gamma_D(x) := \{u \in P(x) | F(x, u, Q(x)) \subset T_D(x)\}.$$

[16]or, the equivalent dual formulation,

$$\forall x \in D\backslash C, \forall p \in N_D(x), \inf_{u \in P(x)} \sup_{v \in Q(x)} \sigma(F(x, u, v), p) \leq 0,$$

where the (regular) normal cone, $N_D(x) := T_D(x)^-$, is the polar cone to the contingent cone $T_D(x)$ and where,

$$\forall p \in X^\star, \sigma(F, p) := \sup_{x \in F} \langle p, x \rangle,$$

is the support function of F.

This theorem is a restatement of Theorems 9.2.14 and 9.2.18 of [11, Aubin, Chapter 9].

5.2 Hamilton–Jacobi–Isaacs Variational Inequalities

Let us recall that the contingent epiderivative $D_\uparrow \mathbf{v}(t,x)$ of $\mathbf{v}$ at (t,x) satisfies the property:

$$\mathcal{E}p(D_\uparrow \mathbf{v}(t,x)) = T_{\mathcal{E}p(\mathbf{v})}(t,x,\mathbf{v}(t,x)).$$

Since the λ-guaranteed viable-capture basin is closed under Lipschitz equations, then its southern border, which is the λ-guaranteed valuation function,

$$V^\sharp_{(\mathbf{b},\mathbf{c})_\lambda}(T,x) := \inf_{\widetilde{u}\in\widetilde{\mathcal{P}}_\lambda} \sup_{(x(\cdot),v(\cdot))\in\mathcal{C}_{\widetilde{u}}(x)} \inf_{t\in[0,T]} L_{(\mathbf{b},\mathbf{c})}(t;(x(\cdot),v(\cdot));\widetilde{u})(T,x), \quad (21)$$

is lower semicontinuous and thus, its epigraph coincides with the λ-guaranteed viable-capture basin.

Theorem 5.2. *Let us assume that the dynamical game (18) is Lipschitz and that the cost functions* $\mathbf{b}$ *and* $\mathbf{c}$ *from* $\mathbf{R}_+ \times X$ *to* $\mathbf{R}_+ \cup \{+\infty\}$ *are nontrivial, nonnegative and lower semicontinuous. Then the* λ*-guaranteed valuation function* $V^\sharp_{(\mathbf{b},\mathbf{c})_\lambda}$ *under the dynamical game (15) is the smallest of the nonnegative lower semicontinuous solutions* $\mathbf{v}$ *to the Hamilton–Jacobi–Isaacs contingent inequalities*

$$\begin{cases} i) & \mathbf{b}(t,x) \le \mathbf{v}(t,x) \le \mathbf{c}(t,x), \\ ii) & \text{if } \mathbf{v}(t,x) < \mathbf{c}(t,x), \\ & \inf_{u\in P(x)} \sup_{w\in F(x,u,Q(x))} (D_\uparrow \mathbf{v}(t,x)(-1,w) + \mathbf{l}(x,u,v) \\ & +\mathbf{m}(x,u,v)\mathbf{v}(t,x)) \le 0, \end{cases}$$

such that there exists a λ*-Lipschitz selection* $\widetilde{u}$ *of the guaranteed regulation map* Γ *defined by*

$$\begin{cases} \Gamma(t,x) \\ := \{u \in P(x) \mid \sup_{w\in F(x,u,Q(x))} (D_\uparrow \mathbf{v}(t,x)(-1,w) + \mathbf{l}(x,u,v) \\ \qquad +\mathbf{m}(x,u,v)\mathbf{v}(t,x)) \le 0\}. \end{cases}$$

Proof. It is a consequence of Theorem 5.1 when $K := \mathcal{E}p(\mathbf{b})$, $C := \mathcal{E}p(\mathbf{c})$ and when the dynamical game is the extended dynamical game (18).

Theorem 5.1 states that the λ-guaranteed viable-capture basin

$$[\mathrm{Capt}_{P_\lambda} \mathrm{Abs}_Q](\mathcal{E}p(\mathbf{b}), \mathcal{E}p(\mathbf{c}))$$

under (18) of the epigraph $\mathcal{E}p(\mathbf{c})$ of $\mathbf{c}$ viable in the epigraph $\mathcal{E}p(\mathbf{b})$ of $\mathbf{b}$ is the largest of the closed subsets $\mathcal{U}$ satisfying $\mathcal{E}p(\mathbf{b}) \subset \mathcal{U} \subset \mathcal{E}p(\mathbf{c})$, the tangential conditions

$$\begin{cases} \forall (t,x,y) \in \mathcal{U}\backslash\mathcal{E}p(\mathbf{c}), \exists u \in P(x), \text{ such that } \forall w \in F(x,u,Q(x)), \\ (-1,w,-\mathbf{m}(x,u,v)y - \mathbf{l}(x,u,v)) \in T_{\mathcal{U}}(t,x,y), \end{cases} \quad (22)$$

and such that there exists a λ-Lipschitz selection of the guaranteed regulation map $\Gamma_{\mathcal{U}}$ defined by

$$\Gamma_{\mathcal{U}}(t,x) := \{u \in P(x) | \{-1\} \times F(x,u,v) \times \{-\mathbf{m}(x,u,v)y - \mathbf{l}(x,u,v)\} \\ \cap T_{\mathcal{U}}(t,x,y) \neq \emptyset\}.$$

Let $(t,x) \mapsto \mathbf{v}(t,x)$ be the southern border of $\mathcal{U}$, that satisfies $\mathcal{U} = \mathcal{E}p(\mathbf{v})$ since $\mathcal{U}$ is closed. When $y := \mathbf{v}(t,x)$, the above condition (22) reads:

$$D_{\uparrow}\mathbf{v}(t,x)(-1,w) \leq -\mathbf{m}(x,u,v)\mathbf{v}(t,x) - \mathbf{l}(x,u,v),$$

because

$$T_{\mathcal{U}}(t,x,\mathbf{v}(t,x)) = \mathcal{E}p(D_{\uparrow}\mathbf{v}(t,x)).$$

Conversely, this condition implies the tangential condition (22) for $y := \mathbf{v}(t,x)$ whenever $(t,x,\mathbf{v}(t,x))$ belongs to $\mathcal{U}$. Otherwise, let $(t,x,y) \in \mathcal{U}$ with $y > \mathbf{v}(t,x)$ and set $\lambda := D_{\uparrow}\mathbf{v}(t,x)(-1,w)$.

By definition of $\lambda := D_{\uparrow}\mathbf{v}(t,x)(-1,w)$, there exist sequences $h_n > 0$ converging to 0, w_n converging to w and λ_n converging to λ such that $(t-h_n, x+h_n w_n, \mathbf{v}(t,x)+h_n\lambda_n)$ belongs to $\mathcal{E}p(\mathbf{v})$. Therefore, for $\mu \in \mathbf{R}$ and h_n small enough,

$$(t-h_n, x+h_n w_n, y+h_n\mu) = (t-h_n, x+h_n w_n, \mathbf{v}(t,x)+h_n\lambda_n) \\ + (0,0,y-\mathbf{v}(t,x)+h_n(\mu-\lambda_n)),$$

belongs to $\mathcal{E}p(\mathbf{v})$ because $y - \mathbf{v}(t,x)$ is strictly positive. This implies that $(-1,w,\mu)$ belongs to the contingent cone to the epigraph $\mathcal{U}$ of $\mathbf{v}$ at (t,x,y), so that tangential condition (22) is satisfied with $\mu := -\mathbf{m}(x,u,v)y - \mathbf{l}(x,u,v)$. □

REFERENCES

[1] Allouch, N. and Florenzano, M. (2004) Edgeworth and Walras equilibria of an arbitrage-free economy, *Econ. Theory*, vol. 23, no. 2, pp. 353–370.

[2] Aubin, J.-P. (1979) *Mathematical Methods of Game and Economic Theory*, North-Holland (Studies in Mathematics and its Applications, vol. 7, 619 pages.).

[3] Aubin, J.-P. (1981) Cooperative fuzzy games, *Math. Op. Res.*, vol. 6, 1–13.

[4] Aubin, J.-P. (1981) Locally Lipschitz cooperative games, *J. Math. Economics*, vol. 8, pp. 241–262.

[5] Aubin, J.-P. (1981) A dynamical, pure exchange economy with feedback pricing, *J. Economic Behavior and Organizations*, 2, pp. 95–127.

[6] Aubin, J.-P. (1981) Contingent derivatives of set-valued maps and existence of solutions to nonlinear inclusions and differential inclusions, *Advances in Mathematics, Supplementary Studies*, Ed. Nachbin L., pp. 160–232.

[7] Aubin, J.-P. (1983) *L'Analyse non linéaire et ses motivations économiques*, Masson (English version: *Optima and Equilibria*, (1993, 1998), Springer-Verlag).

[8] Aubin, J.-P. (1986) A viability approach to Lyapunov's second methods, In: *Dynamical Systems*, Eds. A. Kurzhanski and K. Sigmund, Lectures Notes in Economics and Math. Systems, Springer-Verlag, 287, 31–38.

[9] Aubin, J.-P. (1987) Smooth and heavy solutions to control problems, In: *Nonlinear and Convex Analysis*, Eds. B-L. Lin & Simons S., Proceedings in honor of Ky Fan, Lecture Notes in Pure and Applied Mathematics, June 24–26, 1985.

[10] Aubin, J.-P. (1991) *Viability Theory*, Birkhäuser, Boston, Basel, Berlin.

[11] Aubin, J.-P. (1997) *Dynamic Economic Theory: A Viability Approach*, Springer-Verlag.

[12] Aubin, J.-P. (2001) Viability Kernels and Capture Basins of Sets under Differential inclusions, *SIAM J. Control Optimization*, 40, pp. 853–881.

[13] Aubin, J.-P. (2002) Boundary-value problems for systems of Hamilton–Jacobi–Bellman Inclusions with Constraints, *SIAM J. Control Optimization*, vol. 41, no. 2, pp. 425–456.

[14] Aubin, J.-P. (2001) Regulation of the evolution of the architecture of a network by connectionist tensors operating on coalitions of actors, preprint.

[15] Aubin, J.-P. & Catté, F. (2002) Bilateral fixed-points and algebraic properties of viability kernels and capture basins of sets, *Set-Valued Analysis*, vol. 10, no. 4, pp. 379–416.

[16] Aubin, J.-P. & Cellina, A. (1984) *Differential Inclusions*, Springer-Verlag.

[17] Aubin, J.-P. & da Prato, G. (1995) Stochastic Nagumo's viability theorem, *Stochastic Analysis and Applications*, vol. 13, pp. 1–11.

[18] Aubin, J.-P. & da Prato, G. (1998) The viability theorem for stochastic differential inclusions, *Stochastic Analysis and Applications*, 16, pp. 1–15.

[19] Aubin, J.-P., da Prato, G. & Frankowska, H. (2000) Stochastic invariance for differential inclusions, *Set-Valued Analysis* vol. 8, No. 1-2, pp. 181–201.

[20] Aubin, J.-P. & Dordan, O. (1996) Fuzzy systems, viability theory and toll sets, In *Handbook of Fuzzy Systems, Modeling and Control*, Hung Nguyen Ed.. Kluwer, pp. 461–488.

[21] Aubin, J.-P. & Frankowska H. (1990) *Set-Valued Analysis*, Birkhäuser, Boston, Basel, Berlin.

[22] Aubin, J.-P. & Frankowska, H. (1996) The viability kernel algorithm for computing value functions of infinite horizon optimal control problems, *J.Math. Anal. Appl.*, 201, pp. 555–576.

[23] Aubin, J.-P., Louis-Guerin, C. & Zavalloni, M. (1979) Comptabilité entre conduites sociales réelles dans les groupes et les représentations symboliques de ces groupes : un essai de formalisation mathématique, *Math. Sci. Hum.*, 68, pp. 27–61.

[24] Aubin, J.-P., Pujal, D. & Saint-Pierre, P. (2001) Dynamic management of portfolios with transaction costs under tychastic uncertainty, preprint.

[25] Başar, T. & Bernhard, P. (1991) H^{∞}- *optimal control and related minimax design problems. A dynamic game approach*, Birkhäuser, Boston.

[26] Basile A., de Simone, A. & Graziano, M.G. (1996) On the Aubin-like characterization of competitive equilibria in infinite-dimensional economies, *Rivista di Matematica per le Scienze Economiche e Sociali*, 19, pp. 187–213.

[27] Basile, A. (1993) Finitely additive nonatomic coalition production economies: Core-Walras equivalence, *Int. Econ. Rev.*, 34, pp. 993–995.

[28] Basile, A. (1994) Finitely additive correpondences, *Proc. AMS* 121, pp. 883–891.

[29] Basile, A. (1998) On the ranges of additive correspondences, In: *Functional analysis and economic theory*. Based on the special session of the conference on nonlinear analysis and its applications in engineering and economics, Samos, Greece, July 1996, dedicated to Charalambos Aliprantis on the occasion of his 50th birthday. (ed.) Abramovich, Yuri et al., Springer, Berlin, pp. 47–60.

[30] Buckdahn, R., Quincampoix, M. & Rascanu, A. (1997) Propriétés de viabilité pour des équations différentielles stochastiques rétrogrades et applications à des équations aux dérivées partielles, *Comptes-Rendus de l'Académie des Sciences*, Paris, 235, pp. 1159–1162.

[31] Buckdahn, R., Quincampoix, M. & Rascanu, A. (1998) Stochastic viability for backward stochastic differential equations and applications to partial differential equations, Un. Bretagne Occidentale, 01–1998.

[32] Buckdahn, R., Peng, S., Quincampoix, M. & Rainer, C. (1998) Existence of stochastic control under state constraints, *Comptes-Rendus de l'Académie des Sciences*, Paris, 327, pp. 17–22.

[33] Buckdahn, R., Cardaliaguet, P. & Quincampoix, M. (2000) A representation formula for the mean curvature motion, UBO 08–2000.

[34] Cardaliaguet P., Quincampoix M. & Saint-Pierre, P. (1995) Contribution à l'étude des jeux différentiels quantitatifs et qualitatifs avec contrainte sur l'état, *Comptes-Rendus de l'Académie des Sciences*, Paris, 321, pp. 1543–1548.

[35] Cardaliaguet P. (1994) Domaines dicriminants en jeux différentiels, Thèse de l'Université de Paris-Dauphine.

[36] Cardaliaguet P. (1996) A differential game with two players and one target, *SIAM J. on Control and Optimization*, 34, 4, pp. 1441–1460.

[37] Cardaliaguet P. (1997) On the regularity of semi-permeable surfaces in control theory with application to the optimal exit-time problem (Part II), *SIAM J. on Control Optimization*, vol. 35, no. 5, pp. 1638–1652.

[38] Cardaliaguet P. (2000) *Introduction à la théorie des jeux différentiels*, Lecture Notes, Université Paris-Dauphine.

[39] Cardaliaguet, P., Quincampoix, M. & Saint-Pierre, P. (1999) Set-valued numerical methods for optimal control and differential games, In *Stochastic and differential games. Theory and numerical methods*, Annals of the International Society of Dynamical Games, pp. 177–247, Birkhäuser.

[40] da Prato, G. and Frankowska, H. (1994) A stochastic Filippov Theorem, *Stochastic Calculus*, 12, pp. 409–426.

[41] Doss, H. (1977) Liens entre équations différentielles stochastiques et ordinaires, *Ann. Inst. Henri Poincaré, Calcul des Probabilités et Statistique*, 23, pp. 99–125.

[42] Filar, J.A. & Petrosjan, L.A. (2000) Dynamic cooperative games, *International Game Theory Review*, 2, pp. 47–65.

[43] Florenzano, M. (1990) Edgeworth equilibria, fuzzy core and equilibria of a production economy without ordered preferences, *J. Math. Anal. Appl.*, 153, pp. 18–36.

[44] Frankowska, H. (1987) L'équation d'Hamilton–Jacobi contingente, *Comptes-Rendus de l'Académie des Sciences*, PARIS, Série 1, 304, pp. 295–298.

[45] Frankowska, H. (1987) Optimal trajectories associated to a solution of contingent Hamilton–Jacobi equations, *IEEE, 26th, CDC Conference*, Los Angeles, December 9–11.

[46] Frankowska, H. (1989) Optimal trajectories associated to a solution of contingent Hamilton–Jacobi equations, *Applied Mathematics and Optimization*, 19, pp. 291–311.

[47] Frankowska, H. (1989) Hamilton–Jacobi equation: viscosity solutions and generalized gradients, *J. of Math. Analysis and Appl.* 141, pp. 21–26.

[48] Frankowska, H. (1993) Lower semicontinuous solutions of Hamilton–Jacobi–Bellman equation, *SIAM J. on Control Optimization*, vol. 31, no. 1, pp. 257–272.

[49] Gautier, S. & Thibault, L. (1993) Viability for constrained stochastic differential equations, *Differential Integral Equations*, 6, pp. 1395–1414.

[50] Haurie, A. (1975) On some properties of the characteristic function and core of multistage game of coalitions, *IEEE Trans. Automatic Control*, vol. 20, pp. 238–241.

[51] Gaitsgory, V.V. & Leizarowitz, A. (1999) Limit occoputional measures set for a control system and averaging of singularly perturbed control sysytems, *J. Math. Anal, Appl.*, 233, pp. 461–475.

[52] Isaacs, R. (1965) *Differential Games*, Wiley, New York.

[53] Leitmann, G. (1980) Guaranteed avoidance strategies, *Journal of Optimization Theory and Applications*, Vol.32, pp. 569–576.

[54] Mares, M. (2001) *Fuzzy cooperative games. Cooperation with vague expectations*, Physica Verlag.

[55] Mishizaki, I. & Sokawa, M. (2001) *Fuzzy and multiobjective games for conflict resolution*, Physica Verlag.

[56] Peirce, C. (1893) *Evolutionary love*, The Monist.

[57] Petrosjan, L.A. (2004) Dynamic cooperative games, *Advances in Dynamic Games: Applications to Economics, Finance, Optimization, and Stochastic Control*, Birkhäuser, Boston, vol.7.

[58] Petrosjan, L.A. & Zenkevitch, N.A. (1996) *Game Theory*, World Scientific.

[59] Pujal, D. (2000) Valuation et gestion dynamiques de portefeuilles, Thèse de l'Université de Paris-Dauphine.

[60] Quincampoix, M. (1992) Differential inclusions and target problems, *SIAM J. Control Optimization*, 30, pp. 324–335.

[61] Quincampoix, M. (1992) Enveloppes d'invariance pour des inclusions différentielles Lipschitziennes: applications aux problèmes de cibles, *Comptes-Rendus de l'Académie des Sciences*, Paris, 314, pp. 343–347.

[62] Quincampoix, M. (1990) Frontières de domaines d'invariance et de viabilité pour des inclusions différentielles avec contraintes, *Comptes-Rendus de l'Académie des Sciences*, Paris, 311, pp. 411–416.

[63] Quincampoix, M. & Saint-Pierre P. (1995) An algorithm for viability kernels in Hölderian case: Approximation by discrete viability kernels, *J. Math. Syst. Estimation and Control*, pp. 115–120.

[64] Rockafellar, R.T. and Wets, R. (1997) *Variational Analysis*, Springer-Verlag.

[65] Runggaldier, W.J. (2000) Adaptive and robust control peocedures for risk minimization under uncertainty, In: *Optimal Control and Partial Differential Equations*, pp. 511–520, IOS Press.

[66] Saint-Pierre, P. (1994) Approximation of the viability kernel, *Applied Mathematics & Optimisation*, vol. 29, pp. 187–209.

[67] Zabczyk, J. (1996) *Chance and decision: stochastic control in discrete time*, Quaderni, Scuola Normale di Pisa.

[68] Zabczyk, J. (1999) *Stochastic invariance and consistency of financial models*, preprint, Scuola Normale di Pisa.

Normalized Overtaking Nash Equilibrium for a Class of Distributed Parameter Dynamic Games*

Dean A. Carlson†
Mathematical Reviews
416 Fourth Street
P.O. Box 8604
Ann Arbor, MI 48107-8604, USA
dac@ams.org

Abstract

In this paper we investigate the existence and turnpike properties of overtaking Nash equilibria for an infinite horizon dynamic game in which the dynamic constraints are described by linear evolution equations in a Hilbert space.

1 Introduction

The study of infinite horizon dynamic optimization problems has a rich history during the Twentieth Century beginning with the seminal paper of F. Ramsey [7] in 1928. The primary applications of these problems arise in the arena of business and economic modeling. During the sixties, it was discovered that the usual concept of a minimizer was inadequate to deal with problems defined on an unbounded time horizon since the objective functional, typically described by an improper integral, is unbounded. Consequently, a hierarchy of new types of optimality (e.g., overtaking optimality) were invented to treat these problems. From the seventies into the nineties, these problems were investigated extensively and now much is known regarding these optimization models. The interested reader is directed to the monograph of Carlson, Haurie, and Leizarowitz [3] for a detailed survey of these results.

The theory of dynamic games also has an extensive literature beginning with the work of Isaacs and others in the 1940's and 1950's and continuing up to the present for many classes of systems. In contrast, however, the study of the extensions to the weaker types of optimality, now so familiar in the infinite horizon optimization literature is just beginning. In particular, we mention the works of Carlson and Haurie [4], [5] in which the extension of these ideas was successfully achieved in

*This research was supported by the National Science Foundation (INT–NSF-9972023)
†These results were announced at the Third World Congress of the World Federation of Nonlinear Analysis held in Catania, Italy, July 2000. A summary of these results appears in the proceedings of this congress.

both discrete and continuous time for convex Lagrange problems. In these works the existence of an asymptotic steady-state Nash equilibrium is established and conditions are placed on the model to insure that there exists an overtaking open-loop Nash equilibrium which converges to the steady-state Nash equilibrium. In addition, the strict-diagonal convexity concept developed by Rosen [8] has been extended to provide conditions under which the overtaking Nash equilibrium is unique.

The present work extends the work of Carlson and Haurie [4], [5] to the case of distributed parameter control systems in which the state dynamics of each player is governed by a linear evolution equation with distributed control in a Hilbert space. This work builds on the earlier work of Carlson, Haurie, and Jabrane [2] where the infinite horizon optimal control problem is treated.

Our paper is organized as follows. In section 2 we introduce the basic hypotheses and describe the class of models we treat. Section 3 is devoted to a discussion of the normalized steady-state Nash equilibrium problem. In section 4 we discuss the asymptotic turnpike theorem for overtaking Nash equilibria and in section 5 we investigate the existence of an open-loop Nash equilibrium.

2 Basic Model and Hypotheses

We consider a p-player game in which the state of the j-th player over time, $j = 1, 2, \dots, p$, is a continuous function, $x_j : [0, +\infty) \to E_j$, in which E_j is a separable Hilbert space with inner product $\langle \cdot, \cdot \rangle_{E_j}$. Each of these state variables is governed by a coupled linear evolution equation taking the form

$$\dot{x}_j(t) = A_j \mathbf{x}(t) + B_j u_j(t), \quad \text{a.e.} \quad t \geq 0 \tag{1}$$

in which $\mathbf{x}(\cdot) = (x_1(\cdot), x_2(\cdot), \dots, x_p(\cdot))$ denotes the aggregate state variable defined from $[0, +\infty)$ to the Hilbert space $\mathbf{E} = E_1 \times E_2 \times \cdots \times E_p$ with inner product,

$$\langle \cdot, \cdot \rangle_E \doteq \sum_{j=1}^{p} \langle \cdot, \cdot \rangle_{E_j},$$

A_j is a densely defined linear operator from E into E_j, $u_j : [0, +\infty) \to F_j$ is the control used by player j, and B_j is a bounded linear operator from the separable Hilbert space F_j into E_j. We denote the inner product on F_j by $\langle \cdot, \cdot \rangle_{F_j}$ and proceeding as before we let $\mathbf{F} = F_1 \times F_2 \times \cdots \times F_p$ denote the Hilbert space with inner product

$$\langle \cdot, \cdot \rangle_F \doteq \sum_{j=1}^{p} \langle \cdot, \cdot \rangle_{F_j}.$$

We view the above system of equations as a single equation taking the form

$$\dot{\mathbf{x}}(t) = \mathbf{A}\mathbf{x}(t) + \mathbf{B}\mathbf{u}(t), \tag{2}$$

in which $\mathbf{A} : \mathbf{E} \to \mathbf{E}$ is the densely defined linear operator given by $\mathbf{A}\mathbf{x} = (A_1\mathbf{x}, A_2\mathbf{x}, \ldots, A_j\mathbf{x})$ and $\mathbf{B} : \mathbf{F} \to \mathbf{E}$ is the bounded linear operator defined by $\mathbf{B}\mathbf{u} = (B_1 u_1, B_2 u_2, \ldots, B_p u_p)$. We assume that the operator $\mathbf{A}$ is the generator of a C_0-semigroup $\{\mathbf{S}(t) : \quad t \geq 0\}$. In addition to the above notations we also assume that each player's state, x_j, satisfies a fixed initial condition

$$x_j(0) = x_{j,0} \in E_j, \tag{3}$$

and that the aggregate state variable satisfies a fixed state constraint,

$$\mathbf{x}(t) \in \mathbf{X}, \quad t \geq 0, \tag{4}$$

in which $\mathbf{X}$ is a closed, bounded, convex subset of $\mathbf{E}$. Finally we impose a state-dependent constraint on the controls taking the form

$$u_j(t) \in U_j(\mathbf{x}(t)) \quad \text{a.e.} \quad t \geq 0, \tag{5}$$

where $U_j(\cdot)$ is a set-valued mapping from $\mathbf{X} \subset \mathbf{E}$ into F_j which is closed, uniformly bounded, and convex, with a weakly closed graph. Additionally we suppose that

$$U_j(\lambda\mathbf{x} + (1-\lambda)\mathbf{y}) \subset \lambda U_j(\mathbf{x}) + (1-\lambda) U_j(\mathbf{y})$$

holds for each $\mathbf{x}, \mathbf{y} \in \mathbf{X}$ and $\lambda \in [0, 1]$.

These definitions and hypotheses permit us to make the following definitions.

Definition 2.1. We say that the pair of functions $\{\mathbf{x}(\cdot), \mathbf{u}(\cdot)\} : [0, +\infty) \to \mathbf{E} \times \mathbf{F}$ is a trajectory-control pair for the dynamic game if $\mathbf{x}(\cdot)$ is a continuous function, $\mathbf{u}(\cdot) \in L_2([0, T], \mathbf{F})$ for each $T > 0$, and if $\mathbf{x}(\cdot)$ is a mild solution to equation (2) with control $\mathbf{u}(\cdot)$ and fixed initial condition (3) and if the pair satisfies the constraints (4) and (5). That is, for each $t \geq 0$ we have

$$\mathbf{x}(t) = \mathbf{S}(t)\mathbf{x}_0 + \int_0^t \mathbf{S}(t-s)\mathbf{B}\mathbf{u}(s)\, ds, \tag{6}$$

where $\mathbf{x}_0 = (x_{1,0}, x_{2,0}, \ldots x_{p,0})$.

Further, we say a pair of functions $\{x_k(\cdot), u_k(\cdot)\} : [0, +\infty) \to E_k \times F_k$ is a trajectory-control pair for player k if for each player $j = 1, 2, \ldots, p$, $j \neq k$ there is a corresponding pair of functions $\{x_j(\cdot), u_j(\cdot)\} : [0, +\infty) \to E_j \times F_j$ so that the aggregate pair of functions

$$\{\mathbf{x}(\cdot), \mathbf{u}(\cdot)\} = \{(x_1(\cdot), x_2(\cdot), \ldots, x_p(\cdot)), (u_1(\cdot), u_2(\cdot), \ldots, u_p(\cdot))\}$$

constitutes a trajectory-control pair for the dynamic game.

It is well known that the mild solution to the equation (2) is not necessarily differentiable. However it does enjoy the following "mild differential equation"

$$\frac{d}{dt}\langle \mathbf{x}(t), q\rangle_E = \langle \mathbf{x}(t), \mathbf{A}^* q\rangle_E + \langle \mathbf{Bu}(t), q\rangle_E, \quad \text{a.e.} \quad t \geq 0 \quad \text{and} \quad q \in \mathcal{D}(\mathbf{A}^*) \tag{7}$$

$$\lim_{t\to 0^+} \langle \mathbf{x}(t), q\rangle_E = \langle \mathbf{x}_0, q\rangle_E \quad \text{for} \quad q \in \mathcal{D}(\mathbf{A}^*), \tag{8}$$

where $\mathbf{A}^*$ denotes the adjoint operator associated with $\mathbf{A}$ with densely defined domain $\mathcal{D}(\mathbf{A}^*)$.

The objective of the j-th player up to time $T > 0$ is described by a Lagrange-type integral functional which takes the form

$$J_j^T(\mathbf{x}, u_j) = \int_0^T L_j(\mathbf{x}(t), u_j(t))\, dt, \tag{9}$$

in which $L_j : \mathbf{E} \times F_j \to \mathbb{R}$ is assumed to be lower semi-continuous and convex in (x_j, u_j). In addition we impose the following growth condition:

Assumption 2.1. There exists $K_1 > 0$ and $K > 0$ such that for each $j = 1, 2, \ldots, p$ we have

$$\|\mathbf{x}\|_E^2 + \|u_j\|_{F_j}^2 > K_1 \Longrightarrow L_j(\mathbf{x}, u_j) \geq K(\|\mathbf{x}\|_E^2 + \|u_j\|_{F_j}^2). \tag{10}$$

With this notation we have the following definition.

Definition 2.2. We say a trajectory-control pair for the dynamic game, $\{\mathbf{x}(\cdot), \mathbf{u}(\cdot)\}$ is an admissible trajectory-control pair if the maps $t \to L_j(\mathbf{x}(t), u_j(t))$, $j = 1, 2, \ldots p$ are integrable on $[0, T]$ for each $T > 0$. Additionally we say that the pair of functions $\{x_k(\cdot), u_k(\cdot)\}$ is an admissible pair for player k if for each $j = 1, 2, \ldots p, j \neq k$, there exists a corresponding pair of functions $\{x_j(\cdot), u_j(\cdot)\}$ so that the aggregate pair of functions

$$\{\mathbf{x}(\cdot), \mathbf{u}(\cdot)\} = \{(x_1(\cdot), x_2(\cdot), \ldots, x_p(\cdot)), (u_1(\cdot), u_2(\cdot), \ldots, u_p(\cdot))\}$$

is an admissible trajectory-control pair for the dynamic game.

The above describes the class of dynamic games we wish to study. We seek a Nash equilibria for this dynamic game as $T \to +\infty$. However, dictated by experience in infinite horizon optimization problems it is known that generally along any admissible trajectory-control pair the improper integrals

$$\lim_{T\to+\infty} J_j^T(\mathbf{x}, u_j) = \int_0^{+\infty} L_j(\mathbf{x}(t), u_j(t))\, dt,$$

may diverge. As a consequence of this fact, we seek a weaker type of equilibria known as an overtaking Nash equilibria. To define this type of equilibria we introduce the following notation. For $\mathbf{x} = (x_1, x_2, \ldots x_p) \in \mathbf{E}$ and $y \in E_j$ let $[\mathbf{x}^j, y]$ denote the aggregate vector,

$$(x_1, x_2, \ldots, x_{j-1}, y, x_{j+1}, \ldots x_p) \in \mathbf{E}.$$

We now have the following definitions.

Definition 2.3. We say the admissible trajectory-control pair, $\{\mathbf{x}^*, \mathbf{u}^*\}$, is a

(1) **strong Nash equilibrium** if

(a)

$$\lim_{T\to+\infty} J_j^T(\mathbf{x}^*, u_j^*) = \int_0^{+\infty} L_j(\mathbf{x}^*(t), u_j^*(t))\, dt < +\infty, \tag{11}$$

and

(b)

$$\int_0^{+\infty} L_j(\mathbf{x}^*(t), u_j^*(t))\, dt \leq \liminf_{T\to+\infty} \int_0^T L_j([\mathbf{x}^{*j}, y](t), v(t))\, dt \tag{12}$$

holds for each $j = 1, 2, \ldots, p$ and each trajectory-control pair $y : [0, +\infty) \to E_j$ and $v : [0, +\infty) \to F_j$ for which $\{[\mathbf{x}^{*j}, y], [\mathbf{u}^{*j}, v]\}$ is an admissible trajectory-control pair.

(2) an **overtaking Nash equilibrium** if

$$\liminf_{T\to+\infty} \int_0^T \Big(L_j([\mathbf{x}^{*j}, y](t), v(t)) - L_j(\mathbf{x}^*(t), u_j^*(t))\Big)\, dt \geq 0, \tag{13}$$

holds for each $j = 1, 2, \ldots, p$ and each trajectory-control pair $y : [0, +\infty) \to E_j$ and $v : [0, +\infty) \to F_j$ for which $\{[\mathbf{x}^{*j}, y], [\mathbf{u}^{*j}, v]\}$ is an admissible trajectory-control pair.

(3) a **weakly overtaking Nash equilibrium** if

$$\limsup_{T\to+\infty} \int_0^T \Big(L_j([\mathbf{x}^{*j}, y](t), v(t)) - L_j(\mathbf{x}^*(t), u_j^*(t))\Big)\, dt \geq 0, \tag{14}$$

holds for each $j = 1, 2, \ldots, p$ and each trajectory-control pair $y : [0, +\infty) \to E_j$ and $v : [0, +\infty) \to F_j$ for which $\{[\mathbf{x}^{*j}, y], [\mathbf{u}^{*j}, v]\}$ is an admissible trajectory-control pair.

The above definitions are a direct parallel of the same notions found in the infinite horizon optimal control literature. For a complete discussion of these ideas the reader is directed to Carlson, Haurie, and Leizarowitz [3]. For dynamic games these definitions appear in Carlson and Haurie [4] [5] in which analogous questions for ordinary differential control systems were considered.

In the next section we continue with a discussion of an associated steady-state game.

3 The Steady-State Game

In this section we consider an associated steady-state game. This game is a convex game in a Hilbert space. Specifically we consider the problem of finding a Nash equilibrium of the game with objectives

$$L_j(\mathbf{x}, u_j) \quad j = 1, 2, \ldots, p, \tag{15}$$

subject to the constraints

$$0 = \mathbf{A}\mathbf{x} + \mathbf{B}\mathbf{u} \tag{16}$$

$$\mathbf{x} \in \mathbf{X} \tag{17}$$

$$u_j \in U_j(\mathbf{x}) \quad j = 1, 2, \ldots, p. \tag{18}$$

We seek a Nash equilibrium in the Hilbert space $\mathbf{E} \times \mathbf{F}$ as defined below. To describe this equilibrium we introduce the following notation.

Definition 3.1. We say that the pair $(\mathbf{x}^*, \mathbf{u}^*) \in \mathbf{E} \times \mathbf{F}$ is a steady-state Nash equilibrium if the constraints (16)–(18) are satisfied and if for each $j = 1, 2, \ldots, p$ we have

$$L_j(\mathbf{x}^*, u_j^*) \leq L_j([\mathbf{x}^{*j}, y], v) \tag{19}$$

holds for all $(y, v) \in E_j \times F_j$ for which $([\mathbf{x}^{*j}, y], [\mathbf{u}^{*j}, v])$ satisfies the constraints given by (16)–(18).

This is a convex game in a Hilbert space and it is well known, under appropriate constraint qualifications, that if $(\mathbf{x}^*, \mathbf{u}^*)$ is a Nash equilibrium, then for each $j = 1, 2, \ldots, p$ there exists a multiplier $\bar{\mathbf{q}}_\mathbf{j} \in \mathbf{E}$, such that

$$L_j(\mathbf{x}^*, u_j^*) \leq L_j([\mathbf{x}^{*j}, y], v) - \langle [\mathbf{x}^j, y], \mathbf{A}^*\bar{\mathbf{q}}_\mathbf{j}\rangle_E - \langle \mathbf{B}[\mathbf{u}^{*j}, v], \bar{\mathbf{q}}_\mathbf{j}\rangle_E \tag{20}$$

holds for all $y \in E_j$ such that $[\mathbf{x}^{*j}, y] \in \mathbf{X}$, and all $v \in F_j$ such that $[\mathbf{u}^{*j}, v] \in \mathbf{F}$ satisfies $u_k^* \in U_k([\mathbf{x}^{*j}, y])$ for $k \neq j$ and $v \in U_j([\mathbf{x}^{*j}, y])$.

The main drawback to the formulation considered above is that generally there is no relationship between the multiplier, $\bar{\mathbf{q}}_\mathbf{j}$, of player j and any other player. In 1965, Rosen [8] introduced the notion of a normalized equilibrium for convex games in finite dimensions. Recently, in Carlson [6] these ideas have been extended to convex games with strategies in Hilbert spaces. We briefly now describe these ideas.

To begin we let $r_j > 0$, $j = 1, 2, \ldots, p$, be fixed real numbers and define the weighted objective function $\mathcal{L} : \mathbf{E} \times \mathbf{E} \times \mathbf{F} \to \mathbb{R}$ by the formula

$$\mathcal{L}(\mathbf{x}, \mathbf{y}, \mathbf{v}) = \sum_{j=1}^{p} r_j L_j([\mathbf{x}^j, y_j], v_j). \tag{21}$$

Now let $(\mathbf{x}, \mathbf{u}) \in \mathbf{E}$ satisfy the constraints (16)–(18) and consider the optimization problem

$$\text{minimize } \mathcal{L}(\mathbf{x}, \mathbf{y}, \mathbf{v}) \tag{22}$$

over all $(\mathbf{y}, \mathbf{v}) \in \mathbf{E}$ satisfying the constraints (16)–(18). This is a well defined optimization problem that is parameterized by $(\mathbf{x}, \mathbf{u})$. Let Ω denote the set of all pairs $(\mathbf{x}, \mathbf{u}) \in \mathbf{E} \times \mathbf{F}$ which satisfy the constraints (16)–(18) and define the set-valued mapping $\Gamma : \Omega \to 2^{\Omega}$ by

$$\Gamma(\mathbf{x}, \mathbf{u}) \doteq \{(\mathbf{y}, \mathbf{v}) \in \Omega : \quad \mathcal{L}(\mathbf{x}, \mathbf{y}, \mathbf{v}) \leq \mathcal{L}(\mathbf{x}, \mathbf{z}, \mathbf{w}) \quad \text{for all} \quad (\mathbf{z}, \mathbf{w}) \in \Omega\}. \tag{23}$$

Following the analysis developed in Carlson [6] (or in Rosen [8] for finite-dimensional games) it is known that if $(\bar{\mathbf{x}}, \bar{\mathbf{u}})$ is a fixed point of $\Gamma(\cdot, \cdot)$, that is

$$(\bar{\mathbf{x}}, \bar{\mathbf{u}}) \in \Gamma(\bar{\mathbf{x}}, \bar{\mathbf{u}}), \tag{24}$$

then $(\bar{\mathbf{x}}, \bar{\mathbf{u}})$ is a Nash equilibrium for the above convex game. Moreover, since $(\bar{\mathbf{x}}, \bar{\mathbf{u}})$ is a solution of an optimization problem there exists, with appropriate constraint qualifications, a multiplier, $\bar{\mathbf{q}} \in \mathbf{E}$ such that

$$\mathcal{L}(\bar{\mathbf{x}}, \bar{\mathbf{x}}, \bar{\mathbf{u}}) \leq \mathcal{L}(\bar{\mathbf{x}}, \mathbf{y}, \mathbf{v}) - \big(\langle \mathbf{y}, \mathbf{A}^*\bar{\mathbf{q}}\rangle_E + \langle \mathbf{B}\mathbf{v}, \bar{\mathbf{q}}\rangle_E\big) \tag{25}$$

holds for each $\mathbf{y} \in \mathbf{X}$ and $\mathbf{v} = (v_1, v_2, \dots, v_p) \in \mathbf{F}$ such that $v_k \in U_k(\mathbf{y}), k = 1, 2, \dots, p$. More explicitly this becomes,

$$\sum_{k=1}^{p} r_k L_k(\bar{\mathbf{x}}, \bar{u}_k) \leq \sum_{k=1}^{p} r_k L_k([\bar{\mathbf{x}}^k, y_k], v_k) - \big(\langle \mathbf{y}, \mathbf{A}^*\bar{\mathbf{q}}\rangle_E + \langle \mathbf{B}\mathbf{v}, \bar{\mathbf{q}}\rangle_E\big) \tag{26}$$

holds for each $\mathbf{y} \in \mathbf{X}$ and $\mathbf{v} = (v_1, v_2, \dots, v_p) \in \mathbf{F}$ such that $v_k \in U_k(\mathbf{y})$, $k = 1, 2, \dots, p$. Fixing $j = 1, 2, \dots, p$ and taking $y \in E_j$ and $v \in F_j$ such that $([\bar{\mathbf{x}}^j, y], [\bar{\mathbf{u}}^j, v]) \in \Omega$ it is an easy matter to see that inequality (25) reduces to

$$L_j(\bar{\mathbf{x}}, \bar{u}_j) \leq L_j([\bar{\mathbf{x}}^j, y], v) - \frac{1}{r_j}\Big(\langle [\bar{\mathbf{x}}^j, y], \mathbf{A}^*\bar{\mathbf{q}}\rangle_E + \langle \mathbf{B}[\bar{\mathbf{u}}^j, v], \bar{\mathbf{q}}\rangle_E\Big), \tag{27}$$

which holds for all $y \in E_j$ such that $[\mathbf{x}^{*j}, y] \in \mathbf{X}$, and all $v \in F_j$ such that $[\bar{\mathbf{u}}^j, v] \in \mathbf{F}$ satisfies $\bar{u}_k \in U_k([\mathbf{x}^{*j}, y])$ for $k \neq j$ and $v \in U_j([\mathbf{x}^{*j}, y])$. Identifying $(\mathbf{x}^*, \mathbf{u}^*)$ with $(\bar{\mathbf{x}}, \bar{\mathbf{u}})$ and the multipliers $\bar{\mathbf{q}}_{\mathbf{j}} = (1/r_j)\bar{\mathbf{q}}$ for $j = 1, 2, \dots, p$ we see that equation (27) coincides with equation (20). Thus, the multipliers

$$\bar{\mathbf{q}}_{\mathbf{j}} = (1/r_j)\bar{\mathbf{q}}, \quad j = 1, 2, \dots, p$$

are multipliers for the original static game. Multipliers of this form were first introduced in Rosen [8] and are referred to as *normalized multipliers* with the corresponding Nash equilibrium called a *normalized Nash equilibrium*. The existence of normalized multipliers have been studied in Rosen [8] for finite-dimensional convex games and recently these ideas have been extended to convex games with strategies in Hilbert spaces in Carlson [6]. We direct the interested reader to those works for additional information.

With the discussion given above we make the following assumption.

Assumption 3.1. We assume that there exists a set of positive weights $r_j > 0$, a **unique** normalized Nash equilibrium, denoted by $(\bar{\mathbf{x}}, \bar{\mathbf{u}})$, and a vector $\bar{\mathbf{q}} \in \mathbf{E}$ so that,

$$\bar{\mathbf{q}}_{\mathbf{j}} = (1/r_j)\bar{\mathbf{q}}, \quad j = 1, 2, \ldots, p$$

constitutes a set of normalized multipliers associated with $(\bar{\mathbf{x}}, \bar{\mathbf{u}})$.

Remark 3.1. The uniqueness assumption given above is assured by assuming additional regularity on the model. Specifically what is needed is a *strict diagonal convexity assumption*. This condition originated in Rosen [8] and was extended to the Hilbert space setting considered here in Carlson [6].

4 Turnpike Properties

In economic growth modeling, the primary application of infinite horizon dynamic games, the role of the turnpike property is quite familiar. In the setting considered here one expects that equilibria asymptotically approach the unique steady state Nash equilibrium given in Assumption 3.1. As we shall see, this is indeed the case. To present these results we introduce a new integrand $\mathcal{L}_0 : \mathbf{E}\times\mathbf{E}\times\mathbf{F} \to \mathbb{R}\cup\{+\infty\}$ by the formula

$$\mathcal{L}_0(\mathbf{x}, \mathbf{y}, \mathbf{v} = \begin{cases} \mathcal{L}(\mathbf{x}, \mathbf{y}, \mathbf{v}) - \mathcal{L}(\bar{\mathbf{x}}, \bar{\mathbf{x}}, \bar{\mathbf{u}}) & \\ -\langle \mathbf{y}, \mathbf{A}^*\bar{\mathbf{q}}\rangle_E - \langle \mathbf{B}\mathbf{v}, \bar{\mathbf{q}}\rangle_E & \text{if} \quad \mathbf{x} \in \mathbf{X} \text{ and } v_j \in U_j(\mathbf{y}) \\ +\infty & \text{otherwise.} \end{cases} \tag{28}$$

We observe that this function has the same properties as $\mathcal{L}$ and additionally satisfies $\mathcal{L}_0(\bar{\mathbf{x}}, \mathbf{y}, \mathbf{v}) \geq 0$ for all $(\mathbf{y}, \mathbf{v}) \in \mathbf{E} \times \mathbf{F}$. Further it is also clear that we have the growth condition,

$$\|\mathbf{y}\|_E^2 + \|\mathbf{v}\|_F^2 > K_1 \Longrightarrow \mathcal{L}_0(\bar{\mathbf{x}}, \mathbf{y}, \mathbf{v}) \geq K(\|\mathbf{y}\|_E^2 + \|\mathbf{v}\|_F^2). \tag{29}$$

From this we see that the function $(\mathbf{y}, \mathbf{v}) \to \mathcal{L}_0(\bar{\mathbf{x}}, \mathbf{y}, \mathbf{v})$ enjoys the same properties as the function $\mathcal{L}_0(\cdot, \cdot)$ found in Carlson, Haurie, and Jabrane [2] and so we immediately extract the following results.

Lemma 4.1. *If* $\{\mathbf{y}, \mathbf{v}\}$ *is an admissible trajectory-control pair that satisfies*

$$\int_0^{+\infty} \mathcal{L}_0(\bar{\mathbf{x}}, \mathbf{y}(t), \mathbf{v}(t))\, dt < +\infty, \tag{30}$$

then necessarily $\mathbf{y}(\cdot)$ *is bounded and for every fixed* $S > 0$ *there exists a constant* $C(S)$ *such that*

$$\forall t \geq 0 \quad \int_t^{t+S} \|\mathbf{v}(s)\|_F^2 \, ds \leq C(S). \tag{31}$$

Proof. Lemma 1 in Carlson, Haurie, Jabrane [2]. □

Theorem 4.1. *(Weak Turnpike Theorem) Under Assumption 3.1, if* $\{\mathbf{y}, \mathbf{v}\}$ *is an admissible trajectory-control pair that satisfies*

$$\limsup_{T \to +\infty} \int_0^T (\mathcal{L}(\bar{\mathbf{x}}, \mathbf{y}(t), \mathbf{v}(t)) - \mathcal{L}(\bar{\mathbf{x}}, \bar{\mathbf{x}}, \bar{\mathbf{u}})\, dt = \alpha < \infty, \tag{32}$$

then necessarily

$$\frac{1}{T} \int_0^T \mathbf{y}(t)\, dt \longrightarrow_w \bar{\mathbf{x}} \quad \textit{as} \quad T \to +\infty. \tag{33}$$

Here the notation $\longrightarrow_w$ indicates weak convergence in $\mathbf{E}$.

Proof. Theorem 1 in Carlson, Haurie, Jabrane [2]. □

Remark 4.1. This weak turnpike theorem relates the weak convergence of the average state of an admissible trajectory to the unique steady-state normalized Nash equilibrium, $\bar{\mathbf{x}}$. In addition, it follows from the proof of this result that we also have

$$\frac{1}{T} \int_0^T \mathbf{v}(t)\, dt \longrightarrow_w \bar{\mathbf{u}} \quad \text{as} \quad T \to +\infty.$$

(here convergence is in $\mathbf{F}$) as well. For the details we refer the reader to [2].

To obtain a stronger result we let $\mathcal{G}$ be the set defined by

$$\mathcal{G} = \{\mathbf{x} \in \mathbf{E} : \quad \exists \quad \mathbf{u} \in \mathbf{F} \quad \text{such that} \quad \mathcal{L}_0(\bar{\mathbf{x}}, \mathbf{x}, \mathbf{u}) = 0\}, \tag{34}$$

and give the following definition.

Definition 4.1. Let $\mathcal{F}$ denote the family of all trajectories, $\mathbf{x}(\cdot)$ of the linear system (2) for some locally square integrable control $\mathbf{u}(\cdot)$ satisfying

$$\mathbf{x}(t) \in \mathcal{G} \quad \text{a.e.} \quad t \geq 0. \tag{35}$$

We say that $\mathcal{G}$ has the convergence property $\mathcal{C}$ if $\mathbf{x} \longrightarrow_w \bar{\mathbf{x}}$ as $t \to +\infty$ in $\mathbf{E}$, uniformly in $\mathcal{F}$.

With this definition we immediately obtain the following result.

Theorem 4.2. *Under Assumption (3.1), if $\mathcal{G}$ has property $\mathcal{C}$ and if $\{\tilde{\mathbf{x}}, \tilde{\mathbf{u}}\}$ is an admissible trajectory-control pair for which*

$$\int_0^{+\infty} \mathcal{L}_0(\bar{\mathbf{x}}, \tilde{\mathbf{x}}(t), \tilde{\mathbf{u}}(t))\, dt < +\infty, \tag{36}$$

then necessarily $\tilde{\mathbf{x}}(t) \longrightarrow_w \bar{\mathbf{x}}$ as $t \to +\infty$.

Proof. Theorem 2 in Carlson, Haurie, and Jabrane [2]. □

Remark 4.2. If in the definition of convergence property $\mathcal{C}$ we replace weak convergence in **E** by strong convergence in **E** then, we obtain from the above result that $\|\tilde{\mathbf{x}}(t) - \bar{\mathbf{x}}\|_E \to 0$ as $t \to +\infty$.

The results given above are immediate consequences of the results found in Carlson, Jabrane, and Haurie [2] for the "single-player" case. As we will see in the next section, these results are not quite adequate to establish the desired existence results.

5 Existence of an Overtaking Nash Equilibrium

In this section we provide sufficient conditions to insure that the dynamic game we are considering has an overtaking Nash equilibrium as defined in Section 2. In addition to the assumptions required in the previous sections we require two additional assumptions. The first of these is a controllability assumption that is defined as follows:

Assumption 5.1. There exists an $\epsilon_0 > 0$ and an $S > 0$ such that for all $\hat{\mathbf{x}} \in \mathbf{E}$ satisfying

$$|\langle \hat{\mathbf{x}} - \bar{\mathbf{x}}, \bar{\mathbf{q}} \rangle_E| < \epsilon_0,$$

there exists a trajectory-control pair $\{\mathbf{z}, \mathbf{w}\}$ defined on $[0, S]$ satisfying the constraints (4) and (5) that drives $\hat{\mathbf{x}}$ to $\bar{\mathbf{x}}$ in time S. That is, $\mathbf{z}(\cdot) : [0, S] \to \mathbf{E}$ satisfies

$$\begin{aligned}
\mathbf{z}(t) &= \mathbf{S}(t)\hat{\mathbf{x}} + \int_0^t \mathbf{S}(t-s)\mathbf{B}\mathbf{w}(s)\, ds \\
\mathbf{z}(S) &= \bar{\mathbf{x}} \\
\mathbf{z}(t) &\in \mathbf{X} \\
w_j(t) &\in U_j(\mathbf{z}(t)) \quad j = 1, 2, \ldots p.
\end{aligned}$$

Remark 5.1. Controllability issues in infinite dimensional systems are always very difficult to verify. The above assumption, however, is somewhat weaker than the usual controllability results. Further, since $\bar{\mathbf{x}}$ is a steady state, this could possibly be verified through the use of known null controllability results.

The next assumption we require is as follows:

Assumption 5.2. There exists a set of positive weights, $r_j > 0$, such that Assumption (3.1) holds and,

$$\inf \int_0^{+\infty} \mathcal{L}_0(\mathbf{x}(t), \mathbf{y}(t), \mathbf{v}(t))\, dt, \tag{37}$$

where the infimum is taken over all pairs of admissible trajectory-control pairs, $\{\mathbf{x}, \mathbf{u}\}$ and $\{\mathbf{y}, \mathbf{v}\}$ in which we additionally have, $\mathbf{x}(t) \to_w \bar{\mathbf{x}}$ as $t \to +\infty$, is bounded below.

Remark 5.2. This assumption insures that there is a global lower bound on the integral functional

$$\mathcal{K}(\mathbf{x}, \mathbf{y}, \mathbf{v}) = \int_0^{+\infty} \mathcal{L}_0(\mathbf{x}(t), \mathbf{y}(t), \mathbf{v}(t))\, dt$$

over all pairs of admissible trajectory-control pairs $\{\mathbf{x}, \mathbf{u}\}$ and $\{\mathbf{y}, \mathbf{v}\}$ in which we additionally have, $\mathbf{x}(t) \to_w \bar{\mathbf{x}}$ as $t \to +\infty$. Following the development given in Carlson, Haurie, and Jabrane [2] this condition can be satisfied by considering the infinite horizon optimal control problem in which the objective, up to time $T > 0$ is given by

$$\int_0^T \mathcal{L}(\mathbf{x}(t), \mathbf{y}(t), \mathbf{v}(t)) - \langle \mathbf{y}, \mathbf{A}^* \bar{\mathbf{q}} \rangle_E - \langle \mathbf{B}\mathbf{v}, \bar{\mathbf{q}} \rangle_E\, dt$$

over all pairs of admissible-trajectory control pairs $\{\mathbf{x}, \mathbf{u}\}$ and $\{\mathbf{y}, \mathbf{v}\}$. We further observe that in the case of a single-player game, (i.e., the optimal control case considered in [2]), this hypothesis always is satisfied since the corresponding function $\mathcal{L}_0$ is always non-negative by the Lagrange multiplier theorem.

Our final assumption is a turnpike assumption.

Assumption 5.3. For each admissible trajectory $\mathbf{x} : [0, +\infty) \to \mathbf{E}$ satisfying $\mathbf{x}(t) \to_w \bar{\mathbf{x}}$ as $t \to +\infty$ we have that whenever $\{\mathbf{y}, \mathbf{v}\}$ is an admissible trajectory-control pair satisfying

$$\int_0^{+\infty} \mathcal{L}_0(\mathbf{x}(t), \mathbf{y}(t), \mathbf{v}(t))\, dt < +\infty,$$

then necessarily one has $\mathbf{y}(t) \to_w \bar{\mathbf{x}}$ as $t \to +\infty$.

Remark 5.3. This assumption is, of course, stronger than the conclusion of Theorem 4.2. Nevertheless, in the light of this theorem, this assumption seems plausible and as we shall see permits us to achieve our result.

To provide sufficient conditions for an overtaking Nash equilibrium we consider the set, denoted by $\mathcal{S}$, of all admissible trajectories, $\mathbf{x} : [0, +\infty) \to \mathbf{E}$ satisfying $\mathbf{x}(t) \to_w \bar{\mathbf{x}}$ as $t \to +\infty$. We observe that this set is nonempty and convex by hypothesis. Further, since all of the admissible trajectories are uniformly bounded by (4) we may view this set as a subset of the Hilbert space, $\mathcal{X}$ defined as

$$\mathcal{X} = \left\{ \mathbf{x} : [0, +\infty) \to \mathbf{E} : \quad \int_0^{+\infty} e^{-t} \|\mathbf{x}(t)\|_E^2 \, dt < +\infty \right\}$$

with inner product defined by

$$\langle \mathbf{x}, \mathbf{y} \rangle_{\mathcal{X}} = \int_0^{+\infty} e^{-t} \langle \mathbf{x}(t), \mathbf{y}(t) \rangle_E \, dt.$$

Further, we note that if $\mathbf{x}$ is an admissible trajectory it follows that

$$\|\mathbf{x}\|_{\mathcal{X}} = \left(\int_0^{+\infty} e^{-t} \|\mathbf{x}(t)\|_E^2 \, dt \right)^{\frac{1}{2}} \leq M,$$

in which M is an upper bound for all of the admissible trajectories, giving us that $\mathcal{S}$ is a bounded, and hence a relatively weakly compact subset of $\mathcal{X}$. We now define the following set-valued mapping $\mathcal{B} : \mathcal{S} \to 2^{\mathcal{S}} \backslash \{\emptyset\}$ as follows:

Definition 5.1. We say $\mathbf{y}^* \in \mathcal{B}(\mathbf{x})$ if there exists a control $\mathbf{v}^*(\cdot)$ such that $\{\mathbf{y}^*, \mathbf{v}^*\}$ is an admissible trajectory-control pair such that $\mathbf{y}^*(t) \to_w \bar{\mathbf{x}}$ as $t \to +\infty$ and

$$\int_0^{+\infty} \mathcal{L}_0(\mathbf{x}(t), \mathbf{y}^*(t), \mathbf{v}^*(t)) \, dt = \inf \int_0^{+\infty} \mathcal{L}_0(\mathbf{x}(t), \mathbf{y}(t), \mathbf{v}(t)) \, dt < +\infty$$

where the infimum is taken over all pairs of admissible trajectory-control pairs, $\{\mathbf{y}, \mathbf{v}\}$.

Our reason for introducing this set-valued mapping is given in the following result.

Theorem 5.1. *If there exists an element $\mathbf{x}^* \in \mathcal{S}$ such that*

$$\mathbf{x}^* \in \mathcal{B}(\mathbf{x}^*)$$

(i.e., $\mathbf{x}^*$ is a fixed point of the mapping $\mathbf{x} \to \mathcal{B}(\mathbf{x})$), then $\mathbf{x}^*$ is an overtaking Nash equilibrium for the dynamic game.

Proof. To prove this result let the admissible trajectory-control pair $\{\mathbf{x}^*, \mathbf{u}^*\}$ be such that $\mathbf{x}^* \in \mathcal{B}(\mathbf{x}^*)$ and suppose that there exists an index $j = 1, 2, \ldots, p$ and a corresponding pair of functions $\{y_j, v_j\}$ such that $\{[\mathbf{x}^{*j}, y_j], [\mathbf{u}^{*j}, v_j]\}$ is an admissible trajectory-control pair such that for some $\epsilon^* > 0$ there exists a sequence $\{T_k\}_{k=0}^{+\infty}$, $T_k \to +\infty$ as $k \to +\infty$ such that

$$\int_0^{T_k} L_j(\mathbf{x}^*(t), u_j^*(t))\, dt > \int_0^{T_k} L_j([\mathbf{x}^{*j}, y_j](t), v_j(t))\, dt + \epsilon^*.$$

holds for each $k = 1, 2, \ldots$. From this we obtain,

$$\begin{aligned}
&\int_0^{T_k} \mathcal{L}(\mathbf{x}^*(t), [\mathbf{x}^{*j}, y_j](t), [\mathbf{u}^{*j}, v_j](t))\, dt \\
&= \sum_{l=1}^{p} \int_0^{T_k} r_l L_j([\mathbf{x}^{*j}, [\mathbf{x}^{*j}, y_j]_l(t), [\mathbf{u}^{*j}, v_j]_l(t))\, dt \\
&< \int_0^{T_k} \mathcal{L}(\mathbf{x}^*(t), \mathbf{x}^*(t), \mathbf{u}^*(t))\, dt - r_j\epsilon^*.
\end{aligned}$$

Using the definition of $\mathcal{L}_0$ we get,

$$\begin{aligned}
+r_j\epsilon^* &< \int_0^{T_k} \mathcal{L}(\mathbf{x}^*(t), \mathbf{x}^*(t), \mathbf{u}^*(t)) - \mathcal{L}(\mathbf{x}^*(t), [\mathbf{x}^{*j}, y_j](t), [\mathbf{u}^{*j}, v_j](t))\, dt \\
&= \int_0^{T_k} \mathcal{L}_0(\mathbf{x}^*(t), \mathbf{x}^*(t), \mathbf{u}^*(t)) - \mathcal{L}_0(\mathbf{x}^*(t), [\mathbf{x}^{*j}, y_j](t), [\mathbf{u}^{*j}, v_j](t))\, dt \\
&\quad + \int_0^{T_k} [\langle \mathbf{x}^*(t), \mathbf{A}^*\bar{\mathbf{q}}\rangle_E + \langle \mathbf{B}\mathbf{u}^*(t), \bar{\mathbf{q}}\rangle_E] - [\langle [\mathbf{x}^{*j}, y_j](t), \mathbf{A}^*\bar{\mathbf{q}}\rangle_E \\
&\quad + \langle \mathbf{B}[\mathbf{u}^{*j}, v_j](t), \bar{\mathbf{q}}\rangle_E]\, dt \\
&= \int_0^{T_k} \mathcal{L}_0(\mathbf{x}^*(t), \mathbf{x}^*(t), \mathbf{u}^*(t)) - \mathcal{L}_0(\mathbf{x}^*(t), [\mathbf{x}^{*j}, y_j](t), [\mathbf{u}^{*j}, v_j](t))\, dt \\
&\quad + \int_0^{T_k} \frac{d}{dt}\langle (\mathbf{x}^*(t) - [\mathbf{x}^{*j}, y_j](t)), \bar{\mathbf{q}}\rangle\, dt \\
&= \int_0^{T_k} \mathcal{L}_0(\mathbf{x}^*(t), \mathbf{x}^*(t), \mathbf{u}^*(t)) - \mathcal{L}_0(\mathbf{x}^*(t), [\mathbf{x}^{*j}, y_j](t), [\mathbf{u}^{*j}, v_j](t))\, dt \\
&\quad + \langle (\mathbf{x}^*(T_k) - [\mathbf{x}^{*j}, y_j](T_k)), \bar{\mathbf{q}}\rangle.
\end{aligned}$$

We now divide our considerations into two cases. The first is when we have

$$\lim_{k\to+\infty} \int_0^{T_k} \mathcal{L}_0(\mathbf{x}^*(t), [\mathbf{x}^{*j}, y_j](t), [\mathbf{u}^{*j}, v_j](t))\, dt = +\infty.$$

In this case we have, due to the boundedness of all admissible trajectories, that the right side of the above inequalities tends to $-\infty$ as $k \to +\infty$. Clearly this is not possible since the left side is a fixed positive quantity independent of k. Therefore we must assume that

$$\lim_{k\to+\infty}\int_0^{T_k} \mathcal{L}_0(\mathbf{x}^*(t), [\mathbf{x}^{*j}, y_j](t), [\mathbf{u}^{*j}, v_j](t))\, dt < +\infty.$$

In this case, the optimality of $\mathbf{x}^*$ insures that there exists an index $\hat{k}$ such that for all $k \geq \hat{k}$

$$\int_0^{T_k} \mathcal{L}_0(\mathbf{x}^*(t), \mathbf{x}^*(t), \mathbf{u}^*(t)) - \mathcal{L}_0(\mathbf{x}^*(t), [\mathbf{x}^{*j}, y_j](t), [\mathbf{u}^{*j}, v_j](t))\, dt \leq r_j \frac{\epsilon^*}{3}.$$

Further, Assumption 5.3, insures that we can also assume that whenever $k > \hat{k}$ we have

$$\langle(\mathbf{x}^*(T_k) - [\mathbf{x}^{*j}, y_j](T_k)), \bar{\mathbf{q}}\rangle \leq r_j(\epsilon^*/3).$$

Combining these two results gives us

$$r_j\epsilon^* < (2/3)r_j\epsilon^*,$$

an obvious contradiction, so that the desired result follows. □

As a consequence of this last result, we can establish the existence of an overtaking Nash equilibrium by providing conditions under which the set-valued mapping $\mathcal{B}(\cdot)$ has a fixed point. To do this we require the following infinite dimensional generalization of the Kakutani Fixed Point Theorem given by Bohenblust and Karlin [1] which we state here only for the case of a Hilbert space

Theorem 5.2. *Let H be a weakly separable Hilbert space with S a convex, weakly closed set in E. Let $B : S \to 2^S\backslash\{\emptyset\}$ be a set-valued mapping satisfying the following:*

(1) $B(\mathbf{x})$ is convex for each $\mathbf{x} \in S$.
(2) The graph of B, $\{(\mathbf{x}, \mathbf{y}) \in S \times S : \mathbf{y} \in B(\mathbf{x})\}$, is weakly closed in $H \times H$. That is, if $\{\mathbf{x}_n\}$ and $\{\mathbf{y}_n\}$ are two sequences in S such that $\mathbf{x}_n \to \mathbf{x}$, $\mathbf{y}_n \to \mathbf{y}$, weakly in H with $\mathbf{x}_n \in B(\mathbf{y_n})$, then necessarily we have $\mathbf{x} \in B(\mathbf{y})$.
(3) $\bigcup_{\mathbf{x}\in S} B(\mathbf{x})$ is contained in a sequentially weakly compact set T.

Then there exists $\mathbf{x}^* \in S$ such that $\mathbf{x}^* \in B(\mathbf{x}^*)$.

Proof. See Theorem 5 [1]. □

We now are ready to present our main result.

Theorem 5.3. *Under the general hypothesis of Section 2, and Assumptions (2.1)–(5.3) there exists an overtaking Nash equilibrium for the infinite horizon infinite dimensional dynamic game over the class of all admissible trajectory-control pairs provided there exists at least one admissible trajectory-control pair, say* $\{\hat{\mathbf{x}}, \hat{\mathbf{u}}\}$ *satisfying* $\hat{\mathbf{x}}(t) \to_w \bar{\mathbf{x}}$ *as* $t \to +\infty$.

Proof. In view of Theorem 5.1 it suffices to show that the set-valued mapping $\mathcal{B}(\cdot)$ has a fixed point. To do this we will use Theorem 5.2. We begin by observing that, due to our convexity hypothesis, it is an easy matter to see that $\mathcal{B}(\mathbf{x})$ is convex (giving us conditions (1) of Theorem 5.2) and that condition (3) of Theorem 5.2 holds. Therefore, it remains to show that $\mathcal{B}(\mathbf{x}) \neq \emptyset$ for each $\mathbf{x} \in \mathcal{S}$ and that the graph of this set-valued map is weakly closed. To show that for each $\mathbf{x} \in \mathcal{S}$ we have $\mathcal{B}(\mathbf{x}) \neq \emptyset$ we let $\{T_k\}$ be an unbounded, strictly monotone sequence of positive times such that

$$|\langle \mathbf{x}(T_k) - \bar{\mathbf{x}}, \bar{\mathbf{q}} \rangle| < \epsilon_0.$$

As a result of Assumption 5.1, there exists a trajectory-control pair, $\{\mathbf{z}_k, \mathbf{w}_k\}$ defined on $[0, S]$ that drives the state $\mathbf{x}(T_k)$ to $\bar{\mathbf{x}}$. Consequently, we can construct the following sequence of admissible trajectory-control pairs,

$$\{\tilde{\mathbf{x}}_{\mathbf{k}}(t), \tilde{\mathbf{u}}_{\mathbf{k}}(t)\} = \begin{cases} \{\mathbf{x}(t), \mathbf{u}(t)\} & \text{if } 0 \leq t < T_k \\ \{\mathbf{z}_k(t - T_k), \mathbf{w}_k(t - T_k)\} & \text{if } T_k \leq t < T_k + S \\ \{\bar{\mathbf{x}}, \bar{\mathbf{u}}\} & \text{if } T_k + S \leq t \end{cases}.$$

Further it is an easy matter to see that $\tilde{\mathbf{x}}_{\mathbf{k}} \to \mathbf{x}$ as $k \to +\infty$ pointwise in $\mathbf{E}$. Moreover we also observe that since $\mathcal{L}_0(\bar{\mathbf{x}}, \mathbf{y}, \mathbf{v}) \geq 0$, for each fixed $k = 1, 2, \ldots,$ for all $T > T_k$ and admissible trajectory-control pair $\{\mathbf{y}, \mathbf{v}\}$ we have,

$$\begin{aligned} \int_0^T \mathcal{L}_0(\tilde{\mathbf{x}}_{\mathbf{k}}(t), \mathbf{y}(t), \mathbf{v}(t))\, dt &= \int_0^{T_k} \mathcal{L}_0(\tilde{\mathbf{x}}(t), \mathbf{y}(t), \mathbf{v}(t))\, dt \\ &\quad + \int_{T_k}^{T_k+S} \mathcal{L}_0(\mathbf{z}_{\mathbf{k}}(t - T_k), \mathbf{y}(t), \mathbf{v}(t))\, dt \\ &\quad + \int_{T_k+S}^{T} \mathcal{L}_0(\bar{\mathbf{x}}, \mathbf{y}(t), \mathbf{v}(t))\, dt \\ &\leq \int_0^{T_k} \mathcal{L}_0(\tilde{\mathbf{x}}(t), \mathbf{y}(t), \mathbf{v}(t))\, dt \\ &\quad + \int_{T_k}^{T_k+S} \mathcal{L}_0(\mathbf{z}_{\mathbf{k}}(t - T_k), \mathbf{y}(t), \mathbf{v}(t))\, dt \\ &< +\infty. \end{aligned}$$

This, when combined with Assumption 5.2, insures that the problem of minimizing the functional

$$\int_0^{+\infty} \mathcal{L}_0(\tilde{\mathbf{x}}_{\mathbf{k}}(t), \mathbf{y}(t), \mathbf{v}(t))\, dt,$$

is well defined and that its infimum, over all admissible trajectory-control pairs $\{\mathbf{y}, \mathbf{v}\}$, is finite. Thus, as a result of our convexity and growth hypothesis this functional attains its minimum, say at $\{\mathbf{y}_{\mathbf{k}}, \mathbf{v}_{\mathbf{k}}\}$. Moreover we have that for all $T \geq T_k$,

$$\begin{aligned}\int_0^{T} \mathcal{L}_0(\tilde{\mathbf{x}}_{\mathbf{k}}(t), \mathbf{y}_{\mathbf{k}}(t), \mathbf{v}_{\mathbf{k}}(t))\, dt &= \int_0^{T_k} \mathcal{L}_0(\tilde{\mathbf{x}}(t), \mathbf{y}_{\mathbf{k}}(t), \mathbf{v}_{\mathbf{k}}(t))\, dt \\ &\quad + \int_{T_k}^{T_k+S} \mathcal{L}_0(\mathbf{z}_{\mathbf{k}}(t - T_k), \mathbf{y}_{\mathbf{k}}(t), \mathbf{v}_{\mathbf{k}}(t))\, dt \\ &\quad + \int_{T_k+S}^{T} \mathcal{L}_0(\bar{\mathbf{x}}, \mathbf{y}_{\mathbf{k}}(t), \mathbf{v}_{\mathbf{k}}(t))\, dt \\ &< +\infty,\end{aligned}$$

which implies

$$\int_0^{+\infty} \mathcal{L}_0(\bar{\mathbf{x}}, \mathbf{y}_{\mathbf{k}}(t), \mathbf{v}_{\mathbf{k}}(t))\, dt < +\infty.$$

As a consequence of the turnpike theorem, Theorem 4.2, we have

$$\mathbf{y}_{\mathbf{k}}(t) \to_w \bar{\mathbf{x}} \quad \text{as} \quad t \to +\infty.$$

Therefore, we have $\mathbf{y}_{\mathbf{k}} \in \mathcal{B}(\mathbf{x}_{\mathbf{k}})$ for each $k = 1, 2, \ldots$. Now we observe that by hypothesis the pairs $\{\mathbf{y}_{\mathbf{k}}, \mathbf{v}_{\mathbf{k}}\}$ and $\{\tilde{\mathbf{x}}_{\mathbf{k}}, \tilde{\mathbf{u}}_{\mathbf{k}}\}$ are bounded sequences in the Hilbert space $\mathbf{E} \times \mathbf{F}$ and therefore are weakly compact. This means that without loss of generality we can assume that these sequences pointwise converge weakly in $\mathbf{E} \times \mathbf{F}$ to functions $\{\mathbf{y}^*, \mathbf{v}^*\}$ and $\{\mathbf{x}, \mathbf{u}\}$ as $k \to +\infty$. Further, our convexity and growth assumptions insure that the integral functional

$$\{\mathbf{x}, \mathbf{y}, \mathbf{v}\} \to \int_0^{+\infty} \mathcal{L}_0(\mathbf{x}(t), \mathbf{y}(t), \mathbf{v}(t))\, dt$$

is weakly lower semi-continuous. Thus we have,

$$\int_0^{+\infty} \mathcal{L}_0(\mathbf{x}(t), \mathbf{y}^*(t), \mathbf{v}^*(t))\, dt \leq \liminf_{k \to +\infty} \int_0^{+\infty} \mathcal{L}_0(\tilde{\mathbf{x}}_{\mathbf{k}}(t), \mathbf{y}_{\mathbf{k}}(t), \mathbf{v}_{\mathbf{k}}(t))\, dt.$$

Further, since for each k and for any admissible trajectory-control pair $\{\mathbf{y}, \mathbf{v}\}$, we have

$$\int_0^{+\infty} \mathcal{L}_0(\tilde{\mathbf{x}}_\mathbf{k}(t), \mathbf{y}_\mathbf{k}(t), \mathbf{v}_\mathbf{k}(t))\, dt \le \int_0^{+\infty} \mathcal{L}_0(\tilde{\mathbf{x}}_\mathbf{k}(t), \mathbf{y}(t), \mathbf{v}(t))\, dt,$$

it follows that

$$\int_0^{+\infty} \mathcal{L}_0(\mathbf{x}(t), \mathbf{y}^*(t), \mathbf{v}^*(t))\, dt \le \int_0^{+\infty} \mathcal{L}_0(\mathbf{x}(t), \mathbf{y}(t), \mathbf{v}(t))\, dt,$$

implying that $\mathbf{y}^* \in \mathcal{B}(\mathbf{x})$ (i.e., $\mathcal{B}(\mathbf{x}) \neq \emptyset$) as desired. Finally we observe that an argument similar to the previous one insures that the graph of $\mathcal{B}(\cdot)$ is weakly closed. Thus all of the conditions of Theorem 5.2 are satisfied and there exists a fixed point, say $\hat{\mathbf{x}}$, of $\mathcal{B}(\cdot)$. Appealing to Theorem 5.1 gives us the desired conclusion. □

6 A Simple Example

We conclude this paper with an elementary example to which the existence result obtained above is applicable.

In this example we consider a two-player game in which each player has dynamics governed by the following partial differential equation:

$$\frac{\partial x_i}{\partial t} = -\frac{\partial x_i}{\partial y} + u_i, \quad t > 0, \quad y \in [0, h], \quad i = 1, 2$$

with initial condition

$$x_i(0, y) = x_i^0(y), \quad y \in [0, h], \quad i = 1, 2,$$

in which we assume that $x_i^0 : [0, h] \to \mathbb{R}$ is a given initial state with the property that $x_i^0(0) = 0$. We further suppose that the boundary condition,

$$x_i(t, 0) = 0, \quad t \ge 0$$

holds. Additionally we impose the capacity constraints

$$0 \le x_i(t, y) \le \alpha_i, \quad \text{a.e.} \quad t \ge 0; \quad y \in [0, h]$$

and

$$0 \le u_i(t, y) \le \bar{U}_i \quad \text{a.e.} \quad t \ge 0; \quad y \in [0, h].$$

Here we assume that $\alpha_i > 0$, and $\bar{U}_i > 0$ are fixed constants representing limitations on each of the player's state and controls.

The objective of each player is to minimize an accumulated cost which is described by the integral functional, up to time $T > 0$, by

$$\mathcal{J}_i^T(x_1, x_2, u_i) = \int_0^T \left[\int_0^h q_1^i(y) x_i(t, y)^2 + q_2^i(y) x_2(t, y) x_1(t, y) \right.$$
$$\left. + r_i(y) u_i(t, y)^2 \, dy \right] dt.$$

Here we assume that $q_1^i : [0, h] \to [0, +\infty)$, $q_2^i : [0, h] \to [0, +\infty)$, and $r_i : [0, h] \to [0, +\infty)$ are continuous functions for $i = 1, 2$.

To put this example into the framework considered here we let $E_i = L^2([0, h]; \mathbb{R})$, with the usual norm and inner product, and view the state variable for player i as a map $x_i : [0, +\infty) \to E_i$ (i.e., $x_i(t)(y) = x_i(t, y)$. Similarly we let $F_i = L^2([0, h]; \mathbb{R})$ (again with the usual inner product and norm) and assume the control of each player is a map $u_i : [0, \infty) \to F_i$. The dynamics of each player may now be expressed as

$$\dot{x}_i(t) = A_i x_i(t) + u_i(t),$$

where A_i is the operator with domain

$$\mathcal{D}(A_i) = \{x_i \in E_i : \quad \frac{\partial x_i}{\partial y} \in E_i \quad \text{and} \quad x_i(0) = 0\}$$

defined by

$$(A_i x_i)(y) = \frac{\partial}{\partial y} x_i(y).$$

It is a standard exercise to show that the operators A_i are closed, densely defined and generate a strongly continuous semigroup. Further due to the fact that the dynamics are decoupled these facts carry over to the aggregate operator $\mathbf{A} = (A_1, A_2)$ with domain $\mathcal{D}(\mathbf{A}) = \mathcal{D}(A_1) \times \mathcal{D}(A_2)$. The details can be extracted from a similar example described in Carlson, Haurie, and Leizarowitz [3]. Further we note the we may take

$$\mathbf{X} = \{(x_1, x_2) \in E_1 \times E_2 : (x_1(y), x_2(y)) \in [0, \alpha_1] \times [0, \alpha_2] \quad \text{a.e.} \quad y \in [0, h]\}$$

and

$$U_i(\mathbf{x}) = \{(u_i(\cdot) \in F_i : u_i(y) \in [0, \hat{U}_i] \quad \text{a.e.} \quad y \in [0, h]\}.$$

With regards to the objectives of each player we note that we have $L_i : E_1 \times E_2 \times F_i \to \mathbb{R}$ is given by

$$L_i(x_1, x_2, u_i) = \int_0^h q_1^i(y) x_i(y)^2 + q_2^i(y) x_2(y) x_i(y) + r_i(y) u_i(y)^2 \, dy \, dt.$$

The relevant convexity assumptions for these integrands is now insured by assuming that the functions $f_i : [0, \alpha_1] \times [0, \alpha_2] \times [0, \hat{U}_i] \times [0, h] \to \mathbb{R}$, given by

$$f_i(y, x_1, x_2, u_i) = q_1^i(y)x_i^2 + q_2^i(y)x_2x_1 + r_i(y)u_i^2$$

are convex in the (x_i, u_i) arguments. The positivity assumptions on the functions $q_1^i(\cdot)$ and $r_i(\cdot)$ are sufficient to insure that this condition is satisfied.

It remains to demonstrate that the rest of the hypothesis are satisfied. We first observe that due to the nonnegativity of each of the terms in L_i it is an easy matter to see that the growth hypotheses are satisfied. Further, it is also easy to see that one solution to the steady state problem is $(\bar{\mathbf{x}}, \bar{\mathbf{u}}) = (\mathbf{0}, \mathbf{0})$. The uniqueness of this steady state Nash equilibrium is obtained by requiring a strict diagonal convexity hypothesis to hold. This condition may be realized by assuming that the functions q_1^i and r_i are strictly positive on $[0, h]$. For $\bar{\mathbf{q}}$ we may take $\bar{\mathbf{q}} = \mathbf{0}$ so that we now have both Assumptions 1 and 2 satisfied. Further, Assumption 4 is also satisfied since the objective of each player is nonnegative. It remains to show that Assumptions 3 and 5 are satisfied. To see this we observe that for a given fixed initial condition, $\mathbf{x}_0 = (x_{1,0}, x_{2,0}) : [0, h] \to \mathbb{R}^2$ any admissible control $\mathbf{u} = (u_1, u_2) : [0, +\infty) \times [0, h] \to \mathbb{R}^2$ generates a trajectory, for $i = 1, 2$, on $[0, +\infty) \times [0, h]$ given by (see Carlson, Haurie, Leizarowitz [3] for the details of this representation):

$$x_i(t, y) = \begin{cases} x_{i,0}(y - t) + \int_0^t u_i(s, y - t + s)\, ds & \text{for} \quad 0 \le t \le y \\ 0 & \text{otherwise.} \end{cases}$$

Thus when $t > h$ the admissible trajectory satisfies $x_i(t, y) \equiv 0$. This insures that both Assumption 3 and Assumption 5 are satisfied. With assumptions 1 through 5 all satisfied we may now conclude that the above problem has an overtaking open-loop Nash equilibrium.

REFERENCES

[1] Bohenblust H. F. and Karlin S., *On a theorem of ville*, Contributions to the Theory of Games, Vol. 1 (H. W. Kuhn and A. W. Tucker, eds.), Princeton University Press, Princeton, New Jersey, 1950, pp. 155–160.

[2] Carlson D. A., Haurie A., and Jabrane A., *Existence of overtaking solutions to infinite dimensional control problems on unbounded intervals*, SIAM Journal on Control and Optimization **25** (1987), 1517–1541.

[3] Carlson D. A., Haurie A. B., and Leizarowitz A., *Infinite horizon optimal control; deterministic and stochastic systems*, 2nd ed., Springer-Verlag, New York, New York, 1991.

[4] Carlson D. A. and Haurie A., *A turnpike theory for infinite horizon open-loop differential games with decoupled controls*, New Trends in Dynamic Games and Applications (G. J. Olsder, ed.), Annals of the International Society of Dynamic Games, Birkhäuser, Boston, 1995, pp. 353–376.

[5] Carlson D. A. and Haurie A., *A turnpike theory for infinite horizon open-loop competitive processes*, SIAM Journal on Control and Optimization **34** (1996), no. 4, 1405–1419.

[6] Carlson D. A., *Existence and uniqueness in convex games with strategies in Hilbert spaces*, Annals of the International Society of Dynamic Games (E. Altman, ed.), Birkhäuser, Boston, 2000.

[7] Ramsey F., *A mathematical theory of saving*, Economic Journal **38** (1928), 543–549.

[8] Rosen J. B., *Existence and uniqueness of equilibrium points for concave n-person games*, Econometrica **33** (1965), no. 3, 520–534.

Cooperative Differential Games

Leon A. Petrosjan
St. Petersburgh State University
Bibliotecfhnaya 2
Petrodvorets 199504
St. Petersburg, Russia

Abstract

In this paper the definition of cooperative game in characteristic function form is given. The notions of optimality principle and solution concepts based on it are introduced. The new concept of "imputation distribution procedure" (IDP) is defined and connected with the basic definitions of time-consistency and strong time-consistency. Sufficient conditions of the existence of time-consistent solutions are derived. For a large class of games where these conditions cannot be satisfied, the regularization procedure is developed and new c.f. constructed. The "regularized" core is defined and its strong time-consistency proved.

1 Introduction

In n-person differential games as in classical simultaneous game theory different solution concepts are used. The most common approach is a non-cooperative setting where as a solution the Nash Equilibrium is considered, and also other solutions such as Stackelberg solution, Nash bargaining scheme, Kalai–Smorodinski solution, are frequently used. At the same time not much attention is given to the problem of time-consistency of the solution considered in each specific case. This may follow from the fact that in most cases the Nash Equilibrium turns out to be time-consistent, but not always as was shown in [11]. The time-consistency of other above mentioned solutions take place in exceptional cases.

The problem becomes more serious when cooperative differential games are considered. Usually in cooperative settings players agree to use such controls (strategies) which maximize the sum of their payoffs. As a result the game then develops along the cooperative trajectory (conditionally optimal trajectory). The corresponding maximal total payoff satisfies Bellman's Equation and thus is time-consistent. But the values of characteristic function for each subcoalition of players naturally do not satisfy this property along a conditionally optimal trajectory. The characteristic function plays a key role in construction of solution concepts in cooperative game theory. And impossibility to satisfy Bellman's Equation for values of characteristic function for subcoalitions implies the time-inconsistency of cooperative solution concepts. This was seen first in papers [2], [3], and in the papers [10], [5], [9] it was proposed to introduce a special rule of distribution of the players gain under cooperative behavior over a time interval in such a way that time-consistency of the solution could be restored in given sense.

In this paper we formalize the notion of time-consistency and strong time-consistency for cooperative differential games, propose the regularization method which makes it possible to restore classical simultaneous solution concepts so that they became useful in dynamic game theory. We prove theorems concerning strong time-consistency of regularized solutions and give a constructive method of computing such solutions.

At the same time we do not include latest results published in [1], where the evolution of characteristic function over time resulting as a solution of controlled differential equations is considered.

2 Definition of Cooperative Differential Game in Characteristic Function Form

We will investigate n-person differential games starting from an initial state $x_0 \in R^n$ at moment $t_0 \in R^1$ and prescribed duration $T - t_0$ where $T > t_0$ is a finite number. To underline the dependence of the game from the initial state x_0 and duration $T - t_0$ we will denote this game by $\Gamma(x_0, T - t_0)$. The motion equations have the form

$$\dot{x} = f(x, u_1, \dots, u_n), \quad x \in R^n, \; u_i \in U_i, \; i = 1, \dots, n, \tag{1}$$

$$x(t_0) = x_0 \quad T < \infty, \; t_0, \in R^1, \; T \in R^1. \tag{2}$$

Here $u_i \in U_i$ is the control variable of player i and U_i is a compact set. The payoff function of player i is defined in the following way:

$$K_i(x_0, T - t_0; u_1, \dots, u_n) = \int_{t_0}^{T} h_i(x(t))dt + H_i(x(T)), \tag{3}$$

$$h_i > 0, \; H_i(x) > 0, \; i = 1, \dots, n,$$

where $h_i(x)$, $H_i(x)$ are given continuous functions and $x(t)$ is the trajectory realized in the situation $(u_1, \dots, u_n)$ from the initial state x_0. In the cooperative differential game we consider only open-loop strategies $u_i = u_i(t)$, $t \in [t_0, T]$, $i = 1, \dots, n$ of players.

Consider the cooperative form of the game $\Gamma(x_0, T - t_0)$. In this formalization we suppose that the players before starting the game agree to play $u_1^*, \dots, u_n^*$ such that the corresponding trajectory $x^*(t)$ maximizes the sum of the payoffs

$$\max_u \sum_{i=1}^{n} K_i(x_0, T - t_0; u_1, \dots, u_n) = \sum_{i=1}^{n} K_i(x_0, T - t_0; u_1^*, \dots, u_n^*)$$

$$= \sum_{i=1}^{n} \left[\int_{t_0}^{T} h_i(x^*(t))dt + H_i(x^*(T)) \right] = v(N; x_0, T - t_0),$$

where $x_0 \in R^n$, $T - t_0 \in R^1$ are initial conditions of the game $\Gamma(x_0, T - t_0)$ and N is the set of all players in $\Gamma(x_0, T - t_0)$. The trajectory $x^*(t)$ is called *conditionally optimal*. To define the cooperative game one has to introduce the characteristic function. We will do this in a classical way. Consider a zero-sum game defined over the structure of the game $\Gamma(x_0, T - t_0)$ between the coalition S as first player and the coalition $N \setminus S$ as second player, and suppose that the payoff of S is equal to the sum of payoffs of players from S. Denote this game as $\Gamma_S(x_0, T - t_0)$. Suppose that the value $v(S; x_0, T - t_0)$ of such game exists (the existence of the value of zero-sum differential games is proved under very general conditions).The characteristic function is defined for each $S \subset N$ as value $v(S; x_0, T - t_0)$ of $\Gamma_S(x_0, T - t_0)$. From the definition of $v(S; x_0, T - t_0)$ it follows that $v(S; x_0, T - t_0)$ is superadditive (see [13], [14]). It follows from the superadditivity condition that it is advantageous for the players to form a maximal coalition N and obtain a maximal total payoff $v(N; x_0, T - t_0)$ that is possible in the game. Purposefully, the quantity $v(S; x_0, T - t_0)$ $(S \neq N)$ is equal to a maximal guaranteed payoff of the coalition S obtained irrespective of the behavior of other players, even though the latter form a coalition $N \setminus S$ against S.

Note that the positiveness of payoff functions K_i, $i = 1, \dots, n$ implies that of characteristic function. From the superadditivity of v it follows that

$$v(S'; x_0, T - t_0) \geq v(S; x_0, T - t_0)$$

for any $S,\ S' \subset N$ such that $S \subset S'$, i.e. the superadditivity of the function v in S implies that this function is monotone in S.

The pair $\langle N, v(S; x_0, T - t_0)\rangle$, where N is the set of players, and v the characteristic function, is called the *cooperative differential game in the form of characteristic function* v. For short, it will be denoted by $\Gamma_v(x_0, T - t_0)$.

Various methods for "equitable" allocation of the total profit among players are treated as solutions in cooperative games. The set of such allocations satisfying an optimality principle is called a solution of the cooperative game (in the sense of this optimality principle). We will now define solutions of the game $\Gamma_v(N; x_0, T - t_0)$.

Denote by ξ_i a share of the player $i \in N$ in the total gain $v(N; x_0, T - t_0)$.

Definition 2.1. The vector $\xi = (\xi_1, \dots, \xi_n)$, whose components satisfy the conditions:

1. $\xi_i \geq v(\{i\}; x_0, T - t_0), \quad i \in N,$
2. $\sum_{i \in N} \xi_i = v(N; x_0, T - t_0),$

is called an imputation in the game $\Gamma_v(x_0, T - t_0)$.

Denote the set of all imputations in $\Gamma_v(x_0, T - t_0)$ by $L_v(x_0, T - t_0)$.

Under the solution of $\Gamma_v(x_0, T - t_0)$ we will understand a subset $W_v(x_0, T - t_0) \subset L_v(x_0, T - t_0)$ of imputation set which satisfies additional "optimality" conditions.

The equity of the allocation $\xi = (\xi_1, \ldots, \xi_n)$ representing an imputation is that each player receives at least maximal guaranteed payoff and the entire maximal payoff is distributed evenly without a remainder.

3 Principle of Time-Consistency (Dynamic Stability)

Formalization of the notion of optimal behavior constitutes one of the fundamental problems in the theory of n-person games. At present, for the various classes of games, different solution concepts are constructed. Recall that the players' behavior (strategies in noncooperative games or imputations in cooperative games) satisfying some given optimality principle is called a solution of the game in the sense of this principle and must possess two properties. On the one hand, it must be feasible under conditions of the game where it is applied. On the other hand, it must adequately reflect the conceptual notion of optimality providing special features of the class of games for which it is defined.

In dynamic games, one more requirement is naturally added to the mentioned requirements, viz. the purposefulness and feasibility of an optimality principle are to be preserved throughout the game. This requirement is called the *time-consistency of a solution of the game (dynamic stability).*

The time-consistency of a solution of differential game is the property that, when the game proceeds along a "conditionally optimal" trajectory, at each instant of time the players are to be guided by the same optimality principle, and hence do not have any ground for deviation from the previously adopted "optimal" behavior throughout the game. When the time-consistency is betrayed, at some instant of time there are conditions under which the continuation of the initial behavior becomes non-optimal and hence an initially chosen solution proves to be unfeasible.

Assume that at the start of the game the players adopt an optimality principle and construct a solution based on it (an imputation set satisfying the chosen principle of optimality, say the core, nucleolus, *NM*–solution etc.). From the definition of cooperative game it follows that the evolution of the game is to be along the trajectory providing a maximal total payoff for the players. When moving along this "conditionally optimal" trajectory, the players pass through subgames with current initial states and current duration. In due course, not only the conditions of the game and the players' opportunities, but even the players' interests may change. Therefore, at an instant t the initially optimal solution of the current game may not exist or satisfy players at this instant of time. Then, at the instant t, players will have no ground to keep to the initially chosen "conditionally optimal" trajectory. The latter exactly means the time-inconsistency of the chosen optimality principle and, as a result, the instability of the motion itself.

We now focus our attention on time-consistent solutions in the cooperative differential games.

Let an optimality principle be chosen in the game $\Gamma_v(x_0, T - t_0)$. The solution of this game constructed in the initial state $x(t_0) = x_0$ based on the chosen principle of optimality is denoted by $W_v(x_0, T - t_0)$. The set $W_v(x_0, T - t_0)$ is a subset of the imputation set $L_v(x_0, T - t_0)$ in the game $\Gamma_v(x_0, T - t_0)$. Assume that $W_v(x_0, T - t_0) \neq \emptyset$. Let $x^*(t)$, $t \in [t_0, T]$ be the conditionally optimal trajectory.

The definition suggests that along the conditionally optimal trajectory players obtain the largest total payoff. For simplicity, we assume henceforth that such a trajectory exists. In the absence of the conditionally optimal trajectory, we may introduce the notion of "ϵ-conditionally optimal trajectory" and carry out the necessary constructions with an accuracy ϵ.

We will now consider the behavior of the set $W_v(x_0, T - t_0)$ along the conditionally optimal trajectory $x^*(t)$. Towards this end, in each current state $x^*(t)$ current subgame $\Gamma_v(x^*(t), T - t)$ is defined as follows. In the state $x^*(t)$, we define the characteristic function $v(S; x^*(t), T - t)$ as the value of the zero-sum differential game $\Gamma_S(x^*(t), T - t)$ between coalitions S and $N \setminus S$ from the initial state $x^*(t)$ and duration $T - t$ (as was done already for the game $\Gamma(x_0, T - t_0)$).

The *current cooperative subgame* $\Gamma_v(x^*(t), T - t)$ is defined as $\langle N, v(S, x^*(t), T - t)\rangle$. The imputation set in the game $\Gamma_v(x^*(t), T - t)$ is of the form:

$$L_v(x^*(t), T - t) = \left\{\xi \in R^n \mid \xi_i \geq v(\{i\}; x^*(t), T - t),\ i = 1, \dots, n;\right.$$
$$\left.\sum_{i \in N} \xi_i = v(N; x^*(t), T - t)\right\},$$

where

$$v(N; x^*(t), T - t) = v(N; x_0, T - t_0) - \int_{t_0}^{t} \sum_{i \in N} h_i(x^*(\tau))d\tau.$$

The quantity

$$\int_{t_0}^{t} \sum_{i \in N} h_i(x^*(\tau))d\tau$$

is interpreted as the total gain of the players on the time interval $[t_0, t]$ when the motion is carried out along the trajectory $x^*(t)$.

Consider the family of current games

$$\{\Gamma_v(x^*(t), T - t) = \langle N, v(S; x^*(t), T - t)\rangle, t_0 \leq t \leq T\},$$

determined along the conditionally optimal trajectory $x^*(t)$ and their solutions $W_v(x^*(t), T - t) \subset L_v(x^*(t), T - t)$ generated by the same principle of optimality as the initial solution $W_v(x_0, T - t_0)$.

It is obvious that the set $W_v(x^*(T), 0)$ is a solution of terminal game $\Gamma_v(x^*(T), 0)$ and is composed of the only imputation $H(x^*(T)) = \{H_i(x^*(T)), i = 1, \dots, n\}$, where $H_i(x^*(T))$ is the terminal part of player i's payoff along the trajectory $x^*(t)$.

4 Time-Consistency of the Solution

Let the conditionally optimal trajectory $x^*(t)$ be such that $W_v(x^*(t), T - t) \neq \emptyset$, $t_0 \leq t \leq T$. If this condition is not satisfied, it is impossible for players to adhere to the chosen principle of optimality, since at the very first instant t, when $W_v(x^*(t), T - t) = \emptyset$, the players have no possibility to follow this principle. Assume that in the initial state x_0 the players agree upon the imputation $\xi^0 \in W_v(x_0, T - t_0)$. This means that in the state x_0 the players agree upon such allocation of the total maximal gain that (when the game terminates at the instant T) the share of the ith player is equal to ξ_i^0, i.e. the ith component of the imputation ξ^0. Suppose the player i's payoff (his share) on the time interval $[t_0, t]$ is $\xi_i(x^*(t))$. Then, on the remaining time interval $[t, T]$ according to the ξ^0 he has to receive the gain $\eta_i^t = \xi_i^0 - \xi_i(x^*(t))$. For the original agreement (the imputation ξ^0) to remain in force at the instant t, it is essential that the vector $\eta^t = (\eta_1^t, \dots, \eta_n^t)$ belongs to the set $W_v(x^*(t), T - t)$, i.e. a solution of the current subgame $\Gamma_v(x^*(t), T - t)$. If such a condition is satisfied at each instant of time $t \in [t_0, T]$ along the trajectory $x^*(t)$, then the imputation ξ^0 is realized. Such is the conceptual meaning of the time-consistency of the imputation.

Along the trajectory $x^*(t)$ on the time interval $[t, T]$, $t_0 \leq t \leq T$, the coalition N obtains the payoff

$$v(N; x^*(t), T - t) = \sum_{i \in N} \left[\int_t^T h_i(x^*(\tau)) d\tau + H_i(x^*(T)) \right].$$

Then the difference

$$v(N; x_0, T - t_0) - v(N; x^*(t), T - t) = \int_{t_0}^t \sum_{i \in N} h_i(x^*(\tau)) d\tau$$

is equal to the payoff the coalition N obtains on the time interval $[t_0, t]$. The share of the ith player in this payoff, considering the transferability of payoffs, may be represented as

$$\gamma_i(t) = \int_{t_0}^t \beta_i(\tau) \sum_{i=1}^n h_i(x^*(\tau)) d\tau = \gamma_i(x^*(t), \beta), \tag{4}$$

where $\beta_i(\tau)$ is the $[t_0, T]$ integrable function satisfying the condition

$$\sum_{i=1}^n \beta_i(\tau) = 1, \;\; \beta_i(\tau) \geq 0, \;\; t_0 \leq \tau \leq T, \;\; (i = 1, \dots, n), \tag{5}$$

From (4) we necessarily get

$$\frac{d\gamma_i}{dt} = \beta_i(t) \sum_{i \in N} h_i(x^*(t)).$$

This quantity may be interpreted as an instantaneous gain of the player i at the moment t. Hence it is clear the vector $\beta(t) = (\beta_1(t), \ldots, \beta_n(t))$ prescribes distribution of the total gain among the members of coalition N. By properly choosing $\beta(t)$, the players can ensure the desirable outcome, i.e. to regulate the players' gain receipt with respect to time, so that at each instant $t \in [t_0, T]$ there will be no objection against realization of the original agreement (the imputation ξ^0).

Definition 4.1. The imputation $\xi^0 \in W_v(x_0, T - t_0)$ is called time-consistent in the game $\Gamma_v(x_0, T - t_0)$ if the following conditions are satisfied:

1. there exists a conditionally optimal trajectory $x^*(t)$ along which $W_v(x^*(t), T - t) \neq \emptyset, \quad t_0 \leq t \leq T$,
2. there exists such $[t_0, T]$ integrable vector function $\beta(t) = (\beta_1(t), \ldots, \beta_n(t))$ that for each $t_0 \leq t \leq T, \quad \beta_i(t) \geq 0, \quad \sum_{i=1}^{n} \beta_i(t) = 1$ and

$$\xi^0 \in \bigcap_{t_0 \leq t \leq T} [\gamma(x^*(t), \beta) \oplus W_v(x^*(t), T - t)], \tag{6}$$

where $\gamma(x^*(t), \beta) = (\gamma_1(x^*(t), \beta), \ldots, \gamma_n(x^*(t), \beta))$, and $W_v(x^*(t), T-t)$ is a solution of the current game $\Gamma_v(x^*(t), T - t)$.

The sum $\oplus$ in the above definition has the following meaning: for $\eta \in R^n$ and $A \subset R^n$ $\eta \oplus A = \{\eta + a \mid a \in A\}$.

The cooperative differential game $\Gamma_v(x_0, T-t_0)$ with side payments has a time-consistent solution $W_v(x_0, T-t_0)$ if all of the imputations $\xi \in W_v(x_0, T-t_0)$ are time-consistent.

The conditionally optimal trajectory along which there exists a time-consistent solution of the game $\Gamma_v(x_0, T - t_0)$ is called an *optimal trajectory*.

From the definition of time-consistency at the instant $t = T$ we have $\xi^0 \in \gamma(x^*(T), \beta) \oplus W_v(\overline{x}(T), 0)$, where $W_v(x^*(T), 0)$ is a solution of the current game $\Gamma_v(x^*(T), 0)$ which occurs at the last moment $t = T$ on the trajectory $x^*(t)$, $t \in [t_0, T]$, has a duration 0 and is made up of the only imputation $\xi^T = H(x^*(T)) = \{H_i(x^*(T))\}$, the imputation ξ^0 may be represented as $\xi^0 = \gamma(x^*(T), \beta) + H(x^*(T))$ or

$$\xi^0 = \int_{t_0}^{T} \beta(\tau) \sum_{i \in N} h_i(x^*(\tau)) d\tau + H(x^*(T)).$$

The time-consistent imputation $\xi^0 \in W_v(x_0, T - t_0)$ may be realized as follows. From (6) at any instant $t_0 \leq t \leq T$ we have

$$\xi^0 \in [\gamma(x^*(t), \beta) \oplus W_v(x^*(t), T - t)], \tag{7}$$

where

$$\gamma(x^*(t),\beta) = \int_{t_0}^{t} \beta(\tau)\sum_{i\in N} h_i(x^*(\tau))d\tau$$

is the payoff vector on the time interval $[t_0, t]$, the player i's share in the gain on the same interval being

$$\gamma_i(x^*(t),\beta) = \int_{t_0}^{t} \beta_i(\tau)\sum_{i\in N} h_i(x^*(\tau))d\tau.$$

When the game proceeds along the optimal trajectory, the players on each time interval $[t_0, t]$ share the total gain

$$\int_{t_0}^{t} \sum_{i\in N} h_i(x^*(\tau))d\tau$$

among themselves, and

$$\xi^0 - \gamma(x^*(t),\beta) \in W_v(x^*(t), T-t) \tag{8}$$

so that the inclusion (8) is satisfied. Furthermore, (8) implies the existence of such a vector $\xi^t \in W_v(\overline{x}(t), T-t)$ that $\xi^0 = \gamma(x^*(t),\beta)+\xi^t$. That is, in the description of the above method of choosing $\beta(\tau)$, the vector of the gains to be obtained by the players at the remaining stage of the game

$$\xi^t = \xi^0 - \gamma(x^*(t),\beta) = \int_{t}^{T} \beta_i(\tau)h(x^*(\tau))d\tau + H(x^*(T))$$

belongs to the set $W_v(x^*(t), T-t)$.

We also have

$$\xi^0 = \int_{t_0}^{T} \beta_i(\tau)\sum_{i\in N} h(x^*(\tau))d\tau + H(x^*(T)).$$

The vector function

$$\alpha_i(\tau) = \beta_i(\tau)\sum_{i\in N} h(x^*(\tau))d\tau, i \in N$$

is called the imputation distribution procedure (IDP)

In general, it is fairly easy to see that there may exist an infinite number of vectors $\beta(\tau)$ satisfying conditions (4), (5). Therefore the sharing method proposed here seems to lack true uniqueness. However, for any vector $\beta(\tau)$ satisfying conditions (4)–(5) at each time instant $t_0 \leq t \leq T$ the players are guided by the imputation $\xi^t \in W_v(x^*(t), T-t)$ and the same optimality principle throughout the game, and hence have no reason to violate the previously achieved agreement.

Let us make the following additional assumption.

A. The vector $\xi^t \in W_v(x^*(t), T-t)$ may be chosen as a continuously differentiable monotonic nonincreasing function of the argument t.

Show that by properly choosing $\beta(t)$ we may always ensure time-consistency of the imputation $\xi^0 \in W_v(x_0, T-t_0)$ under assumption A and the first condition of the definition (i.e. along the conditionally optimal trajectory at each time instant $t_0 \le t \le T$ $W_v(x^*(t), T-t) \neq \oslash$).

Choose $\xi^t \in W_v(x^*(t), T-t)$ to be a continuously differentiable nonincreasing function of t, $t_0 \le t \le T$. Construct the difference $\xi^0 - \xi^t = \gamma(t)$, so that we get $\xi^t + \gamma(t) \in W_v(x_0, T-t_0)$. Let $\beta(t) = (\beta_1(t), \dots, \beta_n(t))$ be the $[t_0, T]$ integrable vector function satisfying conditions (4), (5). Instead of writing $\gamma(x^*(t), \beta)$ we will write for simplicity $\gamma(t)$. Rewriting (4) in vector form we get

$$\int_{t_0}^{t} \beta(\tau) \sum_{i \in N} h_i(x^*(\tau)) d\tau = \gamma(t):$$

differentiating with respect to t we get the following expression for $\beta(t)$

$$\beta(t) = \frac{1}{\sum_{i \in N} h_i(x^*(t))} \cdot \frac{d\gamma(t)}{dt} = -\frac{1}{\sum_{i \in N} h_i(x^*(t))} \cdot \frac{d\xi^t}{dt}. \tag{9}$$

Here the last expression follows from the equality

$$\xi^0 = \gamma(t) + \xi^t.$$

Make sure that for such $\beta(t)$ the condition (5) is satisfied. Indeed,

$$\sum_{i \in N} \beta_i(t) = -\frac{\sum_{i \in N} (d\xi_i^t/dt)}{\sum_{i \in N} h_i(x^*(t))} = -\frac{(d/dt)v(N; x^*(t), T-t)}{\sum_{i \in N} h_i(x^*(t))}$$

$$= -\frac{(d/dt)\left[\sum_{i \in N} \left(\int_t^T h_i(x^*(\tau))d\tau + H_i(x^*(T))\right)\right]}{\sum_{i \in N} h_i(x^*(t))}$$

$$= \frac{\sum_{i \in N} h_i(x^*(t))}{\sum_{i \in N} h_i(x^*(t))} = 1$$

(since $\sum_{i \in N} \xi_i^t = v(N; x^*(t), T-t)$).

From condition (3) we have $h_i, H_i > 0$, $i \in N$, and since $d\xi^t/dt \le 0$, then $\beta_i(\tau) \ge 0$.

We set the following theorem.

Theorem 4.1. *If the assumption A is satisfied and*

$$W(x^*(t), T-t) \neq \emptyset, \; t \in [t_0, T], \tag{10}$$

solution $W(x_0, T-t_0)$ *is time-consistent.*

Theoretically, the main problem is to study conditions imposed on the vector function $\beta(t)$ in order to insure the time-consistency of specific forms of solutions $W_v(x_0, T - t_0)$ in various classes of games.

Consider the new concept of strongly time-consistency and define time-consistent solutions for cooperative games with terminal payoffs.

5 Strongly Time-Consistent Solutions

For the time consistent imputation $\xi^0 \in W_v(x_0, T - t_0)$, as follows from the definition, for $t_0 \le t \le T$ there exists such $[t_0, T]$ integrable vector function $\beta(t)$ and imputation ξ^t (generally not unique) from the solution $W_v(x^*(t), T-t)$ of the current game $\Gamma_v(x^*(t), T - t)$ so that $\xi^0 = \gamma(x^*(t), \beta) + \xi^t$. The conditions of time-consistency do not affect the imputations from the set $W_v(x^*(t), T-t)$ which fail to satisfy this equation. Furthermore, of interest is the case when any imputation from current solution $W_v(x^*(t), T - t)$ may provide a "good" continuation for the original agreement, i.e. for a time-consistent imputation $\xi^0 \in W_v(x_0, T - t_0)$ at any instant $t_0 \le t \le T$ and for every $\xi^t \in W_v(x^*(t), T - t)$ the condition $\gamma(x^*(t), \beta) + \xi^t \in W_v(x_0, T - t_0)$, where $\gamma(x^*(T), \beta) + H(x^*(T)) = \xi^0$, is satisfied. By slightly strengthening this requirement, we obtain a qualitatively new time-consistency concept of the solution $W_v(x_0, T-t_0)$ of the game $\Gamma_v(x_0, T-t_0)$ and call it a strongly time-consistent.

Definition 5.1. The imputation $\xi^0 \in W_v(x_0, T - t_0)$ is called strongly time-consistent (STC) in the game $\Gamma_v(x_0, T - t_0)$, if the following conditions are satisfied:

1. the imputation ξ^0 is time-consistent;
2. for any $t_0 \le t_1 \le t_2 \le T$ and $\beta^0(t)$ corresponding to the imputation ξ^0 we have,

$$\gamma(x^*(t_2), \beta^0)) \oplus W_v(x^*(t_2), T - t_2) \subset \gamma(x^*(t_1), \beta^0)) \oplus W_v(x^*(t_1), T - t_1). \tag{12}$$

The cooperative differential game $\Gamma_v(x_0, T - t_0)$ with side payments has a strongly time-consistent solution $W_v(x_0, T - t_0)$ if all the imputations from $W_v(x_0, T - t_0)$ are strongly time-consistent.

6 Terminal Payoffs

In (3), let $h_i \equiv 0, i = 1, \dots, n$. The cooperative differential game with terminal payoffs is denoted by the same symbol $\Gamma_v(x_0, T - t_0)$. In such games the payoffs are made when the game terminates.

Theorem 6.1. *In the cooperative differential game $\Gamma_v(x_0, T - t_0)$ with terminal payoffs $H_i(x(T))$, $i = 1, \ldots, n$, only the vector $H(x^*) = \{H_i(x^*), i = 1, \ldots, n\}$ whose components are equal to the players' payoffs at the end point of the conditionally optimal trajectory may be time-consistent.*

Proof. It follows from the time-consistency of the imputation $\xi^0 \in W_v(x_0, T - t_0)$ that

$$\xi^0 \in \bigcap_{t_0 \le t \le T} W_v(x^*(t), T - t).$$

But since the current game $\Gamma_v(x^*(T), 0)$ is of zero duration, then therein $L_v(x^*(T), 0) = W_v(x^*(T), 0) = H(x^*(T)) = H(x^*)$. Hence

$$\bigcap_{t_0 \le t \le T} W_v(x^*(t), T - t) = H(x^*(T)),$$

i.e. $\xi^0 = H(x^*(T))$ and there are no other imputations. □

Theorem 6.2. *For the existence of the time-consistent solution in the game with terminal payoff it is necessary and sufficient that for all $t_0 \le t \le T$*

$$H(x^*(T)) \in W_v(x^*(t), T - t),$$

where $H(x^*(T))$ is the players' payoff vector at the end point of the conditionally optimal trajectory $x^*(t)$, with $W_v(x^*(t), T - t)$, $t_0 \le t \le T$ being the solutions of the current games along the conditionally optimal trajectory generated by the chosen principle of optimality.

This theorem is a corollary of the previous one.

Thus, if in the game with terminal payoffs there is a time-consistent imputation, then the players in the initial state x_0 have to agree upon realization of the vector (imputation) $H(x^*) \in W_v(x_0, T - t_0)$ and, with the motion along the optimal trajectory $x^*(t)$, at each time instant $t_0 \le t \le T$ this imputation $H(x^*)$ belongs to the solution of the current games $\Gamma_v(x^*(t), T - t)$.

As the Theorem shows, in the game with terminal payoffs only a unique imputation from the set $W_v(x_0, T - t_0)$ may be time-consistent, which is a highly improbable event since this means, that imputation $H(x^*(T))$ belongs to the solutions of all subgames along the conditionally optimal trajectory. Therefore, in such games there is no point in discussing both the time-consistency of the solution $W_v(x_0, T - t_0)$ as a whole and its strong time-consistency.

7 Regularization

For some economic applications it is necessary that the instantaneous gain of player i at the moment t, which by properly choosing $\beta(t)$ regulates the ith player's gain receipt with respect to time

$$\beta_i(t) \sum_{i \in N} h_i(x^*(t)) = \alpha_i(t),$$

is nonnegative (IDP, $\alpha_i \geq 0$). Unfortunately this condition cannot be always guaranteed. In the same time we shall purpose a new characteristic function (c.f.) based on the classical one defined earlier, such that solution defined in games with this new c.f. would be strongly time-consistent and would guarantee nonnegative instantaneous gain of player i at each current moment t.

Let $v(S; x^*(t), T - t,)\quad S \in N$ be the c. f. defined in subgame $\Gamma(x^*(t), T - t))$ in section 1.2 using the classical maxmin approach.

We suppose that the function $V(S; x^*(t), T - t)$ for every fixed $S \subset N$ is continuous on the time interval $[t_0, T]$ (it is always true if the c.f. is defined as the value of the associated zero-sum game played between the coalitions S and $N \setminus S$). For the function $V(N; x^*(t), T - t)\quad (S = N)$ the Bellman's equation along $x^*(t)$ is satisfied, i.e.

$$V(N; x_0(t), T - t_0) = \int_{t_0}^{t} \sum_{i=1}^{n} h_i(x^*(\tau))d\tau + V(N; x^*(t), T - t). \tag{12}$$

We get from (12),

$$V^{'}(N; x^*(t), T - t) = -\left[\sum_{i=1}^{n} h_i(x^*(\tau))\right].$$

Define the new "regularized" function $\overline{v}(S; x_0, T - t_0)$, $S \subset N$ by formula

$$\overline{v}(S; x_0, T - t_0) = -\int_{t_0}^{T} v(S; x^*(\tau), T - \tau)\frac{v^{'}(N; x^*(\tau), T - \tau)}{v(N; x^*(\tau), T - \tau)}d\tau. \tag{13}$$

And in the same manner for $t \in [t_0, T]$

$$\overline{v}(S; x^*(t), T - t) = -\int_{t}^{T} v(S; x^*(\tau), T - \tau)\frac{v^{'}(N; x^*(\tau), T - \tau)}{v(N; x^*(\tau), T - \tau)}d\tau. \tag{14}$$

It can be proved that $\overline{v}$ is superadditive and $\overline{v}(N; x^*(t), T - t) = v(N; x^*(t), T - t)$

Denote the set of imputations defined by characteristic functions $v(S; x_0(t), T - t_0)$, $\overline{v}(S; x^*(t), T - t)$, $t \in [t_0, T]$ by $L(x^*(t), T - t)$ and $\overline{L}(x^*(t), T - t)$ correspondingly. Let $\xi(t) \in L(x^*(t), T - t)$ be an integrable selector, $t \in [t_0, T]$ ($\xi(t)$ is bounded, and if it is measurable on $[t_0, T]$, then it is integrable), define

$$\overline{\xi} = -\int_{t_0}^{T} \xi(t)\frac{v'(N; x^*(t), T - t)}{v(N; x^*(t), T - t)}dt, \tag{15}$$

$$\overline{\xi}(t) = -\int_{t}^{T} \xi(\tau)\frac{v'(N; x^*(\tau), T - \tau)}{v(N; x^*(\tau), T - \tau)}d\tau, \tag{16}$$

$t \in [t_0, T]$.

Definition 7.1. The set $\overline{L}(x_0, T - t_0)$ consists of vectors defined by (15) for all possible integrable selectors $\xi(t)$, $t \in [t_0, T]$ with values in $L(x^*(t), T - t)$.

Let $\xi \in \overline{L}(x_0, T - t_0)$ and the functions $\alpha_i(t)$, $i = 1, \dots, n$, $t \in [t_0, T]$ satisfy the condition

$$\int_{t_0}^{T} \alpha_i(t)dt = \overline{\xi}_i, \quad \alpha_i \geq 0. \tag{17}$$

The vector function $\alpha(t) = \{\alpha_i(t)\}$ defined by the formula (17) is called "imputation distribution procedure" (IDP) (see section 1.4). Define

$$\int_{t_0}^{\Theta} \alpha_i(t)dt = \overline{\xi_i}(\Theta), i = 1, \dots, n.$$

The following formula connects α_i and β_i (see section 1.4)

$$\alpha_i(t) = \beta_i(t) \sum_{i \in N} h_i(x^*(t)).$$

Let $\overline{C}(x_0, T - t_0) \subset \overline{L}(x_0, T - t_0)$ be any of the known classical optimality principles from the cooperative game theory (core, nucleolus, NM-solution, Shapley value or any other OP). Consider $\overline{C}(x_0, T - t_0)$ as an optimality principle in $\Gamma(x_0, T - t_0)$. In the same manner let $\overline{C}(x^*(t), T - t)$ be an optimality principle in $\Gamma(x^*(t), T - t)$, $t \in [t_0, T]$.

The STC of the optimality principle means that if an imputation $\xi \in C(x_0, T - t_0)$ and an IDP $\alpha(t) = \{\alpha_i(t)\}$ of ξ are selected, then after getting payoff by the players, on the time interval $[t_0, \Theta]$

$$\xi_i(\Theta) = \int_{t_0}^{\Theta} \alpha_i(t)dt, i = 1, \dots, n,$$

the optimal income (in the sense of the optimality principle $C(x_*(\Theta), T - \Theta)$) on the time interval $[\Theta, T]$ in the subgame $\Gamma(x^*(\Theta), T - \Theta)$ together with $\xi(\Theta)$ constitutes the imputation belonging to the OP in the original game $\Gamma(x_0, T - t_0)$. The condition is stronger than time-consistency, which means only that the part of the previously considered "optimal" imputation belongs to the solution in the corresponding current subgame $\Gamma(x^*(\Theta), T - \Theta)$.

Suppose $\overline{C}(x_0, T - t_0) = \overline{L}(x_0, T - t_0)$ and $\overline{C}(x^*(t), T - t) = \overline{L}(x^*(t), T - t)$, then

$$\overline{L}(x_0, T - t_0) \supset \overline{\xi}(\Theta) + \overline{L}(x^*(\Theta), T - \Theta)$$

for all $\Theta \in [t_0, T]$ and this implies that the set of all imputations $\overline{L}(x_0, T - t)$ if considered as solution in $\Gamma(x_0, T - t_0)$ is strongly time consistent.

Suppose that the set $\overline{C}(x_0, T - t_0)$ consists of a unique imputation – the Shapley value. In this case from time consistency the strong time-consistency follows immediately.

Suppose now that $C(x_0, T - t_0) \subset L(x_0, T - t_0), C(x^*(t), T - t) \subset L(x^*(t), T - t), t \in [t_0, T]$ are cores of $\Gamma(x_0, T - t_0)$ and correspondingly of subgames $\Gamma(x^*(t), T - t)$.

We suppose that the sets $C(x^*(t), T - t)$, $t \in [t_0, T]$ are nonempty. Let $\widehat{C}(x_0, T - t_0)$ and $\widehat{C}(x^*(t), T - t), t \in [t_0, T]$ be the sets of all possible vectors $\overline{\xi}, \overline{\xi}(t)$ from (15), (16) and $\xi(t) \in C(x^*(t), T - t)$, $t \in [t_0, T]$. And let $\overline{C}(x_0, T - t_0)$ and $\overline{C}(*x(t), T - t), t \in [t_0, T]$ be cores of $\Gamma(x_0, T - t_0)$, $\Gamma(x^*(t), T - t)$ defined for c.f. $\overline{v}(S; x_0, T - t_0)$, $\overline{v}(S; x^*(t), T - t)$.

Proposition 7.1. *The following inclusions hold*

$$\widehat{C}(x_0, T - t_0) \subset \overline{C}(x_0, T - t_0), \tag{18}$$

$$\widehat{C}(x^*(t), T - t) \subset \overline{C}(x^*(t), T - t), \tag{19}$$

$t \in [t_0, T]$.

Proof. The necessary and sufficient condition for imputation $\overline{\xi}$ to belong to the core $\overline{C}(x_0, T - t_0)$ is the condition

$$\sum_{i \in S} \overline{\xi}_i \geq \overline{v}(S; x_0, T - t_0), \; S \subset N.$$

If $\overline{\xi} \in \widehat{C}(x_0, T - t_0)$, then

$$\overline{\xi} = -\int_{t_0}^{T} \xi(t) \frac{v'(N; x^*(t), T - t)}{v(N; x^*(t), T - t)} dt,$$

where $\xi(t) \in C(x^*(t), T - t)$. Thus

$$\sum_{i \in S} \xi_i(t) \geq v(S; x^*(t), T - t), \;\; S \subset N, \;\; t \in [t_0, T].$$

And we get

$$\sum_{i \in S} \overline{\xi}_i = -\int_{t_0}^{T} \sum_{i \in S} \xi(t) \frac{v'(N; x^*, T - t)}{v(N; x^*, T - t)} dt$$

$$\geq -\int_{t_0}^{T} v(S; x^*(t), T - t) \frac{v'(N; x^*(t), T - t)}{v(N; x^*, T - t)} dt = \overline{v}(S; x_0, T - t_0).$$

The inclusion (18) is proved similarly.

Define a new solution in $\Gamma(x_0, T-t_0)$ as $\widehat{C}(x_0, T-t_0)$ which we will call "regularized" subcore. $\widehat{C}(x_0, T-t_0)$ is always time-consistent and strongly time-consistent

$$\widehat{C}(x_0, T-t_0) \supset \int_{t_0}^{t} \xi(\tau)\frac{v'(N; x^*(\tau), T-\tau)}{v(N; x^*(\tau), T-\tau)}d\tau \oplus \widehat{C}(x^*(t), T-t), t \in [t_0, T].$$

Here under $a \oplus A$, where $a \in R^n$, $A \subset R^n$ the set of all vector $a+b,\ b \in A$ is understood. The quantity

$$\alpha_i = -\xi(t)\frac{v'(N; x^*(t), T-t)}{v(N; x^*(t), T-t)} \geq 0$$

is IDP and is nonnegative. □

8 Applications to Associated Problems

Cooperative differential games with random duration. The dynamics of the games is defined by (1) from initial state x_0. But the duration of the game is a random variable with distribution function $F(t)$,

$$\int_{t_0}^{\infty} dF(t) = 1. \tag{20}$$

The payoff has the form:

$$K_i(x_0; u_1, \ldots, u_n) = \int_{t_0}^{\infty}\int_{t_0}^{t} h_i(x(\tau))d\tau dF(t), i \in N.$$

Denote this game by $\Gamma(x_0)$. The c.f. in $\Gamma(x_0)$ is introduced in the same manner as in section 1.2. Denote by $L(x_0)$ the set of imputations and by $C(x_0) \subset L(x_0)$ any solution concept in $\Gamma(x_0)$. For each $\xi \in C(x_0)$ consider a vector function $\alpha(t) = \{\alpha_i(t)\}$ such that

$$\xi_i = \int_{t_0}^{\infty}\int_{t_0}^{t} \alpha_i(\tau)d\tau dF(t). \tag{21}$$

The vector function $\alpha(\tau),\ \tau \geq t$ is called "imputation distribution procedure" (IDP).

Let $x^*(t), t \geq t_0$ be the "conditionally optimal" trajectory in $\Gamma(x_0)$ maximizing the sum of players' payoffs. Consider subgames $\Gamma(x^*(t))$ along $x^*(t)$, and corresponding solutions $C(x^*(t))$. Note that the payoff function in a subgame $\Gamma(x^*(t))$ is defined as

$$K_i(x^*(t); u_1, \ldots, u_n) = \frac{1}{1-F(t)}\int_{t}^{\infty}\int_{t}^{\Theta} h_i(x(\tau))d\tau dF(\Theta), i \in N. \tag{22}$$

Denote by

$$\gamma_i(\Theta) = \int_{t_0}^{\Theta} \int_{t_0}^{t} \alpha_i(\tau) d\tau dF(t), \, i \in N$$

the payoff allocated on the time interval $[t_0, \Theta]$. Then the time-consistency condition will have the form:

Definition 8.1. The solution $C(x_0)$ is called time-consistent if, for each $\overline{\xi} \in C(x_0)$, there exists such IDP $\alpha_i(\tau), \ \tau \geq t_0, \alpha_i(\tau) \geq 0$, that

$$\overline{\xi}_i = \gamma_i(\Theta) + (1 - F(\Theta)) \left(\int_{t_0}^{\Theta} \alpha_i(\tau) d\tau + \overline{\xi}_i^{\Theta} \right), \tag{23}$$

where $\overline{\xi}^{\Theta} = \{\overline{\xi}_i^{\Theta}\} \in C(x^*(\Theta))$.

If $F(\Theta)$ is differentiable, and $\overline{\xi}^{\Theta}$ is a differentiable selector of form $C(x^*(\Theta))$ we get the following formula for $\alpha_i(\Theta)$

$$\alpha_i(\Theta) = \frac{F^{'}(\Theta)}{1 - F(\Theta)} \overline{\xi}_i^{\Theta} - (\overline{\xi}_i^{\Theta})^{'}, \ i \in N. \tag{24}$$

It is clear that the nonnegativeness of $\alpha_i(\Theta)$ can be guaranteed, if we can choose $\overline{\xi}_i^{\Theta}$ as a nonincreasing function of Θ ($\Theta \geq t_0$). If this is impossible then the regularization procedure is proposed similar to the one considered in section 1.7.

An analogous approach can be used for differential cooperative games with uncertain payoffs.

Consider now games with discounted payoffs $\Gamma(x_0)$. Suppose the payoff function is defined as:

$$K_i(x_0; u_1, \ldots, u_n) = \int_{t_0}^{\infty} e^{-\rho(t-t_0)} h_i(x(t)) dt, \ i \in N, \tag{25}$$

then considering the game $\Gamma(x_0)$ and the family of subgames $\Gamma(x^*(t)), t \geq t_0$ along the "conditionally optimal trajectory" with payoffs

$$K_i(x_0; u_1, \ldots, u_n) = \int_{t}^{\infty} e^{-\rho(\tau-t)} h_i(x(\tau)) dt, \ i \in N,$$

we can introduce the sets of imputations $L(x_0), \ L(x^*(t)), \ t \geq t_0$ and the sets of solutions $C(x_0) \subset L(x_0), C(x^*(t)) \subset L(x^*(t)), \ t \geq t_0$ of the corresponding subgames.

Let $\overline{\xi} \in C(x_0)$, introduce IDP $\alpha_i(t)$ as any function, satisfying

$$\overline{\xi}_i = \int_{t_0}^{\infty} e^{-\rho(t-t_0)} \alpha_i(t) dt, \ i \in N. \tag{26}$$

The solution $C(x_0)$ is called time-consistent if for every $\overline{\xi} \in C(x_0)$ there exist such IDP $\alpha(t)$, $\alpha(t) \geq 0$, that

$$\overline{\xi}_i = \int_{t_0}^{\Theta} e^{-\rho(t-t_0)}\alpha_i(t)dt + e^{-\rho(\Theta-t_0)}\overline{\xi}_i^{\Theta}, \; i \in N,$$

where $\overline{\xi} \in C(x^*(\Theta))$.

If ξ^{Θ} is a differentiable selector then we get the following formula for IDP $\alpha(t)$

$$\alpha_i(t) = \rho\overline{\xi}_i - (\overline{\xi}_i^{\Theta})'.$$

In the paper [12] the case when $C(x_0)$ coincides with the Shapley value is considered. For a special environmental problem of pollution cost reduction with discounted payoffs, the time-consistent Shapley value allocation on the time interval $[t_0, \infty)$, which is the IDP for the Shapley value, is computed [12].

In the case when non negativity of α cannot be guaranteed, a regularization procedure similar to the one introduced in section 1.7 is proposed (see[4]).

REFERENCES

[1] Filar, J., Petrosjan, L.A., Dynamic Cooperative Games, *International Game Theory Review, Vol. 2, No. 1* 42–65, 2000.

[2] Haurie, A., On Some Properties of the Characteristic Function and Core of Multistage Game of Coalitions, *IEEE Transaction on Automatic Control* 236–241, April 1975.

[3] Strotz, K. H., Myopia and Inconsistency in Dynamic Utility Maximization , *Review of Economic Studies, Vol 23,* 1955-1956.

[4] Petrosjan, L.A., The Shapley Value for Differential Games, *New Trends in Dynamic Games and Applications*, Geert Olsder ed., Birkhauser, 409–417, 1995.

[5] Petrosjan, L.A., *Differential Games of Pursuit*, World Scientific, London 1993.

[6] Bellman, R.E., *Dynamic Programming*, Princeton University Press, Princeton, NJ, 1957.

[7] Haurie, A., A Note on Nonzero-Sum Differential Games with Bargaining Solution, *Journal of Optimization Theory and Application, Vol 18.1* 31–39, 1976.

[8] Kaitala, V., Pohjola, M., Optimal Recovery of a Shared Resource Stock. A Differential Game Model with Efficient Memory Equilibria, *Natural Resource Modeling, Vol. 3.1* 191–199, 1988.

[9] Petrosjan, L.A., Danilov N.N., *Cooperative Differential Games*, Tomsk University Press 276, 1985.

[10] Petrosjan, L.A., Danilov N.N., *Stability of Solution in Nonzero-Sum Differential Games with Transferable Payoffs*, Vestnik of Leningrad University, No. 1 52–59, 1979.

[11] Petrosjan, L.A., On the Time-Consistency of the Nash Equilibrium in Multistage Games with Discount Payoffs, *Applied Mathematics and Mechanics (ZAMM), Vol. 76 Supplement 3* 535–536, 1996.

[12] Petrosjan, L.A., Zaccour, G., Time-Consistent Shapley Value Allocation of Pollution Cost Reduction, *Journal of Economic Dynamics and Control, Vol. 27*, 381–398, 2003.

[13] McKinsey, *Introduction to the Theory of Games*, McGraw-Hill, NY, 1952.

[14] Owen, G. *Game Theory*, W.B. Saunders Co., Philadelphia, 1968.

PART III
Stopping Games

Selection by Committee

Thomas S. Ferguson
Department of Mathematics
University of California at Los Angeles
Los Angeles, CA 90095-1361, USA
tom@math.ucla.edu

Abstract

The many-player game of selling an asset, introduced by Sakaguchi and extended to monotone voting procedures by Yasuda, Nakagami and Kurano, is reviewed. Conditions for a unique equilibrium among stationary threshold strategies are given.

Key words. selling an asset game, monotone voting procedures, threshold strategies.

AMS Subject Classifications. 60G40, 90 G 39, 90D45.

1 Introduction

The theory of committee decisions is not well developed. Without some treatment of committee decision problems, the Bayesian philosophical viewpoint is not complete. We review an analysis of the problem of selection of a candidate by a committee. This is essentially a game-theoretic version with many players of the problem of selling an asset. A committee is charged with the duty of selecting a candidate for a position. Each member of the committee has his own way of viewing a candidate's worth, which may be related or somewhat opposed to the viewpoints of the other members. Candidates appear sequentially.

The study of this problem was initiated by Sakaguchi [2] and continued in greater depth in Sakaguchi [3,4], Kurano et al. [1], Yasuda et al. [6] and Szajowski and Yasuda [5]. Here we look at the problem of uniqueness of perfect equilibria. This problem arises in the simplest of settings. It may be considered as a study of the equilibrium equations of Sakaguchi.

2 The Problem

Let m denote the number of committee members, referred to below as players, and let M denote the set of all players $M = \{1, \ldots, m\}$. The players sequentially observe i.i.d. m-dimensional vectors, $\mathbf{X}_1, \mathbf{X}_2, \ldots$, from a known distribution $F(\mathbf{x})$ with finite second moments, $E(|\mathbf{X}|^2) < \infty$. After each observation, the players vote on whether or not to accept the present observation (stop) or to continue observing. The players have possibly differing costs of observation. Let $\mathbf{c} = (c_1, \dots, c_m)$ denote the m-vector of costs, with player i paying c_i for each observation. If at stage n, the players vote to accept the present observation, $\mathbf{X}_n = (X_{n1}, \dots, X_{nm})$, then the payoff to player i is $X_{ni} - nc_i$ for $i = 1, \dots, m$.

The voting decision is made according to the rules of a simple game. A coalition is a subset of the players. Let $\mathcal{C} = \{C : C \subset M\}$ denote the class of all coalitions. A simple game is defined by giving the characteristic function, $\phi(C)$, which maps coalitions $C \in \mathcal{C}$ into the set $\{0, 1\}$. Coalitions C for which $\phi(C) = 1$ are called winning coalitions, and those for which $\phi(C) = 0$ are called losing coalitions. Let $\mathcal{W} = \{C : \phi(C) = 1\}$ denote the class of winning coalitions and $\mathcal{L} = \{C : \phi(C) = 0\}$ denote the class of losing coalitions. These are assumed to satisfy the properties, (1) $M \in \mathcal{W}$, (2) $\emptyset \in \mathcal{L}$ and (3) monotonicity: $T \subset S \in \mathcal{L}$ implies $T \in \mathcal{L}$; namely, subsets of losing coalitions are losing and supersets of winning coalitions are winning.

A strategy for each player is a voting rule at each stage based on all past information. As in Sakaguchi, we restrict attention to stationary threshold rules. For player i, this rule is determined by a number a_i, with the understanding that at stage n player i votes to accept $\mathbf{X}_n$ if $X_{ni} > a_i$.

For a given observation, $\mathbf{X} = (X_1, \dots, X_m)$, we denote the stopping set by $A = A_{a_1,\dots,a_m}(\mathbf{X})$. Thus,

$$A = \sum_{C \in \mathcal{W}} \left[\bigcap_{i \in C} \{X_i \geq a_i\} \right] \left[\bigcap_{i \in C^c} \{X_i < a_i\} \right]. \tag{1}$$

Let $V_i = V_i(a_1, \dots, a_m)$ denote player i's expected return from such a joint strategy. If $P(A) = P(A(\mathbf{X})) = 0$, then play never stops and $V_i(A) = -\infty$ for all i. If $P(A) > 0$, we may compute V_i using the optimality equation,

$$V_i = -c_i + E[X_i I(A)] + V_i(1 - P(A)),$$

which we may solve to find

$$V_i = (E[X_i I(A)] - c_i)/P(A) \qquad \text{for } i = 1, \dots, m. \tag{2}$$

We analyze this as a non-cooperative game. We assume that $E|\mathbf{X}|^2 < \infty$. The weaker assumption that $E|\mathbf{X}| < \infty$ may be made but it requires the restriction that the jointly chosen stopping rule be such that the expectation of negative parts of all the returns be greater than $-\infty$. This seems to require cooperation of the players. With the stronger assumption, all stopping rules are allowed, even rules that do not stop with probability one.

We seek Nash equilibria, that is, we seek vectors $(a_1, \dots, a_m)$ such that

$$V_i(a_1, \dots, a_{i-1}, a_i, a_{i+1}, \dots, a_m) = \sup_a V_i(a_1, \dots, a_{i-1}, a, a_{i+1}, \dots, a_m) \tag{3}$$

for all $i = 1, \ldots, m$. There may be many equilibria with $P(A) = 0$ where all players receive $-\infty$. (For example when unanimity is required, if any two players refuse to stop, this is an equilibrium no matter what the other players do.) This is the worst of all possible equilibria, and it is not a (trembling hand) perfect equilibria. Such equilibria are not interesting, so we restrict attention only to equilibria for which $P(A) > 0$.

3 The Equilibrium Equations

Several results from the theory of the house-hunting problem carry over to this more general formulation.

Theorem 3.1. *If $\mathbf{a} = (a_1, \ldots, a_m)$ is an equilibrium vector with finite equilibrium payoff $\mathbf{V} = (V_1, \ldots, V_m)$, then $\mathbf{V}$ is an equilibrium vector with the same equilibrium payoff.*

We would like to say that in any equilibrium, $\mathbf{a} = \mathbf{V}$. But there may exist an $\epsilon > 0$ such that $P(V_i - \epsilon < X_i < V_i + \epsilon) = 0$ in which case, any a_i in the interval $(V_i - \epsilon, V_i + \epsilon)$ gives the same equilibrium payoff vector. All we can say is that we may choose a_i equal to V_i.

Theorem 3.2. *At any equilibrium with $P(A) > 0$ and $a_i = V_i$ for $i = 1, \ldots, m$, we have*

$$E[(X_k - a_k)I(A)] = c_k \qquad \text{for } k = 1, \ldots, m. \tag{4}$$

Equations (4) are necessary and sufficient for an equilibrium with $P(A) > 0$. All such equilibria are perfect.

Moreover, if $P(A) > 0$, the left side of (4) is continuous in a_k with the other a_i, $i \neq k$ fixed, and strictly decreasing from ∞ to 0 and possibly to $-\infty$. This may be seen using (1),

$$\begin{aligned} E[(X_k - a_k)I(A)] = &\sum_{\substack{C \in \mathcal{W} \\ k \in C \\ C \setminus \{k\} \notin \mathcal{W}}} E\left[(X_k - a_k) \prod_{i \in C} I(X_i > a_i) \prod_{i \notin C} I(X_i \leq a_i)\right] \\ &+ \sum_{\substack{C \in \mathcal{W} \\ k \notin C}} E\left[(X_k - a_k) \prod_{i \in C} I(X_i > a_i) \prod_{\substack{i \notin C \\ i \neq k}} I(X_i \leq a_i)\right]. \end{aligned} \tag{5}$$

In any case, for each k there will exist a unique solution for a_k of equation (4) with the other a_i, $i \neq k$, fixed.

The existence of equilibria for general simple games has been proved by Yasuda et al. [6]. Here is an example to show that there may be more than one equilibrium even if the X_i are independent. We take $m = 2$ and X_1 and X_2 independent, with X_1 being uniform on the interval (0, 1), and X_2 taking the value 0 with probability 5/6 and the value 1 with probability 1/6. For both players the cost of observation is 1/8. The voting game is taken to be unanimity; both players must agree on the candidate. There are two perfect equilibria in threshold strategies. In the first, Player 2 votes to accept every candidate, and Player 1 only votes for candidates such that $X_1 > 1/2$. The equilibrium payoff is $(V_1, V_2) = (1/2, -1/12)$. In the second equilibria, Player 1 votes to accept every candidate, and Player 2 only votes for candidates such that $X_2 = 1$. This has equilibrium payoff $(V_1, V_2) = (-1/4, 1/4)$. Clearly Player 1 prefers the first equilibrium and Player 2 the second.

4 Uniqueness of Equilibria in the Independent, Unanimous Consent Case

We assume from here on that the X_i are independent, and that unanimous consent is required to accept a candidate. We give a simple condition for the existence of a unique equilibrium in threshold strategies. Under the unanimous consent voting rule, the set A of (1) becomes simply

$$A = \bigcap_{i \in M} \{X_i > a_i\}, \tag{6}$$

and in the independent case, the equilibrium conditions (4) become

$$E[(X_k - a_k)|X_k > a_k]P(A) = c_k \qquad \text{for } k = 1, \ldots, m. \tag{7}$$

Definition 4.1. A random variable X is said to have strictly decreasing residual expectation if $E[(X - a)|X > a]$ is a strictly decreasing function of a from ∞ when $a = -\infty$, to 0 when a is equal to the right support point of X.

Without the word "strictly", this is called "New better than used in expectation". Note that when X has decreasing residual expectation, then X is a continuous random variable. This is because if X gives positive mass to a point x_0, then $E[(X - a)|X > a]$ has a jump at the point x_0. More generally, one can show that X has decreasing density on its support, and its support is an interval (extending possibly to $+\infty$ but not to $-\infty$).

Theorem 4.1. *In the unanimous consent case, if the X_i are independent and have strictly decreasing residual expectation, then there exists a unique equilibrium in threshold strategies.*

Proof. *(Sketch of Proof)*
From (7), all threshold equilibria must satisfy

$$\frac{1}{c_k} E(X_k - a_k | X_k > a_k) = \frac{1}{P(A)} \qquad \text{for } k = 1, \ldots, m. \tag{8}$$

In particular, this means that all $E(X_k - a_k | X_k > a_k)/c_k$ are equal. From the hypothesis of strictly decreasing residual expectation, we may find for each θ sufficiently large and each $k = 1, \ldots, m$, a unique number $a_k(\theta)$ such that

$$\frac{1}{c_k} E(X_k - a_k(\theta) | X_k > a_k(\theta)) = \theta.$$

As θ decreases, each $a_k(\theta)$ increases strictly and continuously, until one or perhaps several of the $a_k(\theta)$ reach infinity. But as this occurs, $P(A(\theta))$ will decrease strictly and continuously to zero. Therefore there exists a unique value θ_0 such that $\theta_0 = 1/P(A(\theta_0))$. □

5 The Exponential Case

The exponential distribution is on the boundary of the set of distributions with decreasing residual expectation since the residual expectation is constant on the support. Suppose all the X_i have exponential distributions. Since the utilities of the players are determined only up to location and scale, we take without loss of generality all the distributions to have support $(0, \infty)$ and to have mean 1. Then

$$E(X_k - a_k | X_k > a_k) = \begin{cases} 1 & \text{if } a_k \geq 0, \\ 1 - a_k & \text{if } a_k < 0. \end{cases}$$

Suppose without loss of generality that the c_i are arranged in non-decreasing order. If there is a unique smallest c_i, that is if $c_1 < c_2$, then there is a unique solution to equations (7), and it has the property that a_2 through a_n are negative. If $c_1 < 1$, then a_1 is determined from (7) by

$$1 = \frac{c_1}{P(X_1 > a_1)} \qquad \text{or} \qquad a_1 = \log\left(\frac{1}{c_1}\right)$$

and for $i = 2, \ldots, n$,

$$1 - a_i = \frac{c_i}{P(X_1 > a_1)} \qquad \text{or} \qquad a_i = -\frac{c_i - c_1}{c_1}.$$

In this equilibrium, we see a well-known phenomenon. In committee decisions, the person who values time the least has a strong advantage. This is the committee member who is most willing to sit and discuss at length small details until the other members who have more useful ways of spending their time give in. In this example, Player 1 dominates the committee; all the other members accept the first candidate that is satisfactory to Player 1, who uses an optimal strategy as if the other players were not there. In equilibrium, this committee has the structure of a dictatorship.

If $c_1 > 1$ in this example, then all players accept the first candidate to appear. This is agreeable to all committee members, who are using an optimal strategy as if the other players were not there.

If all c_k are equal and less than 1, any set of a_i such that $\prod_{i=1}^{n}(1 - F_i(a_i))$ is equal to the common value of the c_k is in equilibrium. We see that without the condition that the distributions have strictly decreasing residual expectation, there may be a continuum of equilibria.

This phenomenon is not specific to the exponential distribution. If the X_i are i.i.d. with non-decreasing residual expectation on its support, and if $c_1 < \min\{c_2, \dots, c_m\}$, then there is an equilibrium with Player I using an optimal strategy as if the other players were not there, and the other players accepting any candidate agreeable to Player 1.

In the exponential case this equilibrium is the unique perfect equilibrium. For other distributions with non-decreasing residual expectation, this equilibrium may not be unique. Take for example the inverse power law with density $f(x) = \alpha x^{-(\alpha+1)}$ on $(1, \infty)$ with $\alpha = 3$ (so that $EX = 3/2$ and $E(X^2) < \infty$). Suppose $m = 2$, $c_1 = 1/12$, and $c_2 = 1/8$. The equilibrium where Player 1 dominates has equilibrium payoff $(\sqrt{6}, -\frac{3}{4}(2 - \sqrt{6})) = (2.4495, -.3471)$. There is another equilibrium where Player 2 dominates that has equilibrium payoff $(5/6, 2)$.

REFERENCES

[1] M. Kurano, M. Yasuda, and J. Nakagami. Multi-variate stopping problem with a majority rule. *J. Oper. Res. Soc. Jap.*, 23:205–223, 1980.

[2] M. Sakaguchi. Optimal stopping in sampling from a bivariate distribution. *J. Oper. Res. Soc. Jap.*, 16(3):186–200, 1973.

[3] M. Sakaguchi. A bilateral sequential game for sums of bivariate random variables. *J. Oper. Res. Soc. Jap.*, 21(4):486–507, 1978.

[4] M. Sakaguchi. When to stop: randomly appearing bivariate target values. *J. Oper. Res. Soc. Jap.*, 21:45–58, 1978.

[5] K. Szajowski and M. Yasuda. Voting procedure on stopping games of Markov chain. In Shunji Osaki, Anthony H. Christer and Lyn C. Thomas, editors, *UK-Japanese Research Workshop on Stochastic Modelling in Innovative Manufecuring, July 21–22, 1995*, volume 445 of *Lecture Notes in Economics and Mathematical Systems*, pages 68–80. Moller Centre, Churchill College, Univ.

Cambridge, UK, Springer, 1996. Springer Lecture Notes in Economics and Mathematical Systems.

[6] M. Yasuda, J. Nakagami, and M. Kurano. Multi-variate stopping problem with a monotone rule. *J. Oper. Res. Soc. Jap.*, 25:334–350, 1982.

Stopping Game Problem for Dynamic Fuzzy Systems

Yuji Yoshida
Faculty of Economics and Business Administration
The University of Kitakyushu
Kitakyushu 802-8577, Japan
yoshida@kitakyu-u.ac.jp

Masami Yasuda
Department of Mathematics and Informatics
Chiba University, Inage-ku
Chiba 263-8522, Japan;
yasuda@math.s.chiba-u.ac.jp

Masami Kurano
Department of Mathematics
Chiba University
Inage-ku, Chiba 263-8522, Japan
kurano@faculty.chiba-u.jp

Jun-ichi Nakagami
Department of Mathematics and Informatics
Chiba University
Inage-ku, Chiba 263-8522, Japan
nakagami@math.s.chiba-u.ac.jp

Abstract

A stopping game problem is formulated by cooperating with fuzzy stopping time in a decision environment. The dynamic fuzzy system is a fuzzification version of a deterministic dynamic system and the move of the game is a fuzzy relation connecting between two fuzzy states. We define a fuzzy stopping time using several degrees of levels and instances under a monotonicity property, then an "expectation" of the terminal fuzzy state via the stopping time. By inducing a scalarization function (a linear ranking function) as a payoff for the game problem we will evaluate the expectation of the terminal fuzzy state. In particular, a two-person zero-sum game is considered in case its state space is a fuzzy set and a payoff is ordered in a sense of the fuzzy max order. For both players, our aim is to find the equilibrium point of a payoff function. The approach depends on the interval analysis, that is, manipulating a class of sets arising from α-cut of fuzzy sets. We construct an equilibrium fuzzy stopping time under some conditions.

1 Introduction

Optimal stopping on stochastic processes contributes to an essential problem in the sequential decision problem. It is a simple and interesting one because its decision has only two forms, that is, stop or continue. Their applications by many authors are well-known to various fields, economics, engineering etc. The game version of the stopping problem was originated by Dynkin [2] and then Neveu [3] whose pioneering work is named *Dynkin Game*. See [13,11,5] for more references. The results are described clearly and are very attractive, however, we cannot avoid an uncertainty modeling the real problem.

On the other hand the fuzzy theory was founded by Zadeh [17] and then there have been many papers on applications and modeling for extending the results of the classical systems. For example, the fuzzy random variable was studied by Puri and Ralescu in [9].

Here we will discuss a stopping problem concerned with dynamical fuzzy systems. The main discussion is regarding the following two points: One is to define a game value in the zero-sum matrix game under fuzzification and the next is to formulate a fuzzy stopping game using this fuzzy game value for the sequence.

Dynamic fuzzy systems [6] are an extension of Markov decision processes induced by fuzzy configuration. That is, the transition law depending on the state and the action corresponds to a fuzzy relational equation.

In Section 1 the formulation of the fuzzy dynamic system (FDS) and the fuzzy stopping time (FST) are described in order to define a composition of FDS and FST. This base process corresponds to each player's payoff and will be evaluated using the following scalarization called a linear ranking function. In Section 2 we define a game value of a matrix whose elements are fuzzy numbers. A fuzzy stopping model is formulated in Section 3, provided by the previous notions. Also the equilibrium strategy of the model and its game value are obtained under a suitable assumption.

1.1 Preliminaries on Fuzzy Sets

A brief sketch of the notation using here is given as follows: A fuzzy set $\widetilde{a} = \widetilde{a}(x) : x \in \mathbb{R} \to [0, 1]$ on $\mathbb{R}$ is normal, upper semi-continuous, fuzzy convex such that $\widetilde{a}(\lambda x + (1 - \lambda)y) \geq \min\{\widetilde{a}(x), \widetilde{a}(y)\}$ for $\lambda \in [0, 1]$. It may be called a *fuzzy number* instead of fuzzy set because we are considering the real number of the space $\mathbb{R}$. The operation of sum $+$ and a scalar product $\cdot$ for fuzzy sets are defined by $(\widetilde{a} + \widetilde{b})(x) := \sup_{x=x_1+x_2}\{\widetilde{a}(x_1) \wedge \widetilde{b}(x_2)\}$ and $(\lambda \cdot \widetilde{a})(x) := \widetilde{a}(x/\lambda)$ if $\lambda > 0, := \mathbf{1}_{\{0\}}(x)$ if $\lambda = 0$, where $\wedge = \min$ and $\mathbf{1}_{\{\cdot\}}$ means the characteristic function. The α-cut ($\alpha \in [0, 1]$) of the fuzzy set $\widetilde{a}$ is denoted by

$$\widetilde{a}_\alpha := \{x \in E \mid \widetilde{a}(x) \geq \alpha\} \ (\alpha > 0) \tag{1}$$

and $\widetilde{a}_0 := \mathrm{cl}\{x \in E \mid \widetilde{a}(x) > 0\}$, where 'cl' denotes the closure of a set. We frequently use the α-cut in order to define the model and analyze an existence of strategies.

If the operations $+, \cdot$ for any non-empty closed intervals A, B in $\mathbb{R}$ are defined as $A + B := \{x + y \mid x \in A, y \in B\}$, $\lambda \cdot A := \{\lambda x \mid x \in A\}$ and especially $A + \emptyset = \emptyset + A := A$ and $\lambda \cdot \emptyset := \emptyset$, then the following two properties are known.

(1) (Interchanges) Interchanging the operation for fuzzy sets and α-cut is useful in the following discussion. $(\widetilde{a}+\widetilde{b})_\alpha = \widetilde{a}_\alpha + \widetilde{b}_\alpha$ and $(\lambda \cdot \widetilde{a})_\alpha = \lambda \cdot \widetilde{a}_\alpha$ $(\alpha \in [0, 1])$ holds.
(2) (Relation between α-cuts and a fuzzy set) The relation between a fuzzy set and its α-cut $\widetilde{a}(x) = \sup_\alpha \{\alpha \wedge \mathbf{1}_{\widetilde{a}_\alpha}(x)\}$, $x \in \mathbb{R}$, holds.

The following construction of a fuzzy set from a family of subsets was given by Zadeh [17]. So when the family of subsets is given, it can be constructed as a fuzzy set provided the condition are satisfied.

Proposition 1.1. *(Representation Theorem) For a given family of $\{M_\alpha\}$ in $\mathbb{R}$, if (i) $\alpha \le \beta \Rightarrow M_\alpha \supset M_\beta$, (ii) $\alpha_n \uparrow \alpha \Rightarrow M_\alpha = \bigcap_n M_{\alpha_n}$, then there exists a fuzzy set $\widetilde{M}(x)$ such that*

$$\widetilde{M}(x) = \sup_\alpha \{\alpha \wedge \mathbf{1}_{M_\alpha}(x)\} \tag{2}$$

for $x \in \mathbb{R}$.

There are special fuzzy sets, whose α-cuts become closed intervals as follows:

(i) Interval case: There exist two real numbers such that

$$\widetilde{a}_\alpha = \{x \in \mathbb{R} \mid \widetilde{a}_\alpha^- \le x \le \widetilde{a}_\alpha^+\},$$

where $\widetilde{a}_\alpha^+ = [\widetilde{a}]_\alpha^+ = \sup\{x \mid \widetilde{a}(x) \ge \alpha\}$ and $\widetilde{a}_\alpha^- = [\widetilde{a}]_\alpha^- = \inf\{x \mid \widetilde{a}(x) \ge \alpha\}$.
(ii) Triangular-type symmetric L-fuzzy number: (a) $L(x) = L(-x)$, (b) $L(x) = 1 \iff x = 0$, (c) $L(x) \downarrow 0$ $(x \nearrow \infty)$, (d) its support is finite.

For an example, if

$$\widetilde{a}(x) := \begin{cases} L((m-x)/k) & \text{if, } x \ge m \\ L((x-m)/k) & \text{otherwise,} \end{cases}$$

where $L(x) := \max\{1 - |x|, 0\}$, then its α-cut equals to

$$\widetilde{a}_\alpha = \{x \in \mathbb{R} \mid m - (1-\alpha)k \le x \le m + (1-\alpha)k\}.$$

The number m is called a center and k is a spread.

1.2 Fuzzy Dynamic System (FDS)

A *fuzzy dynamic system* is a sequence of fuzzy states generated by the pair of an initial state $\widetilde{s}(x)$ and a convex fuzzy relation $\widetilde{q}(x, y)$. The state space herein is $\mathbb{R} = (-\infty, \infty)$ of real numbers and an initial fuzzy set is a fuzzy number $\widetilde{s} = \widetilde{s}(x)$, $x \in \mathbb{R}$. These are assumed to be given.

Then a finite sequence $\{\widetilde{s}_t; t = 1, 2, \dots, N\}$ is generated by the fuzzy transition law Q recursively as

$$\begin{aligned} \widetilde{s}_1 &:= \widetilde{s}, \\ \widetilde{s}_{t+1} &:= Q(\widetilde{s}_t) := \sup_{x \in \mathbb{R}} \{\widetilde{s}(x) \wedge \widetilde{q}(x, y)\}, \end{aligned} \tag{3}$$

where $\widetilde{q} = \widetilde{q}(x, y) : \mathbb{R} \times \mathbb{R} \to [0, 1]$ is a convex fuzzy relation, that is, a fuzzy number defined by two variables, which satisfies

$$\widetilde{q}(\lambda x_1 + (1 - \lambda)x_2, \lambda y_1 + (1 - \lambda)y_2) \geq \widetilde{q}(x_1, y_1) \wedge \widetilde{q}(x_2, y_2) \tag{4}$$

for $x_1, x_2, y_1, y_2 \in \mathbb{R}$ and $\lambda \in [0, 1]$.

We call the sequence $\{\widetilde{s}_t, t = 1, 2, \dots, N\}$ a fuzzy dynamic system.

1.3 Fuzzy Stopping Time (FST)

A *fuzzy stopping time* $\tilde{\sigma} = \tilde{\sigma}(t)$ on a time-index set $t \in \{1, 2, \dots, N\}$ is

(a) a fuzzy number (set) on $\{1, 2, \dots, N\}$,

(b) and non-increasing, that is,

$$\tilde{\sigma}(t) \geq \tilde{\sigma}(s) \quad \text{for} \quad 1 \leq t < s \leq N \tag{5}$$

with $\tilde{\sigma}(1) = 1$. The interpretation of a fuzzy stopping time $\tilde{\sigma}$ means a degree of continuity, explicitly as the next three kinds

- $\tilde{\sigma}(t) = 1$ is to continue at time t,
- $\tilde{\sigma}(t) = \alpha$ $(0 < \alpha < 1)$ is an intensity of degree for continuity with level α and degree for stopping with level $1 - \alpha$,
- $\tilde{\sigma}(t) = 0$ is to stop at time t.

Since the real value $\tilde{\sigma}(t)$ for $t \in \{1, 2, \dots N\}$ should decrease with time, we impose the requirement of (b). By the definition of α-cut for $\tilde{\sigma}$, clearly

$$\tilde{\sigma}_\alpha = \{1, 2, \dots, \sigma_\alpha\} \tag{6}$$

provided that $\sigma_\alpha := \max\{t \in N \mid \tilde{\sigma}(t) \geq \alpha\}$ for $0 < \alpha$ and $\sigma_0 := \text{cl}\,\{t \in N \mid \tilde{\sigma}(t) > 0\}$ so that $\tilde{\sigma}_\alpha (0 \leq \alpha \leq 1)$ is a connected subset of $\{1, 2, \dots, N\}$.

1.4 Composition of *FDS* and *FST*

Now we consider, by using α-cut and then a representation theorem, a composition of $\{\widetilde{s}_t\}$, $\{\tilde{\sigma}(t)\}$ for $t = 1, 2, \ldots, N$ which are the fuzzy dynamic system (FDS) and the fuzzy stopping time(FST) respectively.

Because α-cut of a fuzzy set $\widetilde{s}$ on $\mathbb{R}$ equals a closed interval, denoted by the superscript $()^-$, $()^+$,

$$\widetilde{s}_\alpha = [\widetilde{s}_\alpha^-, \widetilde{s}_\alpha^+].$$

Conversely, if a family $[\widetilde{s}_\alpha^-, \widetilde{s}_\alpha^+]$, $0 \le \alpha \le 1$, of bounded closed sub-intervals in $\mathbb{R}$ is given, we can construct a fuzzy number $\widetilde{s} = \widetilde{s}(x)$, $x \in \mathbb{R}$ by

$$\widetilde{s}(x) := \sup_{\alpha \in [0,1]} \{\alpha \wedge \mathbf{1}_{[\widetilde{s}_\alpha^-, \widetilde{s}_\alpha^+]}(x)\}, \; x \in \mathbb{R}.$$

Definition 1.2. A composed fuzzy system $\widetilde{s}_{\tilde{\sigma}} = \widetilde{s}_{\tilde{\sigma}}(x)$, $x \in \mathbb{R}$ for a pair of a fuzzy dynamic system and a fuzzy stopping time $(\widetilde{s}_t, \tilde{\sigma}(t))$, $t = 1, 2, \ldots, N$, is defined in the following two steps:

Step 1. For each α, if α-cut of a fuzzy stopping time $\tilde{\sigma}$ is $\{1, 2, \ldots, t\}$, i.e. $\tilde{\sigma}_\alpha = \{1, 2, \ldots, t\}$, then we define α-cut of a composed fuzzy system $\widetilde{s}_{\tilde{\sigma}}$ by

$$(\widetilde{s}_{\tilde{\sigma}})_\alpha := \widetilde{s}_{\tilde{\sigma},\alpha} := \widetilde{s}_{t,\alpha} = [\widetilde{s}_{t,\alpha}^-, \widetilde{s}_{t,\alpha}^+]. \tag{7}$$

Step 2. By letting $S_\alpha := \widetilde{s}_{\tilde{\sigma},\alpha}$, $\alpha \in [0, 1]$, the presentation theorem is applied to a family of S_α:

$$\widetilde{s}_{\tilde{\sigma}}(x) := \sup_{\alpha \in [0,1]} \{\alpha \wedge \mathbf{1}_{S_\alpha}(x)\}, \; x \in \mathbb{R}. \tag{8}$$

Thus a composed fuzzy system $\widetilde{s}_{\tilde{\sigma}}$ of FDS $\{\widetilde{s}_t; t = 1, 2, \ldots, N\}$ and FST $\{\tilde{\sigma}(t); t = 1, 2, \ldots, N\}$ is obtained.

2 Game Value for Fuzzy Matrix Game

The usual sequential decision problems consist of several decisions but the simplest one is two cases, that is, Stopping problem – 2-decision (*Stop, Conti*). Considering the straightforward version in a two-person zero sum game for players PL_I(max), PL_{II}(min), one may adapt the following problem: When either of the player declares "stop", then the system stops and each can get rewards. In case of this stop rule the next three cases occur,

$$\text{Decision table of } (PL_I, PL_{II}) = \begin{pmatrix} (Stop, Stop) & (Stop, Conti) \\ (Conti, Stop) & Next \end{pmatrix}$$

and the corresponding value of the matrix game (zero sum) is

$$v_t = val \begin{pmatrix} r_{(S,S)} & r_{(S,C)} \\ r_{(C,S)} & v_{t+1} \end{pmatrix}, \quad t = 1, 2, \ldots$$

where v_t is a payoff at time t and val means a value of the matrix provided their payoffs $r_{(\cdot,\cdot)}$ are given.

From now we will consider the fuzzy version of this sequential decision problem. First define a value of matrix whose elements are fuzzy numbers.

2.1 Fuzzy Game Value

Definition 2.1. For a matrix $\widetilde{A}$ with each (i, j) element $\widetilde{a}_{ij} = \widetilde{a}_{ij}(x)$, $x \in \mathbb{R}$, is a fuzzy number, define

$$\widetilde{val}(\widetilde{A}) = \widetilde{val}(\widetilde{A})(x) = \widetilde{val} \begin{pmatrix} \widetilde{a}_{11} & \widetilde{a}_{12} \\ \widetilde{a}_{21} & \widetilde{a}_{22} \end{pmatrix} (x) \tag{9}$$

as a map $x \in \mathbb{R} \longmapsto [0, 1]$ in the following steps.

Step 1. For $0 \leq \alpha \leq 1$, let the α-cut of each element $\left(\widetilde{a}_{ij}\right)_\alpha := [a_{ij}^-, a_{ij}^+]$ and define the α-cut of the matrix by

$$\widetilde{A}_\alpha := \left((\widetilde{a}_{ij})_\alpha\right) = \begin{pmatrix} [a_{11}^-, a_{11}^+] & [a_{12}^-, a_{12}^+] \\ [a_{21}^-, a_{21}^+] & [a_{22}^-, a_{22}^+] \end{pmatrix}.$$

Step 2. For each i, j, two real numbers a_{ij}^-, $a_{ij}^+ \in (\widetilde{a}_{ij})_\alpha$, consider each value of the matrix game

$$val\left(a_{ij}^-\right) \quad \text{and} \quad val\left(a_{ij}^+\right)$$

as $\min_j \max_i \left(a_{ij}^\pm\right) = \max_i \min_j \left(a_{ij}^\pm\right)$ in the usual sense.

Step 3. Using these values, define a closed interval, denoted by $val(\widetilde{A}_\alpha)$, as

$$val(\widetilde{A}_\alpha) := \left[val\left(a_{ij}^-\right), val\left(a_{ij}^+\right)\right] = \left[val \begin{pmatrix} a_{11}^- & a_{12}^- \\ a_{21}^- & a_{22}^- \end{pmatrix}, val \begin{pmatrix} a_{11}^+ & a_{12}^+ \\ a_{21}^+ & a_{22}^+ \end{pmatrix}\right]$$

for each α.

Step 4. Construct a fuzzy set (number) $\widetilde{val}(\widetilde{A})$ on $\mathbb{R}$ from a family of $\{val(\widetilde{A}_\alpha); 0 \leq \alpha \leq 1\}$ as

$$\widetilde{val}(\widetilde{A})(x) := \sup_{\alpha \in [0,1]} \{\alpha \wedge \mathbf{1}_{val(\widetilde{A}_\alpha)}(x)\}$$

for $x \in \mathbb{R}$.

Proposition 2.1. *From the definition of* $\widetilde{val}$, *it holds that*

$$\left(\widetilde{val}(\widetilde{A})\right)_\alpha = val(\widetilde{A}_\alpha). \tag{10}$$

Proof. From the definition of the α-cut of $\widetilde{val}$, the result is immediately obtained. □

A *fuzzy max order* (Ramík and Řimánek [10]), between fuzzy numbers is a partial order which defined by the order for interval of α-cut in fuzzy numbers.

Definition 2.2. For two fuzzy numbers in $\mathbb{R}$, $\widetilde{b} \preceq \widetilde{a}$ if and only if the next inequality $\widetilde{b}_\alpha^- \leq \widetilde{a}_\alpha^-$ and $\widetilde{b}_\alpha^+ \leq \widetilde{a}_\alpha^+$ hold in both for all α.

Proposition 2.2. *(Special case with order of elements) If* $\widetilde{A} = \begin{pmatrix} \widetilde{a} & \widetilde{b} \\ \widetilde{c} & \widetilde{d} \end{pmatrix}$ *with*

$$\widetilde{b} \preceq \widetilde{a} \preceq \widetilde{c} \tag{11}$$

where $\preceq$ *means a fuzzy max order, then*

$$\widetilde{val}(\widetilde{A}) = \begin{cases} \widetilde{b} & \text{if } \ \widetilde{d} \preceq \widetilde{b} \\ \widetilde{d} & \text{if } \ \widetilde{b} \preceq \widetilde{d} \preceq \widetilde{c} \\ \widetilde{c} & \text{if } \ \widetilde{c} \preceq \widetilde{d}. \end{cases} \tag{12}$$

Proof. The proof depends on the usual case that a matrix game with each element is a real number. Because of assumption (11) the game has a pure strategy and the element "$\widetilde{a}$" / (*Stop,Stop*) does not become an equilibrium. So this result is extended easily to the interval version and this fuzzy value. □

2.2 A Linear Ranking Function (Scalarization)

In this section we will discuss the evaluation for a fuzzy set. This leads us to define an objective function and an equilibrium strategy for each player.

The first is to consider a function of the scalarization (evaluation) from an interval to a real number. Let a map g from an interval in $\mathbb{R}$ to $\mathbb{R}$ which satisfies

(i) $g(A + B) = g(A) + g(B)$,
(ii) $g(\lambda A) = \lambda g(A)$, $\lambda > 0$,
(iii) $A = [a_1, a_2] \Rightarrow a_1 \leq g([a_1, a_2]) \leq a_2$.

This map is called a *linear ranking function* (Fortemps and Roubens [4]) and it is adapted to the correspondence from the α-cut of a fuzzy number to its reduced-scalar.

Lemma 2.1. *The following three assertions are equivalent.*

(a) A map g is a linear ranking function.

(b) (Affine property) For $\lambda \geq 0,\ \mu,$

$$g(\lambda[0,1]+\mu) = \lambda g([0,1]) + \mu.$$

(c) By letting $k := g([0,1])$,

$$g([a_1, a_2]) = a_1(1-k) + a_2 k.$$

Lemma 2.2. *If* $\widetilde{a} \preceq \widetilde{b}$, *then* $\|\widetilde{a}\|_g \leq \|\widetilde{b}\|_g$ *where*

$$\|\widetilde{a}\|_g := \int_0^1 g(\widetilde{a}_\alpha)\, d\alpha. \tag{13}$$

3 A Fuzzy Stopping Game and Equilibrium Strategies

Our stopping model for the zero-sum case is based on FDS in Section 1.2 and the linear ranking function in Section 2.2 for its evaluation in order to define an objective function and discuss an equilibrium strategy. The data are generated sequentially to define the model by fuzzy transition laws Q, FDS $\{\widetilde{r}^t_{(\cdot,\cdot)};\ t = 1, 2, \ldots, N\}$. A stopping strategy is a pair of FSTs $\tilde{\sigma}_I$, $\tilde{\sigma}_{II}$ defined by (5). Associated with a stopping strategy, we consider a payoff function and thus an equilibrium of the integral (16).

(1) Initial fuzzy state defined on $\mathbb{R}$:

$$\widetilde{r}_{(S,S)},\ \widetilde{r}_{(C,S)},\ \widetilde{r}_{(S,C)}$$

where $\widetilde{r}_{(\cdot,\cdot)} = \widetilde{r}_{(\cdot,\cdot)}(x)$, $x \in \mathbb{R}$.

(2) Fuzzy translation law:

$$Q_{(S,S)},\ Q_{(C,S)},\ Q_{(S,C)}$$

where fuzzy translation laws $Q_{(\cdot,\cdot)}$ are generated by some convex fuzzy relations $\widetilde{q}_{(\cdot,\cdot)}(x, y),\ x, y \in \mathbb{R}$.

(3) FDS $\{\widetilde{r}^t_{(\cdot,\cdot)};\ t = 1, 2, \ldots, N\}$ generated by Q similar to (3):

$$\widetilde{r}^t_{(S,S)},\ \widetilde{r}^t_{(C,S)},\ \widetilde{r}^t_{(S,C)}$$

where

$$\begin{cases} \widetilde{r}^t_{(\cdot,\cdot)} & := Q_{(\cdot,\cdot)}\widetilde{r}^{t-1}_{(\cdot,\cdot)}, \quad t = 2, 3, \ldots, N, \\ \widetilde{r}^1_{(\cdot,\cdot)} & := \widetilde{r}_{(\cdot,\cdot)}. \end{cases} \tag{14}$$

(4) Stopping strategy: Two FSTs $\tilde{\sigma}_I$, $\tilde{\sigma}_{II}$ for each player defined by (5).

(5) Payoff function: $\widetilde{R}(\tilde{\sigma}_I, \tilde{\sigma}_{II})$ whose α-cuts are

$$\widetilde{R}(\tilde{\sigma}_I, \tilde{\sigma}_{II})_\alpha := \begin{cases} \widetilde{r}^t_{(C,S),\alpha} & \text{if} \quad t = \tilde{\sigma}_{I,\alpha} \leq \tilde{\sigma}_{II,\alpha}, \\ \widetilde{r}^t_{(S,C),\alpha} & \text{if} \quad t = \tilde{\sigma}_{II,\alpha} \leq \tilde{\sigma}_{I,\alpha}, \\ \widetilde{r}^t_{(S,S),\alpha} & \text{if} \quad t = \tilde{\sigma}_{I,\alpha} = \tilde{\sigma}_{II,\alpha}. \end{cases} \tag{15}$$

(6) Objective functions:

$$\min_{\tilde{\sigma}_{II}} \max_{\tilde{\sigma}_I} \|\widetilde{R}(\tilde{\sigma}_I, \tilde{\sigma}_{II})\|_g \text{ and } \max_{\tilde{\sigma}_I} \min_{\tilde{\sigma}_{II}} \|\widetilde{R}(\tilde{\sigma}_I, \tilde{\sigma}_{II})\|_g$$

where

$$\|\widetilde{R}(\tilde{\sigma}_I, \tilde{\sigma}_{II})\|_g := \int_0^1 g(\widetilde{R}(\tilde{\sigma}_I, \tilde{\sigma}_{II})_\alpha) \mathrm{d}\alpha \tag{16}$$

with a given linear ranking function g similar to (13).

Thus we have defined a stopping game problem for dynamic fuzzy systems and we shall look for the equilibrium of (16). In order to avoid analytical difficulty, the game values for the zero-sum matrix are restricted within only pure strategies, that is, this is enough to consider the class of FST. Refer to [13]. Explicitly we need the following assumption, which is assumed in several papers [3,5] as

Assumption 3.1. (Dynkin Game) For each t,

$$\widetilde{r}^t_{(S,C)} \preceq \widetilde{r}^t_{(S,S)} \preceq \widetilde{r}^t_{(C,S)}. \tag{17}$$

Consider the next in backward induction

$$\widetilde{v}_t = \widetilde{val} \begin{pmatrix} \widetilde{r}^t_{(S,S)} & \widetilde{r}^t_{(S,C)} \\ \widetilde{r}^t_{(C,S)} & \widetilde{v}_{t+1} \end{pmatrix}, \qquad t = N-1, \ldots, 2, 1, \tag{18}$$

by $\widetilde{val}$ in a fuzzy sense of (9) and

$$\widetilde{v}_N := \widetilde{r}^N_{(S,S)}.$$

Lemma 3.1. *(i) For $t = N-1, \ldots, 2, 1$, and each α,*

$$g(\widetilde{v}_{t,\alpha}) = val \begin{pmatrix} g(\widetilde{r}^t_{(S,S),\alpha}) & g(\widetilde{r}^t_{(S,C),\alpha}) \\ g(\widetilde{r}^t_{(C,S),\alpha}) & g(\widetilde{v}_{t+1,\alpha}) \end{pmatrix} \tag{19}$$

by the scalarization. Here val means in the normal usage.
(ii) For $t = N-1, \ldots, 2, 1$,

$$\|\widetilde{v}_t\|_g = val \begin{pmatrix} \|\widetilde{r}^t_{(S,S)}\|_g & \|\widetilde{r}^t_{(S,C)}\|_g \\ \|\widetilde{r}^t_{(C,S)}\|_g & \|\widetilde{v}_{t+1}\|_g \end{pmatrix}$$

$$= \begin{cases} \|\widetilde{r}^t_{(S,C)}\|_g & \text{if} \quad \|\widetilde{v}_{t+1}\|_g \leq \|\widetilde{r}^t_{(S,C)}\|_g \\ \|\widetilde{v}_{t+1}\|_g & \text{if} \quad \|\widetilde{r}^t_{(S,C)}\|_g \leq \|\widetilde{v}_{t+1}\|_g \leq \|\widetilde{r}^t_{(C,S)}\|_g \\ \|\widetilde{r}^t_{(C,S)}\|_g & \text{if} \quad \|\widetilde{r}^t_{(C,S)}\|_g \leq \|\widetilde{v}_{t+1}\|_g. \end{cases} \tag{20}$$

Definition 3.1. For each α,

$$\begin{aligned} \sigma^*_{I,\alpha} &:= \inf\{1 \le t \le N \mid g(\widetilde{v}_{t,\alpha}) \le g(\widetilde{r}^t_{(S,C),\alpha})\}, \\ \sigma^*_{II,\alpha} &:= \inf\{1 \le t \le N \mid g(\widetilde{v}_{t,\alpha}) \ge g(\widetilde{r}^t_{(C,S),\alpha})\}, \end{aligned} \tag{21}$$

and define

$$\begin{aligned} \tilde{\sigma}^*_I = \tilde{\sigma}^*_I(t) &:= \sup_{\alpha\in[0,1]} \{\alpha \wedge \mathbf{1}_{[0,\sigma^*_{I,\alpha}]}(t)\}, \\ \tilde{\sigma}^*_{II} = \tilde{\sigma}^*_{II}(t) &:= \sup_{\alpha\in[0,1]} \{\alpha \wedge \mathbf{1}_{[0,\sigma^*_{II,\alpha}]}(t)\} \end{aligned} \tag{22}$$

for $t = 1, 2, \dots, N$.

The next assumption is too technical. However we have to induce the class of strategy under the fuzzy configuration as a class of FST that needs to be well defined as (5).

Assumption 3.2. (Regularity of strategy) Each epoch of $\sigma^*_{I,\alpha}, \sigma^*_{II,\alpha}$ in (21) decrease monotonically in $\alpha \in [0, 1]$.

Theorem 3.1. *Under Assumptions 3.1 and 3.2,*

$$\sup_{\tilde{\sigma}_I} \inf_{\tilde{\sigma}_{II}} \|\widetilde{R}(\tilde{\sigma}_I, \tilde{\sigma}_{II})\|_g = \inf_{\tilde{\sigma}_{II}} \sup_{\tilde{\sigma}_I} \|\widetilde{R}(\tilde{\sigma}_I, \tilde{\sigma}_{II})\|_g \tag{23}$$

holds and its equilibrium strategy $\tilde{\sigma}^*_{I,\alpha}, \tilde{\sigma}^*_{II,\alpha}$ for each player satisfies

$$\|\widetilde{R}(\tilde{\sigma}^*_I, \tilde{\sigma}^*_{II})\|_g = \|\widetilde{v}_1\|_g. \tag{24}$$

Proof. The proof is immediately obtained from the previous assumptions and lemmas. □

Remark 3.1. Here we do not show a concrete example, however, there are many examples in the crisp case, that is, $\alpha = 1$. Thus the assumptions are satisfied in the ordinary case.

Remark 3.2. An infinitely planned horizon case, should be considered a fixed point concerned with the fuzzy relational equation (18) of $\widetilde{val}$. Details are not discussed in this paper.

REFERENCES

[1] Aubin J. P., *Mathematical Methods of Game and Economic Theory* (North-Holland, Amsterdam, 1979).

[2] Dynkin E. B., Game variant of a problem and optimal stopping *Soviet Math. Dokl.* **10** (1969) 270–274.

[3] Neveu N., *Discrete-parameter Martingales* (North-Holland, Amsterdam, 1975).

[4] Fortemps P. and M. Roubens, Ranking and defuzzification methods based on area compensation, *Fuzzy Sets and Systems* **82** (1996) 319–330.

[5] Karatzas I., A pathwise approach to Dynkin game, *IMS Lecture Notes-Monograph Series* **30** (1996) 115–125.

[6] Kurano M., M. Yasuda, J. Nakagami and Y. Yoshida, A limit theorem in some dynamic fuzzy systems, *Fuzzy Sets and Systems* **51** (1992) 83–88.

[7] Kurano M., M. Yasuda, J. Nakagami and Y. Yoshida, An approach to stopping problems of a dynamic fuzzy system, preprint.

[8] Kurano M., M. Yasuda, J. Nakagami and Y. Yoshida, Ordering of fuzzy sets – A brief survey and new results, *J. Oper. Res. Soc. of Japan* **43** (2000) 138–148.

[9] Puri M. L. and D.A. Ralescu, Fuzzy random variables, *J. Math. Anal. Appl.* **114** (1986) 409–422.

[10] Ramík J. and J. Řimánek, Inequality relation between fuzzy numbers and its use in fuzzy optimization, *Fuzzy Sets and Systems* **16** (1985) 123–138.

[11] Szajowski K. and M. Yasuda, Voting procedure on stopping games of Markov Chain, *Springer LN in Eco. and Math System* **445** (1997) 68–80.

[12] Wang G. and Y. Zhang, The theory of fuzzy stochastic processes, *Fuzzy Sets and Systems* **51** (1992) 161–178.

[13] Yasuda M., On a randomized strategy in Neveu's stopping problem, *Stochastic Processes and their Applications* **21** (1985) 159–166.

[14] Yoshida Y., M. Yasuda, J. Nakagami and M. Kurano, A monotone fuzzy stopping time for dynamic fuzzy systems, *Bull. Infor. Cyber. Res. Ass. Stat. Sci., Kyushu University* **31** (1999) 91–99.

[15] Yoshida Y., Markov chains with a transition possibility measure and fuzzy dynamic programming, *Fuzzy Sets and Systems* **66** (1994) 39–57.

[16] Yoshida Y., M. Yasuda, J. Nakagami and M. Kurano, Optimal stopping problems of a stochastic and fuzzy system, *J.Math.Anal.Appli.*, **246**(2000) 135–149.

[17] Zadeh L. A., Fuzzy sets, *Inform. and Control* **8** (1965) 338–353.

On Randomized Stopping Games

Elżbieta Z. Ferenstein
Faculty of Mathematics and Information Science
Warsaw University of Technology
00-661 Warsaw, Poland
efer@mini.pw.edu.pl
and
Polish-Japanese Institute of Information Technologies
Koszykowa 86, 02-008 Warsaw, Poland
elzbieta.ferenstein@pjwstk.edu.pl

Abstract

The paper is concerned with two-person nonzero-sum stopping games in which pairs of randomized stopping times are game strategies. For a general form of reward functions, existence of Nash equilibrium strategies is proved under some restrictions for three types of games: quasi-finite-horizon, random-horizon and infinite-horizon games.

1 Introduction and Preliminaries

Let $(\Omega, \mathcal{F}, P)$ be a probability space and $\{\mathcal{F}_n, n \in N\}$ an increasing sequence of sub-σ-fields of $\mathcal{F}$, $N = \{1, 2, \dots\}$. There are given two sequences of trivariate random variables $\{(X_n^i, Y_n^i, W_n^i), n \in N\}$, $i = 1, 2$, defined on $(\Omega, \mathcal{F}, P)$ and $\{\mathcal{F}_n, n \in N\}$ adapted. They represent players' rewards associated with their appropriate decisions.

We will need the conditions

(A1) $E\Big(\sup\limits_{n \in N} |X_n^i|\Big) < \infty, \; E\Big(\sup\limits_{n \in N} |Y_n^i|\Big) < \infty,$

(A2) $E\Big(\sup\limits_{n \in N} |W_n^i|\Big) < \infty$, for $i = 1, 2$.

Let M denote the set of stopping times τ with respect to $\{\mathcal{F}_n, n \in \overline{N}\}$, $\overline{N} = N \cup \{\infty\}$, $\mathcal{F}_\infty = \sigma(\{\mathcal{F}_n, n \in N\})$, i.e. $\tau : \Omega \to \overline{N}$ and $\{\omega \in \Omega : \tau(\omega) = n\} \in \mathcal{F}_n$, for any $n \in \overline{N}$. Ohtsubo (1987, 1991) considered the following two-person nonzero-sum non-cooperative stopping game. There are two players who choose stopping times $\tau_1 \in M$ and $\tau_2 \in M$ as their strategies respectively. Then, the reward for the player i, $i = 1, 2$, is given by

$$\begin{aligned} Z^i(\tau_1, \tau_2) = {} & X_{\tau_i}^i I_{\{\tau_i < \tau_j\}} + Y_{\tau_j}^i I_{\{\tau_j < \tau_i\}} + \\ & + W_{\tau_i}^i I_{\{\tau_i = \tau_j < \infty\}} + \limsup_{n \to \infty} W_n^i I_{\{\tau_1 = \tau_2 = \infty\}}, \end{aligned} \tag{1}$$

where I_A denotes the indicator function of the set A in $\mathcal{F}$. The aim of each of the players is to make his mean reward as large as possible. So, they look for a Nash equilibrium strategy $(\widehat{\tau}_1, \widehat{\tau}_2) \in M \times M$ of the game $\mathcal{G} = (M \times M, V^1, V^2)$ presented in a normal form, where players' payoff functions V^i are their mean rewards

$$V^i(\tau_1, \tau_2) = E(Z^i(\tau_1, \tau_2)), \qquad i = 1, 2. \tag{2}$$

Thus, for any strategy (τ_1, τ_2) from $M \times M$ we have

$$V^1(\widehat{\tau}_1, \widehat{\tau}_2) \geq V^1(\tau_1, \widehat{\tau}_2) \quad \text{and} \quad V^2(\widehat{\tau}_1, \widehat{\tau}_2) \geq V^2(\widehat{\tau}_1, \tau_2).$$

The game $\mathcal{G}$ is a generalization of the Dynkin stopping problem (1969) and its Neveu modification (1975). Dynkin considered a zero-sum stopping game with some constraints on stopping times: the player 1 (player 2) could choose only stopping times with odd (even) values. Many variations of Dynkin's game were investigated in a series of papers. A broad survey of stopping games is given in Nowak and Szajowski (1999). In general, without some special assumptions on reward sequences $\{(X_n^i, Y_n^i, W_n^i), n \in N\}$, $i = 1, 2$, the game $\mathcal{G}$ may not have a Nash equilibrium strategy. For instance, let us consider the following conditions, for $i = 1, 2$, $n \in N$,

(C3) $X_n^i \geq W_n^i \geq Y_n^i$, a.s.,

(C4) $X_n^i \leq W_n^i \leq Y_n^i$, $X_n^i < Y_n^i$, a.s.,

(C5) $A_n^i \subset A_n^{i+1}$, where $A_n^i = \{\omega \in \Omega : X_n^i \geq E(X_{n+1}^i | \mathcal{F}_n)\}$,

(C6) $X_n^1 = X_n^2 = W_n^1$, $Y_n^1 = Y_n^2 = W_n^2$, a.s.

Ohtsubo (1987, 1991) obtained existence of a Nash equilibrium strategy for each of the below cases:

(a) Assumptions (A1) and (C3).
(b) Assumptions (A1), (C4) and (C5).
(c) Assumptions $(A1)$, $(C4)$ and an assumption on the existence of a sequence of bivariate random variables satisfying some kind of a Bellman's type equation in the two-person game situation.

Ferenstein (1992, 1993, 2001) obtained existence of an equilibrium strategy under Assumptions (A1), (A2) and (C6) which is a constraint on a structure of players' rewards (and an additional assumption that subsequent rewards are functions of a homogeneous Markov chain). Existence results for the finite-horizon game with reward sequences (C6) were established in Enns and Ferenstein (1987).

Bobecka and Ferenstein (2001) proved that one may weaken the assumptions by Ohtsubo (1987) to obtain existence of a Nash solution. Namely, it suffices to assume (C3) and for $n \in N$, $i = 1, 2$,

$$E|X_n^i| < \infty, \quad E|Y_n^i| < \infty \quad \text{and} \quad E\left(\sup_{n \in N} (X_n^i)^+\right) < \infty,$$

instead of the strong Assumption (A1).

Without assumptions of type (C3)–(C6), one may get existence theorems for a larger class of game strategies, namely randomized stopping times. For such strategies, Yasuda (1986) obtained existence of a Nash equilibrium of a zero-sum game assuming additionally that rewards are discounted in time. Recently, Rosenberg et al. (2001) proved that one may skip discounting and that still the zero-sum randomized stopping game has a value under the single integrability Conditions (A1), (A2). In this paper we will obtain existence theorems for nonzero-sum randomized stopping games of three types: some generalization of finite-horizon games (called quasi-finite-horizon games), games with random horizons, and infinite-horizon games with reward sequences approaching zero as n tends to infinity. In a quasi-finite-horizon case we assume that players' strategies are suitable subclasses of randomized stopping times. In all above cases we assume that $\mathcal{F}_n$, $n \in N$, are generated by at most countable sets of events. To prove our theorems we will use a general result of game theory, namely the following theorem (Nash (1951), Fan (1966)).

Theorem 1.1. *Suppose that for a two-person non-zero-sum game* $\mathcal{G} = (S_1 \times S_2, H_1, H_2)$ *the following assumptions are fulfilled:*

(i) S_1, S_2 *are nonempty compact convex subsets of a linear separated topological space* E.
(ii) Functions $H_i : S_1 \times S_2 \to R$, $i = 1, 2$, *are continuous.*
(iii) $H_1(\cdot, s_2)$, $H_2(s_1, \cdot)$ *are quasi-concave, for any* $s_j \in S_j$, $j = 1, 2$.

Then, the game $\mathcal{G}$ *has a Nash equilibrium strategy.*

2 The Model and Main Theorems

For randomized stopping games investigated in this paper we admit two representations of players' strategies: randomized stopping strategies or randomized stopping times. We assume in the sequel, without loss of generality, that the underlying probability space $(\Omega, \mathcal{F}, P)$ on which all considered random variables are defined is rich enough to allow randomization.

Denote by Λ the set of randomized stopping times. Let us recall the notion of randomized stopping time as in Chow et al. (1971), similarly in Ferguson (1967). Namely, $\tau \in \Lambda$ iff there exists a filtration $\{\mathcal{U}_n\}_{n \in \overline{N}}$ such that the conditions (a)–(c) below are satisfied.

(a) $\mathcal{F}_n \subset \mathcal{U}_n$, for $n \in \overline{N}$,
(b) $P(A \mid \mathcal{F}_n) = P(A \mid \mathcal{U}_n)$, a.s., $A \in \mathcal{F}_\infty$, $n \in \overline{N}$,
(c) τ is a stopping time with respect to $\{\mathcal{U}_n\}_{n \in \overline{N}}$.

Players observe subsequent events in $\mathcal{F}_n$, $n \in N$, and their decisions either to quit the game or to continue it are independent. Thus, let the set of game strategies $\Lambda \times \Lambda$ contain pairs of randomized stopping times τ_1, τ_2 which are conditionally independent given $\mathcal{F}_n$, for any $n \in N$. Under the strategy (τ_1, τ_2) the payoff for the player i is $V^i(\tau_1, \tau_2)$, defined by (1) and (2), $i = 1, 2$.

Let S be the set of randomized stopping strategies, i.e. the set of $\{\mathcal{F}_n\}_{n\in\overline{N}}$ adapted random sequences $s = \{p_n\}_{n\in\overline{N}}$ such that

$$0 \le p_n \le 1 \quad \text{and} \quad \sum_{n\in\overline{N}} p_n = 1, \quad \text{a.s.} \tag{3}$$

For an randomized stopping strategy $s = \{p_n\}_{n\in\overline{N}}$, p_n is to be interpreted as conditional probability that stopping occurs at time n, given the observations in $\mathcal{F}_n$.

Let $(s_1, s_2) \in S \times S$, $s_i = \{p_n^i\}_{n\in\overline{N}}$, $i = 1, 2$. Then, let us define the player's i payoff as follows

$$H_i(s_1, s_2) = E\Bigg(\sum_{n\in N} (X_n^i p_n^i (1 - p_1^j - \ldots - p_n^j) +$$
$$+ Y_n^i p_n^j (1 - p_1^i - \ldots - p_n^i)) + \sum_{n\in\overline{N}} W_n^i p_n^i p_n^j \Bigg), \qquad i \neq j, \tag{4}$$

where $W_\infty^i = \limsup_{n\to\infty} W_n^i I_{\{\tau_1=\tau_2=\infty\}}$.

Proposition 2.1. (i) *For $(s_1, s_2) \in S \times S$ there exists $(\tau_1, \tau_2) \in \Lambda \times \Lambda$ such that*

$$H_i(s_1, s_2) = V^i(\tau_1, \tau_2), \qquad i = 1, 2. \tag{5}$$

(ii) *For $(\tau_1, \tau_2) \in \Lambda \times \Lambda$ there exists $(s_1, s_2) \in S \times S$ such that the equality (5) is fulfilled.*

Proof. (i) Construction of randomized stopping times τ_1, τ_2 corresponding to s_1, s_2 is presented in Yasuda (1986) (also, similarly in Irle (1995) and Rosenberg et al. (2001)). Let us suppose, without loss of generality, that there are independent random sequences $\{A_n^1\}_{n\in N}$, $\{A_n^2\}_{n\in N}$ of i.i.d. random variables determined on $(\Omega, \mathcal{F}, P)$ with uniform distributions on [0,1] and independent on $\mathcal{F}_\infty$. Let $\mathcal{U}_n$ be the σ-field generated by $\mathcal{F}_n, A_1^1, \ldots, A_n^1, A_1^2, \ldots, A_n^2$. For a strategy $s_i = \{p_n^i\}_{n\in\overline{N}} \in S$ let us define $\widetilde{p}_n^i = p_n^i / \sum_{k\ge n} p_k^i$, $n \in N$, $i = 1, 2$, and stopping times τ_i with respect to the filtration $\{\mathcal{U}_n,\ n \in \overline{N}\}$ as follows

$$\tau_i = \inf\{n \ge 1 : A_n^i \le \widetilde{p}_n^i\}.$$

Then, $\tau_i, i = 1, 2$, are randomized stopping times since (a)–(c) are easily fulfilled. For any $n \in N$ we have, a.s.,

$$P(\tau_i = n \mid \mathcal{F}_\infty) = P(\tau_i = n \mid \mathcal{F}_n) = p_n^i.$$

Hence, from (1), one may calculate

$$V^i(\tau_1, \tau_2) = E(Z^i(\tau_1, \tau_2)) = H_i(s_1, s_2).$$

(ii) Let $(\tau_1, \tau_2) \in \Lambda \times \Lambda$. For $i = 1, 2$, define $s_i = \{p_n^i\}_{n\in\overline{N}}$ so that

$$p_n^i = P(\tau_i = n|\mathcal{F}_n), \text{ a.s.}, \qquad n \in N.$$

Now, (3) is satisfied so $s_i \in S$. $\{\mathcal{F}_n\}_{n\in\overline{N}}$ conditional independence of τ_1, τ_2 gives us (5). □

Because of Proposition 2.1 we may analyze games with players' strategies either from $\Lambda \times \Lambda$ or $S \times S$ depending on the convenience. First, we will examine the game for which probabilities of stopping at n, for sufficiently large n, are bounded from above. It may be called a quasi-finite-horizon randomized stopping game since it is natural to introduce the following definition.

Definition 2.1. Let $r = \{r_n\}_{n\in\overline{N}}$ be a sequence of random variables $\{\mathcal{F}_n\}_{n\in\overline{N}}$ adapted with finite second moments and such that $E\left(\sum_{n=1}^{\infty} r_n^2\right) < \infty$. Let, for some given natural L, the subset of randomized stopping strategies $S_r^L \subset S$ be as follows

$$S_r^L = \{s = \{p_n\}_{n\in\overline{N}} \in S : p_n \le r_n, \text{ a.s.}, \quad \text{for} \quad n \ge L\}.$$

The game $(S_r^L \times S_r^L, H_1, H_2)$ is called quasi-finite-horizon randomized stopping game.

Existence of a Nash equilibrium of a quasi-finite-horizon randomized stopping game is obtained below under the following assumption on the observed filtration.

(A3) For any $n \in N$, $\mathcal{F}_n$ is generated by at most countable set of events $\{B_1^n, \dots, B_{k_n}^n\}$ from $\mathcal{F}$, $k_n \le \infty$.

Theorem 2.1. *Suppose that Assumptions (A1), (A2), (A3) are fulfilled. Then, the game $(S_r^L \times S_r^L, H_1, H_2)$ has a Nash equilibrium strategy.*

Similar game to the quasi-finite-horizon one is a random-horizon game for which we have the following result

Theorem 2.2. *Suppose that Assumptions (A1), (A2), (A3) are fulfilled. Let K be an nonnegative integer valued random variable, independent on $\{\mathcal{F}_n\}_{n\in N}$, and $E(K) < \infty$. Suppose that players' rewards and payoffs, for $(\tau_1, \tau_2) \in \Lambda \times \Lambda$, $i = 1, 2$, are defined as follows*

$$Z_K^i(\tau_1, \tau_2) = X_{\tau_i}^i I_{\{\tau_i < \tau_j, \tau_i \le K\}} + Y_{\tau_j}^i I_{\{\tau_j < \tau_i, \tau_j \le K\}} + W_{\tau_i}^i I_{\{\tau_i = \tau_j \le K\}},$$

$$V_K^i(\tau_1, \tau_2) = E(Z_K^i(\tau_1, \tau_2)).$$

Then, the game $(\Lambda \times \Lambda, V_K^1, V_K^2)$ has a Nash equilibrium strategy.

For an infinite-horizon case we have

Theorem 2.3. *Suppose that Assumptions* (A1), (A2), (A3) *are satisfied and the sequences $\{X_n^i\}_{n\in N}$, $\{Y_n^i\}_{n\in N}$, $\{W_n^i\}_{n\in N}$ tend to 0 as $n \to \infty$, in probability, for $i = 1, 2$. Then, the game $(\Lambda \times \Lambda, V^1, V^2)$ has a Nash equilibrium strategy.*

3 Proofs of the Theorems

In the proofs of our theorems we will show that the assumptions of Theorem 1.1 are fulfilled.

Let E denote the space of random sequences $\{\xi_n\}_{n\in\overline{N}}$ which are $\{\mathcal{F}_n\}_{n\in\overline{N}}$ adapted and have finite expectations of sums of squares of their elements: $E\left(\sum_{n\in\overline{N}}\xi_n^2\right) < \infty$. E equipped with the inner product $\langle\xi,\eta\rangle = \sum_{n\in\overline{N}}\xi_n\eta_n$ is a Hilbert space.

Note that randomized stopping strategy sets introduced in the previous section satisfy inclusions $S_r^L \subset S \subset E$. First, we will state results needed in the proof of Theorem 2.2.

Lemma 3.1. *Let Assumption (A3) be satisfied. Then, S_r^L is compact in E.*

Proof. Let $r = \{r_n\}_{n\in\overline{N}} \in E$, L be fixed natural. S_r^L is a bounded subset of E since for $p = \{p_n\}_{n\in\overline{N}} \in S_r^L$ we have

$$\|p\|^2 = E\left(\sum_{n\in\overline{N}} p_n^2\right) \le E\left(\sum_{n\in\overline{N}} p_n\right) = 1.$$

Moreover, it will occur easy to show that S_r^L is closed. Let $p^k = \{p_n^k\}_{n\in\overline{N}} \in S_r^L$, $k = 1, 2, \ldots$, and $p = \{p_n\}_{n\in\overline{N}} \in E$ be such that

$$\|p^k - p\|^2 = E\left(\sum_{n\in\overline{N}} (p_n^k - p_n)^2\right) \to 0 \quad \text{as} \quad k \to \infty.$$

Hence, there exists subsequence of naturals $\{k_j\}$ such that $\sum_{n\in\overline{N}}(p_n^{k_j} - p_n)^2 \to 0$ as $j \to \infty$, a.s., so for any n we have $p_n^{k_j} \to p_n$ as $j \to \infty$, a.s. Therefore, p_n is $\mathcal{F}_n$-measurable, $0 \le p_n \le 1$ and $p_n \le r_n$ for $n \ge L$, a.s. To prove that $p \in S_r^L$ it remains to show that $\sum_{n\in\overline{N}} p_n = 1$, a.s. For any j and M we may write

$$p_1^{k_j} + \ldots + p_M^{k_j} + \sum_{M<n\in\overline{N}} p_n^{k_j} = 1,$$

which gives us, a.s.,

$$p_1 + \ldots + p_M + \limsup_{j\to\infty} \sum_{M<n\in\overline{N}} p_n^{k_j} = 1$$

and in consequence $\sum_{n\in\overline{N}} p_n = 1$ since the limit above is $\sum_{M<n\in\overline{N}} p_n$ for $0 \le p_n^{k_j} \le r_n, n \ge L, \sum_{n\in\overline{N}} r_n < \infty$, a.s., and M may be taken arbitrarily large.

Now, let us note that S_r^L is weakly compact in E since it is a bounded closed and convex subset of E.

To show that S_r^L is compact it is sufficient to prove that weak convergence implies the convergence is norm in E. So, suppose that $p^k = \{p_n^k\}_{n\in\overline{N}} \in S_r^L$ and p^k converges weakly to $p = \{p_n\} \in S_r^L$ as $k \to \infty$. Thus, for any $h \in E$ we have

$$\langle h, p^k - p\rangle \to 0 \quad \text{as} \quad k \to \infty, \tag{6}$$

since E is Hilbert space. Moreover, note that because of Assumption (A3), for any k and n there exists the sequence of reals $\alpha_{j,n}^k$, $j = 1, \ldots, k_n$, such that we have, a.s.,

$$p_n^k - p_n = \sum_{j=1}^{k_n} \alpha_{j,n}^k I_{B_j^n}. \tag{7}$$

Thus, from (6) and (7), for any n and $j \le k_n$, we have $\alpha_{j,n}^k \to 0$ as $k \to \infty$. Now, let us write, for any M,

$$\|p^k - p\|^2 = \sum_{n=1}^{M}\sum_{j=1}^{k_n}(\alpha_{j,n}^k)^2 P(B_j^n) + E\left(\sum_{M<n\in\overline{N}} (p_n^k - p_n)^2\right).$$

Hence, $\|p^k - p\|^2 \to 0$, as $k \to \infty$, since the expectation on the right side of the above equality may be arbitrarily small if one takes sufficiently large M for $p^k \in S_r^L$, while the first term on the right side tends to 0 as $k \to \infty$ thanks to the dominated convergence theorem. □

Lemma 3.2. *Let $\{X_n\}_{n\in\overline{N}}$ be a sequence of random variables $\{\mathcal{F}_n\}_{n\in\overline{N}}$ adapted such that $E(\sup_{n\in N} |X_n|) < \infty$. Let $p = \{p_n\}_{n\in\overline{N}}$, $q = \{q_n\}_{n\in\overline{N}} \in S_r^L$ and let*

$$J_1(p, q) = E\left(\sum_{n\in\overline{N}} X_n p_n q_n\right),$$

$$J_2(p, q) = E\left(\sum_{n\in\overline{N}} X_n p_n (1 - q_1 - \ldots - q_n)\right).$$

Then, the functions $J_i : S_r^L \times S_r^L \to R$, $i = 1, 2$, are continuous.

Proof. Let $p^k = \{p_n^k\}_{n\in\overline{N}}$, $q^k = \{q_n^k\}_{n\in\overline{N}} \in S_r^L$, $k = 1, 2, \ldots$ be sequences converging to p, q in S_r^L, respectively, as $k \to \infty$. We will show that $J_1(p^k, q^k) \to J_1(p, q)$ as $k \to \infty$. For any natural M let us rewrite $J_1(p^k, q^k)$ as follows

$$J_1(p^k, q^k) = E\left(\sum_{n=1}^{M} X_n p_n^k q_n^k\right) + E\left(\sum_{M<n\in\overline{N}} X_n p_n^k q_n^k\right). \tag{8}$$

Now, to obtain the required convergence it is sufficient to apply the dominated convergence theorem to both terms on the right side of (8). Namely, using the lemma assumptions it is easy to see that we have, for any M and k,

$$\lim_{k\to\infty} E\left(\sum_{n=1}^{M} X_n p_n^k q_n^k\right) = E\left(\sum_{n=1}^{M} X_n p_n q_n\right),$$

and

$$\lim_{M\to\infty}\left|E\left(\sum_{M<n\in\overline{N}} X_n p_n^k q_n^k\right)\right| \le \lim_{M\to\infty} E\left(\sum_{M<n\in\overline{N}} |X_n| r_n^2\right) = 0,$$

which together with (8) gives us continuity of $J_1(\cdot,\cdot)$. The same reasoning assures continuity of $J_2(\cdot,\cdot)$. □

Proof of Theorem 2.1. It is sufficient to note that the assumptions of Theorem 1.1 are satisfied under the assumptions of Theorem 2.1. Namely, (i) is fulfilled since nonempty strategy sets $S_1 = S_2 = S_r^L$ are convex and compact in the light of Lemma 3.1, (ii) is the consequence of Formula (4) and Lemma 3.2. (iii) is straightforward since, for any $p, q \in S_r^L$, the functions $H_1(\cdot, q) : S_r^L \to R$ and $H_2(p,\cdot) : S_r^L \to R$ are linear, so they are quasi-concave, i.e. the sets $\{p \in S_r^L : H_1(p,q) > t\}$ and $\{q \in S_r^L : H_2(p,q) > t\}$ are convex for any real t.

Proof of Theorem 2.2. Let $(\tau_1, \tau_2) \in \Lambda \times \Lambda$ and $s_i = \{p_n^i\}_{n\in\overline{N}} \in S$, $i = 1, 2$, be such that we have, for any $n \in \overline{N}$, a.s.,

$$p_n^i = P(\tau_i = n|\mathcal{F}_n).$$

Let us denote $\overline{G}_n = P(K > n)$, $n = 1, 2, \ldots$

Then, one may obtain

$$\begin{aligned} V_K^i(\tau_1,\tau_2) &= E(X_{\tau_i}^i I_{\{\tau_i<\tau_j,\tau_i\le K\}} + Y_{\tau_j}^i I_{\{\tau_j<\tau_i,\tau_j\le K\}} + W_{\tau_i}^i I_{\{\tau_i=\tau_j\le K\}}) \\ &= E\Bigg(\sum_{n\in N}(\overline{G}_{n-1} X_n^i p_n^i (1 - p_1^j - \ldots - p_n^j) \\ &\quad + \overline{G}_{n-1} Y_n^i p_n^j (1 - p_1^i - \ldots - p_n^i) + \overline{G}_{n-1} W_n^i p_n^i p_n^j\Bigg), \quad i \ne j. \end{aligned} \tag{9}$$

Thus, $V_K^i(\tau_1,\tau_2) := H_i^K(s_1,s_2)$. Hence, according to Proposition 2.1, it is sufficient to show that the game $(S \times S, H_1^K, H_2^K)$ has a Nash equilibrium strategy. Similarly, as in the proof of Lemma 3.1 we obtain that S is weakly compact in E and that the functions $H_1(\cdot, s_2)$, $H_2(s_1,\cdot)$ are quasi-concave, for any $s_j \in S_j$, $j = 1, 2$.

To complete the proof it is sufficient to show that the payoff functions $H_i^K : S \times S \to R$, $i = 1, 2$, are weakly continuous since it assures the assumptions of Theorem 1.1 are fulfilled. Hence, because of the form of H_i^K, given in (9), it remains to prove that the functions $J_l(p, q)$, $p, q \in S$, $l = 1, 2$, below are weakly continuous:

$$J_1(p, q) = E\left(\sum_{n\in N} \overline{G}_{n-1} X_n p_n q_n\right),$$

$$J_2(p, q) = E\left(\sum_{n\in N} \overline{G}_{n-1} X_n p_n (1 - q_1 - \ldots - q_n)\right),$$

where $p = \{p_n\}_{n\in\overline{N}}$, $q = \{q_n\}_{n\in\overline{N}}$, $\{X_n\}_{n\in N}$ is a sequence of univariate random variables $\{\mathcal{F}_n\}_{n\in\overline{N}}$ adapted such that $E(\sup_{n\in N} |X_n|) < \infty$.

Let $(p^k, q^k) \in S \times S$, $k = 1, 2, \ldots$, be a sequence converging weakly to $(p, q) \in S \times S$. We will show that $J_1(p^k, q^k) \to J_1(p, q)$ as $k \to \infty$. For any natural M let us rewrite $J_1(p^k, q^k)$ as follows

$$J_1(p^k, q^k) = E\left(\sum_{n=1}^{M} \overline{G}_{n-1} X_n p_n^k q_n^k\right) + E\left(\sum_{M<n\in N} \overline{G}_{n-1} X_n p_n^k q_n^k\right). \tag{10}$$

Note that using the dominated convergence theorem we obtain, for any k,

$$\lim_{M\to\infty} \left| E\left(\sum_{M<n\in N} \overline{G}_{n-1} X_n p_n^k q_n^k\right)\right| \le \lim_{M\to\infty} E\left(\sup_{n\in N} |X_n| \sum_{M<n\in N} \overline{G}_{n-1}\right) = 0, \tag{11}$$

since $\sum_{n\in N} \overline{G}_n = E(K) < \infty$. Hence, it is sufficient to show the equality below

$$\lim_{k\to\infty} E\left(\sum_{n=1}^{M} \overline{G}_{n-1} X_n p_n^k q_n^k\right) = E\left(\sum_{n=1}^{M} \overline{G}_{n-1} X_n p_n q_n\right). \tag{12}$$

Now, let us note that according to Assumption (A3) we have (7) and, similarly, for any natural n there exists the sequence of reals $\beta_{j,n}^k$, $j = 1, \ldots, k_n$, such that we have, a.s.,

$$q_n^k - q_n = \sum_{j=1}^{k_n} \beta_{j,n}^k I_{B_j^n}, \tag{13}$$

and $\beta_{j,n}^k \to 0$ as $k \to \infty$, for any $j \le k_n$, in view of the weak convergence of the sequence $\{q^k\}$. Thus, the expectation on the left hand side of (12) may be rewritten, after simple algebra, as follows

$$E\left(\sum_{n=1}^{M}\sum_{j=1}^{k_n}\overline{G}_{n-1}X_n\alpha_{j,n}^k\beta_{j,n}^k I_{B_j^n}\right)+E\left(\sum_{n=1}^{M}\overline{G}_{n-1}X_n q_n(p_n^k-p_n)\right)$$

$$+E\left(\sum_{n=1}^{M}\overline{G}_{n-1}X_n p_n q_n^k\right).$$

Now, (12) is fulfilled since in the above formula the last term tends to the right side of (12) and the first two terms tend to 0 as $k\to\infty$, since the sequences $\{p^k\}$ and $\{q^k\}$ converge weakly to p and q respectively, as $k\to\infty$, and one may apply the dominated convergence theorem.

Weak continuity of J_1 is the consequence of (10), (11) and (12). In a similar way we prove that J_2 is weakly continuous. The proof is completed.

Proof of Theorem 2.3. The proof follows the lines of the proof of Theorem 2.2.

REFERENCES

[1] Bobecka K. and Ferenstein E.Z., On nonzero-sum stopping game related to discrete risk processes, *Control and Cybernetics*, 30, 2001, 339–354.

[2] Chow Y.S., Robbins H. and Siegmund D., *Great Expectations: The Theory of Optimal Stopping*, Houghton Mifflin Co., Boston, 1971.

[3] Dynkin E.B., Game variant of a problem on optimal stopping, *Soviet Math. Dokl.,* vol. 10, pp. 270–274, 1969.

[4] Enns E.G. and Ferenstein E.Z., On a multi-person time-sequential game with priorities, *Sequential Analysis*, 6, 239–256, 1987.

[5] Fan K., Applications of a Theorem concerning sets with convex sections, *Math. Annalen*, 163, 189–203, 1966.

[6] Ferenstein E.Z., Two-person non-zero-sum game with priorities, in T.S. Ferguson and S.M. Samuels (eds.) *Contemporary Mathematics*, 125, 119–133, 1992.

[7] Ferenstein E.Z., A variation of the Dynkin's stopping game, *Mathematica Japonica*, 38, 2, 371–379, 1993.

[8] Ferenstein E.Z., On some kind of the Dynkin's stopping game, *Demonstratio Mathematica*, 34, 1, 191–197, 2001.

[9] Ferguson T.S, *Mathematical Statistics. A Decision Theoretic Approach*, Academic Press, New York 1967.

[10] Irle A., Games of stopping with infinite horizon, *ZOR—Math. Methods. Oper. Res.* 42, 3, 345–359, 1995.

[11] Nash J., Non-cooperative games, *Ann. Math.*, 54, 286–295, 1951.

[12] Neveu J., *Discrete Parameter Martingales*, Amsterdam: North-Holland, 1975.

[13] Nowak A.S. and Szajowski K., Nonzero-sum stochastic games, *Annals of the International Society of Dynamic Games*, 4, 297–343, 1999.

[14] Ohtsubo Y., A nonzero-sum extension of Dynkin's stopping problem, *Mathematics of Operations Research*, 12, 277–296, 1987.

[15] Ohtsubo Y., On a discrete-time non-zero-sum Dynkin problem with monotonicity, *J. Appl. Probab.*, 28, 466–472, 1991.

[16] Rosenberg D., Solan E. and Vieille N. Stopping games with randomized strategies. *Probab. Theor. Relat. Fields*, 119, 433–451, 2001.

[17] Yasuda M., On a randomized strategy in Neveu's stopping problem, *Stochastic Processes and their Applications*, 21, 159–166, 1986.

Stopping Games – Recent Results

Eilon Solan
Department of Managerial Economics and Decision Sciences
Kellogg School of Management
Northwestern University
e-solan@kellogg.northwestern.edu
and
School of Mathematical Sciences
Tel Aviv University
Tel Aviv 69978, Israel
eilons@post.tau.ac.il

Nicolas Vieille
Département Finance et Economie
HEC School of Management (HEC)
78 Jouy-en-Josas, France
vieille@hec.fr

Abstract

We survey recent results on the existence of the value in zero-sum stopping games with discrete and continuous time, and on the existence of ε-equilibria in nonzero-sum games with discrete time.

1 Introduction

Stopping games have been introduced by Dynkin [4] as a generalization of optimal stopping problems, and later used in several models in economics and management science, such as optimal equipment replacement, job search, consumer purchase behavior, research and development (see Mamer [11] and the references therein), and the analysis of strategic exit (see Ghemawat and Nalebuff [7] or Fine and Li [5]).

The basic setting is as follows. The game is defined by two processes a and b, defined on a probability space $(\Omega, \mathcal{A}, \mathbf{P})$, endowed with a filtration $\mathbf{F}$. Two players are allowed to stop at any time. The payoff to player 1 is given by one of the two processes depending on who stopped first. Formally, the two players choose stopping times σ and τ and player 1 receives from player 2 the amount

$$\mathbf{E}\left[a_\sigma \mathbf{1}_{\sigma<\tau} + b_\tau \mathbf{1}_{\tau\leq\sigma,\tau<+\infty}\right].$$

Much work has been devoted to the study of this zero-sum game, both in a discrete-time and in a continuous-time framework. In discrete time, Dynkin [4] proved the existence of the value under an assumption that at any stage only one of the two players is allowed to stop, and Neveu [13] proved the existence of the value under the assumption $a \leq b$. After those seminal contributions, most of the literature focused on continuous-time games, in the context of the general theory of stochastic processes. Bismut [1] proved the existence of the value under the assumption $a \leq b$ and an assumption known as Mokobodsky's hypothesis (in addition, several regularity assumptions are needed). The latter assumption was later removed, see e.g. Lepeltier and Maingueneau [10]. Some authors also worked in the diffusion case, see e.g. Cvitanić and Karatzas [3]. Finally, most work involves a symmetrized payoff function

$$\gamma(\sigma, \tau) = \mathbf{E}\left[a_\sigma \mathbf{1}_{\sigma<\tau} + b_\tau \mathbf{1}_{\tau<\sigma} + c_\sigma \mathbf{1}_{\sigma=\tau<+\infty}\right],$$

where c is a third given process, under the assumption $a \leq c \leq b$. This list of references is by no means exhaustive.

Comparatively few studies deal with nonzero-sum (two-player) stopping games. For such games, a, b, c are $\mathbf{R}^2$-valued processes, and the i-th coordinate is the payoff to player i, see e.g. Mamer [11], Morimoto [12], Hideo [8], and Ohtsubo [14],[15].

When the players are restricted to stopping times, the value need not exist in general, even if the processes are nonrandom and constant. For instance, for the stopping game in discrete time defined by $a_n = b_n = 1$ and $c_n = 0$, one has

$$\sup_\sigma \inf_\tau \gamma(\sigma, \tau) = 0 \quad \text{while} \quad \inf_\tau \sup_\sigma \gamma(\sigma, \tau) = 1.$$

The purpose of this paper is to survey recent work on stopping games that aim at obtaining the existence of the value under no-order conditions on the processes a, b and c, by suitably convexifying the set of strategies of the players.

The paper is organized as follows. Section 2 contains a brief discussion of the appropriate convexification. Sections 3 and 4 present results on zero-sum games respectively, for discrete time and continuous time models. In both cases, the proof is sketched in the simple case of deterministic payoff functions. Finally, Section 5 discusses a result for two-player nonzero-sum stopping games with deterministic payoff functions.

2 Randomized Stopping Times

This section contains a brief discussion of the proper way of convexifying the set of stopping times. A more extensive treatment can be found in Rosenberg *et al.*[16], or Touzi and Vieille [22]. We follow the logic of behavior strategies and enlarge the set of strategies by allowing a player to stop, at any stage, with positive probability.

In discrete time, this leads to the following notion. A strategy (of player 1) is a $\mathbf{F}$-adapted process $\mathbf{x} = (x_n)$ with values in $[0, 1]$. x_n is to be interpreted as the probability that player 1 stops the game at stage n, conditional on the game being still alive at that stage. In computing the payoff induced by a pair of strategies $(\mathbf{x}, \mathbf{y})$, one assumes that the randomizations performed by the players in the various stages are mutually independent, and independent from the payoff processes. Thus, a strategy $\mathbf{x}$ that corresponds to the stopping time σ is

$$x_n = \begin{cases} 0 & \text{on } \sigma > n, \\ 1 & \text{on } \sigma \leq n. \end{cases}$$

In continuous time, this leads to the following notion. A strategy (of player 1) is a right-continuous, non-decreasing adapted process (F_t) with $F_t \in [0, 1]$ for each t. Here, F_t may be interpreted as the probability that player 1 will have stopped before time t (including t). Thus, the strategy that corresponds to a stopping time σ is the process (F_t) defined as $F_t = \mathbf{1}_{\sigma \leq t}$. The payoff associated with the two strategies (F_t) and (G_t) can be written as

$$\gamma(F, G) = \mathbf{E}\left[\int_{[0,\infty)} a(1-G)dF + \int_{[0,\infty)} b(1-F)dG + \sum_{0 \leq t < \infty} c_t \Delta F_t \Delta G_t\right],$$

where $\Delta F_t = F_t - F_{t-}$ is the jump of F at time t.

The alternative standard way of convexifying the set of stopping times is to follow the logic of mixed strategies and to define a strategy as, loosely speaking, a probability distribution over stopping times. The equivalence between mixed strategies and behavior strategies holds under fairly general assumptions, see Kuhn [9]. For a discussion specific to the case of stopping games and to the above notions, see Touzi and Vieille [22].

3 Zero-Sum Games in Discrete Time

We deal here with zero-sum stopping games in discrete time. We first describe the setup and the result in a precise way, then give an overview of the proof.

Let $(\Omega, \mathcal{A}, \mathbf{P})$ be a probability space, and $(\mathcal{F}_n)$ be a filtration over $(\Omega, \mathcal{A}, \mathbf{P})$ (the information available at stage n). Let (a_n), (b_n), (c_n) be adapted processes, defined over $(\Omega, \mathcal{A}, \mathbf{P})$. We assume

$$\sup_n |a_n|, \sup_n |b_n|, \sup_n |c_n| \in L^1(\mathbf{P}). \tag{1}$$

By properly enlarging the probability space $(\Omega, \mathcal{A}, \mathbf{P})$, one can assume w.l.o.g. that it supports a double sequence $(X_n, Y_n)_{n=0}^{\infty}$ of *iid* variables, uniformly distributed over $[0, 1]$, such that, for each n: (i) (X_n, Y_n) is independent of the process $(a_k, b_k, c_k)_k$; (ii) (X_n, Y_n) is $\mathcal{F}_{n+1}$-measurable, and independent of $\mathcal{F}_n$. X_n and Y_n are used by the players in their randomizations.

Define the stopping game as follows. A *strategy* for player 1 (*resp.* player 2) is a [0, 1]-valued adapted process $\mathbf{x} = (x_n)$ (*resp.* $\mathbf{y} = (y_n)$). Given strategies $(\mathbf{x}, \mathbf{y})$, define the stopping stages of players 1 and 2 by $\theta_1 = \inf\{n \geq 0, X_n \leq x_n\}$, $\theta_2 = \inf\{n \geq 0, Y_n \leq y_n\}$, and set

$$\theta = \min(\theta_1, \theta_2). \tag{2}$$

Thus, θ is the stage at which the game stops.

We set

$$r(\mathbf{x}, \mathbf{y}) = a_{\theta_1} 1_{\theta_1 < \theta_2} + b_{\theta_2} 1_{\theta_2 < \theta_1} + c_{\theta_1} 1_{\theta_1 = \theta_2 < +\infty}.$$

The payoff of the game is $\gamma(\mathbf{x}, \mathbf{y}) = \mathbf{E}(r(\mathbf{x}, \mathbf{y}))$.

The game has a value $v \in \mathbf{R}$ if

$$v = \sup_{\mathbf{x}} \inf_{\mathbf{y}} \gamma(\mathbf{x}, \mathbf{y}) = \inf_{\mathbf{y}} \sup_{\mathbf{x}} \gamma(\mathbf{x}, \mathbf{y}).$$

It is convenient to introduce discounted payoffs, as in Yasuda [23]. For $\lambda \in (0, 1]$, the λ-discounted evaluation is given by

$$\gamma_\lambda(\mathbf{x}, \mathbf{y}) = \lambda \mathbf{E}\left[(1 - \lambda)^{\theta + 1} r(\mathbf{x}, \mathbf{y})\right]. \tag{3}$$

The game has a λ-discounted value $v_\lambda \in \mathbf{R}$ if

$$v_\lambda = \sup_{\mathbf{x}} \inf_{\mathbf{y}} \gamma_\lambda(\mathbf{x}, \mathbf{y}) = \inf_{\mathbf{y}} \sup_{\mathbf{x}} \gamma_\lambda(\mathbf{x}, \mathbf{y}).$$

Yasuda [23] proved the existence of the discounted value v_λ, by adapting Shapley's [17] argument.

Theorem 3.1. *(Rosenberg et al. [16]) Every stopping game such that (1) holds has a value v. Moreover, $v = \lim_{\lambda \to 0} v_\lambda$.*

We sketch below the proof in the deterministic case; that is, the payoff at each stage n depends only on n. Thus, a_n, b_n, c_n are real numbers. The proof for the general case builds upon the ideas described below.

We denote by G_n the game that starts at stage n. G_n is similar to the original game, but players are restricted to the use of strategies that stop before stage n with probability 0. In particular, G_0 coincides with G.

We denote by $v_n(\lambda)$ the λ-discounted value of G_n, for each $n \in \mathbf{N}$ and $\lambda \in (0, 1]$. Define $v_n(0) = \limsup_{\lambda \to 0} v_n(\lambda)$. We shall prove that $\sup_{\mathbf{x}} \inf_{\mathbf{y}} \gamma(\mathbf{x}, \mathbf{y}) \geq v_0(0)$. By exchanging the roles of the two players, one immediately deduces

$$\inf_{\mathbf{y}} \sup_{\mathbf{x}} \gamma(\mathbf{x}, \mathbf{y}) \leq \liminf_{\lambda \to 0} v_0(\lambda),$$

which implies both conclusions of the theorem, since

$$\sup_{\mathbf{x}} \inf_{\mathbf{y}} \gamma(\mathbf{x}, \mathbf{y}) \le \inf_{\mathbf{y}} \sup_{\mathbf{x}} \gamma(\mathbf{x}, \mathbf{y}).$$

As usual, $v_n(\lambda)$ satisfies a recursive equation (dynamic programming principle), that here takes the form

$$v_n(\lambda) = (1-\lambda) \sup_{x\in[0,1]} \inf_{y\in[0,1]} \{(1-x)(1-y)\, v_{n+1}(\lambda) \\ +x(1-y)\, a_n + y(1-x)\, b_n + xy\, c_n\} \tag{4}$$

By possibly taking a subsequence (λ_p) that converges to zero, we may assume that $v_n(0) = \lim_{\lambda\to 0} v_n(\lambda)$. We let $x_n(\lambda)$ achieve the supremum in (4), and let $x_n(0)$ denote any limit point of $(x_n(\lambda))$ as λ goes to zero. Thus, one has for each $y \in [0,1]$ and every $\lambda \ge 0$,

$$(1-\lambda)(x_n(\lambda)y\, v_{n+1}(\lambda) + x_n(\lambda)(1-y)\, a_n + (1-y)x_n(\lambda)\, b_n \\ + (1-x_n(\lambda))(1-y)\, c_n) \ge v_n(\lambda). \tag{5}$$

This can be rephrased by introducing the process $(\widetilde{v}_n(\lambda))_n$ defined, for $\lambda \ge 0$, by

$$\widetilde{v}_n(\lambda) = \begin{cases} v_n(\lambda) & n \le \theta, \\ r_\theta & n > \theta, \end{cases}$$

where r_θ is equal to a_θ, b_θ or c_θ depending on whether $\theta_1 < \theta_2, \theta_1 > \theta_2$ or $\theta_1 = \theta_2 < +\infty$. Note that $\widetilde{v}_n(\lambda)$ depends on $(\mathbf{x}, \mathbf{y})$, through the value of θ. Though $(v_n(\lambda))$ is a sequence of numbers, $(\widetilde{v}_n(\lambda))$ is a process, the randomness being caused by the random choices of the players. Whenever useful, we shall write $\widetilde{v}_n^{\mathbf{x},\mathbf{y}}(\lambda)$ to emphasize which strategies are used.

Inequality (5) can be rephrased as follows: for every choice of strategy $\mathbf{y}$, and for each $\lambda \ge 0$, the process $((1-\lambda)^{\min(n,\theta)}\widetilde{v}_n(\lambda))$ is a submartingale, provided that player 1 uses the strategy $\mathbf{x}_\lambda := (x_n(\lambda))$.

In particular, the process $(\widetilde{v}_n(0))$ is a submartingale under the pair of strategies $(\mathbf{x}_0, \mathbf{0})$, where $\mathbf{0}$ is the strategy of player 2 that never stops. Thus, it converges, $\mathbf{P}$-a.s., to some random variable $\widetilde{v}_\infty$.

We now split the discussion in two parts. Assume first that, under the pair $(\mathbf{x}_0, \mathbf{0})$, θ is $\mathbf{P}$-a.s. finite. In that case, θ is also $\mathbf{P}$-a.s. finite for $(\mathbf{x}_0, \mathbf{y})$, whatever be $\mathbf{y}$. Thus, for each $\mathbf{y}$, the limit $\widetilde{v}_\infty$ coincides with r_θ. The submartingale property of $\widetilde{v}_n(0)$ then implies that $\gamma(\mathbf{x}_0, \mathbf{y}) = \mathbf{E}\left[\widetilde{v}_\infty^{\mathbf{x}_0,\mathbf{y}}\right] \ge v_0(0)$. Since $\mathbf{y}$ is arbitrary, $\inf_{\mathbf{y}} \gamma(\mathbf{x}_0, \mathbf{y}) \ge v_0(0)$.

Assume now that $\theta = +\infty$ with positive probability, and let $\varepsilon > 0$ be given. On the event $\{\theta = +\infty\}$, $\widetilde{v}_n(0) = v_n(0)$ for each $n \in \mathbf{N}$. Therefore, the sequence of numbers $(v_n(0))$ is convergent, say to v_∞. If $v_\infty \le 2\varepsilon$,

$$\gamma(\mathbf{x}_0, \mathbf{y}) = \mathbf{E}\left[r_\theta \mathbf{1}_{\theta<+\infty}\right] \ge \mathbf{E}\left[\widetilde{v}_\infty^{\mathbf{x}_0,\mathbf{y}}\right] - \varepsilon \ge v_0(0) - 2\varepsilon,$$

hence $\inf_{\mathbf{y}} \gamma(\mathbf{x}_0, \mathbf{y}) \ge v_0(0) - 2\varepsilon$.

The tricky case is when $v_\infty > 2\varepsilon$. We choose $N \in \mathbf{N}$ such that $|v_n(0) - v_\infty| \le \varepsilon$ for each $n \ge N$. We define a strategy of player 1 as follows. We first construct two sequences (λ_p, s_p) (possibly of finite length) by the following recursive device:

- Choose $\lambda_1 \in (0, 1]$ such that $v_N(\lambda_1) > v_N(0) - \varepsilon^2$; set
$$s_1 = \inf\left\{n \ge N, v_n(\lambda_1) \le \varepsilon^2\right\}.$$
- For $p \ge 1$, choose $\lambda_{p+1} \in (0, 1]$ such that $v_{s_p}(\lambda_{p+1}) \ge v_{s_p}(0) - \varepsilon^2$ and set $s_{p+1} = \inf\left\{n \ge s_p, v_n(\lambda_{p+1}) \le \varepsilon^2\right\}$.

We let $\overline{\mathbf{x}}$ be the strategy that coincides with $\mathbf{x}_0$ up to stage $s_0 = N$, and with $\mathbf{x}_{\lambda_{p+1}}$ from stage s_p up to stage s_{p+1}. Consider what may happen between the two stages s_p and s_{p+1}, assuming that the game was not stopped before stage s_p. Whatever be the strategy $\mathbf{y}$ used by player 2, the process $(1 - \lambda_{p+1})^{\min(n,\theta)}\widetilde{v}_n(\lambda_{p+1})$ is a submartingale between s_p and s_{p+1}. Thus, $\widetilde{v}_n(\lambda_{p+1})$ increases on average by a factor of $1/(1 - \lambda_{p+1})$ from one stage to the following, prior to $\min(\theta, s_{p+1})$. Since $v_n(\lambda_{p+1}) \ge \varepsilon - \varepsilon^2$ for $n \le \min(\theta, s_{p+1})$ and all quantities are bounded, it must be that $\min(\theta, s_{p+1})$ is finite.

Since $v_{s_{p+1}}(\lambda_{p+1}) \le \varepsilon^2$, the probability that $\theta < s_{p+1}$ (conditioned on $\theta \ge s_p$) is bounded away from zero, and the payoff to player 1, conditioned on $\theta \le s_{p+1}$, is at least $v_{s_p}(\lambda_{p+1}) \ge v_\infty - \varepsilon - \varepsilon^2$.

These observations imply that $\theta < +\infty$ $\mathbf{P}$-a.s. under $(\overline{\mathbf{x}}, \mathbf{y})$, for each $\mathbf{y}$, and $\gamma(\overline{\mathbf{x}}, \mathbf{y}) \ge v_\infty - 2\varepsilon \ge v_N - 2\varepsilon$. Since $(\widetilde{v}_n(0))$ is a submartingale under $(\mathbf{x}_0, \mathbf{y})$, this implies that $\gamma(\overline{\mathbf{x}}, \mathbf{y}) \ge v_0(0) - 2\varepsilon$.

4 Zero-Sum Games in Continuous Time

We deal here with zero-sum stopping games in continuous time and with finite horizon. Let $(\Omega, \mathcal{F}, \mathbf{P})$ be a complete probability space, and $T > 0$ a fixed terminal time. Let $a = \{a_t, 0 \le t \le T\}$, $b = \{b_t, 0 \le t \le T\}$ and $c = \{c_t, 0 \le t \le T\}$ be real-valued processes, satisfying the integrability condition

$$E\left[\sup_t |a_t| + \sup_t |b_t| + \sup_t |c_t|\right] < +\infty\ , \tag{6}$$

and we denote by $\mathbf{F}$ the $\mathbf{P}$-augmentation of the filtration generated by the processes a, b and c.

We denote by $\mathcal{V}^+$ the set of adapted, right-continuous, non-decreasing processes, with values in $[0, 1]$. For $F, G \in \mathcal{V}^+$, we set

$$\gamma(F, G) = \mathbf{E}\left[\int_{[0,\infty)} a(1-G)dF + \int_{[0,\infty)} b(1-F)dG + \sum_{0\le t<\infty} c_t \Delta F_t \Delta G_t\right].$$

Theorem 4.1. *(Touzi and Vieille [22]) Assume that: (i) (a, b, c) satisfy the integrability condition (6), (ii) the game has a finite horizon T, (iii) a and b are (càdlàg) semimartingales with trajectories continuous at T, (iv) $c \leq b$. Then*

$$\sup_{F \in \mathcal{V}^+} \inf_{G \in \mathcal{V}^+} \gamma(F, G) = \inf_{G \in \mathcal{V}^+} \sup_{F \in \mathcal{V}^+} \gamma(F, G),$$

i.e., the stopping game has a value.

Note that the assumptions on the processes are not symmetric on the two players. The basic idea of the proof is to apply Sion's [19] minmax Theorem to the payoff function $\gamma : \mathcal{V}^+ \times \mathcal{V}^+ \to R$. Define

$$\mathcal{S} = \left\{ (F_t), \mathbf{E}\left[\int_0^T F_t^2 dt \right] < +\infty \right\}.$$

The set $\mathcal{S}$ is a Hilbert space when endowed with the scalar product $\mathbf{E}\left[\int_0^T F_t G_t dt \right]$, and $\mathcal{V}^+$ is a subset of $\mathcal{S}$, compact for the weak topology $\sigma(\mathcal{S}, \mathcal{S})$. However, Sion's Theorem does not apply directly since the payoff function γ does not have enough continuity properties.

This difficulty is circumvented by applying the Sion's Theorem to restricted strategy spaces. Define

$$\mathcal{V}_1 = \left\{ (F_t) \in \mathcal{V}^+, (F_t) \text{ has continuous trajectories, } \mathbf{P}\text{-a.s.} \right\}$$

and

$$\mathcal{V}_2 = \left\{ (G_t) \in \mathcal{V}^+, G_T = 1 \text{ on } \{b_T < 0 < a_T\} \text{ and } \Delta G_T = 0 \text{ on } \{b_T > 0\} \right\},$$

and let $\overline{\mathcal{S}} = \left\{ (F_t), \mathbf{E}\left[\int_0^T F_t^2 dt + F_T^2 \right] < +\infty \right\}$. The space $\overline{\mathcal{S}}$ is a Hilbert space when endowed with the scalar product $\mathbf{E}\left[\int_0^T F_t G_t dt + F_T G_T \right]$. One can check that $\mathcal{V}_2$ is compact for the weak topology $\sigma(\overline{\mathcal{S}}, \overline{\mathcal{S}})$. Moreover, γ is separately continuous on $\mathcal{V}_1 \times \mathcal{V}_2$ for the strong topology. Hence, by Sion's Theorem

$$\sup_{\mathcal{V}_1} \inf_{\mathcal{V}_2} \gamma(F, G) = \inf_{\mathcal{V}_2} \sup_{\mathcal{V}_1} \gamma(F, G).$$

The restriction on player 1's strategies is imposed in order to have continuity of γ. The restriction on player 2's strategies is imposed in such a way that restricting the strategy spaces to $\mathcal{V}_1$ and $\mathcal{V}_2$ respectively entails no loss for the players:

$$\sup_{\mathcal{V}_1} \inf_{\mathcal{V}_2} \gamma(F, G) = \sup_{\mathcal{V}_1} \inf_{\mathcal{V}^+} \gamma(F, G), \text{ and } \inf_{\mathcal{V}_2} \sup_{\mathcal{V}_1} \gamma(F, G) = \inf_{\mathcal{V}_2} \sup_{\mathcal{V}^+} \gamma(F, G). \tag{7}$$

Let us discuss these two equalities in the non-random case. Thus, a, b and c are right-continuous functions defined on $[0, T]$, and continuous at T. The proof in the general case is obtained by elaborating upon the ideas that follow.

The first equality is immediate: let $(F_t) \in \mathcal{V}_1$, and $(G_t) \in \mathcal{V}^+$ be given. Thus, $F : [0, T] \to [0, 1]$ is a continuous, non-decreasing function, while $G : [0, T] \to [0, 1]$ is a right-continuous, non-decreasing function. Let $\widetilde{G} : [0, T] \to [0, 1]$ be the function that agrees with G on $[0, T)$, and where $\widetilde{G}_T$ is equal to G_{T-} or to 1 depending whether $b_T > 0$ or $b_T \leq 0$. Plainly, $\widetilde{G}$ belongs to $\mathcal{V}_2$. On the other hand,

$$\gamma(F, \widetilde{G}) - \gamma(F, G) = \begin{cases} (1 - F_T)b_T(1 - G_T) & \text{if } Y_T \leq 0, \\ -(1 - F_T)b_T \Delta G_T & \text{if } Y_T > 0. \end{cases}$$

In both cases, it is non-positive, which establishes the first equality.

As for the second equality in (7), let $G \in \mathcal{V}_2$ and $F \in \mathcal{V}^+$ be given. By using an argument similar to the one of the previous paragraph, we may assume that F is continuous at T in case $a_T > 0, b_T \geq 0$. Let (F^n) be a sequence of continuous non-decreasing functions such that $F^n_t \to F_{t-}$ for each $t \in [0, T]$. Using the assumption $c \leq b$, one can check that $\limsup \gamma(F^n, G) \geq \gamma(F, G)$, which yields the second equality.

5 Non Zero-Sum Games in Discrete Time

We conclude by presenting a result on two-player non zero-sum stopping games in discrete time. We show how a simple application of Ramsey's Theorem, combined with a result of Flesch *et al.* [6], imply the existence of an ε-equilibrium in the *deterministic* case.

We let here (a_n), (b_n) and (c_n) be three bounded sequences in $\mathbf{R}^2$, and let ρ be a uniform bound on the payoffs. The setup of Section 3 then reduces to the following. A strategy of player 1 is a sequence $\mathbf{x} = (x_n)$ in $[0, 1]$, where x_n is the probability that player 1 will choose to stop in stage n, if the game was not stopped before.

The payoff of the game is defined to be

$$\gamma(\mathbf{x}, \mathbf{y}) = \mathbf{E}\left[r(\mathbf{x}, \mathbf{y})\right],$$

as in Section 3, except that here $\gamma(\mathbf{x}, \mathbf{y}) \in \mathbf{R}^2$.

Theorem 5.1. *(Shmaya et al. [18]) For each $\varepsilon > 0$, the stopping game has an ε-equilibrium $(\mathbf{x}^*, \mathbf{y}^*)$; that is, a pair of strategies that satisfies:*

$$\gamma^1(\mathbf{x}, \mathbf{y}^*) \leq \gamma^1(\mathbf{x}^*, \mathbf{y}^*) + \varepsilon \text{ and } \gamma^2(\mathbf{x}^*, \mathbf{y}) \leq \gamma^2(\mathbf{x}^*, \mathbf{y}^*) + \varepsilon \text{ for each } \mathbf{x}, \mathbf{y}.$$

Note that if payoffs are not uniformly bounded, an ε-equilibrium need not exist. (As a counter example, take the game defined by $a^i_n = b^i_n = n - 1, c^i_n = n - 2$ for every $i = 1, 2$, and every $n \in \mathbf{N}$.) Moreover, even when payoffs are bounded a

0-equilibrium need not exist (e.g. $a_n^i = b_n^i = n/(n+1)$, $c_n^i = (n-1)/n$ for every $i = 1, 2$, and every $n \in \mathbf{N}$).

Fix $\varepsilon > 0$ once and for all, and an ε-discretization Z of the set $[-\rho, \rho]^2$; that is, Z is a finite set such that for every $u \in [-\rho, \rho]^2$ there is $z \in Z$ with $\|z - u\| < \varepsilon$.

For every two positive integers $k < l$ we define a *periodic* stopping game $G(k, l)$ as follows:

$$a_n(k, l) = a_{k+(n \bmod l)},\ b_n(k, l) = b_{k+(n \bmod l)} \text{ and } c_n(k, l) = c_{k+(n \bmod l)}.$$

This is "the game that starts at stage k and restarts at stage l." We denote by $\gamma_{k,l}(\mathbf{x}, \mathbf{y})$ the payoff function in the game $G(k, l)$.

The game $G(k, l)$ may be analyzed as a particular stochastic game $\Gamma(k, l)$ with absorbing states. To see this, assume for example $k = 0$ and $l = 2$, and consider the stochastic game $\Gamma(0, 2)$ described by the matrix

	b_0 *	b_1 *
a_0 *	c_0 *	a_0 *
a_1 *	b_0 *	c_1 *

In this game player 1 chooses a row and player 2 chooses a column. The first line corresponds to the pure strategy *never stop*, the second and third ones to the pure strategies *stop in stage 0* and *stop in stage 1*. The columns are to be interpreted symmetrically for player 2. The meaning of an asterisked entry is that the game moves to an absorbing state (i.e., ends) as soon as such an entry is played.

Thus, each stage of $\Gamma(0, 2)$ corresponds to two stages (a period) of $G(0, 2)$.

Using Flesch *et al.* [6], the game $\Gamma(k, l)$ has a stationary ε-equilibrium, or equivalently, for each $\varepsilon > 0$, the game $G(k, l)$ has a periodic ε-equilibrium $(\mathbf{x}(k, l), \mathbf{y}(k, l))$, with period $l - k$.

For each $k < l$, we choose $z(k, l) \in Z$ such that

$$\|\gamma(\mathbf{x}(k, l), \mathbf{y}(k, l)) - z(k, l)\| < \varepsilon.$$

For every pair of non-negative integers we attached an element in Z – a color. By Ramsey's Theorem (see, e.g., Bollobás [2]) there is an infinite set $K \subseteq \mathbf{N} \cup \{0\}$ and $z \in Z$ such that $z(k, l) = z$ for every $k, l \in K$, $k < l$.

In particular, there exists an increasing sequence of non-negative integers $k_1 < k_2 < \cdots$ such that for every $j \in \mathbf{N}$, $z(k_j, k_{j+1}) = z$.

We define a profile $(\mathbf{x}, \mathbf{y})$ from stage k_1 on by concatenating the profiles

$$(\mathbf{x}(k_i, k_{i+1}), \mathbf{y}(k_i, k_{i+1})) :$$

$(\mathbf{x}, \mathbf{y})$ coincides with $(\mathbf{x}(k_i, k_{i+1}), \mathbf{y}(k_i, k_{i+1}))$ from stage k_i up to stage $k_{i+1} - 1$. To complete the construction before stage k_1, we recall that every finite-stage game has an equilibrium. The profile $(\mathbf{x}, \mathbf{y})$ coincides between stages 0 and $k_1 - 1$ with an equilibrium in the k_1-stage game, whose payoffs are $(a_n, b_n, c_n)_{n<k_1}$ if the play is stopped prior to stage k_1, and is z otherwise. There is no difficulty in proving that $(\mathbf{x}, \mathbf{y})$ is an ε-equilibrium of the stopping game, starting from stage k_1.[1] Moreover, the expected payoff, conditioned that the game was not stopped before stage k_1, is, up to ε, z. It follows that $(\mathbf{x}, \mathbf{y})$ is an ε-equilibrium of the stopping game.

Extensions to more than two players are limited. We actually proved that if every periodic deterministic game admits an ε-equilibrium, then every non-periodic deterministic game admits an ε-equilibrium. Using the technique of Solan [20] instead of that of Flesch *et al.* [6], one can prove that every *three*-player periodic deterministic game admits an ε-equilibrium (though it needs not be periodic). One can then prove that every three-player deterministic stopping game admits an ε-equilibrium.

Unfortunately, it is currently not known whether every n-player periodic deterministic stopping game admits an ε-equilibrium, for $n \geq 4$. For more details, see Solan and Vieille [21].

To generalize this proof to general two-player stopping games, one needs to generalize Ramsey's Theorem to a stochastic setup, and to show that a concatenation of ε-equilibria in periodic non-deterministic games yields an ε'-equilibrium, for some $\varepsilon' > 0$ that goes to 0 as ε goes to 0. Whereas Ramsey's Theorem can be generalized to a stochastic setup, it is not clear yet how to achieve the second goal.

REFERENCES

[1] Bismut J. M. Sur un problème de Dynkin. *Z. Warsch. V. Geb.*, 39: 31–53, 1977.

[2] Bollobás B. *Modern graph theory*. Springer, 1998.

[3] Cvitanic J. and Karatzas I. Backward stochastic differential equations with reflection and Dynkin games. *Ann. Probab.*, 24: 2024–2056, 1996.

[4] Dynkin E. B. Game variant of a problem on optimal stopping. *Soviet Math. Dokl.*, 10: 270–274, 1969.

[5] Fine C. H. and Li L. Equilibrium exit in stochastically declining industries. *Games and Economic Behavior*, 1: 40–59, 1989.

[6] Flesch J., Thuijsman F., and Vrieze O. J. Recursive repeated games with absorbing states. *Mathematics of Operations Research*, 21: 1016–1022, 1996.

[7] Ghemawat P. and Nalebuff B. Exit. *RAND J. Econ.*, 16: 184–194, 1985.

[8] Hideo N. Non-zero-sum stopping games of symmetric Markov processes. *Probab. Theor. Relat. Fields*, 75: 487–497, 1987.

[1]This is true, provided the periodic profiles in $\Gamma(k, l)$ are chosen to satisfy an additionnal property: the probability of absorption in each period is bounded away from 0. We do not elaborate here on this point.

[9] Kuhn H. W. Extensive games and the problem of information. In H.W. Kuhn and A.W. Tucker, editors, *Contributions to the Theory of Games*. Annals of Mathematics Study 28, Princeton University Press, 1953.

[10] Lepeltier J. P. and Mainguenau M. A. Le jeu de Dynkin en théorie générale sans l'hypothèse de Mokobodsky. *Stochastics*, 13: 25–44, 1984.

[11] Mamer J. W. Monotone stopping games. *J. Appl. Probab.*, 24: 386–401, 1987.

[12] Morimoto H. Non zero-sum discrete parameter stochastic games with stopping times. *Probab. Theor. Relat. Fields*, 72: 155–160, 1986.

[13] Neveu J. *Discrete-Parameter Martingales*. North-Holland, Amsterdam, 1975.

[14] Ohtsubo Y. A non zero-sum extension of Dynkin's stopping problem. *Mathematics of Operations Research*, 12: 277–296, 1987.

[15] Ohtsubo Y. On a discrete-time non-zero-sum Dynkin problem with monotonicity. *J. Appl. Probab.*, 28: 466–472, 1991.

[16] Rosenberg D., Solan E., and Vieille N. Stopping games with randomized strategies. *Probab. Theor. Relat. Fields.*, 119: 433–451, 2001.

[17] Shapley L. S. Stochastic games. *Proceedings of the National Academy of Sciences of the U.S.A.*, 39: 1095–1100, 1953.

[18] Shmaya E., Solan E., and Vieille N. An application of Ramsey's theorem to stopping games. *Games and Economic Behavior*, 42 (2003), 300–306.

[19] Sion M. On general minimax theorems. *Pacific Journal of Mathematics*, 8: 171–176, 1958.

[20] Solan E. Three-player absorbing games. *Mathematics of Operations Research*, 24: 669–698, 1999.

[21] Solan E., and Vieille N. Quitting games. *Mathematics of Operations Research*, 26: 265–285, 2001.

[22] Touzi N., and Vieille N. Continuous-time Dynkin games with mixed strategies. *SIAM J. Contol. Optim.* 41: 1073–1088, 2002.

[23] Yasuda M. On a randomized strategy in Neveu's stopping problem. *Stochast. Process. Appl.*, 21: 159–166, 1986.

Dynkin's Games with Randomized Optimal Stopping Rules

Victor Domansky*
St. Petersburg Institute for Economics and Mathematics
Russian Academy of Science
St. Petersburg, Russia
doman@emi.spb.su

Abstract

We consider stopping games for Markov chains in the formulation introduced by Dynkin [6]. Two players observe a Markov sequence and may stop it at any stage. When the chain is stopped the game terminates and Player 1 receives from Player 2 a sum depending on the player who stopped the chain and on its current state. If the game continues infinitely, then Player 1 gets "the payoff at infinity" depending on the "limiting" behavior of the chain trajectory.

We describe the structure of solutions for a class of stopping games with a countable state space and nonnegative payoffs. The payoff is equal to zero if Player 1 stops the chain but not Player 2. These solutions require using of randomized stopping rules. We study an extent of dependence of the value for these games on the "the payoff at infinity". It turns out that this extent is determined with the "limiting" behavior of payoffs and with the transition structure of the chain.

1 Introduction

Stopping games are generalizations of optimal stopping problems for stochastic processes (see e.g. [15]) for the case when several decision makers with different goals are involved.

We consider the game formulation of stopping problem for a homogeneous Markov chain going back to Dynkin [6]. Two players observe the Markov sequence $x_n, n = 0, 1, \ldots,$ with the state space X. Four real-valued functions

$$a_{11},\ a_{12},\ a_{21},\ c : X \to R$$

are given over X.

Each player may stop the chain at any stage n. When the chain is stopped the game is finished and Player 1 wins from Player 2 the sum, depending on the player who stopped the chain and on its current state x. The payoff $a_{11}(x)$ corresponds to simultaneous stopping at the state x. The payoffs $a_{12}(x)$ and $a_{21}(x)$ correspond to

*This study was supported by grant 01-06-80279 of the Russian Foundation of Basic Research which is gratefully acknowledged.

stopping by Player 1 and Player 2 only. If the game continues infinitely Player 1 gets $\underline{\lim}_{n\to\infty} c(x_n)$.

We denote by $\Gamma_c(x)$ the game, depending on the initial state x and on the limit payoff c, distinguishing the latter parameter. Below we give some reasons for it.

Strategies of players for the game $\Gamma_c(x)$ are randomized stopping times τ and σ for the chain. Let T and Σ be the sets of strategies of Players 1 and 2 correspondingly.

Denote the expected gain of Player 1 for the game $\Gamma_c(x)$ under strategies $\tau \in T$, $\sigma \in \Sigma$ by $K_c(\tau, \sigma|x)$:

$$\begin{aligned} K_c(\tau, \sigma|x) = \mathbf{E}_x(&\mathbf{I}_{[\tau=\sigma<\infty]} a_{11}(x_\tau) + \mathbf{I}_{[\tau<\sigma]} a_{12}(x_\tau) \\ &+ \mathbf{I}_{[\tau>\sigma]} a_{21}(x_\sigma) + \mathbf{I}_{[\tau=\sigma=\infty]} \underline{\lim}_{n\to\infty} c(x_n)). \end{aligned}$$

At any stage of the game, players should compare their gains resulting from immediate stopping with their expected gains in the case of continuation of observations under the assumption that both of them use their optimal strategies later on. Dynamic programming considerations imply that the upper and the lower value functions $\overline{v}_c(x)$ and $\underline{v}_c(x)$ of the game $\Gamma_c(x)$ should satisfy the optimality equation that is a game analogue of the Wald–Bellman equation:

$$v_c(x) = \mathrm{val}\begin{Vmatrix} a_{11}(x) & a_{12}(x) \\ a_{21}(x) & a_{22}(x, v_c) \end{Vmatrix} = \mathrm{val}[a_{ij}(x, v_c)]. \tag{1}$$

Here, for integrable function u over X,

$$a_{22}(x, u) = \mathbf{P}_x u = \mathbf{E}[u(x_1)|x_0 = x],$$

$a_{ij}(x, u) = a_{ij}(x)$ for all pairs $(i, j) \neq (2, 2)$. We denote the value of a matrix game with the payoff matrix $[a_{ij}]$ by $\mathrm{val}[a_{ij}]$.

Thus, the value of a stopping game is a fixed point of the operator T, functioning over the set of integrable functions according to formula

$$Tu(x) = \mathrm{val}[a_{ij}(x, u)].$$

In general, such fixed point is not unique and its choice depends on "the payoff at infinity" c. Notice that the function c is not taken into account with the equation (1) by itself.

Algorithms for finding the value $v_c(x)$ are based on iterations of the operator T, applied to a properly chosen initial function $w_c^0(x)$, depending on "the payoff at infinity" c. Thus, the crucial point of the analysis of stopping games under consideration turns to be the determining domains of attraction for the operator T. These domains depend both on the payoff structure and on the transition structure of the chain.

Generally, one should employ mixed strategies to resolve the equation (1). Mixed behavior strategies determine randomized stopping rules.

In Dynkin's model, at each stage there is only one player who may stop the chain. There is a partition $X = X_1 \cup X_2$ of the state space X such that Player 1 can stop at the stage n if $x_n \in X_1$ and Player 2 can stop if $x_n \in X_2$. If the game continues infinitely Player 1 gets zero. This model can be reduced to the described model putting $a_{21}(x) = \infty$ for $x \in X_1$ and $a_{12}(x) = -\infty$ for $x \in X_2$.

Later, the class of stopping games was extended by allowing the players to stop simultaneously, replacing the discrete time by the continuous time etc. See [11], [7], [2], [1], [9]. These works suppose that the payoffs satisfy inequalities

$$a_{12}(x) \leq a_{11}(x) \leq a_{21}(x). \tag{2}$$

Inequalities (2) result in solvability of optimality equations with use of pure strategies only. Namely, for any function u the matrices $[a_{ij}(x, u)]$ have saddle points either at (2, 1), or at (1, 2), or at (2, 2). So, there is no need to use randomized stopping rules.

In the works of Yasuda [18] and Rosenberg, Solan, Vieille [13], the assumption (2) is dropped, but, as before, if the game continues infinitely Player 1 gets zero. It allows obtaining the values and the ε-optimal strategies of players for these games as the limits of values and ε-optimal strategies for discounted games.

Here, we drop both the assumptions (2) and the assumption $c = 0$. We investigate solvability of optimality equations by means of mixed strategies and the impact of "the payoff at infinity" c on the choice of solution.

We describe solutions for a special class of stopping games $\Gamma_c(x)$ with a countable state space. The payoffs are nonnegative with $a_{12}(x) = 0$. Thus, the predetermined stopping at any fixed stage yields Player 1 zero gain. Moreover, any strategy stopping the chain brings him zero, since Player 2 can refrain from stopping for arbitrary long time risking the "payoff at infinity" c only. To lose less than c Player 2 must stop the chain with positive probability.

In Section 2, we examine the games with zero payoffs $a_{21}(x)$ and $a_{12}(x)$. We analyze the structure of randomized stopping rules and give sufficient conditions for the value of such game to be equal to zero.

In Section 3, we continue to analyze the games with zero payoffs $a_{21}(x)$ and $a_{12}(x)$. We prove that the value of such game is more than zero if all states of the chain are transient and the potential of the function $a_{11}(x)^{-1}$ is finite. We give the bounds for values and for optimal randomized stopping times by means of these potentials.

In Section 4, we examine the games with positive payoffs $a_{21}(x)$. For these games we obtain bounds for values and for optimal randomized stopping times using the results of Section 2.

In Section 5, we consider games $\Gamma_c(x)$ for the trivial chain with deterministic transitions $p(x, x + 1) = 1$. In this case the estimates given in Sections 2 and 3 provide the exact values of the games $\Gamma_c(x)$. Several numerical examples illustrate the obtained results.

For conclusion, it should be noted that there are various other approaches to game-theoretic setting of stopping problems. According to one of them Player 1 selects a stopping time and Player 2 picks a distribution of the process (see [14], [8],). Another approach assumes that players independently observe different stochastic processes. The players' payoffs depend on the whole combination of states where the processes are stopped (see e.g. [3], [12], [10]).

Another setting of stopping problem with two players sequentially observing the same process assumes that the payoff function depends on the states chosen by both players, and the player who chooses the state later is informed of the choice of his opponent (see e.g. [17]). This game may be reduced to the above described setting.

2 Games with Zero Payoffs a_{21} and a_{12}. The Case $v_c = 0$

We consider stopping games $\Gamma_c(x)$ with a countable state space X. The transition probabilities of Markov chain $x_n, n = 0, 1, \ldots,$ are given with the infinite transition matrix $P = [p(x, y)], x, y \in X$.

In this section, we assume that payoffs satisfy the following conditions:

$$a_{21}(x) = a_{12}(x) = 0, \quad a_{11}(x) = a(x) > 0, \quad c(x) > 0, \quad \forall x \in X.$$

Under these conditions the optimality equations (1) for the value $v_c(x)$ of the game $\Gamma_c(x)$ can be written as

$$v_c(x) = \frac{a(x) \cdot Pv_c(x)}{a(x) + Pv_c(x)}, \tag{3}$$

where $Pv_c(x) = \sum_{y \in X} p(x, y) \cdot v_c(y)$.

Let L^+ be the class of nonnegative functions f over X such that

$$\mathbf{E}_x \left[\sup_n f(x_n) \right] < \infty, \quad \forall x \in X. \tag{4}$$

Suppose $c \in L^+$. It follows that

$$\hat{c}(x) = \mathbf{E}_x \underline{\lim}_{n \to \infty} c(x_n) < \infty, \quad \forall x \in X.$$

The function $\hat{c}(x)$ is a P-harmonic function, i.e., $P\hat{c}(x) = \mathbf{E_x \hat{c}(x_1)} = \mathbf{\hat{c}(x)}$. Therefore, $\lim_{n\to\infty} \hat{c}(x_n)$ exists $\mathbf{P}_x$-a.s., and the equality $\lim_{n\to\infty} \hat{c}(x_n) = \underline{\lim}_{n\to\infty} c(x_n)$ holds. Thus, $\underline{\lim}_{n\to\infty} c(x_n)$ can be replaced with $\lim_{n\to\infty} \hat{c}(x_n)$. Further, we assume that either c is a P-harmonic function, satisfying $\mathbf{E}_x \lim_{n\to\infty} c(x_n) < \infty$ for all $x \in X$, or $c = \infty$ for all $x \in X$.

Any solution v_c of the equation (3) satisfies the inequalities $0 \le v_c \le Pv_c$, i.e., v_c is a nonnegative P-subharmonic function. Evidently, $v_c(x) \le c(x)$, since the strategy $\sigma(x) = 0$ for all $x \in X$ ensures Player 2 the loss $c(x)$.

Let τ be a stationary strategy, given by function $\tau(x)$. Here $\tau(x)$ is the conditional probability to stop in the case of falling into x:

$$\tau(x) = \mathbf{P}(\tau = n | \tau \geq n; x_n = x).$$

Consider the Markov chain X^τ with state space $X \cup \{d\}$ and with transition probabilities

$$p_\tau(x, y) = (1 - \tau(x))p(x, y), \qquad p_\tau(x, d) = \tau(x), \qquad p_\tau(d, d) = 1,$$

resulting from the initial chain if the strategy τ is used. Falling into the absorbing state d may be regarded as a break of the chain, corresponding to stopping of the initial chain by the strategy τ. We put $a(d) = c(d) = 0$. For $x, y \in X$, we have

$$p_\tau^{(n)}(x, y) = \mathbf{P}_x(\{\tau \geq n\} \cap \{x_n = y\}),$$

where $p_\tau^{(n)}(x, y)$ is the element of n-step transition matrix P_τ^n. It follows that, for the strategy n of Player 2, prescribing to stop at the step n,

$$\begin{aligned} K_c(\tau, n|x) &= \sum_{y \in X} \mathbf{P}_x(\{\tau = n\} \cap \{x_n = y\})a(y) \\ &= \sum_{y \in X} \mathbf{P}_x(\{\tau \geq n\} \cap \{x_n = y\})\mathbf{P}(\tau = n | \tau \geq n; x_n = y)a(y) \\ &= \sum_{y \in X} p_\tau^{(n)}(x, y)\tau(y)a(y) = \mathbf{E}_x^\tau[\tau(x_n)a(x_n)], \end{aligned}$$

where $\mathbf{E}_x^\tau$ is the expectation with respect to the chain X^τ with $x_0 = x$.

Similarly, for the strategy ∞, given by $\sigma_0(x) = 0$, for all $x \in X$,

$$K_c(\tau, \infty | x) = \mathbf{E}_x^\tau \left[\lim_{n \to \infty} c(x_n) \right],$$

and, if the strategy τ stops the chain, i.e. if $\mathbf{P}_x^\tau\{\lim_{n\to\infty} x_n = d\} = 1$, then $K_c(\tau, \infty|x) = 0$.

It yields that, for any strategy $\sigma \in \Sigma$, the payoff function $K_c(\tau, \sigma|x)$ can be written as

$$K_c(\tau, \sigma|x) = \mathbf{E}_x^\tau[\tau(x_\sigma)a(x_\sigma) \cdot I_{[\sigma<\infty]} + \lim_{n\to\infty} c(x_n) \cdot I_{[\sigma=\infty]}].$$

Therefore, the best answer of Player 2 to the strategy τ is the solution of the minimizing stopping problem for the chain X^τ with the current payoff $f(x) = \tau(x)a(x)$ for $x \in X$, $f(d) = 0$, and with the "payoff at infinity" $\lim_{n\to\infty} c(x_n)$ (recall that c is P-harmonic function). Let $W_c^\tau(x)$ be the value of this stopping problem. Note that if the strategy τ stops the chain, then the best answer is ∞ and $W_c^\tau(x) = 0$ for any c.

Recall that L^+ is the class of positive functions satisfying (4). It is known (see e.g. [15]) that if $f, c \in L^+$, then the value of the minimizing stopping problem with the current payoff f and with the P-harmonic "payoff at infinity" c is equal to the maximal P^τ-subharmonic minorant of the function $\min(f(x), c(x))$.

In the same way, the best answer of Player 1 to the strategy σ is the solution of the maximizing stopping problem for the chain X^σ with the current payoff $f(x) = \sigma(x)a(x)$ for $x \in X$, $f(d) = 0$, and with the P-harmonic "payoff at infinity" $c(x)$. Let $W_c^\sigma(x)$ be the value of this stopping problem. Observe that if the strategy σ stops the chain, then the best answer of Player 1 should stop the chain as well.

If both functions f and c are bounded from below, then the value of the maximizing stopping problem with the current payoff f and with the P-harmonic "payoff at infinity" c is equal to the minimal P^σ-superharmonic majorant of the function $\max(f(x), c(x))$.

Proposition 2.1. *Let R be the set of all recurrent states. Then, for any state x such that R is accessible from x $\mathbf{P}_x$-a.s., and for any function c, the value $v_c(x)$ is equal to zero.*

Proof. For any $\varepsilon > 0$ the strategy $\sigma(y) = \varepsilon \cdot a(y)^{-1}$ for $y \in R$, $\sigma(x) = 0$ for all $x \notin R$ stops the chain. It guarantees Player 2 the loss not exceeding ε against any strategy of Player 1, stopping the chain, and the loss zero against the strategy "not to stop at all". □

It follows that for the value of the game $v_c(x)$ to be positive the chain should remain in the set $X - R$ of transient states with positive probability.

Further, we suppose that all states of the chain are transient. Consequently, for any initial state x, the expectation of sojourn time at any state y is finite, i.e., the potential matrix

$$G(x, y) = \sum_{n=0}^{\infty} p^{(n)}(x, y)$$

is finite for all $x, y \in X$. Here $p^{(n)}(x, y)$ is the element of n-step transition matrix P^n, and P^0 is an identity matrix.

Let $S(x)$ denote the potential of the function $a(y)^{-1}$ for the initial state x:

$$S(x) = \sum_{y \in X} G(x, y)a(y)^{-1}. \tag{5}$$

Proposition 2.2. *If $S(x) = \infty$, then, for any payoff at infinity c, the value $v_c(x)$ is equal to zero.*

Proof. For any $\varepsilon > 0$ the strategy $\sigma(y) = \varepsilon \cdot a(y)^{-1}$ for all $y \in X$ stops the chain. It provides Player 2 a loss not exceeding ε against any strategy of Player 1, stopping the chain, and a loss equal to zero against the strategy "not to stop at all". □

3 Games with Zero Payoffs a_{21} and a_{12}. The Case $v_c > 0$

It follows from Proposition 2.2 that, for the value of the game to be positive, the function $S(x)$ given by (5) should be finite for all $x \in X$. Further, in this section, we show that, in this case, for any limiting payoff c such that $\mathbf{E}_x(\lim_{n\to\infty} c(x_n))^{-1} < \infty$, the game has the value $v_c(x)$ not equal to zero.

Consider the superharmonic function

$$u_c(x) = S(x) + h_c(x),$$

where the harmonic function $h_c(x)$ is defined as

$$h_c(x) = \mathbf{E}_x\left(\lim_{n\to\infty} c(x_n)\right)^{-1}.$$

Observe that for any $c > 0$, including $c = \infty$, the function $u_c(x)$ satisfies the equation

$$u_c(x) = a(x)^{-1} + \sum_{y\in X} p(x, y)u_c(y). \tag{6}$$

As $\lim_{n\to\infty} S(x_n) = 0$ because S is potential, the function $u_c(x)$ satisfies the following "boundary condition at infinity":

$$\lim_{n\to\infty} u_c(x_n) = \left(\lim_{n\to\infty} c(x_n)\right)^{-1}. \tag{7}$$

Assuming that the value $v_c(x) > 0$, we may rewrite optimality equations (3) as

$$v_c(x)^{-1} = a(x)^{-1} + \left(\sum_{y\in X} p(x, y)v_c(y)\right)^{-1}. \tag{8}$$

Notice that equations (6) and (8) are rather similar.

Let $w_c(x) = (u_c(x))^{-1}$. Taking into account (7), we obtain

$$\lim_{n\to\infty} w_c(x_n) = \lim_{n\to\infty} c(x_n). \tag{9}$$

Consider the stationary strategy τ_c of Player 1 given by

$$\tau_c(x) = w_c(x) \cdot a(x)^{-1}, \qquad \forall x \in X.$$

Proposition 3.1. *If $c < \infty$, then*

$$\inf_{\sigma} K_c(\tau_c, \sigma|x) = w_c(x),$$

i.e., the strategy τ_c guarantees Player 1 the payoff $w_c(x)$ in the game $\Gamma_c(x)$.

Proof. Let us show that, for all $x \in X$, the inequalities

$$w_c(x) \le (1 - \tau_c(x)) \sum_{y\in X} p(x, y)w_c(y) = \sum_{y\in X} p_{\tau_c}(x, y)w_c(y) \tag{10}$$

hold, i.e., that $w_c(x)$ is a subharmonic function with respect to the chain X^{τ_c}.

In fact, taking into account (6), we obtain

$$1-\tau_c(x)=(u_c(x))^{-1}\cdot\sum_{y\in X}p(x,y)u_c(y)=w_c(x)\cdot\sum_{y\in X}p(x,y)w_c(y)^{-1}.$$

Consequently,

$$\begin{aligned}(1-\tau_c(x))\sum_{y\in X}p(x,y)w_c(y)&=w_c(x)\cdot\sum_{y\in X}p(x,y)w_c(y)^{-1}\sum_{y\in X}p(x,y)w_c(y)\\&\geq w_c(x)\cdot\sum_{y\in X}p(x,y)w_c(y)^{-1}\\&\quad\times\left(\sum_{y\in X}p(x,y)w_c(y)^{-1}\right)^{-1}=w_c(x).\end{aligned}$$

Consider the minimizing stopping problem for the chain X^{τ_c} with the current payoff $\tau_c(x)a(x)=w_c(x)$ for $x\in X$, and with the "payoff at infinity" $\lim_{n\to\infty}c(x_n)$.

It follows from (9) that $\lim_{n\to\infty}c(x_n)$ can be replaced by $\lim_{n\to\infty}w(x_n)$. As $w_c(x)<c(x)$, it follows that $w_c\in L^+$. The function $w_c(x)$ is its own maximal subharmonic minorant. The value of this minimizing stopping problem is equal to $w_c(x)$, and the best answer of Player 2 to the strategy τ_c is "to stop immediately". Consequently, the strategy τ_c provides the gain not less than $w_c(x)$ against any strategy of Player 2. □

Remark. It follows from (10) that the best answer of Player 1 to the strategy $\sigma_c(x)=w_c(x)\cdot a(x)^{-1}$ is "not to stop at all", resulting in

$$K_c(\infty,\sigma_c|x)=\mathbf{E}_x^{\tau}\left[\lim_{n\to\infty}w_c(x_n)\right].$$

Here $w_c(x)\leq K_c(\infty,\sigma_c|x)\leq c(x)$.

Proposition 3.2. *Let $S(x)<\infty$ and $0<\mathbf{E}_x(\lim_{n\to\infty}c(x_n))^{-1}<\infty$ for all $x\in X$; then there exists a solution v_c of equation (3), satisfying boundary conditions (9); the function $v_c(x)$ is defined by*

$$v_c(x)=\lim_{n\to\infty}T^{(n)}w_c(x),$$

where the operator Tu for $u:X\to R$ is given by

$$Tu(x)=\frac{a(x)\cdot\sum_{y\in X}p(x,y)\cdot u(y)}{a(x)+\sum_{y\in X}p(x,y)\cdot u(y)}.$$

Proof. As follows from Proposition 3.1

$$w_c(x)=\min_{\Sigma}K_c(\tau_c,\sigma|x),$$

and therefore the sequence $T^{(n)}w_c(x)$ represents the sequential results of improvement of the strategy τ_c. It follows that the sequence $T^{(n)}w_c(x)$ is non-decreasing and monotonically converges to $v_c(x)$. The function $v_c(x)$ is a solution of equation (3) satisfying boundary conditions (9). □

Remark. It can be shown that T is a contracting operator over the set of functions, satisfying boundary conditions (9), in the metrizable topology of pointwise convergence. It follows that

$$\lim_{n\to\infty} T^{(n)}w_c(x) = \lim_{n\to\infty} T^{(n)}c(x) = v_c(x),$$

and the function $v_c(x)$ is a unique solution of equation (3) satisfying boundary conditions (9).

Theorem 3.1. *Let $S(x) < \infty$ and $0 < \mathbf{E}_x(\lim_{n\to\infty} c(x_n))^{-1} < \infty$ for all $x \in X$; then the game $\Gamma_c(x)$ has the value $v_c(x)$, the stationary strategies τ_c^* and σ_c^* given by*

$$\tau_c^*(x) = \sigma_c^*(x) = v_c(x) \cdot a(x)^{-1}, \qquad \forall x \in X,$$

are optimal strategies.

Proof. It is easy to verify that $v_c(x)$ is bounded harmonic function with respect to the chain $X^{\tau_c^*}$ with sub-stochastic transition probabilities

$$p_{\tau_c^*}(x, y) = (1 - \tau_c^*(x))p(x, y),$$

resulting from the initial chain if the strategy τ_c^* is used. It follows that

$$v_c(x) = \min_{\sigma\in\Sigma} K_c(\tau_c^*, \sigma|x).$$

Analogously, $v_c(x)$ is a bounded harmonic function with respect to the chain $X^{\sigma_c^*}$, and, consequently,

$$v_c(x) = \max_{\tau\in T} K_c(\tau, \sigma_c^*|x).$$

□

Theorem 3.2. *Let $S(x) < \infty$ for all $x \in X$; then the game $\Gamma_\infty(x)$ with the payoff at infinity $c(x) = \infty$ has the value $v_\infty(x)$, satisfying the inequalities*

$$a(x) > v_\infty(x) > S(x)^{-1} = w_\infty(x);$$

the stationary strategy σ_∞^* given by

$$\sigma_\infty^*(x) = v_\infty(x) \cdot a(x)^{-1}, \qquad \forall x \in X,$$

is optimal; Player 1 has only ε-optimal strategies.

Proof. Analogously to the proof of Theorem 3.1, $v_\infty(x)$ is a bounded from below harmonic function with respect to the chain $X_{\sigma^*_\infty}$, and, consequently,

$$v_\infty(x) = \max_{\tau \in T} K_\infty(\tau, \sigma^*_\infty | x).$$

On the other hand, as the function $v_\infty(x)$ is not bounded from above, the symmetric result is incorrect. The strategy τ^*_∞ stops the chain. Consequently, the strategy $\sigma(x) = 0$ of Player 2, prescribing not to stop the chain at all, provides him the loss zero against τ^*_∞. Nevertheless, the strategy τ^*_c for sufficiently large c provides Player 1 the gain arbitrary close to $v_\infty(x)$. □

4 Games with Positive Payoffs $a_{11}(x)$ and $a_{21}(x)$ and with Zero Payoff $a_{12}(x)$

Here we analyze the games $\Gamma_c(x)$ with payoffs satisfying the following conditions:

$$a_{11}(x) = a(x) > a_{21}(x) = b(x) \geq 0, \qquad a_{12}(x) = 0.$$

For the games under consideration, the optimality equation (3) takes the form

$$v_c(x) = \frac{a(x) \sum_{y \in X} p(x, y) v_c(y)}{a(x) - b(x) + \sum_{y \in X} p(x, y) v_c(y)} \quad \text{if} \quad b(x) \leq \sum_{y \in X} p(x, y) v_c(y)$$

and

$$v_c(x) = \sum_{y \in X} p(x, y) v_c(y) \quad \text{if} \quad b(x) > \sum_{y \in X} p(x, y) v_c(y).$$

It follows that $0 \leq v_c \leq P v_c$, i.e., v_c is a nonnegative P-subharmonic function. Evidently, $v_c(x) \leq c(x)$, since the strategy $\sigma(x) = 0$ for all $x \in X$ ensures Player 2 the loss $c(x)$. Consequently, there is a unique Riesz decomposition $v_c(x) = h_c(x) - \pi_c(x)$, where $h_c(x)$ is a harmonic function,

$$h_c(x) = \lim_{n \to \infty} P^n v_c(x) = \mathbf{E}_x \lim_{n \to \infty} v_c(x_n),$$

satisfying $h_c(x) \leq c(x)$. The function $\pi_c(x)$ is a potential of a nonnegative function $P v_c(x) - v_c(x)$,

$$\pi_c(x) = G(x, y)(P v_c(y) - v_c(y)).$$

It satisfies $\lim_{n \to \infty} P^n \pi_c(x) = 0$.

Below we formulate several assertions allowing the estimation of the value $v_c(x)$ of the game $\Gamma_c(x)$ on the basis of results of the preceding section. The proofs are omitted, being rather close to the proofs of the preceding section.

Let $R_\alpha \subset X$ be a recurrent class of states of the chain. Then the harmonic function $c(x)$ is equal to a constant c_α for all states $x \in R_\alpha$.

Proposition 4.1. *For any state $x \in R_\alpha$,*

$$v_c(x) = \min(c_\alpha, \inf_{R_\alpha} b(x)).$$

This proposition reduces the analysis of these games to the case, where all states of the chain are transient and the potential matrix $G(x, y) = \sum_{n=0}^{\infty} p^{(n)}(x, y)$ is finite for all $x, y \in X$.

In this case, along with the game $\Gamma_c(x)$ we can consider the game $\Gamma_c^0(x)$ with the payoff b equal to zero. If the potential Ga^{-1} of the function $a(y)^{-1}$ is finite and this game satisfies the conditions of Theorem 3.1, then we obtain $v_c^0(x) \le v_c(x) \le c(x)$, where $v_c^0(x)$ is the value of the game $\Gamma_c^0(x)$. In this case, for any sufficiently large limiting payoff c, the game has the value $v_c(x)$ not equal to zero and satisfying the "boundary condition at infinity" $\lim_{n\to\infty} v_c(x_n) = \lim_{n\to\infty} c(x_n)$.

Let $W_c^-(x)$ be the value of the minimizing stopping problem with the current payoff $b(x)$ and the "payoff at infinity" $c(x)$:

$$W_c^-(x) = \inf_{\sigma\in\Sigma} \mathbf{E_x}[\mathbf{b}(\mathbf{x}_\sigma) \cdot \mathbf{I}_{[\sigma<\infty]} + \lim_{\mathbf{n}\to\infty} \mathbf{c}(\mathbf{x_n}) \cdot \mathbf{I}_{[\sigma=\infty]}].$$

The function $W_c^-(x)$ is a P-subharmonic function. It satisfies the equation

$$W_c^-(x) = \min(b(x), c(x), PW_c^-(x)),$$

and the following "boundary condition at infinity":

$$\lim_{n\to\infty} W_c^-(x_n) = \min(\underline{\lim}_{n\to\infty} b(x_n), \lim_{n\to\infty} c(x_n)).$$

Thus, if $\underline{\lim}_{n\to\infty} b(x_n) > \lim_{n\to\infty} c(x_n)$, then the strategy $\sigma(x) = 0$ becomes the optimal answer of Player 2 to the strategy $\tau(x) = 0$. Consequently, $\lim_{n\to\infty} v_c(x_n) = \lim_{n\to\infty} c(x_n)$.

In the converse case, if $\underline{\lim}_{n\to\infty} b(x_n) \le \lim_{n\to\infty} c(x_n)$, then the following problem arises: Is it possible for Player 2 to stop the chain and simultaneously ensure himself a loss not exceeding c?

Define the strategy σ_c by

$$\sigma_c(x) = c(x) \cdot a(x)^{-1} \quad \text{if} \quad b(x) \le c(x) < a(x),$$

$\sigma_c(x) = 1$ if $a(x) \le c(x)$, and $\sigma_c(x) = 0$ if $b(x) > c(x)$.

Proposition 4.2. *The strategy σ_c stops the chain ($\mathbf{P_x}$-a.s.) iff the potential $S_c(x)$ of the function $\sigma_c(y)$ for the initial state x*

$$S_c(x) = \sum_{y\in X} G(x, y)\sigma_c(y)$$

is infinite.

The maximal harmonic minorant $(c \wedge d)$ of two harmonic functions c and d is the harmonic function given by

$$(c \wedge d)(x) = \mathbf{E}_x \lim_{n\to\infty} \min(c(x_n), d(x_n)) = \lim_{n\to\infty} P^n \min(c, d)(x).$$

Similarly, the minimal harmonic majorant $(c \vee d)$ of two harmonic functions c and d is the harmonic function given by

$$(c \vee d)(x) = \mathbf{E}_x \lim_{n\to\infty} \max(c(x_n), d(x_n)) = \lim_{n\to\infty} P^n \max(c, d)(x).$$

Proposition 4.3. *If for a certain function d the strategy σ_d stops the chain, then, for any function c,*

$$v_c(x) = v_{(c\wedge d)}(x) \le (c \wedge d)(x).$$

If the strategies σ_c and σ_d stop the chain, then the strategy $\sigma_{(d\wedge c)}$ also stops the chain.

Let C^+ be the set of all harmonic functions c such that the strategy σ_c stops the chain. Since the set of harmonic functions possesses a lattice structure, we can define the greatest lower bound c^+ of the set C^+. The harmonic function $c^+(x)$ satisfies $0 \le c^+(x) \le \infty$.

Proposition 4.4. *The function c^+ is a unique harmonic function such that for any $\varepsilon > 0$ the strategy $\sigma_{(1+\varepsilon)c^+}$ stops the chain, and the strategy $\sigma_{(1-\varepsilon)c^+}$ does not stop the chain.*

It follows from Propositions 4.3 and 4.4 that it is sufficient to consider the games $\Gamma_c(x)$ with $c \le c^+$.

Proposition 4.5. *If the strategy σ_c does not stop the chain, then the value $v_c(x)$ of the game $\Gamma_c(x)$ satisfies the inequality*

$$v_c(x) \ge c(x) \cdot (S_c(x) + 1)^{-1} = w_c(x).$$

Theorem 4.1. *For any $c \le c^+$ the value of the game $\Gamma_c(x)$ is given by*

$$v_c(x) = \lim_{n\to\infty} T^{(n)} c(x).$$

If the strategy σ_c does not stop the chain, then the value $v_c(x)$ of the game $\Gamma_c(x)$ satisfies

$$v_c(x) = \lim_{n\to\infty} T^{(n)} w_c(x),$$

the optimal stationary strategies are given by

$$\tau_c^*(x) = v_c(x) \cdot \left(\sum_{y\in X} p(x, y) v_c(y) \right)^{-1}, \quad \sigma_c^*(x) = v_c(x) \cdot a(x)^{-1}, \quad \forall x \in X.$$

If the strategy σ_c stops the chain, then each of the players may have no optimal strategies.

5 Games with Deterministic Transitions

Here we consider the games $\Gamma_c(x)$ for the chain with the state space being the set of positive integers $X = \{1, 2, \dots\}$ and with trivial transition probabilities $p(x, x+1) = 1$. For this case we can determine the exact values of the games $\Gamma_c(x)$. For more details see [5].

In this case harmonic functions are constants $c \in (0, \infty)$, the potential matrix is $G(x, y) = 1$ for $x \leq y$, $G(x, y) = 0$ for $x > y$, and the potential of $a(y)^{-1}$ is

$$S(x) = \sum_{y=x}^{\infty} a(y)^{-1}.$$

Example 5.1. Let the payoffs be $a(y) = y(y+1)$, $b(y) = 0$. We have

$$S(x) = \sum_{y=x}^{\infty} \frac{1}{y(y+1)} = \frac{1}{x}.$$

For $c \in (0, \infty)$, the value of the game

$$v_c(x) = \frac{xc}{x+c}.$$

The optimal strategies are given by

$$\tau_c(y) = \sigma_c(y) = \frac{c}{(y+1)(y+c)}, \qquad \text{for } y \geq x.$$

In particular,

$$v_1(x) = \frac{x}{x+1},$$

$$\tau_1(y) = \sigma_1(y) = \frac{1}{(y+1)^2}, \qquad \text{for } y \geq x.$$

Therefore, we get for the probability of non-stop

$$\mathbf{P}_x(\tau_1 = \infty) = \mathbf{P}_x(\sigma_1 = \infty) = \prod_{y=x}^{\infty} \left(1 - \frac{1}{(y+1)^2}\right) = \frac{x}{x+1} = v_1(x).$$

For $c = \infty$, the value of the game

$$v_\infty(x) = x.$$

The optimal strategies are given by

$$\tau_\infty(y) = \sigma_\infty(y) = \frac{1}{y+1}, \qquad \text{for } y \geq x.$$

So, the probability of non-stop is

$$\mathbf{P}_x(\tau_\infty = \infty) = \mathbf{P}_x(\sigma_\infty = \infty) = \prod_{y=x}^{\infty} \frac{y}{y+1} = 0.$$

Thus, the strategy τ_∞ does not guarantee Player 1 the gain $v_\infty(x) = x$. However, the strategy τ_c for sufficiently large c is an ε-optimal strategy.

Example 5.2. Let

$$a(x) = x, \qquad b(x) = 0, \text{ for } x = k(k+1)\,, \quad b(x) = 1, \text{ for } x \neq k(k+1)\,,$$

where k is a positive integer.

For this game $c^+ = 1$. If $c \leq 1$, then the game is equivalent to the previous one. For $k(k+1) \leq x < (k+1)(k+2)$ the value of the game $\Gamma_c(x)$ is equal to $ck/(c+k)$, the optimal strategies are given by

$$\tau_c(y) = \sigma_c(y) = 0, \qquad \text{for } y \neq i(i+1)$$

$$\tau_c(y) = \sigma_c(y) = \frac{c}{(i+1)(i+c)}, \qquad \text{for } y = i(i+1) \geq x.$$

For any $c \geq 1$ the value of the game does not depend on c and is less than 1.

Example 5.3. Consider the game $\Gamma_c(x)$ with

$$a(x) = x, \qquad b(x) = \frac{x}{x+1}.$$

For this game $c^+ = 1$. If $c < 1$, then for $x \geq c/(1-c)$ the value of the game $\Gamma_c(x)$ is equal to c, the optimal strategies are given by

$$\tau_c(y) = \sigma_c(y) = 0, \qquad \text{for } y \geq c/(1-c).$$

The value and the mixed optimal strategies for $x < c/(1-c)$ can be found by means of backward induction.

For any $c \geq 1$, the value $v_c(x)$ of the game $\Gamma_c(x)$ is equal to

$$\left(x \cdot \sum_{y=x}^{\infty} \frac{1}{y^2}\right)^{-1}.$$

Observe that

$$v_c(x+1) > \frac{x}{x+1} = b(x),$$

i.e. the matrix of optimality equation does not have saddle point (2, 2).

Example 5.4. Let

$$a(x) = x(x+1), \qquad b(x) = \frac{x}{(x+1)}.$$

For this game $c^+ = \infty$. If $c < 1$, then for $x \geq c/(1-c)$ the value of the game $\Gamma_c(x)$ is equal to c, the optimal strategies are given by

$$\tau_c(y) = \sigma_c(y) = 0, \qquad \text{for } y \geq c/(1-c).$$

The value and the mixed optimal strategies for $x < c/(1-c)$ can be found by means of backward induction.

For $c \geq 1$, the value of the game $\Gamma_c(x)$ is given by $v_c(x) = (x+1)/x$ $(\sum_{y=x}^{\infty}(1/y^2)+(1/c))^{-1}$. Observe that $v_c(x+1) \geq v_1(x+1) = (x+2)/(x+1)$ $(\sum_{y=x+1}^{\infty}(1/y^2)+1)^{-1} > x/(x+1) = b(x)$, i.e. the matrix of optimality equation in fact does not have saddle point (2, 2).

Acknowledgement

I am grateful to referees for instructive and helpful comments and for bibliographical refinements.

REFERENCES

[1] Alario-Lazaret M., Lepeltier J. P., Marshal B., Dynkin Games, - Stochastic Differential Systems (Bad Honnef), 23–32, (1982), *Lecture Notes in Control and Information Sciences,* **43**, *Springer-Verlag*.

[2] Bismuth J. M., Sur un problem de Dynkin, *Z.Warsch. V. Geb.*, **39** (1977), 31–53.

[3] Domansky V., On some game problems of optimal stopping for a sequence of sums of random variables. In: *Modern Directions in Game Theory.* (1976), Vilnius: Mokslas, 86–93.

[4] Domansky V., Game problems of optimal stopping for Markov chains. *Survey of seminars on probab. theory and math. statistics, LOMI Academy of Sciences URSS*, (1977). *Probability Theory and its Appl.*, **v. 23, i.4** (1978), 863–865.

[5] Domansky V, Randomized optimal stopping rules for a class of stopping games, *Probability Theory and its Appl.*, **v. 46, i.4** (2001).

[6] Dynkin E., A game variant of optimal stopping problem, *Dokl. Acad. Nauk*, **v. 185, N1** (1969), 16–19.

[7] Elbakidze N., Constructing value and optimal policies for game stopping problem for Markov process, *Probability Theory and its Appl.*, **v. 21, i.1** (1976), 164–169.

[8] Irle A., Games of stopping with infinite horizon, *ZOR - Mathematical Methods of Operations Research*, **v. 42** (1995), 345–359.

[9] Lepeltier J. P., Maingueneau M. A., Le jeu de Dynkin en theorie generale sans l'hypothese de Mokobodski, *Stochastics*, **13** (1984), 25–44.

[10] Mazalov V., Kochetov E., Games with optimal stopping of random walks, *Probability Theory and its Appl.*, **v. 42, i.4** (1997), 820–826.

[11] Neveu J., *Discrete-Parameter Martingales*, North-Holland, Amsterdam, (1975).

[12] Presman E., Sonin I., Game problems of optimal stopping, Existence and uniqueness of equilibrium points. In: V. Arkin, ed. *Probabilistic Problems for Economics Control*, (1977), Moscow: Nauka, 115–144.

[13] Rosenberg D., Solan E., and Vieille N. Stopping games with randomized strategies. *Probab. Theor. Relat. Fields.*, vol. 119, (2001) 433–451.

[14] Sakaguchi M., Sequential deception games. *Math. Japonica*, **v. 37, N5** (1992) 813–826.

[15] Shirjaev A., *Statistical Sequential Analysis*. Moscow, Nauka, (1976).

[16] Stettner L., On general zero-sum stochastic games with optimal stopping, *Probab. and Math. Stat.*, **v. 3** (1982), 103–112.

[17] Szajowski K., Optimal stopping of a discrete Markov process by two decision makers. *SIAM J. Control and Optimization*, **v. 33, No 5** (1995), 1392–1410.

[18] Yasuda M., On a randomized strategy in Neveu's stopping problem, *Stochastic Processes Appl.*, **21** (1985), 159–166.

[19] Yasuda M., Explicit optimal value for Dynkin's stopping game, *Math. Comput. Modelling*, **v. 22, No 10–12** (1995), 313–324.

Modified Strategies in a Competitive Best Choice Problem with Random Priority

Zdzisław Porosiński
Institute of Mathematics
Wrocław University of Technology
50-370 Wrocław, Poland
porosin@im.pwr.wroc.pl

Abstract

A zero-sum game version of the full-information best choice problem is considered. Two players observe sequentially a stream of *iid* random variables (objects) from a known continuous distribution appearing according to some renewal process with the object of choosing the largest one. The horizon of observation is a positive random variable independent of objects. The observation of the random variables is imperfect and the players are informed only whether the object is greater than or less than some levels specified by both of them. Each player can choose at most one object. If both want to accept the same object, a random assignment mechanism is used. If some Player accepts an object, the other Player can change his level and continues the game alone. A similar game with discrete time and random number of objects is considered as a dual problem. The normal form of the game is derived. For the Poisson stream and the exponential horizon the value of the game and the form of the equilibrium strategy are obtained. In discrete-time case a game with geometric number of objects is completely solved.

1 Introduction

The following zero-sum game version of the continuous-time full-information best choice problem is considered. Two players observe sequentially a stream of objects being *iid* random variables from a known continuous distribution appearing according to some renewal process. The aim is to choose the best (the largest) object and this decision about choosing must be made before a moment T, which is a positive random variable independent of objects. The random variables cannot be perfectly observed. Each time a random variable is sampled the sampler is informed only whether it is greater than or less than some level he specified. Each Player can choose at most one object. Neither recall nor uncertainty of selection is allowed. After each sampling, players take a decision for acceptance or rejection of the object. If both want to accept the same object, priority is given to the player specified by a random assignment mechanism. At the moment when one Player has accepted an object, the other one can change his level and continues the game alone.

A class of suitable strategies and a gain function for the problem is constructed. The natural case of the Poisson renewal process with parameter λ and exponentially distributed T with parameter μ is examined in detail. The game value and the optimal strategy depends on λ and μ by $p = \mu/(\mu+\lambda)$ only.

The results of the paper extend those obtained in the paper by Porosiński & Szajowski [4], which considered a similar problem where the priority was given to a specified player.

2 The Priority Game with Imperfect Observation

Let $\xi_1, \xi_2, \xi_3, \ldots$ be a sequence of *iid* random variables from a common known continuous distribution F defined on a probability space $(\Omega, \mathcal{F}, \mathbf{P})$. ξ's appear according to some renewal process and ρ_n stand for the length of the time interval between ξ_{n-1} and ξ_n (for convenience it is assumed $\xi_0 = 0$ by definition), i.e. $\rho_1, \rho_2, \rho_3, \ldots$ are *iid* positive random variables with a continuous distribution G. A positive random variable T with a distribution H represents the moment, when the observation is terminated. All ξ's, ρ's and T are independent.

The sequence of random variables is sequentially sampled one by one by two decision makers (players). However, the observation is imperfect and the exact realized values are not known. Players specify only their level of sensitivity (impressionability) and they are able to know whether the observed random variable is greater than or less than their prescribed levels, chosen individually. After ξ_n is observed, players have to accept or reject the object. If only one player wants to accept ξ_n, he gets the object. If both wants to accept, a random mechanism chooses one of them to benefit and they know this. One can say that Players have random priority in accepting a realization. When one player accepts an object, then the other investigates the sequence of future realizations, having the opportunity of accepting one of them. After the first acceptance, the player, who has not yes accepted an object, can change his level of sensitivity (i.e. can modify his strategy). Neither recall nor uncertainty of selection is allowed. The aim of the players is to choose *the best object* (the maximal one).

The problem is modeled as a two-person zero-sum game. The continuous-time zero-sum game related to the full-information best choice problem with imperfect observation, has been solved by Porosiński & Szajowski [2]. Since changes in the specified levels determining the strategies of the players appear during the games presented here, then the structure of the strategy sets and the form of the gain functions are different from those considered by Porosiński & Szajowski [2].

Let

$$S_n = \rho_1 + \ldots + \rho_n,\ n = 1, 2, \ldots,\ S_0 = 0, \tag{1}$$

$$N(t) = \max\{n \geq 0 : S_n \leq t\},\ t \geq 0. \tag{2}$$

So S_n is the waiting time of the n-th object and $N(t)$ is the total number of ξ's that appear up to time t. At the moment when ξ_n is observed, all previous values of ξ's and ρ's are known and also it is known whether the moment T follows or not, i.e. the σ-field of information is

$$\mathcal{F}_n = \sigma\{\xi_1, \ldots, \xi_n, \rho_1, \ldots, \rho_n, \chi_{\{T>S_1\}}, \ldots, \chi_{\{T>S_n\}}\},\ n = 1, 2, \ldots,$$

where χ_A stands for the indicator function of the event A.

Let $\mathcal{S}$ be the set of stopping times with respect to $\{\mathcal{F}_n\}_{n=0}^{\infty}$. Since the observation is imperfect,

$$\mathcal{S}^0 = \{\tau \in \mathcal{S} : \tau = \inf\{n \le N(T) : \xi_n \ge x\}, x \in R\}$$

is the class of strategies for the one-person decision problem. This set of strategies is not appropriate for the two-person game, since the behavior of the player, who remains in the game after the first acceptance, depends on the information available when the first acceptance was made. Moreover, in the case when both Players want to choose the same object, priority is given by a *lottery* described by a random variable η with uniform distribution on [0, 1] and a number $\pi \in [0, 1]$, i.e. Player 1 benefits if $\eta \le \pi$, otherwise Player 2 gets the object.

The following modification of the strategy set solves this problem (see Szajowski [5]). Let

$$\mathcal{S}_k^0 = \{\tau \in \mathcal{S}^0 : \tau \ge k\}$$

$\left(\text{and } \mathcal{S}^0 = \mathcal{S}_0^0\right)$. Define the strategy set

$$\mathcal{T}_i = \left\{\left(\sigma_0^i, \left\{\sigma_n^i\right\}\right) : \sigma_0^i \in \mathcal{S}^0,\ \sigma_n^i \in \mathcal{S}_{n+1}^0 \text{ for } n \ge 1\right\},$$

$i = 1, 2$, for Player 1 and 2 respectively. When Player 1 chooses $\mathsf{x} \in \mathcal{T}_1$ and Player 2 chooses $\mathsf{y} \in \mathcal{T}_2$ then the effective stopping times

$$\tau_1 = \sigma_0^1 \chi_{\left\{\sigma_0^1 < \sigma_0^2\right\} \cup \left\{\sigma_0^1 = \sigma_0^2, \eta \le \pi\right\}} + \sigma_{\sigma_0^2}^1 \chi_{\left\{\sigma_0^1 > \sigma_0^2\right\} \cup \left\{\sigma_0^1 = \sigma_0^2, \eta > \pi\right\}},$$

$$\tau_2 = \sigma_0^2 \chi_{\left\{\sigma_0^1 > \sigma_0^2\right\} \cup \left\{\sigma_0^1 = \sigma_0^2, \eta > \pi\right\}} + \sigma_{\sigma_0^1}^2 \chi_{\left\{\sigma_0^1 \le \sigma_0^2\right\} \cup \left\{\sigma_0^1 = \sigma_0^2, \eta \le \pi\right\}}$$

are defined for both of them.

Since the aim is to choose the best object, let

$$f(\mathsf{x}, \mathsf{y}) = \mathbf{P}(\xi_{\tau_1} = \max\{\xi_1, \ldots, \xi_{N(T)}\}) - \mathbf{P}(\xi_{\tau_2} = \max\{\xi_1, \ldots, \xi_{N(T)}\})$$

be the payoff function.

From now on, without loss of generality, we assume that the observed random variables come from the standard uniform distribution (to study the general case, put $F(\xi_n)$ instead of ξ_n).

Definition 2.1. A pair $(\mathsf{x}^*, \mathsf{y}^*) \in \mathcal{T}_1 \times \mathcal{T}_2$ is an equilibrium point in the considered game, if for every $\mathsf{x} \in \mathcal{T}_1$ and $\mathsf{y} \in \mathcal{T}_2$ we have $f(\mathsf{x}, \mathsf{y}^*) \le f(\mathsf{x}^*, \mathsf{y}^*) \le f(\mathsf{x}^*, \mathsf{y})$.

In fact, Players choose their levels of sensitivity at the beginning of the game and they keep these strategies up to the first acceptance. If Player 1 accepts an object first, at that moment Player 2 may change his strategy by choosing a new level. He knows only the strategy of Player 1 (the exact value of chosen object is unknown) and he uses this information when he decides which level to choose. Similarly, if Player 2 accepts an object first, then Player 1 changes his strategy based on his knowledge about decision process so far.

Let Player 1 and Player 2 choose the levels $x \in [0, 1]$ and $y \in [0, 1]$ respectively. Player 1 gets $+1$ when he accepts the first $\xi_s \geq x$ (if $\xi_s < y$ or $\xi_s \geq y$ and the lottery chooses Player 1) and all further objects are less than ξ_s or when Player 2 accepts the first $\xi_s \geq y$ (if $\xi_s < x$ and Player 1 changes his level to $x_s = x_s(y, x)$ or $\xi_s \geq x$, the lottery chooses Player 2 and Player 1 chooses new level $x_s = x_s(x \vee y, 1)$) and the first, after ξ_s, object $\xi_t \geq x_s$ is also greater than ξ_s and there is no object greater than ξ_t later. Since this is a zero-sum game, Player 1 gets -1 when Player 2 gets $+1$ (and, in the description ot winning events for Player 1, x is interchanged with y and "Player 1" with "Player 2"). In other cases, the payoff for Player 1 is 0.

Since the conditional distribution of ξ, given $\xi \in [a, b]$, is uniform on $[a, b]$, the level $x_s(a, b)$ $[y_s(a, b)]$ is the optimal strategy in one-person best choice problem for Player 1 [Player 2] given $N(T) \geq s$, when the opponent has chosen an object $\xi_s \in [a, b)$ according to the uniform distribution. In this way, the strategies $\mathsf{x} = (x, \{x_n\}) \in \mathcal{T}_1$, $\mathsf{y} = (y, \{y_n\}) \in \mathcal{T}_2$ are constructed and the problem is reduced to the zero-sum game on the unit square ($\mathcal{S}^0$ is equivalent to the interval [0, 1]), with the payoff function being the expected value of Player 1's gain.

3 Equivalent Game on Unit Square

The payoff function $f(\mathsf{x}, \mathsf{y})$ for the strategies $\mathsf{x} = (x, \{x_n\}) \in \mathcal{T}_1$, $\mathsf{y} = (y, \{y_n\}) \in \mathcal{T}_2$ can be written as a function of x and y in the form

$$f(x, y, \pi) = \sum_{n=1}^{\infty} f_n(x, y, \pi)\, P(N(T) = n), \tag{3}$$

where $f_n(x, y, \pi)$ denotes the payoff to Player 1, when the levels (x, y) are chosen and $N(T) = n$.

Then for $x < y$ we have

$$\begin{aligned} f_n(x, y, \pi) = {} & \sum_{s=1}^{n} x^{s-1} \int_x^y \xi_s^{n-s}\, d\xi_s + \pi \sum_{s=1}^{n} x^{s-1} \int_y^1 \xi_s^{n-s}\, d\xi_s \\ & + (1-\pi) \sum_{t=2}^{n} \sum_{s=1}^{t-1} x^{s-1} \int_y^1 x_{sy1}^{t-s-1}\, d\xi_s \int_{x_{sy1} \vee \xi_s}^1 \xi_t^{n-t}\, d\xi_t \\ & - \sum_{t=2}^{n} \sum_{s=1}^{t-1} x^{s-1} \int_x^y y_{sxy}^{t-s-1}\, d\xi_s \int_{y_{sxy} \vee \xi_s}^1 \xi_t^{n-t}\, d\xi_t \end{aligned}$$

$$- (1-\pi)\sum_{s=1}^{n} x^{s-1}\int_y^1 \xi_s^{n-s}\,d\xi_s$$

$$- \pi\sum_{t=2}^{n}\sum_{s=1}^{t-1} x^{s-1}\int_y^1 y_{sy1}^{t-s-1}\,d\xi_s\int_{y_{sy1}\vee\xi_s}^1 \xi_t^{n-t}\,d\xi_t \tag{4a}$$

and for $x \geq y$

$$f_n(x,y,\pi) = \sum_{s=1}^{n-1}\sum_{t=s+1}^{n} y^{s-1}\int_y^x x_{syx}^{t-s-1}\,d\xi_s\int_{x_{syx}\vee\xi_s}^1 \xi_t^{n-t}\,d\xi_t$$

$$+ \pi\sum_{s=1}^{n} y^{s-1}\int_x^1 \xi_s^{n-s}\,d\xi_s$$

$$+ (1-\pi)\sum_{s=1}^{n-1}\sum_{t=s+1}^{n} y^{s-1}\int_x^1 x_{sx1}^{t-s-1}\,d\xi_s\int_{x_{sx1}\vee\xi_s}^1 \xi_t^{n-t}\,d\xi_t$$

$$- \sum_{s=1}^{n} y^{s-1}\int_y^x \xi_s^{n-s}\,d\xi_s - (1-\pi)\sum_{s=1}^{n} y^{s-1}\int_x^1 \xi_s^{n-s}\,d\xi_s$$

$$- \pi\sum_{s=1}^{n-1}\sum_{t=s+1}^{n} y^{s-1}\int_x^1 y_{sx1}^{t-s-1}\,d\xi_s\int_{y_{sx1}\vee\xi_s}^1 \xi_t^{n-t}\,d\xi_t, \tag{4b}$$

where $x_{sab} = x_s(a,b)$, $y_{sab} = y_s(a,b)$.

Based on the distributions of T and the process $N(t)$, the distribution of the total number of objects can be found

$$P(N(T)=n) = \int_0^\infty P(S_n \leq t, S_{n+1} > t)dH(t) \tag{5}$$

$$= \int_0^\infty dH(t)\int_0^t P(\rho_{n+1} > t-s)dG^{*n}(s),$$

where G^{*n} stands for the distribution of S_n.

Due to the form of the payoff function, it is very difficult to obtain the optimal levels (x, y) explicitly, even if the distributions of G and H are fixed. Nevertheless, in the natural case considered below, the solution has a very simple form.

4 A Poisson Stream of Objects

Let G be exponential with parameter λ. Thus $(N(t))_{t\in[0,+\infty)}$ is the Poisson process with parameter λ. Moreover, let T have an exponential distribution with parameter μ. In this case the probability that exactly N objects appear up to time T, given by (5), can be calculated as

$$P(N(T)=n) = \int_0^\infty\left(\int_0^t e^{-\lambda(t-s)}\frac{\lambda^n}{(n-1)!}s^{n-1}e^{-\lambda s}ds\right)\mu e^{-\mu t}dt$$

$$= \frac{\mu\lambda^n}{(\lambda+\mu)^{n+1}}.$$

In this case, the levels $x_s(a,b) \stackrel{\text{df}}{=} x_{ab}$, $y_s(a,b) \stackrel{\text{df}}{=} y_{ab}$ do not depend on the moment s, because the distribution of N given $N \geq s$ is geometric with parameter $p \stackrel{\text{df}}{=} \mu/(\lambda+\mu)$ for each s. This enables us to change the order of summation in (3). After summing and integrating the payoff function $f(x, y, \pi)$ is transformed to a function $\bar{f}(s, t, \pi)$ of new coordinate variables $s = p/(1-qx)$, $t = p/(1-qy)$ (this transformation $[0,1] \times [0,1]$ onto $[p,1] \times [p,1]$ preserves monotonicity) of the form

$$\bar{f}(s,t,\pi) \tag{6}$$
$$= \begin{cases} s\ (\bar{g}(s,t,t_{st}) + \pi\bar{g}(t,1,t_{t1}) - (1-\pi)\bar{g}(t,1,s_{t1})) & \text{if } s < t, \\ t\ (-\bar{g}(t,s,s_{ts}) + \pi\bar{g}(s,1,t_{s1}) - (1-\pi)\bar{g}(s,1,s_{s1}))) & \text{if } s \geq t, \end{cases}$$

where $s_{s1}, s_{ts}, s_{t1}, t_{s1}, t_{st}, t_{t1}$ are images of $x_{x1}, x_{yx}, x_{y1}, y_{x1}, y_{xy}, y_{y1}$ respectively, and

$$\bar{g}(A,B,u)$$
$$= \begin{cases} \ln B - \ln A - u\,((1+\ln B)/B - (1+\ln A)/A) & \text{if } p \leq u < A, \\ \ln B - \ln A - u(1+\ln B)/B + u\ln u/A + 1 & \text{if } A \leq u < B, \\ \ln B - \ln A + u(1/A - 1/B)\ln u & \text{if } B \leq u \leq 1. \end{cases}$$

Since Player 1 [Player 2] wants to choose his levels s_{s1}, s_{ts}, s_{t1}, $[t_{s1}, t_{st}, t_{t1}]$ to maximize [minimize] $\bar{f}(s,t,\pi)$, we have to find the $u^* = u^*(A,B)$ for which $\min_{u\in[p,1]} \bar{g}(A,B,u)$ is achieved. Such an optimal u^* has the following form (cf Porosiński & Szajowski [3]): $u^* = \exp(-1 + A(1+\ln B)/B)$ if $B \geq e^{-1}$ and $u^* = e^{-1}$ if $B < e^{-1}$.

In this way the game is transformed to a zero-sum game on $[p,1] \times [p,1]$, with the gain function, obtained from (6) by putting $s_{s1} = t_{s1} = u^*(s,1)$, $s_{ts} = u^*(t,s)$, $t_{st} = u^*(s,t)$, $s_{t1} = t_{t1} = u^*(t,1)$, of the following form

$$\bar{f}(s,t,\pi)$$
$$= \begin{cases} s\left(\ln t - \ln s - \left(\frac{1}{s} - \frac{1}{t}\right)e^{-1} + (2\pi-1)\left(1 - \ln t - \frac{1}{t}e^{t-1}\right)\right) \\ \qquad\qquad\qquad\qquad\qquad\qquad \text{if } s < t < e^{-1}, \\ s\left(\ln t - \ln s + 1 - \frac{1}{s}e^{-1+s(1+\ln t)/t} + (2\pi-1)\left(1 - \ln t - \frac{1}{t}e^{t-1}\right)\right) \\ \qquad\qquad\qquad\qquad\qquad\qquad \text{if } t \geq s \vee e^{-1}, \\ t\left(\ln t - \ln s - \left(\frac{1}{s} - \frac{1}{t}\right)e^{-1} + (2\pi-1)\left(1 - \ln s - \frac{1}{s}e^{s-1}\right)\right) \\ \qquad\qquad\qquad\qquad\qquad\qquad \text{if } t \leq s < e^{-1}, \\ t\left(\ln t - \ln s - 1 + \frac{1}{t}e^{-1+t(1+\ln s)/s} + (2\pi-1)\left(1 - \ln s - \frac{1}{s}e^{s-1}\right)\right) \\ \qquad\qquad\qquad\qquad\qquad\qquad \text{if } s \geq t \vee e^{-1}. \end{cases}$$

Since the relation $\bar{f}(s, t, \pi) + \bar{f}(t, s, 1 - \pi) = 0$ follows, we restrict our considerations to $\pi \in [0.5, 1]$ and from now on we shorten notation, writing $\bar{f}(s, t)$ instead of $\bar{f}(s, t, \pi)$.

The existence and form of equilibrium for such a game can be found in Parthasarathy & Raghavan [1].

Table 1: The solution of the game with random priority π for the Poisson stream of objects with parameter λ, the exponential horizon with parameter μ and $\mu/(\lambda + \mu) \le t^*$

π	s^*	t^*	$v(\pi)$
0.5	0.3679	0.3679	0
0.6	0.3627	0.3292	0.0387
0.75	0.3579	0.2785	0.0894
0.9	0.3548	0.2356	0.1323
1	0.3533	0.2107	0.1571

Theorem 4.1. *Let $F(x, y)$ be continuous function on the unit square $[0, 1] \times [0, 1]$ and let $F(x, y)$ be concave in x for each y. Then the zero-sum game $\Gamma = ([0, 1], [0, 1], F(x, y))$ has an equilibrium of the form $(I_a, \beta I_c + (1 - \beta)I_d)$ for some $0 \le a, c, d, \beta \le 1$, where I_a stands for the distribution concentrated at a.*

Let us consider the function $\bar{f}(s, t)$ on the unit square. It is continuous in both variables, concave on s and has (as a function of s) a unique maximum for every fixed t. Equating to zero the partial derivatives of the function $\bar{f}(s, t)$ we find the point (s^*, t^*) in a triangle $t < s < e^{-1}$, which is the unique solution of the equations

$$e^{-1} - s + (2\pi - 1)(e^{s-1} - s - se^{s-1}) = 0,$$
$$e^{-1} - s - s(\ln t - \ln s) + (2\pi - 1)(e^{s-1} - s + s \ln s) = 0. \quad (7)$$

At the point (s^*, t^*) there is a saddle point of $\bar{f}(s, t)$ on the unit square. Accordingly, the pair of strategies (s^*, t^*) fulfils the minimax conditions $\min_t \bar{f}(s^*, t) = \max_s \min_t \bar{f}(s, t) \overset{\text{df}}{=} v(\pi)$, $\max_s \bar{f}(s, t^*) = \min_t \max_s \bar{f}(s, t) = v(\pi)$. Player 1 has an optimal pure strategy $s^* \in (0, e^{-1}]$ and Player 2 has an optimal pure strategy $t^* = (s^*)^{2\pi} \exp\left((2\pi - 1)e^{s^*-1}\right)$, where $t^* < s^*$. The game value is $v(\pi) = \bar{f}(s^*, t^*) = (s^* - t^*)e^{-1}/s^* + (2\pi - 1)t^*$. This solution on $[0, 1] \times [0, 1]$ is also valid on $[p, 1] \times [p, 1]$ for $p \le t^*$ because only the domain of the gain function $\bar{f}(s, t)$ is dependent on p (the form given by (6) is independent of p). For $p > t^*$ the players have to modify their strategies according to the properties of $\bar{f}(s, t)$ on $[p, 1] \times [p, 1]$.

For the main problem, based on the auxiliary game, we can formulate the following solution.

Proposition 4.1. *Let* $\pi \in [0.5, 1]$ *and let* (s^*, t^*) *be a unique solution of (7). For the Poisson stream of objects with parameter* λ *and the exponential horizon with parameter* μ, *if* $\mu/(\lambda + \mu) \leq t^*$, *there exists a solution of the game having the following form. The optimal level for Player 1 is* $x^* = 1 - (\mu(1 - s^*))/\lambda s^*$ *and the optimal level for Player 2 is* $y^* = 1 - (\mu(1 - t^*))/\lambda t^*$. *The game value* $v(\pi) = (s^* - t^*)e^{-1}/s^* + (2\pi - 1)t^*$ *is independent of* λ *and* μ.

The *small* p case seems to be *natural*. Small $\mu/(\lambda + \mu)$ yields the expected number of objects $EN = \lambda/\mu$ to be big enough. The condition $p \leq t^*$ is equivalent to $EN \geq 1/t^* - 1$, and this lower bound is rather small, e.g. 1.72, 2.04, 3.01 and 3.75 for π equal to 0.5, 0.6, 0.85 and 1 respectively. The problem for $\pi = 1$ considered here should be treated as an asymptotic case, but, if there is no random priority, the lottery mechanizm is needless and the problem is better modeled in Porosiński & Szajowski [4].

REFERENCES

[1] Parthasarathy T. & Raghavan T., Equilibria of continuous two-person games, *Pacific Journal of Mathematics*, **57** (1975), 265–270.

[2] Porosiński Z. & Szajowski K., On continuous-time two person full-information best choice problem with imperfect observation, *Sankyā*, **58** (1996), Series A, 186–193.

[3] Porosiński Z. & Szajowski K., Full-information best choice problem with random starting point, *Math. Japonica*, **52** (2000), 57–63.

[4] Porosiński Z. & Szajowski K., Modified strategies in two person best choice problem with imperfect observation, *Math. Japonica*, **52** (2000), 103–112.

[5] Szajowski K., Double stopping by two decision makers, *Adv. Appl. Probab.*, **25** (1993), 438–452.

Bilateral Approach to the Secretary Problem

David Ramsey
Institute of Mathematics
Wrocław University of Technology
50-370 Wrocław, Poland
ramsey@im.pwr.wroc.pl

Krzysztof Szajowski
Institute of Mathematics
Wrocław University of Technology
50-370 Wrocław, Poland
szajow@im.pwr.wroc.pl

Abstract

A mathematical model of competitive selection of the applicants for a post is considered. There are N applicants with similar qualifications on an interview list. The applicants come in random order and their salary demands are distinct. Two managers, I and II, interview them one at a time. The aim of the manager is to obtain the applicant who demands minimal salary. When both managers want to accept the same candidate, then some rule of assignment to one of the managers is applied. Any candidate hired by a manager will accept the offer with some given probability. A candidate can be hired only at the moment of his appearance and can be accepted at that moment. At each moment n one candidate is presented. The considered problem is a generalization of the best choice problem with uncertain employment and its game version with priority or random priority. The general stopping game model is constructed. The algorithms of construction of the game value and the equilibrium strategies are given. An example is solved.

Key words. optimal stopping problem, game variant, Markov process, random priority, secretary problem

AMS Subject Classifications. 60G40, 62L15, 90D15.

1 Introduction

This paper deals with the mathematical model of competitive selection of the applicants for a post. There are N applicants of similar qualification on an interview list. The applicants come in a random order and their salary demands are distinct. Two managers, called Player 1 and Player 2, interview them one at a time. The aim of the manager is to obtain the applicant who demands minimal salary. When both managers want to accept the same candidate, then some rule of assignment to one of the managers is applied. Any candidate hired by the manager will accept the offer of job with some given probability. A candidate can be hired only at the moment of his appearance and can be accepted at that moment. At each moment n one candidate is presented. The considered problem is related both to the uncertain employment considered by [12] and to the competitive optimal stopping problem with priority (see [4]) or more generally with random priority of the players (see [7], [14]).

Let us formulate the optimal stopping problem with uncertain employment considered by Smith [12] (see also [15]) in a rigorous way. Let a homogeneous Markov process $(X_n, \mathcal{F}_n, \mathbf{P}_x)_{n=0}^N$ be defined on probability space $(\Omega, \mathcal{F}, \mathbf{P})$ with fixed state space $(\mathbb{E}, \mathcal{B})$. Define the gain function $f : \mathbb{E} \to \Re$. Let $\mathfrak{M}^N$ be a set of sequences $\bar{\mu} = \{\mu_n\}_{n=0}^N$ of $\{0, 1\}$-valued random variables such that μ_n is $\mathcal{F}_n$-measurable for every n. Let $\{\eta_n\}_{n=0}^N$ be a sequence of i.i.d. r.v. with the uniform distribution on $[0, 1]$, independent of $\{X_n\}_{n=0}^N$ and $\bar{\mu}$ and let $\alpha = \{\alpha_n\}_{n=0}^N$ be the sequence of real numbers, $\alpha_n \in [0, 1]$. Define $\tau_\alpha(\bar{\mu}) = \inf\{n \geq 0 : \mu_n = 1, \eta_n \leq \alpha_n\}$. In the optimal stopping problem with uncertain employment the aim is to find $\bar{\mu}^*$ such that

$$E_x f(X_{\tau_\alpha(\bar{\mu}^*)}) = \sup_{\bar{\mu} \in \mathfrak{M}^N} E_x f(X_{\tau_\alpha(\bar{\mu})}) \text{ for all } x \in \mathbb{E}$$

and to determine the function $v(x) = E_x f(X_{\tau_\alpha(\bar{\mu}^*)})$. We can look at the above problem as a problem of one decision-maker who wants to accept, on the basis of sequential observation, the most profitable state of the Markov process which appears in the realization but the solicited state is available with some probability only. The availability is unknown before solicitation. If the decision-maker has made an unsuccessful stop he can choose any next state under the same rules. The availability is described by the sequence α.

In a bilateral approach each player can get at most one of the states from the realization of the Markov chain. Since there is only one random sequence $\{X_n\}_{n=0}^N$ in a trial, at each instant n only one player can obtain a realization x_n of X_n. Both players together can accept at most two objects. The problem of assigning the objects to the players when both want to accept the same one can be solved in many ways. In [2] Dynkin assumed that for odd n Player 1 can choose x_n and for even n Player 2 can choose. Other authors solve the problem by more or less arbitrary definition of the payoff function. Sakaguchi [11] considered some version of the bilateral sequential games related to the no-information secretary problem with uncertain employment. The paper investigated the two-person non-zero-sum games with one or two sets of N objects under the conditions of the secretary problem. In the case of one set of objects it can happen that both players attempt to accept the same object. In this case players have half success which is taken into account in the payoff function. Another approach assumes a priority for one decision-maker (see papers by Sakaguchi [10], Enns & Ferenstein [3], Radzik & Szajowski [6], Ravindran & Szajowski [9]) or the random priority (the paper by Fushimi [5], Radzik & Szajowski [7] and Szajowski [14]).

The model of competitive choice of the required object with the uncertain employment and random priority has been formulated and preliminary results have been obtained by Szajowski [13]. At each moment n the state of the Markov process x_n is presented to both players. If the players have not already made an acceptance there are the following possibilities. If only one of them would like to accept the state then he tries to take it. In this moment a random mechanism assigns availability to the state (which can depend on the player and the moment of decision n).

Model A. This is the approach which has been considered by the authors in [8].

(i) If both of them are interested in this state then at first the random device chooses the player who will first solicit the state. The availability of the state is similar to the situation when only one player want to take it.

(ii) If a state is not available for the player chosen by the random device then the observed state at moment n is lost as in the case when both players reject it. The next state in the sequence is interviewed.

Model B. The model differs from **Model A** only in the case when both players would like to accept the same state. So that point (i) is the same.

If the random device chooses Player 1 and the state is not available for him (lottery decides about that) then the observed state at moment n is solicited by Player 2. The state is available for him as in the situation when only Player 2 tries to take it (the random experiment decides about it). If the state is not available then it is lost and the next state in the sequence is interviewed.

Model C. The model differs from **Model A** and **B** in the case when both players would like to accept the same state. This model admits that if the state is not available for the player chosen by the device then the other player is able to solicit the state.

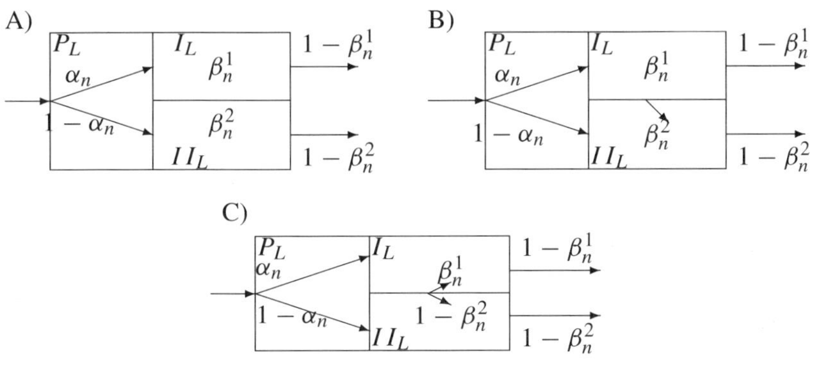

Figure 1: The schemes of decision processes

Fig. 1 presents the scheme of the decision process in each model. The lottery P_L assigns the priority to the players. The random devices I_L and II_L describe availability of the state to Player 1 and Player 2 respectively. In **Model B** there is a door between I_L and II_L which can be opened from the room I_L. In **Model C** the door handles are from both sides.

This paper deals with the extended model described in the point **Model C**. In Section 2 the formal description of a two-step random assignment is given. The algorithms solution of the game related to the model described in Section 2 is presented in Section 3. The examples are solved in Section 4.

2 Two-Step Random Assignment

Let $(X_n, \mathcal{F}_n, \mathbf{P}_x)_{n=0}^N$ be a homogeneous Markov chain defined on a probability space $(\Omega, \mathcal{F}, \mathbf{P})$ with state space $\mathbb{E}\times\mathbb{E}$ and let $f_1 : \mathbb{E}\times\mathbb{E} \to \Re$ and $f_2 : \mathbb{E}\times\mathbb{E} \to \Re$ be $\mathcal{B} \times \mathcal{B}$ real-valued measurable functions. The horizon N is finite. Player i $(i \in \{1, 2\})$ observes the Markov chain and tries to maximize his payoff defined by the function f_i. Each realization x_n of X_n can be accepted by at most one player and neither player can accept more than one realization of the chain. It is assumed there is a lottery, which decides which player has priority when both players wish to accept the same realization. Also, it is assumed that if a player wishes to accept a realization x_n of X_n and has priority, then that player obtains that realization with some probability that is strictly non-zero and strictly less than one (i.e. uncertain employment). If a player has not accepted any of the previous realizations at stage n, then he has two options. The first is to solicit the observed state of the process, the second is to reject it. Once a player has accepted one of the realizations, then he no longer takes part in the game.

If both players wish to accept the same realization, then the lottery chooses which player has priority. Let $(\{\epsilon_n\}_{n=0}^N, \{\alpha_n\}_{n=0}^N)$ be the description of the lottery, where ϵ_i, $i = 0, 1, \dots, N$ are a sequence of i.i.d r.v.s from the [0, 1] uniform distribution and the α_i, $i = 0, 1, \dots, N$ are real numbers, $\alpha_i \in [0, 1]$. When both players wish to accept the same realization x_n of X_n, then Player 1 has priority if $\epsilon_n \leq \alpha_n$, otherwise Player 2 has priority. Similarly, the lottery $(\{\eta_n^i\}_{n=0}^N, \{\beta_n^i\}_{n=0}^N)$ describes the availability of the nth realization of the chain to the ith player. When only Player 1 (Player 2) accepts state x (y) then Player 1 obtains $g_1(x) = \sup_{y\in\mathbb{E}} f_1(x, y)$ $(g_2(y) = \inf_{x\in\mathbb{E}} f_1(x, y))$ by assumption. Similarly, when only Player 1 (Player 2) accepts state x (y) then Player 2 obtains $g_3(x) = \inf_{y\in\mathbb{E}} f_2(x, y)$ $(g_4(x) = \sup_{x\in\mathbb{E}} f_2(x, y))$. If neither player accepts a realization, then they both gain 0.

Let Ω^N be the aggregation of sequences $\overline{\sigma} = \{\omega_n\}_{n=0}^N$ of $\{0, 1\}$-valued random variables such that ω_n is F_n-measurable, $n = 0, 1 \dots, N$. If a player uses $\overline{\sigma}$, then $\sigma_n = 1$ means that he declares willingness to accept the realization x_n of X_n. If $\sigma_n = 0$, then the player is not interested in accepting the realization x_n. Denote $\Omega_k^N = \{\overline{\sigma} : \sigma_0 = 0, \sigma_1 = 0, \dots, \sigma_{k-1} = 0\}$. Let Λ_k^N and Γ_k^N be copies of Ω_k^N $(\Omega^N = \Omega_0^N)$. One can define the sets of strategies $\overline{\Lambda}^N = \{(\overline{\lambda}, \{\overline{\sigma}_n^1\}) : \lambda \in \Lambda^N, \overline{\sigma}_n^1 \in \Lambda_{n+1}^N \forall n\}$ and $\overline{\Gamma}^N = \{(\overline{\gamma}, \{\overline{\sigma}_n^2\}) : \overline{\gamma} \in \Gamma^N, \overline{\sigma}_n^2 \in \Gamma_{n+1}^N \forall n\}$ for Players 1 and 2 respectively. The strategies $\overline{\lambda}$ and $\overline{\gamma}$ are applied by Player 1 and Player 2 respectively, until the first of the two players has obtained one of the realizations of the Markov chain. After that point the other player, Player i say, continues alone using strategy $\overline{\sigma}_n^i$, $i = 1, 2$.

Let $E_x f_1^+(X_n) < \infty$, $E_x f_1^-(X_n) < \infty$, $E_x f_2^+(X_m) < \infty$ and $E_x f_2^-(X_m) < \infty$ for $n, m = 0, 1, \dots, N$ and $x \in \mathbb{E}$. Let $\psi \in \overline{\Lambda}^N$ and $\tau \in \overline{\Gamma}^N$. Based on the strategies ψ and τ used by Player 1 and Player 2 respectively, the definition of the lotteries and the type of model used, the expected gains $\overline{R}_{1,\bullet}(x, \psi, \tau)$ and $\overline{R}_{2,\bullet}(x, \psi, \tau)$ for Player 1 and Player 2 respectively can be obtained. In this way the form of the game $(\overline{\Lambda}^N, \overline{\Gamma}^N, \overline{R}_{1,\bullet}(x, \psi, \tau), \overline{R}_{2,\bullet}(x, \psi, \tau))$ is defined. This game is denoted by $G^\bullet$. For zero sum games the normal form of the game can be simply defined by $(\overline{\Lambda}^N, \overline{\Gamma}^N, \overline{R}_{1,\bullet}(x, \psi, \tau))$ since $\overline{R}_{1,\bullet}(x, \psi, \tau) = -\overline{R}_{2,\bullet}(x, \psi, \tau)$. The three models considered in the introduction are presented in the following section for both zero sum and non-zero sum games.

Definition 2.1. *The pair (ψ^*, τ^*) is an equilibrium point in the game $G^\bullet$ if for every $x \in \mathbb{E}$, $\psi \in \overline{\Lambda}^N$ and $\tau \in \overline{\Gamma}^N$ the following two inequalities hold*

$$\overline{R}_{1,\bullet}(x, \psi, \tau^*) \le \overline{R}_{1,\bullet}(x, \psi^*, \tau^*), \tag{1}$$

$$\overline{R}_{2,\bullet}(x, \psi^*, \tau) \le \overline{R}_{2,\bullet}(x, \psi^*, \tau^*). \tag{2}$$

In the particular case of zero-sum games, these conditions simplify to

$$\overline{R}_{1,\bullet}(x, \psi, \tau^*) \le \overline{R}_{1,\bullet}(x, \psi^*, \tau^*) \le \overline{R}_{1,\bullet}(x, \psi^*, \tau). \tag{3}$$

The aim is to construct equilibrium pairs (ψ^*, τ^*). After one of the players accepts realization x_n at time n, the other player tries to maximize his gain without any disturbance from the player choosing first, as in the optimal stopping problem with uncertain employment (see Smith [12]). Thus, if neither player has accepted a realization up to stage n, the players must take into account the potential danger from a future decision of the opponent, in order to decide whether or not to accept the realization x_n of X_n. In order to do this, they consider some auxiliary game $G_a^\bullet$.

Let $\psi = (\overline{\lambda}, \{\overline{\sigma}_n^1\})$ and $\tau = (\overline{\gamma}, \{\overline{\sigma}_n^2\})$. Define $s_0(x, y) = \beta_N^2 f_2(x, y) + (1 - \beta_N^2) g_3(x)$, $S_0(x, y) = \beta_N^1 f_1(x, y) + (1 - \beta_N^1) g_2(y)$ and

$$s_n(x, y) = \sup_{t \in \Gamma_{N-n}^N} E_y f_2(x, X_{\sigma(\tau, \beta^2)}) \tag{4}$$

$$S_n(x, y) = \sup_{s \in \Lambda_{N-n}^N} E_x f_1(X_{\sigma(\psi, \beta^1)}, y) \tag{5}$$

for all $x, y \in \mathbb{E}$, $n = 1, 2 \dots, N$, where $\sigma(\psi, \beta^1) = \inf\{0 \le n \le N : \sigma_n^1 = 1, \eta_n^1 \le \beta_n^1\}$ and $\sigma(\tau, \beta^2) = \inf\{0 \le n \le N : \sigma_n^2 = 1, \eta_n^2 \le \beta_n^2\}$. By backward induction (see Bellman [1]), the functions $s_n(x, y)$ can be constructed as $s_n(x, y) =$

$\max\{\beta_n^2 f_2(x,y)+(1-\beta_n^2)T_2 s_{n-1}(x,y), T_2 s_{n-1}(x,y)\}$ and the functions $S_n(x,y)$ has the form $S_n(x,y) = \max\{\beta_n^1 f_1(x,y) + (1-\beta_n^1)T_1 S_{n-1}(x,y), T_1 S_{n-1}(x,y)\}$ respectively, where $T_1 f(x,y) = E_y f(x, X_1)$ and $T_2 f(x,y) = E_x f(X_1, x)$. The operations minimum, maximum, T_2 and T_1 all preserve measurability. Hence $s_n(x,y)$ and $S_n(x,y)$ are $B \otimes B$ measurable. If Player 1 has obtained x at moment n and Player 2 has not yet obtained any realization, then the expected gain of Player 2 is given by $h_2(n,x)$ ($i \in \{1,2\}$), where

$$h_2(n,x) = E_x s_{N-n-1}(x, X_1) \tag{6}$$

for $n = 0, 1, \ldots, N-1$ and $h_2(N,x) = g_3(x)$. Let the future expected reward of Player 1 in such a case be denoted $h_1(n,x)$. If the game is a zero-sum game, then $h_1(n,x) = -h_2(n,x)$.

When Player 2 is the first player to obtain a realization at time n, then the expected gain of Player 1 is given by $H_1(n,x)$, where

$$H_1(n,x) = E_x S_{N-n-1}(X_1, x) \tag{7}$$

for $n = 0, 1, \ldots, N-1$ and $H_1(N,x) = g_2(x)$. Let the future expected reward of Player 2 in such a case be denoted by $H_2(n,x)$. If the game is a zero-sum game, then $H_2(n,x) = -H_1(n,x)$.

Based upon the solutions of the optimization problems when a player remains alone in the decision process, we can consider such an auxiliary game $G_a^{\bullet}$. The form of this game depends on the model determining what happens when both players wish to accept the same state.

3 The Extended Model

Assume that the model deciding the priority assignment is Model C, as given in the introduction. The game related to Model C will be denoted G^C. The sets of strategies available to Player 1 and Player 2 are $\overline{\Lambda}^N$ and $\overline{\Gamma}^N$ respectively. For $\psi = (\overline{\lambda}, \{\overline{\sigma}_n^1\}) \in \overline{\Lambda}^N$ and $\tau = (\overline{\gamma}, \{\overline{\sigma}_n^2\}) \in \overline{\Gamma}^N$, we define the following random variables

$$\begin{aligned}
\lambda_{\alpha,\beta^1,\beta^2}(\psi,\tau) = \inf\{0 \le n \le N : &(\lambda_n = 1, \gamma_n = 1, \epsilon_n \le \alpha_n, \eta_n^1 \le \beta_n^1) \\
&\text{or } (\lambda_n = 1, \gamma_n = 0, \eta_n^1 \le \beta_n^1) \\
&\text{or } (\lambda_n = 1, \gamma_n = 1, \epsilon_n > \alpha_n, \eta_n^2 > \beta_n^2, \eta_n^1 \le \beta_n^1)\}, \\
\gamma_{\alpha,\beta^1,\beta^2}(\psi,\tau) = \inf\{0 \le n \le N : &(\lambda_n = 1, \gamma_n = 1, \epsilon_n > \alpha_n, \eta_n^2 \le \beta_n^2) \\
&\text{or } (\lambda_n = 0, \gamma_n = 1, \eta_n^2 \le \beta_n^2) \\
&\text{or } (\lambda_n = 1, \gamma_n = 1, \epsilon_n \le \alpha_n, \eta_n^1 > \beta_n^1, \eta_n^2 \le \beta_n^2)\}.
\end{aligned}$$

Let

$$\rho_1(\psi,\tau) = \lambda_{\alpha,\beta^1,\beta^2}(\psi,\tau)\mathbb{I}_{\{\lambda_{\alpha,\beta^1,\beta^2}(\psi,\tau)<\gamma_{\alpha,\beta^1,\beta^2}(\psi,\tau)\}}$$
$$+\sigma_{\gamma_{\alpha,\beta^1,\beta^2}}(\psi,\beta^1)\mathbb{I}_{\{\lambda_{\alpha,\beta^1,\beta^2}(\psi,\tau)>\gamma_{\alpha,\beta^1,\beta^2}(\psi,\tau)\}}$$

and

$$\rho_2(\psi,\tau) = \gamma_{\alpha,\beta^1,\beta^2}(\psi,\tau)\mathbb{I}_{\{\lambda_{\alpha,\beta^1,\beta^2}(\psi,\tau)>\gamma_{\alpha,\beta^1,\beta^2}(\psi,\tau)\}}$$
$$+\sigma_{\lambda_{\alpha,\beta^1,\beta^2}}(\tau,\beta^2)\mathbb{I}_{\{\lambda_{\alpha,\beta^1,\beta^2}(\psi,\tau)<\gamma_{\alpha,\beta^1,\beta^2}(\psi,\tau)\}}.$$

We have

$$\overline{R}_{1,C}(x,\psi,\tau) = E_x f_1(X_{\rho_1}(\psi,\tau), X_{\rho_2}(\psi,\tau)),$$
$$\overline{R}_{2,C}(x,\psi,\tau) = E_x f_2(X_{\rho_1}(\psi,\tau), X_{\rho_2}(\psi,\tau)).$$

In the auxiliary game G_a^C, the sets of strategies available to Player 1 and Player 2 are Λ^N and Γ^N respectively. For $\overline{\lambda} \in \Lambda^N$ and $\overline{\gamma} \in \Gamma^N$ we define the random variables

$$\begin{aligned}
\overline{\lambda}_{\alpha,\beta^1,\beta^2}(\overline{\lambda},\overline{\gamma}) = \inf\{0 \le n \le N : &(\lambda_n = 1, \gamma_n = 1, \epsilon_n \le \alpha_n, \eta_n^1 \le \beta_n^1)\\
&\text{or } (\lambda_n = 1, \gamma_n = 0, \eta_n^1 \le \beta_n^1)\\
&\text{or } (\lambda_n = 1, \gamma_n = 1, \epsilon_n > \alpha_n, \eta_n^2 > \beta_n^2, \eta_n^1 \le \beta_n^1)\},\\
\overline{\gamma}_{\alpha,\beta^1,\beta^2}(\overline{\lambda},\overline{\gamma}) = \inf\{0 \le n \le N : &(\lambda_n = 1, \gamma_n = 1, \epsilon_n > \alpha_n, \eta_n^2 \le \beta_n^2)\\
&\text{or } (\lambda_n = 0, \gamma_n = 1, \eta_n^2 \le \beta_n^2)\\
&\text{or } (\lambda_n = 1, \gamma_n = 1, \epsilon_n \le \alpha_n, \eta_n^1 > \beta_n^1, \eta_n^2 \le \beta_n^2)\}.
\end{aligned}$$

As long as $\overline{\lambda}_{\alpha,\beta^1,\beta^2}(\overline{\lambda},\overline{\gamma}) \le N$ or $\overline{\gamma}_{\alpha,\beta^1,\beta^2}(\overline{\lambda},\overline{\gamma}) \le N$, the payoff function for the i-th player is defined as follows

$$\begin{aligned}
r_i(\overline{\lambda}_{\alpha,\beta^1,\beta^2}(\overline{\lambda},\overline{\gamma}), \overline{\gamma}_{\alpha,\beta^1,\beta^2}(\overline{\lambda},\overline{\gamma})) = {} & h_i(\overline{\lambda}_{\alpha,\beta^1,\beta^2}(\overline{\lambda},\overline{\gamma}), X_{\overline{\lambda}_{\alpha,\beta^1,\beta^2}(\overline{\lambda},\overline{\gamma})})\\
&\times \mathbb{I}_{\{\overline{\lambda}_{\alpha,\beta^1,\beta^2}(\overline{\lambda},\overline{\gamma})<\overline{\gamma}_{\alpha,\beta^1,\beta^2}(\overline{\lambda},\overline{\gamma})\}}\\
&+ H_i(\overline{\gamma}_{\alpha,\beta^1,\beta^2}(\overline{\lambda},\overline{\gamma}), X_{\overline{\gamma}_{\alpha,\beta^1,\beta^2}(\overline{\lambda},\overline{\gamma})})\\
&\times \mathbb{I}_{\{\overline{\lambda}_{\alpha,\beta^1,\beta^2}(\overline{\lambda},\overline{\gamma})\ge\overline{\gamma}_{\alpha,\beta^1,\beta^2}(\overline{\lambda},\overline{\gamma})\}}
\end{aligned} \tag{8}$$

otherwise the payoff to each player is 0.

Firstly, we consider zero sum games. As a solution to such a game, we look for an equilibrium pair $(\overline{\lambda}^*, \overline{\gamma}^*)$ such that

$$\begin{aligned} R(x, \overline{\lambda}_{\alpha,\beta^1,\beta^2}(\overline{\lambda}, \overline{\gamma}^*), \overline{\gamma}_{\alpha,\beta^1,\beta^2}(\overline{\lambda}, \overline{\gamma}^*)) &\leq R(x, \overline{\lambda}_{\alpha,\beta^1,\beta^2}(\overline{\lambda}^*, \overline{\gamma}^*), \overline{\gamma}_{\alpha,\beta^1,\beta^2}(\overline{\lambda}^*, \overline{\gamma}^*)) \\ &\leq R(x, \overline{\lambda}_{\alpha,\beta^1,\beta^2}(\overline{\lambda}^*, \overline{\gamma}), \overline{\gamma}_{\alpha,\beta^1,\beta^2}(\overline{\lambda}^*, \overline{\gamma})) \end{aligned} \tag{9}$$

for all $x \in \mathbb{E}$, where

$$R(x, \overline{\lambda}_{\alpha,\beta^1,\beta^2}(\overline{\lambda}, \overline{\gamma}), \overline{\gamma}_{\alpha,\beta^1,\beta^2}(\overline{\lambda}, \overline{\gamma})) = E_x r_1(\overline{\lambda}_{\alpha,\beta^1,\beta^2}(\overline{\lambda}, \overline{\gamma}), \overline{\gamma}_{\alpha,\beta^1,\beta^2}(\overline{\lambda}, \overline{\gamma})).$$

As in Model A, we can define a sequence $v_n(x)$, $n = 0, 1, \ldots, N+1$ on $\mathbb{E}$ by setting $v_{N+1}(x) = 0$ and

$$v_n(x) = \mathrm{val} \begin{bmatrix} \begin{array}{c} \alpha_n(\beta_n^1 h_1(n,x) + (1-\beta_n^1) g(n,x,\beta_n^2)) \\ +(1-\alpha_n)(\beta_n^2 H_1(n,x) + (1-\beta_n^2) G(n,x,\beta_n^1)) \end{array} & G(n,x,\beta_n^1) \\ g(n,x,\beta_n^2) & T v_{n+1}(x) \end{bmatrix} \tag{10}$$

for $n = 0, 1, \ldots, N$, where $G(n, x, \beta_n^1) = \beta_n^1 h_1(n, x) + (1 - \beta_n^1) T v_{n+1}(x)$ and $g(n, x, \beta_n^2) = \beta_n^2 H_1(n, x) + (1 - \beta_n^2) T v_{n+1}(x)$. By subtracting $T v_{n+1}(x)$ from each entry above, it can be seen that the game above is equivalent to a game with matrix A, where

$$A = \begin{bmatrix} \mathfrak{a}_{ss} & \mathfrak{a}_{sc} \\ \mathfrak{a}_{cs} & \mathfrak{a}_{cc} \end{bmatrix} = \begin{bmatrix} \begin{array}{c} \alpha(a + (1-\beta)b) \\ +(1-\alpha)(b + (1-\gamma)a) \end{array} & a \\ b & 0 \end{bmatrix} \tag{11}$$

where $a, b, \alpha, \beta, \gamma$ are real numbers and $\alpha, \beta, \gamma \in [0, 1]$. By direct checking we obtain

Lemma 3.1. *The two-person zero-sum game with payoff matrix A given above has an equilibrium point (ϵ, δ) in pure strategies, where*

$$(\epsilon, \delta) = \begin{cases} (s, s) & \text{if } (1 - (1-\alpha)\gamma)a \geq \alpha\beta b \cap (1 - \alpha\beta)b \leq (1-\alpha)\gamma a, \\ (s, f) & \text{if } a \geq 0 \cap (1 - \alpha\beta)b > (1-\alpha)\gamma a, \\ (f, s) & \text{if } b \leq 0 \cap (1 - (1-\alpha)\gamma)a < \alpha\beta b, \\ (f, f) & \text{if } a < 0 \cap b > 0. \end{cases} \tag{12}$$

Denote

$$
\begin{aligned}
A_n^{ss} &= \{x \in \mathbb{E} \,:\, (1-(1-\alpha_n)\beta_n^2)(h_1(n,x) - Tv_{n+1}(x)) \\
&\quad \geq \alpha_n\beta_n^2(H_1(n,x) - Tv_{n+1}(x)), (1-\alpha\beta_n^1)(H_1(n,x) - Tv_{n+1}(x)) \\
&\quad \leq (1-\alpha_n)\beta_n^1(h_1(n,x) - Tv_{n+1}(x))\} \qquad (13)\\
A_n^{sf} &= \{x \in \mathbb{E} \,:\, h_1(n,x) \geq Tv_{n+1}(x), (1-\alpha_n\beta_n^1)(H_1(n,x) - Tv_{n+1}(x)) \\
&\quad > (1-\alpha_n)\beta_n^1(h_1(n,x) - Tv_{n+1}(x))\} \qquad (14)\\
A_n^{fs} &= \{x \in \mathbb{E} \,:\, H_1(n,x) \leq Tv_{n+1}(x), (1-(1-\alpha_n)\beta_n^2)(h_1(n,x) - Tv_{n+1}(x)) \\
&\quad < \alpha_n\beta_n^2(H_1(n,x) - Tv_{n+1}(x))\} \qquad (15)
\end{aligned}
$$

and

$$A_n^{ff} = \mathbb{E} \setminus (A_n^{ss} \cup A_n^{sf} \cup A_n^{fs}) \qquad (16)$$

By the definition of the sets A_n^{ss}, A_n^{sf}, $A_n^{fs} \in \mathcal{B}$ and Lemma 3.1 we have

$$
\begin{aligned}
v_n(x) = {} & [\alpha_n(\beta_n^1(h_1(n,x) - Tv_{n+1}(x)) \\
& + (1-\beta_n^1)\beta_n^2(H_1(n,x) - Tv_{n+1}(x))) \\
& + (1-\alpha_n)(\beta_n^2(H_1(n,x) - Tv_{n+1}(x)) \\
& + (1-\beta_n^2)\beta_n^1(h_1(n,x) - Tv_{n+1}(x)))]\mathbb{I}_{A_n^{ss}}(x) \\
& + \beta_n^1(h_1(n,x) - Tv_{n+1}(x))\mathbb{I}_{A_n^{sf}}(x) \\
& + \beta_n^2(H_1(n,x) - Tv_{n+1}(x))\mathbb{I}_{A_n^{sf}}(x) + Tv_{n+1}(x). \qquad (17)
\end{aligned}
$$

Define

$$\lambda_n^* = \begin{cases} 1 & \text{if } X_n \in A_n^{ss} \cup A_n^{sf}, \\ 0 & \text{otherwise.} \end{cases} \qquad (18)$$

$$\gamma_n^* = \begin{cases} 1 & \text{if } X_n \in A_n^{ss} \cup A_n^{fs}, \\ 0 & \text{otherwise.} \end{cases} \qquad (19)$$

The stopping times λ_n^* and γ_n^* are defined by Equations (18) and (19) with the appropriate $A_n^\bullet$ given by Equations (13)–(16).

Theorem 3.1. *Game G_A^C with payoff function (8) and sets of strategies Λ^N and Γ^N available to Player 1 and Player 2 respectively, has an equilibrium pair (λ^*, γ^*) defined by Equations (18) and (19), based on (13)–(16). The value of the game to Player 1 is $v_0(x)$.*

Now we construct an equilibrium pair (ψ^*, τ^*) for game G^C. Let $(\overline{\lambda}^*, \overline{\gamma}^*)$ be an equilibrium point in G_a^C.

Define (see [12] and [15])

$$\sigma_{n,m}^{1^*} = \begin{cases} 1 & \text{if } S_{N-m}(X_m, X_n) = f(X_m, X_n), \\ 0 & \text{if } S_{N-m}(X_m, X_n) > f(X_m, X_n) \end{cases} \tag{20}$$

$$\sigma_{n,m}^{2^*} = \begin{cases} 1 & \text{if } s_{N-m}(X_n, X_m) = f(X_n, X_m), \\ 0 & \text{if } s_{N-m}(X_n, X_m) > f(X_n, X_m) \end{cases} \tag{21}$$

Theorem 3.2. *Game G^C has a solution. The equilibrium point is (ψ^*, τ^*), such that $\psi^* = (\overline{\lambda}^*, \{\overline{\sigma}_n^{1^*}\})$ and $\tau^* = (\overline{\gamma}^*, \{\overline{\sigma}_n^{2^*}\})$. $(\overline{\lambda}^*, \overline{\gamma}^*)$ is an equilibrium point in G_a^C and the strategies $\{\overline{\sigma}_n^{1^*}\}$ and $\{\overline{\sigma}_n^{2^*}\}$ are defined by Equations (20) and (21) respectively. The value of the game is $v_0(x)$, where $v_n(x)$ is given by Equation (17).*

Now we consider non-zero sum games. In this case we must search for an equilibrium pair such that

$$\begin{aligned} &R_1(x, \overline{\lambda}_{\alpha,\beta^1,\beta^2}(\overline{\lambda}, \overline{\gamma}^*), \overline{\gamma}_{\alpha,\beta^1,\beta^2}(\overline{\lambda}, \overline{\gamma}^*)) \\ &\quad \leq R_1(x, \overline{\lambda}_{\alpha,\beta^1,\beta^2}(\overline{\lambda}^*, \overline{\gamma}^*), \overline{\gamma}_{\alpha,\beta^1,\beta^2}(\overline{\lambda}^*, \overline{\gamma}^*)), \\ &R_2(x, \overline{\lambda}_{\alpha,\beta^1,\beta^2}(\overline{\lambda}^*, \overline{\gamma}), \overline{\gamma}_{\alpha,\beta^1,\beta^2}(\overline{\lambda}^*, \overline{\gamma})) \\ &\quad \leq R_2(x, \overline{\lambda}_{\alpha,\beta^1,\beta^2}(\overline{\lambda}^*, \overline{\gamma}^*), \overline{\gamma}_{\alpha,\beta^1,\beta^2}(\overline{\lambda}^*, \overline{\gamma}^*)). \end{aligned}$$

Let $v_{1,n}(x)$ ($v_{2,n}(x)$) be the value of this game to the first (second) player on observing the realization x_n. The payoff matrix for player 1 is of the same form as the matrix given in Equation (10), except that $v_{1,\bullet}(x)$ replaces $v_\bullet(x)$. $a, b, \alpha, \beta, \gamma$ are defined as before from the matrix given in Equation (11). The payoff matrix for the second player has the form

$$\begin{bmatrix} \alpha_n(\beta_n^1 h_2(n,x) + (1-\beta_n^1) g(n,x,\beta_n^2)) & g(n,x,\beta_n^2) \\ +(1-\alpha_n)(\beta_n^2 H_2(n,x) + (1-\beta_n^2) G(n, x\beta_n^1)) & \\ G(n,x,\beta_n^1) & Tv_{2,n+1}(x) \end{bmatrix}. \tag{22}$$

Subtracting $Tv_{n+1}(x)$ this matrix is equivalent to one of the form

$$A = \begin{bmatrix} \mathfrak{a}_{ss} & \mathfrak{a}_{sc} \\ \mathfrak{a}_{cs} & \mathfrak{a}_{cc} \end{bmatrix} = \begin{bmatrix} \alpha(a_2 + (1-\beta) b_2) & b_2 \\ +(1-\alpha)(b_2 + (1-\gamma) a_2) & \\ a_2 & 0 \end{bmatrix}. \tag{23}$$

By direct checking we have

Lemma 3.2. *The two-person game with payoff matrices given by (11) and (23) has an equilibrium point in pure strategies given by* (ϵ, δ), *where*

$$(\epsilon, \delta) = \begin{cases} (s, s) & \text{if } (1-(1-\alpha)\gamma)a \geq \alpha\beta b \cap (1-\alpha\beta)b_2 \geq \gamma a_2(1-\alpha), \\ (s, f) & \text{if } a \geq 0 \cap (1-\alpha\beta)b_2 < \gamma a_2(1-\alpha), \\ (f, s) & \text{if } (1-(1-\alpha)\gamma)a < \alpha\beta b \cap b_2 \geq 0, \\ (f, f) & \text{if } a < 0 \cap b_2 < 0. \end{cases} \tag{24}$$

There is not necessarily a unique pure equilibrium in the game.

4 Example

In all the games considered we assume that an applicant accepts a job offer from Player i with probability r_i. If both players wish to accept an applicant, then Player 1 has priority with probability p, otherwise Player 2 has priority. If an applicant rejects an offer from the player with priority, that applicant then accepts the offer from the other player with the appropriate probability. The aim of each player is to employ the best applicant. Thus, the players should only accept applicants, who are the best seen so far (such applicants will be henceforth known as candidates). We obtain asymptotic results for a large number N of applicants. Let t be the proportion of applicants already seen. t will be referred to as the time.

In order to find the equilibrium strategies in the game, we first need to calculate the optimal strategy of a lone searcher. Let $U_i(t)$ be the probability that Player i obtains the best candidate, given that he/she is searching alone at time t. A player should accept a candidate at time t, iff $t \geq U_i(t)$. Smith [12] shows that

$$U_i(t) = \begin{cases} \frac{r_i}{1-r_i}(t^{r_i} - t) & t_i \leq t \leq 1, \\ t_i & 0 \leq t < t_i, \end{cases}$$

where $t_i = r_i^{1/(1-r_i)}$ satisfies $t_i = U_i(t_i)$. Player i's optimal strategy is to accept a candidate, iff $t \geq t_i$.

Example 4.1. Zero-sum game model

In this case it is assumed that a player's payoff is 1 if he/she obtains the best candidate, – 1 if the other player obtains the best candidate and 0 otherwise. Define k_i to be the probability that Player i obtains a candidate when both players wish to accept that candidate. It follows that $k_1 = r_1[p + (1-p)(1-r_2)]$ and $k_2 = r_2[(1-p) + p(1-r_1)]$. Define k_3 to be the probability that neither player obtains a candidate, when both players wish to accept a candidate. Hence, $k_3 = (1-r_1)(1-r_2)$. Let $w(t)$ be the expected value of the game to Player 1 when both of the players are still searching at time t. Thus $w(0)$ is the value of the game to Player 1. The payoff matrix on the appearance of a candidate for this game is given by

$$\begin{pmatrix} k_1[t - U_2(t)] + k_2[U_1(t) - t] + k_3 w(t) & r_1[t - U_2(t)] + (1 - r_1)w(t) \\ r_2[U_1(t) - t] + (1 - r_2)w(t) & w(t) \end{pmatrix}.$$

Rows 1 and 2 (Columns 1 and 2) give the appropriate payoffs when Player 1 (Player 2) accepts and rejects a candidate respectively. The game is solved by recursion. For large t both of the players accept a candidate at a Nash equilibrium. From the form of the payoff matrix, both players accepting a candidate forms a Nash equilibrium when the following inequalities are satisfied

$$\begin{aligned} r_2[U_1(t) - t] + (1 - r_2)w(t) &\le k_1[t - U_2(t)] + k_2[U_1(t) - t] + k_3 w(t) \\ &\le r_1[t - U_2(t)] + (1 - r_1)w(t). \end{aligned}$$

Suppose it is stable for both players to accept a candidate if $t \ge t_{2,2}$. Considering the distribution of the arrival time of the next candidate, it can be shown that

$$w(t) = \int_t^1 \frac{t}{s^2}\{k_1[s - U_2(s)] + k_2[U_1(s) - s] + k_3 w(s)\}ds.$$

Dividing by t and differentiating

$$w'(t) - \frac{(1 - k_3)w(t)}{t} = \frac{k_1}{t^2}[U_2(t) - t] + \frac{k_2}{t^2}[t - U_1(t)].$$

Together with the boundary condition $w(1) = 0$, this gives

$$w(t) = C_1 t^{1-k_3} + C_2 t + C_3 t^{r_1} - C_4 t^{r_2},$$

where $C_3 = k_2 r_1 / r_2(1 - r_1)^2$, $C_4 = k_1 r_2 / r_1 (1 - r_2)^2$ and

$$C_1 = \frac{(1 - k_3)[k_1 r_2(1 - r_1) - k_2 r_1(1 - r_2)]}{r_1 r_2 (1 - r_1)^2 (1 - r_2)^2},$$

$$C_2 = \frac{k_2(1 - r_2) - k_1(1 - r_1)}{(1 - r_1)^2 (1 - r_2)^2}.$$

In the case $r_1 = r_2 = r$ this simplifies to

$$w(t) = \frac{r^2(2p - 1)}{(1 - r)^3}[(2 - r)t^{r(2-r)} - t - (1 - r)t^r]. \tag{25}$$

In this case (here $t_2 = t_1$), from the symmetry of the game it suffices to consider $p \ge 0.5$. Intuitively, for $p > 0.5$ Player 1 should be the more choosy of the two players. Hence, in this case we look for a Nash equilibrium of the form

$$(\phi^*, \tau^*) = \begin{cases} (a, a) & t \geq t_{2,2}, \\ (r, a) & t_{2,1} \leq t < t_{2,2}, \\ (r, r) & t < t_{2,1}. \end{cases}$$

From the arguments presented above, it follows that $t_{2,2}$ satisfies

$$(1 + (2p - 1)r)[t_{2,2} - U_1(t_{2,2})] = (1 - r)w(t_{2,2}). \tag{26}$$

It follows from Equation (25) that $w(t) > 0$ for $t \in [t_{2,2}, 1)$. Hence, it can be seen that for $p > 0.5$, $t_{2,2} > t_1$. For $p = 0.5$, $w(t) = 0$ on this interval and hence $t_{2,2} = t_1$. In this particular case it is simple to show that for $t < t_1$ both players reject a candidate at a Nash equilibrium. In the more general case, the relation between $t_{2,2}$ and the optimal thresholds for a lone searcher are not so clear and so, henceforth, results are given only in the case $r_1 = r_2$. However, the method of solution in the general case is similar.

It can be shown that for $p > 0.5$ and $t_{2,1} < t < t_{2,2}$

$$w'(t) - \frac{pw(t)}{t} = \frac{p}{t}[t - U_1(t)].$$

It should be noted here that $U_1(t)$ changes form at $t = t_1$. Considering the payoff matrix $t_{2,1}$ satisfies $w(t_{2,1}) = U(t_{2,1}) - t_{2,1}$. For $t_1 \leq t \leq t_{2,2}$

$$W(t) = C_5 t^p - \frac{pt^p \ln(t)}{1 - p} + \frac{pt}{(1 - p)^2},$$

where C_5 is calculated from the boundary condition at $t_{2,2}$. Since $w(t) > 0$ on this interval, it follows that $t_{2,1} < t_1$. On the interval $[t_{2,1}, t_1]$, we have

$$w(t) = C_6 t^p + t_1 + \frac{pt}{1 - p},$$

where C_6 is calculated from the boundary condition at t_1. For $t \leq t_{2,1}$ the value function $w(t)$ is constant. Table 1 gives results for $p = 1$ (Player 1 always has priority) and various values of r.

Table 1: Numerical solution of the bilateral selection problem

r	$t_{2,1}$	$t_{2,2}$	$w(0)$
0.5	0.2139	0.2710	0.0571
0.6	0.2342	0.2995	0.0652
0.7	0.2512	0.3229	0.0716
0.8	0.2654	0.3419	0.0764
0.9	0.2771	0.3568	0.0797
0.95	0.2821	0.3628	0.0807

REFERENCES

[1] Bellman R. *Dynamic Programming*. Princeton Press, 1957.

[2] Dynkin E. B. Game variant of a problem on optimal stopping. *Soviet Math. Dokl.*, 10: 270–274, 1969.

[3] Enns E. G. and Ferenstein E. The horse game. *J. Oper. Res. Soc. Jap.*, 28: 51–62, 1985.

[4] Ferenstein E. Z. Two-person non-zero-sum games with priorities. In Thomas S. Ferguson and Stephen M. Samuels, editors, *Strategies for Sequential Search and Selection in Real Time, Proceedings of the AMS-IMS-SIAM Join Summer Research Conferences held June 21–27, 1990*, volume 125 of *Contemporary Mathematics*, pages 119–133, University of Massachusetts at Amherst, 1992.

[5] Fushimi M. The secretary problem in a competitive situation. *J. Oper. Res. Soc. Jap.*, 24: 350–358, 1981.

[6] Radzik T. and Szajowski K. On some sequential game. *Pure and Appl. Math. Sci*, 28: 51–63, 1988.

[7] Radzik T. and Szajowski K. Sequential games with random priority. *Sequential Analysis*, 9(4): 361–377, 1990.

[8] Ramsey D. and Szajowski K. Random assignment and uncertain employment in optimal stopping of Markov processes. *Game Theory and Appl.*, 7: 147–157, 2002.

[9] Ravindran G. and Szajowski K. Non-zero sum game with priority as Dynkin's game. *Math. Japonica*, 37(3): 401–413, 1992.

[10] Sakaguchi M. Bilateral sequential games related to the no-information secretary problem. *Math. Japonica*, 29: 961–974, 1984.

[11] Sakaguchi M. Non-zero-sum games for some generalized secretary problems. *Math. Japonica*, 30: 585–603, 1985.

[12] Smith M. H. A secretary problem with uncertain employment. *J. Appl. Probab.*, 12: 620–624, 1975.

[13] Szajowski K. Uncertain employment in competitive best choice problems. In K.Ano, editor, *International Conference on Stochastic Models and Optimal Stopping, Nagoya 19-21.12.1994*, pages 1–12, Nagoya, Japan, 1994. Faculty of Business Administration, Nanzan University, Nanzan University.

[14] Szajowski K. Optimal stopping of a discrete Markov processes by two decision makers. *SIAM J. Control and Optimization*, 33(5): 1392–1410, 1995.

[15] Yasuda M. On a stopping problem involving refusal and forced stopping. *J. Appl. Probab.*, 20: 71–81, 1983.

Optimal Stopping Games where Players have Weighted Privilege

Minoru Sakaguchi
Professor Emeritus, Osaka University,
Toyonaka, Osaka 560-0002, Japan
smf@mc.kcom.ne.jp

Abstract

A non-zero-sum n-stage game version of a full-information best-choice problem under expected net value (ENV) maximization is analyzed and the solutions are obtained in some special cases of 2-person and 3-person games. The essential feature contained in this multistage game is the fact that the players have their own weights by which at each stage one player's desired decision is preferred to the opponent's one by drawing a lottery.

Key words. Repeated game, Equilibrium value, Optimality equation.

AMS Subject Classifications. 60G40, 90 G 39, 90D45.

1 A Two-Person Optimal Stopping Game

A non-zero-sum game version of the discrete-time, full-information best-choice problem under ENV-maximization is considered in this section. We first state the problem as follows:

(1°) There are two players I and II (sometimes, they are denoted by players 1 and 2) and a sequence of n nonnegative *iid* r.v.s. $\{X_i\}_{i=1}^n$ with a common cdf $F(x)$. Both players observe this sequence sequentially one by one.

(2°) Observing each X_i, both players select simultaneously and independently, either to accept (A) or to reject (R) the observation of X_i. If I–II choice pair is A-A, then player I (II) accepts to receive X_i with probability $w^1 (w^2 \equiv 1 - w^1)$, $\frac{1}{2} \leqq w^1 \leqq 1$, and drops out from the play thereafter. The remained player continues his one-person game. If I–II choice is A-R (R-A), then I (II) accepts X_i and drops out and his opponent continues the remaining one-person game. If I–II choice is R-R, then X_i is rejected and the players face the next X_{i+1}.

(3°) The aim of each player in the game is to determine his acceptance strategy under which he maximizes his expected net value.

Define state $(x|n)$ to mean that (1) both players remain in the game, (2) there remains n r.v.s to be observed and the players currently face the first observation $X_1 = x$. Player i's strategy, $i = 1, 2$, in state $(x|n)$ is to choose A with probability

$\varphi^i(x,n) \in [0, 1]$, and R with probability $\overline{\varphi}^i(x,n) \equiv 1 - \varphi^i(x, n)$. Evidently $\varphi^i(x|1) \equiv 1$, for every $x \in (0, \infty)$, $i = 1, 2$.

Let V_n^i be the value of the game for player i in the n-problem. Then we have

$$\begin{aligned} V_n^i = E_F\Big[&(w^i x + w^j U_{n-1})\varphi^i(x, n)\varphi^j(x,n) + x\varphi^i(x, n)\overline{\varphi}^j(x,n) \\ &+U_{n-1}\overline{\varphi}^i(x,n)\varphi^j(x,n) + V_{n-1}^i\overline{\varphi}^i(x,n)\overline{\varphi}^j(x,n)\Big] \\ &(j = 3-i,\ i = 1,2,\ n = 1,2,\dots,\ V_0^1 = V_0^2 \equiv 0). \end{aligned} \tag{1}$$

Here U_{n-1} is the value of the game for the remaining player when his opponent has already dropped out with $n-1$ unobserved r.v.s thereafter. The optimal strategy for the player in this state is evidently to accept (reject) if $x > (<) U_{n-1}$, since the sequence $\{U_n\}$ satisfies the recursion

$$\begin{aligned} U_n &= E_F(X \vee U_{n-1}) \qquad (n = 1, 2, \dots;\ U_0 \equiv 0) \\ &= \frac{1}{2}(1 + {U_{n-1}}^2). \qquad (\text{if } F(x) = x, 0 \leqq x \leqq 1). \end{aligned} \tag{2}$$

Our problem is to find

$$(V_n^1,\ V_n^2) \to \text{Nash eq.}\{(\varphi^1(\cdot, i),\ \varphi^2(\cdot,\ i))\}_{i=1}^n. \tag{3}$$

We hereafter write $\langle w^1, w^2\rangle$ as $\langle w, \overline{w}\rangle$. We consider the problem (1)~(3) as the optimal stopping game described by the Optimality Equation

$$(V_n^1,\ V_n^2) = E_F\{eq.\ val\ M_n(X)\} \tag{4}$$

where

$$M_n(x) = \begin{array}{c|c|c|} & \text{R} & \text{A} \\ \hline \text{R} & V_{n-1}^1,\ V_{n-1}^2 & U_{n-1},\ x \\ \hline \text{A} & x,\ U_{n-1} & wx + \overline{w}U_{n-1},\ \overline{w}x + wU_{n-1} \\ \hline \end{array} \tag{5}$$

is the bimatrix game which the players face in state $(x|n)$, *eq.val* $(\cdot)$ is an equilibrium value and we assume that $M_n(x)$ has a unique equilibrium for every x satisfying $0 < F(x) < 1$.

The problems we consider in this paper belong to a class of best-choice problems combined with sequential games. Recent works related to this area of problems are Enns and Ferenstein [1], [2] , Mazalov [4] and Sakaguchi [5], [7], [8]. A very important and now classical literature in full-information best-choice problems is Gilbert and Mosteller [3]. Also a recent look at optimal stopping games in various phases can be found in Sakaguchi [6].

2 A Special Case in Two-Person Games where $F(x) = x\ (0 \leqq x \leqq 1)$ and $w = \frac{1}{2}$

Considering symmetry in the players of the game the bimatrix (5) becomes

	R	A
R	$V,\ V$	$U,\ x$
A	$x,\ U$	$\frac{1}{2}(x+U),\ \frac{1}{2}(x+U)$

(6)

where subscripts of V_{n-1} and U_{n-1} are omitted for brevity. If we suppose that $0 < V < U < 1$, as commonsense tells us, is satisfied, then the bimatrix game (6), for each $x \in [0,\ 1]$ has the equlibria:

Condition	Common eq. value	Common eq. strategy
$0 \leqq x < V$	V	Choose R
$V < x < U$	$\dfrac{2(U-V)x - UV}{x + U - 2V}$	Randomize R and A with prob. $\dfrac{2z}{1+z}$ for A
$U < x \leqq 1$	$\frac{1}{2}(x+U)$	Choose A

(7)

where $z \equiv (x - V)/(U - V)$. (See Remarks 2.1 and 5.1 that follow).

By using this fact, some algebra based on (4–7), and induction arguments, we prove

Theorem 2.1. *The solution to the optimal stopping game (OSG) described by Optimality Equation (4)–(5), for $F(x) = x\ (0 \leqq x \leqq 1)$ and $w = \frac{1}{2}$, is as follows. The common equilibrium strategy for both player is in state $(x|n)$,*

$$\begin{cases} \text{Choose R}, & \text{if } 0 \leqq x < V_{n-1}, \\ \text{Randomize R and A} & \\ \quad \text{with probability } \dfrac{2z_{n-1}}{1+z_{n-1}} \text{ for A}, & \text{if } V_{n-1} < x < U_{n-1}, \\ \text{Choose A}, & \text{if } U_{n-1} < x \leqq 1, \end{cases}$$

where $z_{n-1} = (x - V_{n-1})/(U_{n-1} - V_{n-1})$ and the equilibrium values are (V_n, V_n), where the sequences $\{U_n\}$ and $\{V_n\}$ are given by the recursions (2) and

$$V_n = \frac{1}{4}(1 + U_{n-1}^2) - (U_{n-1} - V_{n-1})(\lambda U_{n-1} + \overline{\lambda} V_{n-1}), \quad (n \geqq 1,\ U_0 = V_0 = 0), \tag{8}$$

where $\lambda \equiv 2\log 2 - 1 \doteqdot 0.38629$. *The sequences satisfy* $0 < V_n < U_n < 1$, $(n \geqq 1)$, *and* $V_n \uparrow 1$ *as* $n \to \infty$.

Numerical values of V_n, $n = 1(1)12$ are given in Table 5.2 of Section 5.

By using these values the common equilibrium strategy for the equal-weight game in state $(x|n)$, $n = 1(1)12$ are shown in Figure 1. The shaded region means that the player here should randomize the two decision R and A.

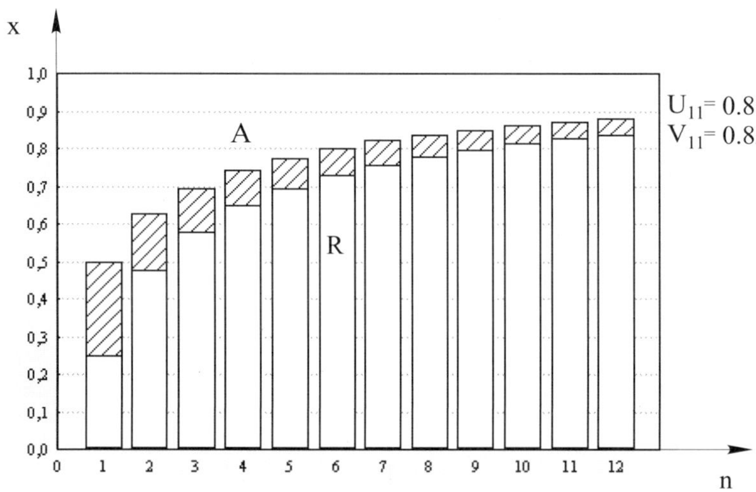

Figure 1: Common eq. strategy for the equal-weight game in state $(x|n)$, $n = 1(1)12$.

Remark 2.1. By (6–7) we can write as $V_n = \int_0^1 h_n(x)\,dx$ where

$$h_n(x) = \begin{cases} V, & \text{if } 0 \leqq x < V, \\ r(x), & \text{if } V < x < U, \\ \frac{1}{2}(x+U), & \text{if } U < x \leqq 1. \end{cases} \tag{9}$$

The integrand is continuous in [0, 1] and concavely increasing in $[V, U]$ (see Figure 2).

3 A Three-Person Optimal Stopping Game

The analysis made in the previous two sections can be extended to three-person games. We state the problem in correspondence to $(1^\circ) \sim (3^\circ)$ in Section 1, as follows:

(1^+) There are three persons I, II, and III. These players have their weights w_1, w_2, and w_3, respectively. Let $1 \geqq w_1 \geqq w_2 \geqq w_3 \geqq 0$ $w_1 + w_2 + w_3 = 1$, and $w_{(i,j)} \equiv w_i/(w_i + w_j)$, $i \neq j$.

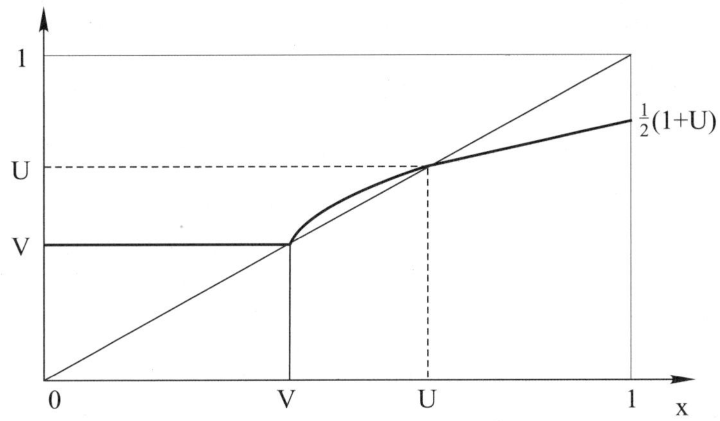

Figure 2: The function $h_n(x)$ in (9).

(2^+) If the three-players' choice is A–A–A, then player I (II, III) accepts X_t with probability w_1 (w_2, w_3) and drops out from the play thereafter. The two players remaining continue their two-person game with their "revised" new weights $\{w_{(i,j)}\}$. If the three players' choice is R–A–A, then II (III) accepts X_t with probability $w_{(2,\,3)}$ ($w_{(3,\,2)}$) dropping out from the game, and the remaining players III (II) and I continue their two-person game with their revised new weights. If the three players' choice is R–R–A, then III accepts X_t and drops out and his opponents I and II continue the remaining two-person game. If the players' choice-triple is R–R–R, then X_t is rejected and the players face the next X_{t+1}. In cases of other four choice-triples A–R–A, A–A–R, R–A–R, and A–R–R, the game is played similarly as mentioned above.

(3^+). The aim of each player is the same as in (3°).

Definition of state $(x|n)$ is the same as in Section 1, with the single difference that there are three players. Let W_n^i, $i = 1, 2, 3$, be the value of the game for player i, for the n-problem.

The statement of the problem in dynamic programming framework is as follows. Denote, by V_n $(w_{(i,\,j)})$, the value for player i in the two-person game against j, with weights $w_{(i,\,j)}$ for i, and $w_{(j,\,i)}$ for j. In state $(x|n)$, players face a trimatrix game with the payoff matrix $M_n(x)$, which is

$$M_n(x) = \begin{cases} \text{R by III} & M_{n,\,R}(x) \\ \text{A by III} & M_{n,\,A}(x) \end{cases} \qquad (10)$$

where $M_{n,\,R}(x) =$

(I) \ (II)	R	A
R	$W^1,\ W^2,\ W^3$	$V(w_{13}),\ x,\ V(w_{31})$
A	$x,\ V(w_{23}),\ V(w_{32})$	$w_{12}x + w_{21}V(w_{13})$, $w_{12}V(w_{23}) + w_{21}x$, $w_{12}V(w_{32}) + w_{21}V(w_{31})$

$M_{n,\,A}(x) =$

	R	A
R	$V(w_{12}),\ V(w_{21}),\ x$	$w_{23}V(w_{13}) + w_{32}V(w_{12}))$, $w_{23}x + w_{32}V(w_{21})$, $w_{23}V(w_{31}) + w_{32}x$
A	$w_{13}x + w_{31}V(w_{12})$, $w_{13}V(w_{23}) + w_{31}V(w_{21})$, $w_{13}V(w_{32}) + w_{31}x$	$w_1x + w_2V(w_{13}) + w_3V(w_{12})$, $w_1V(w_{23}) + w_2x + w_3V(w_{21})$, $w_1V(w_{32}) + w_2V(w_{31}) + w_3x$

In these two matrices the subscript $n-1$ is omitted in W and V. Also $w_{(i,\,j)}$ are rewritten as w_{ij}. The optimality equation is

$$(W_n^1,\ W_n^2,\ W_n^3) = E_F[eq.val M_n(X)] \qquad (n \geqq 1;\ W_0^i = V_0(w_{ij}) = 0,\ \text{for every } i,\ j) \tag{11}$$

provided the equilibrium value of $M_n(x)$ exists uniquely.

4 A Special Case of Three-Person Games where $F(x) = x$ $(0 \leqq x \leqq 1)$ and $w_1 = w_2 = w_3 = \frac{1}{3}$

Let $W_n(V_n)$ be the value of the game for each player in the equal-weight 3-person (2-person), n-problem. The details about $\{V_n\}$ are already given by Theorem 2.1 in Section 2.

Considering symmetry in the role of three players, equation (10) in state $(x|n)$ becomes

$$M_n(x) = \begin{cases} \xrightarrow{\text{R by III}} M_{n,\,R}(x) \\ \xrightarrow{\text{A by III}} M_{n,\,A}(x) \end{cases} \tag{12}$$

where $M_{n,\,R}(x) =$

(II)

		R	A
(I)	R	$W,\ W,\ W$	$V,\ x,\ V$
	A	$x,\ V,\ V$	$\frac{1}{2}(x+V),\ \frac{1}{2}(x+V),\ V$

$M_{n,\,A}(x) =$

	R	A
R	$V,\ V,\ x$	$V,\ \frac{1}{2}(x+V),\ \frac{1}{2}(x+V),$
A	$\frac{1}{2}(x+V),\ V,\ \frac{1}{2}(x+V),$	$\frac{1}{3}(x+2V),\ \frac{1}{3}(x+2V),\ \frac{1}{3}(x+2V),$

In these two matrices the subscripts $n-1$ in W and V are omitted for simplicity. The Optimality Equation is

$$(W_n,\ W_n,\ W_n) = E_F[eq.val M_n(X)] \qquad (n \geqq 1,\ W_0 \equiv 0) \qquad (13)$$

provided that the equilibrium value of $M_n(x)$ exists uniquely for every $x \in (0,\ 1)$ (see Remark 5.1 in Section 5).

We prove

Theorem 4.1. *The solution of the optimal stopping game described by Optimality Equation (10)–(11) for $F(x) = x$ $(0 \leqq x \leqq 1)$ and $w_1 = w_2 = w_3 = \frac{1}{3}$ is the following. The common equilibrium strategy for each player in state $(x|n)$ is*

$$\begin{cases} \textit{Choose } R, & \textit{if } 0 \leqq x < W_{n-1} \\ \textit{Randomize } R \textit{ and } A \textit{ with prob.} & \\ \dfrac{3}{2}\left[1 - \dfrac{1}{2+z_{n-1}} - \dfrac{\sqrt{(1-z_{n-1})(3+z_{n-1})}}{\sqrt{3}(2+z_{n-1})}\right] \textit{for } A, & \textit{if } W_{n-1} < x < V_{n-1} \\ \textit{Choose } A, & \textit{if } V_{n-1} < x \leqq 1, \end{cases}$$

where $z_{n-1} = (x - W_{n-1})/(V_{n-1} - W_{n-1})$, and the common equilibrium value is W_n.

The sequences $\{V_n\}$ and $\{W_n\}$ are determined by the recursions

$$V_n = 2(1-\log 2)(U-V)^2 + UV + \frac{1}{4}(1 + 2U - 3U^2) \qquad (14)$$

$$\begin{aligned} W_n =&\mu(V-W)^2 + VW + \frac{1}{6}(1-V)(1+5V) \\ =&\frac{1}{6}(1+4V+V^2) - (V-W)(\overline{\mu}V + \mu W), \end{aligned} \tag{15}$$

respectively (subscripts $n-1$ in V and W are omitted in the r.h.s.), where

$$\mu \equiv \frac{3}{4} - \frac{\sqrt{3}}{2}\int_0^1 \frac{(1-z)^{3/2}(3+z)^{1/2}}{(2+z)^2}\,dz. \tag{16}$$

Moreover $0 < W_n < V_n < 1$ $(n \geqq 1)$ *and* $W_n \uparrow 1$ *as* $n \to \infty$.

Evaluating the definite integral in (16), we obtain

$$\int_0^1 \frac{(1-z)^{3/2}(3+z)^{1/2}}{(2+z)^2}\,dz \doteqdot 0.16381$$

and hence (16) becomes

$$\mu \doteqdot \frac{3}{4} - \frac{\sqrt{3}}{2}0.16381 \doteqdot 0.60814. \tag{17}$$

It is worth noting that this value of μ is almost equal to $\overline{\lambda} = 2(1-\log 2) \doteqdot 0.61371$ in (8) for the 2-person equal-weight game.

Using this value of μ, we compute $\{W_n\}$ from (15), and the result is given by Table 5.2, for $n = 1(1)12$.

5 Final Remarks

Solutions to the other special games where $F(x) = x$ $(0 \leqq x \leqq 1)$, and $W = \langle 1, 0\rangle$, $\left\langle \frac{1}{2}, \frac{1}{2}, 0\right\rangle$, or $\langle 1, 0, 0\rangle$ are also discussed (although omitted).

We shall give three more remarks. Remark 2.1 is in Section 2.

Remark 5.1. The main results of the present paper are Theorems 2.1 and 4.1. The Optimality Equation (4)–(5) [(12)-(13)] have their meaning only if (5) [(12)] has a unique equilibrium value for every realized value x of the r.v. X.

The payoff matrices (5) and (12) are 2×2 bimatrix, and $2 \times 2 \times 2$ trimatrix, respectively. It is possible that for some x the equilibrium consists of a (lower-dimensional) continuity of pure-strategy equilibrium, and one completely-mixed-strategy equilibrium We have to assume that we adopt the latter when discussing these Optimality Equations.

Remark 5.2. Equilibrium values of the two optimal stopping games derived in Theorems 2.1 and 4.1 are tabulated for $n = 1(1)12$ in Table 1.

Table 1: The equilibrium values of equal-weight games

Based on n	1-person game Eq.(2) U_n	2-person game Eq.(8) V_n	3-person game Eq.(15) W_n
1	0.5000	0.2500	0.1667(= 1/6)
2	0.6250	0.4759	0.3271
3	0.6953	0.5800	0.4643
4	0.7417	0.6468	0.5506
5	0.7751	0.6935	0.6110
6	0.8004	0.7286	0.6561
7	0.8203	0.7561	0.6913
8	0.8364	0.7782	0.7196
9	0.8498	0.7965	0.7429
10	0.8611	0.8119	0.7625
11	0.8707	0.8250	0.7792
12	0.8791	0.8364	0.7936

Remark 5.3. Similar n-stage game versions of full-information and no-information problems under winning-probability maximization can be analyzed in the same way as was done in this paper. The results obtained in this approach will appear in a forthcoming paper.

Acknowledgements

The author thanks Krzysztof Szajowski of Wrocław University of Technology, Poland, for his helpful discussions and comments on the present problem.

REFERENCES

[1] Enns E. G. and Ferenstein E., The horse game, J. Oper. Res. Soc. Japan 28 1985, 51–62.

[2] Enns E. G. and Ferenstein E., On a multi-person time-sequential game with priorities, Sequential Analysis 6 1987, 239–256.

[3] Gilbert J. P. and Mosteller F., Recognizing the maximum of a sequence, J. Am. Stat. Assoc. 61 1966, 35–73.

[4] Mazalov V. V., A game related to optimal stopping of two sequences of independent random variables having different distributions, Math. Japonica 43 1996, 212–218.

[5] Sakaguchi M., Sequential games with priority under expected value maximization. Math. Japonica 36 1991, 545–562.

[6] Sakaguchi M., Optimal stopping games – A review, Math. Japonica 42 1995, 343–351.

[7] Sakaguchi M., Optimal stopping games for bivariate uniform distribution, Math. Japonica 41 1995, 677–687.

[8] Sakaguchi M., A non-zero-sum repeated game – Criminal vs. Police, Math. Japonica 48 1998, 427–436.

Equilibrium in an Arbitration Procedure

Vladimir V. Mazalov
Karelian Research Centre
Petrozavodsk 185 610, Russia
vmazalov@krc.karelia.ru

Anatoliy A. Zabelin
Chita State Pedagogical University
Chita 672090, Russia

Abstract

A bargaining problem with two players, Labor (player L) and Management (player M) is considered. The players must decide the monthly wage paid to L by M. At the beginning players L and M submit their offers x and y. If $x \leq y$ there is an agreement at $(x + y)/2$. If not, the arbiter is called in and he chooses the offer which is nearest to his solution α. The arbiter imposes a solution α which is concentrated in two points with probabilities $p = 1/2$. The equilibrium in the arbitration game among pure and mixed strategies is derived.

1 Introduction

We shall consider a zero-sum game intended to model labor-management negotiations using an arbitration procedure. Imagine that two players: Labor (player L) and Management (player M) bargain on a wage bill which has to be in the range $[0, 1]$ where the current wage bill is normalized at zero, and the known maximum management ability to pay is at 1. Player L is interested maximizing a wage bill as much as possible and the player M has the opposite goal.

At the beginning the players L and M submit their offers x and y respectively, $x, y \in [0, 1]$. If $x \leq y$ there is an agreement at $(x + y)/2$. If not, the arbiter A is called in and he has to choose one of the decisions.

There are different approaches in analyzing the arbitration models [1–6]. In the Nash scheme [1–2] the role of arbiter is passive, he must determine an equitable efficient outcome among bargaining alternatives. Another model was described in [3] in the framework of differential games. Similar competitive decision-making under uncertainty model was analyzed in [5]. In [6] the players have an arbitration committee which is composed of two arbitrators who propose their solutions for the wage at every stage of negotiations.

We consider here the final-offer arbitration procedure [4] which allows the arbiter only to choose one of the two final offers made by the players. We suppose here that the arbiter imposes a solution α which is a random variable being concentrated in two points a and $1-a$ with equal probabilities $1/2$, $0 \le a \le 1$. The arbiter chooses the offer which is nearest to his solution α.

So, we have a zero-sum game determined in unit square where strategies of players L and M are the real numbers $x, y \in [0, 1]$ and payoff function in this game has the form $H(x, y) = EH_\alpha(x, y)$, where

$$H_\alpha(x, y) = \begin{cases} (x+y)/2, & \text{if } x \le y \\ x, & \text{if } x > y, |x-\alpha| < |y-\alpha| \\ y, & \text{if } x > y, |x-\alpha| > |y-\alpha| \\ \alpha, & \text{if } x > y, |x-\alpha| = |y-\alpha| \end{cases} \tag{1}$$

Notice at once that from symmetry it follows that the optimal strategies of players must be symmetric with respect to the midpoint of the interval $[0, 1]$ and the value of the game is equal to $1/2$. Consequently, it is enough to find the optimal strategy only for one of the players, for instance, M.

Below we show that the equilibrium in this game in dependence on value a can be among pure (section 2) and mixed (section 4) strategies.

The solution of this game with α distributed in two points with different probabilities was derived in [7].

2 Equilibrium in Pure Strategies

For enough small values a the equilibrium consists of pure strategies.

Theorem 2.1. *If $a \le \frac{1}{4}$, optimal strategies are $x = 1$ and $y = 0$.*

Proof. Assume the player M is going to play $y = 0$. From the game rules it follows that any offer x of player L from the interval $[0, 2a)$ will be closer to the arbiter solution. For $x > 2a$ the solution a will be closer to the offer of player M and the solution $1-a$ will be closer to the offer of player L. According to (1) we can represent the payoff of player L in the form

$$H(x, y) = \begin{cases} x, & \text{if } x \in [0, 2a) \\ \frac{1}{2}a + \frac{1}{2}2a = \frac{3}{2}a, & \text{if } x = 2a \\ \frac{1}{2}0 + \frac{1}{2}x = \frac{1}{2}x, & \text{if } x \in (2a, 1] \end{cases} \tag{2}$$

The graph of function $H(x, y)$ consists of two straight lines. Its form is shown in Fig. 1.

Evidently, if $a \le \frac{1}{4}$ then $H(x, 0) \le \frac{1}{2}$ for any x. We obtain equality for $x = 1$. It follows immediately from here that the best response of player L is the pure strategy $x = 1$.

According to symmetry $x = 1$, $y = 0$ form an equilibrium in the game. □

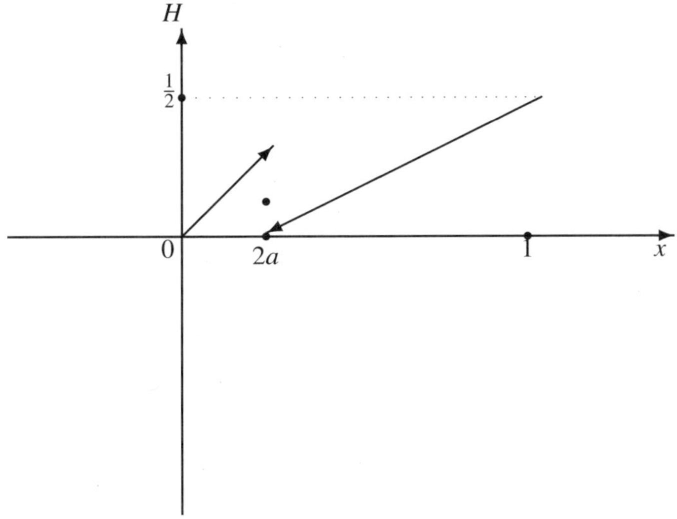

Figure 1: Graph of function $H(x, 0)$

Remark 1. Thus, for small a a player has to offer maximal profitability for himself in the hope that the arbiter's opinion will be lucky for him.

3 Method for Obtaining the Equilibrium Among Mixed Strategies

In case $a > \frac{1}{4}$ there is no equilibrium among pure strategies. Denote by $F(y)$ mixed strategy of player M. Suppose that support of distribution F lies in the left half of the interval $[0, 1]$. It follows from the game rules that the payoff $H(x, F) \leq \frac{1}{2}$ for all $x < \frac{1}{2}$. For $x \geq \frac{1}{2}$ the payoff has form

$$H(x, F) = \frac{1}{2}\left[F(2a - x)x + \int_{2a-x}^{\frac{1}{2}} y dF(y)\right] + \frac{1}{2}\left[F\Big(2(1-a) - x\Big)x + \int_{2(1-a)-x}^{\frac{1}{2}} y dF(y)\right]. \quad (3)$$

We will search a form of distribution function $F(y)$ which satisfies the following conditions (see Fig. 2):

(i) a support of distribution F must be concentrated in the interval $[b_1, b_2] \subset [0, 1/2]$,
(ii) payoff function $H(x, F)$ is constant in the equal length intervals $[c_1, c_2] \subset [1/2, 1]$ and $[d_1, d_2] \subset [1/2, 1]$ with $1/2 \leq c_1 \leq c_2 \leq d_1 \leq d_2$, and its value is equal to $1/2$,

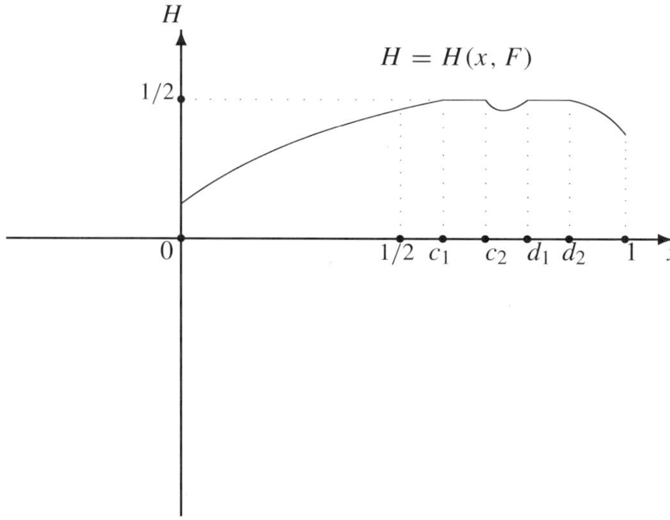

Figure 2: Graph of function $H(x, F)$

(iii) $H(x, F) \leq 1/2$ for other x,
(iv) For all $x \in [c_1, c_2]$ the location of point $2(1-a)-x$ is from the right b_2, and for all $x \in [d_1, d_2]$ the location of point $2a-x$ is from the left b_1. Below we show the existence of such intervals.

For such distribution function (assumptions (i), (iv)) we obtain from (2) for $x \in [c_1, c_2]$

$$H(x, F) = \frac{1}{2}\left[F(2a-x)x + \int_{2a-x}^{\frac{1}{2}} y dF(y)\right] + \frac{1}{2}x, x \in [c_1, c_2]. \quad (4)$$

Using the assumption that $H(x, F)$ is constant in the interval $[c_1, c_2]$ we can determine the analytic form of distribution function $F(y)$. Let us differentiate (3) and equate it to zero.

$$\frac{dH}{dx} = \frac{1}{2}\left[-F'(2a-x)x + F(2a-x) + (2a-x)F'(2a-x)\right] + \frac{1}{2} = 0. \quad (4)$$

Changing the argument in (4) $2a - x = y$ we obtain the differential equation

$$2F'(y)(y-a) = -[F(y)+1], y \in [2a-c_2, 2a-c_1],$$

which solution represents the distribution function $F(y)$ in the interval $[2a - c_2, 2a - c_1]$

$$F(y) = -1 + \frac{C_1}{\sqrt{a-y}}, y \in [2a-c_2, 2a-c_1],$$

where C_1 is a constant value.

Anologously, the function $H(x, F)$ in the interval $[d_1, d_2]$ has the form

$$H(x, F) = \frac{1}{2}\int_0^{\frac{1}{2}} ydF(y) + \frac{1}{2}[F(2(1-a)-x)x + \int_{2(1-a)-x}^{\frac{1}{2}} ydF(y)], x \in [1-b, 1]. \quad (5)$$

From the assumption (ii) in the interval $[d_1, d_2]$ it follows

$$\frac{dH}{dx} = \frac{1}{2}\Big[-F'(2(1-a)-x)x + F(2(1-a)-x) + (2(1-a)-x) F'(2(1-a)-x)\Big] = 0.$$

Changing the argument $2(1-a) - x = y$ we obtain differential equation

$$2F'(y)\,(y-(1-a)) = -F(y), y \in [2(1-a)-d_2, 2(1-a)-d_1].$$

Its solution

$$F(y) = \frac{C_2}{\sqrt{1-a-y}}, y \in [2(1-a)-d_2, 2(1-a)-d_1]. \quad (6)$$

So, we find a distribution function $F(y)$, $y \in \left[0, \frac{1}{2}\right]$ with parameters $c_1, c_2, d_1, d_2, C_1, C_2$ with constant value of payoff function $H(x, F)$ for player L in the intervals $[c_1, c_2]$ and $[d_1, d_2]$. To prove optimality of $F(y)$ for player M it is sufficient to show that $H(x, F) \leq 1/2$ for all x.

4 Optimal Strategies

A decision of the game depends on the value a and is split for two cases. We'll see below that it is represented by the "golden section" of the interval [0, 1] (denote it z). Remember that $z \approx 0.618$ is a solution of the quadratic equation $z^2+z-1=0$.

4.1 Case $\frac{1}{4} < a \leq z^2 \approx 0.382$

Theorem 4.1. *For*

$$\frac{1}{4} < a \leq z^2, \quad (7)$$

optimal strategies in the arbitration game with payoff function (1) have the form

$$G(x) = \begin{cases} 0, & x \in [0, 2a-b] \\ 1 - \frac{1-\sqrt{a}}{\sqrt{x-a}}, & x \in (2a-b, 2a] \\ 2 - \frac{1}{\sqrt{a}}, & x \in (2a, 1-b] \\ 2 - \frac{1/\sqrt{a}-1}{\sqrt{x+a-1}}, & x \in (1-b, 1] \end{cases} \tag{8}$$

$$F(y) = \begin{cases} -1 + \frac{1/\sqrt{a}-1}{\sqrt{a-y}}, & y \in [0, b] \\ -1 + \frac{1}{\sqrt{a}}, & y \in (b, 1-2a] \\ \frac{1-\sqrt{a}}{\sqrt{1-a-y}}, & y \in (1-2a, 1-2a+b] \\ 1, & y \in (1-2a+b, 1] \end{cases} \tag{9}$$

where $b = 2\sqrt{a} - 1$.

Proof. Consider the expression (9). It is easy to show that from condition (7) for a the following inequalities

$$0 < b \leq 1 - 2a \leq 1 - 2a + b \leq 1/2$$

take place. Function $F(y)$ is continuous, non-decreasing and its support is concentrated on the set $[0, b] \bigcup [1-2a, 1-2a+b] \subset [0, 1/2]$.

Moreover, its value at zero

$$F(0) = -1 + \frac{1}{a} - \frac{1}{\sqrt{a}} \geq 0$$

is nonnegative due to assumption $a \leq z^2$.

So, the function $F(y)$ of the form (9) represents a distribution of the strategy of player M. □

Lemma 4.1. *Mean value* $\bar{y}$ *of mixed strategy of the form (9) is equal to b.*

Proof. After simplification

$$\bar{y} = \int_0^1 y dF(y) = \int_0^b y dF(y) + \int_{1-2a}^{1-2a+b} y dF(y)$$

$$= \left(\frac{1}{\sqrt{a}} - 1\right) \int_0^{2\sqrt{a}-1} y d\left(\frac{1}{\sqrt{a-y}}\right) + \left(1 - \sqrt{a}\right) \int_{1-2a}^{2\sqrt{a}-2a} y d\left(\frac{1}{\sqrt{1-a-y}}\right)$$

and integration by parts we obtain

$$\bar{y} = 2\sqrt{a} - 1 = b.$$

Suppose now that player M uses the mixed strategy of the form (9) and find the best response of player L. Notice that $F(y)$ has the form described in the previous section. Consequently, the payoff $H(x, F)$ of player L will be constant in the intervals $[2a - b, 2a) \bigcup [1 - b, 1]$. Let us calculate this value.

Substituting $x = 2a$ in (3) and $x = 1 - b$ in (5) we obtain respectively

$$\begin{aligned} H(2a - 0, F) &= \frac{1}{2}[F(0)2a + \bar{y}] + \frac{1}{2}2a \\ &= \frac{1}{2}\left[\left(-1 + \frac{1}{a} - \frac{1}{\sqrt{a}}\right)2a + 2\sqrt{a} - 1\right] + \frac{1}{2}2a = 1/2, \end{aligned}$$

$$H(1 - b, F) = \frac{1}{2}\bar{y} + \frac{1}{2}\Big[F(1 - 2a + b)(1 - b)\Big] = \frac{1}{2}b + \frac{1}{2}(1 - b) = 1/2.$$

So, the payoff function $H(x, F)$ has constant value in the intervals $[2a - b, 2a]$ and $[1 - b, 1]$ and is equal to $1/2$.

Show now that the value of function $H(x, F)$ for other $x \in \left[\frac{1}{2}, 1\right]$ is not larger than $\frac{1}{2}$. For $x \in (2a, 1 - b)$ it follows from (2) that

$$H(x, F) = \frac{1}{2}\bar{y} + \frac{1}{2}x$$

is an increasing function of x with maximal value $H(1 - b, F) = 1/2$.

For $x \in \left[\frac{1}{2}, 2a - b\right)$ it follows from (2) that

$$H(x, F) = \frac{1}{2}\left[xF(b) + \int_{1-2a}^{1-2a+b} y dF(y)\right] + \frac{1}{2}x$$

is an increasing function also with maximal value at the point $x = 2a - b$, $H(2a - b, F) = 1/2$.

Thus (see Fig. 3), $H(x, F) \leq 1/2$ for all x. It follows from here that for any mixed strategy $G(x)$ of player M his payoff $H(G, F)$ will not be larger than $1/2$. Hence, $F(y)$ of the form (8) represents the optimal strategy of player M.

Notice that the value of function $H(x, F)$ is equal to $1/2$ in the intervals $[2a - b, 2a)$ and $[1 - b, 1]$ which are symmetrically located with respect to the point $1/2$ for the intervals $(1 - 2a, b + 1 - 2a]$ and $[0, b]$. Consequently, these intervals are the support of the optimal distribution $G(x)$.

We noticed earlier the optimal strategy of player L must be symmetric with respect to $1/2$ for the optimal strategy of player M. Formula (8) for $G(x)$ corresponds to precisely this distribution.

Theorem is proven. □

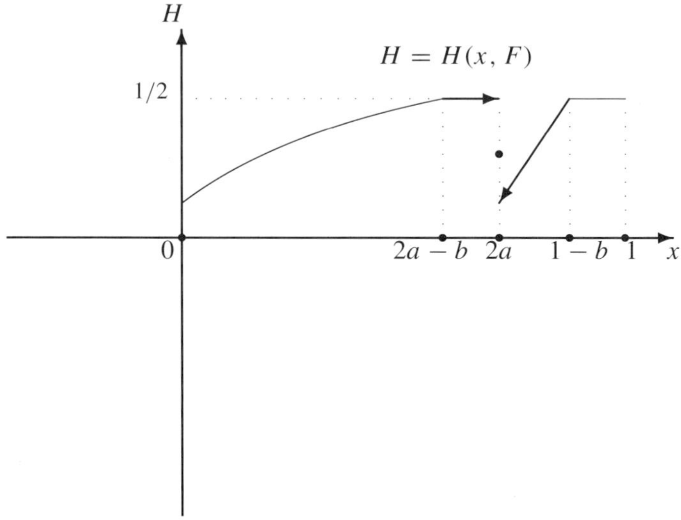

Figure 3: Graph of function $H(x, F)$

4.2 Case $z^2 < a \leq 1/2$

Theorem 4.2. *For*

$$z^2 < a \leq 1/2 \tag{10}$$

optimal strategies in the arbitration game with payoff function (1) have the form

$$G(x) = \begin{cases} 0, & x \in [0, 4a-1-b] \\ 1 - \frac{\sqrt{z(1-2a)}}{\sqrt{x-a}}, & x \in (4a-1-b, 2a-b] \\ 2 - \frac{\sqrt{a-b}}{\sqrt{x+a-1}}, & x \in (2a-b, 1-b] \\ 1, & x \in (1-b, 1] \end{cases} \tag{11}$$

$$F(y) = \begin{cases} 0, & y \in [0, b] \\ -1 + \frac{\sqrt{a-b}}{\sqrt{a-y}}, & y \in (b, b+1-2a] \\ \frac{\sqrt{z(1-2a)}}{\sqrt{1-a-y}}, & y \in (b+1-2a, b+2(1-2a)] \\ 1, & y \in (b+2(1-2a), 1] \end{cases} \tag{12}$$

where $b = a - (1-2a)(1+z)$.

Proof. The correctness of the expression (12) follows from the inequalities

$$0 \leq b \leq b+1-2a \leq b+2(1-2a) \leq 1/2.$$

which are true due to assumption (10). Let us prove the first inequality, others are trivial. The inequality $b \geq 0$ is equivalent to $a \geq (1+z)/(3+2z)$. The latter one is true because $a > z^2$ (assumption) and $z^2 = (1+z)/(3+2z)$ (it is a property of golden section z because $z^2 = 1 - z = (1+z)/(3+2z)$).

Function $F(y)$ of the form (12) is non-decreasing with values in the interval $[0, 1]$, therefore it represents a mixed strategy of player M. □

Lemma 4.2. *Mean value $\bar{y}$ of mixed strategy (12) is equal to $b + 1 - 2a = a - z(1 - 2a)$.*

Proof. can be developed by direct calculations. □

Suppose now that player M uses a mixed strategy of the form (12) and determine the best response of player L.

Consider the payoff $H(x, F)$. Because the distribution F has the form described in the section 3 function, $H(x, F)$ is constant in the interval $(4a - 1 - b, 1 - b]$. It is easy to calculate that this value is equal to $1/2$.

In the interval $[1 - b, 1]$ function $H(x, F)$ has the form

$$H(x, F) = \frac{1}{2}\bar{y} + \frac{1}{2}\left[xF(2(1-a) - x) + \bar{y} - \int\limits_{b}^{2(1-a)-x} y dF(y) \right] =$$

$$= \bar{y} + \frac{1}{2}\left[x\left(\frac{\sqrt{a-b}}{\sqrt{3a-2+x}} - 1\right) - \sqrt{a-b} \int\limits_{b}^{2(1-a)-x} y d\left(\frac{1}{\sqrt{a-y}}\right)\right]$$

After simplification we obtain

$$H(x, F) = \bar{y} + b/2 + (1+z)(1-2a) - x/2 + (1-2a)\frac{\sqrt{a-b}}{\sqrt{3a-2+x}}. \quad (13)$$

Function (13) is decreasing in the interval $[1 - b, 1]$ with maximal value in the point $x = 1 - b$ equal to $1/2$.

In the interval $\left[\frac{1}{2}, 4a - 1 - b\right)$ we have

$$H(x, F) = \frac{1}{2}\left[\int\limits_{b}^{2a-x} y dF(y) + \int\limits_{2a-x}^{b+2(1-2a)} y dF(y) \right] + \frac{1}{2}x =$$

$$\frac{1}{2}\left[x\frac{\sqrt{z(1-2a)}}{\sqrt{1-3a+x}} + \sqrt{z(1-2a)} \int\limits_{2a-x}^{b+2(1-2a)} y d\left(\frac{1}{\sqrt{1-a-y}}\right)\right] =$$

$$\frac{x}{2} - (1-2a)\frac{\sqrt{z(1-2a)}}{\sqrt{1-3a+x}} + z(1-2a) + \frac{b + 2(1-2a)}{2}. \quad (14)$$

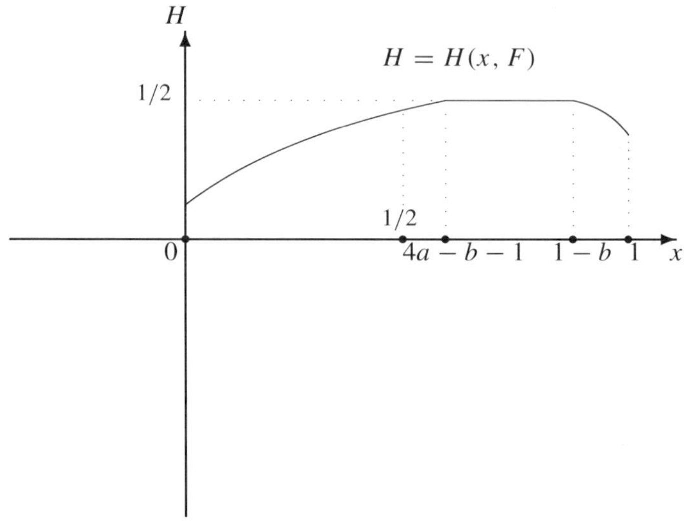

Figure 4: Graph of function $H(x, F)$

Function (14) increases x in the interval $\left[\frac{1}{2}, 4a - 1 - b\right)$ and its maximal value, in the point $x = 4a - 1 - b$ is equal to $1/2$.

Therefore, the graph of the function $H(x, F)$ has the form presented in Fig. 4. Its values are not larger than $1/2$ and are equal to this value in the interval $(4a - 1 - b, 1 - b]$. Thus, we show

$$H(x, F) \leq 1/2, x \in [0, 1].$$

Consequently, for any mixed strategy G of player L

$$H(G, F) \leq 1/2.$$

It follows immediately from here that F of the form (12) is the optimal strategy of player M.

From symmetry of the problem the symmetric, in respect to the point $1/2$ strategy G of the form (11), will be the optimal strategy of player L.

Theorem is proven.

Remark 2. It is interesting to notice that optimal rule in this arbitration game is determined by the "golden section" value. The optimal rule prescribes that player M locate his offers in the interval $[0, 1/2]$ and player L locate his offers in the interval $[1/2, 1]$. So, there are no situations with offers of player M larger than offers of L. Notice also that the mean values of distributions F and G coincide with the middle points of the support intervals.

Acknowledgements

The research was supported by Russian Foundation for Basic Research (project 01-01-00126).

REFERENCES

[1] Nash J., The bargaining problem, *Econometrica*, **18** (1950), 155–162.

[2] Luce R. D., Raiffa H., *Games and Decisions*, Wiley (1957).

[3] Leitman G., Collective bargaining: a differential game, *J. Opt. Th. Appl.* **11** (1973), 405–412.

[4] Chatterjee K., Comparison of arbitration procedures: Models with complete and incomplete information, *IEEE Transactions on Systems, Man, and Cybernetics*, smc-11, no. 2 (1981), 101–109.

[5] Kilgour D. M., Game-theoretic properties of final-offer arbitration, *Group Decision and Negotiation*, 3 (1994), 285–301.

[6] Sakaguchi M., A time-sequential game related to an arbitration procedure, *Math. Japonica* **29**, no. 3 (1984), 491–502.

[7] Mazalov V. V., Zabelin A. A., Discrete final-offer arbitration model, RIMS Seria 1263 *Development of the optimization theory for the dynamic systems and their applications* (2002), Kyoto University Press, 117–130.

PART IV

Applications of Dynamic Games to Economics, Finance and Queuing Theory

Applications of Dynamic Games in Queues

Eitan Altman
INRIA B.P. 93,
2004 route des Lucioles,
06902 Sophia Antipolis, France
Eitan.Altman@sophia.inria.fr

Abstract

Queueing phenomena, along with many related decision problems, are well known to all of us from daily situations. We often need to answer questions concerning whether to queue or not, where to queue, how long to queue etc. In networking applications, both in road traffic as well as in telecommunication networks, individuals or some central controllers are frequently faced with similar questions. These decisions have often to be taken in a randomly changing environment, in a decentralized way, and with partial information. This gives rise to many challenging problems in dynamic games. We shall describe in this overview some generic queueing problems (both static and dynamic) that require game theoretic models and solutions.

1 Introduction

We shall focus in this overview on some sample of decision problems that arise in queueing, and whose solutions requires game theoretic modeling and solution approaches. The main generic problems are

1. *To queue or not to queue?* In a context of networks, this question is related to admission control, i.e. to whether a new connection, or an arriving packet is accepted to the network (or to a specific node of a network).
2. *When to queue?* Individuals who have to queue in order to receive service often face this problem. They may have an idea about the global demand, i.e. the probability distribution of the total number of persons that would arrive and compete for the same service, but the decisions of arrival times of other persons are typically unknown.
3. *Where to queue?* This is one of the most studied problems in networking games. There are several routes between one or more sources and one or more destinations that involve costs or delays. The latter depend on the congestion at each part of the route (i.e. each node or link). The congestion is of course a function of the routing decisions. Different users have to determine their route choices. This type of game is known as the traffic assignment problem in the context of road traffic, and as a networking routing game in the context of telecommunications. We shall present two simple examples of stochastic games in which decisions of where to queue have to be taken dynamically (depending on the time varying state of the system).

4. *How much to queue?* This is perhaps the second most important type of games that occur in telecommunication networks. There are applications that can control the amount or the rate of information sent to the network. For example, file transfers and electronic mail that are sent in the Internet use a dynamic mechanism (called TCP - Transport Control Protocol) to adjust the rate of transmission to the encountered congestion. Real time video and voice applications often use dynamic compression which results not only in controlling the rate of transmission, but also the amount of data to be transmitted. This type of control determines how much data would be queued at different nodes in the network, it influences both the delay (or cost) of that application and contributes to delay (or cost) of other applications.

A key element that appears in many types of networking games and games in queues is the congestion. It has an impact on costs and on delays. In the context of telecommunications, it may further affect other measures of quality of the communications, as perceived by a connection, such as loss probabilities of its packets and the delay variations. Large amounts of losses or large delay variations are quite harmful in real time voice and video applications.

In this paper we first present some samples of problems in each of the above categories. Through these problems we shall identify some features of networking games. We shall study in particular some tendencies and properties of equilibrium policies, such as

- **Existence and uniqueness:** In some games uniqueness of equilibrium will be established, whereas in other cases, counterexamples will be presented to show that multiple equilibria may occur.
- **Structure of equilibria:** Threshold and switching-curves characterize policies that often appear in dynamic games in networking at equilibrium. We shall also present counterexamples where equilibria with such a structure may be expected, but fail to exist.
- **Tendencies between players in equilibria:** We shall encounter games that exhibit the tendency of "avoiding the crowd" as well as games with the opposite tendency of "joining the crowd". Such tendencies are related to dynamic behavior when the system is not at equilibrium. We shall mention, in particular, some recent extensions due to Yao [38] of submodular games (that were initially introduced by Topkis [35]) along with some queueing examples, which provide tools to identify the two types of tendencies.
- **Efficiency of the equilibrium:** Some of the examples will be used to illustrate how the equilibrium performs in terms of payoff as compared to a globally (cooperative) optimal solution that would be obtained if there were a single decision maker.

We briefly mention some references on this topic and other closely related ones. Reference [9] presents a survey on zero-sum stochastic games in queueing problems. Such games model a worst-case type design for the case that some parameters that are unknown can be modeled as if they were controlled by an opponent player. This reference also presents the methodology for handling unbounded costs which are typical for queueing applications, and for obtaining structural results of saddle point policies. An example in which such techniques have been used in non-zero sum games in queues can be found in [5]. Other recent surveys on applications of games in networks can be found in [6,13]. Finally, many games occurring in queueing systems are described in detail in [22] and references therein.

The structure of the paper is as follows. In the next four sections we address models that fall into the categories of whether to queue or not (Section 2), when to queue (Section 3), where to queue (Section 4) and how much to queue (Section 5). All along these sections we address study properties and features of the equilibria. We proceed with two methodological sections that focus on applications of S-modular games in queueing problems, and establish some properties of the values, of policies, and of the convergence to equilibrium.

2 To Queue or not to Queue: The PC-MF Game

We summarize in this section the networking problem from [12] in which individuals that require service have to make the decision whether to be served at some powerful facility that is shared among all the users that are present there, or to be served at a private slower service facility. As illustrated in Fig. 1, the slow facility may represent a personal computer (PC), the fast facility may represent a powerful workstation or a mainframe (MF), and an individual may represent a job that has to be processed.

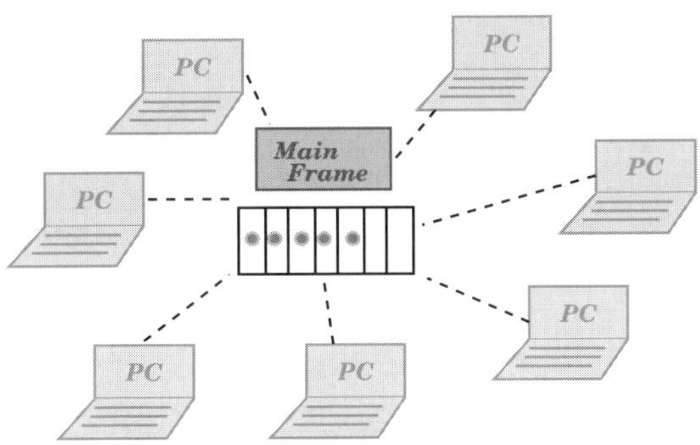

Figure 1: To queue for a fast service at a mainframe or to receive a slow service at a personal PC?

The model. Requests for processing jobs arrive at the system. Inter-arrival times are general i.i.d. (independent and identically distributed) with mean λ^{-1}. Upon arrival of a request, the user connects to MF and observes the load. Based on this information it decides whether to queue or not to queue there. MF shares its computing capacity between all present users there. This is called **Processor Sharing** discipline. The service at MF is exponentially distributed with rate $\mu(x)$. Let x be the number of jobs queued there. Typically $\mu(x) = \mu$ is a constant and the service intensity per customer is μ/x. But the results below hold more generally provided that $\mu(x)/x$ decreases in x [12].

The PCs offer fixed expected service time of θ^{-1}. (Here, θ is assumed to be the same for all PCs, although one can also solve for different θ's, see [15]).

Denote

$X(t)$:= the number of customers at MF at time t,

$T_k,\ k \geq 0$:= arrival time of job C_k, where $0 = T_0 < T_1 < T_2 < \ldots$ ($t = 0$ is the arrival time of the first customer).

The queue-length process $X(t)$ is defined to be left-continuous (thus $X(T_k)$ is the queue length just *prior* to the possible admission of customer C_k to MF).

The kth arriving individual, C_k, must decide at T_k whether to join MF, or to be served at a PC, after observing $X(T_k)$.

Denote u_k:= the strategy for C_k, is the probability of joining MF if $X(T_k) = x$. Let U be the class of such maps, and let $\pi := (u_0, u_1, \ldots)$ denote a multi-strategy of all customers.

Performance measure. Define

w_k:= service duration of customer C_k,

$W_k(x, \pi)$:= expected service time of C_k, given that x customers are present at MF at her arrival.

$V_k(x, \pi)$:= expected service time of C_k if she chose to be served at MF, given that x customers are present at MF at her arrival.

Then

$$W_k(x, \pi) = u_k(x) V_k(x, \pi) + (1 - u_k(x))\theta^{-1}.$$

Observe that V_k depends on π through $\{u_l,\ l > k\}$, the decision rules of *subsequent* customers.

Each customer wishes to minimize her own service time.

To this end, she should evaluate her expected service time at the two queues, namely $V_k(x, \pi)$ and θ^{-1}, and choose the lower one.

Definition 2.1. (**Threshold Policies**) For any $0 \leq q \leq 1$ and integer $L \geq 0$, the decision rule u is an $[L, q]$-threshold rule if

$$u(x) = \begin{cases} 1 & \text{if} \quad x < L \\ q & \text{if} \quad x = L \\ 0 & \text{if} \quad x > L \end{cases} . \tag{1}$$

A customer who employs this rule joins MF if the queue length x is smaller than L, while if $x = L$ she does so with probability q. Otherwise she joins PC.

An $[L, q]$ threshold rule will be denoted by $[g]$ where $g = L + q$. Note that $[L, 1]$ and $[L+1, 0]$ are identical.

Theorem 2.1.

(i) For any equilibrium policy $\pi^* = (u_0, u_1, \ldots)$, each decision rule u_k is a threshold rule.
(ii) A symmetric equilibrium policy $\pi^* = (u^*, u^*, \ldots)$ exists, is unique, and u^* is a threshold rule.

Basic steps of proof:

(i) For every policy π and $k \geq 0$, one shows that $V_k(x, \pi)$ is strictly increasing in x [12].
(ii) Assume that all jobs other than C_k use a threshold policy $[g]$. Then $V_k(x, [g]^\infty)$ can be shown to be:

(1) strictly increasing in g, and
(2) continuous in g.

We note that the best response to a policy g is either a singleton or a whole interval of the form $[i, i+1]$ where i is an integer. The latter occurs when player k is indifferent between a threshold $[i, 0]$ and $[i, 1]$. It then easily follows that a fixed point exists which provides a symmetric threshold equilibrium [12]. ∎

Numerical Examples

The following numerical example with $\theta = 10$, $\mu = 100$ is taken from [12]. In that reference, a recursive procedure is given to compute $V(x, [L, q])$. Figures 2–3 are obtained using that procedure to compute the performance of a threshold x when all other users have threshold $[15, 1]$, for various values of input rates. The value of the threshold that defines the best response to the threshold $[15, 1]$ for a given input rate will then be obtained at the point where $V(x, [15, 1])$ becomes larger than θ^{-1}.

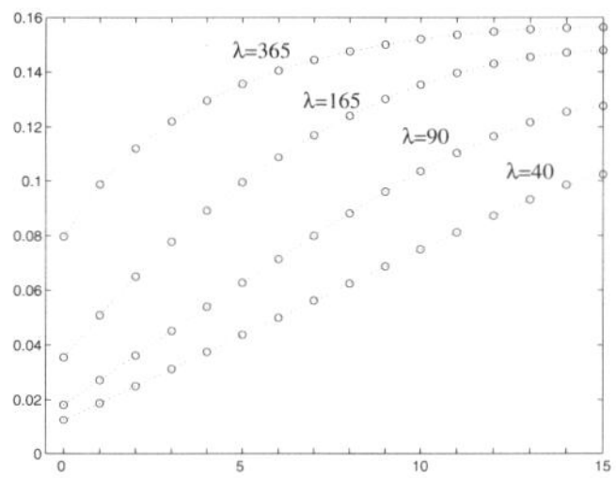

Figure 2: $V(x, [15, 1])$ as a function of x for various λ's

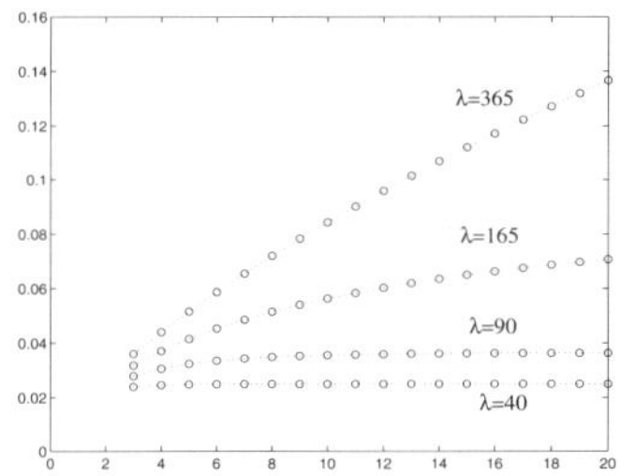

Figure 3: $V(3, [L, 1])$ as a function of L for various λ's

Fig. 3 shows the equilibrium symmetrical threshold that is obtained for this problem. It also compares it to the socially optimal problem (that would be obtained if a single decision maker tried to minimize the expected average waiting time of all customers). We see that the equilibrium threshold is always larger than the one corresponding to the socially optimal threshold. In particular, as the input rate λ becomes very large (overload conditions), the equilibrium threshold will converge to 10 and the socially optimal threshold will converge to 1. The equilibrium threshold of 10 in overload provides an expected waiting time at the MF that equals θ^{-1}. It is ten times larger that the expected waiting time at MF for the socially optimal policy.

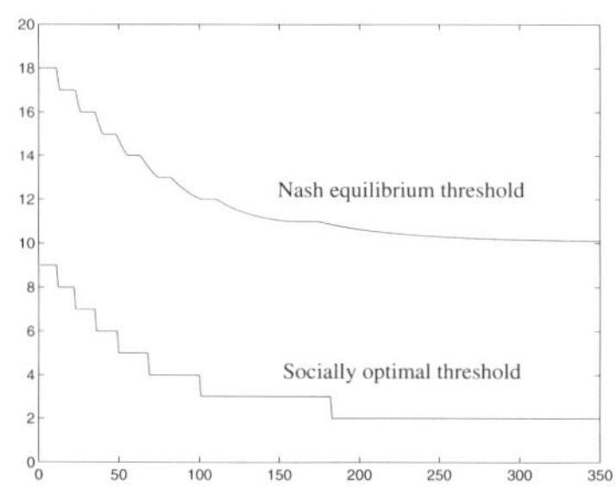

Figure 4: $g^* = L^* + q^*$ as a function of λ

Conclusions. We have observed in this first example that when all other users have a threshold g, the threshold of the best response policy is decreasing in that g. Thus the behavior of a user here goes in the opposite direction as the behavior of the rest of the users. Following the terminology of Hassin and Haviv [20], we call this property "Avoiding the Crowd". It is similar kind of behavior than we would expect in road traffic, where drivers try to avoid rush hours. In future sections we shall also encounter the opposite type of behavior, of "Joining the Crowd". In our example, this behavior implied the uniqueness of the equilibrium.

3 When to Queue?

We present two games that fall into this categories. The first is the problem of when to arrive at a bank. The second is of when to retry to make a phone call. We then briefly mention a third related game that also involves retransmission decisions and that is related to access to a radio channel.

3.1 When to Arrive at the Bank?

The aim of this example is to illustrate that a very common situation we frequently face is in fact a queueing game, where the decision is of when to queue.

We consider the following scenario. A bank opens between 9h00 and 12h00. All customers that arrive before 12h00 are served that day. A random number X

of customers wish to get a service on a given day. Service times are i.i.d. with exponential distribution. The order of service is FCFS (First Come First Served). Each customer wishes to minimize her own expected waiting time, i.e. the expected time elapsed from her arrival at the bank till she gets served.

This problem has been studied in [17]. The authors show that there exists a symmetric equilibrium distribution F of the arrival time with support $[T_0, 12h00]$ for all players, with $T_0 < 9h00$, such that if all customers follow F, then no individual has an incentive to deviate from using the distribution F.

We finally note that if we eliminate the FCFS regime among those who arrive before the bank opens, then $T_0 = 9h00$. This could reduce average waiting time!

3.2 Games in Retrial Queues: When to Retry to Make a Phone Call

We are all used to obtaining occasionally a busy signal when attempting to make phone calls. This signal is either due to the fact that the destination is currently busy with another phone call or to congestion that may cause calls to be blocked. A person whose call is blocked may typically retry calling. The goal of the example, taken from [16,21], is to analyze the individual's choice of time between retrials.

The model: Calls arrive according to a Poisson process with average rate λ. Service rates are i.i.d. with mean τ and finite variance σ^2. Let $S^2 := \tau^2 + \sigma^2$ and $\rho := \lambda\tau$. Between retrials calls are said to be "in orbit". Times between retrials of the ith call in orbit are exponentially distributed with expected value of $1/\theta_i$.

We note that exponentially distributed times are frequently used in models in telecommunications, and in particular in telephony models due to much accumulated experimental results. In particular, it is known that the duration of phone calls is well modeled by the exponential distribution. The exponential distribution is also attractive mathematically since it leads to simple Markovian models due to its memoryless property (i.e. the property $P(X > t + s | X \geq t) = P(X > s)$ for any $t, s > 0$).

We assume that each retrial costs c, and the waiting time costs w per time unit. We begin by presenting the performance of the socially optimal policy [27].

The socially optimal policy. The expected time in orbit when all retrials use the same parameter θ is

$$W = \frac{\rho}{1-\rho}\left(\frac{1}{\theta} + \frac{S^2}{2\tau}\right),$$

so the average cost per call is

$$(w + c\theta)W = \frac{\rho}{1-\rho}\left(\frac{cS^2}{2\tau}\theta + \frac{w}{\theta}\right) + \frac{\rho}{1-\rho}\left(\frac{wS^2}{2\tau} + c\right).$$

This is minimized at

$$\theta^* = \frac{\sqrt{2w\tau/c}}{S}.$$

A remarkable observation is that this rate is independent of the arrival rate!

If θ^* is used, the two terms that depend on θ turn out to be equal: the *waiting* cost and the *retrial* costs coincide.

The game In the game case, Kulkarni [27] computes $g(\theta, \gamma)$, which is the expected waiting time of an individual who retries at rate γ while all the others use retrial rate θ. This allows them to obtain the equilibrium rate:

$$\theta_e = \frac{w\rho + \sqrt{w^2\rho^2 + 16w\tau c(1-\rho)(2-\rho)/S^2}}{4c(1-\rho)}.$$

We observe that θ_e monotonically increases to infinity as λ increases (to $1/\sigma$).

Conclusions. Here are some conclusions from this example.

1. The ratio between the cost at equilibrium and the globally optimal cost tends to infinity. Thus, the non-cooperative nature of the game leads to high inefficiency.
2. The equilibrium retrial rate is larger than the globally optimal retrial rate. They tend to coincide as $\rho \to 0$.
3. Both equilibrium and optimal retrial rates are monotonically decreasing in the variance of the service times.
4. There is a unique equilibrium.

3.3 Access to a Radio Channel: When to Retransmit

Phone calls are not the only example where we find retransmissions. Another example, typical to cellular communications as well as satellite communications, is in the access of packets to a common radio channel. If more than one terminal attempts to transmit a packet at the same time then a collision occurs and all terminals involved in the collision have to retransmit their packets later. A terminal wishes typically to maximize the throughput (i.e. the expected average number of successful transmissions). A further cost may be incurred per each transmission attempt (it could represent the energy consumption). The problem has various formulations as stochastic games, in which each terminal controls the time till it attempts a retransmission of a packet that has collided, or the probability of its retransmission at a given time slot [8,24,28]. In [8], a comparison between a cooperative solution and a non-cooperative solution is presented. It is shown that as the arrival rate of packets increases, the cooperative solution tends to reduce the retransmission probabilities; retransmissions are thus delayed so as to avoid more collisions. Due to this behavior, the throughput is shown to increase with the arrival rate of packets. In the non-cooperative case, a single symmetric equilibrium is obtained in which, in contrast, the retransmission probabilities tend to increase with the arrival rate of packets. Thus, as the arrival rate increases, terminals become more aggressive, and the system throughput decreases. For large arrival rates it decreases to zero.

4 Where to Queue?

We briefly mention some background on the well-studied traffic assignment problem in a general network, and then present two examples in more detail of stochastic games occurring in simple queueing problems.

4.1 The Traffic Assignment Problem

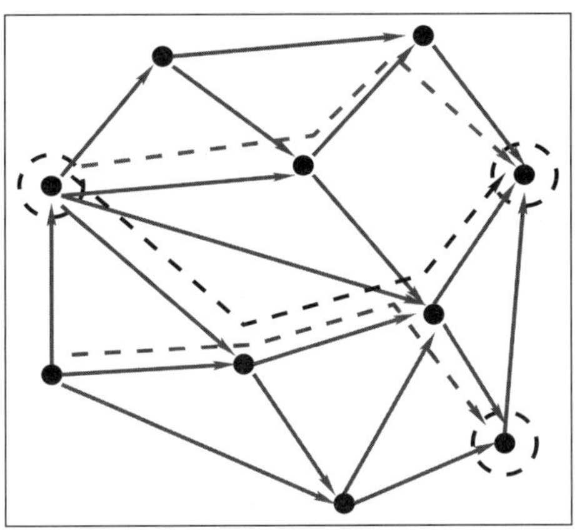

Figure 5: The assignment problem and competitive routing

One of the most studied game problems that arises in networks is the traffic assignment problem [29,30]. The generic problem is to determine which path to take in a network, that may be shared by many types of users having different sources and destinations. The costs, or delays (or other performance measures, such as loss probabilities) for each user may be related to the congestion it suffers between the source and destination, and is a function of choices of routes of other users as well. This problem has been studied already in the fifties by Wardrop [36] in the context of road traffic, where a user corresponds to a single driver. The game has been modeled using a continuum number of players where the influence of a single player (driver) on the cost of others can be neglected. Due to this property, such a player is often called atomless. Existence and uniqueness of the equilibrium (known as Wardrop equilibrium) have been established under quite general conditions, and many efficient computational methods have been developed (see [30] for an overview). This type of equilibrium has also been discovered in the context of mobile telecommunications in [18] and in the context of potential games in [33] (see also [32]). Wardrop principles have also been obtained independently around thirty years before Wardrop in an economic context [31].

An alternative game formulation arises when decisions are not taken by each individual, but instead, by a finite number of big organizations (the players) such as service providers, in the context of telecommunications, or of transportation companies, in the context of road traffic. Such players are often called atomic players. The decision taken by a player is what fraction of its flow to ship over each possible route. Nash equilibrium is then used as the solution concept. Some references in this type of modeling are [29,23], as well as [30] and references therein.

Till now we have not mentioned the relation between the traffic assignment problem and queueing. It turns out that there is a large class of queueing networks for which the average expected delays or the expected queue size satisfy the assumptions of the traffic assignment models studied in the literature. These are known as BCMP queueing networks (known to have a property called "product form", which makes link costs additive). We refer in particular to the papers [25,11] that describe these kinds of networks in the context of the traffic assignment problem.

Interesting extensions to dynamic and differential games context exist for the assignment model. We shall mention [37] and references therein. In this reference, the players have to ship a given amount of flow within a certain period and can decide dynamically at what rate to ship at each instant. A dynamic mixed equilibrium (that combines both atomless and atomic users) is computed using the maximum principle.

4.2 The Gas Station Game

The example we present here illustrates the dynamic routing choices between two paths. When a routing decision is made, the decision maker knows the congestion state of only one of the routes; the congestion state in the second route is unknown to the decision maker. The problem originates from the context of two gas stations on a highway [19]. A driver arriving at the first station sees the number of other cars already queued there and has to decide whether to join that queue, or to proceed to the next gas station. The state of the next gas station (i.e. the number of cars there) is not available when making the decision. The situation is illustrated in Fig. 6. The exact mathematical solution of the model was obtained in [10] and we describe it below.

This problem has natural applications in telecommunication networks: when making routing decisions for packets in a network, the state in a downstream node may become available after a considerable delay, which makes that information irrelevant when taking the routing decisions.

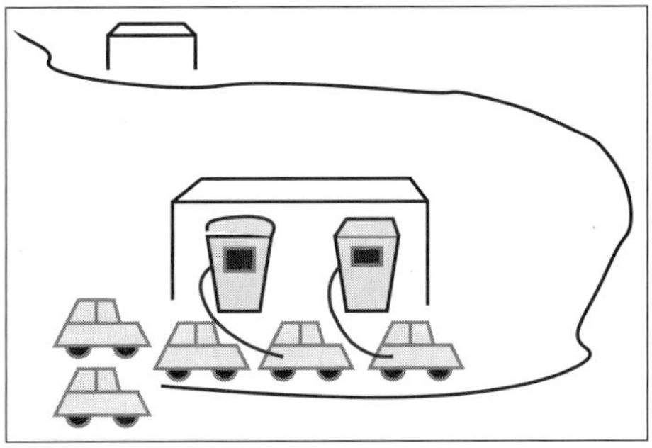

Figure 6: The gas station problem

Although the precise congestion state of the second route is unknown, its probability distribution, which depends on the routing policy, can be computed by the router.

We assume that the times that correspond to the arrival instants of individuals is a Poisson process with rate λ. Each arrival is a player, so there is a countable number of players, each of whom takes one routing decision, of whether to join the first or the second service station. A player eventually leaves the system and does not affect it anymore, once she receives service.

To obtain an equilibrium, we need to compute the joint distribution of the congestion state in both routes as a function of the routing policy.

We restrict to random threshold policies (n, r):

- if the number of packets in the first path is less than or equal to $n - 1$ at the instance of an arrival, the arriving packet is sent to path 1.
- If the number is n then it is routed to path 1 with probability r.
- If the number of packets is greater than n then it is routed to path 2.

The delay in each path is modeled by a state dependent queue:

- Service time at queue i is exponentially distributed with parameter μ_i
- Global inter-arrival times are exponential i.i.d. with parameter λ.

When all arrivals use policy (n, r), the steady state distribution is obtained by solving the steady state probabilities of the continuous time Markov chain [10].

If an arrival finds i customers at queue 1, it computes

$$E_i[X_2] = E\left[X_2 | X_1 = i\right]$$

and takes a routing decision according to whether the following inequality holds:

$$T^{n,r}(i, 1) := \frac{i+1}{\mu_1} \leq \frac{E_i[X_2]+1}{\mu_2} =: T^{n,r}(i, 2).$$

To compute it, the arrival should know the policy (n, r) used by all *previous* arrivals.

If the decisions of the arrival as a function of i coincide with (n, r) then (n, r) is a Nash equilibrium.

The optimal response against $[g] = (n, r)$ is monotonically decreasing in g. This is the **Avoid the Crowd** behavior.

Computing the conditional distributions, one can show [10] that there are parameters $(\mu_1, \mu_2, \lambda, n, r)$ for which the optimal response to (n, r) is indeed a threshold policy.

Denote

$$\rho := \frac{\lambda}{\mu_1}, \qquad s := \frac{\mu_2}{\mu_1}$$

There are other parameters for which the optimal response to (n, r) is a *two-threshold policy* characterized by $t^-(n, \rho, s)$ and $t^+(n, \rho, s)$ as follows.

It is optimal to route a packet to queue 2 **if** $t^-(n, \rho, s) \le X_1 \le t^+(n, \rho, s)$ **and to queue** 1 **otherwise.**

At the boundaries t^- and t^+ routing to queue 1 or randomizing is also optimal if $T^{n,r}(i, 1) = T^{n,r}(i, 2)$. For parameters in which the best response does not have a single threshold, we cannot conclude anymore what is the structure of a Nash equilibrium.

Example [10]. Consider $n = 3,\ r = 1,\ \rho = \lambda/\mu_1 = 1$ and $s = \mu_2/\mu_1 = 0.56$. We plot in Fig. 7 $T^{n,r}(i, 1)$ and $T^{n,r}(i, 2)$ for $i = 0, 1, \ldots, 4$.

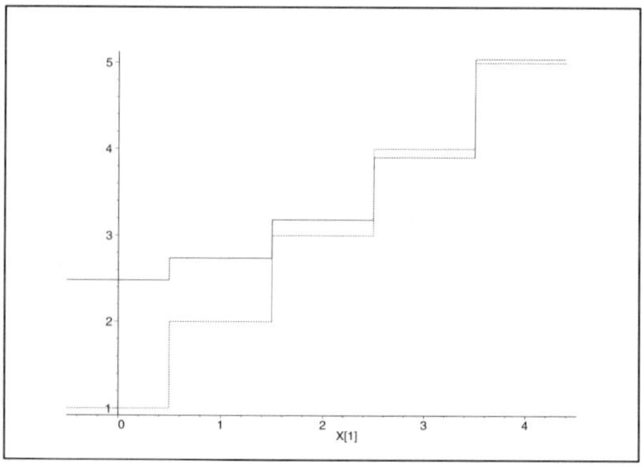

Figure 7: $T^{n,r}(i, 1)$ and $T^{n,r}(i, 2)$

Conclusions: As opposed to the example of the choice of players between a PC and a MF in Sec. 2, we saw in this section an example where for some parameters there may be no threshold type (n, r) equilibria. The form of the Nash equilibrium in these cases remains an open problem. Moreover, even the question of existence of a Nash equilibrium is then an open question.

Yet, in the case of equal service rates in both stations, a threshold equilibrium does exist, and it turns out to have the same type of behavior as in the PC-MF game, i.e. it is unique, and the best response has a tendency of "Avoiding the Crowd" [19].

By actually analyzing (numerically) the equilibrium for equal service rates, it was noted in [19] that when players use the equilibrium strategy, then the revenue of the first station is higher than the second one. Thus the additional information that the users have on the state of the first station produces an extra profit to that station. An interesting open problem is whether the second station can increase its profits by using a different pricing than the first station, so that users will have an extra incentive to go to that station. Determining an optimal pricing is also an interesting problem.

4.3 Where to Queue: Queues With Priority

We present a second queueing problem modeled as a stochastic game with infinitely many individual players due to [2,20]. We assume again that players arrive at the system according to a Poisson process with intensity λ, and have to take a decision of whether to join a low priority (second class) or a high priority (first class) queue, as illustrated in Fig. 8. There is a single server that serves both queues but gives strict priority to first class customers. Thus a customer in the second class queue gets served only when the first class queue is empty. We assume exponentially distributed service time with parameter μ and define $\rho := \lambda/\mu$.

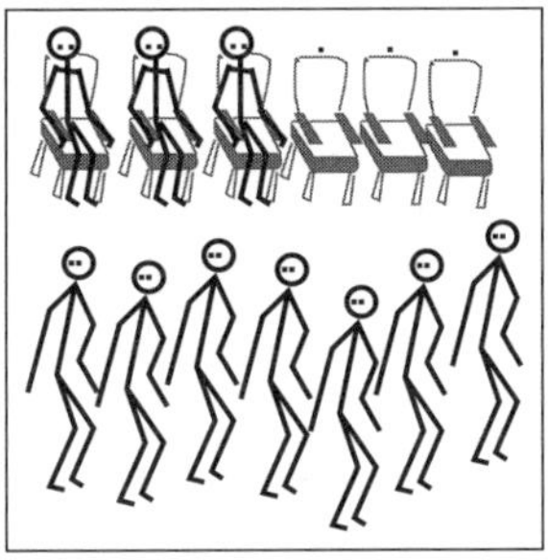

Figure 8: Choice between first and second class priorities

The game model is then as follows:

The actions: Upon arrival, a customer (player) observes the two queues and may purchase the high priority for a payment of an amount θ, or join the low priority queue.

The state: The state is a pair of integers (i, j) corresponding to the number of customers in each queue; i is the number of high priority customers and j, the number of low priority ones.

The analysis of this problem can be considerably simplified by using the following **monotonicity property**, identified in [2]: If for some strategy adopted by everybody, it is optimal for an individual to purchase priority at (i, j), then he must purchase priority at (r, j) for $r > i$.

This implies that the problem has an effective **lower dimensional state space:** It follows that starting at $(0, 0)$ and playing optimally, there is some n such that the only reachable states are

$$(0, j),\ j \leq n, \quad \text{and } (i, n),\ i \geq 1.$$

Indeed, due to monotonicity, if at some state $(0, m)$ it is optimal not to purchase priority, it is also optimal at states $(0, i)$, for $i \leq m$. Let $n - 1$ be the largest such state. Then starting from (0, 0) we go through states $(0, i)$, $i < n$, until $(0, n - 1)$ is reached. At $(0, n)$ it is optimal to purchase priority. We then move to state $(1, n)$.

The low priority queue does not decrease as long as there are high-priority customers. Due to monotonicity, it also does not increase as long as there are high-priority customers since at (i, n), $i \geq 1$ arrivals purchase priority! Therefore we remain at (i, n), as long as $i \geq 1$.

The Equilibrium. Suppose that the customers in the population, except for a given individual, adopt a common threshold policy $[g]$. Then the optimal threshold for the individual is *non-decreasing* in g.

This property is called **"Follow the Crowd" Behavior**

This property clearly implies *Existence* of an equilibrium, that can be obtained by a monotonically best response argument.

However, it turns out that there is no uniqueness of the equilibrium! Indeed, Hassin and Haviv have shown in [20] that there may be up to

$$\left\lfloor \frac{1}{1-\rho} \right\rfloor$$

pure threshold Nash equilibria, as well as other mixed equilibria! They further present numerical examples of multiple equilibria.

We conclude that in this problem we have definitely a different behavior of the equilibria than in the previous stochastic games in which we had threshold equilibria (the PC-MF game and the gas station game).

5 How Much to Queue?

Two main approaches exist for controlling the congestion at the network in high speed communication systems. The first, in which sources control their *transmission rate*, has been adopted by the ATM-forum [1] (an international standardization organism for high speed telecommunication networks) for applications that have flexibility in their requirement for bandwidth (and which then use the so called "Available Bit Rate" service). Although the general approach and signaling mechanisms have been standardized, the actual way to control the sources has been left open and is still an open area of research and development. An alternative approach is the so called "*window based*" flow control which is used at the Internet. An example of a stochastic game based on the latter approach is given in [26].

We present in this section a linear quadratic differential game model for flow control based on the first approach, which has been introduced and analyzed in [7], in which M users control their transmission rate into a single bottleneck queue. Thus, each user has to answer dynamically the question of "how much to queue", see Fig. 9.

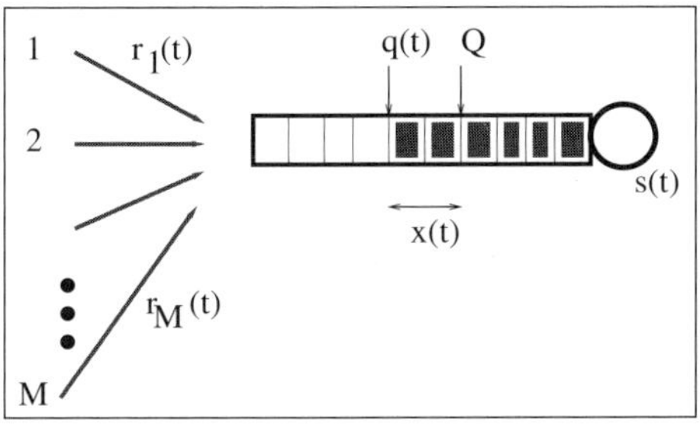

Figure 9: A competitive flow control model

The output rate of the queue (i.e. the server's rate) is given by $s(t)$. The *controlled* input rate of user m is denoted by $r_m(t)$. Let $q(t)$ denote the instantaneous queue length.

It is desirable for telecommunication applications to keep the instantaneous size of the queue close to some target: Q. Indeed, if the queue is too large then information packets might be lost due to overflow, and moreover, packets suffer large delays. If the queue is empty, on the other hand, then the output rate is constrained by the input rate, which might be lower than $s(t)$. We thus may have underutilization of the available service rate and we lose in throughput.

Define $x(t) := q(t) - Q$ to be the shifted instantaneous queue length. It will serve as the state of the queue. We assume that there is some desirable share of bandwidth available for user m, given by $a_m s(t)$, where $\sum_m a_{m=1}^M = 1$.

Define $u_m(t) := r_m(t) - a_m s(t)$ to be the shifted control.

Ignoring the nonlinearity in the dynamics that are due to boundary effects (at an empty queue or at a full queue), we obtain the following idealized dynamics:

$$\frac{dx}{dt} = \sum_{m=1}^{M} (r_m - a_m s) = \sum_{m=1}^{M} u_m, \tag{2}$$

Justification for using the linearized model is presented in [7].

Policies and information. We consider the following class of history dependent policies for all users:

$$u_m(t) = \mu_m(t, x_t), \qquad t \in [0, \infty).$$

μ_m is piecewise continuous in its first argument, piecewise Lipschitz continuous in its second argument. The class of all such policies for user m is $\mathcal{U}_m$.

Objectives. The cost per user is a linear combination of the cost related to deviation from the desirable queue length value Q and deviating from the desirable share of the bandwidth. We consider two versions that differ by some scaling factor:

- N1: the individual cost to be minimized by user m $(m \in \mathcal{M} = \{1, \dots, M\})$ is

$$J_m^{N1}(u) = \int_0^\infty \left(|x(t)|^2 + \frac{1}{c_m} |u_m(t)|^2 \right) dt. \tag{3}$$

- N2: the individual cost to be minimized by controller m $(m \in \mathcal{M})$ is

$$J_m^{N2}(u) = \int_0^\infty \left(\frac{1}{M} |x(t)|^2 + \frac{1}{c_m} |u_m(t)|^2 \right) dt. \tag{4}$$

In case N2 the "effort" for keeping the deviations of the queue length from the desired value is split equally between the users.

Nash equilibria. We seek a multi-policy $\mu^* := (\mu_1^*, \dots, \mu_M^*)$ such that no user has an incentive to deviate from, i.e.

$$J_m^{N1}(\mu^*) = \inf_{\mu_m \in \mathcal{U}_m} J_m^{N1}([\mu_m | \mu_{-m}^*]) \tag{5}$$

where $[\mu_m | \mu_{-m}^*]$ is the policy obtained when for each $j \neq m$, player j uses policy μ_j^*, and player m uses μ_m. We define similarly the problem with the cost J^{N_2}.

Main results. We shall show that the flow control game has a simple computable equilibrium and value. We further show a uniqueness result.

The equilibrium: For case Ni $(i = 1, 2)$, there exists an equilibrium given by

$$\mu_{Ni,m}^*(x) = -\beta_m^{Ni} x \ , \ m = 1, \dots, M,$$

where β_m^{Ni} is given by

$$\beta_m^{N1} = \overline{\beta}^{(N1)} - \sqrt{\overline{\beta}^{(N1)2} - c_m}$$

where $\overline{\beta}^{(N1)} := \sum_{m=1}^M \beta_m^{N1}$, $i = 1, 2$, are the unique solutions of

$$\overline{\beta}^{(N1)} = \frac{1}{M-1} \sum_{m=1}^{M} \sqrt{(\overline{\beta}^{(N1)})^2 - c_m}$$

and for the case N2:

$$\beta_m^{N2} = \overline{\beta}^{(N2)} - \sqrt{\overline{\beta}^{(N2)2} - \frac{c_m}{M}},$$

where $\overline{\beta}^{(N2)} := \sum_{m=1}^M \beta_m^{N2}$, $i = 1, 2$, are the unique solutions of

$$\overline{\beta}^{(N2)} = \frac{1}{M-1} \sum_{m=1}^{M} \sqrt{(\overline{\beta}^{(N2)})^2 - \frac{c_m}{M}} = \frac{\overline{\beta}^{(N1)}}{\sqrt{M}}.$$

Moreover,

$$\beta_m^{N1} = \beta_m^{N2} \sqrt{M}.$$

For each case, this is the unique equilibrium among stationary policies and is time-consistent.

The value. The costs accruing to user m, under the two Nash equilibria above, are given by

$$J_m^{N1}(\mu_{N1}^*) = \frac{\beta_m^{N1}}{c_m} x^2$$

and

$$J_m^{N2}(\mu_{N2}^*) = \frac{\beta_m^{N2}}{c_m} x^2 = \frac{1}{\sqrt{M}} J_m^{N1}(u_{N1}^*).$$

The symmetric case. In the case $c_m = c_j =: c$ for all $m, j \in \mathcal{M}$ we obtain:

$$\beta_m^{N1} = \sqrt{\frac{c}{2M-1}}, \qquad \text{and} \qquad \beta_m^{N2} = \sqrt{\frac{c}{M(2M-1)}}, \qquad \forall m \in \mathcal{M};$$

The case of $M = 2$. We assume general values of c_m's. we have for $m = 1, 2,\ j \neq m$,

$$\beta_m^{N1} = \left[-\frac{2c_j - c_m}{3} + 2\frac{\sqrt{c_1^2 - c_1 c_2 + c_2^2}}{3} \right]^{1/2}, \qquad \beta_m^{N2} = \frac{\beta_m^{N1}}{\sqrt{2}}.$$

If moreover, $c_1 = c_2 = c$, then $\beta_m^{N1} = \sqrt{c/3}$, $\beta_m^{N2} = \sqrt{c/6}$.

We now sketch the proof for N1. The one for N2 is similar.

Proof for (N1). Choose a candidate solution

$$u_m^*(x) = -\beta_m x\ ,\ m = 1, \dots, M, \quad \text{where } \beta_m = \overline{\beta} - \sqrt{\overline{\beta}^2 - c_m}$$

where $\overline{\beta} := \sum_{m=1}^M \beta_m$, is the unique solution of

$$\overline{\beta} = \frac{1}{M-1} \sum_{m=1}^M \sqrt{\overline{\beta}^2 - c_m}.$$

Fix u_j for $j \neq m$. Player m is faced with a linear quadratic optimal control problem with the dynamics

$$dx/dt = u_m - \beta_{-m} x,\ \beta_{-m} = \sum_{j \neq m} \beta_j$$

and cost $J_m^{N1}(u)$ that is strictly convex in u_m. Her optimal response is $u_m = -c_m P_m x$, where P_m is the unique positive solution of the Riccati equation

$$-2\beta_{-m} P_m - P_m^2 c_m + 1 = 0. \tag{6}$$

Denoting $\beta'_m = c_m P_m$, we obtain from (6)

$$\beta'_m = f_m(\beta_{-m}) := -\beta_{-m} + \sqrt{\beta_{-m}^2 + c_m}.$$

u is in equilibrium if and only if $\beta' = \beta$, or

$$\overline{\beta}^2 = \beta_{-m}^2 + c_m. \tag{7}$$

Hence

$$\beta_m = \overline{\beta} - \sqrt{\overline{\beta}^2 - c_m}.$$

Summing over $m \in \mathcal{M}$ we obtain

$$\Delta := \overline{\beta} - \frac{1}{M-1} \sum_{m=1}^{M} \sqrt{\overline{\beta}^2 - c_m} = 0.$$

Uniqueness follows since

- Δ is strictly decreasing in $\overline{\beta}$ over the interval $[\max_m \sqrt{c_m}, \infty)$,
- it is positive at $\overline{\beta} = \max_m \sqrt{c_m}$, and
- it tends to $-\infty$ as $\beta \to \infty$.

5.1 Greedy Decentralized Algorithms

The Nash equilibrium requires some coordination, i.e. a (non-binding) agreement according to which all users follow the Nash equilibrium. Moreover, to compute the equilibrium, a player needs to have the knowledge of other individual utilities (c_m). Both the coordination as well as the knowledge of others' utilities are restrictive and non-realistic assumptions. This motivated the authors in [7] to propose several greedy decentralized "best response" algorithms.

A greedy "best response" algorithm is defined by the following four conditions [14]:

- Each user updates from time to time their policy by computing the best response against the most recently announced policies of the other users.
- The time between updates is sufficiently large, so that the control problem faced by a user when they update their policy is well approximated by the original infinite horizon problem.
- The order of updates is arbitrary, but each user performs updates infinitely often.
- When the nth update occurs, a subset $K_n \subset \{1, \dots, M\}$ of users simultaneously update their policies.

Proposed algorithms

- **Parallel update algorithm (PUA):** $K_n = \{1, \ldots, M\}$ for all n.
- **Round robin algorithm (RRA):** K_n is a singleton for all n and equals $(n + k)\bmod M + 1$, where k is an arbitrary integer.
- **Asynchronous algorithm (AA):** K_n is a singleton for all n and is chosen at random.

The initial policy used by each user is linear.

$\beta^{(n)}$:= value of the linear coefficient defining the policies corresponding to the end of the nth iteration.

The optimal response at each step n:

$$\beta_m^{(n)} = \begin{cases} f_m(\beta_{-m}^{(n-1)}) & \text{if } m \in K_n \\ \beta_m^{(n-1)} & \text{otherwise,} \end{cases} \tag{8}$$

where

$$f_m(\beta_{-m}) := -\beta_{-m} + \sqrt{\beta_{-m}^2 + c_m}. \tag{9}$$

Convergence results. We briefly mention the convergence results obtained in [7].

Theorem 5.1. *Consider PUA.*
(i.a) Let $\beta_k^{(1)} = 0$ for all k. Then $\beta_k^{(2n)}$ monotonically decrease in n and $\beta_k^{(2n+1)}$ monotonically increase in n, for every player k, and thus, the following limits exist: $\hat{\beta}_k := \lim_{n\to\infty} \beta_k^{(2n)}$, $\tilde{\beta}_k := \lim_{n\to\infty} \beta_k^{(2n+1)}$.
(i.b) Assume that $\hat{\beta}_k = \tilde{\beta}_k$ (defined as above, with $\beta_k^{(1)} = 0$ for all k). Consider now a different initial condition satisfying either $\beta_k^{(1)} \le \beta_k$ for all k (where β_k is the unique Nash), or $\beta_k^{(1)} \ge \beta_k$ for all k. Then for all k, $\lim_{n\to\infty} \beta_k^{(n)} = \beta_k$.

Global convergence. A global convergence result is obtained for $M = 2$:

Theorem 5.2. *If*

- *(ii.a) $M = 2$, and either $\beta_k^{(1)} \le \beta_k$ for all k, or $\beta_k^{(1)} \ge \beta_k$ for all k; or if*
- *(ii.b) $\beta_k^{(1)}$ and $c := c_k$ are the same for all k, then $\beta^{(n)}$ converges to the unique equilibrium β^*.*

Local convergence.

Theorem 5.3. *For arbitrary c_k, there exists some neighborhood V of the unique equilibrium β^* such that if $\beta_k^{(1)} \in V$ then $\beta^{(n)}$ converges to the unique equilibrium β^*.*

Numerical examples. Some examples from [7] on the convergence of the greedy algorithm are presented below. The parallel updates are seen to converge slowest with oscillating behavior.

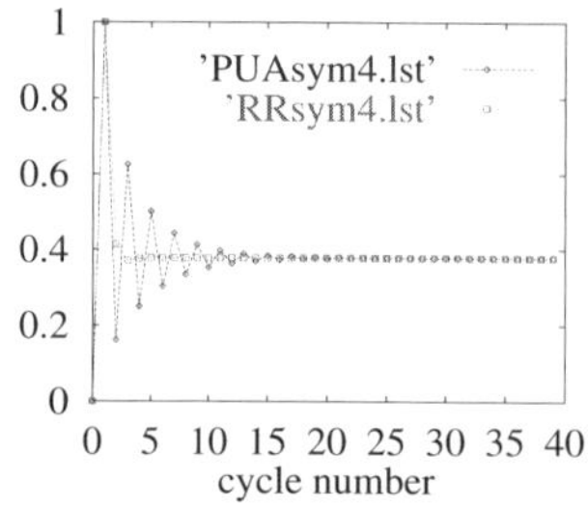

Figure 10: PUA versus RRA for $M = 4$

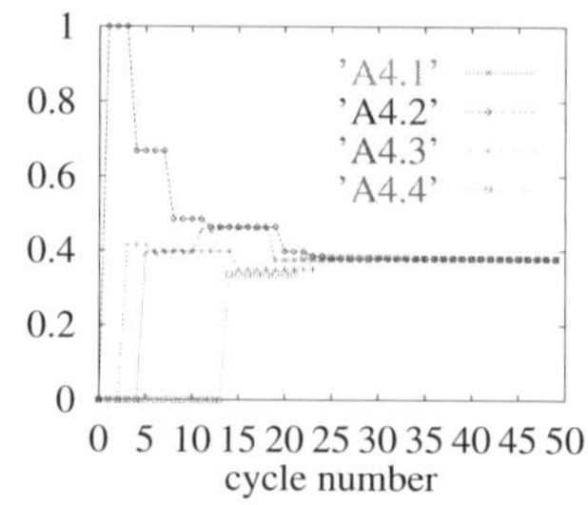

Figure 11: AA for $M = 4$

6 S-Modular Games

We have seen in several examples in previous sections that equilibria best response strategies have either the "Join the Crowd" property or the "Avoid the Crowd" property, which typically leads to threshold equilibria policies. These properties also turn out to be useful when we seek to obtain convergence to equilibria from a non-equilibrium initial point. These issues will be presented in this section within the framework of S-modular games due to Yao [38] who extends the notion of sub-modular games introduced by Topkis [35].

6.1 Model, Definitions and Assumptions

General models are developed in [35,38] for games where the strategy space S_i of player i is a compact sub-lattice of $R^{\overline{m}}$. By sub-lattice we mean that it has the property that for any two elements x, y that are contained in S_i, $\min(x, y)$ (denoted by $x \wedge y$) and $\max(x, y)$ (denoted by $x \vee y$) are also contained there (by $max(x, y)$ we mean the componentwise max, and similarly with the min). We describe below the main results for the case that $\overline{m} = 1$.

Definition 6.1. The utility f_i for player i is super-modular if and only if

$$f_i(x \wedge y) + f_i(x \vee y) \geq f_i(x) + f_i(y).$$

It is sub-modular if the opposite inequality holds.

If f_i is twice differentiable then super-modularity is equivalent to

$$\frac{\partial^2 f_i(x)}{\partial x_1 \partial x_2} \geq 0.$$

Monotonicity of maximizers. The following important property was shown to hold in [35]. Let f be a supermodular function. Then the maximizer with respect to x_i is increasing in x_j, $j \neq i$.

More precisely, define the best response

$$BR_1^*(x_2) = argmax_{x_1} f(x_1, x_2);$$

if there is more than one argmax above we shall always limit ourselves to the smallest one (or always limit ourselves to the largest one). Then $x_2 \leq x_2'$ implies $BR_1^*(x_2) \leq BR_1^*(x_2')$. This monotonicity property holds also for non-independent policy sets such as (10), provided that they satisfy the ascending property (defined below).

Definition 6.2. (**Monotonicity of sublattices**) Let A and B be sub-lattices. We say that $A \prec B$ if for any $a \in A$ and $b \in B$, $a \wedge b \in A$ and $a \vee b \in B$.

Next, we introduce some properties on the policy spaces.

Consider two players. We allow S_i to depend on x_j

$$S_i = S_i(x_j), \quad i, j = 1, 2, \quad i \neq j. \tag{10}$$

Monotonicity of policy sets. We assume

$$x_j \leq x_j' \Longrightarrow S_i(x_j) \prec S_i(x_j').$$

This is called the **Ascending Property**. We define similarly the **Descending Property**.

Lower semi-continuity of policies. We say that the point to set map $S_i(\cdot)$ is lower semi continuous if for any $x_j^k \to x_j^*$ and $x_i^* \in S_i(x_j^*)$ $(j \neq i)$, there exist $\{x_i^k\}$ s.t. $x_i^k \in S_i(x_j^k)$ for each k, and $x_i^k \to x_i^*$.

6.2 Existence of Equilibria and Round Robin Algorithms

Consider an n-player game. Yao [38, Algorithm 1] and Topkis [35, algorithm I] consider a greedy round robin scheme where, at some infinite strictly increasing sequence of time instants T_k, players update their strategies each using the best response to the strategies of the others. Player l updates at times T_k with $k = mn+l$, $m = 1, 2, 3, \ldots$.

Assume lower semi-continuity and compactness of the strategy sets. Under these conditions, super-modularity together with the ascending property implies monotonic convergence of the payoffs to an equilibrium [38]. The monotonicity is in the same direction for all players: the sequences of strategies for each player either all increase or all decrease.

The same type of result is also obtained in [38, Thm. 2.3] with sub-modularity instead of super-modularity for the case of two players, where the ascending property is replaced by the descending property. The monotonic convergence of the round robin policies still holds but is in the opposite direction: the sequence of responses of one player increases to his/her equilibrium strategy, while the ones of the other player decrease.

In both cases, there need not be a unique equilibrium.

Yao [38] further extends these results to cases of costs (or utilities) that are sub-modular in some components and super-modular in others. The notion of s-modularity is used to describe either sub-modularity or super-modularity. Another extension in [38] is to vector policies (i.e. a strategy of a player is in a sublattice of R^k).

Next we present several examples for games in queues where S-modularity can be used. The first two examples are due to Yao [38].

6.3 Example of Super-Modularity: Queues in Tandem

Consider a set of queues in tandem. Each queue has a server whose speed is controlled. The utility of each server rewards the throughput and penalizes the delay. Under appropriate conditions, it is then shown in [38] that the players have compatible incentives: if one speeds up, the other also wants to speed up.

More precisely, consider two queues in tandem with i.i.d. exponentially distributed service times with parameters $\mu_i, i = 1, 2$. Let $\mu_i \le u$ for some constant u. Server one has an infinite source of input jobs There is an infinite buffer between servers 1 and 2. **The throughput** is given by $\mu_1 \wedge \mu_2$.

The expected number of jobs in the buffer is given [38] by

$$\frac{\mu_1}{\mu_2 - \mu_1}$$

when $\mu_1 < \mu_2$, and is otherwise infinite.

Let

- $p_i(\mu_1 \wedge \mu_2)$ be the profit of server i,
- $c_i(\mu_i)$ be the operating cost,
- $g(\cdot)$ be the inventory cost.

The utilities of the players are defined as

$$f_1(\mu_1, \mu_2) := p_1(\mu_1 \wedge \mu_2) - c_1(\mu_1) - g\left(\frac{\mu_1}{\mu_2 - \mu_1}\right)$$

$$f_2(\mu_1, \mu_2) := p_2(\mu_1 \wedge \mu_2) - c_2(\mu_2) - g\left(\frac{\mu_1}{\mu_2 - \mu_1}\right).$$

The strategy spaces are given by

$$S_1(\mu_2) = \{\mu_1 : 0 \le \mu_1 \le \mu_2\},$$

$$S_2(\mu_1) = \{\mu_2 : \mu_1 \le \mu_2 \le u\}.$$

It is shown in [38] that if g is convex increasing then f_i are super-modular. So we can apply the results of the previous subsection, and obtain (1) the property of "joining the crowd" of the best response policies, (2) existence of an equilibrium, (3) convergence to equilibrium of some round robin dynamic update schemes.

6.4 Example of Sub-Modularity: Flow Control

We consider now an example for sub-modularity. There is a single queueing centre with two input streams of Poisson arrivals with rates λ_1 and λ_2. The rates of the streams are controlled by 2 players.

The queueing center consists of c servers and no buffers. Each server has one unit of service rate.

When all servers are occupied, an arrival is blocked and lost.

The blocking probability is given by the Erlang loss formula:

$$B(\lambda) = \frac{\lambda^c}{c!}\left[\sum_{k=0}^{c}\frac{\lambda^k}{k!}\right]^{-1}$$

where $\lambda = \lambda_1 + \lambda_2$.

Suppose user i maximizes

$$f_i = r_i(\lambda_i) - c_i(\lambda B(\lambda)).$$

c_i is assumed to be convex increasing. $\lambda B(\lambda)$ is the total loss rate.

Then it is easy to check [38] that f_i are sub-modular.

Two different settings can be assumed for the strategy sets. In the first, the available set for player i consists of $\lambda_i \leq \overline{\lambda}$. Alternatively, we may consider that the strategy sets of the players depend on each other and the sum of input rates has to be bounded: $\lambda \leq \overline{\lambda}$. Then S_i satisfy the descending property. We can thus apply again the results of subsection 6.2.

6.5 A Flow Versus Service Control

Exponentially inter-arrivals as we used in previous examples are quite appealing to handle mathematically, and they can model sporadic arrivals, or alternatively, information packets that arrive one after the other but whose size can be approximated by an exponential random variable. In this example we consider, in contrast, a constant time T between arrivals of packets, which can be used for modeling ATM (Asynchronous Transfer Mode) networks in which information packets have a fixed size.

We consider a single node with a periodic arrival process, in which the first player controls the constant time period $T \in [\underline{T}, \overline{T}]$ between any two consecutive arrivals. We consider a single server with no buffer. The service time distribution is exponential with a parameter $\mu \in [\underline{\mu}, \overline{\mu}]$, which is controlled by the second player. If an arrival finds the server busy then it is lost. The loss probability is given by

$$P_l = \exp(-\mu T),$$

which is simply the probability that the random service time of (the previous) customer is greater than the constant T.

The transmission rate of packets is T^{-1}, but since a fraction P_l is lost then the goodput (the actual rate of packets that are transmitted successfully) is

$$G = \frac{1}{T}(1 - \exp(-\mu T)).$$

We assume that the utility of the first player is the goodput plus some function of the input rate T^{-1}. The server earns a reward that is also proportional to the goodput, and has some extra operation costs which is a function of the service rate μ. In other words,

$$J_1(T, \mu) = \frac{1}{T}(1 - \exp(-\mu T)) + f(T^{-1}),$$
$$J_2(T, \mu) = \frac{1}{T}(1 - \exp(-\mu T)) + g(\mu).$$

We then have for $i = 1, 2$

$$\frac{\partial J_i{}^2}{\partial T \partial \mu} = -\mu \exp(-\mu T) \leq 0.$$

We conclude that the cost is sub-modular.

7 Applications of S-Modularity to Stochastic Games

In the previous examples of S-modularity, we used that notion in static games and applied it directly to the policies. An alternative use of (a stronger version of) S-modularity is in applying it to dynamic programming operators, which allows us to obtain structural results for equilibrium policies. We shall restrict ourselves in this section to zero-sum games in which player one maximizes some cost and player two seeks to minimize it.

Before introducing the stochastic game setting, we first introduce the structural results in a setting of a set of matrix games. We shall then explain in subsection 7.2 how this has been used to solve in [3,4] a stochastic zero-sum game in which both flow and service are controlled. We then solve in the subsequent subsection a more involved stochastic game in which there are several sources of flow. We conclude with a discussion.

7.1 Monotonic Policies are Matrix Games

Let $\mathbf{X} = \{0, 1, 2, \ldots, L\}$ where L may be finite or infinite. Consider a set of matrix games $\{S(x, a, b)\}_{a,b}$ indexed by $x \in \mathbf{X}$ where $a \in \mathbf{A}(x)$ and $b \in \mathbf{B}(x)$ are the finite set of actions of players 1 and 2 respectively. We assume that $\mathbf{A}(x)$ is monotonically increasing in x and that $\mathbf{B}(x)$ is monotone decreasing in x. If game x is played and the players choose actions a and b, then player 2 pays to player 1 $S(x, a, b)$.

We introduce the following assumptions:

J1 (strong sub-modularity): For any $b^1 > b^2$ and any a^1 and a^2,

$$S(x, a^1, b^2) - S(x, a^1, b^1) - [S(x-1, a^2, b^2) \\ - S(x-1, a^2, b^1)] > 0, \quad 0 < x \leq L.$$

J2 (strong super-modularity): For any $a^1 > a^2$ and any b^1 and b^2,

$$S(x, a^2, b^1) - S(x, a^1, b^1) - [S(x-1, a^2, b^2) \\ - S(x-1, a^1, b^2)] < 0, \quad 0 < x \leq L.$$

u_x (resp. v_x) will denote a randomized strategy for player 1 and player 2, respectively, at game x. A policy $u = \{u_x\}, x \in \mathbf{X}$ for player 1 is a sequence of randomized strategies for all possible games. A policy for player 2 is defined similarly.

We now describe the type of monotonicity of optimal policies that will occur in the above problems. Let $v : \mathbf{X} \to \mathcal{M}(\mathbf{B})$. Denote $b_x^{sup}(v) :=$ the greatest b in the support of u_x, i.e. the greatest $b \in \mathbf{B}$ that is chosen by u with positive probability in state x. Denote $b_x^{inf}(u) :=$ the smallest b in the support of u_x. We say that a decision rule u_t at time t is **strongly monotonically decreasing** if for any $x \in \mathbf{X}$ and y with $y < x$, $b_y^{inf}(u_t) \geq b_x^{sup}(u_t)$. The analogous definitions hold naturally for player 1.

Note that as a direct consequence from the definition of strongly monotonic policies we have

Lemma 7.1. *If u is strongly monotone then it randomizes in at most $|\mathbf{A}| - 1$ games. If v is strongly monotone then it randomizes in at most $|\mathbf{B}| - 1$ games.*

Denote val$S(x)$:= the value of the matrix game x, and $S(x, u, v)$:= the expected cost that player 2 pays to player 1 at game x when mixed strategies u and v are used.

We now present the main structural results for the saddle point policies for the games $x \in \mathbf{X}$.

Theorem 7.1. *Choose any saddle point policies u and v for players 1 and 2, respectively.*

(i) Assume J1. Then u is strongly monotone.
(ii) Assume J2. Then v is strongly monotone.

Proof. Assume J1. Let v be any saddle point policy for player 2. Then for any $b^2 > b^1$, and any policy u, we have by J1

$$\Delta(u, x) := S(x, u_x, b^2) - S(x, u_x, b^1) - [S(x-1, u_{x-1}, b^2) - S(x-1, u_{x-1}, b^1)] > 0.$$

We show that this implies that $b_x^{sup}(v) \le b_{x-1}^{inf}(v)$. Suppose $\Delta(x, u)$ is positive for all u, but the latter does not hold, i.e. $b_x^{sup}(v) > b_{x-1}^{inf}(v)$. Set $b^1 = b_{x-1}^{inf}(v)$ and $b_2 = b_x^{sup}(v)$. Let u be any saddle point policy for player 1. It then immediately follows (for more details see e.g. Lemma 3.1 (i) in [3]) that

$$\text{val}S(x) = S(x, u_x, b_2) \le S(x, u_x, b_1).$$

Hence since $\Delta(x, u)$ is strictly positive, we have

$$S(x-1, u_{x-1}, b_2) < S(x-1, u_{x-1}, b_1) = \text{val}S(x-1)$$

(the last inequality is again straightforward and more details can be found in Lemma 3.1 (i) in [3]). But this contradicts the definition of the value of the game $S(x-1)$. This establishes (1) by contradiction. The proof of (ii) is symmetric. □

Corollary 7.1. *(i) Assume that J1 holds. Assume that* $\mathbf{A} = \{a^1, a^2\}$ *where* $a^1 < a^2$. *Any saddle point policy* u^* *for player 1 satisfies the following. There exists* $m_u \in \mathbf{X}$ *such that*

$$u_x^* = \begin{cases} (1,0) & \text{if } x > m_u \\ (q_u, \bar{q}_u) & \text{if } x = m_u \\ (0,1) & \text{if } x < m_u \end{cases} \tag{11}$$

where $q_u \in [0, 1]$ *is some constant. Moreover, any saddle point* v_x^* *for player 2 needs no randomizations in any game except for (perhaps)* $x = m_u$.
(ii) Assume that $\mathbf{B} = \{b^1, b^2\}$ *where* $b^2 > b^1$. *Any saddle point policy* v^* *for player 2 satisfies the following. Then there exists* $m_v \in \mathbf{X}$ *such that*

$$v_x^* = \begin{cases} (1,0) & \text{if } x > m_v \\ (q_v, \bar{q}_v) & \text{if } x = m_v \\ (0,1) & \text{if } x < m_v \end{cases} \tag{12}$$

where $q_v \in [0, 1]$ *is some constant u. Moreover,* u_x^* *needs no randomizations in any game except for (perhaps)* $x = m_v$.

Remark 7.1. It follows from Corollary 7.1 that if $\mathbf{A} = \{a^1, a^2\}$ and $\mathbf{B} = \{b^1, b^2\}$ and if $m_u \ne m_v$ then no randomizations are needed at any state by both u^* and v^*. If $m_u = m_v$ then randomization may be needed only at $x = m_u = m_v$.

7.2 Stochastic Game Between the Control of Flow and the Control of Service

We now briefly present the stochastic game in [3,4] consisting of a queue with a single input flow controlled by one player, whose adversary controls the service. We shall emphasize here the type of results and the solution approach. In the next subsection we shall present in more detail a model with several input flows and a controlled server.

Considered is a discrete-time single-server queue with a buffer of size $L \leq \infty$. We shall use below x to denote the state, which refers to the number of customers in the queue. We assume that at most one customer may join the system in a time slot. This possible arrival is assumed to occur immediately after the beginning of the time slot. The state corresponds to the number of customers in the queue at the beginning of a time slot. Let a_{min} and a_{max} be two real numbers satisfying $0 < a_{min} \leq a_{max} < 1$. At the end of the slot, if the queue is non-empty and if the action of the server is a, then the service of a customer is successfully completed with probability $a \in \mathbf{A}$ where $\mathbf{A}$ is a finite subset of $[a_{min}, a_{max}]$. If the service fails the customer remains in the queue, and if it succeeds then the customer leaves the system.

Let b_{min}, b_{max} be two real numbers satisfying $0 \leq b_{min} \leq b_{max} < 1$. At the beginning of each time slot, if the state is x then the flow controller chooses an action b from a finite set $\mathbf{B}(x) \subset [b_{min}, b_{max}]$. In this case, the probability of having one arrival during this time slot is equal to b. If the buffer is finite ($L < \infty$) we assume that $0 \in \mathbf{B}(x)$ for all x; moreover, when the buffer is full, no arrivals are possible ($\mathbf{B}(L) = \{0\}$). In all states other than L we assume that the available actions are the same, and we denote them by $\mathbf{B}(x) = \mathbf{B}$.

We assume that a customer that enters an empty system may leave the system (with probability a, when action a is used) at the end of this same time slot. The transition law P is:

$$P_{xaby} := \begin{cases} \bar{b}a, & \text{if } L \geq x \geq 1,\ y = x - 1; \\ ba + \bar{b}\bar{a}, & \text{if } L \geq x \geq 1,\ y = x; \\ b\bar{a}, & \text{if } L > x \geq 0,\ y = x + 1; \\ 1 - b\bar{a}, & \text{if } y = x = 0; \end{cases}$$

(for any number $\chi \in [0, 1]$, $\bar{\chi} := 1 - \chi$). We define an immediate payoff

$$c(x, a, b) := h(x) + \theta(a) + \zeta(b) \tag{13}$$

for all $x \in \mathbf{X}$, $a \in \mathbf{A}$ and $b \in \mathbf{B}$. We assume that $h(x)$ is a real-valued *increasing convex* function on $\mathbf{X}$ which is polynomially bounded, θ is a real function on $\mathbf{A}$ and ζ is a real function on $\mathbf{B}$. It is natural (although not necessary for the structural results) to assume that θ is increasing in a and $\theta \geq 0$ whereas ζ is decreasing in b and $\zeta \leq 0$. h can be interpreted as a function which gives the holding cost rates, ζ

as a reward function related to the acceptance of an incoming customer, and θ as a cost function per quality of service.

We thus consider a flow control problem, where the real controller is the one that chooses actions b, and is playing in an unknown environment of service conditions, so that the first player, which we call "nature" represents an imaginary service controller.

Let U (V) be the class of profiles of player 1 (player 2 resp.). A profile $u \in U$ $(v \in V)$ is a sequence $u = (u_0, u_1, \dots)$ $(v = (v_0, v_1, \dots)$ resp.) where u_n (resp. v_n) is a conditional probability on $\mathbf{A}$ (resp. $\mathbf{B}$) given the history of all states and actions of both players as well as the current state. Thus each player is assumed to have the information of all last actions of both players as well as the current and past states of the system. Both players know the action sets, the immediate cost c, the initial state and the transition probabilities P. A special class of profiles are the stationary policies, which are those we defined in the previous subsection: the choice of a mixed action is then only a function of the current state.

Let u be a policy of player 1 and v a policy of player 2. Let ξ be a fixed number in $[0, 1)$. Define the discounted cost:

$$V_\xi(x, u, v) := E^{u,v}\left[\sum_{t=0}^{\infty} \xi^t \, c(X_t, A_t, B_t) \mid X_0 = x\right], \tag{14}$$

where X_t, A_t and B_t are the stochastic processes describing the state and actions at time t. Define the following problem $(\mathbf{Q}_\xi)$: Find u, v that achieve

$$V_\xi(x) := \sup_{u \in U} \inf_{v \in V} V_\xi(x, u, v), \qquad \forall x \in \mathbf{X}. \tag{15}$$

It is known [3,4] that there exist stationary saddle point policies (u^*, v^*) for the players, i.e. stationary policies that satisfy

$$\begin{aligned}\sup_{u \in U} \inf_{v \in V} V_\xi(x, u, v) &= \inf_{v \in V} \sup_{u \in U} V_\xi(x, u, v) = \sup_{u \in U} V_\xi(x, u, v^*)\\ &= \inf_{v \in V} V_\xi(x, u^*, v) = V_\xi(x, u^*, v^*).\end{aligned}$$

$V_\xi(x)$ is called the ξ-discounted *value* of the stochastic game. This value is further known to be finite. (The existence of stationary saddle point policies and the finiteness of the value also holds in the expected average cost, yet for the case of an infinite buffer, some stability conditions have to be imposed (see [3,4]).

In the discounted stochastic game, the value function is known to satisfy the dynamic programming equation (due to Shapley [34]), which says that for every state x, the value function $V_\xi(x)$ evaluated at that state is equal to the value of some *matrix game* $S(x, a, b)$ (whose entries depend on x, on the discount factor, the immediate costs, the transition probabilities and the value functions at some other states). Moreover, the stationary policies u and v for which u_x and v_x are saddle points of the game $S(x)$ are optimal policies for the original stochastic games. This representation also holds for the case of infinite state space (see [4,9]). It is shown in [3] using dynamic programming arguments that the set $S(x)$ of matrix games satisfies properties J1 and J2 of the previous subsection. This allows us to conclude that any stationary policies obtained from the saddle point of the matrix games $S(x)$ are strongly monotone, as described in the previous subsection, and are saddle point strategies for the stochastic game. This structure also carries on to the expected average cost criterion (see [9]). More details are given in the next subsection for a more general problem.

7.3 The Case of Several Input Flows

We now extend our previous model to allow, in addition to the controlled flow and the controlled service, additional non-controlled flows that join the same queue. They can be viewed as higher priority flows since they are not subject to flow control and packets of these flows cannot be rejected. The probability that the total number of packets brought by all the non-controlled flows together equals i is given by p_i, $i \geq 0$. We assume that $p_0 > 0$ (this is a necessary condition for the queue to be stable). All other parameters are the same as in the previous subsection. We shall only consider here the case of an infinite buffer.

Define for $x \in \mathbf{X}$, $f : \mathbf{X} \cup \{-1\} \to R$,

$$S(x, a, b, f) := c(x, a, b) + \xi\Big(\bar{a}bf(x+1) + (ab + \bar{a}\bar{b})f(x) + a\bar{b}f(x-1)\Big),$$

$$g(f, x) := \sum_{i=0}^{\infty} p_i f(x+i), \qquad x \in \mathbf{X} \cup \{-1\}.$$

The dynamic programming equation for this problem for the discounted cost is given by

$$V_\xi(x) = \text{val}_{a,b} S(x, a, b, g(V_\xi)), \quad x \in \mathbf{X}$$

where we define (with some abuse of notation) $V_\xi(-1) = V_\xi(0)$.

We shall say that $f : \mathbf{X} \to R$ satisfies assumption:

WC (weakly convex) if for all $x \in \mathbf{X}$,

$$f(x+2) - f(x+1) \geq f(x+1) - f(x). \tag{16}$$

SC(x) (strongly convex) if for x given,

$$f(x+2) - f(x+1) > f(x+1) - f(x). \tag{17}$$

MI if $f(x)$ is monotonically increasing in x, i.e. for any $x \in \mathbf{X}$,

$$f(x+1) \geq f(x) \tag{18}$$

We further use the above properties (with some abuse of notation) for the class of policies $f : \mathbf{X} \cup \{-1\} \to R$ that satisfy $f(-1) = f(0)$.

Let U^* be the set of stationary policies for the service controller such that $u \in U^*$ if and only if for any $x \in \mathbf{X}$, u_x is optimal for player 1 in the matrix game $S(x, V_\xi)$. Let V^* be the set of stationary policies for the flow controller such that $v \in V^*$ if and only if for any $x \in \mathbf{X}$, v_x is optimal for player 2 in the matrix game $S(x, V_\xi)$. Any pair (u, v) such that $u \in U^*$ and $v \in V^*$ is optimal for problem $\mathbf{Q}_\xi$ (see e.g. [9]).

We now present the main results.

Theorem 7.2. *Consider the discounted cost. If the holding cost h satisfies* **MI**, **WC** *and either $h(1) > h(0)$ or* **SC(0)** *then*

(i) *The value function V_ξ is strictly convex and increasing,*
(ii) *V_ξ satisfies assumptions J1 and J2,*
(iii) *Any of the optimal policies $u \in U^*$ and $v \in V^*$ are strongly monotone.*

Proof. We proceed using value iteration to show that value function V_ξ is weakly convex and increasing. We note that the function which is zero for all x satisfies these properties. Assume that a function $f : \mathbf{X} \to R$ satisfies these properties and so it also satisfies them when extending the range of f to $\mathbf{X} \cup \{-1\}$ and setting $f(-1) := f(0)$. It clearly follows that $g(f)$ also satisfies these properties. One can then use exactly the same proof as in Lemma 4.3 in [3] (replacing f by $g(f)$) to show that

$$T_\xi(f, x) := \mathrm{val}(S(x, g(f))$$

satisfies these properties. It then follows by induction that for all integers n, $T_\xi^n(0, x)$ satisfies the properties. These properties pass to the limit as $n \to \infty$, and this limit equals the value of the game (see e.g. [9]): $V_\xi(x) = \lim_{n\to\infty} T_\xi^n(0, x)$. Hence V_ξ is increasing and weakly convex. We can now proceed as in the proof of Lemma 4.3 in [3] to conclude that V_ξ is in fact strictly convex so

(i) is established.
(ii) The proof of this part follows as in the proof of Theorem 4.1 in [3].
(iii) follows from (ii) by applying Theorem 7.1. □

7.4 Discussion

We obtained structural results for a zero-sum stochastic game using a stronger version of S-modularity notions, and applying them to the matrix games that appear in the dynamic programming equation. This allowed us to show that both players in the stochastic game have strongly monotonic saddle point policies. In the case that each player has only two actions, the saddle point policies are of a threshold type. Such threshold policies can be parameterized by a single real number, as we saw in Section 2. One could then ask the question whether S-modularity can be applied directly to these strategies. In other words, if both players use threshold policies, is the corresponding discounted or average cost S-modular? We have not been able to answer that question. First, even when restricting to threshold policies, the corresponding costs do not have a simple expression as in the examples in the previous section, so one cannot check easily the S-modularity. In fact, the power of the approach we use in this section is that we do not need to compute the cost corresponding to each pair of threshold policies.

An alternative way to show that there exist equilibria within threshold policies in an N-player game would be to show that the optimal response of a player when all other players restrict to some threshold policy, is also a threshold policy, and then use some simple fixed point arguments. We used a similar approach in previous sections, such as Section 2. Yet the setting there was quite different, there were infinitely many players, each taking one action, then staying for some time in the system without taking further decisions, and then disappearing from the system. In this section, in contrast, we consider a finite number of "permanent players" that control continuously the flow or the service. In this context of two permanent players, we have not been able to show that a best response to a threshold policy is also a threshold policy, and whether it is true remains an open problem. This problem made it particularly hard to extend our problem to non zero-sum games (for example, in which there are two players that send flow to the system and their flow is controlled by two non-cooperative players). Yet, there are some examples of stochastic non-zero sum games with a finite number of permanent players in which it is possible to follow that approach and show that the best response to threshold policies of other players is also a threshold policy, and that there exist equilibria among threshold policies (see [5]).

REFERENCES

[1] ATM Forum Technical Committee, *Traffic Management Specification*, Version 4.0, af-tm-0056, April 1996.

[2] Adiri I. and Yechiali U., "Optimal priority purchasing and pricing decisions in nonmonopoly and monopoly queues", *Operations Research*, 1974.

[3] Altman E., "Monotonicity of optimal policies in a zero sum game: a flow control model", *Advances in Dynamic Games and Applications*, **1**, 269–286, 1994.

[4] Altman E., "Flow control using the theory of zero-sum Markov games", *IEEE Trans. Automatic Control*, **39**, no. 4, 814–818, 1994.

[5] Altman E., "Non zero-sum stochastic games in admission, service and routing control in queueing systems", *Queueing Systems*, **23**, 259–279, 1996.

[6] Altman E., Boulogne T., El Azouzi R., Jiménez T. and Wynter L., "A survey on networking games", submitted to *Telecommunication Systems*, Nov. 2000. Available on http://www-sop.inria.fr/mistral/personnel/Eitan.Altman/ntkgame.html.

[7] Altman E. and Başar T, "Multi-user rate-based flow control", *IEEE Trans. on Communications*, **46**, no. 2, 940–949, 1998.

[8] Altman E., El-Azouzi R., and Jiménez T., "Slotted Aloha as a stochastic game with partial information", In *Proceedings of WiOpt'03*, Sophia-Antipolis, France, pp. 3–5, March 2003.

[9] Altman E. and Hordijk A., "Zero-sum Markov games and worst-case optimal control of queueing systems", invited paper, *Queueing Systems*, **21**, special issue on optimization of queueing systems, Ed. S. Stidham, pp. 415–447, 1995.

[10] Altman E., Jiménez T., Nunez-Queija R and Yechiali U., "Queueing analysis for optimal routing with partial information", *Stochastic Models*, **20**, no. 2, 149–172, 2004.

[11] Altman E. and Kameda H., "Equilibria for multiclass routing in multi-agent networks", *40th IEEE Conference on Decision and Control*, Orlando, Florida, U.S.A., Dec. 2001.

[12] Altman E. and Shimkin N., "Individual equilibrium and learning in processor sharing systems", *Operations Research*, **46**, 776–784, 1998.

[13] Altman E. and Wynter L., "Equilibrium, games, and pricing in transportation and telecommunication networks", **4**, no. 1, 7–21.

[14] Başar T. and Olsder G. J., Dynamic Noncooperative Game Theory, *SIAM Classics in Applied Mathematics*, 1999.

[15] Ben-Shahar I. and Orda A. and Shimkin N., "Dynamic service sharing with heterogeneous preferences", *QUESTA*, **35**, nos. 1–4, 83–103, 2000.

[16] Elcan A., "Optimal customer return rate for an M/M/1 queueing system with retrials", *Probability in the Engineering and Informational Sciences*, **8**, 521–539, 1994.

[17] Glazer A. and Hassin R., "?/M/1: On the equilibrium distribution of customer arrivals", *European Journal of Operations Research*, **13**, 146–150, 1983.

[18] Gupta P. and Kumar P. R., "A system and traffic dependent adaptive routing algorithm for ad hoc networks", In *Proceedings of the 36th IEEE Conference on Decision and Control*, 2375–2380, San Diego, USA, Dec. 1997.

[19] Hassin R., "On the advantage of being the first server", *Management Science*, **42**, 618–623, 1996.

[20] Hassin R. and Haviv M., "Equilibrium threshold strategies: the case of queues with priorities", *Operations Research*, **45**, 966–973, 1997.

[21] Hassin R. and Haviv M., "On optimal and equilibrium retrial rates in a busy system", *Probability in the Engineering and Informational Sciences*, **10**, 223–227, 1996.

[22] Hassin R. and Haviv M., *To queue or not to queue: equilibrium behavior in queueing systems*, International series in Operations Research and Management Science, Volume 59, Kluwer Academic Publishers, Boston, 2002.

[23] Haurie A. and Marcotte P., "On the relationship between Nash-Cournot and Wardrop equilibria", *Networks*, **15**, 295–308, 1985.

[24] Jin Y. and Kesidis G., "Equilibria of a noncooperative game for heterogeneous users of an ALOHA network", *IEEE Commun. Letters*, **6**, no. 7, 282–284, 2002.

[25] Kameda H. and Zhang Y., "Uniqueness of the solution for optimal static routing in open BCMP queueing networks", *Mathematical and Computer Modeling*, **22**, 119–130, 1995.

[26] Korilis Y. A. and Lazar A., "On the existence of equilibria in noncooperative optimal flow control", *Journal of the ACM*, **42**, no. 3, 584–613, 1995.

[27] Kulkarni V. G., "On queueing systems with retrials", *Journal of Applied Probability*, **20**, 380–389, 1983.

[28] MacKenzie A. B. and Wicker S. B., "Selfish users in Aloha: A game theoretic approach", *Proceedings of the Fall 2001 IEEE Vehicular Technology Conference*, 2001.

[29] Orda A., Rom R. and Shimkin N., Competitive routing in multi-user communication networks, *IEEE/ACM Transactions on Networking*, **1**, no. 5, 510–520, 1993.

[30] Patriksson M., *The traffic assignment problem: models and methods*, VSP BV, P.O. Box 346, 3700 AH Zeist, The Netherlands, 1994.

[31] Pigou A. C., *The economics of welfare*, McMillan & Co., London, 1920.

[32] Rosenthal R. W., "A class of games possessing pure strategy Nash equilibria", *Int. J. Game Theory*, **2**, 65–67, 1973.

[33] Sandholm W. H., "Potential games with continuous player sets", *Journal of Economic Theory*, **97**, 81–108, 2001.

[34] Shapley L. S.,"Stochastic games", *Proceeding of the National Academy of Sciences USA*, **39**, 1095–1100, 1953.

[35] Topkis D., "Equilibrium points in nonzero-sum n-person submodular games", *SIAM J. Control and Optimization*, **17**, 773–787, Nov. 1979.

[36] Wardrop J. G, "Some theoretical aspects of road traffic research communication networks", *Proc. Inst. Civ. Eng.*, **1**, no. 2, 325–378, 1952.

[37] Wie B. W., "A differential game approach to the dynamic mixed behavior traffic network equilibrium problem", *European Journal of Operational Research*, **83**, 117–136, 1995.

[38] Yao D. D., "S-Modular games, with queuing applications", *Queuing Systems: Theory and Applications*, **21**, 449–475, 1995.

Equilibria for Multiclass Routing Problems in Multi-Agent Networks

Eitan Altman
INRIA, 2004 route des Lucioles
06902 Sophia Antipolis, France
Eitan.Altman@sophia.inria.fr

Hisao Kameda
Institute of Information Science and Electronics
University of Tsukuba, Tsukuba Science City
Ibaraki 305-8573, Japan
kameda@is.tsukuba.ac.jp

Abstract

We study optimal static routing problems in open multiclass networks with state-independent arrival and service rates. Our goal is to study the uniqueness of optimal routing under different scenarios. We consider first the overall optimal policy, that is the routing policy whereby the overall mean cost of a job is minimized. We then consider an individually optimal policy whereby jobs are routed so that each job may feel that its own expected cost is minimized if it knows the mean cost for each path. This is related to the Wardrop equilibrium concept in a multiclass framework. We finally study the case of class optimization, in which each of several classes of jobs tries to minimize the averaged cost per job within that class; this is related to the Nash equilibrium concept. For all three settings, we show that the routing decisions at optimum need not be unique, but that the utilizations in some large class of links are uniquely determined.

Key words. Routing, Networks, Game theory

1 Introduction

We consider the problem of optimal routing in networks. Much previous work has been devoted to the routing problem in which at each node one may take new routing decisions. We consider a more general framework in which the sources have to decide how to route their traffic between different existing paths. (These two problems coincide in the case where the set of paths equals the set of all possible sequences of consecutive directed links which originate at the source and end at the destination.) In the ATM (Asynchronous Transfer Mode [26])[1] environment, this

[1]ATM is one of the leading architectures for high speed networks, which allows one to integrate data, voice and video applications in a single network

problem arises when we wish to decide on how to route traffic on a given existing set of virtual paths or virtual connections. Our framework thus allows us to handle routing both in a packet-switching as well as in a circuit-switching environment. We consider three different frameworks:

(i) The overall optimization criterion, where a single controller makes the routing decisions. Extensive literature exists for this approach, both in telecommunication applications as well as in load balancing for distributed computer systems [5,11,12,14,28].
(ii) The individual optimality, in which each routed individual chooses its own path so as to minimize its own cost. An individual is assumed to have an infinitesimally small impact on the load in the network and thus on costs of other individuals. This framework has been extensively investigated in transportation science, see [6,10,23], and was also considered in the context of telecommunication [13] and in distributed computing [11–13]. The suitable optimization concept for this setting is of the Wardrop equilibrium [29]; it is defined as a set of routing decisions for all individuals such that a path is followed by an individual if and only if it has the lowest cost for that individual. Individual optimality is the most natural concept in networks which implement Bellman–Ford type algorithms on a packet base; in such networks, packets are routed along the shortest path.[2]
(iii) The class optimization; a class may correspond to all the traffic generated by a big organization. It may represent a service provider in a telecommunication network in case that it is the service providers that take the routing decisions for their subscribers. A class contains a large amount of individuals and has a non-negligible impact on the load in the network. Each class wishes to minimize the average cost per individual, averaged over all individuals within that class; there is thus a single entity in each class which takes the routing decisions for all individuals of that class. The suitable optimization concept for this approach is that of Nash equilibrium [10]; it is defined as a set of routing decisions for different classes such that no class can decrease its own cost by unilaterally deviating from its decision. This approach was used in telecommunication applications in [22,15,19], in load balancing problems in distributed computer systems [17,11,12] and in transportation science in [10].

Except for some exceptions [1,3,7], optimal solutions for the different frameworks may lead to quite different performances. In particular, the well-known

[2]In [9], for example, a shortest path adaptive decentralised routing protocol is proposed for an ad-hoc network. The authors justify this approach by saying that it does not suffice to minimize the average delay of a packet, since one has to consider in addition the resequencing delays due to the fact that packets may follow different paths. By using shortest paths, it is likely that the sojourn times of different packets do not vary much, so that resequencing delays are minimized.

Braess paradox [4] shows that in the individual optimization, adding a link to the network may result in a new equilibrium with larger delays to all users. The same phenomenon may occur also in class optimality [17,15], but cannot occur in the case of overall optimization (in both the class and individual optimal solutions, traffic is sent over the new link, whereas this link is unused by the overall optimal solution). Another important difference between the different solution concepts is that for convex increasing link costs, the solutions are known to be unique (in terms of link utilization) for the individual and the overall optimization, under very general topologies for the single class case (we extend this result to the case of several classes). However, for the class optimization problem we know of simple counter examples (see [22] for a 4-node example) where the equilibrium is not unique.

An optimization problem does not necessarily have a unique solution. If they are not unique, it is necessary to make clear the range and characteristics of the solutions, in particular, when we calculate numerically the optimal solutions and when we intend to analyse the effects of the system parameters on the optimal solutions. [13] already studied the first two approaches (overall and individual optimization) and characterised the uniqueness for a particular cost structure, that of open BCMP queueing networks [2,18][3] with state-independent arrival and service rates. We extend here these results to a fairly general cost function. We also extend substantially the results obtained in [22] for the uniqueness of class optimization.

In Section 2 we provide the notation and some assumptions used in this paper. In Section 3 we obtain the overall optimal solution, and discuss the uniqueness of the overall optimal solution. In Section 4 we show similar results on the uniqueness of the individually optimal solution. Some results on uniqueness for class optimization are presented in Section 5. Numerical examples are presented in Section 6, and the paper ends with a concluding section 7.

2 Notation and Assumptions

We consider an open network model that consists of a set $\mathcal{M}$ containing M links. We assume that in the network there are pairs of origin and destination points. We call the pair of one origin and one destination points an *O-D pair*. The unit entity that is routed through the network is called a *job*. Each job arrives at one of the origin points and departs from one of the destination points. The origin and destination points of a job are determined when the job arrives in the network.

Jobs are classified into J different classes. For the sake of simplicity, we assume that jobs do not change their class while passing through the network. A class k job may have one of several different origin–destination pairs. Such a class may represent all the users of a given service provider in a context of telecommunications, provided that they control the routes of the traffic of their subscribers. In the

[3]More details on these networks are given later on. The name BCMP was given owing to the initials of its authors [2]

context of road traffic it may represent the set of vehicles of a given type, such as buses, or trucks, or bicycles etc. With this in mind, it is natural to expect that jobs of different classes are faced with different types of routing decisions.

A class k job with the O-D pair (o, d) originates at node o and has as destination node d which it reaches through a series of links, which we refer to as a *path*, and then leaves the system. We assume that links are class-dependent directional, i.e. for each class, there is a given direction in which the flow can be sent.

In many previous papers [22,15], routing could be done at each node. In this paper we follow the more general approach in which a job of class k with O-D pair (o, d) has to choose one of a given finite set of paths (see also [13,23]). We call this set the *paths* of job class k O-D pair (o, d).

We assume that we can choose the job flow rate of each *path* in order to achieve a performance objective. A path may be a given sequence of links that connect the origin and destination nodes. In that case, it may correspond to a virtual connection in ATM. We allow, however, a path to be some more general object. It may contain a number of sub-paths; we assume however that once the job flow rate of a path is given, the job flow rate of each sub-path in the path is fully determined (and is not the object of a control decision). That is, the relative flow rate of each sub-path in the same path is governed by some fixed transfer proportions (or probabilities) between the links.

For example, one may consider paths that include noisy links, where lost packets have to be retransmitted locally over the link. Thus, some given proportion of the traffic in this path use a direct sub-path (no losses) whereas others have to loop (this model losses and retransmissions). Another example of a path containing several sub-paths is a network in which switches route arriving traffic in some fixed proportions between outgoing links (sub-paths); if this proportion is not controlled by the entity that takes routing decisions for the class, then the resulting routes from the outgoing links are still considered a single path.

The solution of a routing problem is characterised by the chosen values of job flow rates of all paths. Below, the rate of flows will be real numbers, and methods related to convex optimization will be used. Our model is therefore appropriate for applications in which we may ignore the discrete nature of packets and/or sessions, thus avoiding the high complexity of discrete optimization methodologies. A job can correspond to an IP packet in a network through which a large number of packets flow (and in which the route of each packet may be different). There are cases, as in ATM, in which all packets of a connection have to follow the same route. Our model will still be useful provided that we need to choose the route of a large number of sessions.

Notation Regarding the Network:

$D^{(k)}$ = Set of O-D pairs for class k jobs.
$\Pi_d^{(k)}$ = Set of paths that class k jobs of O-D pair $d \in D^{(k)}$ flow through.

$\Pi^{(k)}$ = Set of all paths for class k jobs, i.e., $\Pi^{(k)} = \bigcup_{d \in D^{(k)}} \Pi_d^{(k)}$.

$$\gamma_{pd}^{kk'} = \begin{cases} 1 & \text{if } p \in \Pi_d^{(k')} \text{ and } k = k', \\ 0 & \text{otherwise.} \end{cases}$$

Notation Regarding Arrivals to the Network and Flow Rates:

$\phi_d^{(k)}$ = Rate at which class k jobs join O-D pair $d \in D^{(k)}$.
$\phi^{(k)}$ = Total job arrival rate of class k jobs, i.e., $\phi^{(k)} = \sum_{d \in D^{(k)}} \phi_d^{(k)}$.

Φ = System-wide total job arrival rate, i.e., $\Phi = \sum_{k=1}^{J} \phi^{(k)}$.

$x_p^{(k)}$ = Rate at which class k jobs flow through path p.
δ_{lp} = Percentage of the rate $x_p^{(k)}$ that pass through link l, for $p \in \Pi^{(k)}$.
$\lambda_l^{(k)}$ = Rate at which class k jobs visit link l, $\lambda_l^{(k)} = \sum_{p \in \Pi^{(k)}} \delta_{lp} x_p^{(k)}$.

$\phi_d^{(k)}$, $\phi^{(k)}$, Φ and δ_{lp} are given constants (and not decision variables).

Notation Regarding Service and Performance Values in the Open Network:

$\mu_l^{(k)}$ = Constant denoting the service rate of class k jobs at link l.
$\rho_l^{(k)}$ = $\lambda_l^{(k)}/\mu_l^{(k)}$. Utilization of link l for class k jobs.

$\rho_l = \sum_{k=1}^{J} \rho_l^{(k)}$. Total utilization of link l.[4]

$\hat{T}_l^{(k)}$ = Mean cost of class k jobs at link l.
$T_l(\rho_l)$ = Weighted cost per unit flow in link l.
$T_p^{(k)}$ = Average class k cost of path p, $p \in \Pi^{(k)}$, $k = 1, 2, \dots, J$.[5]
Δ = Overall mean cost of a job (averaged over all classes).
$\Delta^{(k)}$ = Overall mean cost of a job of class k.

Notation Regarding Vectors and Matrices:

$\boldsymbol{\rho} = [\rho_1, \rho_2, \dots, \rho_M]^{\mathrm{T}}$ where T means 'transpose'. We call this the utilization vector.
$\boldsymbol{\lambda} = [\lambda_1^{(1)}, \lambda_2^{(1)}, \dots, \lambda_M^{(1)}, \dots, \lambda_1^{(k)}, \lambda_2^{(k)}, \dots, \lambda_M^{(k)}, \dots,]^{\mathrm{T}}$, i.e, the vector of total flows over all links.
$\boldsymbol{\phi} = [\phi_1^{(1)}, \phi_2^{(1)}, \dots, \phi_1^{(2)}, \phi_2^{(2)}, \dots]^{\mathrm{T}}$, i.e., the arrival rate vector.

[4]Each term in the sum is positive even if the directions of flows are not the same.
[5]$T_l^{(k)}$, T_l and $\hat{T}_l^{(k)}$ are defined at the end of the section (Assumption B1)

$\mathbf{x} = [x_1^{(1)}, x_2^{(1)}, \dots, x_1^{(2)}, x_2^{(2)}, \dots]^{\mathrm{T}}$, i.e., the path flow rate vector.

$\boldsymbol{\alpha} = [\alpha_1^{(1)}, \alpha_2^{(1)}, \dots, \alpha_1^{(2)}, \alpha_2^{(2)}, \dots]^{\mathrm{T}}$, i.e., the vector whose elements are $\alpha_d^{(k)}$, $d \in D^{(k)}$, $k = 1, 2, \dots, J$; the elements $\alpha_d^{(k)}$ correspond to some Lagrange multipliers.

$\boldsymbol{\xi} = [\xi_1^{(1)}, \xi_2^{(1)}, \dots, \xi_1^{(2)}, \xi_2^{(2)}, \dots]^{\mathrm{T}}$, i.e., the vector whose elements are $\xi_l^{(k)}$, $l \in \mathcal{M}$, $k = 1, 2, \dots, J$; the elements $\xi_l^{(k)}$ correspond to some Lagrange multipliers.

$\mathbf{T} = [T_1^{(1)}, T_2^{(1)}, \dots, T_1^{(2)}, T_2^{(2)}, \dots]^{\mathrm{T}}$, i.e., the vector whose elements are $T_p^{(k)}$, $p \in \Pi^{(k)}$, $k = 1, 2, \dots, J$.

$\mathbf{x}^{(k)} = [x_1^{(k)}, x_2^{(k)}, \dots]^{\mathrm{T}}$, i.e., the path flow rate vector for class k jobs. $\boldsymbol{\phi}^{(k)}$, $\boldsymbol{\alpha}^{(k)}$, and $\mathbf{T}^{(k)}$ are defined similarly.

$\mathbf{x}^{-k} = [x_1^{(1)}, x_2^{(1)}, \dots, x_1^{(k-1)}, \dots, x_1^{(k+1)}, \dots]^{\mathrm{T}}$, i.e., the path flow rate vector for jobs of the classes other than class k.

$$\boldsymbol{\Gamma} = \begin{bmatrix} \gamma_{11}^{11} & \gamma_{12}^{11} & \cdots & \gamma_{11}^{12} & \gamma_{12}^{12} & \cdots \\ \gamma_{21}^{11} & \gamma_{22}^{11} & \cdots & \gamma_{21}^{12} & \gamma_{22}^{12} & \cdots \\ \vdots & & \ddots & \vdots & & \ddots \\ \gamma_{11}^{21} & \gamma_{12}^{21} & \cdots & \gamma_{11}^{22} & \gamma_{12}^{22} & \cdots \\ \gamma_{21}^{21} & \gamma_{22}^{21} & \cdots & \gamma_{21}^{22} & \gamma_{22}^{22} & \cdots \\ \vdots & & \ddots & \vdots & & \ddots \end{bmatrix}$$ i.e., the incident matrix whose (i, j) element is $\gamma_{pd}^{kk'}$, $p \in \Pi^{(k)}$, $k = 1, 2, \dots, J$, $d \in D^{(k')}$, $k' = 1, 2, \dots, J$, where $i = p + \sum_{\kappa=1}^{k-1} |\Pi^{(\kappa)}|$ and $j = d + \sum_{\kappa=1}^{k'-1} |D^{(\kappa)}|$.

$\mathbf{x} \cdot \mathbf{y} = \sum_i x_i y_i$, i.e., the inner product of vectors $\mathbf{x} = [x_1, x_2, \dots]^{\mathrm{T}}$ and $\mathbf{y} = [y_1, y_2, \dots]^{\mathrm{T}}$.

We make the following assumptions on the cost:

B1: The cost over a path is given by as a weighted sum of link-by-link costs over the path: associated with each link $l \in \mathcal{M}$ there is a cost $T_l(\rho_l)$ per flow unit, that depends on the utilization of the link (the function T_l does not depend on the class k!). There is further a class dependent weight factor $\mu_l^{(k)}$ per link l. Thus, the cost per unit flow of class k on link l is $\hat{T}_l^{(k)} = T_l/\mu_l^{(k)}$. Thus the average cost per unit flow of class k job that passes through path $p \in \Pi^{(k)}$ is

$$T_p^{(k)} = \sum_{l \in \mathcal{M}} \delta_{lp} \hat{T}_l^{(k)} = \sum_{l \in \mathcal{M}} \frac{\delta_{lp}}{\mu_l^{(k)}} T_l(\rho_l). \tag{1}$$

(For examples of such costs, see [13].)

B2: $T_l : [0, \infty) \to [0, \infty]$, and $T_l(0)$ is finite.

B3: The set $\mathcal{M}$ is composed of two disjoint sets of links:

(i) $\mathcal{M}_{\mathcal{I}}$, for which $T_l(\rho_l)$ are convex and strictly increasing (in the interval where they are finite),
(ii) $\mathcal{M}_{\mathcal{C}}$, for which $T_l(\rho_l) = T_l$ are constant (independent of ρ_l).

B4: $T_l(\rho_l)$ are continuous. Moreover, they are continuously differentiable whenever they are finite.

Assumptions $B1$–$B4$ cover in particular the cost that is mostly used in networking games in telecommunications, which is the expected queueing delay in the so-called BCMP queueing networks [2,18] with state-independent arrival and service rates. These include open networks (networks in which all arrivals eventually leave the network) in which jobs arrive at nodes according to independent Poisson processes, and in which each node of the network is represented by (i) a queue which has independent exponentially distributed service times and FIFO (first in first out) service order, or (ii) a queue which has a generally (independent) distributed service time and a processor sharing service or discipline.

Denote $\boldsymbol{\rho}_U = \boldsymbol{\rho}_U(\mathbf{x}) = \boldsymbol{\rho}|_{\rho_l=0, l\in\mathcal{M}_{\mathcal{C}}}$. This is the same as $\boldsymbol{\rho}$ except that $\rho_l = 0$ for all $l \in \mathcal{M}_{\mathcal{C}}$.

The overall mean cost of a job, Δ, can be written as

$$\Delta = \sum_{k=1}^{J} \sum_{p\in\Pi^{(k)}} \frac{x_p^{(k)}}{\Phi} T_p^{(k)} = \frac{1}{\Phi} \sum_{l\in\mathcal{M}} \rho_l T_l(\rho_l).$$

The mean cost of a job of class k, $\Delta^{(k)}$, can be written as

$$\Delta^{(k)} = \sum_{p\in\Pi^{(k)}} \frac{x_p^{(k)}}{\phi^{(k)}} T_p^{(k)} = \frac{1}{\phi^{(k)}} \sum_{l\in\mathcal{M}} \rho_l^{(k)} T_l(\rho_l).$$

Note that the following conditions should be satisfied for each $k = 1, 2, \ldots, J$,

$$\sum_{p\in\Pi_d^{(k)}} x_p^{(k)} = \phi_d^{(k)}, \quad d \in D^{(k)}, \tag{2}$$

$$x_p^{(k)} \geq 0, \quad p \in \Pi^{(k)}. \tag{3}$$

We can express (2) as

$$\sum_{k'=1}^{J} \sum_{p\in\Pi^{(k')}} \gamma_{pd}^{k'k} x_p^{(k')} = \phi_d^{(k)}, \quad d \in D^{(k)},$$

$$\text{or, equivalently,} \qquad \boldsymbol{\Gamma}^{\mathrm{T}}\mathbf{x} = \boldsymbol{\phi}. \tag{4}$$

Remark. Our model includes those discussed for the static routing problems of communications networks [5,8,14]. It also includes those of the load balancing problems of distributed computer systems such as given in [12,20,27,28]. From condition B3 it is easy to see that $T_l^{(k)}$ is a convex function of $\lambda_l^{(k)}, l \in \mathcal{M}_\mathcal{I}, k = 1, 2, \ldots, J$. It follows that $T_l^{(k)}$ is also convex with respect to $\mathbf{x}$.

3 Overall Optimal Solution

By the overall optimal policy we mean the policy whereby routing is determined so as to minimize the overall mean cost of a job. The problem of minimizing the overall mean cost is stated as follows:

$$\text{minimize:} \quad \Delta \quad = \quad \frac{1}{\Phi} \sum_{l \in \mathcal{M}} \rho_l T_l(\rho_l) \tag{5}$$

with respect to $\mathbf{x}$ subject to

$$\mathbf{\Gamma}^{\mathrm{T}} \mathbf{x} = \boldsymbol{\phi}, \quad \mathbf{x} \geq 0, \tag{6}$$

where $\rho_l = \sum_{k=1}^{J} \lambda_l^{(k)} / \mu_l^{(k)}$ and $\lambda_l^{(k)} = \sum_{p \in \Pi^{(k)}} \delta_{lp} x_p^{(k)}$. Note that (6) are the same as (2) and (3), respectively. We call the above problem the *overall optimization problem*, and its solution the *overall optimal solution*.

Define

$t_p^{(k)} \quad = \partial(\Phi\Delta)/\partial x_p^{(k)}$, i.e., class k marginal cost of path p, $p \in \Pi^{(k)}, k = 1, 2, \ldots, J$.

$\mathbf{t}$ $=[t_1^{(1)}, t_2^{(1)}, \ldots, t_1^{(2)}, t_2^{(2)}, \ldots]^{\mathrm{T}}$ is the gradient vector of the function $\Phi\Delta$, i.e., the vector whose elements are $t_p^{(k)}$, $p \in \Pi^{(k)}, k = 1, 2, \ldots, J$.

Lemma 3.1. $\mathbf{x}$ *is an optimal solution of the problem* (5) *if and only if* $\mathbf{x}$ *satisfies the following conditions. There exist Lagrange multipliers* $\boldsymbol{\alpha}$ *such that*

$$[\mathbf{t}(\mathbf{x}) - \mathbf{\Gamma}\boldsymbol{\alpha}] \cdot \mathbf{x} = 0, \tag{7}$$

$$\mathbf{t}(\mathbf{x}) - \mathbf{\Gamma}\boldsymbol{\alpha} \geq 0, \tag{8}$$

$$\mathbf{\Gamma}^{\mathrm{T}} \mathbf{x} - \boldsymbol{\phi} = 0, \tag{9}$$

$$\mathbf{x} \geq 0. \tag{10}$$

Proof. Since the objective function (5) is convex and the feasible region of its constraints is a convex set, any local solution of the problem is a global solution point. So by applying Karush–Kuhn–Tucker's Theorem [25], we obtain that $\mathbf{x}$ is an optimal solution of problems (5)–(6) if and only if $\mathbf{x}$ satisfies the following conditions. There exists Lagrange multipliers α such that (7)–(10) hold. □

Lemma 3.2. $\bar{\mathbf{x}}$ *is an optimal solution of the problem* (5) *if and only if*

$$\begin{aligned} \mathbf{t}(\bar{\mathbf{x}}) \cdot (\mathbf{x} - \bar{\mathbf{x}}) &\geq 0, \quad \textit{for all} \quad \mathbf{x} \\ \textit{such that } \boldsymbol{\Gamma}^{\mathrm{T}}\mathbf{x} &= \boldsymbol{\phi} \quad \textit{and} \quad \mathbf{x} \geq 0. \end{aligned} \tag{11}$$

Proof. (11) holds for some $\bar{\mathbf{x}}$ if and only if $\bar{\mathbf{x}}$ is the solution of the following linear program (where the decision variables are $\mathbf{x}$):

$$\min \mathbf{t}(\bar{\mathbf{x}}) \cdot \mathbf{x} \text{ subject to } \boldsymbol{\Gamma}^{\mathrm{T}}\mathbf{x} = \boldsymbol{\phi}, \mathbf{x} \geq 0$$

with $\bar{\mathbf{x}}$ fixed. $\mathbf{x}$ is an optimal solution of the linear program if and only if $\mathbf{x}$ satisfies the Kuhn–Tucker conditions (see e.g. pp 158–165 in [25]) for the Lagrangian

$$L(\mathbf{x}, \boldsymbol{\alpha}^*) = \mathbf{t}(\bar{\mathbf{x}}) \cdot \mathbf{x} + \boldsymbol{\alpha}^* \cdot (\boldsymbol{\phi} - \boldsymbol{\Gamma}^{\mathrm{T}}\mathbf{x}). \tag{12}$$

The Kuhn–Tucker conditions are

$$\frac{\partial L}{\partial \mathbf{x}} = \mathbf{t}(\bar{\mathbf{x}}) - \boldsymbol{\Gamma}\boldsymbol{\alpha}^* \geq 0, \tag{13}$$

$$\frac{\partial L}{\partial \mathbf{x}} \cdot \mathbf{x} = [\mathbf{t}(\bar{\mathbf{x}}) - \boldsymbol{\Gamma}\boldsymbol{\alpha}^*] \cdot \mathbf{x} = 0, \tag{14}$$

$$\frac{\partial L}{\partial \boldsymbol{\alpha}^*} = \boldsymbol{\phi} - \boldsymbol{\Gamma}^{\mathrm{T}}\mathbf{x} = 0, \tag{15}$$

$$\mathbf{x} \geq 0. \tag{16}$$

That is, the relation (11) (i.e., the statement that $\bar{\mathbf{x}}$ is a solution of the above linear program) is equivalent to the set of relations in Lemma 3.1 for some (finite) Lagrange multiplier $\boldsymbol{\alpha}^*$ (for the finiteness, see Cor. on 5.1 p. 165 in [25]). □

From condition B1 we see that Δ depends only on the utilization of each link, ρ_l, which results from the path flow rate matrix. It is possible, therefore, that different values of the path flow rate matrix result in the same utilization of each link and the same minimum mean cost.

We define below the concept of monotonicity of vector-valued functions with vector-valued arguments.

Definition. Let $\mathbf{F}(\bullet)$ be a vector-valued function that is defined on a domain $S \subseteq R^n$ and that has values $\mathbf{F}(\mathbf{x})$ in R^n. This function is *monotonic* in S if for every pair $\mathbf{x}, \mathbf{y} \in S$

$$(\mathbf{x} - \mathbf{y}) \cdot [\mathbf{F}(\mathbf{x}) - \mathbf{F}(\mathbf{y})] \geq 0.$$

It is *strictly monotone* if, for every pair $\mathbf{x}, \mathbf{y} \in S$ with $\mathbf{x} \neq \mathbf{y}$,

$$(\mathbf{x} - \mathbf{y}) \cdot [\mathbf{F}(\mathbf{x}) - \mathbf{F}(\mathbf{y})] > 0.$$

We need the following property:

Lemma 3.3. *Assume B1–B4, and let $l \in \mathcal{M}_{\mathcal{I}}$. Then*

(i) $T_l(\rho_l)$ is finite if and only if its derivative $T_l'(\rho_l)$ is finite.
(ii) If $T_l'(\rho_l)$ is infinite then for any $\mathbf{x}$ for which the load on link l is ρ_l, the corresponding cost $\Delta(\mathbf{x})$ is infinite.

Proof. (i) Due to the convexity of T_l, we have

$$T_l(\rho_l) = \int_0^{\rho_l} T_l'(\zeta_l)d\zeta_l \le \rho_l T_l'(\rho_l).$$

By B2, if $T_l(\rho_l) = \infty$ then $\rho_l > 0$, which implies by the latter equation that $T_l'(\rho_l)$ is infinite.
For the converse, assume that $T_l(\rho_l)$ is finite. Then by continuity, there exists some $\epsilon > 0$ such that $T_l(\rho_l + \epsilon)$ is finite. Since T_l is convex, $T_l'(\rho_l) \le \epsilon^{-1}(T_l(\rho_l + \epsilon) - T_l(\rho_l))$ and is thus finite as well.

(ii) If $T_l'(\rho_l)$ is infinite then by (i), $T_l(\rho_l)$ is infinite; moreover, $\rho_l > 0$ by assumption B2, so that $\Delta(\rho) = \infty$ (by (5)). □

For the function $\mathbf{t}(\mathbf{x})$ we have the following.

Lemma 3.4. *Assume B1–B4. Whenever finite, $\mathbf{t}(\mathbf{x})$ is monotone but is not strictly monotone, i.e., for arbitrary $\mathbf{x}$ and $\mathbf{x}'$ ($\mathbf{x} \neq \mathbf{x}'$), if $\Delta(\mathbf{x})$ or $\Delta(\mathbf{x}')$ is finite then*

$$(\mathbf{x} - \mathbf{x}') \cdot [\mathbf{t}(\mathbf{x}) - \mathbf{t}(\mathbf{x}')] > 0 \quad if \quad \boldsymbol{\rho}_U \neq \boldsymbol{\rho}_U', \tag{17}$$

$$= 0 \quad if \quad \boldsymbol{\rho}_U = \boldsymbol{\rho}_U', \tag{18}$$

where $\boldsymbol{\rho}_U = \boldsymbol{\rho}_U(\mathbf{x})$ and $\boldsymbol{\rho}_U' := \boldsymbol{\rho}_U(\mathbf{x}')$ are the utilization vectors that $\mathbf{x}$ and $\mathbf{x}'$ result in, respectively.

Proof. Assume that $\boldsymbol{\rho}_U \neq \boldsymbol{\rho}_U'$. Then

$$\begin{aligned}
&(\mathbf{x} - \mathbf{x}') \cdot [\mathbf{t}(\mathbf{x}) - \mathbf{t}(\mathbf{x}')] \\
&= \sum_{k=1}^{J} \sum_{p \in \Pi^{(k)}} (x_p^{(k)} - x_p^{'(k)})[t_p^{(k)}(\mathbf{x}) - t_p^{(k)}(\mathbf{x}')] \\
&= \sum_{k=1}^{J} \sum_{p \in \Pi^{(k)}} \sum_{l \in \mathcal{M}} (x_p^{(k)} - x_p^{'(k)}) \\
&\quad \times \frac{\delta_{lp}}{\mu_l^{(k)}} \left\{ \left[T_l(\rho_l) - T_l(\rho_l') \right] + \left[\rho_l \frac{dT_l(\rho_l)}{d\rho_l} - \rho_l' \frac{dT_l(\rho_l')}{d\rho_l'} \right] \right\} \\
&= \sum_{l \in \mathcal{M}_{\mathcal{I}}} (\rho_l - \rho_l') \left\{ \left[T_l(\rho_l) - T_l(\rho_l') \right] + \left[\rho_l \frac{dT_l(\rho_l)}{d\rho_l} - \rho_l' \frac{dT_l(\rho_l')}{d\rho_l'} \right] \right\} > 0
\end{aligned}$$

(The second equality above follows from (1). The last inequality follows from the strict monotonicity of $T_l(\rho_l)$, as well as the fact that its derivative is increasing in ρ_l, and the derivative remains increasing when multiplied by ρ_l. Due to Lemma 3.3, if $\Delta(\rho)$ is finite then $T_l'(\rho_l)$ is finite for all links $l \in \mathcal{M}$ (and similarly for $\Delta(\rho')$). The last inequality follows since by condition $B3$, $T_l(\rho_l)$ are strictly monotone and $\rho_l dT_l(\rho_l)/d\rho_l$ are increasing for $l \in \mathcal{M}_\mathcal{I}$. Therefore we have the relations (17) and (18). □

Theorem 3.1. *Assume $B1$–$B4$ and that there exists some finite feasible solution. Then the utilization in each link $k \in \mathcal{M}_\mathcal{I}$ is uniquely determined and is the same for all overall optimal solutions.*

Proof. Suppose that the overall optimal policy has two distinct solutions $\hat{\mathbf{x}} =$ and $\tilde{\mathbf{x}}$, which result in the utilization vectors $\hat{\boldsymbol{\rho}}_U := \boldsymbol{\rho}_U(\hat{\mathbf{x}})$ and $\tilde{\boldsymbol{\rho}}_U := \boldsymbol{\rho}_U(\tilde{\mathbf{x}})$, respectively, and $\hat{\boldsymbol{\rho}}_U \neq \tilde{\boldsymbol{\rho}}_U$. Then we have from Lemma 3.2, $\mathbf{t}(\hat{\mathbf{x}}) \cdot (\tilde{\mathbf{x}} - \hat{\mathbf{x}}) \geq 0$, $\mathbf{t}(\tilde{\mathbf{x}}) \cdot (\hat{\mathbf{x}} - \tilde{\mathbf{x}}) \geq 0$. Hence

$$(\hat{\mathbf{x}} - \tilde{\mathbf{x}}) \cdot \left[\mathbf{t}(\hat{\mathbf{x}}) - \mathbf{t}(\tilde{\mathbf{x}}) \right] \leq 0.$$

From Lemma 3.4 we have

$$(\hat{\mathbf{x}} - \tilde{\mathbf{x}}) \cdot \left[\mathbf{t}(\hat{\mathbf{x}}) - \mathbf{t}(\tilde{\mathbf{x}}) \right] > 0,$$

since $\hat{\boldsymbol{\rho}}_U \neq \tilde{\boldsymbol{\rho}}_U$. This leads to a contradiction. That is, if there exist two distinct optimal solutions, the utilization vectors of both the solutions must be the same. Note that the utilization of link $l \in \mathcal{M}_C$ is considered always zero. Naturally, in that case, $\sum_{l \in \mathcal{M}_C} \rho_l$ must be unique but each of ρ_l, $l \in \mathcal{M}_C$, need not be unique. □

Note that even when the utilization in each link is unique, the overall optimal solution may not be unique. This is due to the fact that T depends only on $\boldsymbol{\rho}$ (see (5)) (thus if $\mathbf{x}$ is overall optimal then any solution $\mathbf{x}'$ that gives rise to the same value of $\boldsymbol{\rho}$ will be optimal as well). In Section 5 of [13] there is an example of the cases where more than one optimal solution exists.

Now let us consider the range of the optimal solutions. From the above, we obtain the following relations that characterize the range of the optimal solutions.

$$\sum_{k=1}^{J} \sum_{p \in \Pi^{(k)}} \delta_{lp} \frac{x_p^{(k)}}{\mu_l^{(k)}} = \rho_l, \quad l \in \mathcal{M}_\mathcal{I}, \tag{19}$$

$$\sum_{l \in \mathcal{M}_C} \sum_{k=1}^{J} \sum_{p \in \Pi^{(k)}} \delta_{lp} \frac{x_p^{(k)}}{\mu_l^{(k)}} = \sum_{l \in \mathcal{M}_C} \rho_l,$$

and for $k = 1, 2, \dots, J$,

$$\sum_{p \in \Pi_d^{(k)}} x_p^{(k)} = \phi_d^{(k)}, \; d \in D^{(k)}, \tag{20}$$

$$x_p^{(k)} \geq 0, \; p \in \Pi^{(k)}, \tag{21}$$

where the value of each ρ_l is what an optimal solution $\mathbf{x}$ results in. From the relations (19)–(21) we see that optimal path flow rates belong to a convex polyhedron. Then we have the following proposition about the uniqueness of the optimal solutions.

Corollary 3.1. *The overall optimal solution is unique if and only if the total number of elements in* $\mathbf{x}$ *does not exceed the number of linearly independent equations in the set of linear equations* (19)–(20).

4 Individually Optimal Solution

By the individually optimal policy we mean that jobs are scheduled so that each job may feel that its own mean cost is minimum if it knows the mean cost $T_p^{(k)}(\mathbf{x})$ of each path of the O-D pair d, $p \in \Pi_d^{(k)}$, $k = 1, \dots, J$. By the *individual optimization problem* we mean the problem of obtaining the routing decision that achieves the objective of the individually optimal policy. We call the solution of the individual optimization problem the *individually optimal solution* or the *equilibrium*. In the equilibrium, no user has any incentive to make a unilateral decision to change its route. Wardrop [29] considered this equilibrium for a transportation network and defined it through two principles: a policy is equilibrium if for each individual of a class, the delay along paths which are actually used between the source and destination are (i) the same, and (ii) they are smaller than or equal to the delays along paths not used. It is well known that the solution of the Wardrop equilibrium can be obtained by a single mathematical problem that is obtained by a transformation of the cost [23]. This is related to the fact that this is a special case of *potential games* (with a continuum of players), see e.g. [21,24]. We shall obtain a similar solution approach through a mathematical program for our setting as well.

We assume that there is a routing decision and that $\mathbf{x}$ is the path flow rate matrix which results from the routing decision. The individually optimal policy requires that a class k job of O-D pair d should follow a path $\hat{p}$ that satisfies

$$T_{\hat{p}}^{(k)}(\mathbf{x}) = \min_{p \in \Pi_d^{(k)}} T_p^{(k)}(\mathbf{x}) \tag{22}$$

for all $d \in D^{(k)}$, $k = 1, 2, \dots, J$. If a routing decision satisfies the above condition, we say the routing decision realizes the individually optimal policy.

Definition. The path flow rate vector **x** is said to satisfy the equilibrium conditions for a multi-class open network if the following relations are satisfied for all $d \in D^{(k)},\ k = 1, 2, \ldots, J$,

$$T_p^{(k)}(\mathbf{x}) \geq A_d^{(k)},\ x_p^{(k)} = 0, \tag{23}$$

$$T_p^{(k)}(\mathbf{x}) = A_d^{(k)},\ x_p^{(k)} > 0, \tag{24}$$

$$\sum_{p \in \Pi_d^{(k)}} x_p^{(k)} = \phi_d^{(k)}, \tag{25}$$

$$x_p^{(k)} \geq 0,\quad p \in \Pi_d^{(k)}, \tag{26}$$

$$\text{where } A_d^{(k)} = \min_{p \in \Pi_d^{(k)}} T_p^{(k)}(\mathbf{x}),\quad d \in D^{(k)}, k = 1, \ldots, J.$$

Note that (23)–(26) are identical to the relations

$$[\mathbf{T}(\mathbf{x}) - \mathbf{\Gamma A}] \cdot \mathbf{x} = 0, \tag{27}$$

$$\mathbf{T}(\mathbf{x}) - \mathbf{\Gamma A} \geq 0, \tag{28}$$

$$\mathbf{\Gamma}^{\mathrm{T}}\mathbf{x} - \boldsymbol{\phi} = 0, \tag{29}$$

$$\mathbf{x} \geq 0, \tag{30}$$

where $\mathbf{A} = [A_1^{(1)}, A_2^{(1)}, \ldots, A_1^{(2)}, A_2^{(2)}, \ldots]^{\mathrm{T}}$, i.e., the vector whose elements are $A_d^{(k)},\ d \in D^{(k)},\ k = 1, 2, \ldots, J$. The above definition is the natural extension of the notion of Wardrop [29] equilibrium to our setting.

Theorem 4.1. *Assume $B1$–$B4$. There exists an individually optimal solution* **x** *which satisfies the relations* (27)–(30).

Proof. Define $\tilde{T}(\mathbf{x})$ by

$$\tilde{T}(\mathbf{x}) = \frac{1}{\Phi}\left[\sum_{l \in \mathcal{M}_{\mathcal{I}}} \int_0^{\rho_l} T_l(s)ds + \sum_{l \in \mathcal{M}_C} \rho_l T_l\right],$$

(where we recall that $\rho_l = \sum_{k=1}^{J} \lambda_l^{(k)}/\mu_l^{(k)}$ and $\lambda_l^{(k)} = \sum_{p \in \Pi^{(k)}} \delta_{lp} x_p^{(k)}$). Note that $\tilde{T}(\mathbf{x})$ is a convex increasing function of **x**. Then by (1),

$$T_p^{(k)}(\mathbf{x}) = \frac{\partial}{\partial x_p^{(k)}}(\Phi \tilde{T}(\mathbf{x})).$$

Introduce the following convex nonlinear program:

minimize $\tilde{T}(\mathbf{x})$ with respect to **x** s.t. (29)–(30).

The Kuhn–Tucker conditions are the same as (27)–(30). Therefore, the program should have an optimal solution which must satisfy relations (27)–(30). □

We can express the individually optimal solution in the variational inequality form by using the same method as that for the overall optimal solution as follows.

Corollary 4.1. *Assuming $B1$–$B4$, $\bar{\mathbf{x}}$ is an individually optimal solution if and only if it is feasible and*

$$\mathbf{T}(\bar{\mathbf{x}}) \cdot (\mathbf{x} - \bar{\mathbf{x}}) \geq 0, \quad \textit{for all} \quad \mathbf{x}$$
$$\textit{such that } \boldsymbol{\Gamma}^{\mathrm{T}}\mathbf{x} = \boldsymbol{\phi} \quad \textit{and} \quad \mathbf{x} \geq 0.$$

Proof. Similar to the proof of Lemma 3.2. □

Lemma 4.1. *Assume $B1$–$B4$. Whenever finite, the function $\mathbf{T}(\mathbf{x})$ is monotone but is not strictly monotone. That is, for arbitrary $\mathbf{x}$ and $\mathbf{x}'$ ($\mathbf{x} \neq \mathbf{x}'$), if $\mathbf{T}(\mathbf{x})$ are finite or $\mathbf{T}(\mathbf{x}')$ are finite then*

$$(\mathbf{x} - \mathbf{x}') \cdot [\mathbf{T}(\mathbf{x}) - \mathbf{T}(\mathbf{x}')] > 0 \quad \textit{if} \quad \boldsymbol{\rho}_U \neq \boldsymbol{\rho}'_U, \tag{31}$$
$$= 0 \quad \textit{if} \quad \boldsymbol{\rho}_U = \boldsymbol{\rho}'_U \tag{32}$$

where $\boldsymbol{\rho}_U$ and $\boldsymbol{\rho}'_U$ are the utilization vectors that $\mathbf{x}$ and $\mathbf{x}'$ result in respectively.

Proof. This Lemma can be proved in the same way as the Lemma 3.4. Assume that $\boldsymbol{\rho}_U \neq \boldsymbol{\rho}'_U$. Then

$$(\mathbf{x} - \mathbf{x}') \cdot [\mathbf{T}(\mathbf{x}) - \mathbf{T}(\mathbf{x}')]$$
$$= \sum_{k=1}^{J} \sum_{p \in \Pi^{(k)}} \sum_{l \in \mathcal{M}_{\mathcal{I}}} (x_p^{(k)} - x_p^{'(k)}) \times \frac{\delta_{lp}}{\mu_l^{(k)}} (T_l(\rho_l) - T_l(\rho'_l))$$
$$= \sum_{l \in \mathcal{M}_{\mathcal{I}}} (\rho_l - \rho'_l)(T_l(\rho_l) - T_l(\rho'_l)) > 0$$

The last inequality follows since by condition B3, $T_l(\rho_l)$ are strictly monotone for $l \in \mathcal{M}_{\mathcal{I}}$. Therefore we have the relations (31) and (32). □

Theorem 4.2. *Assume $B1$–$B4$. Then all equilibria, for which all users have finite cost, have the same utilization on links $l \in \mathcal{M}_{\mathcal{I}}$.*

Proof. We can prove this theorem in the same way as Theorem 3.1. □

Here again, individually optimal solution may not be unique. The range of the individually optimal solutions (related to finite costs) is given by the same set of relations as (19)–(21) but with possibly different values of ρ_l, $l = 1, 2, \ldots, M$.

Next, we illustrate that the uniqueness of the utilization is indeed restricted to equilibria with finite cost. Consider the following network. There are 4 nodes:

$\{1, 2, 3, 4\}$ and 1 class. The set of links is $\{(12), (13), (24), (34), (23)\}$. There is an amount of flow of $\phi = \phi^{(1)} = 1$ to ship between the source node 1 and the destination node 4. The cost per link is given by

$$T_l(\rho_l) = \frac{1}{1 - \rho_l}.$$

The strategy in which all the flow goes along the path $(1 \to 2 \to 3 \to 4)$ is individually optimal. Indeed, given that all users follow this path, no individual can decrease his/her cost by choosing another path. This gives rise to infinite cost for all individuals.

However, there exists another individual optimal strategy: to route half of the flow along the path $1 \to 2 \to 4$ and the other half through the path $1 \to 3 \to 4$. This is the unique equilibrium that has finite cost for all users.

5 Class Optimal Solution

Our purpose in this section is to present equivalent characterizations of the class optimal solution and to extend the known uniqueness results to more general assumptions.

The question of uniqueness for the class optimal solution has only been treated for some special cases [1,16,22]. A counterexample in [22] shows that different class optimal solutions may exist, with different utilizations.

The following assumption will be made throughout:

G: If not all classes have finite cost, then at least one of the classes which has infinite cost can change its own flow to make this cost finite.

5.1 Problem Formulation

By the class optimal policy we mean that jobs are scheduled so that the expected cost of each class may be minimum under the condition that the scheduling decisions on jobs of the other classes are given and fixed. By the *class optimization problem* we mean the problem of obtaining the routing decision $\mathbf{x}$ that achieves the objective of the class optimal policy. We call the solution of the class optimization problem the *class optimal solution* or the *Nash equilibrium.* In the Nash equilibrium, no class has any incentive to make a unilateral decision to change the decision on the routes of the jobs of the class.

Assumption G above implies that in any Nash equilibrium, all classes have finite costs.

We assume that there is a routing decision and that $\mathbf{x}$ is the path flow rate matrix which results from the routing decision. The class optimal policy requires that

$$\Delta^{(k)}(\mathbf{x}^{(k)}, \mathbf{x}^{-k}) = \min_{\mathbf{x}'^{(k)}} \Delta^{(k)}(\mathbf{x}'^{(k)}, \mathbf{x}^{-k}) \tag{33}$$

for all $k = 1, 2, \ldots, J$ ($\Delta^{(k)}(\mathbf{x}'^{(k)}, \mathbf{x}^{-k})$ is the overall mean cost of a job of class k given that other classes use flow rate $\mathbf{x}^{-k}$, and class k uses $\mathbf{x}'^{(k)}$). If $\mathbf{x}$ satisfies the above condition we say that it realizes the class optimal policy.

The problem of minimizing the mean cost for jobs of class k is stated as follows:

$$\text{minimize:}\ \Delta^{(k)} = \frac{1}{\phi^{(k)}} \sum_{l \in \mathcal{M}} \rho_l^{(k)} T_l(\rho_l) \tag{34}$$

with respect to $\mathbf{x}^{(k)}$ with $\mathbf{x}^{-k}$ being fixed subject to

$$\mathbf{\Gamma}^{\mathrm{T}}\mathbf{x} = \boldsymbol{\phi}, \mathbf{x} \geq 0$$

(where we recall that $\rho_l^{(k)} = \lambda_l^{(k)}/\mu_l^{(k)}$, $\rho_l = \sum_{k=1}^{J} \rho_l^{(k)}$ and $\lambda_l^{(k)} = \sum_{p \in \Pi^{(k)}} \delta_{lp} x_p^{(k)}$).

5.2 Variational Inequalities and Kuhn–Tucker Conditions

As in the previous sections we can get the Kuhn–Tucker conditions and the variational inequalities form by using the same reasoning as before. First we define

$\tilde{t}_p^{(k)} = \partial(\phi^P(k)\Delta^{(k)})/\partial x_p^{(k)}$, i.e., class k marginal class-cost of path p, $p \in \Pi^{(k)}$, $k = 1, 2, \ldots, J$.

$\tilde{\mathbf{t}} = [\tilde{t}_1^{(1)}, \tilde{t}_2^{(1)}, \ldots, \tilde{t}_1^{(2)}, \tilde{t}_2^{(2)}, \ldots]^{\mathrm{T}}$, i.e., the vector whose elements are $\tilde{t}_p^{(k)}$, $p \in \Pi^{(k)}$, $k = 1, 2, \ldots, J$.

$\tilde{\mathbf{t}}^{(k)} = [\tilde{t}_1^{(k)}, \tilde{t}_2^{(k)}, \ldots]^{\mathrm{T}}$.

Lemma 5.1. *$\mathbf{x}$ is an optimal solution of the problem* (33) *if and only if $\mathbf{x}$ is feasible and it satisfies the following conditions*

$$[\tilde{\mathbf{t}}(\mathbf{x}) - \mathbf{\Gamma}\boldsymbol{\alpha}] \cdot \mathbf{x} = 0, \tag{35}$$

$$\tilde{\mathbf{t}}(\mathbf{x}) - \mathbf{\Gamma}\boldsymbol{\alpha} \geq 0, \tag{36}$$

$$\mathbf{\Gamma}^{\mathrm{T}}\mathbf{x} - \boldsymbol{\phi} = 0, \tag{37}$$

$$\mathbf{x} \geq 0. \tag{38}$$

Proof. Since the objective function (34) is convex and the feasible region of its constraints is a convex set, any local solution of the problem is a global solution point. To obtain the optimal solution of player k given the policies $\mathbf{x}^{-k}$ of other players, we construct the Lagrangian function for (34), $k = 1, 2, \ldots, J$,

$$L^{(k)}(\mathbf{x}, \boldsymbol{\alpha}^{(k)}) = \Phi^{(k)}\Delta^{(k)} + \boldsymbol{\alpha}^{(k)} \cdot (\boldsymbol{\phi}^{(k)} - \mathbf{\Gamma}^{(k)\mathrm{T}}\mathbf{x}^{(k)}).$$

where $\mathbf{\Gamma}^{(k)}$ = incident matrix whose (p, d) element is γ_{pd}^{kk}, $p \in \Pi^{(k)}$, $d \in D^{(k)}$, $k = 1, 2, \dots, J$.

By the Kuhn–Tucker theorem $\mathbf{x}$ is an optimal solution if and only if there exists some (finite) $\boldsymbol{\alpha}$ and the following relations hold for $k = 1, 2, \dots, J$

$$\frac{\partial L^{(k)}}{\partial \mathbf{x}^{(k)}} = \tilde{\mathbf{t}}^{(k)}(\mathbf{x}) - \mathbf{\Gamma}^{(k)}\boldsymbol{\alpha}^{(k)} \geq 0, \tag{39}$$

$$\frac{\partial L^{(k)}}{\partial \mathbf{x}^{(k)}} \cdot \mathbf{x}^{(k)} = [\tilde{\mathbf{t}}^{(k)}(\mathbf{x}) - \mathbf{\Gamma}^{(k)}\boldsymbol{\alpha}^{(k)}] \cdot \mathbf{x}^{(k)} = 0, \tag{40}$$

$$\frac{\partial L^{(k)}}{\partial \boldsymbol{\alpha}^{(k)}} = \boldsymbol{\phi}^{(k)} - {\mathbf{\Gamma}^{(k)}}^{\mathrm{T}}\mathbf{x}^{(k)} = 0, \tag{41}$$

$$\mathbf{x}^{(k)} \geq 0, \tag{42}$$

where $(\partial L^{(k)})/(\partial \mathbf{x})$ denotes the vector whose elements are $(\partial L^{(k)})/(\partial x_p^{(k)})$, $p \in \Pi^{(k)}$, $k = 1, 2, \dots, J$ (for the finiteness, see [25] Cor. 5.1). We see that relations (39)–(42) for $k = 1, 2, \dots, J$ are the same as relations (35)–(38). □

We can express the class optimal solution in the variational inequality form in the same way as the overall optimal solution as follows.

Corollary 5.1. *Assume $B1$–$B4$. $\bar{\mathbf{x}}$ is a class optimal solution if and only if it is feasible and*

$$\tilde{\mathbf{t}}(\bar{\mathbf{x}}) \cdot (\mathbf{x} - \bar{\mathbf{x}}) \geq 0, \textit{ for all } \mathbf{x}$$
$$\textit{such that } \mathbf{\Gamma}^{\mathrm{T}}\mathbf{x} = \boldsymbol{\phi} \textit{ and } \mathbf{x} \geq 0.$$

Proof. Similar to the proof of Lemma 3.2. □

5.3 All Positive Flows

We make the following assumptions:

- $\mu_l^{(k)}$ can be represented as $a^{(k)}\mu_l$, and $0 < \mu_l^{(k)}$ is finite.
- At each node, each class may re-route all the flow that it sends through that node to any of the out-going links of that node. Thus the set of paths for class k equals the set of all possible sequences of consecutive directed links which originate at a source s and end at the destination d, $sd \in D^{(k)}$.
- The rate of traffic of class k that enters the network at node v is given by $\phi_v^{(k)}$; if this quantity is negative this means that the traffic of class k leaves node v at a rate of $|\phi_v^{(k)}|$. We assume that $\sum_v \phi_v^{(k)} = 0$.

For each node u and class k, denote by $In(u, k)$ the set of its in-going links, and denote by $Out(u, k)$ the set of its out-going links.

Due to the second assumption, we may work directly with the decision variables $\lambda_l^{(k)}$ instead of working with the path flows. For each node v we can then replace (4) by:

$$\sum_{l \in Out(v,k)} \lambda_l^{(k)} = \sum_{l \in In(v,k)} \lambda_l^{(k)} + \phi_v^{(k)}$$

We shall use the Kuhn–Tucker condition. To do so, we define the Lagrangian

$$L^{(k)}(\boldsymbol{\lambda}, \boldsymbol{\xi}^{(k)}) = \sum_{l \in \mathcal{M}} \rho_l^{(k)} T_l -$$

$$\sum_u \xi_u^{(k)} \left[\sum_{l \in Out(u,k)} \lambda_l^{(k)} - \sum_{l \in In(u,k)} \lambda_l^{(k)} - \phi_u^{(k)} \right].$$

Here, $\boldsymbol{\xi}^{(k)} = [\xi_1^{(k)}, \xi_2^{(k)}, \ldots, \xi_M^{(k)}]^{\mathrm{T}}$ is the vector of Lagrange multipliers for class k.

An assignment $\boldsymbol{\lambda}^*$ is class-optimal if and only if the following Kuhn–Tucker conditions hold. There exists some $\boldsymbol{\xi}^{(k)} = [\xi_u^{(k)}]$ such that

$$\frac{\partial L^{(k)}(\boldsymbol{\lambda}^*, \boldsymbol{\xi}^{(k)})}{\partial \lambda_l^{(k)}} \geq 0, \tag{43}$$

$$\frac{\partial L^{(k)}(\boldsymbol{\lambda}^*, \boldsymbol{\xi}^{(k)})}{\partial \lambda_l^{(k)}} = 0 \text{ if, } \lambda_l^{(k)} > 0; \tag{44}$$

$$\lambda_l^{(k)} \geq 0, \quad \sum_{l \in Out(v,k)} \lambda_l^{(k)} = \sum_{l \in In(v,k)} \lambda_l^{(k)} + \phi_v^{(k)}.$$

Define

$$K_l^{(k)}(\rho_l^{(k)}, \rho_l) = \frac{\partial \rho_l^{(k)} T_l(\rho_l)}{\partial \lambda_l^{(k)}}.$$

Then

$$K_l^{(k)}(\rho_l^{(k)}, \rho_l) = \frac{1}{\mu_l^{(k)}} \left(\rho_l^{(k)} \frac{\partial T_l(\rho_l)}{\partial \rho_l} + T_l(\rho_l) \right)$$

Conditions (43)–(44) can be rewritten as

$$K_l^{(k)}(\rho_l^{(k)}, \rho_l) \geq \xi_u^{(k)} - \xi_v^{(k)}, \tag{45}$$

with equality if $\lambda_l^{(k)} > 0$ and $l = (u, v)$. Note that condition B3 implies that $K_l^{(k)}(\rho_l^{(k)}, \rho_l)$ is strictly monotonically increasing in both arguments.

Lemma 5.2. *Assume $B1$–$B4$. Assume that $\boldsymbol{\lambda}$ and $\hat{\boldsymbol{\lambda}}$ are two class-optimal solutions with finite costs. If $\rho_l = \hat{\rho}_l$ for all links l of type $\mathcal{M}_{\mathcal{I}}$ (see assumption B3) then $\lambda_l^{(k)} = \hat{\lambda}_l^{(k)}$, $k = 1, \ldots, J$.*

Proof. Assume that under the assumptions of the Lemma the conclusions do not hold. Then there exist some $l \in \mathcal{M}_I$ and some k such that

$$\hat{\lambda}_l^{(k)} > \lambda_l^{(k)}. \tag{46}$$

We now construct another network with the same nodes and links as the original one, with the flow on a link l between two points u and v given by $|\hat{\lambda}_l^{(k)} - \lambda_l^{(k)}|$, its direction is (uv) if and only if $\hat{\lambda}_{uv}^{(k)} - \lambda_{uv}^{(k)} > 0$ and is otherwise (vu). In this network there are no inputs and outputs. It follows from (46) that the network contains a cycle $\mathcal{C}$ with strictly positive flow.

We now consider any link $(uv) \in \mathcal{C}$. Then in the original network either (uv) is the direction of the flow of class k and$\hat{\lambda}_{(uv)}^{(k)} > \lambda_{(uv)}^{(k)}$, or the direction is (vu) and $\hat{\lambda}_{(vu)}^{(k)} < \lambda_{(vu)}^{(k)}$. In the first case we have by Kuhn–Tucker conditions

$$\hat{\xi}_u^{(k)} - \hat{\xi}_v^{(k)} = K_{(uv)}^{(k)}(\hat{\rho}_{(uv)}^{(k)}, \hat{\rho}_{(uv)}) \geq K_{(uv)}^{(k)}(\rho_{(uv)}^{(k)}, \rho_{(uv)}) = \xi_u^{(k)} - \xi_v^{(k)}. \tag{47}$$

In the second case, we have

$$\xi_v^{(k)} - \xi_u^{(k)} = K_{(vu)}^{(k)}(\rho_{(vu)}^{(k)}, \rho_{(vu)}) \geq K_{(vu)}^{(k)}(\hat{\rho}_{(vu)}^{(k)}, \hat{\rho}_{(vu)}) = \hat{\xi}_v^{(k)} - \hat{\xi}_u^{(k)}. \tag{48}$$

Due to the strict monotonicity of K for $l \in \mathcal{M}_I$, there is at least one link in $\mathcal{C}$ for which a strict inequality holds in the corresponding inequality among (47) and (48). This implies that

$$0 = \sum_{i=1}^{|\mathcal{C}|} (\hat{\xi}_i^{(k)} - \hat{\xi}_{i-1}^{(k)}) > \sum_{i=1}^{|\mathcal{C}|} (\xi_i^{(k)} - \xi_{i-1}^{(k)}) = 0$$

which is a contradiction. Thus the Lemma is established. □

Theorem 5.1. *Assume $B1$–$B4$. Denote by $\mathcal{M}_1(\boldsymbol{\lambda})$ the sets of links l such that $\lambda_l^{(k)} > 0, \forall k = 1, \ldots, J$ for an assignment $\boldsymbol{\lambda}$.*
Assume that $\boldsymbol{\lambda}$ and $\hat{\lambda}$ are two class-optimal solutions with finite costs for all players.
Assume that $\lambda_l^{(k)} = 0, \forall k, \forall l \notin \mathcal{M}_1(\boldsymbol{\lambda})$, $\hat{\lambda}_l^{(k)} = 0, \forall k, \forall l \notin \mathcal{M}_1(\hat{\boldsymbol{\lambda}})$.
Then $\lambda_l^{(k)} = \hat{\lambda}_l^{(k)}$ for all $l \in \mathcal{M}_{\mathcal{I}}$ (see assumption B3 for the definition of $\mathcal{M}_{\mathcal{I}}$).

Proof. Denote $\xi_u = \sum_{k=1}^{J} a^{(k)} \xi_u^{(k)}$, where $a^{(k)}$ is defined in the beginning of Subsection 5.3, and

$$S_l(\rho_l) = \sum_k \mu_l^{(k)} K_l^{(k)}(\rho_l^{(k)}, \rho_l) = \rho_l \frac{\partial T_l(\rho_l)}{\partial \rho_l} + J T_l(\rho_l).$$

Note that the assumption that costs are finite and Lemma 3.3 imply that $S_l(\rho_l)$ are finite and Assumption B3 implies that $S_l(\rho_l)$ is strictly monotone.

Let $\hat{\boldsymbol{\xi}}$ denote the vector of the Lagrange multipliers corresponding to $\hat{\boldsymbol{\lambda}}$. (45) implies that

$$\mu_{uv}^{-1} S_{uv}(\rho_{uv}) \geq \xi_u - \xi_v, \tag{49}$$

with equality for $(u, v) \in \mathcal{M}_1(\boldsymbol{\lambda})$. A similar relation holds for $\hat{\boldsymbol{\lambda}}$. We obtain that

$$\begin{aligned} 0 &\leq \sum_{(u,v)\in\mathcal{M}} (\rho_{uv} - \hat{\rho}_{uv})(S_{uv}(\rho_{uv}) - S_{uv}(\hat{\rho}_{uv})) \qquad (50) \\ &\leq \sum_{(u,v)\in\mathcal{M}} \mu_{uv}(\rho_{uv} - \hat{\rho}_{uv})\Big((\xi_u - \hat{\xi}_u) - (\xi_v - \hat{\xi}_v)\Big) = 0. \end{aligned}$$

The first inequality follows from the strict monotonicity of $S_l(\rho_l)$ for $l \in \mathcal{M}_{\mathcal{I}}$; for $l \in \mathcal{M}_{\mathcal{C}}$ this relation is trivial. The second inequality holds in fact for each pair u, v (and not just for the sum). Indeed, for $(u, v) \in \mathcal{M}_1(\boldsymbol{\lambda}) \cap \mathcal{M}_1(\hat{\boldsymbol{\lambda}})$ this relation holds with equality due to (49). This is also the case for $(u, v) \notin \mathcal{M}_1(\boldsymbol{\lambda}) \cup \mathcal{M}_1(\hat{\boldsymbol{\lambda}})$, since in that case $\rho_{uv} = \hat{\rho}_{uv} = 0$. Consider next the case $(u, v) \in \mathcal{M}_1(\boldsymbol{\lambda})$, $(u, v) \notin \mathcal{M}_1(\hat{\lambda})$. Then we have

$$\begin{aligned} &(\rho_{uv} - \hat{\rho}_{uv})(S_{uv}(\rho_{uv}) - S_{uv}(\hat{\rho}_{uv})) \\ &\quad = \rho_{uv}(S_{uv}(\rho_{uv}) - S_{uv}(\hat{\rho}_{uv})) \leq \mu_{uv}\rho_{uv}(\Big(\xi_u - \hat{\xi}_u) - (\xi_v - \hat{\xi}_v)\Big). \end{aligned}$$

A symmetric argument establishes the case $(u, v) \in \mathcal{M}_1(\hat{\boldsymbol{\lambda}})$, $(u, v) \notin \mathcal{M}_1(\boldsymbol{\lambda})$. We finally establish the last equality in (50).

$$\begin{aligned} &\sum_{(u,v)\in\mathcal{M}} \mu_{uv}(\rho_{uv} - \hat{\rho}_{uv})\Big((\xi_u - \hat{\xi}_u) - (\xi_v - \hat{\xi}_v)\Big) \\ &\quad = \sum_j (\xi_j - \hat{\xi}_j) \sum_w (\rho_{jw} - \hat{\rho}_{jw})\mu_{jw} - \sum_j (\xi_j - \hat{\xi}_j) \sum_w (\rho_{wj} - \hat{\rho}_{wj})\mu_{wj} \\ &\quad = \sum_{k=1}^{J} \Big(\sum_j (\xi_j - \hat{\xi}_j) \sum_w (\rho_{jw}^{(k)} - \hat{\rho}_{jw}^{(k)})\mu_{jw} \\ &\quad - \sum_j (\xi_j - \hat{\xi}_j) \sum_w (\rho_{wj}^{(k)} - \hat{\rho}_{wj}^{(k)})\mu_{wj} \Big) \\ &\quad = \sum_{k=1}^{J} \frac{1}{a^{(k)}} \Big[\sum_j (\xi_j - \hat{\xi}_j) \Big(\sum_{l\in Out(j,k)} (\lambda_l^{(k)} - \hat{\lambda}_l^{(k)}) - \sum_{l\in In(j,k)} (\lambda_l^{(k)} - \hat{\lambda}_l^{(k)}) \Big) \Big] \\ &\quad = 0 \end{aligned}$$

(we used the fact that the sum of $\phi_v^{(k)}$ over all nodes v is equal to zero so that the difference of ingoing and outgoing lambda's is also zero). We conclude from (50) that $\rho_l = \hat{\rho}_l$ for all links in $\mathcal{M}_{\mathcal{I}}$. The proof follows from Lemma 5.2. □

Remark 5.1. Theorem 5.1 and its proof are substantial extensions of [22] who considered the special case where μ_l^i do not depend on l and i, where there is a *single source-destination* pair which is the same for all users (all paths and all classes), and where $\mathcal{M}_1(\boldsymbol{\lambda}) = \mathcal{M}_1(\hat{\boldsymbol{\lambda}})$. Moreover, the costs of all links are assumed in [22] to be strictly increasing.

Next, we present an example of load balancing [17] that occurs in distributed computing, in which different classes have different sources and where our uniqueness result may apply.

Example 5.1. There are two processors and a single communication means that connects them. Nodes are numbered 1 and 2. We associate a class with each node (and thus have two players in the game). Node i has the external arrival of jobs to process with rate ϕ_i and it has to decide what fraction of the arriving jobs would be processed locally and what fraction should be forwarded to the other node. Delay is incurred at each node (processing delay) as well as in the communication bus (communication delay), and the goal of each class is to minimize the average delay of jobs of that class. The delay at each network element (nodes and communication bus) is an increasing function of the total job rates that use that element (thus the decisions of one class also influence the cost for the other class). This load balancing problem can be modeled as a network game that consists of three nodes and three links:

- Nodes: s_1, s_2, d, where s_i is the source of jobs of class i, and d is a common destination.
- Links: $s_i d$, $i = 1, 2$ represent the processor i, and $s_1 s_2$ represents the communication bus.

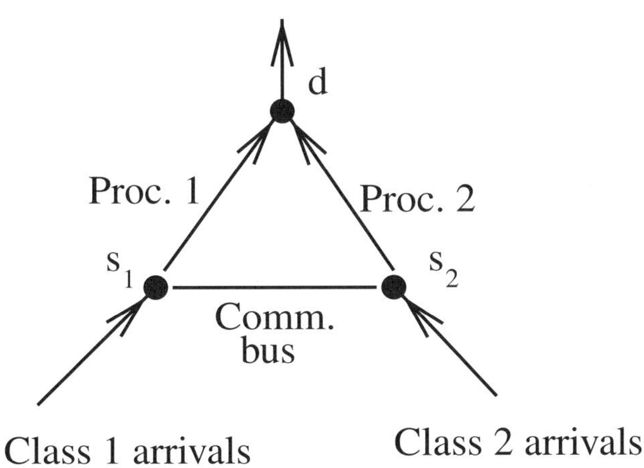

Figure 1: A network representation of the load balancing problem

- Paths: Class i has two paths, $s_i \to d$ (corresponding to local processing) and path $s_i \to s_j \to d$, that corresponds to forwarding jobs to the other processor.

This network model is depicted in Fig. 1. We conclude that for the above problem, there is at most one equilibrium (under the appropriate assumptions on the delay functions) at which each class splits its arrival flows: a fraction is processed locally and a fraction is forwarded. Numerical examples can be found in [17] (in which the problem of the uniqueness of the equilibrium is not addressed).

6 Numerical Examples

Consider a simple example of a network composed of two parallel links $\mathcal{M} = \{a, b\}$ and J identical classes: Each link can be identified with a path. We consider for simplicity $\mu_a^{(k)} = 1, \mu_b^{(k)} = 2, k = 1, \dots, J$. Consider an M/M/1 type cost, i.e.

$$\hat{T}_l = \frac{1/\mu_l^{(k)}}{1-\rho_l},$$

$l = a, b, k = 1, \dots, J$ (T_l is infinite for $\rho_l \geq 1$). Let $\phi^k = 2/J$. We note that

$$\rho_a = \frac{\sum_j \lambda_a^{(j)}}{1} = 2 - \sum_j \lambda_b^{(j)} = 2(1-\rho_b).$$

Hence, $\rho_a < 1$ implies that $\rho_b > 0.5$. We have:

$$J\Delta = \sum_{l=a,b} \rho_l T_l(\rho_l) = \sum_{l=a,b} \frac{\rho_l}{1-\rho_l}$$
$$= \frac{2(1-\rho_b)}{2\rho_b - 1} + \frac{\rho_b}{1-\rho_b}.$$

The overall optimal solution is obtained at $\rho_b^* = \sqrt{1/2}$, which gives $\rho_a^* = 2 - \sqrt{2}$ and $\Delta(\rho) = (2\sqrt{2}+1)/J$.

In order to obtain the individual optimization, we note that

$$T_l^{(k)} = \frac{1/\mu_l}{1-\rho_l}, \quad k = 1, \dots, J.$$

This gives

$$T_a^{(k)} = \frac{1}{2\rho_b - 1}, \quad T_b^{(k)} = \frac{1}{2(1-\rho_b)}, \quad k = 1, \dots, J.$$

The individual optimum is obtained at $\bar{\rho}_a = 1/4$, $\bar{\rho}_b = 3/4$, which gives delays along the two links of $T_a^{(k)} = T_b^{(k)} = 2$.

In both cases the solution in terms of the ρ_l's is unique. Any choice of rates $x_l^{(k)}$ that gives the corresponding ρ_l is optimal, and it is clearly not unique. For example, if $J = 2$,

$$x_l^{(k)} = \mu_l \rho_l^* / 2, \ l = a, b, \ k = 1, 2$$

is an overall optimal solution and

$$x_l^{(k)} = \mu_l \bar{\rho}_l / 2, \ l = a, b, k = 1, 2$$

is an individually optimal solution. Another overall optimal solution is

$$x_a^{(1)} = \rho_a^{(1)} = \rho_a^*, x_b^{(2)} = 1 - x_a^{(1)}, x_a^{(2)} = 0, x_b^{(2)} = 1,$$

and another individually optimal solution is

$$x_a^{(1)} = \rho_a^{(1)} = \bar{\rho}_a, x_b^{(2)} = 1 - x_a^{(1)}, x_a^{(2)} = 0, x_b^{(2)} = 1.$$

Unlike the overall and individually optimal solutions, the class optimal solution for this problem is indeed unique, as has been shown in [22].

7 Concluding Remarks and Perspectives

We studied *multiclass* static routing problems with several types of optimization concepts in networks: the overall optimization, individual optimization and class optimization. The routing problem is of the type studied in [13], where one has to determine the assignment of the flow rates among different paths. This setting is more general than the one in which routing decisions may be taken at each node [15,22]; it is of special importance to telecommunications networks, in particular ATM networks, in which users have to route their traffic through predetermined virtual paths. We established the uniqueness of the utilization under the optimal solutions for different types of optimization problems.

Our flow allocation model is a simplification of the most general ones expected to be encountered in actual communication networks. In particular, we considered a single cost per decision maker which is based on additive link costs. This model covers costs such as expected delays, but may fall short of covering other types of costs such as loss probabilities or call rejection rates.

We should mention that in practice network conditions may change frequently; this means that one should update the routing decisions from time to time. We believe that our static optimization could be a starting point for the design of future distributed adaptive routing protocols (see e.g. [9]).

REFERENCES

[1] Altman E., Başar E., Jiménez T. and Shimkin N., "Competitive routing in networks with polynomial cost", *IEEE Trans. on Automatic Control*, **47**, 92–96, 2002.

[2] Baskett F., Chandy K. M., Muntz R. R. and Palacios F., "Open, closed, and mixed networks of queues with different classes of customers", *J. ACM*. **22**, 248–260, 1975.

[3] Bennett L. D., "The existence of equivalent mathematical programs for certain mixed equilibrium traffic assignment problems", *European J. of Operations Research*, **71**, 177–187, 1993.

[4] Braess D., "Über ein Paradoxen aus der Verkehrsplanung", *Unternehmensforschung* **12**, 258–268, 1968.

[5] Cantor D. G. and Gerla M., "Optimal routing in a packet-switched computer network," *IEEE Trans. Computers* **C-23**, 1062–1069, 1973.

[6] Dafermos S., "The traffic assignment problem for multiclass-user transportation networks", *Transportation Sci.* **6**, 73–87, 1972.

[7] Dafermos S. and Sparrow F. T., "The traffic assignment problem for a general network", *Journal of Research of the National Bureau of Standards-B. Math. Sci.*, **73B**, No. 2, 91–118, 1969.

[8] Fratta L., Gerla M. and Kleinrock L., "The flow deviation method—an approach to the store-and-forward communication network design", *Networks* **3**, 97–133, 1973.

[9] Gupta P. and Kumar P. R., "A system and traffic dependent adaptive routing algorithm for ad hoc networks", *Proceedings of the 37th IEEE Conference on Decision and Control*, Tampa, Florida, USA, Dec. 1998.

[10] Haurie A. and Marcott P., "On the relationship between Nash-Cournot and Wardrop equilibria", *Networks* **15**, 295–308, 1985.

[11] Kameda H., Kozawa T. and Li J., "Anomalous relations among various performance objectives in distributed computer systems", *Proc. World Congress on Systems Simulation*, 459–465, Sept, 1997.

[12] Kameda H., Li J., Kim C. and Zhang Y., *Optimal Load Balancing in Distributed Computer Systems*, Springer, London, 1997.

[13] Kameda H. and Zhang Y., "Uniqueness of the solution for optimal static routing in open BCMP queueing networks", *Math. Comput. Modeling* **22**, No. 10–12, 119–130, 1995.

[14] Kleinrock L., *Queueing Systems, Vol 2*, Wiley, New York, 1976.

[15] Korilis, Y. A., Lazar, A. A. and Orda, A., "Architecting noncooperative networks", *IEEE Journal of Selected Areas of Communications*, **13**, 1241–1251, 1995.

[16] Kameda H., Altman E. and Kozawa T., "A case where a paradox like Braess's occurs in the Nash equilibrium but does not occur in the Wardrop equilibrium – a situation of load balancing in distributed computer systems", *Proc. of the 38th IEEE Conference on Decision and Control*, Phoenix, Arizona, USA, Dec. 1999.

[17] Kameda H., Altman E., Kozawa T. and Hosokawa Y., "Braess-like paradoxes of Nash equilibria for load balancing in distributed computer systems", *IEEE Trans. on Automatic Control*, **45**, No. 9, 1687–1691, 2000.

[18] Kelly F. P., *Reversibility and Stochastic Networks*, John Wiley & Sons, New York (1979).

[19] La R. J. and Anantharam V., "Optimal routing control: game theoretic approach", *Proc. of the 36th IEEE Conference on Decision and Control*, San Diego, California, Dec. 1997.

[20] Li J. and Kameda H., "Load balancing problems for multiclass jobs in distributed/parallel computer systems," *IEEE Trans. Comp.* **47**, No. 3, 1998, pp. 322–1998.

[21] Monderer D. and Shapley L. S., "Potential games", *Games and Econ. Behavior*, **14**, 124–143, 1996.

[22] Orda A., Rom R. and Shimkin N., "Competitive routing in multi-user environments", *IEEE/ACM Trans. on Networking*, **1**, No. 5, 510–521, 1993.

[23] Patriksson M. *The Traffic Assignment Problem: Models and Methods* VSP BV, P.O. Box 346, 3700 AH Zeist, The Netherlands, 1994.

[24] Sandholm W. H., "Potential games with continuous player sets", *Journal of Economic Theory*, **97**, No. 1, 81–108, 2001.

[25] Shapiro J. F., *Mathematical Programming, Structures and Algorithms*, J. Wiley, New York, 1979.

[26] Stallings W., *High Speed Networks: TCP/IP and ATM design principles*. Upper Saddle River, NJ, USA: Prentice-Hall, Inc., 1998.

[27] Tantawi A. N. and Towsley D., "A general model for optimal static load balancing in star network configurations", *Proc. of PERFORMANCE' 84*, North-Holland, New York, 277–291 (1984).

[28] Tantawi A. N. and Towsley D., "Optimal static load balancing in distributed computer systems", *J. ACM* **32**, 445–465, 1985.

[29] Wardrop J. G., "Some theoretic aspects of road traffic research", *Proc. Inst. Civ. Eng.* Part 2, 1, 325–378, 1952.

Endogenous Shocks and Evolutionary Strategy: Application to a Three-Players Game

Ekkehard C. Ernst*
Directorate General Economics
European Central Bank
60311 Frankfurt, Germany
ekkehard.ernst@ecb.int

Bruno Amable
Faculté des Sciences Economies
Université Paris X – Nanterre
France
Bruno.Amable@cepremap.ens.fr

Stefano Palombarini
Faculté des Sciences Economies
Université Paris VIII
France
stefano.palombarini@cepremap.cnrs.fr

Abstract

An evolutionary game with three players – trade unions, financial investors and firms – is presented where each player has a short-term and a long-term maximizing strategy at hand. The short-term strategy maximizes current payoffs without taking into account benefits from future cooperation while long-term strategies depend on the cooperative behavior of the other players. We first determine equilibria arising in the static game and determine under which conditions long-term cooperation may emerge. We then endogenize the stochastic environment, making it subject to the strategies selected and show how additional equilibria and strategy cycles arise in an evolutionary set-up.

1 Introduction

Evolutionary game theory has gained an important place in economic modeling through the last two decades. The possibility of analyzing strategic interactions in

*The authors wish to thank participants at the 9th International Symposium for Dynamic Games, 2000, and an anonymous referee for very helpful comments. Any remaining errors are ours. The views expressed in this paper are those of the authors, and do not necessarily represent those of the ECB, the University of Nanterre or the University Paris VIII.

a low-rationality framework and the elimination of non-strict Nash equilibria has increasingly attracted economists to follow this approach. New developments in stochastic evolutionary game theory (SEGT) – first suggested by Foster and Young [5] – have moreover allowed to narrow even further the possible outcomes – at least in the very long-run.

Especially when analyzing the emergence and change of institutional arrangements in societies, this last approach has now gained some authority: social customs, patterns of interaction and adoption of standards have been analyzed successfully by SEGT. Researchers in the social sciences often face situations where the adoption of one particular strategy by one player increases the marginal returns of this – or some related – strategy for other players: a case for strategic complementarities. In this situation, coordination problems easily arise and SEGT allows the singling out of the equilibrium that is characterized by risk dominance, a concept that makes sense when allowing for very long-run developments of strategic configurations.

However, as has been put forward by Bergin and Lipman [2], the results these models generate do not in general survive small modifications of the underlying stochastic processes. In particular, state-dependent stochastic processes will not necessarily allow selection of the risk-dominant equilibrium in the long-run.

Starting from this insight, the following paper proposes an application of a three-player coordination problem to analyze the strategic interactions between trade unions, banks and firms. In particular, by making the stochastic process by which the game's payoffs are selected completely endogenous to the strategic configurations we can go even beyond the initial result of Bergin and Lipman's paper, showing that even the underlying Nash equilibria no longer need to exist: consequently, no (stochastically) evolutionary stable strategies will arise.

In the model presented here, a coordination problem between trade unions, financial intermediaries and firms arises due to a multi-dimensional strategy space. Firms have access to a long- and a short-term technology where the productivity of the long-term technology crucially depends on a stable labor input: as soon as the firm is obliged to lay-off parts of its workforce, the technology does not yield any profits any more. The short-term technology is more flexible and allows an easy rotation of the workforce which can be used as a strategic weapon against wage demands.

Workers, on the other hand, have the possibility of joining a trade union and bargaining over wages (only) or making use of an additional (costly) instrument to increase their employment stability: in our setting we suggest that this second element is an investment in their human capital that helps to reduce the idiosyncratic risk at the plant level.

Lastly, financial investors have the choice to invest directly through the stock market or to join (or form) a bank. The first choice gives them full access to the dividend flow whereas in the second case they have to bear intermediation costs.

Both strategies have different impacts on the risk distribution and the availability of information in the economy.

The marginal return to each type of strategy (either long or short term) is increased when at least one of the other players adopts the same strategy type (and even more so when both remaining players do). Hence, a coordination problem arises with equilibria that can be Pareto-ranked.

This three-players game may be solvable by using the theory of global games[1]; in these games, the risk dominant equilibrium is selected. However, the analysis of the macroeconomy suggests that interdependencies between microeconomic actors and production pools may have considerable impact on the risk structure of the economy and hence on the strategic choices that institutional actors (and firms) undertake. More importantly, the endogenous nature of the stochastic process may even affect the Nash characteristic of some of these equilibria and hence will no longer allow the application of global games or SEGT theorems.

Introducing institutional questions like these into economic modeling has gained importance in economics in recent years, especially in the literature on economic growth. Authors have become aware that the underlying mechanisms (externalities, complementarities) of these models may be influenced by factors which had not been included in standard economics until then. Moreover, historical accounts like Altvater, Hoffmann and Semmler [1] or others, suggest that it is institutional change as much as the mere existence of institutions that may influence the long-run evolution of a country's economic performance.

The paper is organized as follows: In the next two parts we analyze endogenous labor market relations and the corresponding technology choices and explore the resulting strategic game between firms, trade unions and financial investors. In section four we open the analysis for macroeconomic relations and show how and under what conditions institutional cycles can arise. A final section concludes the discussion.

2 The Stochastic Environment

The economy is composed of N production poles each containing a given number n of employees. The employees may be able to switch between these poles while they are "off" in order to find the optimal (i.e. corresponding the best to their preferences) pole with respect to the wage contract proposed at that pole. At each point in time, one of these poles is "on", i.e. organizes a game between workers, managers and its financiers in order to figure out the wage and debt contract which will be utilized in the following period where the pole is "off" (in the sense of the game play, not in the sense of production).

Once the contract has been determined, the pole starts production; if no contract agreement could be found, the pole does not produce anything until the next time

[1]There exists an important resemblance between global games and SEGT ([3,8]).

it plays the game. Moreover, we assume time consistency, i.e. that the players are committed to the strategies they decided on before the production process.

Production underlies idiosyncratic shocks u_{it} with $u_{it} \sim G_i(u_{it}|h_i, T_i)$ supposed to be stationnary and $(\partial G_i/\partial h_i) < 0$, $(\partial^2 G_i/\partial h_i \partial T_i) < 0$ where h_i denotes the workers' human capital and T_i the firm's technology choice to be determined later; human capital improves the average realization of the shock but less so for less specific technologies (with lower T_i). Moreover, idiosyncratic shocks are uncorrelated among pools, i.e. $cov(u_i, u_j) = 0 \, \forall \, i \neq j$.

The firm faces a demand function depending on aggregate income, i.e. aggregate wages and profits in each period. If the pools are not interconnected then the overall uncertainty simply equals the productivity uncertainty u_{it}. However, if pools are interconnected, then a single pool faces a collective risk through aggregate demand. Given that pools are symmetric this can be written as $\eta_{it} = (1/N) \sum_{j \neq i} \Pr(\varepsilon_{jt} \geq \underline{\varepsilon}_j) \cdot \varepsilon_{jt}$, where $\varepsilon_{it} = (1/N)u_{it} + \eta_{it}$ stands for the overall risk a firm faces and $\Pr(\varepsilon_{it} \geq \underline{\varepsilon}_j)$ for the probability that pool j reaches a certain minimum state $\underline{\varepsilon}_j$ (which will be determined later on). Cumulative distribution functions are given by $F_i(\varepsilon_{it})$ and $H_i(\eta_{it})$, both depending on the distribution functions of all idiosyncratic shocks.

After contracts have been determined, production starts yielding expected outputs y_i^e until the shock reaches some minimum state $\underline{\varepsilon}_i$ below which financial investors are no longer willing to keep their engagement at pool i. In this case, the firm will be liquidated. The exit probability can be determined as:

$$s_i = 1 - \Pr(\varepsilon_{it} \geq \underline{\varepsilon}_i) = \int_0^{\underline{\varepsilon}_i} dF_i(\varepsilon_{it}). \tag{1}$$

3 Firms, Banks and Workers: Technological Choice

In a first step, we concentrate on the microeconomic relations only, leaving out pool interdependency and focusing only on idiosyncratic shocks. We will come back to the aggregate shock when we look at the macroeconomic links. In the following we therefore drop pool indices i.

3.1 Technology and Profits

Technology

The firm has to make a technology choice $T \in [0, 1]$ with technologies ranging from completely unspecific ($T = 0$) to completely specific ($T = 1$). The more specific a technology is, the lower will be its resale value in case of liquidation V^L, $V^L = V(T)$, $V' < 0$. To be precise, we want to assume that $V^L(0) = V^{\max} > 0$ and $V^L(1) = 0$. The installed technology yields the expected output $y^e \equiv Ey = p(h)y(T)$ where $y' > 0$ and $p = \int_0^{\infty} u \cdot dG(u|\, h, T)$ and $p'(h) > 0$, i.e. the impact of human capital, h, on p will be more important the more specific the technology is.

The lower the degree of specificity, the less important will be specific human capital investment to use the technology; therefore the workforce can easily be replaced. In case of wage negotiations firms could then use outside labor to replace parts of the insider force, increasing therefore its bargaining position. This will reduce the probability for workers to find a job covered by a collective agreement by $q = q(T, \xi)$ with $q_T < 0$, $q_\xi > 0$, $q(0, \xi) = \overline{q}(\xi) \in (0, 1)$, $q(1, \xi) = 0$, i.e. $q(T, \xi)$ measures the reduction in workers' bargaining power due to the availability of outside labor. Exogenous factors (e.g. legislation, immigration) may restrict the maximum amount of this outside labor pool that can be employed – even in the case of a completely un-specific technology – to a value $\overline{q}(\xi)$ less than one. In particular, we consider ξ as being *net* immigration [2] and analyze in the following how the equilibrium behavior changes with this parameter.

Profits

In each period the firm expects to earn a return depending on its technology choice, $y^e = p(h)y(T)$, spending w to hire workers and facing a probability $s(\kappa)$ to be liquidated due to weak performance and impatient financial investors. Denoting r the interest rate, the firm's Bellman equation can be set up as:

$$r\pi = p(h)y(T) - w + s(\kappa)[V^L(T) - \pi] + \dot{\pi}$$

which can be rewritten at the steady state (where $\dot{\pi} = 0$) as:

$$\pi = \frac{p(h)y(T) + s(\kappa)V^L(T) - w}{r + s(\kappa)}. \tag{2}$$

As can be easily seen from this formula, all the cross derivatives between the strategic variables are positive leading to strategic complementarities:

$$\frac{\partial^2 \pi}{\partial h \partial T} > 0; \; \frac{\partial^2 \pi}{\partial h \partial \kappa} > 0; \; \frac{\partial^2 \pi}{\partial T \partial \kappa} > 0. \tag{3}$$

3.2 Trade Unions and Wage Bargaining

In order to determine the wage bargaining position for workers we have to consider their maximization problem. Workers will obtain W when being employed and covered by a collective wage agreement. With probability $s(\kappa)$ the firm is liquidated

[2] Net immigration is influenced by a variety of factors including costs of immigration, perceived benefits abroad and at home, etc. which will not be taken up here in detail.

and they lose the job, obtaining their unemployment value U benefiting from some outside opportunity R. Workers can spend η to invest in their human capital h; this investment will be lost when unemployed but can be recovered on a job. Being unemployed, they will find a new unionized job with probability $\theta - q(T, \xi)$. Therefore, the two states in which the worker can be, have the following values:

$$
\begin{aligned}
rW &= w + s(\kappa)(U - W) + \dot{W}, \\
rU &= R - \eta h + (\theta - q(T, \xi))(W - U) + \dot{U}.
\end{aligned}
$$

Again placing ourselves at the steady state, the wage to be earned from being employed can be written as:

$$
\begin{aligned}
r(W - U) &= w + \eta h - R + (s(\kappa) + \theta - q(T, \xi))(U - W) \\
\Leftrightarrow W - U &= \frac{w + \eta h - R}{r + s(\kappa) + \theta - q(T, \xi)}.
\end{aligned}
\tag{4}
$$

We are assuming a Nash bargaining game with the firm's fallback position (i.e. in the case no agreement can be found) to be normalized to zero and union's bargaining power σ. Both bargaining parties therefore have to select a wage w^b such as to maximize:

$$
w^b = \arg\max \pi^{\sigma} (W - U)^{1-\sigma}.
$$

The optimal solution to this problem has to satisfy the following first-order condition:

$$
(1 - \sigma)\pi = \sigma(W - U)
$$

which can be rewritten as – using (2) and (4):

$$
(1 - \sigma)\frac{p(h)y(T) + s(\kappa)V^L(T) - w^b}{r + s(\kappa)} = \sigma \frac{w^b + \eta h - R}{r + s(\kappa) + \theta - q(T, \xi)}
$$

which can be used to calculate the bargained wage as:

$$
w^b = \frac{\begin{array}{l}(1 - \sigma)(p(h)y(T) + s(\kappa)V^L(T))(r + s(\kappa) + \theta - q(T, \xi)) \\ +\sigma(R - \eta h)(r + s(\kappa))\end{array}}{r + s(\kappa) + (1 - \sigma)(\theta - q(T, \xi))}
$$

With the impact of human capital on the distribution of the outcome of the production process being low for short-run technologies, investment in human capital will depress the bargained wage for short-term horizon production processes. Moreover, whenever firms can use outside labor, $q(T, \xi)$, it will have a dampening effect on wage demands.

3.3 The Financial Relation

The financial relation is characterized by the dispersion of ownership $\kappa \in [0, 1]$ (the higher the κ the more concentrated is the ownership). If the firm is in financial distress then the building of a coalition to refinance it will be the more difficult the more the financial contractors are dispersed, i.e. the lower is κ (see [7], pp 295–299). The more finance is concentrated in the hands of a couple of banks the higher the chance for the firm to survive even in case of low realization.

Banks are assumed to be ready to restructure the enterprise and to refinance even projects with (temporarily) low expected outcomes in order to obtain credit repayments. This ability makes them accept a profit level (temporarily) lower than that required by the stock market. In this way they secure the investment even in the presence of an adverse demand or production shock. Investors on the stock market, however, are not easily able to form a stable coalition to refinance a failing project or they may not dispose of sufficient liquidity to do it ([4]); therefore they decide to liquidate the firm.

Specifically we want to suppose that survival is guaranteed if there is only bank finance involved, $\kappa = 1 \Rightarrow s(\kappa) = 0$ (i.e. $\underline{\varepsilon} = 0$). For completely dispersed ownership, survival becomes more difficult, i.e. $\kappa = 0 \Rightarrow s(\kappa) \gg 0$ (i.e. $\underline{\varepsilon} \gg 0$). According to the technology and liquidity value assumptions, the following holds given the union's strategy:

$$\pi^e(T = 1, \kappa = 1) > \pi^e(T = 0, \kappa = 1) \text{ and}$$
$$\pi^e(T = 0, \kappa = 0) > \pi^e(T = 1, \kappa = 0).$$

Furthermore we want to make the assumption that the use of the short-term technology is efficient, i.e. $\pi^e > V^L(0)$ which can be rewritten as $\pi^e(T = 0, \kappa = 1) > \pi^e(T = 0, \kappa = 0)$.

In order to determine the resulting financial relationship that emerges from the game, we have to consider the return accruing to financial investors as being a function of the financial relation they choose. The returns for the financial investor depending on his financial relation choice and on the technology T used by the firm are supposed to be:

$$\Psi(T, \kappa = 1) = r \cdot \pi^e(T, \kappa = 1) - c^B$$
$$\Psi(T, \kappa = 0) = r \cdot \pi^e(T, \kappa = 0)$$

where c^B stands for the organization costs when forming or joining a bank and is – for the moment – considered to be exogenously given as the other pools will not interfere with the pay-offs of the playing pool.

In the following figure, the different stages of the game and the moments of bargaining have been summarized:

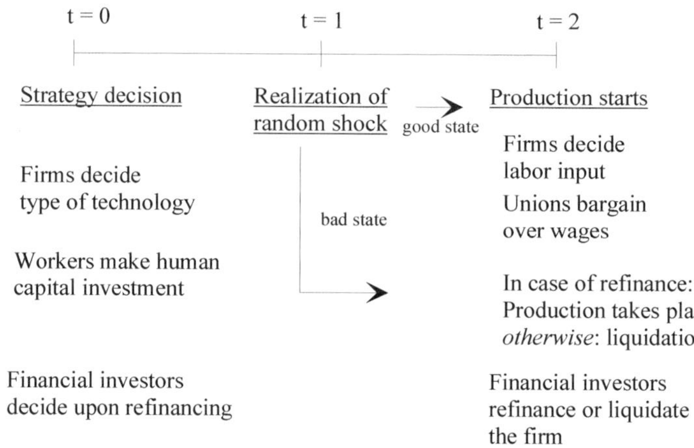

Figure 1: Timing of the game

3.4 The Nash Game

Having defined the various pay-offs we can now determine the equilibria of this three-players game. Firms have the choice over technologies, T, while workers choose the amount of human capital, h, they are ready to secure in the labor relation which in turn determines the time horizon of trade unions. Financial investors decide upon the degree, κ, to which they are ready to reschedule debt and to save failing firms from bankruptcy. Forming a bank comes at a cost as the restructuring of failing investment projects will necessitate time and money to proceed. Moreover, the information produced by the firm through the stock market is no longer available; the bank has to go through costly monitoring in order to obtain the necessary information.

In order to simplify the representation, we are only considering binary choices such that $T \in \{0, 1\}$, $h \in \{0, 1\}$ and $\kappa \in \{0, 1\}$ with the eight strategy combination denominated as (**S**, s, FS), (**S**, s, FL), (**L**, s, FS), (**L**, s, FL), (**S**, l, FS), (**S**, l, FL), (**L**, l, FS), (**L**, l, FL). In the following game, firms choose rows, trade unions choose columns while financial investors choose matrices (see Table 1, p. 378).

Given the structure of the game, the following proposition can be easily verified:

Proposition 3.1. *Given the above hypotheses concerning the technology choices and profit functions and*

(i) banking costs lie in the interval:

$$p(h)y(1) - w^b > \frac{r(r+\bar{s})c^B}{\delta\bar{s}} > p(h)y(0) - w^b - rV^L(0) \ \forall w^b, h \quad (5)$$

with $\bar{s} = s(0)$,

(ii) and human capital investment has the following characteristics:

$$\frac{\eta(s(\kappa)+\sigma r)}{(1-\sigma)(r+\theta-\overline{q}(\xi))} > y(0)(p(1)-p(0)) \quad \textit{and}$$

$$\frac{\eta\sigma r}{(r+\theta)(1-\sigma)} < y(1)(p(1)-p(0)) \tag{6}$$

*there exist two Nash equilibria in the above game: (**S**, s, FS) and (**L**, l, FL). The game is therefore a coordination game.*

Proof. See appendix. □

Remark 3.1. Given that $\overline{q}(\xi) \in (0, 1)$

$$\frac{s(\kappa)+\sigma r}{r+\theta-\overline{q}(\xi)} > \frac{\sigma r}{r+\theta}$$

this second condition is not trivial, the coordination game structure therefore only emerges for human capital to have a sensibly different impact on the productivity of the two technologies. Moreover, notice that the first part of (6) depends on the availability of outside labor (which in turn may be influenced by net immigration). Therefore, when $\exists\overline{\xi}$ such that

$$\frac{\eta(s(\kappa)+\sigma r)}{(1-\sigma)(r+\theta-\overline{q}(\xi))} < y(0)(p(1)-p(0))\forall\xi < \overline{\xi}$$

then only the long-term strategy will emerge.

Hence, the strategic complementarities that exist between the three decision variables T, κ and h (see (3)) create a coordination problem. In order to determine the equilibrium emerging in the long-run when the players face a problem of strategic uncertainty, the theory of global games can be used to find conditions under which either the long-run or the short-run equilibrium will be chosen. In general, low banking costs, high profitability of the long-run technology and low education costs will be beneficial for the (**L**, *l*, *FL*)-equilibrium to emerge. However, in this contribution we want to concentrate on the possibility of endogenous stochastic processes as they emerge out of a macroeconomic demand spillover.

4 Demand Shocks and Fluctuations

In SEGT the stochastic environment under which behavior selection takes place is usually considered to be exogenously given although it has been shown that the results obtained that way are crucially dependent on the underlying assumptions about the stochastic processes ([2]). This is justified by the fact that boundedly

Table 1: Strategic game between trade unions, firms and financial investors.

Financial Investors	*Short term (FS)*		*Long term (FL)*	
Firms, Trade Unions	Short term (s)	Long term (l)	Short term (s)	Long term (l)
Short term (S)	$\delta\pi^e(T=0,\kappa=0)$ $W_1(T=0,h=0)$ $\pi^e(T=0,\kappa=0)$	$\delta\pi^e(T=0,\kappa=0)$ $W_2(T=0,h=1)$ $\pi^e(T=0,\kappa=0)$	$\delta\pi^e(T=0,\kappa=1)-c^B$ $W_3(T=0,h=0)$ $\pi^e(T=0,\kappa=1)$	$\delta\pi^e(T=0,\kappa=1)-c^B$ $W_4(T=0,h=1)$ $\pi^e(T=0,\kappa=1)$
Long term (L)	$\delta\pi^e(T=1,\kappa=0)$ $W_1(T=1,h=0)$ $\pi^e(T=1,\kappa=0)$	$\delta\pi^e(T=1,\kappa=0)$ $W_2(T=1,h=1)$ $\pi^e(T=1,\kappa=0)$	$\delta\pi^e(T=1,\kappa=1)-c^B$ $W_3(T=1,h=0)$ $\pi^e(T=1,\kappa=1)$	$\delta\pi^e(T=1,\kappa=1)-c^B$ $W_4(T=1,h=1)$ $\pi^e(T=1,\kappa=1)$

rational agents select strategies randomly and hence the payoff matrix remains stable, independently of the state of the game. However, as soon as the pay-off matrix integrates random elements itself, this hypothesis cannot be kept any more – as can be easily seen from the stochastic structure of a global game.

In the set-up of the model thus far, only one production pool is supposed to be 'on' at each round, determining strategies and hence supply and demand on the micro level. In a world determined by division of labor, however, demand cannot be locally satisfied but has to flow between different pools of the play grid. Letting all pools play at all times but only one pool updating its contractual environment would create exactly this kind of interdependency. In this situation, however, the shock no longer depends only on the idiosyncratic element at the firm level but also depends on the realization of all the other shocks in the economy determining aggregate demands.

In order to introduce this new element we will use the stochastic variable as defined by (1). This variable is partly determined by the survival of all of the other production pools:

$$\varepsilon_{it} = \frac{1}{N}u_{it} + \eta_{it} = \frac{1}{N}u_{it} + \frac{1}{N}\sum_{j\neq i}\Pr(\varepsilon_{jt} \geq \underline{\varepsilon}_j)\cdot\varepsilon_{jt} \tag{7}$$

which is additive separable in shocks of pool i and shocks of all the other pools and shows cumulative density function $F_i(\varepsilon_{it})$. This will have an immediate impact on the realization of the expected output:

$$p_i(h_i) = \int_0^{\infty} \varepsilon_i \cdot dF(\varepsilon_i|h_i, T_i).$$

Here h_i and T_i only have an indirect effect on $F_i(\varepsilon_i)$ through their effect on the distribution of G. Their effect on the expected ouput is therefore dampened by the pool interdependency.

In a similar way, the liquidation probability for the nth pool is as before:

$$s_i = \int_0^{\underline{\varepsilon}_i} dF_i(\varepsilon_{it}).$$

Notice that again as in the preceding (idiosyncratic) case the minimum state of survival $\underline{\varepsilon}_i$ is a function of κ_i such that $s_i = s_i(\kappa_i)$, $s_i' < 0$. However, while $\underline{\varepsilon}_i$ is a function of κ_i alone, due to the interdependencies the distribution function, F_i is also a function of the financial relations of all the other pools:

$$\frac{\partial F_i}{\partial \kappa_i} > 0 \ \forall i \in \{1, .., N\}.$$

Therefore the survival probability depends also on the strategy equilibrium of the other pools:

$$s_i = s_i(\kappa_1, \ldots, \kappa_N), \ \frac{\partial s_i}{\partial \kappa_j} < 0 \ \forall j \in \{1, 2, .., N\}.$$

Moreover, from (7) it follows that increasing the number of pools playing $\kappa = 1$ also increases the survival probability of each individual pool; taking expectations we obtain:

$$E(\varepsilon_{it}) = \frac{1}{N} E(u_{it}) + \frac{1}{N} \sum_{j \neq i} \Pr(\varepsilon_{jt} \geq \underline{\varepsilon}_j) \cdot E(\varepsilon_{jt}).$$

Knowing that $\Pr(\varepsilon_{jt} \geq \underline{\varepsilon}_j | \kappa_j = 1, \kappa_{\neg j}) > \Pr(\varepsilon_{jt} \geq \underline{\varepsilon}_j | \kappa_j = 0, \kappa_{\neg j})$ and given a sufficiently high number of pools, N, the survival probability can be approximated by a continuous relationship.

$$k = |\boldsymbol{\kappa}| : s_i = s_i(k) \text{ with } \frac{\partial s_i}{\partial k} < 0 \text{ as } \frac{\partial E(\varepsilon_{it})}{\partial k} > 0 \ \forall k \in [0, 1]. \tag{8}$$

This is not the only way by which the characteristics of the macroeconomy influence the strategic decisions on the microeconomic level. Given the equilibrium strategies of all pools an information set $\Phi(\boldsymbol{\kappa})$ is accessible which contains the (stock market) prices of those firms that are public; this set is important in order to make interferences on the idiosyncratic shock. As long as ε_{it} contains private information which is only revealed during the play one important condition for informational efficiency is publicly (costless) observable prices. In a stationary world (which ours is) past (stock market) prices will then allow making inferences as to whether or not the firm is likely to suffer from liquidation in the next play. However, as banks learn about the private information through costly monitoring and restructuring of the firm, they are not likely to make this information public. Therefore it is plausible to assume that:

$$\Phi(\boldsymbol{\kappa}_1) \subset \Phi(\boldsymbol{\kappa}_2) \ \forall |\boldsymbol{\kappa}_1| < |\boldsymbol{\kappa}_2|.$$

However, due to the symmetry of firms, the smaller the information set the less interference one can draw about one particular firm and therefore the higher has to be the engagement of banks in monitoring their particular corporate client. The banking costs c^B can therefore no longer be thought of as exogenously given. Instead, they are likely to go up the fewer public firms there are:

$$c_i^B = c_i^B(\kappa_1, \ldots, \kappa_N), \ \frac{\partial c_i^B}{\partial \kappa_j} < 0 \ \forall j \in \{1, 2, .., N\}.$$

These results concerning linkages between pools can now be used to draw conclusions about new equilibrium behavior. In the following, two cases will be distinguished.

4.1 Opposition Between Unions and Stock Markets

Even without knowing the exact distribution of the summary variable ε_{it} we can draw conclusions on equilibria obtained in proposition (3.1). Here, we consider that the second condition of that proposition holds and hence that a coordination problem may exist. We therefore do not consider a change of the equilibrium structure due to different configurations of human capital distributions throughout the economy.

It is obvious that under the new specification of pool interdependency, satisfaction of the first condition of proposition (3.1) also depends on the equilibria obtained in the other pools; the stochastic term is not longer state-independent and the Nash character of the equilibrium may break down any time the equilibrium distribution over the whole range of production pools changes.

Proposition 4.1. *Suppose that condition (6) holds. Then:*

1. *Suppose that* $\delta\bar{s}(\boldsymbol{\kappa}')(p(1)y(1) - w^b) < r(r + \bar{s}(\boldsymbol{\kappa}'))c^B(\boldsymbol{\kappa}')$ *with* $|\boldsymbol{\kappa}'| \leq |\mathbf{1}|$ *and* $\kappa_i = 1$ *for production pool* i*. Then the long-term equilibrium (**L**, l, FL) is no longer Nash for all* $\boldsymbol{\kappa} \geq \boldsymbol{\kappa}'$.
2. *Suppose that* $(r + \bar{s}(\boldsymbol{\kappa}''))rc^B(\boldsymbol{\kappa}'') > \delta\bar{s}(\boldsymbol{\kappa}'')(p(h)y(0) - w^b - rV^L(0))$ *with* $|\boldsymbol{\kappa}''| \leq |\mathbf{1}|$ *and* $\kappa_i = 0$ *for production pool* i*. Then the short-term equilibrium (**S**, s, FS) is no longer Nash for all* $\boldsymbol{\kappa} \leq \boldsymbol{\kappa}''$.
3. *Suppose that* $\boldsymbol{\kappa}'' > \boldsymbol{\kappa}'$ *and that* $|\boldsymbol{\kappa}'| < W_L/(W_S + W_L) < |\boldsymbol{\kappa}''|$ *where* W_S, W_L *are measures for the workers' pay-off in the short-run and long-run equilibrium respectively. Then an interior equilibrium exists which is a saddle point.*

Proof. See appendix. □

The interior equilibrium that is hence obtained in proposition (4.1) is unstable and gives rise to a new situation where, depending on the initial conditions, the economy will end up in a situation where one of the players chooses the long-run strategy, while the others opt for the short-run strategy, following the fact that both (**L**, *l*, *FL*) and (**S**, *s*, *FS*) are repelling. Simulation 2 represents a number of trajectories that lead to either of the opposition equilibria (see Table 2, p. 389).

More important, though, than the question of whether this interior equilibrium exists is to ask what gives rise to its existence. As we have argued in the beginning of this section, the important fact of pool interdependency and aggregate uncertainty has rarely been treated in the literature on evolutionary games. This is largely due to the fact that the source of the uncertainty has been attached to the individual non-rationality. However, in games with random payoff matrices it does not make sense to treat uncertainty as idiosyncratic and exogenous to the individual decision process. Banking activity will provide (intertemporal) insurance for the production process, and in that way set incentives for trade unions and firms to opt for the long-term strategy. However, the banking process is costly in terms of information provision and refinancing. Given a sufficiently smooth economic development, workers will have an incentive to free-ride on the insurance provided by the rest of the pools in the economy and opt for short-run gains by reducing their human capital investment and bargaining for higher wages.

Conversely it is true that an economy largely dominated by stock market finance will have more efficient information processing. Private information will become public through stock market prices giving the right investment information for financial investors. Consequently, when a sufficient number of workers guarantee through their human capital investment a smooth functioning of the economy, the individual investor will weigh banking costs of reduced information processing capacities higher than increased stabilizing of the economy. The investor will therefore opt for a more arm's length financial relation with the firm, guaranteeing a higher pay-off from the investment.

4.2 Union-Led Institutional Cycle

Besides the effect of pool interdependency on the best reply strategy set of financial investors, workers may face a similar spillover. In the following, we want to analyze the consequences when the interdependency effects affect the best reply sets of one of the other player populations, i.e. for instance when the composition of the equilibrium set of financial investors affect best reply strategies of workers.

Going back to condition (6) one sees that the crucial factor that this condition holds is that the long-run strategy better insures the worker facing a long-run strategy by firms. However, in this new set-up an argument similar to the one before prevails: the union that decides to switch to a short-term strategy can free-ride on the stabilized economy due to high banking activity. On the other hand, in an equilibrium situation where (almost) all pools are coordinated on short-run strategies, unions may be willing to not use their bargaining power and to get at least parts of the benefits of an insurance strategy. Defining $\Theta \equiv [(1-\sigma)(r+\theta)/\eta](p(1)-p(0))$ and $\gamma \equiv \sigma(r-\overline{q}(\xi))+\overline{q}(\xi)$ we can prove the following proposition.

Proposition 4.2. *Suppose that there exist* $\boldsymbol{\kappa}', \boldsymbol{\kappa}'' \in [\mathbf{0}, \mathbf{1}]$ *such that*

$$\begin{aligned} W(T=1,\boldsymbol{\kappa},h=0) &> W(T=1,\boldsymbol{\kappa},h=1) \\ &\Leftrightarrow s(\boldsymbol{\kappa}) > \Theta y(1) - \sigma r \forall |\boldsymbol{\kappa}| \geq |\boldsymbol{\kappa}'| \quad \textit{and} \\ W(T=0,\boldsymbol{\kappa},h=0) &< W(T=0,\boldsymbol{\kappa},h=1) \\ &\Leftrightarrow s(\boldsymbol{\kappa}) < \Theta y(0) - \gamma \forall |\boldsymbol{\kappa}| \leq |\boldsymbol{\kappa}''|, \end{aligned}$$

then one equilibrium exists in the unit square $[0, 1]^2$. *It is a center.*

Proof. See appendix. □

The public good character of a stabilized economy therefore gives rise to a cyclical behavior of strategy choice by banks and trade unions. The costly long-run strategy does not pay off enough with firm's high survival rates such that it is better for unions to fully use their bargaining power for increased wages. As more and more unions follow this strategy switch, financial investors do not get enough to justify their banking investment and try to raise their financial income by directly investing in firms. However, this destabilizes the economic output path and volatility raises again, pushing unions to reconsider their short-term strategic position in order to switch again. It is therefore justified to speak of an institutional-economic cycle as the characteristics of both the institutional setting and the economic performance vary over the course of the game play.

Finally, getting back to the earlier remark (3.1), it can be shown that the existence of this oscillating behavior depends on the availability of outside labor as stated by the following corollary.

Corollary 4.1. *Provided that* $\{\xi | s(\boldsymbol{\kappa}) > \Theta y(0) - \sigma r - \overline{q}(\xi)(1-\sigma)\} \neq \emptyset \forall \boldsymbol{\kappa}$ *holds, then with increasing availability of outside labor a saddle point emerges and the (**S**, s, FS)-position attracts all flows.*

Proof. See appendix. □

Hence, with increasing immigration (or other forms of outside labor such as a generally higher level of unemployment), the short-term strategy (**S**, *s*, *FS*) will become an attractor and the interior equilibrium will become saddle-point unstable. With increasing availability of labor the amplitude of the cycle dampens and then breaks down when the critical point is reached as can be seen from the simulation 3 (see Table 3, p. 389).

5 Conclusion

The objective of the preceding article has been to show how in a three-player situation with long-term and short-term strategies a strategic complementarity can bring about a coordination game situation with multiple equilibria. In this game, each player – by investing in a relation-specific asset – increases the value of the relation and therefore the marginal incentives for the two remaining players to adopt a similar strategy. With sufficiently high human capital or banking costs, however, the long-term strategy is only profitable if strategy coordination is prevalent.

Moreover, we demonstrated that the resulting coordination equilibria are likely to be unstable – and hence no longer Nash – once one takes the macroeconomic externality into account that is created by the stabilizing effect of the long-term strategy. Abandoning the assumption of local autarky of pools playing, demand is aggregated on the macroeconomic level creating a sectoral correlation of otherwise idiosyncratic shocks. When the resulting collective risk is taken into account – which is justified from the standpoint of these microeconomic considerations – an analysis of the replicator dynamics shows that saddle-point or cyclical equilibria can arise where financial investors and trade unions switch in a cyclical manner between their long- and short-term strategies.

We have argued that this result qualifies in an important way the recent developments of stochastic evolutionary game theory, where shocks are considered to be idiosyncratic to individual players. These games – as well as global games which are isomorphic – select the risk-dominant equilibrium. However, as soon as the underlying stochastic process has to be considered endogenous to the economy – as in our case – one can no longer assume state independency of random shocks, a necessary condition for SGET to yield the risk-dominance result. Under conditions exposed in the last proposition, a Nash equilibrium need not exist even in mixed strategies.

Interesting extensions of the suggested model would be to calibrate a complete macroeconomic model on data concerning the long-run economic and institutional evolution of particular countries and to see whether or not this would reproduce characteristics of their long-run business cycle behavior. One other question which has not been addressed here is the impact of the union's strategic choice on the growth rate. An increasing claim on the national output may reduce incentives to innovative activity and hence reduce growth. A by-product would be a negative relationship between growth and economic volatility, a well-known stylized fact.

6 Appendix

6.1 Proof of Proposition 3.1

In order for a coordination game between (**S**, s, FS) and (**L**, l, FL) to exist we must have for financial investors (see game 1):

$$\Psi_1 > \Psi_2 \text{ and } \Psi_3 < \Psi_4 \tag{9}$$

with

$$\Psi_1(T=0,\kappa=0,h) = \delta\pi^e(T=0,\kappa=0)$$

$$= \delta\frac{p(h)y(0)+\bar{s}V^L(T)-w^b}{r+\bar{s}} \tag{10}$$

$$\Psi_2(T=0,\kappa=1,h) = \delta\pi^e(T=0,\kappa=1) - c^B$$

$$= \delta\frac{p(h)y(0)-w^b}{r} - c^B \tag{11}$$

$$\Psi_3(T=1,\kappa=0,h) = \delta\pi^e(T=1,\kappa=0) = \delta\frac{p(h)y(1)-w^b}{r+\bar{s}} \tag{12}$$

$$\Psi_4(T=1,\kappa=1,h) = \delta\pi^e(T=1,\kappa=1) - c^B$$

$$= \delta\frac{p(h)y(1)-w^b}{r} - c^B \tag{13}$$

Substituting (10)–(13) into (9) leads to condition (5).

Given that (5) holds, the conditions for trade unions boil down to the following two inequalities:

$$W_1 > W_2 \text{ and } W_3 < W_4 \tag{14}$$

where $W_1, \ldots, W_4$ represent job values under different strategic choices with

$$W_1(T = 0, \kappa = 0, h = 0)$$
$$= \frac{(1-\sigma)(p(0)y(0) + \overline{s}\overline{V})(r + \theta - \overline{q}) + (r\sigma + \overline{s})R}{r[r + \overline{s} + (\theta - \overline{q})(1 - \sigma)]} \tag{15}$$

$$W_2(T = 0, \kappa = 0, h = 1)$$
$$= \frac{(1-\sigma)(p(1)y(0) + \overline{s}\overline{V})(r + \theta - \overline{q}) + (r\sigma + \overline{s})(R - \eta)}{r[r + \overline{s} + (\theta - \overline{q})(1 - \sigma)]} \tag{16}$$

$$W_3(T = 1, \kappa = 1, h = 0)$$
$$= \frac{(1-\sigma)p(0)y(1)(r + \theta) + \sigma Rr}{r[r + (1 - \sigma)\theta]} \tag{17}$$

$$W_4(T = 1, \kappa = 1, h = 1)$$
$$= \frac{(1-\sigma)p(1)y(1)(r + \theta) + \sigma(R - \eta)r}{r[r + (1 - \sigma)\theta]} \tag{18}$$

where $\overline{s} = s(0)$, $\overline{q} = q(0)$ and $\overline{V} = V^L(0)$. Substituting (15)–(18) into (14) leads to condition (6). □

6.2 Proof of Propositions 4.1 and 4.2

In order to reduce the complexity of the dynamic model we want to concentrate on the shifts in financial investors' and trade unions' strategy. Firms are supposed to always play the best reply strategy and therefore the kind of technology they adopt on the individual level is not subject to dynamic lags. However, as trade unions and financial investors base their strategic decision on the whole population as in a 'playing the field' model ([10], pp 72–73) they use a statistical model to determine their best reply. Moreover, we will make extensive use of the continuity of the existing probability and the shock with respect to the distribution of strategies (see (8); this is obviously only justified in a large population (large N).

Let us first define the following two variables:

$k : \boldsymbol{\kappa} \mapsto [0, 1]$	percentage of bank-based financed firms
ω	percentage of unions following a wage maximizing strategy.

In the following dynamic analysis we suppose that the growth rate of the number of players using the same pure strategy depends positively on the excess payoff over the average payoff in its player population. Using Taylor's formulation of the n-population replicator dynamics this is written as (see [9]):

$$\dot{x}_{ih} = [u_i(e_i^l, x_{-i}) - u_i(x)]x_{il}$$

with $i \in \{1, .., n\}$, pure strategy $l \in S_i$, S_i: strategy set for player population i and population state x. For the two-player game and with payoff matrix A_i having zeros off the diagonal (like in a coordination game) this can be rewritten as:

$$\dot{x}_{il} = [a_{1i}x_j - a_{2i}(1 - x_j)]x_i(1 - x_i),\ j \neq i.$$

With this simplified form of the replicator dynamics we are then able to prove the propositions in this section.

Opposition Between Unions and Stock Markets

Given the assumption about the firm's behavior the game resembles a coordination game between financial investors and workers. As the workers payoff matrix is not supposed to depend on the stochastic distribution (given the strategy choices of the two other players, the workers will have the same best reply strategy independently of $\boldsymbol{\kappa}$) we can represent their payoff matrix as:

$$W = \begin{pmatrix} W_S & 0 \\ 0 & W_L \end{pmatrix}$$

where $W_S \equiv W_1 - W_2$, $W_L \equiv W_4 - W_3$. Columns represent worker's strategy and rows the financial investor's strategies (which corresponds to the firm's technology choice given the assumption about its best replies). The payoff matrix for financial investors is a little different:

$$F = \begin{pmatrix} F_S \cdot (k - k'') & 0 \\ 0 & F_L \cdot (k' - k) \end{pmatrix}$$

where $k' = |\boldsymbol{\kappa}'|$ and $k'' = |\boldsymbol{\kappa}''|$ and $F_S \equiv \Psi_1 - \Psi_2$, $F_L \equiv \Psi_4 - \Psi_3$. Given the continuity of $s(\kappa)$ with respect to κ (see (8)), the two conditions of proposition 4.1 can be translated by giving financial investors a negative return at the short-term equilibrium for $k < k''$ and a negative return for the long-run equilibrium for $k > k'$. For our argument, only the relative payoff evolution is relevant here.

The resulting replicator dynamics is then written as:

$$\dot{k} = [F_L \cdot (k' - k)(1 - \omega) - F_S \cdot (k - k'')\omega] \cdot k(1 - k)$$

$$\dot{\omega} = [W_S \cdot k - W_L \cdot (1 - k)] \cdot \omega(1 - \omega)$$

with isoclines $\dot{k} = 0 \Rightarrow \widetilde{\omega} = [F_L \cdot (k' - k)]/[F_L \cdot (k' - k) + F_S \cdot (k - k'')]$ and $\dot{\omega} = 0 \Rightarrow \widetilde{k} = W_L/[W_S + W_L]$. For $k' < \widetilde{k} < k''$ the two lines cut and an interior stationary point $(\widetilde{k}, \widetilde{\omega})$ exists. The Jacobian for this point is

$$\boldsymbol{J} = \begin{pmatrix} -(F_S \cdot \widetilde{\omega} + F_L(1 - \widetilde{\omega}))\widetilde{k}(1 - \widetilde{k}) & [F_S \cdot (k'' - \widetilde{k}) - F_L \cdot (k' - \widetilde{k})] \cdot \widetilde{k}(1 - \widetilde{k}) \\ (W_S + W_L) \cdot \widetilde{\omega}(1 - \widetilde{\omega}) & 0 \end{pmatrix}$$

with trace $TR = J_{11} < 0$ and determinant $Det = -J_{21}J_{12}$. Moreover we have $J_{21} > 0$, $J_{12} > 0$ for $F_L/F_S > (1 - k'')/(1 - k')$. Provided that $k'' > k'$ and – by definition – $F_L > F_S$, this last condition will be satisfied and hence the interior equilibrium will be saddle-point unstable. □

Union-Led Cycles

Under repercussions of financial investors' strategies on workers' best replies the workers' payoff matrix has to be modified. We suggest analyzing the simplest case where investors' best replies do not change due to composition effects; their payoff matrix remains unchanged. When the conditions of the proposition hold, the workers' payoff matrix therefore resembles:

$$W = \begin{pmatrix} W_S(k - k'') & 0 \\ 0 & W_L(k' - k) \end{pmatrix}$$

where $k' = |\kappa'|$ and $k'' = |\kappa''|$ and W_S and W_L are defined as above. Using a similar argument as above, given the continuity of $s(\kappa)$ with respect to κ (see (8)), the two conditions of proposition 4.2 can be translated by giving workers a negative payoff at the short-term equilibrium for $k < k''$ and a negative payoff for the long-run equilibrium for $k > k'$. Again, for our argument, only the relative payoff evolution is relevant here.

The payoff matrix for financial investors is now standard again:

$$F = \begin{pmatrix} F_S & 0 \\ 0 & F_L \end{pmatrix}$$

with F_S and F_L as defined earlier.

The resulting replicator dynamic system is then writen as:

$$\dot{k} = [F_L \cdot (1-\omega) - F_S\omega] \cdot k(1-k)$$
$$\dot{\omega} = [W_S(k-k'')k + W_L(k-k')(1-k)] \cdot \omega(1-\omega)$$

with isoclines

$$\dot{k} = 0 \Rightarrow \widetilde{\omega} = \frac{F_L}{F_S + F_L} \text{ and}$$

$$\dot{\omega} = 0 \Rightarrow \widetilde{k}_{1/2} = \frac{1}{2(W_L - W_s)}$$
$$\left[W_L(1+k') - W_S k'' \pm \sqrt{(W_L(1+k') - W_S k'')^2 - 4(W_L - W_S)W_L k'}\right].$$

As can be easily checked, the upper cutting point $(\widetilde{k}_2, \widetilde{\omega})$ lies outside the unit square for all $W_L > W_S$ and $k'' < k'$. The Jacobian for the remaining equilibrium inside the play grid of the new system now is:

$$\boldsymbol{J}_1 = \begin{pmatrix} 0 & [F_S + F_L] \cdot \widetilde{k}_1(1-\widetilde{k}_1) \\ 2(W_L - W_s)\widetilde{k}_1 - W_L(1+k') + W_S k'' & 0 \end{pmatrix}$$

with trace $TR = 0$ and determinant $Det = -J_{21}J_{12}$. As $J_{12} > 0$ and $J_{21} < 0$, the determinant will be positive leading to two complex eigenvalues. Given that $TR = 0$, the equilibrium will be a center.

With increasing net immigration, ξ, the short-term strategy may become dominant for firms and financial investors, whatever the strategic choice of unions. Using W_1 and W_2 this means $W_1(\overline{\xi}) > W_2(\overline{\xi})\forall\boldsymbol{\kappa}$. Moreover, one can easily show that $W_S(\xi) \equiv W_1(\xi) - W_2(\xi)$ is increasing with ξ. Therefore when $\exists\overline{\xi}$ such that $q(\overline{\xi}) < 1 \wedge W_1(\overline{\xi}) > W_2(\overline{\xi})\forall\boldsymbol{\kappa}$, the above worker payoff matrix changes and rather looks like

$$W = \begin{pmatrix} W_S & 0 \\ 0 & W_L(k'-k) \end{pmatrix}$$

where no $k'' \in [0, 1]$ exists such that $W_1 < W_2$. Carrying out similar calculations as before with this modified payoff matrix, one concludes that the Jacobian for the equilibrium inside the play grid has now a negative determinant, making the equilibrium a saddle point. Moreover, given that the high equilibrium is still unstable for $k > k'$, only the short-run equilibrium will be stable. □

6.3 Simulations

The simulations are presented in Tables 2 and 3.

Table 2: Banks' and Unions' codynamics

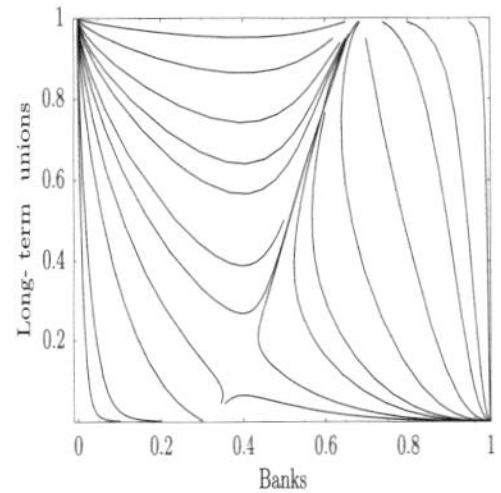

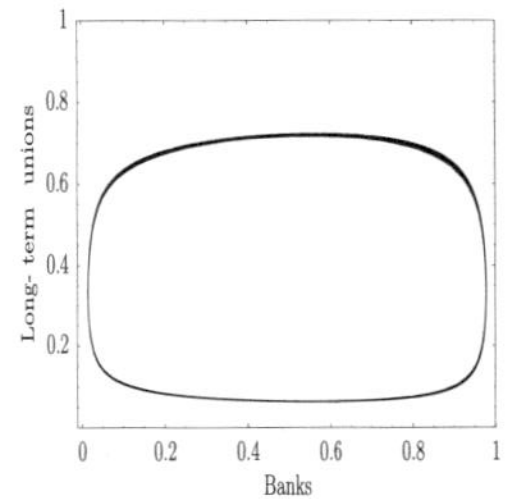

The above simulations have been run using the following system of differential equations:

$$\dot{k} = [F_L \cdot w(t)(k(t) - k'') - F_S(k' - k(t)) \cdot (1 - w(t)) \cdot k(t)(1 - k(t))]$$

$$\dot{w} = [W_L \cdot k(t) - W_S \cdot (1 - k(t))] \cdot w(t)(1 - w(t))$$

where $W_S = 2$, $W_L = 3$, $F_S = 2$, $F_L = 4$, $k' = 0.3$, $k'' = 0.7$.

The above simulation has been run using the following system of differential equations:

$$\dot{k} = [F_L \cdot w(t) - F_S(1 - w(t))] \cdot k(t)(1 - k(t))$$

$$\dot{w} = [W_L \cdot k(t) - W_S \cdot (1 - k(t))] \cdot w(t)(1 - w(t))$$

where $W_S = 2$, $W_L = 3$, $F_S = 2$, $F_L = 4$.

Table 3: Increasing availability of outside labor

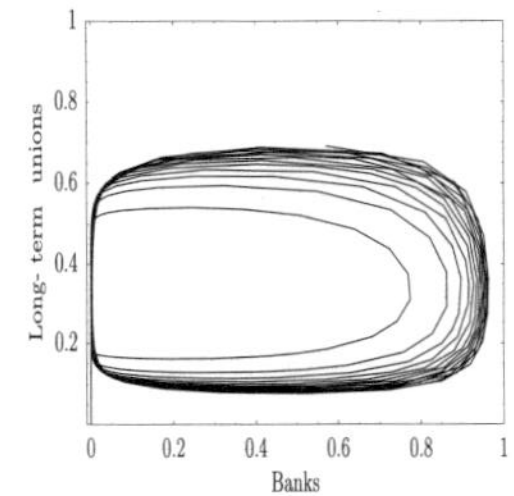

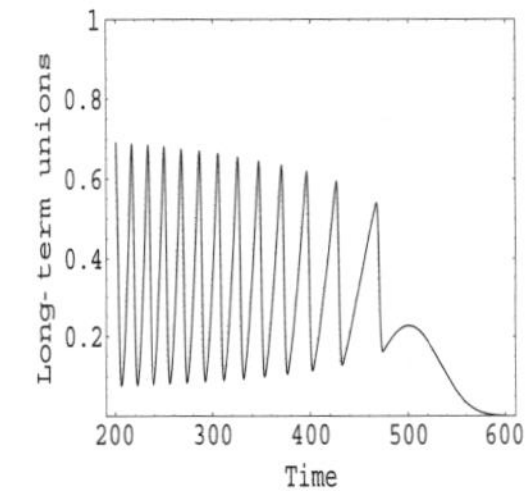

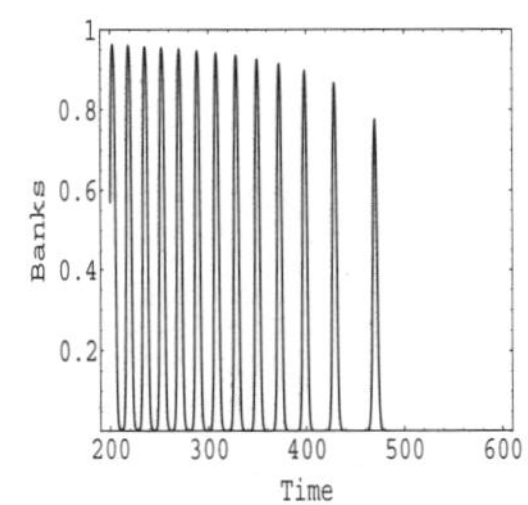

The simulation on this page has been run using the following system of differential equations:

$$\dot{k} = [F_L \cdot w(t) - F_S(1 - w(t))] \cdot k(t)(1 - k(t))$$

$$\dot{w} = [W_L \cdot k(t)(k'' - k(t)) - W_S \cdot (k(t) - k'(1 - 0.002 \cdot t)) (1 - k(t))] \cdot w(t)(1 - w(t))$$

where $W_S = 2$, $W_L = 3$, $F_S = 2$, $F_L = 4$, $k' = 0.3$, $k'' = 0.7$.

REFERENCES

[1] Altvater E., Hoffmann J. and Semmler W., *Vom Wirtschaftswunder zur Wirtschaftskrise* (Olle & Wolter, Berlin, 1982).

[2] Bergin J. and Lipman B. L., Evolution with state-dependent mutations, *Econometrica* **64/4** (1996) 943–956.

[3] Carlsson H. and van Damme E., Global games and equilibrium selection, *Econometrica* **61** (1993) 989–1018.

[4] Dewatripont M. and Maskin E., Credit and efficiency in centralized and decentralized economies, *Review of Economic Studies* **62** (1995) 541–555.

[5] Foster D. P. and Young H. P., Stochastic evolutionary game dynamics, *Theoretical Population Biology* **38** (1990) 219–232.

[6] Friedman D., Evolutionary games in economics, *Econometrica* **59/3** (1991) 637–666.

[7] Greenbaum S. I. and Thakor A. V., *Contemporary Financial Intermediation*, (Dryden Press, Fort Worth, 1995).

[8] Osano H., An evolutionary model of corporate governance and employment contracts, *Journal of the Japanese and International Economics* **11** (1997) 403–436.

[9] Taylor P., Evolutionary stable strategies with two types of players, *Journal of Applied Probability* **16** (1979) 76–83.

[10] Young H. P., *Individual Strategy and Social Structure* (Princeton University Press, Princeton, 1998).

Robust Control Approach to Option Pricing, Including Transaction Costs

Pierre Bernhard
University of Nice (UNSA-CNRS)
06903 Sophia Antipolis, France
Pierre.Bernhard@essi.fr

Abstract

We adopt the robust control, or game theoretic, approach of [5] to option pricing. In this approach, uncertainty is described by a restricted *set* of possible price trajectories, without endowing this set with any probability measure. We seek a hedge against every possible price trajectory.

In the absence of transaction costs, the continuous trading theory leads to a very simple differential game, but to an uninteresting financial result, as the hedging strategy obtained lacks robustness to the unmodeled transaction costs. (A feature avoided by the classical Black and Scholes theory through the use of unbounded variation cost trajectories. See [5].)

We therefore introduce transaction costs into the model. We examine first the continuous time model. Its mathematical complexity makes it beyond a complete solution at this time, but the partial results obtained do point to a robust strategy, and as a matter of fact justify the second part of the paper.

In that second part, we examine the discrete time theory, deemed closer to a realistic trading strategy. We introduce transaction costs into the model from the outset and derive a pricing equation, which can be seen as a discretization of the quasi variational inequality of the continuous time theory. The discrete time theory is well suited to a numerical solution. We give some numerical results. In the particular case where the transaction costs are null, we recover our theory of [5], and in particular the Cox, Ross and Rubinstein formula when the contingent claim is a convex function of the terminal price of the underlying security.

1 Introduction

We consider the classical problem of pricing a contingent claim based upon an underlying stock of curent price $S(t)$, and defined by its terminal value, or *payoff*, $M(S(T))$ at *exercise time* T. In the case of a (European) call, we have $M(s) = [s - K]_+ = \max\{s - K, 0\}$ for a given *striking price* K.

This problem is classically solved by Black and Scholes'theory [6] in the continuous trading framework, and approached by the theory of Cox, Ross and Rubinstein [8] in the limiting vanishing step size case for the discrete trading, discrete time theory. The fundamental device of these theories, due to Merton, is to construct a *replicating* portfolio, made up of the underlying stock and riskless bonds, and a self-financed trading strategy, or *hedging strategy*, that together yield the same payoff as the contingent claim to be priced. However, the classical theory of Black and Scholes, based upon the "geometric diffusion" market model, is known to have the major weakness that transaction costs cannot be taken into account in any meaningful way. See [15].

In [5], we proposed a robust control approach to that same idea, that we quickly review hereafter. The distinctive feature of our theory is in our market model. We forgo any stochastic description of the underlying stock price. Instead, we assume that we know hard bounds on the possible (relative) variation rate of the stock price. And we seek to manage our portfolio through self-financed trading in such a way as to do at least as well as the option, in terms of final value, on all possible price histories, leading to a minimax control problem.

A very similar approach has been taken independantly and simultaneously with our research by J.-P. Aubin, D. Pujal and coworkers, see [13,2], using their tools of viability theory. A game theoretic approach is also used by [12] in connection with transaction costs, but to investigate a different problem of optimizing these costs from the viewpoint of the banker. Essentially the same market model as ours has been proposed in [14], where they give it the name we shall use of "interval model".

In the absence of transaction costs, the continuous trading theory leads to a simple differential game. However, the solution of that game yields the so-called "parity value" for the option, something rather far from observed prices on the market. Correlatively, the bang-bang hedging strategy obtained, that we call the "naive strategy", lacks robustness to the unmodeled transaction costs, in particular if the underlying stock price fluctuates close to the money resulting in a perpetual dillema for the trader. (In [5] we argued that its inherent robustness is the main reason to prefer the Black and Scholes strategy and option price. This is done at the price of adopting, for the underlying stock price, trajectories of unbounded variation and known quadratic relative variation—the volatility. Whether this is realistic in a world where stock prices are updated at discrete time instants is a matter of debate.)

We propose then to include transaction costs into our differential game model. This leads to a three-dimensional impulse control game that has up to now resisted our attempts to solve it via classical means.[1] It displays, however, at least one feature of robustness against fluctuating stock prices: the fact that no trading should occur during a final time interval, after a final jump in portfolio composition. This gives a strong hint as how the solution can be approached by a discrete time theory, with a step-size function of the transaction costs and of the maximum relative variation rate hypothesized for the underlying stock.

Therefore, in a second part, we investigate the discrete time theory. The theory we obtain is well suited to a numerical solution, particularily so in the convex

[1] At the time of the revision of this paper we are close to a solution in terms of characteristics. It displays an interesting new type of singularity.

case where we are able to show that it preserves the convexity of the option price with respect to the current underlying stock's price, yielding a simplification into the computation. Also, not surprisingly, the discrete time pricing equation can be seen as a finite differences approximation of the continuous time equation. But the continuous time theory is far from developed to the point where a rigorous convergence proof would be feasible.

In the case where the transaction costs vanish, we recover our theory of [5]. Hence, if furthermore the contingent claim's payoff, is a convex function of the stock price (e.g. for a simple European call), it is strongly reminiscent of the theory of Cox, Ross and Rubinstein [8], to which it gives a normative value even for a non-vanishing step size. Otherwise, it is shown to give a higher equilibrium price to the option than the previous theory. (If one identifies our $(1-\alpha)$ and $(1+\beta)$ with their d and u.)

2 Continuous Time Theory

2.1 The Models

2.1.1 Market Model

In that market, we have a riskless security, called *bonds*, evolving at a known constant rate, which in fact sets the lending and borrowing rate on that market. Let this rate be denoted ρ. The exercise time of the option considered is T. And let

$$R(t) = \mathrm{e}^{\rho(t-T)}$$

be either the value of a bond, or the end-time factor in our market.

We denote $S(t)$ the underlying stock price at time t. We let:

Definition 2.1. The set Ω of *admissible* price histories is defined by two positive numbers $\tilde{\alpha}$ and $\tilde{\beta}$, and is the set of all absolutely continuous time functions $t \mapsto S(t)$ such that at every instant where it is differentiable, it satisfies the inequalities

$$-\tilde{\alpha} \le \frac{\dot{S}}{S} \le \tilde{\beta} \tag{1}$$

or, equivalently that between any two instants of time $t_1 < t_2$,

$$\mathrm{e}^{-\tilde{\alpha}(t_2-t_1)} S(t_1) \le S(t_2) \le \mathrm{e}^{\tilde{\beta}(t_2-t_1)} S(t_1).$$

We choose to represent that hypothesis in a system theoretic fashion:

$$\dot{S} = \tilde{\tau} S, \tag{2}$$

where the time function $t \mapsto \tilde{\tau}(t) \in [-\tilde{\alpha}, \tilde{\beta}]$ is assumed to be measurable, and represents an a priori unknown disturbance.

We shall use the notations $\alpha = \tilde{\alpha} + \rho$ and $\beta = \tilde{\beta} - \rho$. The positive numbers ρ, α and β describe the market model and are assumed known.

Although this presentation is meant to emphasize the fact that there are no probabilities involved, and that our "disturbance" $\tilde{\tau}$ is just a mathematical device, a renaming of the quantity that we have assumed to be bounded, it may be useful to relate this form with one more reminiscent of the classical geometric diffusion model. Let $\mu = (\tilde{\beta} - \tilde{\alpha})/2$ and $\sigma = (\tilde{\beta} + \tilde{\alpha})/2$. Let also $\nu = (\tilde{\tau} - \mu)/\sigma$. Now, all stock price trajectories can be represented by the system

$$\dot{S} = (\mu + \sigma\nu)S \tag{3}$$

where ν is any measurable time function $t \mapsto \nu(t)$ satisfying $|\nu(t)| \leq 1,\ \forall t$. In a sense, this is a "normalized" disturbance, and therefore σ is a measure of the volatility of the stock considered.

2.1.2 Portfolio Model

We form a portfolio made up of x shares of the underlying stock, and y riskless bonds. The value or *worth* $\tilde{w}$ of this portfolio at any time instant is thus

$$\tilde{w}(t) = x(t)S(t) + y(t)R(t).$$

We aim to precisely define what is a self-financed hedging strategy.

Let us investigate how a self-financed trading strategy behaves in the presence of transaction costs, since this is our objective. Assume these costs are proportional to the amount traded, not necessarily with the same proportionality ratio for the two comodities considered. Let us call c_0 the trading cost ratio for the riskless bond, and c_1 for the underlying stock. Each transaction should finance those costs. Therefore, let $\mathrm{d}x$ be the variation in x at a stock price of S, and $\mathrm{d}y$ the variation in y at a bond value R, we should have

$$\mathrm{d}x S + c_1|\mathrm{d}x|S + \mathrm{d}y R + c_0|\mathrm{d}y|R = 0. \tag{4}$$

We therefore let:

Definition 2.2.

a. A *dynamic portfolio* (or simply *portfolio*) is a pair of bounded variation time functions $(x(\cdot), y(\cdot))$ defined over $[0, T]$.
b. A dynamic portfolio is said to be *self financed* if it satisfies (4) (in the sense of Stieltjes calculus).

The costs c_0 and c_1 are assumed small, may be of the order of a few percent. Introduce

$$\varepsilon = \mathrm{sign}(\mathrm{d}x), \qquad C_\varepsilon := \varepsilon\frac{c_0 + c_1}{1 - \varepsilon c_0} \tag{5}$$

Proposition 2.1. *A self-financed dynamic portfolio is entirely defined by its intial composition $(x(0), y(0))$ and the bounded variation time function $x(\cdot)$. The time function $y(\cdot)$ and its worth $\tilde{w}(\cdot)$ can be reconstructed through integration of the differential relations*

$$dy = -\frac{(1+\varepsilon c_1)}{(1-\varepsilon c_0)}\frac{S}{R}dx \tag{6}$$

and

$$d\tilde{w} = \rho\tilde{w}dt + (\tilde{\tau} - \rho)xSdt - C_\varepsilon Sdx. \tag{7}$$

Proof. Because c_0 and c_1 are smaller than one, it easily follows that $\mathrm{d}x$ and $\mathrm{d}y$ should have opposite signs. Recall that $\varepsilon = \mathrm{sign}(\mathrm{d}x)$. It therefore comes

$$(1+\varepsilon c_1)\mathrm{d}xS + (1-\varepsilon c_0)\mathrm{d}yR = 0,$$

hence (6). Further more, the classical fact that $y\mathrm{d}R = \rho yR\mathrm{d}t = \rho(\tilde{w} - xS)\mathrm{d}t$, yields (7). □

Notice that in (7), the last term is always negative, and represents the loss in portfolio value due to the trading costs. Of course, the case without transactions costs can be recovered by letting $c_0 = c_1 = C_\varepsilon = 0$ in the above theory.

We shall let, for short, $C_{+1} = C^+$ and $C_{-1} = C^-$ (a negative number). It is worthwile to examine two extreme cases:

The case $c_0 = 0$. If the riskless bond is, say money, and trading in that commodity is free, then we simply have $C_\varepsilon = \varepsilon c$, the transaction cost on the stock.

The case $c_0 = c_1 = c$. If both transaction costs are equal, an interesting feature that shows up is that

$$1 + C^+ = \frac{1+c}{1-c} = \frac{1}{1+C^-}.$$

More generally, we may notice the folowing fact:

Proposition 2.2. *Whenever $c_0 \le c_1$, one has*

$$1 < 1 + C^+ \le \frac{1}{1+C^-}$$

2.2 Hedging Strategies

2.2.1 Variations of x

We would like to let $x(\cdot)$ be our control. But there are costs associated with its *variations*. Hence, in a classical system theoretic fashion, we are led to consider its derivative as the control. However, another difficulty shows up, since we want also to allow discontinuities in x. We would therefore need a theory of impulse control in differential games. It is worthwhile to write the Isaacs quasi-variational inequality that this formally leads to. But the theory of this inequation is not available at this time.

We shall in some respect get around that difficulty with the following approximating device. We shall let the monetary flux be our control:

$$\dot{x}S = \tilde{\xi}, \qquad \text{or} \qquad S\mathrm{d}x = \tilde{\xi}\mathrm{d}t, \qquad \tilde{\xi} \in [-\widetilde{X}, +\widetilde{X}], \tag{8}$$

where we shall take $\widetilde{X}$ to be a (very) large positive number, and investigate the limit of the solution found as $\widetilde{X} \to \infty$. Whether this limit is the solution of an impulsive problem is a rather technical question, the more so here that we shall deal with a differential game, not a mere control problem. The tools introduced in a recent article [1] seem appropriate to attempt an extension to game problems. We shall not be concerned with that problem here.

2.2.2 Trading Strategies

We shall let our control $\tilde{\xi}$ be a function of time, $S(t)$, and $x(t)$ if necessary. We need to chose this function in such a way that the induced differential equations have a solution. We therefore let

Definition 2.3. An admissible *trading strategy* is a function $\tilde{\varphi} : [0, T] \times \mathbb{R}^+ \times \mathbb{R} \to \mathbb{R}$ such that the differential equation (8) with $\tilde{\xi} = \tilde{\varphi}(t, S(t), x(t))$ has a unique solution $x(\cdot)$ for every $x(0)$ and admissible time function $S(\cdot) \in \Omega$.

As a result, an admissible trading strategy, together with an initial portfolio $(x(0), y(0))$ yields a well-defined dynamic portfolio.

Our aim is described by the following definition:

Definition 2.4. At an initial market price $S(0)$ given,

a. An initial portfolio and an admissible trading strategy $\tilde{\varphi}$ constitute a *hedge* at $S(0)$ if they insure that

$$\forall S(\cdot) \in \Omega \text{ with } S(0) \text{ given}, \quad \tilde{w}(T) \geq M(S(T)). \tag{9}$$

b. The strategy $\tilde{\varphi}$ is a *hedging strategy* for the initial portfolio $(x(0), y(0))$ at $S(0)$ if together they constitute a hedge.
c. An initial portfolio is said to be *hedging* at $S(0)$ if there exists a related hedging strategy for it.

Finally, the relation with pricing is as follows:

Definition 2.5. The *equilibrium price* of the contingent claim $(T, M(\cdot))$ at $S(0)$ is the least worth $\tilde{w}(0) = y(0)R(0)$ of all hedging initial portfolios of the form $(0, y(0))$.

This last definition stems from the following remark. Let an initial hedging portfolio be given as $(x(0), y(0))$. Let $\varepsilon = \text{sign}(x(0))$, and let us assume that $x(0)$ and $y(0)$ have different signs, as will be the case for efficient hedging portfolios in the case of simple european options. The cost of creating it, or its *price* is

$$\begin{aligned} P(x(0), y(0)) &= (1 + \varepsilon c_1)x(0)S(0) + (1 - \varepsilon c_0)y(0)R(0) \\ &= (1 - \varepsilon c_0)[\tilde{w}(0) + C_\varepsilon x(0)S(0)]. \end{aligned}$$

In the notations, of the next paragraph, this leads to define the price of the hedge as $P = (1 - \varepsilon c_0)R(0)\min_v[W(0, u(0), v) + C_\varepsilon v]$. On the other hand, if our theory allows for a jump in x and y at initial time, satisfying (4), then it follows from (the same reasoning as that leading to) equation (22) that indeed, $R(0)W(0, u, 0) = R(0)\min_v[W(0, u, v) + C_\varepsilon v]$. Since ε will be the same for all efficient hedging portfolios ($+1$ or -1 in the case of a call or a put respectively), in comparing the worth of hedging portfolios, we may neglect the factor $(1 - \varepsilon c_0)$, which disappears altogether if we assume that our original wealth was invested in bonds.

2.3 End-time Values

It is convenient to transform everything in end-time values. We shall let

$$\begin{aligned} &u = \frac{S}{R}, \quad v = \frac{xS}{R}, \quad w = \frac{\tilde{w}}{R}, \\ &\tau = \tilde{\tau} - \rho, \quad -\alpha = -\tilde{\alpha} - \rho, \quad \beta = \tilde{\beta} - \rho, \\ &\xi = \frac{\tilde{\xi}}{R}, \quad X = \frac{\widetilde{X}}{R}. \end{aligned}$$

Notice that S, x and y are readily recovered from u, v, and w with the help of

$$x = \frac{v}{u}, \qquad y = w - v. \tag{10}$$

With these notations, the dynamics of the market and portfolio become

$$\dot{u} = \tau u, \tag{11}$$

$$\dot{v} = \tau v + \xi, \tag{12}$$

$$\dot{w} = \tau v - C_\varepsilon \xi. \tag{13}$$

$$\tau \in [-\alpha, \beta], \qquad \xi \in [-X, +X]. \tag{14}$$

Our objective is to find the cheapest hedging portfolio and corresponding hedging strategy

$$\xi = \varphi(t, u, v) \tag{15}$$

2.4 Mathematical Analysis of the Problem

The aim (9) of a hedging strategy φ can be written as

$$\forall\tau(\cdot) \in [-\alpha, \beta], \qquad M(u(T)) - w(T) \le 0,$$

where one remembers that $u(T)$ is a function of $u(0)$ and τ, and $w(T)$ a function of $v(0)$, $w(0)$, and both $\tau(\cdot)$ and φ. Obviously, this is equivalent to

$$\sup_{\tau(\cdot)}[M(u(T)) - w(T)] \le 0.$$

And for a given $u(0)$, $v(0)$, $w(0)$, there exists a hedging strategy if (and only if provided that the min below exists)

$$\min_{\varphi}\sup_{\tau(\cdot)}[M(u(T)) - w(T)] \le 0. \tag{16}$$

Hence, in a typical "robust control" fashion, we face a minimax control problem or dynamic game problem.

Now, notice that w does not appear in the right hand side of the dynamics (11), (12), (13). Hence, we may integrate (13) in

$$w(t) = w(0) + \int_0^t (\tau(s)v(s) - C_\varepsilon\xi(s))\mathrm{d}s.$$

The relation (16) can therefore be rewritten

$$\min_{\varphi}\sup_{\tau(\cdot)}[M(u(T)) - \int_0^T (\tau(t)v(t) - C_\varepsilon\xi(t))\mathrm{d}t - w(0)] \le 0.$$

Now, $u(t)$, $v(t)$ and hence $\xi(t) = \varphi(t, u(t), v(t))$ are independent of $w(0)$. Hence, the above relation is satisfied provided that

$$w(0) \ge \min_{\varphi}\sup_{\tau(\cdot)}[M(u(T)) - \int_0^T (\tau(t)v(t) - C_\varepsilon\xi(t))\mathrm{d}t].$$

We are thus led to the investigation of the function

$$W(t, u(t), v(t)) = \min_{\varphi}\sup_{\tau(\cdot)}[M(u(T)) - \int_t^T (\tau(s)v(s) - C_\varepsilon\xi(s))\,\mathrm{d}s], \tag{17}$$

and define the *price* of the contingent claim investigated as $W(0, u(0), 0)$.

We introduce the Isaacs equation of this game:

$$\begin{aligned}
&\frac{\partial W}{\partial t} + \min_{\xi}\sup_{\tau\in[-\alpha,\beta]}\left\{\tau\left[\frac{\partial W}{\partial u}u + \left(\frac{\partial W}{\partial v} - 1\right)v\right] + \left(\frac{\partial W}{\partial v} + C_\varepsilon\right)\xi\right\} = 0,\\
&W(T, u, v) = M(u).
\end{aligned} \tag{18}$$

(Notice that the function between braces in the r.h.s. above is not differentiable in ξ because of the definition of C_ε, involving $\varepsilon = \operatorname{sign}(\xi)$.)

If we adopt the approximation device of restricting ξ to a finite interval $[-X, X]$, then the $\min_\xi$ in (18) above should be restricted accordingly. And we get

Theorem 2.3. *If (18) has a solution W, then $e^{-\rho T} W(0, u(0), 0)$ is the approximated equilibrium price of the contingent claim investigated.*

Proof. From standard differential games theory (see [3,7]), W is indeed the min-sup in (17). Hence the worth of the cheapest hedging initial portfolio with $x(0) = 0$, hence $v(0) = 0$, for a given $u(0)$ is $w(0) = W(0, u(0), 0)$. Going back to the original variables $\tilde{w}(0) = \mathrm{e}^{-\rho T} w(0)$ yields the result. □

2.5 No Transaction Costs

We first consider the case without transaction costs, i.e. $C_\varepsilon = 0$. This was investigated in more detail in [5], but we stress here a game theoretic analysis which was not discussed there.

Here, we need not keep xS or v as a state variable as it can be changed instantly at no cost. We therefore remain with two state variables (plus time):

$$\dot{u} = \tau u,$$
$$\dot{w} = \tau v.$$

The control variables are $\tau \in [-\alpha, \beta]$ and $v \in \mathbb{R}$. A trading strategy in this context will be a function $\varphi : [0, T] \times \mathbb{R}^+ \to \mathbb{R}$ giving $v(t) = \varphi(t, u(t))$. The problem at hand is to find states controllable by v to the set $w(T) \geq M(u(T))$ against any control of τ.

The above analysis simplifies in

$$w(T) = w(0) + \int_0^T \tau(t) v(t) \mathrm{d}t.$$

and

$$w(0) = \min_{\varphi} \sup_{\tau(\cdot)} \left[M(u(T)) - \int_0^T \tau(t) v(t) \mathrm{d}t \right],$$

which thus provides the equilibrium option price sought.

Therefore, our pricing equation is Isaac's equation for this game:

$$\frac{\partial W}{\partial t} + \min_{v} \max_{\tau} \left[\tau \left(\frac{\partial W}{\partial u} u - v \right) \right] = 0, \quad W(T, u) = M(u). \tag{19}$$

Theorem 2.4. *In the absence of transaction costs, the equilibrium price of the contingent claim investigated is the so-called* parity value

$$e^{-\rho T} M(e^{\rho T} S(0)).$$

The corresonding hedging strategy is given by

$$x(t) = \frac{dM}{ds}(e^{\rho(T-t)} S(t)).$$

Proof. For any v, the $\max_\tau$ in (19) is non negative. Hence it is maximized by the choice $v = (\partial W/\partial u)u$, and Isaac's equation is reduced to $\partial W/\partial t = 0$. Hence its solution is $W(t,u) = M(u)$. The results follow. □

We see that we recover the classical fact that the optimal x is the sensitivity of the option's value.

2.5.1 European Calls

Let us examine the case of a European call. Then $M = [u - K]_+$ is not differentiable. Yet it is easy to see that $W = M$ is indeed the viscosity solution of Isaac's equation, since at $u = K$, we do have that for any p between 0 and 1,

$$\min_v \max_\tau [\tau(pu - v)] = 0.$$

However, this is not the last word about the non-differentiability of W.

The corresponding strategy is the "naive strategy":

$$v = \begin{cases} 0 & \text{if} \quad u \le K, \\ u & \text{if} \quad u \ge K. \end{cases}$$

The strategy at $u = K$ is better analyzed in terms of the semi-permeability of the manifold $w = [u - K]_+$. It is readily apparent that if we do not want the corner to leak, we need that both $\dot{w} \ge 0$ and $\dot{w} \ge \dot{u}$, which requires that $v = 0$ if $\tau < 0$ (to abide by the first constraint) and $v = 1$ if $\tau > 0$ (to abide by the second.) Hence the hedging strategy is a bang-bang function of the *sign of the variation* of the underlying stock price; A very undesirable feature.

Moreover, if there are transaction costs, and if the prices oscillate around the parity value, this will induce constant large buy and sell decisions which will cost much and ruin the hedging strategy.

2.6 Non-zero Transaction Costs: A Partial Solution

To be more specific, we consider the case of a European call with striking price K, where $M(s) = \max\{0, s - K\}$.

2.6.1 Three-D Impulsive Control Formulation

In this paragraph, we investigate the problem allowing for instantaneous trading of a finite amount of securities, hence jumps in x and y at times chosen by the trader, but still in accordance with (4).

We let the trader choose instants of time t_k and trading amounts ξ_k with signs ε_k, and augment the dynamic equations with the jump conditions

$$v(t_k^+) = v(t_k^-) + \xi_k, \tag{20}$$

$$w(t_k^+) = w(t_k^-) - C_{\varepsilon_k}\xi_k. \tag{21}$$

The definition of a trading strategy for this paragraph is therefore as follows:

Definition 2.6. A trading strategy is defined by

a. a measurable function $\xi(\cdot) : [0, T] \to \mathbb{R}$ called the *continuous part*,
b. an *impulsive part* made of
   - a finite increasing sequence of time instants $\{t_k\}$
   - a sequence of corresponding numbers $\{\xi_k\}$,

The corresponding dynamic portfolio is given by the equations (12) (20), and the worth of the portfolio can be computed from (13) (21).

We denote symbolically by φ a feedback rule that lets one decide whether to make a jump and of how much, and also compute the continuous part of the trading strategy, knowing past and present u's and v's.

As previously noticed, w does not enter the right hand side. Using the equality

$$w(T) = w(0) + \int_0^T (\tau v(t) - C_\varepsilon \xi(t))\mathrm{d}t - \sum_k C_{\varepsilon_k}\xi_k,$$

We therefore end up with the dynamics (11), (12), (20) and the problem is to find

$$W(0, u(0), v(0)) = \min_{\varphi} \sup_{\tau(\cdot)} \left[M(u(T)) - \int_0^T (\tau v(t) - C_\varepsilon \xi(t))\mathrm{d}t + \sum_k C_{\varepsilon_k}\xi_k \right].$$

Determination of this impulsive minimax is beyond the scope of the current theory. One should refer to the theory of impulse control, as developed in [4]. However, added difficulties arise. On the one hand, this is a game not a control problem. On the other hand, this is a deterministic problem, so that the corresponding PDE is first order, and one would need to extend to quasi-variational inequalities the technique of viscosity solutions. Moreover, the quasivariational inequality (QVI) is further degenerate due to the fact that the second term in the brace in (22) is nonpositive.[2]

[2] In that respect, we would have a more classical impulse control problem if the transaction costs were chosen affine, with a fixed part added to the proportional part.

Yet, it is interesting to write the quasi-variational inequality that is formally associated with this impulsive game:

$$0 = \min\left\{\frac{\partial W}{\partial t} + \max_{\tau\in[-\alpha,\beta]} \tau\left[\frac{\partial W}{\partial u}u + \left(\frac{\partial W}{\partial v} - 1\right)v\right],\right.$$
$$\left.\min_{\xi}[W(t,u,v+\xi) - W(t,u,v) + C_\varepsilon \xi]\right\} \quad (22)$$

with furthermore $\partial W/\partial v \in [-C^+, -C^-]$ whenever the first minimum is obtained by the first term in the brace, in order for $(\partial W/\partial v + C_\varepsilon)\xi$ to have a minimum in ξ, which is then reached at $\xi = 0$.

It might be possible to construct the solution of this QVI. We *conjecture* that the solution leads to hedging strategy involving, for any realistic initial condition, an initial jump in v followed by a "coasting" period where ξ takes intermediate values depending on the variations of u, followed by a final period with $\xi = 0$ as the next paragraph shows, hence, providing a non-trivial hedging strategy for option pricing with transaction costs, a feat known to be impossible with the classical theory, see [15].

2.6.2 Four-D Non-impulsive Analysis

We now turn to the approximation device consisting in bounding $|\xi|$ by a very large number X that we shall let go to infinity.

We also turn back to the 3-state formulation, considering the qualitative problem of driving the final state to the set $\{w(T) - M(u(T)) \geq 0\}$. The barrier of this problem is the graph of the function W of the 2-state variable formulation. But the geometric intuition of semi-permeable surfaces will help here. Notice also that, to gain in intuition (we like to think of the hedging strategy as maximizing the value of the portfolio), we have changed the sign of the terminal term. We make use of the Isaacs–Breakwell theory. The reader unfamiliar with that theory could as well turn directly to the subsection 2.6.3.

Hamiltonian set-up. In terms of differential games, we must construct a "barrier" separating states that can be driven to the desired set at time T against all disturbances from those for which at least one disturbance function exists that will prevent the aim being reached. Although we shall only sketch the mathematical details, we shall make free use of the theory. See, e.g. [10,11].

Because we are in fixed end-time, the state space of this game is four-dimensional: (t, u, v, w). Let (n, p, q, r) be a semi-permeable normal. It satisfies

$$n + \max_{\xi}\min_{\tau}\{[pu + (q + r)v]\tau + (q - C_\varepsilon)\xi\} = 0,$$

and the controls on barrier trajectories are given by

$$\tau = \begin{cases} -\alpha & \text{if} \quad pu + (q+r)v > 0, \\ \beta & \text{if} \quad pu + (q+r)v < 0, \end{cases} \qquad \xi = \begin{cases} X & \text{if} \quad q > C^+, \\ 0 & \text{if} \quad C^- < q < C^+, \\ -X & \text{if} \quad q < C^-. \end{cases}$$

Furthermore, on a smooth part of a barrier, along the barrier trajectories the semi-permeable normal satisfies the adjoint equations:

$$\begin{aligned} \dot{n} &= 0, \\ \dot{p} &= -\tau p, \\ \dot{q} &= -(1+q)\tau, \\ \dot{r} &= 0. \end{aligned}$$

Barrier sheet towards $u < K$. Let us construct the natural barrier arriving on the part $u < K$, $w = 0$ of the target set boundary, that we parametrize with $u(T) = s$, $v(T) = \chi$. We get

$$\begin{aligned} u(T) &= s \le K, & p(T) &= 0, \\ v(T) &= \chi, & q(T) &= 0, \\ w(T) &= 0, & r(T) &= 1. \end{aligned}$$

So, at final time, we get $\tau = -\alpha$ and $\xi = 0$. The equations integrate backwards in

$$\begin{aligned} u(t) &= s\mathrm{e}^{\alpha(T-t)}, & p(t) &= 0, \\ v(t) &= \chi \mathrm{e}^{\alpha(T-t)}, & q(t) &= \mathrm{e}^{-\alpha(T-t)} - 1, \\ w(t) &= \chi(\mathrm{e}^{\alpha(T-t)} - 1), & r(t) &= 1. \end{aligned}$$

This solution is not valid before the time t_α when q crosses the value C^-, i.e.

$$T - t_\alpha = \frac{1}{\alpha} \ln\left(\frac{1}{1+C^-}\right).$$

Prior to t_α, one has $q < C^-$, and therefore, if that trajectory is still part of a barrier, $\xi = -X$. In view of the fact that we are interested in the case $X \to \infty$, this means a negative jump in v, i.e. in x, the underlying stock content of our portfolio.

Regular barrier sheet towards $u(T) > K$. We now consider the natural barrier towards $u(T) > K$, $w(T) = u(T) - K$. We again parametrize that boundary by $u(T) = s$ and $v(t) = \chi$. We get now:

$$\begin{aligned} u(T) &= s \geq K, & p(T) &= -1,\\ v(T) &= \chi, & q(T) &= 0,\\ w(T) &= s - K, & r(T) &= 1. \end{aligned}$$

The corresponding value of τ at time T depends on the sign of $s - \chi$. Let us first consider the case $s > \chi$. We have then at time T and just before $\tau = \beta$, and still $\xi = 0$. The differential equations integrate backwards in

$$\begin{aligned} u(t) &= s\mathrm{e}^{-\beta(T-t)}, & p(t) &= -\mathrm{e}^{\beta(T-t)},\\ v(t) &= \chi\mathrm{e}^{-\beta(T-t)}, & q(t) &= \mathrm{e}^{\beta(T-t)} - 1,\\ w(t) &= \chi(\mathrm{e}^{-\beta(T-t)} - 1) + s - K, & r(t) &= 1. \end{aligned}$$

This solution is not valid before the time t_β when q crosses the value C^+:

$$T - t_\beta = \frac{1}{\beta}\ln(1 + C^+). \tag{23}$$

Prior to that time, and again if the trajectories were part of a continuing barrier, we would have $\xi = X$, which indicates a positive jump in v, hence in x.

Singular barrier sheet towards $u(T) > K$. A particular case arises if we consider the case $s = \chi$. Then it is readily apparent that $pu + (q + r)v$ remains null along any time interval $[t, T]$ on which $\xi = 0$. Then, any τ satisfies the semi-permeability condition, and so does $\xi = 0$ as long as

$$\ln(1 + C^-) \leq \int_t^T \tau(\theta)\mathrm{d}\theta \leq \ln(1 + C^+).$$

Along these trajectories, $u = v \in [s/(1+C^+), s/(1+C^-)] \cap [s\mathrm{e}^{-\beta(T-t)}, s\mathrm{e}^{\alpha(T-t)}]$, and $w = u - K$.

Thus, for each s we have two free parameters: t and $\int_t^T \tau\,\mathrm{d}\theta$, yet this constitutes only a 2-D manifold, because all are embedded into the 2-D manifold $u = v = w + K$, t arbitrary.

One of these trajectories for each s is obtained with $\tau = \beta$. It is the "last" trajectory of the sheet we would construct with $\chi \leq s$. It will come as no surprise to the reader that the trajectories constructed with $\chi > s$, i.e. in our original variables $x > 1$, will play no role in the solution.

Intersection. These two three-dimensional sheets intersect along a 2-D edge, that joins continuously with the above singular 2-D manifold, and that we can parametrize with $h = T - t$ and $u \in [K\exp(-\beta h), K\exp(\alpha h)]$ as

$$v = \hat{v}(h, u) := \frac{u\mathrm{e}^{\beta h} - K}{\mathrm{e}^{\beta h} - \mathrm{e}^{-\alpha h}}, \qquad w = \hat{w}(h, u) := (1 - \mathrm{e}^{-\alpha h})\hat{v}(h, u).$$

Notice that because of Proposition 2.2, and assuming that $\alpha \leq \beta$ (maximum rate of decrease of a stock price not larger than the maximum rate of increase), $t_\alpha < t_\beta$. Hence the above intersection only holds over the time interval $[t_\beta, T]$, because before, the sheet towards $u > K$ is missing.

A careful study of the intersection shows that it is a τ-dispersal line, i.e. that the trader must watch the evolution of the stock price and adapt to it.

Composite barrier. These semi-permeable surfaces define a composite natural barrier that can as usual be described as (the graph of) a function $w = W(t, u, v) = \inf\{w \mid (t, u, v, w) \text{ is hedgeable}\}$. We get here $h = T - t$ and

$$W(t, u, v) = \begin{cases} (1 - \mathrm{e}^{-\alpha h})v & \text{if} \quad v \geq \hat{v}(h, u), \\ (1 - \mathrm{e}^{\beta h})v + u\mathrm{e}^{\beta h} - K & \text{if} \quad v \leq \hat{v}(h, u). \end{cases}$$

We may further notice that if $u \leq K \exp(-\beta h)$, we always are in the first case above, and if $u \geq K \exp(\alpha h)$, taking into account the fact that we consider only the cases $v \leq u$, we always are in the second case.

2.6.3 Interpretation of the Results

As a cue to interpreting the geometry of the barrier in the state space, notice that if a point (t, u, v, w) is "hedgeable", then any point (t, u, v, w') with $w' > w$ is also admissible. Therefore for a given (t, u), say, we should look for a barrier point with those coordinates and the lowest possible w as a limiting admissible state, and therefore an equilibrium price for the call.

The main point we have shown is that the terminal part of the play leads to $\xi = 0$ as an optimal behavior, i.e. to a constant x, no trading is necessary during the final $T - t_\beta$ time interval. Therefore the main weakness of the "naive" strategy of [5], which was a risk of constant and costly trading, is avoided. The whole idea to include the transaction costs into the model was aimed at that result.

If $u < K\mathrm{e}^{-\beta(T-t)}$, then according to our model, the call is and shall remain out of the money. The value of the call is 0. As a matter of fact, the only relevant barrier is our sheet towards $u < K$. For a given u and t it is intersected at minimum w by $\chi = 0$ and indeed yields $w = 0$.

If $u > K\mathrm{e}^{-\beta(T-t)}$, the call may end up in the money. If moreover $u > K\mathrm{e}^{\alpha(T-t)}$, then it will surely do so. The intersection of the two sheets of the composite barrier is at $v = u$, $w = u - K$. One should have (at least) one share of the stock, and may have borrowed an amount $KR(t)$, worth K at exercise time.

In between, and for the last instant of times, the limiting v and w are just such that with no trading, if the stock goes down at maximum rate, we shall end up with $w = 0$, and if it goes up at maximum rate, we shall have just $w = u - K$. The corresponding equilibrium value for the call is a linear function of u (with $h = T - t$ the maturity):

$$w = \frac{1 - e^{-\alpha h}}{e^{\beta h} - e^{-\alpha h}} (u e^{\beta h} - K).$$

It coincides with one step of the discrete time theory hereafter (see also [5]), and therefore more prominently of [8]. As a matter of fact, since we have found that the optimal behavior was to let $T - t_\beta$ time pass without trading, therefore without incurring trading costs, we find one step of the discrete time theory with that step size.

We therefore suggest exploiting this result by using a discrete time strategy with that step size. This is the minimum time it takes for the stock to increase by a relative amount $-C^-$, i.e. of the order of (but slightly less than) $c_0 + c_1$.

2.6.4 Closing Costs

Before we investigate the discrete time theory, we must make a digression on closing costs.

If there are trading costs, it is not equivalent to ending up with no stock and no debt or with, say s worth of stock, and as much in debt, as there is a cost to selling the stock and using the proceeds to repay the debt.

The target set at exercise time should therefore be changed to reflect that fact. Let $\eta = \text{sign}(v)$. The correct target set for $u(T) < K$ is then

$$w + C_{-\eta} v \geq 0.$$

(If we know that only positive v's will be used, we may simply set $w + C^- v \geq 0$, but the above form is useful for the theoretical analysis.) In the case $u(T) > K$, there are two possible ways of comparing a portfolio and the option. Either we decide to liquidate any position, and compare situations with $v = 0$, or we want to bring our portfolio to a position similar to that just after exercising the call, i.e. with $v = u$. Both methods do not lead to the same conclusions. The first one has the advantage of leading to a continuous target set for the portfolio. Incidentally, one should then exercise the call only if the net proceeds after liquidating the position are positive, i.e. if $(1 + C^-)u(T) - K > 0$. And the target set then reads

$$w + C_{-\eta} v \geq [(1 + C^-)u - K]_+.$$

We forgo the mathematical analysis of this case, as it is at this time less advanced than the previous one. The analysis seems to point to a delay without trading at the end of length

$$T - t_\beta = \frac{1}{\beta} \ln \frac{1 + C^+}{1 + C^-}, \tag{24}$$

hence roughly twice as long as in the case without closure costs.

3 Discrete Time

We turn to the discrete time theory. In [5], we argued that this is a more realistic theory as a trader is likely to pay attention to a given portfolio a finite number of times per day. Here, however, we have another justification, arising from the continuous time theory itself, where we have seen that introducing transaction costs automatically leads to optimal hedging strategies made of jumps in the contents of the portfolio, with no trading at least after the last jump.[3] This suggests using the characteristic step size (23) or (24) above. This is typically of the order of one third to one half day. The present theory can also be exploited for other step sizes.

3.1 The Model

3.1.1 Market Model

We let now the time t be an integer, i.e. we take the step size Δt as our unit of time, so that t is now an integer ranging from 0 to a given positive integer T. Let also the price of a unit riskless bound be

$$R(t) = (1+\rho)^{(t-T)}$$

so that ρ in this section is $e^{\rho\Delta t} - 1$ of the continuous time theory. Likewise, concerning the underlying stock price S we let:

Definition 3.1. The set Ω of *admissible* price histories is defined by two positive numbers $\tilde{\alpha}$ and $\tilde{\beta}$ and is the set of all sequences $\{S(t)\}$, $t \in \{1, \dots, T\}$ such that

$$(1-\tilde{\alpha})S(t) \le S(t+1) \le (1+\tilde{\beta})S(t) \tag{25}$$

We choose to define

$$\tilde{\tau}_t := \frac{S(t+1) - S(t)}{S(t)},$$

so that the above definition also reads

$$S(t+1) = (1+\tilde{\tau}_t)S(t), \qquad \tilde{\tau}_t \in [-\tilde{\alpha}, \tilde{\beta}]. \tag{26}$$

Notice that the $\tilde{\alpha}$ and $\tilde{\beta}$ of this section are related to those of the previous section via the same relation as ρ.

3.1.2 Portfolio Model

We call x_t and y_t the number of shares in the portfolio immediately *before* the transactions at time t, and $\tilde{w}_t = x_t S(t) + y_t R(t)$ the corresponding value of the portfolio. We shall make use of $\tilde{w}_t^+ = x_{t+1}S(t) + y_{t+1}R(t)$, the value of the portfolio immediately *after* the trading at time t, and likewise for v and w below. Let also $dx(t) = x_{t+1} - x_t$ and $dy(t) = y_{t+1} - y_t$. We therefore have

[3] And affine transaction costs would lead to purely impulsive strategies anyhow.

Definition 3.2.

a. A *dynamic portfolio* is a pair of sequences $(\{x_t\}, \{y_t\})$, defined over $t \in \{1, \dots, T\}$.
b. A dynamic portfolio is said to be *self financed* if it satisfies (as equation (4)):

$$dx(t)S(t) + c_1|dx(t)|S(t) + dy(t)R(t) + c_0|dy(t)|R(t) = 0. \tag{27}$$

We choose $\tilde{\xi}_t$, the amount in S traded at time t as our control, so that

$$(x_{t+1} - x_t)S(t) = \tilde{\xi}_t.$$

Let $\varepsilon = \text{sign}(\tilde{\xi})$. The same reasoning as in the continuous time case may be used to conclude that self-financing of the strategy imposes the following:

Proposition 3.1. *A self-financed dynamic portfolio satisfies*

$$(y_{t+1} - y_t)R(t) = -\frac{1 + \varepsilon c_1}{1 - \varepsilon c_0}\tilde{\xi}_t.$$

and

$$\tilde{w}_{t+1} = (1 + \rho)(\tilde{w}_t - C_\varepsilon \tilde{\xi}_t) + (\tilde{\tau}_t - \rho)(x_t S(t) + \tilde{\xi}_t).$$

Introduce, as in the continuous time theory, the end-time values

$$u_t = \frac{S(t)}{R(t)}, \quad v_t = \frac{x_t S(t)}{R(t)}, \quad w_t = \frac{\tilde{w}_t}{R(t)}, \quad \xi_t = \frac{\tilde{\xi}_t}{R(t)},$$

as above,

$$v_t^+ = \frac{x_{t+1} S(t)}{R(t)}, \quad w_t^+ = \frac{\tilde{w}_t^+}{R(t)},$$

and let

$$-\alpha := \frac{-\tilde{\alpha} - \rho}{1 + \rho} \le \tau_t := \frac{\tilde{\tau}_t - \rho}{1 + \rho} \le \beta := \frac{\tilde{\beta} - \rho}{1 + \rho}, \tag{28}$$

After some simple calculations, the discrete time market and portfolio model then read

$$\begin{aligned} & & u_{t+1} &= (1 + \tau_t)u_t, \\ v_t^+ &= v_t + \xi_t, & v_{t+1} &= (1 + \tau_t)v_t^+, \\ w_t^+ &= w_t - C_\varepsilon \xi_t, & w_{t+1} &= w_t^+ + \tau_t v_t^+. \end{aligned} \tag{29}$$

We shall consider feedback trading strategies:

Definition 3.3. A trading strategy is a sequence of functions $\varphi_t : \mathbb{R}^+ \times \mathbb{R} \to \mathbb{R}$. The self-financed dynamic portfolio generated from an initial portfolio (x_0, y_0) by a price history $\{S_t\} \in \Omega$ and the trading strategy $\{\varphi_t\}$ is the pair of sequences generated by (29) with, for all $t \in \{0, 1, \dots, T-1\} : \xi_t = \varphi_t(u_t, v_t)$.

We now state our objective: finding hedging strategies and the equilibrium price of the contingent claim.

Definition 3.4. At a given market price $S(0)$,

a. An initial portfolio (x_0, y_0) and a trading strategy constitute a *hedge* at $S(0)$ if, for any $\{S_t\} \in \Omega$, with $S(0)$ given, together they yield $w_T \geq M(u_T)$.
b. The corresponding trading strategy is then called a *hedging strategy*.
c. An initial portfolio (x_0, y_0) is said *hedging* at $S(0)$ if there exists a corresponding hedging strategy.

And finally

Definition 3.5. The *equilibrium price* of the contingent claim investigated at $S(0)$ is the least worth $w_0 = y_0 R(0)$ of all hedging portfolios of the form $(0, y_0)$.

Remark 3.1. The definition *a)* above may be slightly modified to reflect the preferred notion of hedge in the presence of closing costs (e.g., judged at T^+ imposing $v_T^+ \geq u_T$, or alternatively $v_T^+ = 0$ and $w_T^+ \geq [(1 - c_1)u - K]_+$). See the section on continuous trading for further hindsight into these definitions.

3.2 Dynamic Programming

Let $\mathcal{A}_t$ be the set of states (u_t, v_t, w_t) from which there exists a trading strategy $\xi_k = \varphi_k(u_k, v_k)$, $k \geq t$ that, for any possible future sequence $\{\tau_k\}$, drives the portfolio to an admissible state at time T, i.e. such that $w_T \geq M(u_T)$. It is clear that if $(u, v, w) \in \mathcal{A}_t$, then any (u, v, w') with $w' > w$ will also be in $\mathcal{A}_t$. We may thus characterize the set $\mathcal{A}_t$ as the epigraph of its *floor function*

$$W_t(u, v) = \min \left\{ w \left| \begin{pmatrix} u \\ v \\ w \end{pmatrix} \in \mathcal{A}_t \right. \right\} \quad \text{so that} \quad \begin{pmatrix} u \\ v \\ w \end{pmatrix} \in \mathcal{A}_t \Leftrightarrow w \geq W_t(u, v).$$

It is convenient to perform the classical dynamic programming construction, leading to the Isaacs equation, in two steps. Let first $\mathcal{A}_t^+$ be the set of states (u_t, v_t^+, w_t^+) at t^+ that will be driven to $\mathcal{A}_{t+1}$ by any τ, and W_t^+ the corresponding floor function. Thus,

$$\begin{pmatrix} u \\ v^+ \\ w^+ \end{pmatrix} \in \mathcal{A}_t^+ \Longleftrightarrow \forall \tau \in [-\alpha, \beta], \quad \begin{pmatrix} (1+\tau)u \\ (1+\tau)v^+ \\ w^+ + \tau v^+ \end{pmatrix} \in \mathcal{A}_{t+1}.$$

or equivalently

$$w^+ \geq W_t^+(u, v^+) \iff \forall \tau \in [-\alpha, \beta],\, w^+ + \tau v^+ \geq W_{t+1}((1+\tau)u, (1+\tau)v^+).$$

Thus, we have

$$W_t^+(u, v^+) = \max_{\tau \in [-\alpha, \beta]} [W_{t+1}((1+\tau)u, (1+\tau)v^+) - \tau v^+]. \tag{30}$$

Now, $\mathcal{A}_t$ is the set of all states (u, v, w) that can be sent by an appropriate control ξ into a (u, v^+, w^+) in $\mathcal{A}_{t+1}$, hence,

$$w \geq W_t(u, v) \iff \exists \xi : w - C_\varepsilon \xi \geq W_t^+(u, v + \xi).$$

Therefore

$$W_t(u, v) = \min_\xi [W_t^+(u, v + \xi) + C_\varepsilon \xi] \tag{31}$$

It is useful to give the form taken by the recursion merging the two steps from t_{t+1} to t^+ and from t^+ to t into a single Isaacs equation:

$$W_t(u, v) = \min_\xi \max_{\tau \in [-\alpha, \beta]} [W_{t+1}((1+\tau)u, (1+\tau)(v + \xi)) - \tau(v + \xi) + C_\varepsilon \xi]. \tag{32}$$

to be initialized with

$$W_T(u, v) = M(u) \tag{33}$$

We have thus proved the following result:

Theorem 3.1. *If equations (32), (33) have a solution $W_t(u, v)$, the equilibrium price of the contingent claim at $S(0)$ is $(1+\rho)^{-T} W_0((1+\rho)^T S(0), 0)$.*

(The coefficients $(1+\rho)^{-T}$ and $(1+\rho)^T$ are there to come back in the original variables, as opposed to their end-time values.)

Equations (30) and (31), or equivalently (32) and (33), also provide a constructive algorithm to numerically compute the equilibrium price. We discuss that matter at the end of Subsection 3.4.

3.3 Limiting Cases

3.3.1 Zero Transaction Costs

The case with no transaction costs corresponds here to $C_\varepsilon = 0$. Then, in (32) ξ only appears in combination with v as $(v + \xi)$ (i.e. v^+), which can therefore be taken as our mute maximization variable. If moreover the final value W_T does not depend on v, then the r.h.s. above never depends on v either, leading to a function $W_t(u)$:

$$W_t(u) = \min_{v^+} \max_{\tau\in[-\alpha,\beta]} [W_{t+1}((1+\tau)u) - \tau v^+].$$

We shall argue below that in the case of simple European options, the maximum in τ is reached at an end point of the admissible interval. As a consequence, this maximum is minimum when $\tau = -\alpha$ and $\tau = \beta$ yield the same value, i.e.

$$W_{t+1}((1-\alpha)u) + \alpha v^+ = W_{t+1}((1+\beta)u) - \beta v^+$$

leading to

$$v^+(u) = \frac{W_{t+1}((1+\beta)u) - W_{t+1}((1-\alpha)u)}{\alpha+\beta}$$

and thus

$$W_t(u) = \frac{\alpha}{\alpha+\beta} W_{t+1}((1+\beta)u) + \frac{\beta}{\alpha+\beta} W_{t+1}((1-\alpha)u). \tag{34}$$

These are exactly the equations obtained by Cox, Ross and Rubinstein [8]. The reference [5] develops in some more detail the reason why the two theories seem to coincide. The (big ?) difference, though, is that the theory of Cox, Ross, and Rubinstein is based upon a market model which is *not* realistic for finite step sizes, and is only meant to be meaningful in the limit as the step size goes to zero. Here we have a normative theory even for finite step sizes. However, with the same historical data, we shall be led to a model with a larger volatility $\sigma = (\alpha+\beta)/2$ than in their approach.

3.3.2 Vanishing Step Size

It is interesting to investigate the limiting case of our theory, with transaction costs, when the step size goes to zero in the above recursion. Thus we replace the step size "one" of the above theory by h. We choose to modelize stock price histories of *bounded variation*, as opposed to the classical Black and Scholes model. (See [5] for more details.) Thus replace τ by $\tau_h = h\tau \in [-h\alpha, h\beta]$.

Equation (32) now reads

$$W_t(u,v) = \min_{\xi} \max_{\tau\in[-\alpha,\beta]} [W_{t+h}((1+h\tau)u, (1+h\tau)(v+\xi)) - h\tau(v+\xi) + C_\varepsilon \xi]$$

The following analysis is formal. We strongly conjecture that it can be made precise, though at a rather high mathematical price. Rewrite the above equation as

$$0 = \min_{\xi} \max_{\tau\in[-\alpha,\beta]} \Big\{ W_{t+h}((1+h\tau)u, (1+h\tau)v) - W_t(u,v) - h\tau(v+\xi) + \\ W_{t+h}((1+h\tau)u, (1+h\tau)(v+\xi)) - W_{t+h}((1+h\tau)u, (1+h\tau)v) + C_\varepsilon \xi \Big\}. \tag{35}$$

The first line in the above display always goes to zero as $h \to 0$. Now, two situations may arise as $h \to 0$.

Either the minimum in ξ is attained for a non zero ξ. Nevertheless, the second line must also go to zero, since the sum does. Therefore, in the limit we must have

$$\min_{\xi}[W_t(u, v + \xi) - W_t(u, v) + C_\varepsilon \xi] = 0,$$

where we recognize the second term of (22). In this case, placing $\xi = 0$ will yield a positive r.h.s. But the second line in the display (35) is zero for $\xi = 0$. Thus the first is positive, and the term above is the minimum of the two lines.

Or the minimum in ξ is reached at $\xi = 0$. This means that the second line in the display would be positive for non zero ξ's. And the first line reads, dividing through by the positive h,

$$\max_{\tau \in [-\alpha, \beta]} \left[\frac{1}{h} \Big(W_{t+h}((1 + h\tau)u, (1 + h\tau)v) - W_t(u, v) \Big) - \tau v \right] = 0.$$

In the limit as h goes to zero we recognize the first line of (22).

Altogether, we see that we end up with the quasi-variational inequality (QVI) (22) of the continuous time theory. Thus, equation (32) can be seen as an "upwind" finite difference scheme for the continuous QVI, strongly suggesting that the solution of the discrete trading problem converges to that of the continuous trading one as the step size goes to zero.

3.4 The Convex Case

As is the case without transaction costs, convexity of the evaluation function M at terminal time is preserved by the recursion and helps in the computations. Let us state the main fact:

Theorem 3.2.

a. *The functions W_t and W_t^+ generated by the recursion (30), (31), (33) are convex in v for each u.*
b. *If furthermore the function M is convex, then they are jointly convex in (u, v).*

Proof. We provide the proof of the second statement. The first one goes along the same lines, just simpler.

Notice first that M being convex (in u), W_T is jointly convex in (u, v). Assume that $W_{t+1}(u, v)$ is convex in (u, v). So is $W_{t+1}((1 + \tau)u, (1 + \tau)v) - \tau v$. And therefore, according to (30), W_t^+ is the maximum of a family of convex functions, thus convex.

Assume therefore that W_t^+ is convex in (u, v). Introduce the extended function

$$\Gamma(\eta, \xi) = \begin{cases} +\infty & \text{if} \quad \eta \neq 0, \\ -C_{-\varepsilon}\xi & \text{if} \quad \eta = 0. \end{cases}$$

It is convex in (ξ, η). Now, (31) reads

$$W_t(u, v) = \min_{\eta,\xi}[W_t^+(u - \eta, v - \xi) + \Gamma(\eta, \xi)].$$

Hence W_t is the inf convolution of two convex functions, therefore it is convex. The theorem follows by induction. □

Beyond its theoretical significance, —there are deep reasons to expect the value of a call to be convex, at least in u (see, e.g. [9])— this fact has an important computational consequence. Let us first emphasize the following fact:

Corollary 3.1. *If M is convex, the function $\tau \mapsto [W_{t+1}((1+\tau)u, (1+\tau)v) - \tau v]$ is convex.*

As a consequence, the maximum in τ in (30) is necessarily reached at an end point of the segment $[-\alpha, \beta]$. Computationally, this means that the maximization is reduced to comparing two values, a significant simplification. The practical consequence is that using the recurrence relation to compute a pricing is very fast for a convex terminal value. We typically had run times of 6 seconds per time step on a 500 MHz PC, with u and v discretized in 200 steps each, a golden search in ξ, and $P1$ finite elements interpolation of the function W_t.

This will not be so for a digital call, say. The maximization in τ then has to be done via an exhaustive search. Yet, since we preserve the convexity in v, the minimization in ξ can still be performed via an efficient algorithm.

Remark 3.2. The various ways of taking closing costs into account usually preserve the convexity of M.

3.5 Partial Solution for a European Call

If we take $M(s) = \max\{0, s - K\}$, we can perform "by hand" the first steps of the recursion (32). We find the following facts.

- The optimal choice for $\xi(T)$ is always zero: there is no incentive to perform a portfolio readjustment at final time since this has a cost, and buys us nothing in this formulation without closure costs.
- If $u_t \geq (1 - \alpha)^{t-T} K$, one finds that the recursion reaches a fixed point

  $$W_t(u, v) = [(1 + C^+)(1 + \beta) - 1](u - v) + u - K.$$

  The optimal trading strategy is always to jump to $v = u$, i.e. own one share of the underlying stock.
- If $u_t \leq (1 + \beta)^{t-T} K$, the situation is slightly more subtle, at least for large trading costs. As a matter of fact, if $\alpha + C^- < 0$, for the last time steps, where $(1 - \alpha)^{T-t} > 1 + C^-$, the optimal hedging strategy is $\xi_t = 0$, hence do not trade. This is again a feature of the robustness of this theory against small variations in u. For earlier t's, the optimal hedging strategy is to jump to $v = 0$, and the lowest value of a replicating portfolio is

$$W_t(u, v) = [1 - (1 + C^-)(1 - \alpha)]v.$$

That is, one needs to have at least $(1 + C^-)(1 - \alpha)v$ worth of riskless bonds to pay for the trading in of the stock at hand at the next time instant. (Remember that v_t is the value of the stock in the portfolio *before* the trading at time t.)

- For u between these two limits, the value is piecewise affine in u and v. We have shown in the previous subsection that the "worst" market evolution, i.e. the one that dimensions the necessary portfolio, is always an extreme value, which makes the numerical solution very fast (only $\tau = -\alpha$ and $\tau = \beta$ have to be compared).
- In the case where $-\alpha = C^-$, which corresponds to the critical time of the continuous time theory, further degeneracies appear in the minimization in ξ in the recursion (32), making $\xi = 0$ a possible optimal hedging strategy for a larger region of the (u, v) space.
- Adding a closure cost does not change the above results much.

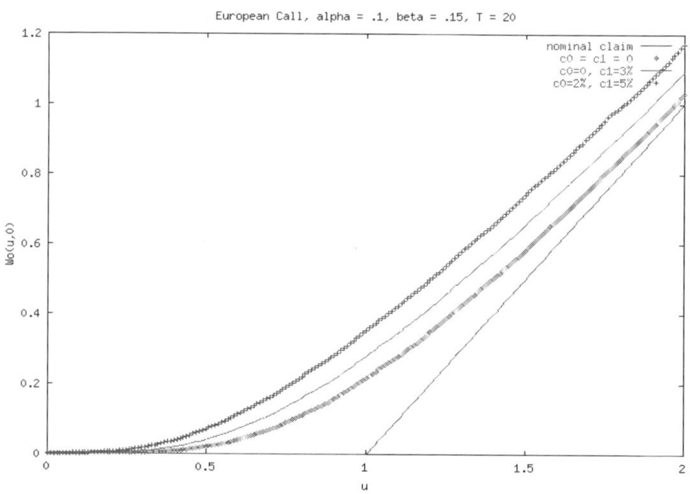

Figure 1: Equilibrium prices for various transaction costs, 20 time steps

4 Conclusions

Our objective was to introduce transaction costs in our non-stochastic theory of option pricing, not so much to have a more realistic theory, although this may be of interest, but mainly because we speculated that doing so would alleviate the bad feature of the previous continuous trading theory, which was found to give a naive strategy much too sensitive to trading costs (which had been neglected) in case the stock price oscillates around the present-value of the striking cost.

This aim seems to be indeed achieved, since at least for a (small) final time interval, no trading occurs. Yet at this time, we do not have a complete solution of the continuous time problem.

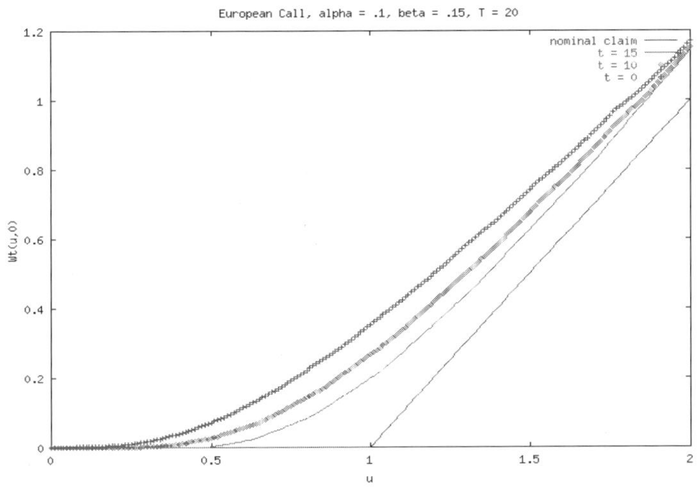

Figure 2: Equilibrium prices for various maturities, $c_0 = 2\%$, $c_1 = 5\%$

So we turn to a discrete time scheme, which may be considered as a more realistic formulation anyway. The corresponding problem can easily be solved numerically. The option value it will yield is reminiscent of that of Cox, Ross and Rubinstein in that it is piecewise affine, and coincides with it in the case of simple put or call options with no transaction costs in the theory. We point out that this is now a normative theory even with a finite step size, and not only as the step size vanishes. A particularily meaningful step size is the critical time deduced from the continuous trading strategy. Moreover, we conjecture that the limit of that theory as the step size goes to zero is the solution of the continuous time theory, in effect making our algorithm an efficient approximation scheme for the latter.

At this time, detailed numerical comparisons are being made between this theory and classical stochastic theories and between option pricing with or without trading costs. Even more importantly, we want to see whether this theory accounts for observed option prices on the market. Notice that we have more parameters to adjust than Black and Scholes, say.

Acknowledgements

We acknowledge the excellent work of our graduate students David Mac Audière, Pierre Mazzara, Raphaël Salique, and Virginie Silicio, who performed many numerical computations in relation with the discrete theory. The curves we show in Figures 1 and 2 were computed with their program.

We also wish to acknowledge the kindly worded comments of the anonymous reviewer that made the author aware of serious deficiencies in the original paper, and resulted in extensive re-writing.

REFERENCES

[1] Alvarez O. and Bardi M.: "Viscosity solutions methods for singular perturbations in deterministic and stochastic control", preprint # 2, Dip. Matematica P. e A., Università di Padova, Italy, February 2000.

[2] Aubin J-P., Pujal D., and Saint-Pierre P.: *Dynamic Management of Portfolios with Transaction Costs under Tychastic Uncertainty*, preprint, 2001.

[3] Bardi M. and Capuzzo-Dolcetta I. *Optimal Control & Viscosity solutions of Hamilton-Jacobi-Bellman Equations*, Birkhauser, Cambridge, MA, USA, 1997.

[4] Bensoussan A. and Lions J-L.: *Contrôle impulsionnel et inéquations quasi-variationnelles*, Gauthier-Villars, Paris, 1982.

[5] Bernhard P.: "A robust control approach to option pricing", in M. Salmon (ed.) *Robust Decision Theory and Ambiguity in Finance*, City University Press, London, 2003.

[6] Black F. and Scholes M.: "The pricing of options and corporate liabilities", *Journal of Political Economy*, **81**, pp 637–659, 1973.

[7] Crépey S.: "*Contribution à des méthodes numériques appliquées à la Finance et aux Jeux Différentiels*", thèse de l'École Polytechnique, Palaiseau, 2000.

[8] Cox J. C., Ross S. A., and Rubinstein M.: "Option pricing: a simplified approach", *Journal of Financial Economics*, **7**, pp 229–263, 1979.

[9] Cox J. C. and Rubinstein M.: *Options Markets* Prentice Hall, Englewood Cliffs, NJ, USA, 1985.

[10] Isaacs R.: *Differential Games*, John Wiley, New-York, 1965.

[11] Lewin J.: *Differential Games*, Springer Verlag, London, 1994.

[12] Olsder G-J.: "Control-theoretic thoughts on option pricing", *International Game Theory Review*, **2**, pp 209–228, 2000.

[13] Pujal D.: *Évaluation et gestion dynamiques de portefeuilles*, Thesis, Paris Dauphine University, 2000.

[14] Roorda B., Engwerda J., and Schumacher H.: "Performance of hedging strategies in interval models", preprint, 2000.

[15] Soner H. M., Shreve S. E., and Cvitanic J.: "There is no non trivial hedging strategy for option pricing with transaction costs", *The Annals of Applied Probability*, **5**, pp 327–355, 1995.

S-Adapted Equilibria in Games Played over Event Trees: An Overview

Alain Haurie*
HEC Management Studies
University of Geneva
1211 Geneva 4, Switzerland
alain.haurie@hec.unige.ch

Georges Zaccour†
GERAD
Ecole des HEC Montréal
Montréal, Québec H3T 2A7, Canada
georges.zaccour@hec.ca

Abstract

This paper exposes in voluntarily simple terms the concept of S-adapted equilibrium introduced to represent and compute economic equilibria on stochastic markets. A model of the European gas market, that has been at the origin of the introduction of the concept, is recalled in this paper and the results obtained in 1987, when the contingent equilibrium has been computed for a time horizon extending until 2020, are compared with the observed trend in these markets over the last two decades. The information structure subsumed by this concept of S-adapted strategies is then analyzed, using different paradigms of dynamic games. The paper terminates with some open and intriguing questions related to the time consistency and subgame perfectness of the dynamic equilibrium thus introduced.

1 Introduction

This paper gives a tutorial overview of an equilibrium concept for a class of stochastic games that model random competitive processes where the uncertainty is not influenced by the actions of the agents (players). In [15] such a solution concept for stochastic games has been introduced and characterized under the name of S-adapted equilibrium. A stochastic variational inequality (VI) served to characterize the equilibrium. There has been recently a renewed interest in the application of mathematical programming techniques to the solution of stochastic variational

*The research of this author has been supported by an SNSF grant
†The research of this author has been supported by an NSERC grant.

inequalities [6] and the modeling of oligopolies in random markets [9], [10]. In [6] a motivating example consisting of a model of the European gas market was borrowed from [19] which seems to have been the first application of this concept. This model of the European gas market is documented in [14] and [13] which are not widely available reports. So our first aim is to recall this particular model and to show how the S-adaptation was "invented" to represent energy and resource markets. This has defined a new class of stochastic games where the characteristic feature is for the random disturbances not to be influenced by the players' actions. Our second aim is to analyze in more detail the properties of the S-adapted information structure, using the context of classical games in extensive forms [4], [5], the framework of stochastic variational inequalities and the paradigm of piecewise deterministic differential games [11], respectively.

The paper is organized in two parts and six sections. Part 1 corresponds to section 2 and summarizes the model of the European gas market that has been the origin of the S-adapted strategy concept. We show that an S-adapted strategy has an interpretation in terms of contingent gas contracts with indexation clauses. The second part of the paper deals with a general theory of games that are played on "uncontrolled event trees", i.e. where the uncertainty is not dependent on the players' actions and can be represented by a discrete event process. In section 3 we start from the basics and we interpret the concept of S-adapted strategies in the classical context of games in extensive form. In section 4 we develop the relations that exist between the S-adapted equilibria defined in [14] and the concept of stochastic variational inequalities; we discuss existence and uniqueness of an S-adapted equilibrium. In section 5, we expose the extension of the concept of S-adapted strategies to the context of multistage and of piecewise deterministic games. For multistage games we provide a maximum principle which clearly shows the proximity of the concept with *open-loop equilibria* for deterministic multistage or differential games. In continuous time we relate the S-adapted information structure to the class of piecewise deterministic differential games (PDDG) introduced in [7]. The PDDG formalism has been applied to the analysis of stochastic oligopolies in [11], under the assumption that the random market disturbances are independent of the firms' actions. The *Piecewise-Open-Loop* information structure was used in this model which differs from the S-adapted one. In [9] the S-adapted information structure was applied to the stochastic oligopoly of Ref. [11] and the results, compared numerically, showed a surprising proximity. In the present paper we investigate a little further the links that can be established with the theory of open-loop differential games and, in particular, the necessary optimality conditions in the form of maximum principles. This leads us to the final section 6 where we address the issues of time consistency and of subgame perfectness.

Part I. The Motivating Energy Modeling Study

In 1987, in his PhD thesis [19], G. Zaccour extended the stochastic programming method to a game theoretic context, by introducing a Nash equilibrium in a class of strategies where the decisions of each player were indexed over the nodes of an event tree. This event tree was used to represent the uncontrolled uncertainty characterizing the oil price scenarios. This solution concept has been called *S*-adapted for "sample path adapted". In this first part we revisit this initial model that has had the misfortune of not being published in a widely circulated journal[1] although it has served as the motivating example for a recent publication concerning the solution of stochastic variational inequalities [6].

2 A Model of the European Gas Market

In this section the European gas market model is presented with enough detail to permit an analysis of the scenario simulation results that were obtained and an interpretation in terms of contingent contracts. Furthermore, as the modeling was undertaken almost 20 years ago, we can now compare the simulations obtained from the model with the actual realization in the period 1985–2000. So this is a case where the model can be put to a test against reality.

2.1 The Structure of the European Gas Market

Typically the European gas market is organized as a multilevel structure involving producers, distributors and consumers as schematized in Figure 1.

A modeling exercise has been undertaken in 1985 to represent the evolution of this market over the following four periods indicated in Table 1.

Table 1: The four time periods of the gas market model

1985–1989	1990–1994	1995–1999	2000–2019

The model does not consider the distributors as active players and concentrates on the producers and their production units as indicated in Table 2.

Table 2: The players and their production units

Algeria	→	Arzew, Skikda, Algeria Pipe
Holland	→	Groningen
Norway	→	EKOFISK, Troll
FUSSR	→	USSR1, USSR2

The nine consuming regions are shown in Table 3.

We assume that the market is oligopolistic and that the producers behave as Nash–Cournot equilibrium seekers.

[1]The reason for this delayed publication deserves to be told. The paper, submitted to a famous OR journal, has been lost by the editor in charge...

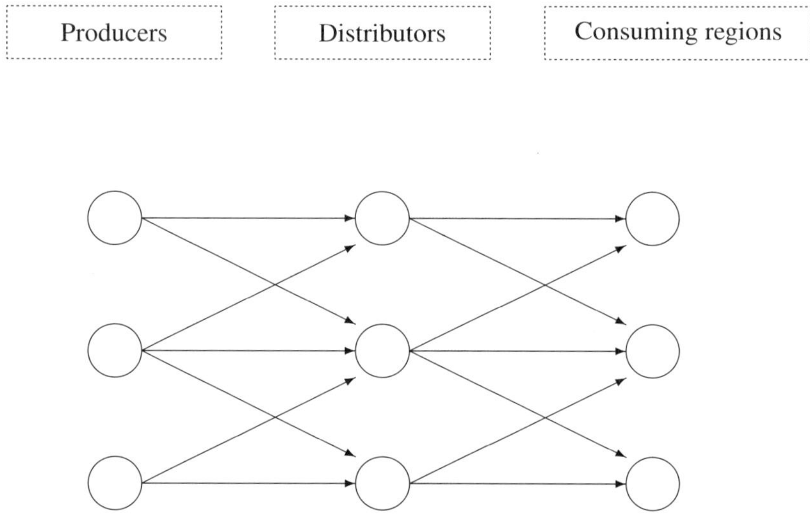

Figure 1: A multilevel market structure

Table 3: The consuming regions

Belgium–Luxembourg	
France North	France South
Italy North	Italy South
Holland	UK
FRC North	FRC South

2.2 Modeling of Gas Contracts

Competition on natural gas markets takes usually the form of negotiation of long-term contracts between producers and distributors. From a producer perspective, the main motivation for long-term contracts is that developing a gas field is an extremely expensive and lengthy venture and therefore a (long-term) contract is a guarantee that a demand exists for this field. From the perspective of a distributing company, a contract is synonymous with a reliable supply at a certain price. In practice, contracts are much more complex arrangements than a simple pair (price, quantity). They usually include number of operational clauses, a price indexation clause, lower and upper bounds on quantity etc.

The price escalation clause in gas contracts is based mainly on the price of oil and the lower bound on quantity represents a take-or-pay clause. Since obviously future price of oil is a random event, a gas contract between a seller and a buyer may take the form of a set of pairs (price, quantity), each one corresponding to a state of the world considered for oil price.

This indexation clause in the contractual terms is translated in the game-theoretic model through the introduction of the *S-adapted* information structure. For achieving this, one represents the uncertainty as an event tree, where the paths correspond to different oil price scenarios. The gas demand law in each consuming region is affected by the random oil price scenario. A player decision vector specifies for each node of the event tree the quantities to be exchanged with the distributors under the price of oil characteristic of this node. The tree structure can be interpreted in contractual terms: For a producer and distributor, the set of pairs (price, quantity) mentioned above is obtained by collecting the results for all nodes in the event tree that correspond to the same period. The lowest quantity in this set can be interpreted as a take-or-pay quantity clause. The highest one provides the upper bound on the volume of gas that can be lifted (annually for instance). Therefore these results correspond to quantities of gas that are contingent to the price of oil.

2.3 A Model of Competition Under Uncertainty

We can provide a mathematical formulation of the model sketched above. Consider a set of time periods $t \in \{0, \dots, T\}$. Let J be the set of consuming regions. In each region $j \in J$ one considers a set of demand laws indexed over a set of scenarios Ω which represent different conjunctural evolutions of the gas market. A scenario $\omega \in \Omega$ is represented as a path in an event tree as indicated in Figure 2. The set $\mathcal{N}^t$ contains the nodes n^t of the event tree at period t. These nodes represent all the different possible histories of the random events up to time t. We say that a scenario passes through node n^t if it has up to time t the history represented by node n^t. The cumulative probability of all the scenarios passing through the node $n^t \in \mathcal{N}^t$ is denoted $\pi(n^t)$. The set of players (producers) is denoted $M = \{1, \dots, m\}$. Each

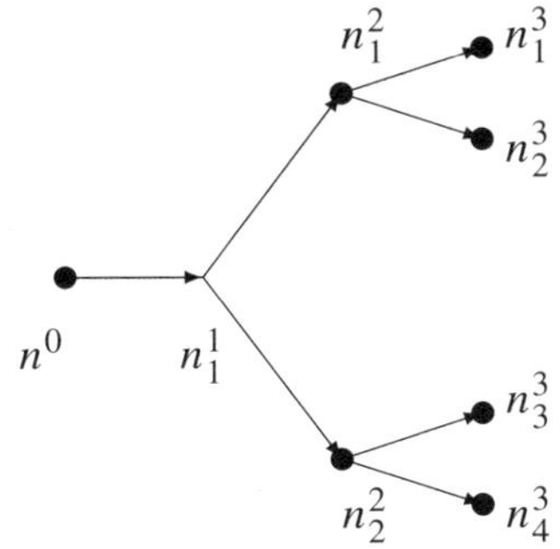

Figure 2: Event tree

player k controls a set U_k of production units. Each unit $\ell \in U_k$ is described by

a. The reserves it exploits $R_\ell(t)$ (with initial condition $R_\ell(0) = R_\ell^0$ given)
b. The production capacity $K_\ell(t)$ (with initial condition $K_\ell(0) = K_\ell^0$ given)

c. The production cost function $G_\ell(q_\ell(t))$, which is defined as a convex twice differentiable monotonically increasing function
d. The investment cost function $\Gamma_\ell(q_\ell(t))$, also defined as a given convex twice differentiable monotonically increasing function.

The players have to choose, at each period, the amount they produce and the investment in capacity expansion. When they make their decisions they know the node of the event tree that has been reached, i.e. they know the history of the gas market. The players decisions are therefore indexed over the nodes of the event tree. These contingent decisions are taken by the different players in order to be in a Nash equilibrium, i.e. each player's contingency plan is the best reply, in terms of expected profits to the contingency plans adopted by the other players.

The stochastic game problem can then be summarized as follows

$$\text{Equil.}\ [\pi_k = \sum_{t=0}^{T} \beta_i^t \Big\{ \sum_{j\in J} \sum_{\ell\in U_k} \sum_{n^t\in\mathcal{N}^t} \pi(n^t) \big[q_{\ell j}(n^t)(P_j(n^t)) \\ - G_\ell(q_{\ell j}(n^t)) - \Gamma_\ell(I_\ell(n^t))\big]\Big\}]_{k\in M} \tag{1}$$

s.t.

$$R_\ell(n^t) = R_\ell(a(n^t)) - q_\ell(a(n^t)) \quad \text{Reserves depletion} \tag{2}$$

$$K_\ell(n^t) = K_\ell(a(n^t)) + I_\ell(a(n^t)) \quad \text{Capacity expansion} \tag{3}$$

$$q_\ell(n^t) \le K_\ell(n^t) \quad \text{Capacity bounds} \tag{4}$$

$$P_j(n^t) = f_j(Q_j(n^t)) \quad \text{Demand laws} \tag{5}$$

$$0 \le I_\ell(a(n^t)),\ q_\ell(n^t),\ R_\ell(n^t) \\ \forall t, n^t \in \mathcal{N}^t, \ell \in U_i, k \in M. \tag{6}$$

In the profit function (1), the term $P_j(n^t)$ represents the clearing market price of gas in consuming region j, given the economic conjuncture represented by the node n^t. This price depends on the total quantity supplied on the market and will be further discussed in the next subsection. In the constraints (2) and (3), we use the notation $a(n^t)$ to denote the unique predecessor of the node n^t in the event tree. These constraints represent *state equations* in the parlance of control theory. As indicated, the investment and reserve commitment decisions taken at the antecedent node $a(n^t)$ determine the state variables $R_\ell(n^t)$ and $K_\ell(n^t)$.

One complements the model with the following specifications:

Production cost functions: We assume that the marginal production cost[2] increases when the production q gets closer to the capacity K

[2]The numerical values for the parameters entering in the definition of the cost and demand functions can be obtained in the report [13] which is available upon request to the authors.

$$G'(q_\ell) = \alpha_\ell + \frac{\gamma_\ell}{K_\ell - q_\ell}.$$

Investment cost functions: We assume constant marginal investment cost with differences among producing units

$$\Gamma'(I_\ell) = \begin{cases} \$\ 15/\text{MMBTU Troll} \\ \$\ 16.9/\text{MMBTU USSR 2} \\ \equiv 0 \text{ Algeria, Holland.} \end{cases}$$

2.4 The Stochastic Demand Law

The natural gas (inverse) demand law in each consuming region defines the price of gas as an affine function of the total quantity put on this market by the producers. Both the slope and intercept at the origin are random parameters depending on the price of oil. The event tree of Figure 3 shows the possible time evolutions of oil price. It shows a starting value of \$ 30/BBL in 1985-89 with a predicted decline to \$ 15/BBL in 1990-1994 followed in 1995-1999 by two possible price increases to \$ 20/BBL or \$ 30/BBL respectively. Further branching occurs after that period leading to prices ranging from \$ 20 to \$ 60/BBL in 2000-2019. At each node the probabilities of each branch is also indicated. Different demand laws are specified for each node.

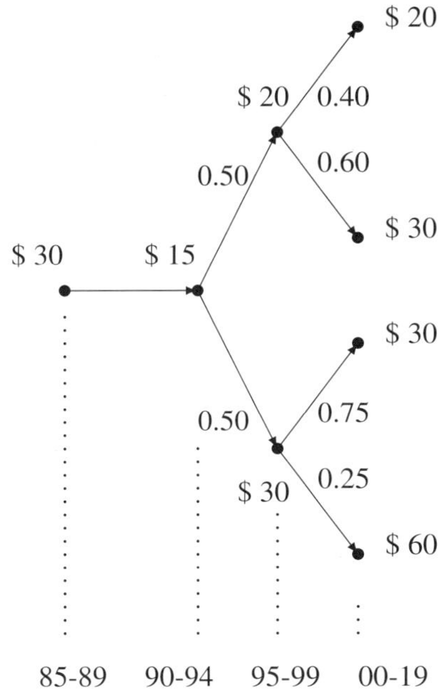

Figure 3: Stochastic oil prices

Table 4: Flows in mtoe (NHV)

	Period	1985–89	1990–94	1995–99		2000–2019			
Source	Destination	$ 30	$ 15	$ 20	$ 30	$ 20	$ 30	$ 30	$ 60
U.K.	U.K.	33.70	35.00	37.00	37.00	38.00	38.00	38.00	38.00
FRG	FRG (North)	10.00	13.70	12.74	11.40	12.40	11.62	11.71	8.39
FRG	FRG (South)	3.70	0.10	1.06	2.40	1.41	2.18	2.09	5.41
France	France (North)	1.86	2.61	3.00	3.00	3.00	3.00	3.00	1.58
France	France (South)	2.94	0.29	0.00	0.00	0.00	0.00	0.00	1.42
Italy	Italy (North)	2.03	0.00	0.04	0.69	0.45	1.06	1.10	1.39
Italy	Italy (South)	8.37	10.00	9.96	9.32	9.55	8.94	8.90	8.61
Alg. pipe	Italy (South)	5.06	4.25	7.55	9.00	9.30	10.82	10.84	9.66
Arzew	France (North)	3.24	2.43	4.48	5.47	5.53	7.00	7.03	6.28
Arzew	Belgium Lux.	3.55	2.79	3.98	4.55	4.61	5.30	5.32	4.52
Skikda	France (South)	5.28	3.97	6.93	8.83	8.18	10.55	10.55	10.25
FURSS1	FRG (South)	13.67	16.01	16.90	15.88	17.88	17.26	17.30	12.82
FURSS1	France (South)	7.92	12.56	15.14	14.14	17.45	16.75	16.74	13.22
FURSS1	Italy (North)	9.72	14.54	17.15	16.38	18.90	18.60	18.62	14.70
FURSS2	U.K.	0.00	0.00	22.38	22.42	22.44	22.51	22.51	22.09
Holland	Holland	22.93	20.96	24.11	23.97	24.35	23.97	23.90	22.08
Holland	FRG (North)	16.66	13.77	13.79	14.60	12.63	12.99	12.90	13.00
Holland	France (North)	3.31	4.04	3.93	3.71	3.50	3.14	3.10	3.52
Holland	Italy (North)	5.34	4.13	5.11	6.25	4.63	5.50	5.41	7.17
Holland	Belgium Lux.	3.86	4.67	4.22	3.88	4.25	3.77	3.75	3.29
Ekofisk	U.K.	9.07	8.70	4.52	4.70	0.00	0.00	0.00	0.00
Ekofisk	FRG (North)	7.06	3.82	8.34	9.50	0.00	0.00	0.00	0.00
Ekofisk	France (North)	1.49	1.23	11.88	1.65	0.00	0.00	0.00	0.00
Ekofisk	Belgium Lux.	2.18	2.58	3.06	2.83	0.27	0.27	0.00	0.00
Troll	U.K.	0.00	0.00	0.00	0.00	9.23	9.10	8.97	7.50
Troll	FRG (North)	0.00	0.00	0.00	0.00	12.34	13.11	13.03	13.99
Troll	France (North)	0.00	0.00	0.00	0.00	3.17	3.07	3.03	3.88
Troll	Belgium Lux.	0.00	0.00	0.00	0.00	3.82	3.47	3.72	3.47

2.5 The Equilibrium Strategies

The computation of the equilibrium strategies, adapted to the event tree describing the uncertainty on demand gave the flows to consuming regions shown in Table 4 and total production decisions shown in Table 7.

The equilibrium strategy is interesting in a period where there are several nodes, like in 1995–99. Depending on the prevailing oil price ($ 20 or $ 30) the investment decision will be different. Consider, for example, the reserve commitments shown in Table 7. In 1995–99, Algeria commits 22.94 *mtoe* or 27.85 *mtoe* of its reserves if the price of oil is $ 20 or $ 30 respectively. The reserve commitment decision is an inter-temporal decision, since like investment for capacity expansion, it determines the evolution of the reserves (another type of nonrenewable capital). In Figure 3 one sees the different situations corresponding to the two nodes. In the $ 20 node the perspective is a branching between $ 20 or $ 30 prices for the future (final) period 2000–2019, with respective probabilities 0.40 and 0.60. The reserve commitment decision is taken via an arbitrage between the immediate profit (in 1995–99) and the value of reserves in the forthcoming period 2000–2019. This future value is obtained as the conditional expectation of the two possible values if the oil price were to be low ($ 20) or moderate ($ 30). Indeed this arbitrage is also made as the best response to the commitments decided by the competitors. In the $ 30 node, of period 1995–99, the perspective is a branching between $ 30 or $ 60 prices for the future (final) period 2000–2019, with respective probabilities 0.75 and 0.25. This is a different perspective, with different immediate profits and future values of reserves, hence the different commitment.

The resulting prices of gas on the different markets are shown in Table 5.

Table 5: Price of gas $ 83/MMBTU on different markets

Period	**1985–89**	**1990–94**	**1995–99**		**2000–2019**			
Oil price	**$30**	**$15**	**$20**	**$30**	**$20**	**$30**	**$30**	**$60**
Holland	6.29	3.26	4.11	5.88	4.06	5.85	5.85	10.60
U.K.	4.72	2.61	2.66	3.50	2.66	3.51	3.515	4.715
Germany (North)	4.35	2.59	3.23	4.41	3.24	4.44	4.45	7.29
Germany (South)	4.35	2.59	3.23	4.42	3.24	4.44	4.45	7.29
France (North)	4.57	2.71	3.11	4.02	3.04	3.95	3.46	6.18
France (South)	4.57	2.66	3.06	4.00	2.97	3.82	3.82	6.18
Italy (North)	4.22	2.66	3.28	4.39	3.35	4.45	4.45	7.33
Italy (South)	4.22	2.71	3.28	4.39	3.25	4.45	4.45	7.33
Belgium-Luxemb.	5.11	3.09	3.71	4.95	3.07	4.93	4.94	8.04

The total gas demand is shown in Table 6. Annual production and reserves at the end of period are in Table 7.

Table 6: Annual total gas demand in mtoe (PCI)

Period	1985–89	1990–94	1995–99		2000–2019			
Oil price	$30	$15	$20	$30	$20	$30	$30	$60
Gas Demand	182.9	182.1	227.3	231.7	247.2	252.0	251.5	236.3

Table 7: Annual production and reserves at the end of period

Period Oil price $83/bbl		1985–89 $30	1990–94 $15	1995–99 $20	 $30	 $20	2000–2019 $30	 $30	 $60
Algeria	(a)	17.14	13.44	22.94	27.85	27.62	33.66	33.74	30.71
	(b)	2745	2678	2564	2539	2011	1890	1864	1925
Holland	(a)	52.09	47.56	51.17	52.40	49.37	49.37	49.06	49.06
	(b)	1481	1243	987	981				
Norway	(a)	19.80	16.32	17.80	18.87	28.82	29.01	28.75	28.85
	(b)	176(1) 1733(2)	94 1733	5 1733	 1733	 1202	 1179	 1199	 1197
FUSSR	(a)	31.31	43.10	71.58	68.83	76.63	75.12	75.16	62.85
	(b)	3280	3065	2707	2721	1174	1205	1218	1464

(1) Ekofisk
(2) Troll
(a) production
(b) reserves

2.6 20 Years After

This model was proposed almost 20 years ago to represent the possible evolutions of market shares on a time horizon extending up to 2020. We have now observed the evolution of the price of oil and we can obtain from the energy statistics the volumes of gas circulating in Europe. We may thus compare the market shares[3] of the different producers "forecast" in 1985 for the next fifteen years and observations. The assumptions made at that date regarding the price of oil, expressed in $1983, were as follows: $30/bbl for the period 1985-1989, $15/bbl for 1990-1994 and two possibilities, $20 and $30, for 1995-1999. During this last period, the price of oil was actually closer to $20/bbl than to $30/bbl and therefore we shall compare the realizations with the "forecast" made under this assumption. The market shares

[3]We could have done this comparison on the basis of quantities instead of market shares. The advantage of this formulation is that all root mean square errors are easily comparable with market shares as unit of measurement.

predicted and the ones observed have been compared[4]. We only report the *Root Mean Square* errors (RMSE) for the different markets in Table 8.

Table 8: Root Mean Square Errors

Holland	0.03113
UK	0.15103
Germany	0.04296
France	0.09296
Italy	0.09296
Belgium–Luxembourg	0.05461

As it can be readily seen, the model predicted quite well in the case of Holland, Germany and Belgium–Luxembourg. The performance is good for France and Italy. In the case of France, the model forecast an important role for Norway and in the case of Italy, the model underestimated the market share of the USSR. The relatively bad performance in the UK case is due to a forecast of an important role of Russian gas during the last period (1995–1999) and the result was no gas from Russia at all.

This comparison exercise shows that the S-adapted information structure, when applied to model stochastic resource markets can provide useful insights. One could also re-run the model for the realized scenario and compute the value of information by comparing the deterministic solution with the uncertain one.

2.7 In Summary

The European gas market modeling exercise has the following original features:

- It models some real life contingent contracts that have indexation clauses triggered by the random evolution of the price of oil;
- In doing so it uses a formalism of a Nash–Cournot game played over an event tree that represents the possible evolutions (scenarios) of the price of oil;
- The equilibrium is characterized as the solution of an extended variational inequality for which we have existence and uniqueness results, as shown in section 4;
- The model has dynamical constraints representing the capacity expansion and the reserve depletion processes; the decision of investment and production (reserve commitment) are therefore similar to control variables that influence the evolution of a state variable. An economic interpretation of the decision taken at each node can be obtained by considering the tradeoffs for the agents between the immediate gain and the impact of the current decision on the conditional expected value of the state variables in the next few periods, as discussed in section 5.
- Finally a comparison of the scenarios simulated and the actual realization shows a good fit.

[4]The fully detailed tables can be obtained from the authors, on request.

Part II. The Theory of Games Played on Uncontrolled Event Trees

In Part 1 we have reviewed an economic model where the players choose contingency plans represented as solutions indexed over the set of nodes of an event tree. In this second part we study the information structure that is subsumed by this indexing of decisions over an event tree. For that we use a variety of different paradigms that are used to study dynamic games.

3 A Formulation of S-Adapted Strategies in the Classical Game Theory Framework

In this section we present S-adapted strategies as a particular type of information structure for a game in extensive form. To simplify the exposition we restrict ourselves to the two-player case.

3.1 A Two-Player Game in Extensive Form

We consider a two-player game with sequential simultaneous moves. Nature intervenes randomly after every decision run consisting of an action chosen by player 1 and an action chosen by player 2. Players cannot observe the action taken by the opponent. Nature selects an event randomly and independently of players' moves. Nature's moves can be sequentially dependent, like *e.g.* in a Markov chain or can be the representation of a more general stochastic process. Nature's moves are known to each player. Players use V-N-M utilities so they strive to maximize their expected payoffs. A run is a sequence of a two-decision move, one by each player, followed by Nature's move. The game is played over T runs. Payoffs (in utils) depend on the terminal node reached after the last run. A run is illustrated in Figure 4 where, as usual, the action nodes are represented by squares, event nodes by circles. The information structure is represented by the dotted lines that connect together all the elements belonging to the same information set. At the beginning of the run, Player 1, knowing that the game is in node s chooses an action. Player 2 chooses her action without knowing the choice made by Player 1, as indicated by the dotted line. Nature then makes a random move, without taking into account the actions chosen by the players (this is also indicated by a dotted line crossing all the nodes, at that decision level). Next run is played similarly, Player 1 knowing the new node s' which is the updated history of Nature's moves, etc...

3.2 Normal Form Representation

We index the runs over a set of time periods $t \in \{0, 1, \ldots, T-1\}$. It is convenient to introduce an initial state of Nature denoted $\xi(0)$. Nature's moves make this state evolve over periods. The Nature state evolution is then represented by a stochastic process $\{\xi(t) : t = 0, 1, \ldots, T\}$. The information structure described above in Figure 4 says that each player has at her disposal at period t the whole history

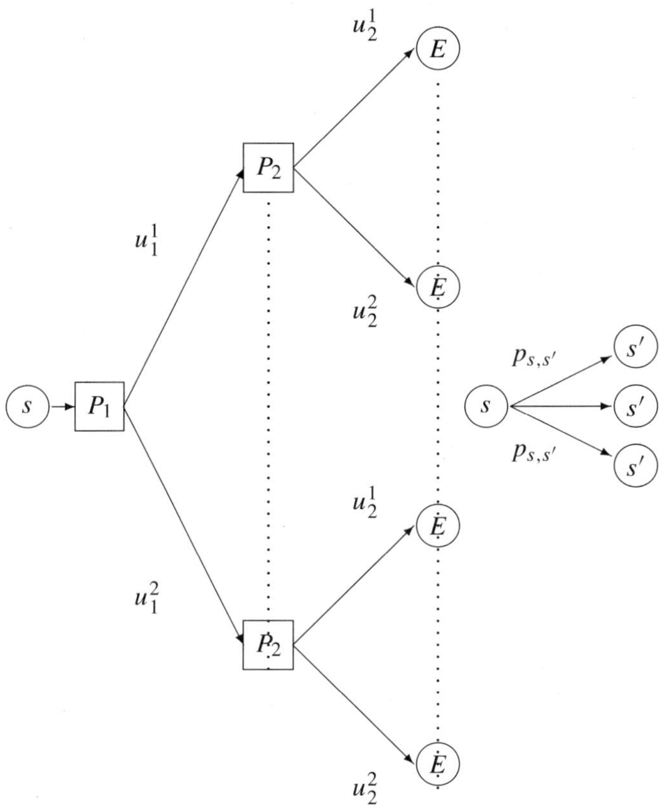

Figure 4: A "run" in the extensive form game

$h^t = \{\xi(0), \ldots, \xi(t)\}$ of the $\xi(\cdot)$ process up to time t. Therefore Player k picks, at time t an action in a finite set U_k^t, using a strategy defined by the mappings

$$\gamma_k^t : h^t \mapsto \gamma_k^t \in \mathcal{P}[U_k^t],\ k = 1, 2, t = 0, 1, \ldots, T-1, \tag{7}$$

where $\mathcal{P}[U_k^t]$ is the simplex of probability measures on the finite set U_k^t. The rewards are defined at terminal nodes $n(T)$ of the game tree. We represent them by two functions of the terminal nodes $G_k(n(T))$, $k = 1, 2$. With this construct, when the players adopt strategies $\gamma_k = \{\gamma_{kt} : t = 0, 1, \ldots, T-1\}$, $k = 1, 2$, respectively, the rewards become random variables $\tilde{G}_k(T)$ with probability laws depending on the strategies pair $\underline{\gamma} = (\gamma_1, \gamma_2)$. So we define the game payoffs by

$$V_k(\gamma_1, \gamma_2) = \mathrm{E}_{\underline{\gamma}}\left[\tilde{G}_k(T)\right], k = 1, 2. \tag{8}$$

We now use the usual definition of an equilibrium

Definition 3.1. An equilibrium is an admissible strategy pair (γ_1^*, γ_2^*) such that

$$V_1(\gamma_1, \gamma_2^*) \leq V_1(\gamma_1^*, \gamma_2^*), \tag{9}$$

$$V_2(\gamma_1^*, \gamma_2) \leq V_2(\gamma_1^*, \gamma_2^*), \tag{10}$$

for any admissible strategy γ_1 or γ_2.

We notice that, with our assumption of finiteness of state and action sets, and using Kuhn's theorem [16], we can formulate this game as a (large scale) bimatrix game. However, if one deals with the modeling of an economic competition, a formulation as a matrix game is not very convenient. First of all the sizes of the matrices tend to be huge. Furthermore the equilibria will be most probably obtained as mixed-strategies with a rather difficult interpretation. This is why in the following section we shall reformulate the model as a stochastic concave game with actions in vector spaces.

4 S-Adapted Strategies in Dynamic Games

In this section we extend the framework in order to describe economic competition "a la Cournot" in a dynamic and random environment. For that purpose we shall introduce a formalism that is akin to *stochastic programming* ([2]). From now on we use the term *period* instead of *run*.

4.1 Strategies as Variables Indexed on an Event Tree

The set of periods is $\mathcal{T} = \{0, 1, \ldots, T\}$. The randomness affecting the competition process is an exogenous stochastic process $(\xi(t) : t \in \mathcal{T})$ represented by an *event tree*. This tree has a root node n_0 in period 0 and has a set of nodes $\mathcal{N}^t$ in period $t = 1, \ldots, T$. Each node $n^t \in \mathcal{N}^t$ represents a possible sample value of the history h^t of the $\xi(\cdot)$ process up to time t. The tree graph structure represents the nesting of information as time periods succeed each other. Introduce the following notations:

- $a(n^t) \in \mathcal{N}^{t-1}$ is the unique predecessor of node $n^t \in \mathcal{N}^t$ in the event tree graph for $t = 1, \ldots, T$;
- $\mathcal{S}(n^t) \in \mathcal{N}^{t+1}$ is the set of possible direct successors of node $n^t \in \mathcal{N}^t$ for $t = 0, \ldots, T-1$;
- a complete path from the root node n_0 to a terminal node n^T is called a *scenario*. Each scenario has a probability and the probabilities of all scenarios sum to 1;
- with each node n^t is associated the "probability of passing through this node" denoted $\pi(n^t)$; it is the sum of the probabilities of all the scenarios that contain this node, indeed $\pi(n_0) = 1$ and $\pi(n^T)$ is equal to the probability of the single scenario that terminates in (leaf) node n^T.

We consider a set $M = \{1, \ldots, m\}$ of players. For each player $k \in M$, we define a set of decision variables indexed over the set of nodes. We call $x_k^{n^t} \in \mathbb{R}^{m_k}$ the decision variables of player k at node n^t.

Remark 4.1. The S-adapted information structure is subsumed by the indexing of the decision variables over the set of nodes in the event tree. Indeed, each node of the event tree is an exhaustive summary of the history of the $\xi(\cdot)$-process. Making the decision variables depend on the nodes in the event tree is therefore equivalent to saying that the decisions are adapted to the history of the $\xi(\cdot)$-process, which is exactly what is meant by S-adapted strategies.

4.2 Rewards and Constraints

4.2.1 Transition and Terminal Rewards

Define, for each node n^t, $t = 1, \ldots .T$, a *transition reward* function

$$L_k^{n^t}(\underline{x}^{n^t}, \underline{x}^{a(n^t)}),$$

where $L_k^{n^t}(\cdot, \cdot)$ is twice continuously differentiable. Notice that we make the rewards depend on the transition from decisions at the preceding period and antecedent node $a(n^t)$ and decisions at the current period and node n^t; therefore it is a transition reward that should be associated with time period $t-1$ and therefore discounted by a factor β_k^{t-1} if a discount factor $\beta_k \in [0, 1]$ were used by Player k. We shall not introduce discount factors in our formalism, for the sake of keeping it as simple as possible.

At each terminal node n^T is defined a terminal reward $\Phi_k^{n^T}(\underline{x}^{n^T})$ which is also supposed to be twice continuously differentiable.

4.2.2 Constraints

We introduce two groups of constraints for each player

$$f_k^{n^t}(x_k^{n^t}) \geq 0, \tag{11}$$

$$g_k^{n^t}(x_k^{n^t}, x_k^{a(n^t)}) \geq 0, \tag{12}$$

where f_k and g_k are mappings from Euclidean spaces to Euclidean spaces that are also supposed to be twice continuously differentiable.

Remark 4.2. These constraints are decoupled for the different players. We could introduce as well constraints that involve ancestors of the current node n^t, like *e.g.* $h_k(x_k^{n^t}, x_k^{a(n^t)}, \ldots, x_k^{a^P(n^t)}) \geq 0$, where $a^\ell(n^t)$ represents the operator that associates with a node n^t its ℓ-period ancestor in the event tree[5]. We shall keep the formulation (11–12) for the sake of simplifying the notations.

4.3 Normal Form Game and S-Adapted Equilibrium

Definition 4.1. An admissible S-adapted strategy for player k, is a vector $\gamma_k = \{x_k^{n^t} : n^t \in \mathcal{N}^t,\ t = 0, \dots, T-1\}$ that satisfies all the constraints (11)–(12). We call Γ_k the set of admissible S-adapted strategies of Player k.

Associated with an admissible S-adapted strategy vector $\underline{\gamma} = \{\gamma_k\}_{k\in M}$ we define the payoffs

$$V_k(\underline{\gamma}) = \sum^{t=1,\dots,T} \sum_{n^t\in\mathcal{N}^t} \pi(n^t) L_k^{n^t}(\underline{x}^{n^t}, \underline{x}^{a(n^t)}) + \sum_{n^T\in\mathcal{N}_T} \pi^{n^T} \Phi_k^{n^T}(\underline{x}^{n^T}), \quad k \in M.$$

The strategy sets Γ_k and the payoff functions $V_k : \Gamma_1 \times \dots \times \Gamma_k \times \dots \times \Gamma_m \to \mathbb{R}$ define a game in normal form. An S-adapted equilibrium is an equilibrium for this particular normal form game.

Definition 4.2. An S-adapted equilibrium is an admissible S-adapted strategy vector $\underline{\gamma}^*$ such that

$$V_k(\underline{\gamma}^*) \le V_k([\underline{\gamma}^{*-k}, \gamma_k]), \quad k = 1, \dots, m, \tag{13}$$

where,

$$[\underline{\gamma}^{*-j}, \gamma_k] = (\gamma_1^*, \dots, \gamma_k, \dots, \gamma_m^*),$$

represents a unilateral deviation by Player k from the equilibrium.

4.4 Concave Game Properties and Variational Inequality Reformulation

The game defined above is a concave game as defined by Rosen [17]. We can therefore adapt the existence and uniqueness theorems proved by Rosen.

4.4.1 Existence and Uniqueness Conditions

Lemma 4.1. *Assume the functions $L_k^{n^t}(\underline{x}^{n^t}, \underline{x}^{a(n^t)})$ and $g_k^{n^t}(x_k^{n^t}, x_k^{a(n^t)})$ are concave in $(x_k^{n^t}, x_k^{a(n^t)})$; assume the functions $f_k^{n^t}(x_k^{n^t})$ are concave in $x_k^{n^t}$ and assume the functions $\Phi_k^{n^T}(\underline{x}^{n^T})$ are concave in $x_k^{n^T}$. Assume that the set of admissible strategies is compact. Then there exists an equilibrium.*

[5]This type of constraints could be used to represent capacity accumulation in technology models. A technology with life τ has, at time t, a capacity which corresponds to the sum of all investments made over the past τ years $(t - \tau + 1, \dots, t)$.

Proof. Using classical results proved in [17] and based on the Kakutani fixed point theorem one easily obtains the existence of equilibria. □

We can also adapt the results proved in [17] to provide conditions under which this equilibrium is unique. It is convenient to introduce the notations $\mathbf{x}_k^{n^t} = (x_k^{n^t}, x_k^{a(n^t)})$ and $\underline{\mathbf{x}}^{n^t} = (\underline{x}^{n^t}, \underline{x}^{a(n^t)})$ respectively. Define now the *pseudo-gradients*

$$\mathcal{G}^{n^t}(\underline{\mathbf{x}}^{n^t}) = \begin{pmatrix} \dfrac{\partial L_1^{n^t}(\underline{\mathbf{x}}^{n^t})}{\partial \mathbf{x}_1^{n^t}} \\ \vdots \\ \dfrac{\partial L_k^{n^t}(\underline{\mathbf{x}}^{n^t})}{\partial \mathbf{x}_k^{n^t}} \\ \vdots \\ \dfrac{\partial L_m^{n^t}(\underline{\mathbf{x}}^{n^t})}{\partial \mathbf{x}_m^{n^t}} \end{pmatrix}, \quad t = 1, \ldots, T, \tag{14}$$

$$\hat{\mathcal{G}}^{n^t}(\underline{x}^{n^T}) = \begin{pmatrix} \dfrac{\partial \Phi_1^{n^t}(\underline{x}^{n^T})}{\partial x_1^{n^T}} \\ \vdots \\ \dfrac{\partial \Phi_k^{n^t}(\underline{\mathbf{x}}^{n^t})}{\partial x_k^{n^T}} \\ \vdots \\ \dfrac{\partial \Phi_m^{n^t}(\underline{\mathbf{x}}^{n^t})}{\partial x_m^{n^T}} \end{pmatrix}, \quad t = 1, \ldots, T, \tag{15}$$

and the Jacobian matrices

$$\mathcal{J}^{n^t}(\underline{\mathbf{x}}^{n^t}) = \frac{\partial \mathcal{G}^{n^t}(\underline{\mathbf{x}}^{n^t})}{\partial \underline{\mathbf{x}}^{n^t}}, \quad t = 1, \ldots, T \tag{16}$$

$$\hat{\mathcal{J}}^{n^t}(\underline{x}^{n^T}) = \frac{\partial \hat{\mathcal{G}}^{n^T}(\underline{x}^{n^T})}{\partial \underline{x}^{n^T}}. \tag{17}$$

Lemma 4.2. *If, for all* $\underline{\mathbf{x}}^{n^t}$ *the matrices* $\mathcal{Q}^{n^t}(\underline{\mathbf{x}}^{n^t}) = \frac{1}{2}[\mathcal{J}^{n^t}(\underline{\mathbf{x}}^{n^t}) + (\mathcal{J}^{n^t}(\underline{\mathbf{x}}^{n^t}))']$ *and* $\hat{\mathcal{Q}}^{n^T}(\underline{x}^{n^T}) = \frac{1}{2}[\mathcal{J}^{n^T}(\underline{x}^{n^T}) + (\mathcal{J}^{n^T}(\underline{x}^{n^T}))']$ *are negative definite, then the equilibrium is unique.*

Proof. This is exactly the setting of Theorem-5 in Rosen's paper [17]. The negative definiteness of the matrices $\mathcal{Q}^{n^t}(\underline{\mathbf{x}}^{n^t})$ and $\hat{\mathcal{Q}}^{n^T}(\underline{x}^{n^T})$ implies the strict diagonal concavity of the function $\sum_{k=1}^{m} V_k(\underline{\gamma})$ which implies uniqueness of the equilibrium. □

Remark 4.3. Another result of Rosen's paper [17] could extend straightforwardly to this stochastic equilibrium framework. It is related to the existence of *normalized equilibria* when there exists a coupled constraint $h(\underline{\mathbf{x}}^{n^t}) \geq 0$ at each node, where h is a concave function satisfying the constraint qualification conditions. We shall not develop further this aspect of the modeling of interacting firms that have to satisfy jointly some global constraints, as it is typically the case in environmental management. We refer to [3], [8], [12] for more developments on dynamic game models with coupled constraints.

4.4.2 Variational Inequality Reformulation

It is well known that Nash-equilibria in concave games can be characterized as solutions of a variational inequality (VI). This type of characterization extends readily to the case of equilibria in the class of S-adapted strategies. This has been done in [6] and, more recently, in [9] and [10]. The VI formulation is useful for the design of efficient numerical methods. The challenge in the solution of stochastic VI's is to deal with the large dimension of the problem, when the event tree grows bigger. In the three references given above, one applies approximation techniques to obtain ε-equilibria using reduced size event trees, obtained through sampling. We shall not develop further this topic and refer the interested reader to these papers for details.

5 A Stochastic Control Formulation

In this section we slightly modify the formalism used previously in order to establish a link with the theory of open-loop multistage or differential games. The particularity of the control formalism is to have state equations that define state variables as consequences of the choice of control variables. The control variables are therefore the independent variables, the state variables are dependent variables. Introducing costate variables, defined as the Lagrange multipliers associated with the state equations, permits the elimination of state variable variations in the expression of the differential of the Lagrangian. Then, the necessary optimality conditions are written in terms of the control variations. This is, in a few words, the idea behind the maximum or extremum principles that have been obtained in open-loop optimal control or differential game Nash-equilibrium problems. We shall now explore the possibility characterizing the S-adapted equilibria through maximum principles.

We shall first obtain such a characterization for a multistage game, with random disturbances represented by an uncontrolled event tree and with a general state equation shared by all players. In a second subsection we shall explore this type of characterization for a piecewise deterministic game, with random Markov disturbances, where each player has a specific dynamics, the linking occurring only at the level of the reward functions that involve the whole state variable vector.

5.1 A Multistage Game Formulation

Consider an event tree with nodes $n^t \in \mathcal{N}^t$, $t = 0, 1, \ldots, T$. Let $\pi(n^t)$ be the probability of passing through node $n^t \in \mathcal{N}^t$, $t = 0, 1, \ldots, T$. A system of state equations is defined over the event tree. Let $X \subset \mathbb{R}^\nu$, with ν a given positive integer, be a state set. For each node $n^t \in \mathcal{N}^t$, $t = 0, 1, \ldots, T$ let $U_k^{n^t} \subset \mathbb{R}^{\mu_k^{n^t}}$, with $\mu_k^{n^t}$ a given positive integer, be the control set of player k. Denote $\underline{U}^{n^t} = U_1^{n^t} \times \cdots \times U_k^{n^t} \times \cdots \times U_m^{n^t}$ the product control sets. Associated with each node n^t we define a transition function $f^{n^t}(\cdot,\cdot) : X \times \underline{U}^{n^t} \mapsto X$. The state equations[6] are given by

$$x(n^t) = f^{a(n^t)}(x(a(n^t)), \underline{u}(a(n^t))), \tag{18}$$

$$\underline{u}(a(n^t)) \in \underline{U}^{a(n^t)}, \quad n^t \in \mathcal{N}^t, t = 1, \ldots, T. \tag{19}$$

According to the state equation (18), at each node n^t, a vector of controls decision $\underline{u}(n^t)$ will determine, in association with the current state $x(n^t)$ the state $x(n')$, for all descendent nodes $n' \in \mathcal{D}(n^t)$. This is illustrated in Figure 5.

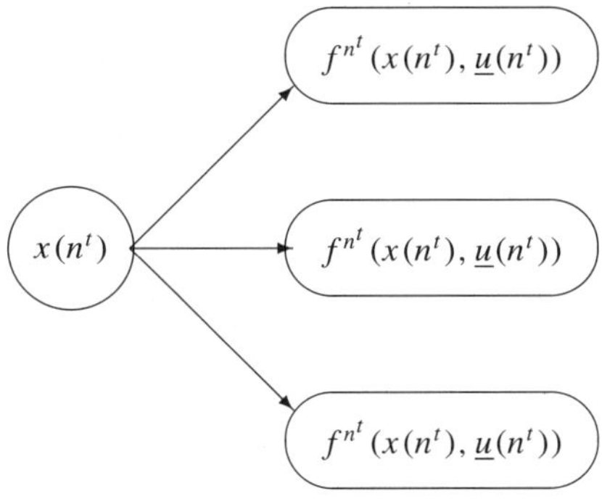

Figure 5: The stochastic state equations

In this state equation formalism, the decisions of the players are the controls $u_k(n^t)$ that are chosen independently. The state variables $x(n^t)$ are determined once the controls have been chosen. The state variables are shared by all the players and

[6]In order to differentiate the treatment of dynamic games from the preceding formulation as concave games, we shall use different notations for the indexing of variables over the event tree nodes. More specifically we shall now use $x(n^t)$ instead of x^{n^t}.

they enter into the definition of their reward functions as shown below in Eq. (20). At each node n^t, $t = 0, \dots, T-1$, the reward to player k is a function of the state and of the controls of all players, given by $\phi_k^{n^t}(x(n^t), \underline{u}(n^t))$. At a terminal node n^T the reward to Player k is given by a function $\Phi_k^{n^T}(x(n^T))$. This defines the following multistage game, where we denote $\mathbf{x} = \{x(n^t) : n^t \in \mathcal{N}^t, t = 0, \dots, T\}$ and $\underline{\mathbf{u}} = \{\underline{u}(n^t) : n^t \in \mathcal{N}^t, t = 0, \dots, T-1\}$ respectively and where $J_k(\mathbf{x}, \underline{\mathbf{u}})$ is the payoff of Player k:

$$J_k(\mathbf{x}, \underline{\mathbf{u}}) = \sum_{t=0}^{T-1} \sum_{n^t \in \mathcal{N}^t} \pi(n^t)\phi(n^t)_k(x(n^t), \underline{u}(n^t)) + \sum_{n^t \in \mathcal{N}_T} \pi^{n^T} \Phi_k^{n^T}(x^{n^T}), \quad k \in M, \tag{20}$$

s.t.

$$x(n^t) = f^{a(n^t)}(x(a(n^t)), \underline{u}(a(n^t))), \tag{21}$$

$$\underline{u}(a(n^t)) \in \underline{U}^{a(n^t)}, \quad n^t \in \mathcal{N}^t, t = 1, \dots, T, \tag{22}$$

$$x(n_0) = x^0 \text{ given.} \tag{23}$$

Definition 5.1. An admissible S-adapted strategy for Player k is defined by a vector $\underline{\mathbf{u}} = \{\underline{u}(n^t) : n^t \in \mathcal{N}^t, t = 0, \dots, T-1\}$. It defines a plan of actions adapted to the history of the random process represented by the event tree.

We can now define a game in normal form, with payoffs $V_k(\underline{\mathbf{u}}, x^0) = J_k(\mathbf{x}, \underline{\mathbf{u}})$, $k \in M$, where $\mathbf{x}$ is obtained from $\underline{\mathbf{u}}$ as the unique solution of the state equations that emanates from the initial state x^0.

Definition 5.2. An S-adapted equilibrium is an admissible S-adapted strategy $\underline{\mathbf{u}}^*$ such that for every player k the following holds

$$V_k(\underline{\mathbf{u}}^*, x^0) \geq V_k([\underline{\mathbf{u}}^{*-k}, \mathbf{u}_k], x^0). \tag{24}$$

It defines a plan of actions adapted to the history of the random process represented by the event tree.

Remark 5.1. In the definition of the S-adapted equilibrium we can notice the close resemblance to the open-loop information structure. The important difference lies essentially in the definition of state equations over an exogenous event tree, and of the controls as vectors indexed over the set of nodes of the event tree.

5.2 Lagrange Multipliers and Maximum Principles

As for open-loop multistage games we can formulate necessary conditions for an S-adapted equilibrium, in the form of a maximum principle.

For each player k, we form the Lagrangian

$$\mathcal{L}_k(x, \underline{u}, \lambda_k) = \phi_k^{n_0}(x^{n_0}, \underline{u}^{n_0}) + \sum_{t=1}^{T-1} \sum_{n^t \in \mathcal{N}^t} \pi(n^t)$$

$$\left\{\phi_k^{n^t}(x(n^t), \underline{u}(n^t)) + \lambda_k(n^t) \bullet (f^{a(n^t)}(x(a(n^t)), \underline{u}(a(n^t))) - x(n^t))\right\}$$

$$+ \sum_{n^t \in \mathcal{N}_T} \pi^{n^T} \left\{\Phi_k^{n^T}(x^{n^T}) + \lambda_k(n^T) \bullet (f^{a(n^T)}(x(a(n^T)), \underline{u}(a(n^T))) - x(n^T))\right\} \tag{25}$$

where the symbol $\bullet$ is used to represent a scalar product. In this expression we have introduced, for each player k, a *costate variable* λ_k^n, also indexed over the set of nodes and having the same dimension as x^n. Then we can define, for each player k and each node $n \in \mathcal{N}^t$, $t = 0, 1, \ldots, T-1$, the pre-Hamiltonian function

$$H_k^n(x(n), \underline{u}(n), \underline{\lambda}_k(\mathcal{S}(n))) = \phi_k^n(x(n), \underline{u}^n) + (\sum_{n' \in \mathcal{S}(n)} \frac{\pi^{n'}}{\pi^n} \lambda_k(n')) \bullet f^n(x(n), \underline{u}(n)), \tag{26}$$

where $\underline{\lambda}_k(\mathcal{S}(n))$ stands for all the vectors $\lambda_k^{n'}$ with $n' \in \mathcal{S}(n)$.

Remark 5.2. The main difference with the usual Hamiltonian of multistage games with the open-loop information structure, lies in the consideration of an average sensitivity vector $(\sum_{n' \in \mathcal{S}(n)} (\pi^{n'}/\pi^n)\lambda_k(n'))$. The interpretation of this average sensitivity is clearer when one notices that the ratio $(\pi^{n'}/\pi^n)$ is the transition probability from node n to node n'.

Theorem 5.1. *Assume that* $\underline{\mathbf{u}}^*$ *is an S-adapted equilibrium at* x^0*, generating the state trajectory* $x^{*(n^t)}$ *over the event tree. Then there exists, for each player k a costate trajectory* $\lambda_k(n^t)$ *such that the following conditions hold, for* $k \in M$*,* $u_k(n^t) = u_k^{*(n^t)}$*,* $x(n^t) = x^{*(n^t)}$

$$0 = \frac{\partial H_k^n}{\partial u_k^n} \quad n \in \mathcal{N}^t,\ t = 0, 1, \ldots, T-1, \tag{27}$$

$$\lambda_k^n = \frac{\partial H_k^n}{\partial x^n} \quad n \in \mathcal{N}^t,\ t = 0, 1, \ldots, T-1, \tag{28}$$

$$\lambda_k^{n^T} = \frac{\partial \Phi_k^{n^T}(x^{n^T})}{\partial x^{n^T}} \quad n^T \in \mathcal{N}_T. \tag{29}$$

Proof. In the Lagrangian expression (25) we regroup the terms that contain x^n.

$$\mathcal{L}_k(x, \underline{u}, \lambda_k) = \sum_{t=0}^{T-1} \sum_{n^t \in \mathcal{N}^t} \pi(n^t) \left\{ H_k(n^t)(x(n^t), \underline{u}(n^t), \underline{\lambda}_k^{S(n^t)}) - \lambda_k(n^t) x(n^t) \right\}$$
$$+ \sum_{n^T \in \mathcal{N}_T} \pi^{n^T} \left\{ \Phi_k^{n^T}(x^{n^T}) - \lambda_k^{n^T} x^{n^T} \right\}. \tag{30}$$

Conditions (26) and (27)–(29) are then obtained by equaling to 0 the partial derivatives of the Lagrangian w.r.t. x^t and u^t. □

5.3 Piecewise Deterministic Differential Games

When one formulates an S-adapted strategy for a continuous time dynamic game, one obtains a structure which is quite close to the class of piecewise deterministic games introduced in [7]. This will be presented below, in the particular case where each player controls her own dynamics.

5.3.1 A Stochastic Systems with Decoupled Controls

We consider a system observed over an infinite time interval $t \in [0, T)$. Player k is characterized at time t by a state variable $x_k(t) \in \mathbb{R}^{m_k}$. We assume that each player k controls the system by selecting an absolutely continuous function $x_k(\cdot)$, also called *trajectory*. Let I be a finite set. For each $i \in I$ a reward rate to Player k is given by the function $L_k^i(\underline{x}(t), \dot{x}_k(t))$ taking value in the extended line $\mathbb{R} \cup \{-\infty\}$. The reward takes a value $-\infty$ when the velocity $\dot{x}_k(t)$ is not admissible. We notice that only the time derivative of $x_k(\cdot)$ enters into the reward function of Player k, whereas the whole vector $\underline{x}(\cdot)$ appears as an argument of the reward function to create the interdependency between all players. We could also define terminal rewards $\Phi_k^i(\underline{x}(T))$ with values in $\mathbb{R}$, but, for the sake of simplicity we shall assume $\Phi_k^i(\underline{x}(T)) \equiv 0$.

Let $\xi(\cdot)$ be a stochastic (uncontrolled) jump process taking values in the set I. We denote $\tau_0 = 0$, τ_ν, $\nu = 1, 2, \ldots$, the time of the ν-th jump, $i_\nu = \xi(\tau_\nu)$, $h_\nu = (\tau_0, i_0, \tau_1, i_1, \ldots, \tau_\nu, i_\nu)$ the history of the $\xi(\cdot)$-process up to jump time τ_ν, and $\tilde{h}(t) = (\xi(s) : s \le t)$ the history of the $\xi(\cdot)$-process up to current time t.

5.3.2 S-Adapted Information Structure and Equilibria

At each jump time τ_ν, each player k, knowing the history h_ν, selects an a.c. function $x_k(t) : [\tau_\nu, T) \to \mathbb{R}^{m_k}$, with $x_k(\tau_\nu) = x_k(\tau_\nu^-)$ that is the trajectory of this player until the next jump occurs at random time $\min\{T, \tau_{\nu+1}\}$.

Even though the $\xi(\cdot)$-process can be very general, we shall concentrate from now on the case of a continuous time Markov chain with constant jump rates

$$q_{ij} = \lim_{dt \to 0} \frac{\mathrm{P}[\xi(t + dt) = j | \xi(t) = i]}{dt}. \tag{31}$$

Associated with a strategy vector $\underline{\gamma}$ we define the payoff to Player k as given by

$$V_k(\underline{\gamma}; i, \underline{x}^0) = \mathrm{E}_{\underline{\gamma}}\left[\int_0^T L_k^{\xi(t)}(\underline{x}(t), \dot{x}_k(t))\, dt | \xi(0) = i, \underline{x}(0) = \underline{x}^0\right]. \qquad (32)$$

As usual an equilibrium at $(i, \underline{x}^0)$ is defined as a strategy vector $\underline{\gamma}^*$ such that for every admissible $[\underline{\gamma}^{*-k}, \gamma_k]$ the following holds

$$V_k(\underline{\gamma}^*; i, \underline{x}^0) \geq V_k([\underline{\gamma}^{*-k}, \gamma_k]; i, \underline{x}^0). \qquad (33)$$

5.3.3 Comparison with Piecewise Open-Loop Equilibria

In [7] an information structure, called *Piecewise Open-Loop* (POL) has been introduced for this type of piecewise deterministic games. In POL strategies, the players observe, at each jump time τ_ν, the ξ-process value $\xi^\nu = \xi(\tau_\nu)$ and the trajectory values $\underline{x}^\nu = \underline{x}(\tau_\nu)$. Then each player k selects an a.c. function $x_k(t) : [\tau_\nu, T) \to \mathbb{R}^{m_k}$, with $x_k(\tau_\nu) = x_k(\tau_\nu^-)$ that is the trajectory of this player until the next jump occurs at random time $\min\{T, \tau_{\nu+1}\}$. In [11] POL equilibria have been computed for an oligopoly model with uncontrolled jump Markov disturbances. The POL information structure leads to a class of Markov strategies defining a Markov process $\chi(\tau_\nu) = (\xi(\tau_\nu), \underline{x}(\tau_\nu))$ observed at successive random times τ_ν. POL equilibria are characterized by a discrete event dynamic programming equation.

The important difference, between S-adapted and POL strategies is that, in the former, the trajectory states are not observed at jump times. In both information structures, the players observe the values taken by the $\xi(\cdot)$ process. In POL the players use Markov strategies that depend only on the current $\chi(\tau_\nu) = (\xi(\tau_\nu), \underline{x}(\tau_\nu))$ observation. In an S-adapted strategy, the player has to recall the whole history h_ν of the $\xi(\cdot)$-process up to time τ_ν. This information is necessary for reconstructing the trajectories realization up to time τ_ν. If the players were restricting their information uniquely to the current $\xi(\tau_\nu)$ value they could not infer the current trajectory state value $\underline{x}(\tau_\nu)$.

5.3.4 A Stochastic Maximum Principle

We consider state and costate variables $\tilde{x}(t, h_\nu)$ and $\tilde{\lambda}_k(t, h_\nu)$ that are indexed over the history of the $\xi(\cdot)$-process up to the last jump time τ_ν and with a running time $t \geq \tau_\nu$. We denote $(t, h_\nu)_\omega$ the history sample value, at time t, that is associated with an elementary realization (scenario) $\omega \in \Omega$.

We build the Lagrangian

$$\begin{aligned}\mathcal{L}_k(\underline{x}, z_k, \lambda_k) = \int_{\omega\in\Omega} dp(\omega) \int_0^T \Big\{ & L_k(\underline{\tilde{x}}((t, h_\nu)_\omega), z_k((t, h_\nu)_\omega)) \\ & + \tilde{\lambda}_k((t, h_\nu)_\omega) \bullet (z_k((t, h_\nu)_\omega) - \dot{\tilde{x}}_k((t, h_\nu)_\omega))\Big\}\, dt. \qquad (34)\end{aligned}$$

After change of order of integration and integration by parts we obtain

$$\mathcal{L}_k(\underline{x}, z_k, \lambda_k) = \int_{\omega\in\Omega} dp(\omega) \Bigg[\int_0^T \Big\{ L_k(\tilde{\underline{x}}((t, h_\nu)_\omega), \tilde{z}_k((t, h_\nu)_\omega)) + \tilde{\lambda}_k((t, h_\nu)_\omega) \bullet (z_k((t, h_\nu)_\omega)) + \dot{\tilde{\lambda}}_k((t, h_\nu)_\omega) \bullet \tilde{x}_k((t, h_\nu)_\omega)) \Big\} dt + [\tilde{\lambda}_k((t, h_\nu)_\omega) \bullet \tilde{x}_k((t, h_\nu)_\omega)]_0^T \Bigg]. \tag{35}$$

If we define the pre-Hamiltonian for Player k at history point $h_\nu(\omega)$ as

$$\tilde{H}_k(\tilde{x}, \tilde{\lambda}_k, \tilde{z}_k) = L_k(\tilde{x}, \tilde{z}_k) + \tilde{\lambda}_k \bullet \tilde{z}_k, \tag{36}$$

the expression (35) becomes

$$\mathcal{L}_k(\underline{x}, z_k, \lambda_k) = \int_{\omega\in\Omega} dp(\omega) \Bigg[\int_0^T \Big\{ \tilde{H}_k(\tilde{\underline{x}}((t, h_\nu)_\omega), \tilde{\lambda}_k((t, h_\nu)_\omega), \tilde{z}_k((t, h_\nu)_\omega)) + \dot{\tilde{\lambda}}_k((t, h_\nu)_\omega) \bullet \tilde{x}_k((t, h_\nu)_\omega)) \Big\}\, dt + [\tilde{\lambda}_k((t, h_\nu)_\omega) \bullet \tilde{x}_k((t, h_\nu)_\omega)]_0^T \Bigg]. \tag{37}$$

If we introduce the Hamiltonian

$$\tilde{\mathcal{H}}_k(\tilde{\underline{x}}, \tilde{\lambda}_k) = \sup_{z_k} \{ L_k(\tilde{\underline{x}}, z_k) + \tilde{\lambda}_k \bullet z_k \} \tag{38}$$

We have now to consider the set of all scenarios ω that share the same history up to time t. Call $h_\nu^{\theta,i}$ the history h_ν with $\tau_\nu = \theta$ and $\xi(\tau_\nu = i)$. If the history of the process is $(t, h_\nu^{\theta,i})$, with $t \geq \theta$, on a small time interval $[t, t+dt)$, we can have either an evolution $(t+dt, h_\nu^{\theta,i})$ with an elementary probability $1 - q^i dt$ or a switch to $(t+dt, h_{\nu+1}^{t,i})$, $j \neq i$, with an elementary probability $q_{ij}dt$. Here $h_{\nu+1}^{t,i}$ stands for the history where one appends the jump time t and the after jump state $\xi(t) = j$ to the history $h_\nu^{\theta,i}$.

We shall still simplify the notations by designing $\tilde{\lambda}^{\nu,\theta,i}$ and $\tilde{x}^{\nu,\theta,i}$ the costate and state variables indexed on history $h_\nu^{\theta,i}$ and $\tilde{\lambda}_k^{\nu+1,t,j}$ the costate variable indexed over the history $h_{\nu+1}^{t,i}$. Through standard reasoning of convex analysis we finally obtain the optimality conditions in the form of a family of coupled Hamiltonian systems

$$\dot{\tilde{x}}^{\nu,\theta,i}(t) \in -\partial_{\lambda_k} \tilde{\mathcal{H}}_k(\tilde{x}^{\nu,\theta,i}(t), \tilde{\lambda}_k^{\nu,\theta,i}(t)), \tag{39}$$

$$\dot{\tilde{\lambda}}^{\nu,\theta,i}(t) \in -\partial_{x_k} \tilde{\mathcal{H}}_k(\tilde{x}^{\nu,\theta,i}(t), \tilde{\lambda}_k^{\nu,\theta,i}(t)) - \sum_{j\in\mathcal{I}-i} q_{ij}(\tilde{\lambda}_k^{\nu+1,t,j}(t) - \tilde{\lambda}_k^{\nu,\theta,i}(t)). \tag{40}$$

Remark 5.3. The family of coupled Hamiltonian systems defined in (39)–(40) is very rich as there is a continuum of possible histories. This characterization of the equilibrium will be useful for understanding the qualitative properties of the equilibrium strategies.

6 The Time Consistency and Subgame Perfectness Issues

6.1 A Reminder

Dynamic games admit equilibria that may have two desirable properties:

Time consistency: This property is verified for open-loop Nash equilibria in deterministic differential games. It says that, along an *equilibrium trajectory*, if the game is stopped at a given time t and then the play resumes from the state $\underline{x}^*(t)$ reached along the trajectory, the continuation of the same open-loop equilibrium will still be an admissible equilibrium for this new game.

Subgame pefectness: This property (see [18]) holds true for feedback Nash equilibria in deterministic or stochastic differential games. It says that, if one has defined an equilibrium strategy $\underline{\gamma}^*$ and that the players have not played according to this equilibrium for some time and a state $x(t)$ has been reached at time t, then resuming the play according to $\underline{\gamma}^*$ from the initial condition $t, \underline{x}(t)$ will still be an equilibrium.

6.2 Properties of S-Adapted Equilibria

6.2.1 Time Consistency

S-adapted equilibria are time consistent. Consider first the case of a multistage game as defined by Eqs. (20)–(23) that is played on a finite event tree. Given an S-adapted strategy defined by the controls $\underline{u}(n^t)$ indexed over the nodes n^t and an initial state $\underline{x}(n^0)$ one can define an *extended trajectory* which associates a state $\underline{x}(n^t)$ with each node n^t. Now, let $\mathbf{\underline{u}}^* = \{\underline{u}^*(n^t) : n^t \in \mathcal{N}^t, t = 0, 1 \ldots, T-1\}$ be an S-adapted equilibrium, with the associated extended trajectory $\mathbf{\underline{x}}^* = \{\underline{x}^*(n^t) : n^t \in \mathcal{N}^t, t = 0, 1 \ldots, T\}$ emanating from $\underline{x}(n^0)$ and let us consider the subgame that starts at a node n^τ, with initial state $\underline{x}^*(n^\tau)$. The S-adapted strategy $\underline{u}^*(\cdot)$ restricted to the nodes that are descendants of n^τ, is still an equilibrium for this subgame. We call $\mathcal{D}[n^\tau]$ the set of all nodes that are descendants of n^τ. The proof of this result is, as usual, a direct application of the *tenet of transition* or *Bellman principle*. If the restricted strategy $\{\underline{u}^*(n^t) : n^t \in \mathcal{D}[n^\tau]\}$ were not an equilibrium, a Player k could improve her payoff by changing unilaterally her strategy on this portion of the event tree. The new strategy for the subgame could be combined with the vector $\{\underline{u}^*(n^t) : n^t \in \mathcal{N} \setminus \mathcal{D}[n^\tau]\}$ to define a new S-adapted strategy for the whole game, where Player k has changed her strategy unilaterally and has improved her payoff. This would contradict the equilibrium property assumed for $\underline{u}^*(\cdot)$.

6.2.2 Subgame Perfectness

S-adapted equilibria are not subgame perfect. The reason is that these strategies are very much akin to open-loop ones for deterministic games. If players do not play correctly at a node n^t, the control vector is $\underline{u}(n^t)$ instead of $\underline{u}^*(n^t)$. Then the trajectory is perturbed and the initial S-adapted strategy is not an equilibrium any more for the subgames that would initiate from the nodes $n' \in \mathcal{S}(n^t)$ with trajectory state $\underline{x}' = f^{n't}(\underline{x}^*(n^t), \underline{u}(n^t))$. In short, an error in play will destroy the equilibrium property for the subsequent subgames.

6.2.3 Extended Time Consistency for PDDGs

For PDDG's, the S-adapted equilibria are time-consistent over a set of sample trajectories that is very rich indeed. Since the set of scenarios is infinite and non-denumerable, it is possible to reach at time t a discrete state $\xi(t) = i$ and any continuous state $\underline{x}(t) \in \mathcal{X}^i(t)$, where $\mathcal{X}^i(t)$ is an infinite non-denumerable set. Assume that the players, after playing a non-equilibrium strategy reach at time t a state $(i, \underline{x})$ where $\underline{x} \in \mathcal{X}^i(t)$. This means that there is a history $h^{t,i}_{\nu+1}$ which would have led to $\underline{x}$ at time t if the players had played the equilibrium. Furthermore, since we assume that the uncontrolled random process is a Markov chain, the future of the process is only conditioned by the current state $\xi(t) = i$. Therefore, for the subgame defined from that point on, the S-adapted equilibrium remains an equilibrium. Therefore, we see that a sort of *local subgame perfectness* is likely to be observed in PDDGS's played under the S-adapted information structure.

7 Conclusion

In this paper we have explored the different facets of the concept of S-adapted equilibrium in a dynamic game. We have recalled that the concept has been introduced to represent the adaptation to random uncontrolled events that take place in the indexation clauses of gas contracts in the energy markets. Recalling the model that had been identified in 1987 to represent the evolution of the European gas market until 2020, we observe that the predicted equilibrium trajectories are reasonably close to those that have been observed in practice. We have then studied the various facets of the information structure subsumed by the S-adapted strategy concept with the help of different paradigms used in the theory of dynamic games. The most basic one is the formulation of a game in extensive form. The S-adapted strategy can be represented simply, using the classical tool of information sets. We then moved to a representation as a concave game, where the S-adapted information structure leads to equilibria that can be computed through mathematical programming techniques; furthermore, existence and uniqueness conditions are readily obtained, by straightforward extension of the classical results of Rosen concerning concave games. Finally, we have explored the control/differential game formalism. In a multistage game context one can formulate a maximum/equilibrium principle that characterizes an S-adapted equilibrium. In a differential game context, with jump Markov disturbances, the maximum principle becomes more "abstract" due to the fact that there is a continuum of possible trajectories for the continuous time

Markov chain. The maximum principles are related to the property of time consistency that the S-adapted equilibria exhibit. The property of subgame perfectness indeed is not verified, although due to the richness of possible trajectories of the jump Markov process, the time consistency property leads to a situation which is close to subgame perfectness in the domain of reachable initial states. However, the full exploration of this property is still to be made.

REFERENCES

[1] Başar T., Time consistency and robustness of equilibria in noncooperative dynamic games in F. van der Ploeg and A. J. de Zeeuw (eds.) *Dynamic Policy Games in Economics* , North Holland, Amsterdam, 1989.

[2] Birge J.R. and Louveaux F., *Introduction to Stochastic Programming* Springer Series in Operations Research. Springer-Verlag, New York, 1997.

[3] Carlson D. and Haurie A., Infinite horizon dynamic games with coupled state constraints, in J.A. filar, V. Gaitsgory and K. Misukami (eds.) Advances in dynamic games and applications, *Annals of the International Society of Dynamic Games*, Vol. 5, pp. 195–212, Birkhäuser, Boston, 2000.

[4] Friedman J.W., *Game Theory with Economic Applications*, Oxford University Press, Oxford, 1986.

[5] Fudenberg D. and Tirole J., *Game Theory*, MIT Press, Cambridge, MA 1991.

[6] Gürkan G., Özge A. Y. and Robinson S. M., Sample-path solution of stochastic variational inequalities, *Mathematical Programming*, 84 (1999), pp. 313–333.

[7] Haurie A., Piecewise deterministic differential games, in T. Baçar and P.Bernhard (eds.) *Differential Games and Applications*, Lecture Notes in Control and Information Sciences, Vol. 119, Springer-Verlag, Berlin, 1989.

[8] Haurie A., Environmental coordination in dynamic oligopolistic markets, *Group Decision and Negotiations*, Vol. 4, pp. 49–67, 1995.

[9] Haurie A. and Moresino F., Computation of S-adapted equilibria in piecewise deterministic games via stochastic programming methods, *Annals of the International Society of Dynamic Games*, Vol. 6, pp. 225–252, Birkhäuser, Cambridge, MA, 2001.

[10] Haurie A. and Moresino F., S-Adapted Oligopoly Equilibria and Approximations in Stochastic Variational Inequalities, *Annals of Operations Research*, to appear.

[11] Haurie A., Roche M., Turnpikes and computation of piecewsie open-loop equilibria in stochastic differential games, *Journal of Economic Dynamics and Control*, Vol. 18, pp. 317–344, 1994.

[12] Haurie A., Zaccour G., Differential game models of global environmental management, in C. Carrara and J.A. Filar eds., Control and game-theoretic models of the environment, *Annals of the International Society of Dynamic Games*, Vol. 2, pp. 3–23, Birkhäuser, Boston, 1995.

[13] Haurie A., Zaccour G., Legrand J. and Smeers Y., *A stochastic dynamic Nash-Cournot model for the European gas market*, *Cahiers du GERAD*, No. G-87–24, October 1987.

[14] Haurie A., Zaccour G., Legrand J. and Smeers Y., Un modèle de Nash Cournot stochastique et dynamique pour le marché européen du gaz, in *Actes du colloque Modélisation et analyse des marchés du gaz naturel*, HEC, Montréal, 1988.

[15] Haurie A., Zaccour G. and Smeers Y., Stochastic equilibrium programming for dynamic oligopolistic markets, *Journal of Optimization Theory and Applications*, vol. 66, No. 2, 243–253, 1990.

[16] Kuhn H.W., Extensive games and the problem of information, in H. W. Kuhn and A. W. Tucker (eds.), *Contributions to the theory of games*, Vol. 2, Annals of Mathematical Studies No 28, Princeton University Press, Princeton, New Jersey, 1953, pp. 193–216.

[17] Rosen J. B., Existence and uniqueness of equilibrium points for concave N-person games, *Econometrica*, vol. 33, 1965, pp. 520–534.

[18] Selten R., Rexamination of the Perfectness Concept for Equilibrium Points in Extensive Games, *International Journal of Game Theory*, Vol. 4, 1975, pp. 25–55.

[19] Zaccour G., *Théorie des jeux et marchés énergétiques: marché européen de gaz naturel et échanges d'électricité*, PhD thesis, HEC, Montréal, 1987.

Existence of Nash Equilibria in Endogenous Rent-Seeking Games

Koji Okuguchi
Gifu Shotoku Gakuen University
Gifu-shi, Gifu-ken 500-8288, Japan
okuguchi@gifu.shotoku.ac.jp

Abstract

The existence of Nash equilibria is investigated for one-stage and two-stage rent-seeking games with endogenous rent which depends on the aggregate expenditure by all rent-seeking agents such as individuals, firms or countries. Under reasonable assumptions on lottery production functions and rent function, both games turn out to have a unique equilibrium. The conditions for the equilibrium aggregate expenditure by all agents to increase in the first stage and for the total rent over two stages to dissipate are derived for the two-stage rent-seeking games.

1 Introduction

Rent-seeking games in which agents such as individuals, firms and countries try to obtain rents or prizes through non-market activities in the form of bribing bureaucrats, politicians, or in more general terms, rent-providers, are widely observed phenomena. Since the pioneering work of Tullock [6], many papers have been written on rent-seeking games, but in most of them rents have been assumed to be fixed at exogenously given levels independent of agents' expenditures to rent-providers. Furthermore, most of the papers on rent-seeking games have analyzed only one-stage or one-shot games in which the games are played only once. If x_i is i-th agent's expenditure for rent-seeking activity, and if x_i is assumed to produce lotteries in the amount of $f_i(x_i)$, where $0 < f_i' < \infty$ for $x_i \in (0, \infty)$, its probability of winning the rent is given by the expression $f_i(x_i)/\Sigma_{j=1}^n f_j(x_j)$, where n is the number of active agents. In general, the lottery production function may take any functional form. However, in most papers on rent-seeking games, it has been assumed that $f_i(x_i) = x_i^\alpha$, where α is a positive constant not necessarily less than one. If $\alpha < 1$, the lottery production function has decreasing returns, and if $\alpha > 1$, increasing returns. In the constant returns case where $\alpha = 1$, i-th agent's probability of winning the rent is equal to the ratio of its expenditure to the total expenditure by all agents. Perez-Castrillo and Verdier [4] have conducted a systematic analysis of the existence of pure Nash equilibrium in a one-stage rent-seeking game with exogenous rent on the basis of winning probability given by a logit function, *i.e.* $x_i^\alpha / \Sigma_{j=1}^n x_j^\alpha$. Okuguchi [2] has proven the existence of a unique pure Nash equilibrium in a similar game without using logit function for winning probability but assuming homogeneous agents, *i.e.* $f_i(x_i) \equiv f(x_i)$ for all i, as well as diminishing returns technology for lottery production. Szidarovszky and Okuguchi [3,5] have generalized his result for the case with nonhomogeneous agents. Chung [1] has first analyzed one-stage rent-seeking game where prize or rent is endogenously determined by the aggregate expenditure by all agents. In his game, any agent's probability of winning the rent is given by a logit function, and the aggregate expenditure by all agents is assumed to have immediate influence on the amount of rent offered by the rent-provider. This, however, is inappropriate when rent-seeking activities extend over more than one period. In multi-stage rent-seeking games agents learn how to influence the rent-provider over time, or the rent-provider learns how much rent to provide over time taking into account agents' expenditures in earlier stages. It may therefore be more appropriate to assume aggregate expenditure by all agents to influence the rent-provider with a time delay.

The plan of this paper is as follows. In Section 2, we will analyze the existence of pure Nash equilibrium in a one-stage rent-seeking game with endogenous rent and without assuming logit function for winning probability. In Section 3, after formulating a two-stage rent-seeking game, where the rent in the first stage is assumed to be exogenously given but that in the second stage is determined by the aggregate expenditure by all agents in the first stage, we will prove the existence of a unique subgame perfect Nash equilibrium. We will also analyze the effects of a change in the number of agents on the first-stage equilibrium aggregate expenditure by all agents. These effects will have bearings on the rent dissipation problem, which will be our concern in Section 4. Section 5 concludes.

2 One-Stage Game with Endogenous Rent

Let n be the number of agents in a rent-seeking game, $f(x_i)$ be the i-th agent's lottery production function, where x_i is its expenditure devoted to rent-seeking activity, and $R\left(\sum_{j=1}^n x_j\right)$ be the rent as a function of the aggregate expenditure by all agents. We have to assume the following.

Assumption 1.

$$\begin{cases} f(0) = 0,\ f'(0) = \infty,\ f'(\infty) = 0, \\ f'(x_i) > 0,\ f''(x_i) < 0,\ x_i \in (0, \infty). \end{cases}$$

Assumption 2.

$$\begin{cases} R(0) = 0,\ R(\infty) = \textit{a finite positive number}, \\ R'(0) = \infty,\ R'(\infty) = 0, \\ R'(X) > 0,\ R''(X) < 0,\ X \equiv \Sigma x_j \in (0, \infty). \end{cases}$$

Since the i-th agent's winning probability for the rent is $f(x_i)/\Sigma_{j=1}^{n} f(x_j)$, its expected net-rent, π_i, is defined by

$$\pi_i = R(\Sigma x_j) f(x_i)/\Sigma f(x_j) - x_i, i = 1, 2, \dots, n, \tag{1}$$

where π_i is assumed to satisfy

Assumption 3.

$$\pi_i(x_1, \dots, x_i, \dots, x_n) = 0 \ \ for \ \ x_i = 0, i = 1, 2, \dots, n.$$

We note that in view of Assumption 2,

$$x_1 = x_2 = \dots = x_n = 0$$

is not an equilibrium. The first-order condition for maximizing π_i with respect to x_i is

$$\begin{aligned} \frac{\partial \pi_i}{\partial x_i} &= \left\{ \left[f'(x_i) \Sigma f(x_j) - f(x_i) f'(x_i) \right] / (\Sigma f(x_j))^2 \right\} R(\Sigma x_j) \\ &\quad + f(x_i) R'(\Sigma x_j)/\Sigma f(x_j) - 1 = 0, i = 1, 2, \dots, n. \end{aligned} \tag{2}$$

It is clear that $x_i \equiv x$ for all i. Rewrite (2) as

$$G(x) = H(x), \tag{3}$$

where

$$\begin{aligned} G(x) &\equiv g(x)(n-1)/n^2, \\ g(x) &\equiv f'(x) R(nx)/f(x), \\ H(x) &\equiv 1 - R'(nx)/n. \end{aligned}$$

In view of the assumption on f, f' and R, we see that $g(x)$ is sufficiently large for a sufficiently small x and $g(\infty) = 0$. Differentiating $g(x)$ we can show that $g' < 0$ if and only if

$$\{f'' f - (f)^2/ff'\} R/R' + n = \{(\sigma_2 - \sigma_1)/\epsilon + 1\} n < 0,$$

where

$$\sigma_1 \equiv x \frac{f'}{f}, \sigma_2 \equiv x \frac{f''}{f'}, \epsilon \equiv X \frac{R'}{R}, X \equiv nx.$$

σ_1 and σ_2 are the elasticities of f and f', respectively, with respect to change in x, and ϵ is the elasticity of the rent with respect to change in the aggregate expenditure by all agents. The immediately above inequality holds under

Assumption 4. $\sigma_1 - \sigma_2 > \varepsilon$.

The function $H(x)$ on the RHS of (3) has the following properties due to Assumption 2.

$$H(0) = 1 - \frac{R'(0)}{n} < 0,$$

$$H(\infty) = 1 - \frac{R'(\infty)}{n} > 0,$$

$$H'(x) = -R''(nx) > 0, x \in (0, \infty).$$

Given n, $G(x)$ is strictly decreasing in x if and only if Assumption 4 holds. It takes a sufficiently large value for a sufficiently small x, and converges to 0 as x approaches ∞. On the other hand, $H(x)$ is strictly increasing. It takes a negative value and a positive one for $x = 0$ and $x = \infty$, respectively. Hence the curves for $G(x)$ and $H(x)$ can be shown as in Figure 1, and they have a unique intersection. Summarizing, we have

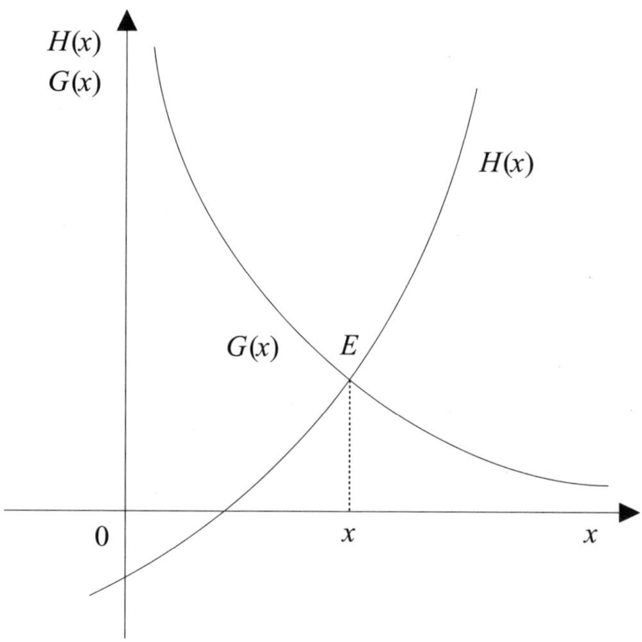

Figure 1: Existence of a unique Nash equilibrium.

Theorem 2.1. *Under Assumptions 1–4, the one-stage endogenous rent-seeking game has a unique symmetric pure Nash equilibrium.*

The lottery production function given by,

$$f(x_i) = x_i^{\alpha}, 0 < \alpha < 1,$$

yields

$$\sigma_1 = \alpha, \sigma_2 = \alpha - 1.$$

Assumption 4 therefore becomes

$$\varepsilon < 1,$$

where ϵ need not necessarily be constant. Hence,

Corollary 2.1. *Under Assumptions 1–3, there exists a unique symmetric pure Nash equilibrium if the elasticity of the rent with respect to change in the aggregate expenditure by all agents is less than one.*

3 Two-Stage Rent-Seeking Game

We now formulate a two-stage rent-seeking game in which the first-stage rent is exogenously given but the second-stage one depends on the aggregate expenditure by all agents in the first stage. Without loss of generality, we normalize the first-stage rent to 1. Let x_{1i} and x_{2i} be the i-th agent's first and second stage expenditures respectively. Let, in addition, $f_i(x_{1i}) = x_{1i}$, $f_i(x_{2i}) = x_{2i}$, and $R(\Sigma x_{1j})$ be *the second-stage rent*. Furthermore, let $\delta = 1/(1+r)$, where r is the rate of interest. The expected present value of the net-rent for the i-th agent over the two stages, $\pi_i \equiv \pi_{1i} + \pi_{2i}$, is given by

$$\pi_i \equiv \frac{x_{1i}}{\Sigma x_{1j}} - x_{1i} + \delta \left\{ \frac{R(\Sigma x_{1j}) x_{2i}}{\Sigma x_{2j}} - x_{2i} \right\}, i = 1, 2, \dots, n. \tag{4}$$

We will now examine the existence of a subgame perfect Nash equilibrium. Given Σx_{1j}, the optimal strategy for the second stage is the solution of the following equation.

$$\frac{\partial \pi_{2i}}{\partial x_{2i}} = \delta \left\{ \frac{R(\Sigma x_{1j})(\Sigma x_{2j} - x_{2i})}{(\Sigma x_{2j})^2} - 1 \right\} = 0, i = 1, 2, \dots, n. \tag{5}$$

It is easily found that the optimal strategy is identical for all agents. Define $X_1 \equiv \Sigma x_{1j}$, and let the second stage optimal strategy as a function of the aggregate expenditure in the first stage be

$$x_{2i} \equiv x_2 \equiv \varphi(X_1) \equiv \{(n-1)/n^2\} R(X_1), i = 1, 2, \dots, n.$$

Clearly, $\varphi'(X_1) > 0$ and $\varphi''(X_1) < 0$. We may rewrite (4) as

$$\pi_i = \frac{x_{1i}}{\Sigma x_{1j}} - x_{1i} + \delta \left\{ \frac{R(\Sigma x_{1j})}{n} - \varphi(\Sigma x_{1j}) \right\}, i = 1, 2, \dots, n. \tag{6}$$

The first-order condition for maximization of this with respect to x_{1i} is

$$\frac{\partial \pi_i}{\partial x_{1i}} = \frac{(X_1 - x_{1i})}{X_1^2} - 1 + \delta \left\{ \frac{R'(X_1)}{n} - \varphi'(X_1) \right\} = 0, i = 1, 2, \ldots, n. \quad (7)$$

The second-order condition is satisfied in view of $R'' < 0$ and the definition of $\varphi(X)$. Clearly, $x_{1i} \equiv x_1$ for all i. Define

$$\psi(X_1) \equiv (n-1)/X_1 + \delta R'(X_1)/n. \quad (8)$$

This expression has the following properties.

$$\psi(X_1) = \infty \text{ for a sufficiently small } X_1,$$
$$\psi(\infty) = 0,$$
$$\psi'(X_1) = -(n-1)/X_1^2 + \delta R''(X_1)/n < 0, X_1 \in (0, \infty).$$

Summing (5) from 1 to n and rearranging, we get the equilibrium condition,

$$\psi(X_1) = n. \quad (9)$$

The downward-sloping curve for $\psi(X_1)$ and the horizontal line with height n intersect uniquely as in Figure 2. This establishes

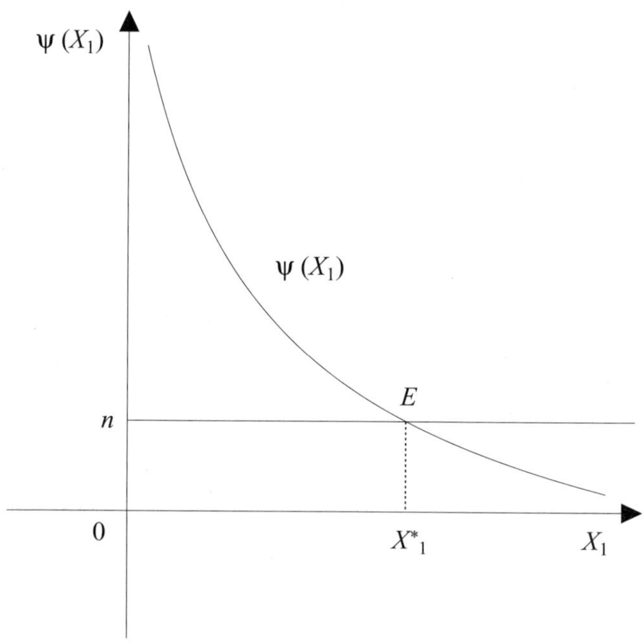

Figure 2: Subgame perfect Nash equilibrium in two-stage rent-seeking game.

Theorem 3.1. *The two-stage rent-seeking game with endogenous rent has a unique symmetric subgame perfect Nash equilibrium.*

We note in passing that since the curve for ψ shifts upwards as δ increases, the equilibrium value X_1^* increases as the interest rate decreases. It is easy to see that X_1^*, the solution of (7), depends on n. To see this dependence more clearly, note that the function $\psi(X_1)$ depends on n, and let the solutions of (9) for n and $(n+1)$ agents be $X_1^*(n)$ and $X_1^*(n+1)$, respectively. Taking into account Figure 2 for the cases of n and $(n+1)$ agents, we can assert that $X_1^*(n+1) \gtreqless X_1^*(n)$ according as

$$\frac{n}{X_1^*(n)} + \delta\frac{R'(X_1^*(n))}{n+1} - \frac{n-1}{X_1^*(n)} - \delta\frac{R'(X_1^*(n))}{n} \gtreqless 1, \tag{10}$$

which, in the light of

$$\delta\frac{R'(X_1^*(n))}{n} = \frac{\big(n - [(n-1)/X_1^*(n)]\big)}{n}$$

is rewritten as

$$\frac{2n - (2n+1)X_1^*(n)}{(n+1)X_1^*(n)} \gtreqless 0. \tag{11}$$

Hence,

$$X_1^*(n+1) \gtreqless X_1^*(n) \text{ according as } X_1^*(n) \gtreqless n/(n+0.5) \equiv X_1^0.$$

Hence, we have proven

Theorem 3.2. *The subgame perfect Nash equilibrium aggregate expenditure in the first-stage increases and decreases in the event of an increase in the number of agents if $X_1^*(n) < X_1^0$ and $X_1^*(n) > X_1^0$ respectively.*

4 Rent Dissipation

In this section we will analyze whether the total rent over two stages will dissipate or not as the number of agents increases. Let the ratio of the present value of total expenditure by all agents over two stages to the present value of the total rent over two stages be

$$\begin{aligned} D(X_1^*) &\equiv (X_1^* + \delta X_2^*)/(1 + \delta R(X_1^*)) \\ &\equiv (X_1^* + \delta n\varphi(X_1^*))/(1 + \delta R(X_1^*)). \end{aligned}$$

Since $X_1^* \leq 1$, $D(X_1^*) < 1$ holds. If D increases as n increases, the rent dissipates. By simple calculation we get

$$D(X_1^*(n+1)) - D(X_1^*(n)) \tag{12}$$

$$= \frac{(X_1^*(n+1) - X_1^*(n)) + \delta X_1^*(n) X_1^*(n+1)\left[\frac{R(X_1^*(n))}{X_1^*(n)} - \frac{R(X_1^*(n+1))}{X_1^*(n+1)}\right]}{(1+\delta R(X_1^*(n)))(1+\delta R(X_1^*(n+1)))}$$

$$+ \frac{\frac{n}{n+1}\delta R(X_1^*(n+1)) - \frac{n-1}{n}\delta R(X_1^*(n)) + \frac{\delta^2}{n(n+1)} R(X_1^*(n)) R(X_1^*(n+1))}{(1+\delta R(X_1^*(n)))(1+\delta R(X_1^*(n+1)))}.$$

If $X_1^*(n+1) > X_1^*(n)$, all four expressions between the braces in the numerator of (12) become positive. The positivity of the expression between the second braces comes from the concavity of the function R. If, on the other hand, $X_1^*(n+1) < X_1^*(n)$, the sign of the right hand side of (12) is indeterminate. We have therefore established

Theorem 4.1. *If $X_1^*(n) < X_1^0$, the rent dissipates as the number of agents increases. In the other case it may or may not dissipate.*

5 Conclusions

In this paper we have analyzed the existence of a unique pure Nash equilibrium in two rent-seeking games. In Section 2, we have proven the existence of a unique pure Nash equilibrium in a one-stage game, where the lottery production function has decreasing returns and where, in addition, rent depends on the aggregate expenditure by all agents. In Section 3, we have formulated and analyzed a two-stage rent-seeking game, where the first-stage rent is fixed but that in the second stage is determined by the agents' total expenditure in the first stage. We have proven that under our assumptions a unique subgame perfect Nash equilibrium exists. The equilibrium aggregate expenditure in the first stage by all agents may increase, decrease or may not change as the number of agents increases. In Section 4, we have derived a sufficient condition for rent dissipation for our two-stage game.

REFERENCES

[1] Chung, T., "Rent-seeking contest when the prize increases with the aggregate efforts", *Public Choice*, **87**(1996), 55–66.

[2] Okuguchi, K., "Decreasing returns and existence of Nash equilibrium in a rent-seeking game", Discussion paper, *Nanzan University* (1996).

[3] Okuguchi, K. and F. Szidarovszky, *The Theory of Oligopoly with Multi-product Firms*, Second rev. ed., Berlin, Heidelberg and New York: Springer-Verlag (1999).

[4] Perez-Castrillo, J.D. and T. Verdier, "A general analysis of rent-seeking games", *Public Choice*, **73**(1992), 335–350.

[5] Szidarovszky, F. and K. Okuguchi, "On the existence and uniqueness of pure Nash equilibrium in rent-seeking games", *Games and Economic Behavior*, **18** (1997), 35–140, also in D. A. Walker (ed.), *Equilibrium III, Some Recent Types of Equilibrium Models*, Gros, U.K and Mass. USA: Edward Elger Publishing, 2000.

[6] Tullock. G., "The welfare cost of tariffs, monopolies and theft", *Western Economic Journal*, **5**(1967), 224–232.

[7] Tullock, G., "Efficient rent-seeking", in J. M. Buchanan, et al. (eds.), *Toward a Theory of the Rent-Seeking Society*(1980), College Station, Texas A&M Press.

A Dynamic Game with Continuum of Players and its Counterpart with Finitely Many Players

Agnieszka Wiszniewska-Matyszkiel*
Institute of Applied Mathematics & Mechanics,
Warsaw University,
02-097 Warsaw, Poland
agnese@mimuw.edu.pl

Abstract

The purpose of this paper is to compare two ways of modelling exploitation of common renewable resource by a large group of players.

1 Introduction

Let us consider the following situation: a large group of players extract a common renewable resource. This resource may be the atmosphere (extraction stands for pollution), oceanic fishery (fishing), a lake or a river (fishing or pollution), or a forest (extracting wood). Large group means "large enough to make one player's influence on the system insignificant – negligible".

Examples from real life show that in such situations players facing lack of restrictions usually do not take into account their influence on the ecosystem and just maximize their instantaneous payoff, which often mean that they extract as much as they can. This is a general rule and there may be only few exemptions, for whom using more "environment-friendly" strategy increases payoff even if they know that their decision has negligible effect on the state of the environment.

Such a situation may lead to destruction of the common resource, disastrous to all its users. It took place in the whaling industry: species after species of whales got almost extinct.

The purpose of this paper is to compare two ways of modeling a situation like those mentioned above by two kinds of games: with a continuum of players and with finitely many players. In both kinds of games the measures on spaces of players are normalized in order that the results could be comparable. By increasing the number of players we do not want to illustrate situations in which additional physical agents join the game and increase the potential maximal aggregate exploitation,

*The research is supported by KBN grants # 1 H02B 015 15 and # 5 H02B 008 20

but the situation in which the decision making process becomes more decentralized but potential maximal aggregate exploitation remains the same (in the case of a rainforest we have the same workers and equipment, but the decision about the level of extraction can be made by one, two, $n \ldots$ players or by so many players that the influence of one player on the ecosystem is negligible: e.g. when each lumberman decides for himself).

We recall some properties of dynamic games with continuum of players. The games considered are with discrete time and such that the instantaneous payoff of every player depends on his own strategy and the aggregate (or mean) strategy of all players, strategies available in one-stage games depend on the state of the system changing in response to the aggregate strategy. The properties obtained are essentially different from properties of games with finitely many players; all of them were proven in Wiszniewska-Matyszkiel [12]. The theoretical results allow us to find all dynamic equilibria in the game with a continuum of players. An equilibrium in the game with finitely many players is calculated directly by using optimization theory, mainly Bellman's equation.

A continuous time analogue of the problem considered in this paper is examined by Wiszniewska-Matyszkiel in [17].

1.1 Games with a Continuum of Players

Games with continuum of players (e.g. Schmeidler [5], Mas-Colell [4], Balder [1], Wieczorek [6], Wieczorek and Wiszniewska [7], Wiszniewska-Matyszkiel [9]) illustrate situations where the number of agents is large enough to make a single agent insignificant – *negligible*. This happens in many real situations: at competitive markets, stock exchange, or while we consider emission of greenhouse gases and similar global effects of exploitation of the common global ecosystem.

Obviously, the insignificance of a single player does not disappear when time is introduced into consideration. Moreover, this phenomenon may occur more visibly: the relation between a dynamic equilibrium and a family of static equilibria corresponding to it, quoted in theorem 3.1 below, is unlikely to hold in a dynamic game with a finite number of players.

Although the general theory of dynamic games with continuum of players is still being developed, there are interesting applications of such games: Wiszniewska-Matyszkiel [8] and [11] concerning models of exploitation of common ecosystems by large groups of players, Karatzas, Shubik and Sudderth [3] and Wiszniewska-Matyszkiel [15] and [18] analyzing dynamic games with continuum of players modeling financial markets and Wiszniewska-Matyszkiel [12] containing an example of dynamic game modeling of presidential elections preceded by a campaign.

This paper appears in a sequence of the author's papers concerning dynamic games with a continuum of players: [13] and [14] developing a general theory of such games and [12], [15] devoted to a certain class of games with discrete time and continuum of players with special focus on applications.

1.2 Example to be Considered

The example we are going to examine in this paper is a relatively simple discrete time dynamic game describing exploitation of a rainforest. It was first considered by Wiszniewska-Matyszkiel [12] in its continuum-of-players form.

This example exhibits essential properties of more general dynamic games (called decomposable large games, as in Wiszniewska-Matyszkiel [12]). Moreover, we show that considering finitely many players, no matter how many, makes modeling the insignificance of a single player's effect on the system impossible.

The example will be formally stated along with the general game. The situation it describes is as follows:

Example 1.1. *Discrete time rainforest*

A group of players extract a rainforest. They can extract some fixed surplus part of the state without damaging the system. If they extract more, they affect the "basis" of regeneration (because of heavy rains the soil erodes immediately, which then makes it unable to sustain trees). Even if they extract less than the surplus, the future state of the system cannot be improved.

This example is going to be formalized in the sequel.

Some surveys on game-theoretic models of exploitation of common ecosystems can be found in e.g. Kaitala [2] or Wiszniewska-Matyszkiel [10] (this survey contains also some information about dynamic games with continuum of players modeling such exploitation) and [16].

2 Definition of Simple Dynamic Games

We shall consider two games, differing from each other by the space of players. The space of players will be denoted by Ω while λ will denote the measure defined on Ω: in the case of a *game with a continuum of players*, the *players* are assumed to form the unit interval ($\Omega = [0, 1]$) with the Lebesgue measure λ, in the other case, there are n players ($\Omega = \{1, \dots, n\}$), all of them of measure $\lambda(i) = 1/n$.

A set $\mathbb{X} \subset \mathbb{R}^m$ will be the *set of* (possible) *states of the system*; in our example $m = 1$ and $\mathbb{X} = [0, M]$.

There is an *initial state* of the system $\bar{x} \in \mathbb{X}$; in our example $\bar{x} > 0$.

A *time set* is either $\mathbb{T} = \{0, 1, 2, \dots\}$ or $\mathbb{T} = \{0, 1, \dots, T\}$, in our example the first case is considered: $\mathbb{T} = \{0, 1, 2, \dots\}$.

The symbol $\mathfrak{X}$ will stand for the set of all functions $X : \mathbb{T} \to \mathbb{X}$, and is called a *set of trajectories*.

Every function $X \in \mathfrak{X}$ such that $X(t_0) = \bar{x}$ will be called a *trajectory of the system starting from* $\bar{x}$ ($X(t)$ will denote the state of the system at time t for this trajectory).

A set $\mathbb{S} \subset \mathbb{R}^k$ will be the *space of* (static) *strategies*; in our example $k = 1$ and $\mathbb{S} = \mathbb{R}_+$.

The set of all functions $D : \mathbb{T} \to \mathbb{S}$ will be denoted by $\mathfrak{S}$.

A nonempty valued correspondence $S : \Omega \times \mathbb{X} \multimap \mathbb{S}$ is a *correspondence of players' strategies*. Instead of $S(\omega, \cdot)$ we shall write S_ω. The set $S_\omega(x)$ is understood as the *set of* (static) *strategies available to player* ω *at* (time) t *and* (state of the system) x. Every $d \in S_\omega(x)$ is *individual static strategy available to player* ω *at* t *and* x. If the trajectory of the system is X, then any function $D \in \mathfrak{S}$ such that $D(t) \in S_\omega(X(t))$ is ω *'s dynamic strategy available to him at* X.

In our example $S_\omega(x) = [0, c \cdot x]$, for some constant $r < c \leq 1 + r$, where the number $r > 0$ denotes the *rate of regeneration of the system* and its meaning will be explained in the sequel.

Any function $\delta : \Omega \to \mathbb{S}$ measurable with respect to the corresponding measures on both sets and for almost every ω fulfilling $\delta(\omega) \in S_\omega(x)$, will be called *static profile available at* (state) x; $\delta(\omega)$ is *player* ω*'s strategy at* (profile) δ. The set of all static profiles will be denoted by Σ.

The number $u_\delta = \int\limits_\Omega \delta(\omega) d\lambda(\omega)$ is called the *aggregate of* δ, where λ denotes the appropriate measure on the set of players: the Lebesgue measure for the continuum-of-players case and the normalized counting measure for finite sets.

The sets of all aggregates of static profiles available at various states constitute a correspondence $Y : \mathbb{X} \overset{\text{onto}}{\multimap} \mathbb{Y}$, called a *correspondence of available profile aggregates*, whose range $\mathbb{Y}$ is called a *set of aggregates*.

The set of all functions $U : \mathbb{T} \to \mathbb{Y}$ will be denoted by $\mathfrak{Y}$. Such functions will be called *aggregate functions*.

The next element of the game is a function $P : \Omega \times \mathbb{S} \times \mathbb{Y} \times \mathbb{X} \to \mathbb{R} \cup \{-\infty\}$, called a *players' instantaneous payoff function*. The function $P(\omega, \cdot, \cdot, \cdot)$, denoted by P_ω, is an *individual instantaneous payoff function* (*instantaneous payoff* for short) *of player* ω. In general, the function P may depend also on time.

In our example $P_\omega(d, u, x) = \ln(d)$ (we also let $\ln 0 = -\infty$).

A *regeneration function* $\varphi : \mathbb{X} \times \mathbb{Y} \to \mathbb{R}^k$ is a function such that the behavior of the system is ruled by the equation

$$X(t+1) = X(t) + \varphi(X(t), U(t))$$

(the *behavior equation*).

In our example $k = 1$ and the regeneration function fulfills $\varphi(x, u) = -\max(0, u - r \cdot x)$ i.e. the behaviour equation has the form

$$X(t+1) = X(t) - \max\left(0, U(t) - r \cdot X(t)\right),$$

where r is the *rate of regeneration of the system*.

Let U be an aggregate function. The trajectory starting from $\bar{x}$ and fulfilling the behavior equation together with U will be called *corresponding to* U, and denoted by X_U. Such a trajectory is obviously unique.

An aggregate function U is *admissible* if for every t, $U(t)$ is available at $X_U(t)$ (i.e. $U(t) \in Y(X_U(t))$).

The next element of the model is a summable function $\Psi : \mathbb{T} \to \mathbb{R}_+ \setminus \{0\}$ called a *discounting function*; in our example $\Psi(t) = (1+\xi)^{-t}$ for some $\xi > 0$.

A function $\Delta : \Omega \times \mathbb{T} \to \mathbb{S}$ measurable as a function of the first argument, is called a *dynamic profile* if for every t the function $\Delta(\cdot, t)$ is a static profile available at the state $X_{U_\Delta}(t)$, where the aggregate function U_Δ is defined by $U_\Delta(t) = u_{\Delta(\cdot,t)}$ and called the *aggregate of the dynamic profile* Δ. The trajectory X_{U_Δ} is called *corresponding to the dynamic profile* Δ and for simplicity denoted by X^Δ.

The *payoff* function of player ω is a function $\Pi_\omega : \mathfrak{S} \times \mathfrak{Y} \to \overline{\mathbb{R}}$ (it is a function of ω's own strategy at a profile and the aggregate of this profile). The payoff is equal to instantaneous payoffs discounted and summed over time:

$$\Pi_\omega(D, U) = \sum_{t \in \mathbb{T}} P_\omega(D(t), U(t), X_U(t))\Psi(t).$$

If a player ω expects some admissible aggregate function U, he maximizes his payoff given U.

3 Results

In this section we shall present some results concerning equilibria in both kinds of games.

A *Nash* (or *Cournot-Nash*) *equilibrium* is a profile such that almost no player has an incentive to change his strategy, unless the remaining players have changed theirs. Here "almost no" denotes almost no according to the measure considered on the set of players. Such formulation encompasses both standard definition of Nash equilibrium for games with finitely many players (since no player is of measure 0) and standard definition of Nash equilibrium in games with continuum of players (like in Schmeidler [5], Balder [1] or Wiszniewska-Matyszkiel [9]).

3.1 Results for Games with a Continuum of Players Proven Earlier

In the case of games with continuum of players like those considered in this paper, we can reformulate the definition of equilibrium as in Wiszniewska-Matyszkiel [12] and consider profiles such that almost no player has an incentive to change his strategy, unless the aggregate of the profile has changed. Obviously, we shall obtain the same equilibria.

An *equilibrium* (or *dynamic equilibrium* to emphasize the difference from static equilibrium which will be defined in the sequel) is a dynamic profile Δ such that

$$\text{for a.e. } \omega,\ \Delta(\omega) \in \operatorname{Argmax} \left\{ \Pi_\omega(D, U_\Delta) \;\middle|\; D \in \mathfrak{S}, D(t) \in S_\omega(X^\Delta(t)) \text{ for all } t \right\},$$

i.e. for a.e. ω and every dynamic strategy D available at X^Δ, $\Pi_\omega(\Delta(\omega), U_\Delta) \geq \Pi_\omega(D, U_\Delta)$.

If U is an admissible aggregate function, then the *set of player ω's dynamic best responses to U* is equal to

$$\mathbf{B}_\omega(U) = \operatorname{Argmax}\{\Pi_\omega(D, U) \mid D \in \mathfrak{S}, D(t) \in S_\omega(X_U(t)) \text{ for all } t\},$$

and the *set of aggregates of dynamic best responses*

$$\overline{\mathbf{B}}(U) = \{U_\Delta \mid \Delta \in \Sigma, \Delta(\omega, \cdot) \in \mathbf{B}_\omega(U) \text{ for a.e. } \omega\}.$$

A dynamic equilibrium is, equivalently, a dynamic profile Δ whose aggregate is a fixed point of the correspondence $\overline{\mathbf{B}}$.

It is worth emphasizing that Nash equilibrium can be defined this way only in the game with continuum of players, since in games with finitely many players each players has non-negligible influence on the aggregate.

In the case of games with continuum of players besides the equilibrium in the dynamic game we shall consider equilibria in static games corresponding to it. Such procedure is possible also in games with finitely many players, but in that case it does not make any sense: there will be no correspondence between static and dynamic equilibria, and static equilibria cannot be used to calculate equilibrium in the dynamic game.

We define as follows.

A game G_x with the same space of players, with player's ω payoff function equal to his instantaneous payoff function $P_\omega(\cdot, \cdot, x)$ and player's ω strategy set equal to $S_\omega(x)$, will be called a *static game at state x corresponding to $\mathfrak{G}$*.

A *static equilibrium at state x* is a static profile δ such that for a.e. ω, $\delta(\omega) \in \operatorname{Argmax}_{d \in S_\omega(x)} P_\omega(d, u_\delta, x)$.

The *set of static best responses of player ω to $u \in \mathbb{Y}$* (representing aggregate of a profile) at state x is defined by

$$B_\omega(u, x) = \in \operatorname*{Argmax}_{d \in S_\omega(x)} P_\omega(d, u, x),$$

and the *set of aggregates of static best responses to u at state x* by

$$\overline{B}(u, x) = \{u_\delta \mid \delta \in \Sigma \text{ and for a.e. } \omega, \delta(\omega) \in B_\omega(u, x)\}.$$

Static equilibrium profile δ may be equivalently defined by the condition

$$u_\delta \in \overline{B}(u_\delta, x),$$

i.e. δ is a profile, whose aggregate is a fixed point of the correspondence $\overline{B}(\cdot, x)$.

As in the case of the definition of dynamic equilibrium, this definition makes sense only for games with continuum of players.

There are some relations between dynamic and static best response sets, and between dynamic and static equilibria. These relations may seem counterintuitive at first since they differ from known results for games with finitely many players, but in fact they are quite simple in the case of dynamic games with discrete time and continuum of players.

The following results were proven in Wiszniewska-Matyszkiel [12]:

Theorem 3.1. *Let* (Ω, λ) *be the unit interval* $[0, 1]$ *with the Lebesgue measure.*

a) For every admissible aggregate function U *and a dynamic strategy* D*, condition 1 implies 2. If, moreover, the payoff* $\Pi_\omega(D, U)$ *is finite, then 1 and 2 are equivalent.*

1. *For every* t *the vector* $D(t)$ *is a static best response of player* ω *to* $U(t)$ *at the state of the system* $X_U(t)$.
2. *The function* D *is a dynamic best response of player* ω *to the aggregate function* U.

b) For every dynamic profile Δ *condition 3 implies 4. If for almost every player the payoff* $\Pi_\omega(\Delta(\omega, \cdot), U_\Delta)$ *is finite, then 3 and 4 are equivalent.*

3. *The static profile* $\Delta(\cdot, t)$ *is for every* t *a static equilibrium at the state of the system* $X^\Delta(t)$.
4. *The dynamic profile* Δ *constitutes a dynamic equilibrium.*

An obvious consequence of theorem 3.1 is the following statement:

Corollary 3.1. *Let* (Ω, λ) *be the unit interval with the Lebesgue measure. If the function* P_ω *is upper semi-continuous and strictly quasi-concave in* d *and the values of the correspondence* S_ω *are compact convex, then for every aggregate function* U *such that the set of admissible payoffs of player* ω *for this* U *is bounded from above and is not equal to* $\{-\infty\}$*, there is exactly one* ω*'s dynamic best response to* U.

For our example we can state the following result.

Proposition 3.1. *Let* (Ω, λ) *be the unit interval with the Lebesgue measure.*

a) If $c = 1 + r$*, then no dynamic profile such that a set of players of positive measure get finite payoffs is an equilibrium, and every dynamic profile yielding the destruction of the system at finite time (i.e.* $\exists \bar{t} \quad \forall t > \bar{t} \quad X^\Delta(t) = 0$*) is an equilibrium.*

b) If $c < 1 + r$*, then the only (up to measure equivalence) equilibrium is a profile* Δ *such that* $\Delta(\omega, t) = c \cdot \bar{x} \cdot (1 + r - c)^t$ *(the only profile fulfilling* $\Delta(\omega, t) = c \cdot X^\Delta(t)$ *for all* ω*,* t*).*

For comparison with the counterpart of this proposition for the n-player case, the proof is quoted.

Proof.

a) Let Δ be a dynamic equilibrium such that a set of players of a positive measure get finite payoffs, U the aggregate of this profile, and X the corresponding trajectory. By theorem 3.1, the only static best response of every player to $U(t)$ at the state $X(t)$ is $d = (1 + r)X(t)$, therefore the aggregate of the resulting static profile is equal to $(1 + r)X(t)$. Therefore $U(\bar{t}) = (1 + r)X(\bar{t})$ for $\bar{t}$ such that $X(\bar{t}) > 0$, which implies that $X(t) = 0$ for $t > \bar{t}$, therefore $S_\omega(X(t)) = \{0\}$. So for $t \geq \bar{t}$ every dynamic profile such that X is the trajectory corresponding to it, fulfills $\Delta(\omega, t) \equiv 0$.

Hence for $t \geq \bar{t}$ the instantaneous payoff $P_\omega(\Delta(\omega, t), U_\Delta(t), X^\Delta(t)) = \ln 0 = -\infty$. Therefore for a.e. player the payoff is equal to $-\infty$.

That means, there exists no equilibrium with payoffs finite for a non-negligible set of players.

Now let Δ be a dynamic profile destructing the system at a finite time, U the statistics of the profile Δ and X the trajectory corresponding to it. The trajectory X fulfills $X(t) = 0$ for $t > \bar{t} \in \mathbb{R}$. Since the only possible strategy of player ω for $t > \bar{t}$ is $d = 0$, his only possible payoff for the aggregate function U is $-\infty$, whatever admissible dynamic strategy he chooses.

Therefore $\mathbf{B}_\omega(U) = \{D \in \mathfrak{S} | \text{ for a.e. } t \;\; D(t) \in S_\omega(X(t))\} \ni \Delta(\omega, \cdot)$, which ends this part of the proof.

b) Let us note that there exists no profile destructing the system at finite time. Therefore best responses for every profile's aggregate yield payoffs greater than $-\infty$. Since for every ω, u, x $B_\omega(u, x) = \{c \cdot x\}$, the only (up to measure equivalence) equilibrium fulfils $\Delta(\omega, t) = c \cdot X^\Delta(t)$ for every ω, t. Therefore $\Delta(\omega, t) = c \cdot \bar{x} \cdot (1 + r - c)^t$ for every ω, t. □

3.2 Games with Finitely Many Players

Now we shall consider the game with a finite number n of players ($\Omega = \{1, \dots, n\}$ with all atoms of measure $\lambda(\{i\}) = 1/n$).

In this case the definition of an equilibrium cannot be simplified as it was in the case of continuum of players.

An *equilibrium* is a dynamic profile Δ such that for a.e. ω,

$$\begin{aligned} \Delta(\omega) \in \operatorname{Argmax} \Big\{ &\Pi_\omega(\widetilde{\Delta}(\omega), U_{\widetilde{\Delta}}) \\ &\big| \, \widetilde{\Delta} - \text{dynamic profile s.t. } \widetilde{\Delta}(\nu) = \Delta(\nu) \text{ for all } \nu \neq \omega \Big\}, \end{aligned}$$

i.e. for a.e. ω and every dynamic profile $\widetilde{\Delta}$ such that $\widetilde{\Delta}(\nu) = \Delta(\nu)$ for all $\nu \neq \omega$, $\Pi_\omega(\Delta(\omega), U_\Delta) \geq \Pi_\omega(\widetilde{\Delta}(\omega), U_{\widetilde{\Delta}})$.

Analogously we can define a dynamic equilibrium by best responses: it is a dynamic profile such that each player's strategy is the best response to the strategies of the remaining players, with best responses defined in the obvious way.

The following proposition states that the players take their impact on the system into account and extract less than in the non-atomic case in order to protect the system.

Proposition 3.1. *Let $c = 1 + r$.*

a) No dynamic profile yielding payoffs equal to $-\infty$ for any player is an equilibrium.

b) A profile Δ defined by the equation $\Delta(\omega, t) = \bar{z}_n \cdot \bar{x} \cdot (1 + r - \bar{z}_n)^t$ for $\bar{z}_n = \max\left(n\xi(1+r)/(1+n\xi), r\right)$ is a dynamic equilibrium (Δ is the only solution to the system of equations $\Delta(\omega, t) = \bar{z}_n \cdot X^{\Delta}(t)$).

In the case of finitely many players we do not have any useful relation between static and dynamic equilibria, and the scheme of the proof will be quite different: the equilibrium will be calculated by definition.

The best response to a profile can be reduced only to the best response to the aggregate of the strategies of the remaining players – which in this case usually differs from the aggregate of the whole profile.

To prove proposition 3.1 we shall need the following lemma stating the structure of the set of best responses of one player (without loss of generality the n-th player) under some assumption about the remaining players' choices. That is we shall calculate the best response of the n-th player, to any profile Δ such that the aggregate of the remaining players' strategies is equal to $C \cdot X^{\Delta}(t)$ for a constant C $\left(\text{i.e. } \frac{1}{n} \sum_{\omega=1}^{n-1} \Delta(\omega, t) = C \cdot X^{\Delta}(t)\right)$ for all t.

Lemma 3.1. *Assume that the profile Δ is such that $\frac{1}{n} \sum_{\omega=1}^{n-1} \Delta(\omega, t) = C \cdot X^{\Delta}(t)$ for all t. Then the only best response of the n-th player is of the fixed rate form*

$$D(t) = \bar{z} \cdot X(t), \text{ where } \bar{z} = \begin{cases} n \cdot (r - C) & \text{if } \frac{n\xi \cdot (1+r-C)}{1+\xi} < n \cdot (r - C), \\ 1 + r & \text{if } \frac{n\xi \cdot (1+r-C)}{1+\xi} > 1 + r, \\ \frac{n\xi \cdot (1+r-C)}{1+\xi} & \text{if neither of above.} \end{cases}$$

and $X(t)$ is the actual state at time t.

Proof. The optimization problem of the n-th player is:

$$\sup_{D \in \mathfrak{D}_C} \sum_{t=0}^{\infty} \frac{\ln(D(t))}{(1+\xi)^t},$$

where the set $\mathfrak{D}_C$ is the set of all controls available to the player while the remaining players extract $C \cdot X$.

Let us denote by Z the function of the ratio $Z(t) = D(t)/X(t)$, where X is the trajectory corresponding to $\frac{1}{n} D + CX$. Obviously, every ratio function Z defines unique control function D. We shall denote by U^Z the aggregate of every profile such that the first $n - 1$ players extract at the aggregate rate C and the n-th player extracts at rate Z.

We can find the optimal control for optimization over a constrained class of controls $\widetilde{\mathfrak{D}}_C = \{D \in \mathfrak{D}_C : Z(t) \equiv z,\ 0 \le z \le 1 + r\}$ (defined by fixed ratio of the state $Z_z(t) \equiv z$) for various initial states and calculate the value function $V_C : \mathbb{X} \to [-\infty, +\infty)$:

$$V_C(\bar{x}) = \sup_{D \in \widetilde{\mathfrak{D}}_C} \sum_{t=0}^{\infty} \frac{\ln(D(t))}{(1+\xi)^t}.$$

This is equivalent to $V_C(\bar{x}) = \sup_{0 \le z \le 1+r} \sum_{t=0}^{\infty} \frac{\ln\left(z \cdot X_{U^{Z_z}}(t)\right)}{(1+\xi)^t}$.

Let us note that if $C \le r - [(1+r)/n]$ then the n-th player cannot affect the system whatever strategy he chooses. Therefore $\bar{z} = 1 + r$ obviously defines his unique optimal control in this case by $D(t) = \bar{z} \cdot X(t)$ and this D is optimal not only over constrained class of controls, but also over the whole class of controls.

Now we shall calculate the optimal $\bar{z}$ in the case where $C > r - [(1+r)/n]$.

The trajectory $X_{U^{Z_z}}$ corresponding to the aggregate function U^{Z_z} (such that $U^{Z_z}(t) = ((1/n)z + C)\, X_{U^{Z_z}}(t))$ fulfills $X_{U^{Z_z}}(t) = \bar{x}\,(1 + r - (1/n)z - C)^t$ for $z \ge \max(0, n(r - C))$ while for $z < \max(0, n(r - C))$ it fulfills $X \equiv \bar{x}$. Therefore

$$\begin{aligned}
\bar{z} &\in \operatorname*{Argmax}_{0 \le z \le 1+r} \sum_{t=0}^{\infty} \frac{\ln\left(z \cdot X_{U^{Z_z}}(t)\right)}{(1+\xi)^t} \\
&= \operatorname*{Argmax}_{\max(0, n\cdot(r-C)) \le z \le 1+r} \sum_{t=0}^{\infty} \frac{\ln\left(z \cdot \bar{x} \cdot \left(1 + r - \frac{1}{n}z - C\right)^t\right)}{(1+\xi)^t} \\
&= \operatorname*{Argmax}_{\max(0, n\cdot(r-C)) \le z \le 1+r} \left((\ln \bar{x} + \ln z) \cdot \sum_{t=0}^{\infty} \frac{1}{(1+\xi)^t} \right. \\
&\qquad \left. + \ln\left(1 + r - \frac{1}{n}z - C\right) \cdot \sum_{t=0}^{\infty} \frac{t}{(1+\xi)^t} \right) \\
&= \operatorname*{Argmax}_{\max(0, n\cdot(r-C)) \le z \le 1+r} \left((\ln \bar{x} + \ln z) \cdot \frac{1+\xi}{\xi} \right. \\
&\qquad \left. + \ln\left(1 + r - \frac{1}{n}z - C\right) \cdot \frac{1+\xi}{\xi^2} \right).
\end{aligned}$$

The necessary condition for optimum in the interval $(\max(0, n \cdot (r - C)), 1 + r)$ is $(1/z) \cdot (1+\xi)/\xi - \frac{1}{n}/(1 + r - \frac{1}{n}z - C) \cdot (1+\xi)/\xi^2 = 0$, which is equivalent to $(1/z) = 1/\left(n\xi \cdot \left(1 + r - \frac{1}{n}z - C\right)\right)$, therefore $\bar{z} = (n\xi \cdot (1 + r - C))/(1 + \xi)$, which is obviously positive.

It is immediate that the maximized function is strictly concave, therefore if $(n\xi \cdot (1+r-C))/(1+\xi) \in [n \cdot (r-C)), 1+r]$, then the maximum is attained at this point.

If $(n\xi \cdot (1+r-C))/(1+\xi) < n \cdot (r-C)$, then the optimal $\bar{z}$ is equal to $n \cdot (r-C)$.

If $(n\xi \cdot (1+r-C))/(1+\xi) > 1+r$, then the optimal $\bar{z}$ is equal to $1+r$.

Therefore the optimal $\bar{z}$ is defined by

$$\bar{z} = \begin{cases} n \cdot (r-C) & \text{if } \frac{n\xi \cdot (1+r-C)}{1+\xi} < n \cdot (r-C), \\ 1+r & \text{if } \frac{n\xi \cdot (1+r-C)}{1+\xi} > 1+r, \\ \frac{n\xi \cdot (1+r-C)}{1+\xi} & \text{if neither of above.} \end{cases}$$

To prove the optimality of the control function we have calculated above, we shall use the *Bellman equation* for the value function W :

$$W(x) = \max_{d \in S_\omega(x)} \left(\ln d + \frac{1}{1+\xi} W\left(x + \varphi\left(x, \frac{1}{n} d + Cx \right) \right) \right),$$

with the transversality condition

$$(1+\xi)^{-t} \cdot W\left(\bar{X}(t)\right) \stackrel{t \to \infty}{\to} 0$$

along the trajectory $\bar{X} = X_U$ corresponding to the aggregate function U defined by $U(t) = (1/n)\bar{D}(t) + C\bar{X}_U(t)$ for $\bar{D}$ such that

$$\bar{D}(t) = \operatorname*{Argmax}_{d \in S_\omega(\bar{X}(t))} \left(\ln d + \frac{1}{1+\xi} W\left(\bar{X}(t) + \varphi\left(\bar{X}(t), \frac{1}{n} d + C\bar{X}(t) \right) \right) \right).$$

This is equivalent to

$$W(x) = \max_{\max(0, n \cdot (r-C)) \le z \le 1+r} \left(\ln(z \cdot x) + \frac{1}{1+\xi} W\left(x \cdot \left(1 + r - \frac{1}{n} z - C \right) \right) \right),$$

with the transversality condition $1/(1+\xi)^t \cdot W\left(\bar{X}(t)\right) \stackrel{t \to \infty}{\to} 0$ along the trajectory $\bar{X} = X_U$ corresponding to the aggregate function U defined by $U(t) = \left((1/n)\bar{Z}(t) + C\right) \bar{X}_U(t)$ for $\bar{Z}$ such that

$$\bar{Z}(t) = \operatorname*{Argmax}_{\max(0, n \cdot (r-C)) \le z \le 1+r} \left(\ln\left(z\bar{X}(t)\right) + \frac{1}{1+\xi} W\left(\bar{X}(t) \cdot \left(1 + r - \frac{1}{n} z - C \right) \right) \right).$$

We shall check whether the value function V_C for the optimization over the constrained set of controls (calculated above) fulfills this equation:

$$V_C(x) = \sup_{\max(0, n \cdot (r-C)) \le z \le 1+r} \ln(zx) + \frac{1}{1+\xi} V_C\left(x \cdot \left(1 + r - \frac{1}{n} z - C \right) \right),$$

which is an easy calculation.

The transversality condition is also easy to check:

$$\frac{1}{(1+\xi)^t}\cdot V_C\left(\bar{X}(t)\right)=\frac{1}{(1+\xi)^t}\cdot V_C\left(\bar{x}\cdot\left(1+r-\frac{1}{n}\bar{z}-C\right)^t\right)$$

$$=\frac{1}{(1+\xi)^t}\left[\frac{1+\xi}{\xi}\left(\ln\left(\bar{x}\cdot\left(1+r-\frac{1}{n}\bar{z}-C\right)^t\right)+\ln\bar{z}\right)\right.$$

$$\left.+\frac{1+\xi}{\xi^2}\ln\left(1+r-\frac{1}{n}\bar{z}-C\right)\right]\overset{t\to\infty}{\to}0.$$

Therefore V_C is the value function for our optimization, which implies that the optimal control to our problem is defined by the fixed ratio $\bar{z}$ of the state.

The last thing to show is that there is at most one optimal control. To prove this, let us assume, conversely, that there exist two optimal controls, $U'\neq U''$. By the strict concavity of the logarithm we have the inequality

$$\sum_{t=0}^{\infty}\frac{\ln\left(\frac{1}{2}U'(t)+\frac{1}{2}U''(t)\right)}{(1+\xi)^t}>\frac{1}{2}\sum_{t=0}^{\infty}\frac{\ln U'(t)}{(1+\xi)^t}+\frac{1}{2}\sum_{t=0}^{\infty}\frac{\ln U''(t)}{(1+\xi)^t},$$

which contradicts optimality of U' and U'' if $U=\frac{1}{2}U'+\frac{1}{2}U''$ defines an admissible control of the n-th player, i.e. for each t $U(t)\leq c\cdot X_U(t)$.

To prove this, we first show that for each t we have the inequality $X_U(t)\geq\frac{1}{2}X_{U'}(t)+\frac{1}{2}X_{U''}(t)$, which can be done simply by induction. This implies that $U(t)=\frac{1}{2}U'(t)+\frac{1}{2}U''(t)\leq\frac{1}{2}\cdot c\cdot X_{U'}(t)+\frac{1}{2}\cdot c\cdot X_{U''}(t)\leq c\cdot X_U(t)$. □

Proof. (of proposition 3.1)

a) Using lemma 3.1 we prove that the payoffs of the players at any equilibrium will be always greater than $-\infty$.

At first assume that each of $n-1$ players always extracts as much as he can: $(1+r)X(t)$. Then the optimal control of the remaining player at time t equals $\bar{z}\cdot X(t)$, where $\bar{z}$ is the constant defining the unique best response of the remaining player, therefore the aggregate function U fulfills $U(t)=((1/n)\cdot\bar{z}+[(n-1)/n](1+r))\cdot X(t)$.

Since $0<\bar{z}=(1+r)\xi/(1+\xi)<1+r$, the system will not be destroyed and payoffs of all players are greater than $-\infty$.

If at some time t the $n-1$ players extract less than $[(n-1)/n](1+r)X(t)$, then, by taking a strategy as above, the player get payoff greater than $-\infty$, therefore for the optimal $\bar{S}$ he also gets payoff greater than $-\infty$, which ends this part of the proof.

b) We use lemma 3.1 and check which $\bar{z}$ defines the best response to $C=[(n-1)/n]\bar{z}$:

First we check for $(n\xi\cdot(1+r-C))/(1+\xi)<n\cdot(r-C)$. Then $z=n\,(r-[(n-1)/n]z)$, i.e. $z=r$. If we substitute this into the inequality, we get that $z=r$ holds whenever $r>n\xi$.

Now we check when $(n\xi \cdot (1+r-C))/(1+\xi) > 1+r$, i.e. $(\xi \cdot (1+r))/(1+\xi) > 1+r$, which is never fulfilled.

Last we check what happens if neither of above inequalities hold. Then $\bar{z} = \frac{n\xi \cdot (1+r-[(n-1)/n]\bar{z})}{1+\xi}$, therefore $\bar{z} = (n\xi \cdot (1+r))/(1+n\xi)$.

When we combine all this results, we get $\bar{z}_n = \max\left([(n\xi \cdot (1+r))/(1+n\xi)], r\right) < 1+r$. □

3.3 The Game with Continuum of Players Versus Games with Finite Number of Players

Let us note that the constants $\bar{z}_n$ defining the players' rate of extraction at symmetric equilibria for the n-player games fulfill: $\bar{z}_n < 1+r$ and $\bar{z}_n \nearrow 1+r$ as n tends to infinity.

This implies that the system will not be destroyed at finite time: the players do not feel negligible and they care for the system, although for a large n single player's influence on the system is small. It is important that such careful behavior is not necessary to avoid a disaster: in our model while at least one player refrains from extracting $1+r$, the remaining ones cannot destroy the system.

The convergence to the "disastrous" limit $1+r$ is what we expect: this is the rate of extraction of the symmetric equilibrium in the case of a continuum of players.

4 Conclusions

In this paper we stated some properties of dynamic games with a continuum of players and discrete time, such that the instantaneous payoff of every player depends on his own strategy and the aggregate strategy, strategies available in one-stage games depend on the state of the system changing in response to the aggregate strategy.

In such a game a player's strategy is a best response to some dynamic aggregate strategy if (and only if whenever the payoff is finite) it is a sequence of best responses to the respective aggregates in one-stage games corresponding to the dynamic game, while a dynamic profile is an equilibrium if (and only if whenever the payoff of almost every player is finite) it is a sequence of equilibria in one-stage games corresponding to the dynamic game. These results, strange at first sight, stem from the phenomenon of insignificance of every single player.

This phenomenon was illustrated by an example of extraction of a common ecosystem presented twice in different ways, as two games differing from each other only by space of players. A continuum of players always destroy the system in equilibrium, while finitely, but arbitrarily many players will not do it.

Therefore games with a continuum of players are better for modeling real situations in which people feel negligible, especially those situations in which a large group of people, each of them feeling insignificant, extract a common ecosystem.

As it should be expected, the rate of extraction in a symmetric equilibrium in the n-players game increases and tends to the rate of extraction in a continuum-of-players game as n increases and tends to infinity.

REFERENCES

[1] Balder E. (1995) *A Unifying Approach to Existence of Nash Equilibria*, International Journal of Game Theory **24**, 79–94

[2] Kaitala V. (1986) *Game Theory Models in Fisheries Management: A Survey*, 252–266, in T. Başar (ed.), 1986, *Dynamic Games and Applications in Economics*, Lecture Notes in Economics and Mathematical Systems, Springer

[3] Karatzas I., Shubik M., Sudderth W. D. (1994) *Construction of Stationary Markov Equilibria in a Strategic Market Game*, Mathematics of Operations Research **19**, 975–1006

[4] Mas-Colell A. (1984) *On the Theorem of Schmeidler*, Journal of Mathematical Economics **13**, 201–206

[5] Schmeidler D. (1973) *Equilibrium Points of Nonatomic Games*, Journal of Statistical Physics **17**, 295–300

[6] Wieczorek A. (1997) *Simple Large Games and Their Applications to Problems with Many Agents*, Report **842**, Institute of Computer Science, Polish Academy of Sciences

[7] Wieczorek A., Wiszniewska A. (Wiszniewska-Matyszkiel) (1999) *A Game-Theoretic Model of Social Adaptation in an Infinite Population*, Applicationes Mathematicae **25**, 417–430

[8] Wiszniewska-Matyszkiel A. (2000) *Dynamic Game with Continuum of Players Modelling "the Tragedy of the Commons"*, Game Theory and Applications **5** (Petrosjan, Mazalov eds.), 162–187

[9] Wiszniewska-Matyszkiel A. (2000) *Existence of Pure Equilibria in Games with Continuum of Players*, Topological Methods in Nonlinear Analysis **16**, 339–349

[10] Wiszniewska-Matyszkiel A. (2001) *Exploitation of Ecosystems and Game Theory I: Deterministic Noncooperative Games* (in Polish), Matematyka Stosowana **2** (43), 11–31

[11] Wiszniewska-Matyszkiel A. (2001) *"The Tragedy of the Commons" Modelled by Large Games*, Annals of the International Society of Dynamic Games **6**, (E. Altman, O. Pourtallier eds.) 323–345

[12] Wiszniewska-Matyszkiel A. (2002) *Discrete Time Dynamic Games with Continuum of Players I: Decomposable Games*, International Game Theory Review **4**, 331–342

[13] Wiszniewska-Matyszkiel A. (2002) *Static and Dynamic Equilibria in Games with Continuum of Players*, Positivity **6**, 433–453

[14] Wiszniewska-Matyszkiel A. (2003) *Static and Dynamic Equilibria in Stochastic Games with Continuum of Players*, Control and Cybernetics **32**, 103–126

[15] Wiszniewska-Matyszkiel A. (2003) *Discrete Time Dynamic Games with Continuum of Players II: Semi-Decomposable Games. Modelling Financial Markets*, International Game Theory Review **5**, 27–40

[16] Wiszniewska-Matyszkiel A. (2003) *Exploitation of Ecosystems and Game Theory II: Cooperation* (in Polish), Matematyka Stosowana **4** (45), 56–77

[17] Wiszniewska-Matyszkiel A. (2003) *Common Resources, Optimality and Taxes in Dynamic Games with Increasing Number of Players*, preprint RW 03-06 (127) Institute of Applied Mathematics and Mechanics, Warsaw University; submitted

[18] Wiszniewska-Matyszkiel A. (2004) *Modelling Stock Exchange by Games with Continuum of Players*, in Polish, submitted

PART V
Numerical Methods and Algorithms for Solving Dynamic Games

Distributed Algorithms for Nash Equilibria of Flow Control Games

Tansu Alpcan
Coordinated Science Laboratory
University of Illinois
Urbana, IL 61801 USA
alpcan@control.csl.uiuc.edu

Tamer Başar
Coordinated Science Laboratory
University of Illinois
Urbana, IL 61801, USA
tbasar@control.csl.uiuc.edu

Abstract

We develop a mathematical model within a game theoretical framework to capture the flow control problem for variable rate traffic at a bottleneck node. In this context, we also address various issues such as pricing and allocation of a single resource among a given number of users. We obtain a distributed, end-to-end flow control using cost functions defined as the difference between particular pricing and utility functions. We prove the existence and uniqueness of a Nash equilibrium for two different utility functions. The paper also discusses three distributed update algorithms, parallel, random and gradient update, which are shown to be globally stable under explicitly derived conditions. The convergence properties and robustness of each algorithm are studied through extensive simulations.

1 Introduction

Today's Internet strives to comply with the demands of a broad range of applications, much different from the original design goals. Implementation of RTT (Real-Time Traffic) for increasingly popular applications like VoIP (Voice over IP) or video conferencing is one such example. Pricing of the network resources and charging the users in a way proportional with their usage is yet another one. There is also an increasing need for mechanisms that ensure fair allocation of network resources among the users. Achieving all these goals is only possible with the introduction of new efficient and real-time implementable congestion and flow control schemes, and we are already seeing efforts towards improving and modifying the flow control mechanisms of TCP. Among many different approaches here, the game theoretic one has been enjoying increasing popularity as it provides a fitting framework to study the underlying network optimization problems [1–4].

Game theory provides a natural framework for developing pricing mechanisms to solve rate control, fairness and even routing problems. Users on the Internet are of completely non-cooperative nature in terms of their demands for bandwidth, and this leads specifically to the use of non-cooperative game theory for flow and congestion control. An appropriate solution concept here is the non-cooperative Nash equilibrium [5]. In this approach, a distributed, non-cooperative network game is defined, where each user tries to minimize a specific cost function by adjusting his flow rate, with the remaining users' flows fixed. An advantage of this approach comes from the fact that it leads to distributed schemes, and not a centralized control for the network, which fits well with today's as well as tomorrow's expected trends of decentralized computing.

Most network games in the literature are focused on elastic, best-effort type traffic [2]. As an example of a study that addresses the flow control problem in a game theoretic framework, we can cite Altman and Başar [1], who show that if an appropriate cost function and pricing mechanism are used, one can find an efficient Nash equilibrium for a multi-user network, which is further stable under different update algorithms. Another game-theoretic study is the one by Korilis and Lazar [6], who investigate the existence of Nash equilibria of the flow control model introduced earlier in [7]. They develop a general approach to study the existence of Nash equilibria by using the concept of best reply correspondence.

We consider in this paper a more general model, with two components. The first pertains to a classical admission control mechanism [8, p.494], where the users are admitted to the network after given a certain QoS (Quality of Service) guarantee in accordance with the available resources of the network. The guaranteed minimum flow rate meets the requirements of the intended RTT application. The second component of the model concerns elastic flow, and it accommodates a wide range of traffic types, from medium to high elasticity. The distributed end-to-end control system is modeled as a network game where users, or players, adjust their excess flows rates, or strategies, according to their individual needs but also taking into account the state of the network. The cost function we adopt for this purpose features, in addition to a relevant pricing function, an inherent feedback mechanism, enabling the users to acquire the basic (essential) information about the state of the network. The non-cooperative game framework provides equilibrium conditions for the system and most importantly, the market structure, where supply and demand for bandwidth determine the allocation of network resources and prices. Fairness is also an important issue, and this is built into the network so that those users who are willing to pay for resources more than others receive a proportionately larger portion of the resources. We model the individual user's demand for bandwidth in terms of two different utility functions: an affine and a logarithmic function. For extensions of the non-cooperative flow control game considered in this study to general network topologies and Internet-style networks including stability and convergence analysis under non-negligible delays, we refer to our subsequent studies [9–12].

In the next section, we provide a description of the proposed model, and in Section 3 we derive a unique Nash equilibrium. In Section 4, we investigate various update algorithms and establish stability conditions. Simulation results are presented in Section 5, which are followed by the concluding remarks of Section 6.

2 The Model and The Cost Function

We consider a bottleneck node in a general network topology, with a certain level, C, of available bandwidth, which is shared by N users or connections. The ith user's flow rate, λ_i, consists of two parts: The guaranteed minimum flow rate, $\lambda_{i,min}$, and the variable excess flow rate, x_i, defined as the difference between the total flow and the minimum flow: $x_i = \lambda_i - \lambda_{i,min}$. The guaranteed flow rate is negotiated between the user and the network at the time of the connection setup and remains constant thereafter. It plays a crucial role in meeting the QoS requirements necessary for real-time traffic types. The problem of giving users guarantees for their requested minimum flows while at the same time preserving the network resources by respecting the bound C on the maximum available bandwidth at the bottleneck node, can be solved through an admission control mechanism.

We assume that during the connection M out of N users request an excess bandwidth on top of their prenegotiated guaranteed minimum flow rates. Hence, we can consider a non-cooperative network game where M users compete for the remaining available bandwidth, m, after all guaranteed minimum flows are subtracted from the total capacity: $m = C - \sum_{i=1}^{N} \lambda_{i,min}$. We note that the excess flow rate of a user is elastic in nature, i.e. it has no QoS guarantees and is bounded above by the total available excess bandwidth, m.

The cost function for each user entering the game is defined as the difference between the pricing and the utility functions. This cost function not only sets the dynamic prices, but also captures the demand of a user for bandwidth. The first term of the cost function, the pricing function, is defined as:

$$P_i(x_i, x_{-i}) = \frac{k_i}{m - (x_i + x_{-i})^+}(x_i)^2, \tag{1}$$

where x_i is the excess bandwidth taken by the ith user, x_{-i} is defined as $x_{-i} := \sum_{j \neq i} x_j$, and $k_i \geq 0$ is the pricing parameter determined by the network. The pricing term not only sets the actual price, but also has the regulatory function of providing the user with feedback about the network status via the denominator term, $m - \sum_j x_j$. This term may also be seen as capturing the delay experienced by the ith user. As the sum of flows of users approach the available capacity, m, the denominator of (1) approaches zero, and hence the price increases without bound. This preserves the network resources by forcing the users to decrease their elastic flows. At the same time, a proportional relationship between demand and price is obtained, which ensures that the prices are set according to market forces. The price per unit flow is chosen to be proportional to the flow rate itself, resulting in the quadratic term, x_i^2. This pricing structure discourages a user from picking up an arbitrary value for the minimum flow rate, $\lambda_{i,min}$, with the intention of minimizing the fixed costs. In addition, it eliminates wide fluctuations in the excess rate, x_i.

The second part of the cost function, the utility function U_i, quantifies the user's utility for having excess bandwidth and captures to some extent the 'human factor'. Although it cannot be exactly known to the network, some statistical estimates can be collected, taking into account the habits of a specific type of a user over a certain period of time. A reasonable assumption is to take it to be strictly concave for elastic flows. Specifically, a logarithmic function can best represent the utility function of the user in this case [13]. Hence, a possible realistic utility function for a user demanding excess flow can be defined in terms of x_i as:

$$U_i(x_i) = \ln(1 + x_i) + d_i, \quad x_i \geq 0 \,\forall i, \tag{2}$$

where d_i is a positive constant and will have no effect on the optimization process. Based on the given pricing and utility functions, the cost function for user i is simply, $J_i = P_i - U_i$. In other words, the flow rate of the ith user results from the interaction between price and demand:

$$J_i(x_i, x_{-i}) = \frac{k_i x_i^2}{m - (x_i + x_{-i})} - \ln(1 + x_i) - d_i, \quad x_i \geq 0 \tag{3}$$

Notice that the excess utility function, and hence the cost function, are defined only in the region where $x_i \geq 0$. In the next section we will show that even if $x_i \leq 0$ is allowed, the Nash equilibrium solution will still be the nonnegative.

A drawback of the realistic utility function above is that it leads to nonlinear reaction functions for the users. Therefore an analytical analysis of the cost function with this utility is very difficult and limited, if not impossible, even though an existence and uniqueness result (on Nash equilibria) could be obtained, as we will do in the next section. In order to make the analysis tractable, and to obtain explicit results, we will use linear utility functions for the users, which lead to a set of linear equations as reaction functions. Accordingly, for tractability we will also adopt as the utility function of user i:

$$\tilde{U}_i(x_i) = a_i x_i + \tilde{d}_i, \quad x_i \geq 0 \,\forall i, \tag{4}$$

where a_i is a positive constant not exceeding 1, and $\tilde{d}_i > 0$. One possible interpretation for the linear utility $\tilde{U}_i$ is that it constitutes a linear approximation to the actual utility function at a given point. In this case, the system is analyzed locally in the vicinity of the chosen point. The parameter a_i is the slope of the utility function at that point, say x_i^0:

$$a_i := \frac{\partial U_i(x_i^0)}{\partial x_i} \Rightarrow 0 < a_i \leq 1, \quad \forall i.$$

The counterpart of the cost function (3) with the linear utility function (4) is

$$\tilde{J}_i(x_i, x_{-i}) = \frac{k_i x_i^2}{m - (x_i + x_{-i})} - a_i x_i - \tilde{d}_i, \quad x_i \geq 0. \tag{5}$$

Another interpretation for the linear utility would come from a worst-case perspective. The constant a_i can be chosen so as to provide an upper bound for marginal utility, or the slope of the logarithmic function at any given point x_i. It will be shown later that given the total flow rates of all other users, x_{-i}, the optimal flow rate of the ith user under linear utility, with $a_i = 1$, is always higher than the one under logarithmic utility. Hence, linear utility (4) with $a_i = 1$ leads to a worst-case flow for the ith user, namely

$$a_i = \max_{x_i \geq 0} \frac{\partial U_i(x_i)}{\partial x_i} \iff a_i = \max_{x_i \geq 0} \frac{1}{1 + x_i} \Rightarrow a_i = 1, \quad \forall i.$$

The parameter $\tilde{d}_i$ in (4) is the same as in (2), and since it is a constant it can be ignored in the subsequent analysis. Notice that the same cost function structure (5) is arrived at in both local and worst-case analyses, with a_i chosen as described above. Combining the worst-case and local analyses in a single step simplifies the problem at hand significantly.

3 Existence and Uniqueness of a Nash Equilibrium

We first prove the existence of a unique Nash equilibrium for the logarithmic utility model, described by the cost function (3). Next, we give a similar result under the linear-utility cost function (5). Additionally, we derive the reaction functions in the linear-utility case and compute the equilibrium point explicitly. We conclude the section with a result (a Proposition) justifying the worst-case linear-utility function analysis.

3.1 Existence and Uniqueness under Logarithmic Utility Functions

The underlying M-player non-cooperative game here is defined in terms of the cost functions $J_i(x_i, x_{-i}), i = 1, \ldots, M$, where J_i is defined by (3), and the constraints

$$x_i \geq 0, \tag{6}$$

$$\sum_{j=1}^{M} x_j \leq m. \tag{7}$$

The first constraint is dictated by the fact that the ith user has requested a flow rate of at least $\lambda_{i,min}$. The second constraint is a physical capacity constraint, which says that the aggregate sum of all flows in a node cannot exceed its total capacity.

Theorem 3.1. *There exists a unique Nash equilibrium in the network game defined by (3) and (6)–(7), which is also an inner solution.*

Proof. Let $x := (x_1, \dots, x_M)'$ be the vector of flow rates of the M players, and X be the subset of $\mathbb{R}^M$ where x belongs in view of the constraints (6)–(7). Then, X is closed and bounded (therefore compact), and is convex, but is not rectangular. On X, each $J_i(x_i, x_{-i})$ is continuous and bounded, except on the hyperplane defined by (7) where it is infinite, and is moreover analytic in that region. Its first derivative with respect to x_i is

$$\frac{\partial J_i(x_i, x_{-i})}{\partial x_i} = \frac{k_i x_i^2 + 2k_i x_i[m - (x_i + x_{-i})]}{[m - (x_i + x_{-i})]^2} - \frac{1}{1 + x_i}, \tag{8}$$

and its second derivative is

$$\frac{\partial^2 J(x_i, x_{-i})}{\partial x_i^2} = \frac{4k_i x_i}{[m - (x_i + x_{-i})]^2} + \frac{2k_i x_i^2}{[m - (x_i + x_{-i})]^3} + \frac{2k_i}{[m - (x_i + x_{-i})]} + \frac{1}{(1 + x_i)^2}, \tag{9}$$

which is clearly seen to be well-defined and positive on X, except on the hyperplane (7). Hence, for an arbitrarily small $\varepsilon > 0$, if we replace (7) with

$$\sum_{j=1}^{M} x_j \le m - \varepsilon, \tag{10}$$

and denote the corresponding constraint set defined by (6) and (10) by X_ε, on X_ε we have a game with a convex, compact nonrectangular action set, and a convex cost function for each player. By a standard theorem of game theory (Theorem 4.4, p. 176 in [5]), it admits a Nash equilibrium. Let us denote this equilibrium solution by x_ε^*, to depict its possible dependence on ε. We now claim that provided that $\varepsilon > 0$ is sufficiently small, x_ε^* does not depend on ε, and hence it provides also a Nash equilibrium solution to the original game on X.

To prove this claim, let us study the optimization problem faced by a generic player, i, in the computation of the Nash equilibrium. This optimization problem is

$$\min_{0 \le x_i \le m_i(\varepsilon)} J_i(x_i, x_{-i}^*(\varepsilon)), \tag{11}$$

where $m_i(\varepsilon) := m - \varepsilon - x_{-i}^*(\varepsilon)$, and $x_{-i}^*(\varepsilon)$ is the total flow from x_ε^* of all players, except the ith. We first note from (8) that $(\partial J_i/\partial x_i)(0, x_{-i}^*(\varepsilon)) < 0$, and hence $x_i^*(\varepsilon)$ cannot be zero, and therefore $x_i^*(\varepsilon) > 0$. We next show that $x_i^*(\varepsilon)$ cannot equal $m_i(\varepsilon)$ either, and hence the solution to (11) has to be an inner solution. To see this, let us evaluate (8) at $x_i = m_i(\varepsilon)$:

$$\frac{\partial J_i(m_i, x_{-i}^*(\varepsilon))}{\partial x_i} = \frac{k_i m_i(m_i + 2\varepsilon)}{\varepsilon^2} - \frac{1}{1 + m_i}. \tag{12}$$

Now, unless $m_i(\varepsilon) = O(\varepsilon)$, the first term above becomes unbounded as $\varepsilon \to 0$, dominating the second term and therefore making (12) positive, which implies that $x_i = m_i(\varepsilon)$ cannot be the solution to (11). Hence, the only way (11) can have a boundary solution is if $m_i(\varepsilon) = O(\varepsilon)$, but since there is only a single constraint, namely (10), we would have $m_j(\varepsilon) = O(\varepsilon)$ for all $j = 1, \dots, M$. Hence the total flow would be $O(\varepsilon)$, which contradicts the initial hypothesis that $\sum_{j=1}^{M} x_j^*(\varepsilon) = m - \varepsilon$. Therefore, for ε sufficiently small, the solution to (11) has to be inner, and since this applies to all players, the Nash equilibrium has to be independent of ε for $\varepsilon > 0$ sufficiently small. As a byproduct, we have also proven the other claim of the theorem, which is that the Nash equilibrium is inner.

We next show the uniqueness of the Nash equilibrium. To preserve notation, let $\partial^2 J(x_i, x_{-i})/\partial x_i^2$ given by (9) be denoted by B_i. Further introduce, for $i, j = 1, \dots, M,\ j \neq i$,

$$\frac{\partial^2 J_i(x_i, x_{-i})}{\partial x_i \partial x_j} = \frac{2k_i x_i}{[m - (x_i + x_{-i})]^2} + \frac{2k_i x_i^2}{[m - (x_i + x_{-i})]^3} := A_{i,j},$$

with both B_i and $A_{i,j}$ defined on X_ε, to avoid singularity on the hyperplane (7). Suppose that there are two Nash equilibria, represented by two flow vectors x^1 and x^0, with elements x_i^0 and x_i^1, respectively. Define the pseudo-gradient vector:

$$g(x) = \begin{pmatrix} \nabla_{x_1} J_1(x_1, x_{-1}) \\ \vdots \\ \nabla_{x_M} J_M(x_M, x_{-M}) \end{pmatrix}.$$

As the Nash equilibrium is necessarily an inner solution, it follows from the first order optimality condition that $g(x^0) = 0$ and $g(x^1) = 0$. Define the flow vector $x(\theta)$ as a convex combination of the two equilibrium points $x^0,\ x^1$:

$$x(\theta) = \theta x^0 + (1 - \theta) x^1,$$

where $0 < \theta < 1$. By differentiating $x(\theta)$ with respect to θ, we obtain

$$\frac{dg(x(\theta))}{d\theta} = G(x(\theta)) \frac{dx(\theta)}{d\theta} = G(x(\theta))(x^1 - x^0), \tag{13}$$

where $G(x)$ is defined as the Jacobian of $g(x)$ with respect to x:

$$G(x) := \begin{pmatrix} B_1 & A_1 & \cdots & A_1 \\ A_2 & B_2 & & A_2 \\ \vdots & & \ddots & \vdots \\ A_M & A_M & \cdots & B_M \end{pmatrix}_{M \times M},$$

where we have used the simpler notation A_i for $A_{i,j}$, since $A_{i,j}$ does not depend on the second index j. Integrating (13) over θ from $\theta = 0$ to $\theta = 1$ we obtain

$$0 = g(x^1) - g(x^0) = \left[\int_0^1 G(x(\theta)) d\theta \right] (x^1 - x^0), \tag{14}$$

where $(x^1 - x^0)$ is a constant flow vector. Let $\overline{B_i(x)} = \int_0^1 B_i(x(\theta))d\theta$ and $\overline{A_i(x)} = \int_0^1 A_i(x(\theta))d\theta$. In view of constraints (6) and (10), we have:

$$B_i(x_i, x_{-i}) > A_{i,j}(x_i, x_{-i}) > 0, \quad \forall i, j.$$

Thus, it follows that $\overline{B_i(x)} > \overline{A_i(x)} > 0$, for any flow vector $x(\theta)$. □

In order to simplify the notation, define the matrix $\mathcal{G}(x^1, x^0)$

$$\mathcal{G}(x^1, x^0) := \int_0^1 G(x(\theta))d\theta = \begin{pmatrix} \overline{B_1} & \overline{A_1} & \cdots & \overline{A_1} \\ \overline{A_2} & \overline{B_2} & & \overline{A_2} \\ \vdots & & \ddots & \vdots \\ \overline{A_M} & \overline{A_M} & \cdots & \overline{B_M} \end{pmatrix}_{M \times M} .$$

Lemma 3.1. *The matrix $\mathcal{G}(x^1, x^0)$ is full rank for any fixed x.*

Proof. The matrix $\mathcal{G}$ is full rank if the only vector y satisfying $\mathcal{G}\, y = 0$ is the null-vector, $y = 0$. Expanding this equation for all i, one obtains

$$\overline{B_i} y_i + \overline{A_i} \sum_{j \neq i} y_i = 0. \tag{15}$$

Rearranging the terms in (15) and solving for y_i:

$$y_i = \frac{-\overline{A_i}}{\overline{B_i} - \overline{A_i}} \sum_{j=1}^{M} y_j, \tag{16}$$

where the term $\overline{A_i}/(\overline{B_i} - \overline{A_i})$ is strictly positive as a result of $\overline{B_i} > \overline{A_i} > 0$. Now summing over i, we get

$$\sum_{i=1}^{M} y_i = - \sum_{i=1}^{M} \frac{\overline{A_i}}{\overline{B_i} - \overline{A_i}} \sum_{j=1}^{M} y_j,$$

which can only hold if $\sum_{j=1}^{M} y_j = 0$, and from (16), $y_i = 0 \; \forall i$. Thus, we conclude that the matrix $\mathcal{G}$ is full rank.

Finally, we rewrite (14) in the following form:

$$0 = \mathcal{G} \cdot [x^1 - x^0]. \tag{17}$$

By Lemma 3.1, the square matrix $\mathcal{G}$ is full rank. Therefore, the only possibility for (17) to hold is $x^1 - x^0 = 0$. Therefore, the Nash equilibrium is unique. □

The lines of the proof above can actually be used to prove uniqueness of Nash equilibrium whenever it exists for more general M-player games. We state and prove this result below, which should be of independent interest.

Corollary 3.1. *Consider a non-cooperative game with M players, with twice differentiable cost functions, $J_i(x_i, x_{-i})$ for player i, and a compact strategy space, X. Let the pseudo-gradient vector be denoted by $g(x)$, i.e.*

$$g(x) := \begin{pmatrix} \nabla_{x_1} J_1(x_1, x_{-1}) \\ \vdots \\ \nabla_{x_M} J_M(x_M, x_{-M}) \end{pmatrix},$$

and the Jacobian of $g(x)$ with respect to x be denoted by $G(x)$. Further define

$$\mathcal{G} := \int_0^1 G(x(\theta))d\theta.$$

If for all strategies that are in the interior of the strategy space, X^o, the matrix $\mathcal{G}$ is full rank, then there is at most one equilibrium in X^o.

Proof. Suppose that there exist two Nash equilibria, represented by two flow vectors x^1 and x^0, with elements x_i^0 and x_i^1, respectively. As we consider only inner solutions, it follows from first-order optimality conditions that $g(x^0) = 0$ and $g(x^1) = 0$. Define the flow vector $x(\theta)$ as a convex combination of the two equilibrium points x^0, x^1:

$$x(\theta) = \theta x^0 + (1 - \theta)x^1,$$

where $0 < \theta < 1$. By differentiating $x(\theta)$ with respect to θ, we obtain

$$\frac{dg(x(\theta))}{d\theta} = G(x(\theta))\frac{dx(\theta)}{d\theta} = G(x(\theta))(x^1 - x^0).$$

Integrating this over θ from $\theta = 0$ to $\theta = 1$ we obtain

$$0 = g(x^1) - g(x^0) = \left[\int_0^1 G(x(\theta))d\theta\right](x^1 - x^0) \equiv \mathcal{G} \cdot [x^1 - x^0]. \tag{18}$$

Since $\mathcal{G}$ is full rank, the only possibility for (18) to hold is $x^1 - x^0 = 0$. Thus, the Nash equilibrium is unique whenever it exists. □

3.2 Existence of Unique Nash Equilibrium under Linear Utility Functions

Here, we show the existence of a unique Nash equilibrium for cost functions with linear utility. Furthermore, by exploiting the linearity of reaction functions, we compute the equilibrium point explicitly. Since we carry out the analysis for a general a_i, it applies not only to the worst-case analysis, but also to the local analysis, where the logarithmic utility function is approximated by a linear function.

Again, each user minimizes his cost function (3), under the constraints given in (6) and (7). First assuming an inner solution, we have for the ith user:

$$\frac{\partial \tilde{J}_i(x_i, x_{-i})}{\partial x_i} = \frac{k_i x_i^2 + 2k_i x_i m - 2k_i x_i (x_i + x_{-i})}{(m - (x_i + x_{-i}))^2} - a_i = 0, \tag{19}$$

which can be solved for x_i, to lead to: [1]

$$x_i = (m - x_{-i}) \left[1 \pm \left(\frac{k_i}{k_i + a_i} \right)^{\frac{1}{2}} \right]$$

The solution with the plus sign is eliminated in view of the constraint $m - x_{-i} \geq x_i$; hence, the only feasible solution is the one with the minus sign:

$$x_i = (m - x_{-i}) \left[1 - \left(\frac{k_i}{k_i + a_i} \right)^{\frac{1}{2}} \right] \equiv m q_i - q_i x_{-i}, \tag{20}$$

where

$$q_i := 1 - \left(\frac{k_i}{k_i + a_i} \right)^{\frac{1}{2}} \tag{21}$$

To complete the derivation we now check the boundary solutions. For the boundary point $x_i = 0$, we observe from (19) that $(\partial \tilde{J}_i(x_i, x_{-i})/\partial x_i) = -a_i$, which means that the user can decrease his cost by increasing x_i. Hence, this cannot be an equilibrium point. For the other boundary points $x_i = m - x_{-i}$ and $x_{-i} = m$, we observe that at these points the cost goes to infinity. As a result, the inner solution is the unique optimal response for the constrained optimization problem of the ith user, for each fixed $x_{-i} < m$. We observe from (20) that the unique optimal flow for the ith user is a linear function of the aggregate flow of all other users. This set of M equations can now be solved for $x_i, \quad i = 1, \ldots, M$. To ease the notation, let $\bar{x} := x_i + x_{-i}$. Then, (20) can be rewritten as

$$x_i = m q_i - q_i (\bar{x} - x_i) \Rightarrow x_i = \frac{q_i}{1 - q_i} m - \frac{q_i}{1 - q_i} \bar{x}.$$

Summing both sides from 1 to M, and letting

$$\lambda := \sum_{i=1}^{M} \frac{q_i}{1 - q_i}, \tag{22}$$

we obtain

$$\bar{x} = \lambda m - \lambda \bar{x} \Rightarrow \bar{x} = \frac{\lambda}{1 + \lambda} m.$$

[1]Here we assume throughout that $x_i \neq m - x_{-i}$, which will be seen shortly not to be an assumption at all.

Note that λ is well defined and positive, since $0 < q_i < 1 \ \forall i$. Hence $\bar{x} < m$, thus satisfying the underlying constraint. Finally, substituting $\bar{x}$ above into the expression for x_i (in terms of $\bar{x}$), yields the following unique solution to (20):

$$x_i^* = \frac{1}{1+\lambda} \frac{q_i}{1-q_i} m, \quad i = 1, \ldots, M. \tag{23}$$

Note that (23) is feasible since it is strictly positive, and $\sum_{i=1}^{M} x_i^* < m$. We summarize this result in the following theorem, whose proof follows from the leading derivation:

Theorem 3.2. *There exists a unique Nash equilibrium in the network game with users having linear utility functions, and it is given by (23), where q_i and λ are given by (21) and (22) respectively.*

We conclude the section with an important proposition, justifying the worst-case analysis based on linear utility functions.

Proposition 3.1. *Given the total flow rates of all users except the ith one, $x_{-i} = \sum_{j \neq i} x_j$, the optimal flow rate of the ith user, $x_{i,nonlin}^{opt}$, having a logarithmic utility and the cost function (3) is less than the optimal rate $x_{i,lin}^{opt}$ obtained when the same user has the linear utility (4) with $a_i = 1$ and cost function (5).*

Proof. The optimal solution of the ith user was already shown to be an inner solution. Differentiating the linear utility cost, $\tilde{J}_i$, given by (5), and the logarithmic utility cost, J_i, given by (3), both with respect to x_i, we obtain:

$$\tilde{J}_i^{'} = P_i^{'} - \tilde{U}_i^{'} = P_i^{'} - 1, \tag{24}$$

$$J_i^{'} = P_i^{'} - U_i^{'} = P_i^{'} - \frac{1}{1+x_i}, \tag{25}$$

where a 'prime' denotes a partial derivative with respect to x_i. The pricing function P_i in (1) is unimodal with a global minimum at $x_i = 0$. Hence, for $x_i \geq 0$, $P_i^{'}$ is a monotonically increasing function passing through the origin. Hence, the point at which (24) is zero is strictly larger than the point at which (25) is zero, that is $x_{i,nonlin}^{opt} < x_{i,lin}^{opt}$. This completes the proof. □

An intuitive explanation of this result lies in the high marginal demand of worst-case utility, $a = 1$. The marginal demand of a user with linear utility is higher than the one with logarithmic utility. The proposition above is based on this difference in demand.

4 Update Algorithms and Stability

In the previous section, it was shown that a unique equilibrium point exists under different cost functions, where each user attains a minimum cost given the equilibrium flow rates of the other users. In a distributed environment, however, each user acts independently and convergence to this equilibrium point does not occur instantaneously. Hence it is important to study the evolution of various iterative processes toward the unique equilibrium. In the literature, there exist various iterative update schemes with different convergence and stability properties [2]. We consider here three asynchronous update schemes relevant to the proposed model: PUA, parallel update algorithm, which is also known as the Jacobi algorithm; RUA, random update algorithm, and GUA, gradient update algorithm, also known as Jacobi over relaxation [14]. For the specific model at hand, individual users do not need to know the specific flow rate of other users, except their sum. This feature is of great importance for possible applications, as it simplifies substantially the information flow within the system.

4.1 Parallel Update Algorithm (PUA)

In PUA, the users optimize their flow rates at each iteration, in discrete time intervals $\ldots, n-1, n, n+1 \ldots$. If the time intervals are chosen to be longer than twice the maximum delay in the transmission of flow information, it is possible to model the system as an ideal, delay-free one. In a system with delays, users update their flows using the available (delayed) information.

One important feature of PUA is that the users are myopic. They optimize their flow rates based on instant costs and parameters, ignoring future implications of their actions. In a delay-free system, this behavior affects convergence rate adversely as it will be seen in the simulations.

For the cost function (3), the players use either nonlinear programming techniques to minimize their cost at each iteration or directly the reaction function. The analytical solution to the optimization problem of the ith user turns out to be the root of the third-order equation:

$$k_i x_i^3 + (2k_i(m - x_{-i}) + k - 1)x_i^2 + (2(k+1)(m - x_{-i}))x_i + [m - x_i]^2 = 0. \tag{26}$$

Only one root of this equation, denoted $\tilde{x}_i$, is feasible: $0 < \tilde{x}_i < m$. The closed-form solution for this root is at the same time the reaction function, which is highly nonlinear in contrast to the linear reaction function given by (20). As the root of (26) involves a complicated expression, we write the nonlinear reaction function of the ith user only symbolically :

$$x_i^{(n+1)} = f(x_{-i}^{(n)}, k_i). \tag{27}$$

Stability and convergence of the system is as important as the existence of a unique equilibrium. In an unstable system, the flow rates may oscillate indefinitely if there is a deviation from equilibrium. Or, if the system does not have the global convergence property, there exists the possibility of not reaching the equilibrium at all through an iteration starting at an arbitrary feasible point.

We now study the convergence of PUA for the linear reaction case. The update function for the ith user is (from (20)):

$$x_i^{(n+1)} = mq_i - q_i x_{-i}^{(n)} \;\; \forall i, n, \tag{28}$$

where q_i was defined in (21). Let $\Delta x_i = x_i - x_i^*$, where x_i^* is the flow rate of the ith user at the Nash equilibrium. Then we have

$$\Delta x_i^{(n+1)} = -q_i \Delta x_{-i}^{(n)}, \;\; \forall i. \tag{29}$$

Let

$$\|\Delta x\| = \max_i |\Delta x_i|,$$

and note that from (29):

$$\|\Delta x^{(n+1)}\| \leq (M-1) \max_i |q_i| \; \|\Delta x^{(n)}\|.$$

Clearly, we have a contraction mapping in (29) if $(M-1)\max_i |q_i| < 1$. Thus, the following sufficient condition ensures the stability of the system with linear utility under the PUA algorithm:

$$|q_i| \leq \frac{1}{M}, \;\; i = 1, \ldots, M. \tag{30}$$

One simple way of meeting this condition is to set $q_i = 1/M, \; i = 1, \ldots, M.$ From (21), (30) translates into the following stability condition on the pricing parameter k_i for each i :

$$k_i \geq \frac{(M-1)^2}{2M-1} a_i. \tag{31}$$

Notice that these apply not only to the analysis in the linear-reaction case, but also to the local analysis of the nonlinear-utility cost function (3). Thus, the system is locally stable and convergent under PUA if the condition above is satisfied. Next, we show that the system not only has local stability and convergence property, but is also globally stable and convergent for the nonlinear-utility cost. Before making a precise statement of this result and proceeding with the proof, we present the following two useful Lemmas.

Lemma 4.1. *For any feasible point $\mathbf{x_o} = \mathbf{x^{(n)}}$ at time (n), let $x_i^{(n+1)}$ be the outcome of the ith user's nonlinear reaction function (27). If $x_i^{(n)} > x_i^*$, where $\mathbf{x}^*$ is the unique equilibrium, then $x_i^{(n+1)} \leq x_i^{(n)}$.*

Proof. Assume that $x_i^{(n+1)} > x_i^{(n)} > x_i^*$. Given the flow rate, $x_i^{(n)}$, of the ith user at the time instant n, we linearize the logarithmic utilility, U_i, given in (2) around this value, and turn it into (4) by defining

$$a_i(x_i^{(n)}) := \frac{\partial U_i(x_i^{(n)})}{\partial x_i} = \frac{1}{1 + x_i^{(n)}}.$$

Thus, the nonlinear reaction function (27) is linearized to (28) at $x_i^{(n)}$. Using the fact that $a_i(x_i^{(n)}) > \partial U_i(x_i^{(n+1)})/\partial x_i$ and following an argument similar to that used in the proof of Proposition 3.1, the resulting flow, $x_{i,lin}^{(n+1)}$, provides an upper bound on $x_i^{(n+1)}$. Combined with the contraction property of the linear reaction function, we obtain:

$$x_i^{(n+1)} < x_{i,lin}^{(n+1)} < x_i^{(n)}. \tag{32}$$

Obviously, (32) contradicts the assumption made, and hence $x_i^{(n+1)} \leq x_i^{(n)}$. □

Lemma 4.2. *For any feasible point* $\mathbf{x_0} = \mathbf{x^{(n)}}$ *at time* (n), *if* $x_i^{(n)} < x_i^*$, *then* $x_i^{(n+1)} \geq x_i^{(n)}$.

Proof. The proof is very similar to that of Lemma 4.1. Suppose that $x_i^{(n+1)} < x_i^{(n)} < x_i^*$. Then, it can be shown that $a_i < \partial U_i(x_i^{(n+1)})/\partial x_i$, and $x_{i,lin}^{(n+1)}$ provides a lower bound on $x_i^{(n+1)}$. Again using the contraction property,

$$x_i^{(n+1)} > x_{i,lin}^{(n+1)} > x_i^{(n)}.$$

As this contradicts the initial hypothesis, $x_i^{(n+1)} \geq x_i^{(n)}$. □

Theorem 4.1. *The PUA is globally convergent and stable for both the linear and logarithmic utility cost functions (3) and (5), under the sufficient condition (31).*

Proof. The convergence result for the linear utility case was obtained above. In principle, it is also possible to derive the global convergence result for the logarithmic utility case, using the same method, but this time the reaction function is (27). This reaction function was obtained as one of the roots of the third-order equation (26), and is highly nonlinear. Hence the method based on reaction functions becomes practically intractable. We will, therefore, make use of the local and worst-case analyses to obtain a global convergence result. As in Lemma 4.1, the nonlinear reaction function (27) can be linearized to (28) at each $x_i^{(n)}$. Also note that the existence of a unique feasible Nash equilibrium, $0 < x_i^* < m$, was already established for the network game.

For any feasible initial point, $\mathbf{x_0}$, we have the following cases for the ith user:

In the first case, $x_i^{(n)} > x_i^*$, where $x_i^{(n)} = x_{i,0}$ is the starting point. According to Lemma 4.1, there are two possibilities: $x_i^* < x_i^{(n+1)} < x_i^{(n)}$ and $x_i^{(n+1)} < x_i^* < x_i^{(n)}$. The former case results in a monotonically decreasing sequence bounded below by x_i^*, whereas the latter case leads to an oscillating sequence around the equilibrium.

In the second case, $x_i^{(n+1)} < x_i^*$, where $x_i^{(n+1)}$ can be considered as the starting point at any time instant $(n+1)$. Again by Lemma 4.2, there are two possibilities: $x_i^* > x_i^{(n+1)} > x_i^{(n)}$ and $x_i^{(n+1)} > x_i^* > x_i^{(n)}$. Similar to the previous case, the former leads to a monotonically increasing sequence bounded above by x_i^*, and the latter results again in an oscillating sequence around the equilibrium. In order to analyze these cases, one can define the relative distance to equilibrium as $\Delta x_i^{(n)} := x_i^{(n)} - x_i^*$.

If $x_i^{(n+1)} > x_i^{(n)}$, then the linearized reaction function at $x_i^{(n)}$ provides an upper bound, $x_{i,linear}^{(n+1)}$, on $x_i^{(n+1)}$, following an argument similar to the one in Proposition 3.1. The fact that $(\partial U_i^{(n)}/\partial x_i) < a_i^{(n)}$ justifies the given bound. Using the contraction property of linearized reaction function (29) and the worst-case bound above, we obtain:

$$\Delta x_{i,linear}^{(n+1)} < \Delta x_i^{(n)} < \Delta x_{i,linear}^{(n)} \quad \forall i. \tag{33}$$

For $x_i^{(n+1)} < x_i^{(n)}$, following a similar argument, it is easy to show that the relation (33) also holds. Therefore, for all possible cases, we have shown that, at each iteration locally linearized flows provide a decreasing upper bound to the iterates of the nonlinear reaction function for the distance to equilibrium. Figure 1 summarizes this discussion graphically.

The flow rate of any i^{th} user converges to the unique Nash equilibrium and the nonlinear system is stable and globally convergent from any feasible initial point $\mathbf{x_0}$. We note that the proof is based on the convergence of the linear system, which is required for the convergence of the nonlinear system. Moreover, condition (30), or equivalently (31), is sufficient for the convergence of the iteration corresponding to linear and nonlinear reaction functions. □

4.2 Random Update Algorithm (RUA)

Random update scheme is a stochastic modification of PUA. The users optimize their flow rates in discrete time intervals and infinitely often, with a predefined probability p_i, $0 < p_i < 1$, for user i. Thus, at each iteration a random set of users among the M update their flow rates. Again, the users are myopic and make instantaneous optimizations. In the limiting case, $p_i = 1$, RUA is the same as PUA. The non-ideal system with delay is also similar to PUA. The users make decisions based on delayed information at the updates, if the round trip delay is longer than the discrete time interval.

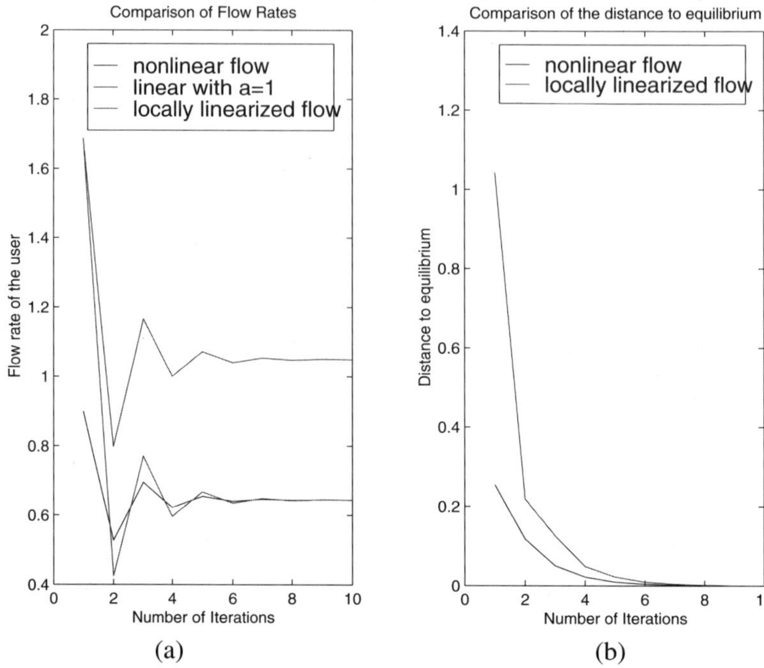

Figure 1: (a) Comparison of nonlinear, locally linearized: $a_i = 1/(1 + x_i)$ and linear worst-case: $a_i = 1$ flow rates, (b) distance to equilibrium, nonlinear and locally linearized.

For the linear-utility case (3) with linear reaction function (20), the update scheme may be formulated for the ith user as follows:

$$x_i^{(n+1)} = \begin{cases} mq_i - x_{-i}^{(n)} q_i, & \text{with probability } p_i, \\ x_i^{(n)}, & \text{with probability } 1 - p_i. \end{cases}$$

Subtracting x_i^* from both sides, we obtain:

$$\triangle x_i^{(n+1)} = \begin{cases} -q_i \triangle x_{-i}^{(n)}, & \text{with probability } p_i, \\ \triangle x_i^{(n)}, & \text{with probability } 1 - p_i. \end{cases}$$

Taking the absolute value of both sides, and then taking expectations, leads to

$$\begin{aligned} E|\triangle x_i^{(n+1)}| &\le p_i q_i E|\triangle x_{-i}^{(n)}| + (1 - p_i) E|\triangle x_i^{(n)}| \\ &\le p_i q_i \sum_{j=1}^{M} E|\triangle x_i^{(n)}| + (1 - p_i(1 + q_i)) E|\triangle x_i^{(n)}|. \end{aligned}$$

Choosing $p_i \geq 1/(1+q_i)$, this expression can further be bounded by

$$E|\Delta x_i^{(n+1)}| \leq p_i q_i \sum_{j=1}^{M} E|\Delta x_j^{(n)}|,$$

and summing over all users we obtain

$$\sum_{i=1}^{M} E|\Delta x_i^{(n+1)}| \leq \left(\sum_{i=1}^{M} p_i q_i\right) \cdot \sum_{i=1}^{M} E|\Delta x_i^{(n)}| \tag{34}$$

If $\sum_{i=1}^{M} p_i q_i < 1$, $\mu^{(n)} := \sum_{i=1}^{M} E|\Delta x_i^{(n)}|$ is a decreasing positive sequence, and hence converges to the only equilibrium state of (34), zero. This implies convergence of each individual term in the summation to zero, which in turn says that $x_i^{(n)} \rightarrow x_i^*, i = 1, \ldots, M$, with probability 1. Notice that the sufficient condition (30) for the stability of PUA also guarantees the stability of RUA for the linear utility case.

The question now comes up as to the choice of p_i that would lead to fastest convergence in (34), which we will call the optimal update probability. Maheswaran and Başar [15] show that in a quadratic system without delay, one can find a fairly tight bound for optimal update probability as number of users goes to infinity, and this bound is $\frac{2}{3}$. Repeating the same analysis for this model and linear cost function leads to an exact update probability $p_{opt} = \frac{2}{3}$, which is optimal for a large number of users.

The stability and convergence results obtained also apply to the local analysis of the nonlinear utility function as in PUA. Hence the nonlinear utility case is locally stable under RUA. Moreover, Lemma 4.1 and Lemma 4.2 are valid for RUA, and hence Theorem 4.1 holds, indicating the global stability of the algorithm.

4.3 Gradient Update Algorithm (GUA)

The gradient update algorithm can be described as a relaxation of PUA. For this scheme, we define a relaxation parameter s_i, $0 < s_i < 1$, for i^{th} user, which determines the step-size the user takes towards the equilibrium solution at each iteration.

For the linear utility case, the algorithm is defined as:

$$x_i^{(n+1)} = x_i^{(n)} + s_i \cdot [(mq_i - x_{-i}^{(n)} q_i) - x_i^{(n)}] \;\; \forall i, n. \tag{35}$$

Different from both PUA and RUA, the users are not myopic in this scheme. Although they seem to choose suboptimal flow rates at each iteration instead of exact optimal solutions, they benefit from this strategy by reaching the equilibrium faster. GUA, despite its deterministic nature like PUA, is actually very similar to RUA in analysis. When compared with PUA, as we observe in simulations, GUA converges faster to Nash equilibrium than PUA in highly loaded delay-free systems, where there is a high demand for scarce resources and users act simultaneously. An intuitive explanation can be made using the fact that equilibrium point is quite dynamic in loaded systems during iterations. In PUA, users update their flows as if it is static, while in GUA, users behave more cautiously and do not rush to the temporary equilibrium point at each iteration. Thus, the wide fluctuations in flow rates, which can be observed in PUA, are avoided in this case. One can interpret the relaxation parameter s_i also as a measure of this caution. Another advantage of GUA, its relative insensitivity to delays in the system, can also be explained with the same reasoning.

A similar but deterministic version of the convergence analysis of RUA for the linear utility function yields the same convergence result as in RUA, except that p_i is now replaced with s_i:

$$\triangle x_i^{(n+1)} = (1 - s_i)\triangle x_i^{(n)} - s_i q_i \sum_{j \neq i}^{M} \triangle x_j^{(n)} \tag{36}$$

$$\Rightarrow |\triangle x_i^{(n+1)}| \leq (1 - s_i)|\triangle x_i^{(n)}| + s_i q_i \sum_{j \neq i}^{M} |\triangle x_j^{(n)}|, \quad \forall i. \tag{37}$$

Choosing $s_i \geq 1/(1 + q_i)$, and imposing the condition $\sum_{i=1}^{M} s_i q_i < 1$, the flow rates of the users converge to the unique equilibrium as in other schemes. Using (35), we obtain:

$$\lim_{n \to \infty} x_i^{(n)} = x_i^* = mq_i - x_{-i}^* q_i, \quad \forall i.$$

The sufficient condition (30) also guarantees the stability of GUA for the linear utility case. Moreover, since GUA is a modification of PUA, it can be shown that Lemma 4.1 and Lemma 4.2 hold for the GUA as well. Thus, the stability results of the RUA are directly applicable to GUA for both the linear and nonlinear reaction functions.

Next, we investigate the possibility of finding an optimal relaxation parameter, s, for the linear utility case, in the sense that it leads to fastest convergence to the equilibrium. In order to simplify the analysis, we assume symmetric users, resulting in $q_i = q = 1/M$, and $s_i = s$, $\forall i$. For the special case of symmetric initial conditions, we obtain from (36):

$$\triangle x_i^{(n+1)} = [1 - s(1 + (M - 1)q)]\triangle x_i^{(n)}.$$

The value of s, leading to fastest convergence in this case is

$$s_{opt} = \frac{1}{1 + (M - 1)/M} \Rightarrow \lim_{M \to \infty} s_{opt} = 0.5, \tag{38}$$

which leads to one-step convergence.

For the general case, however, it is not possible to find a unique optimal value of s, as different starting points for users which result in different $\triangle x_i$ at each iteration, affect the optimal value of s. Using simulations, we conclude that optimal value of s for a delay-free linear system should be in the range $0.5 < s_{opt} < 1$.

The analysis for the linear utility case applies to the nonlinear utility case locally, giving the same local stability and convergence results. One can show that in addition to the local results, global convergence and stability of PUA also apply to GUA. Therefore, GUA converges globally to the unique equilibrium in the nonlinear utility case. As it will be shown in numerical examples, GUA becomes advantageous only under heavy load, and loses its fast convergence property in lightly loaded systems.

5 Numerical Simulations of the Update Schemes

Each update scheme analyzed in the previous section is simulated using MATLAB. The proposed model is tested through extensive simulations for both nonlinear and linear reaction functions. The latter can be considered as either worst-case analysis or local approximation to the nonlinear utility cost. The system is simulated first without delay under all three update schemes: PUA, RUA and GUA. Next, in the second group of simulations, uniformly distributed delays are added to the system for more realistic analysis. The convergence rate is measured as the number of iterations required to reach the unique Nash equilibrium. As a simplification, we assume symmetric users in most cases, where cost parameters like, a, k, q, update probability, p, for RUA, and relaxation parameter, s, for GUA are not user specific. Starting condition for simulations is the origin, i.e. zero initial flow, unless otherwise stated. The following criterion is used as the stopping criterion, where M is the total number of users. $\sum_{i=1}^{M} |x_i^{(n+1)} - x_i^{(n)}| \leq M \cdot \epsilon$. The stopping distance is chosen sufficiently small, $\epsilon = 10^{-5}$, for accuracy in all simulations.

5.1 Simulations for Delay-free Case

The convergence of the update algorithms for different numbers of users, as a crucial parameter, is investigated throughout the analysis. However, we first implemented the basic PUA algorithm with $M = 20$ users with linear reaction functions and $a = 1$ indicating a high demand for bandwidth. $k = 10$ is chosen to ensure stability. From Figure 2(a), we observe the undesirable, wide oscillations in flow rates of users, which is a disadvantage of PUA under a heavily loaded delay-free system. In this case, although the number of users is small, the low value of pricing parameter k loads the system. Absence of delay in the system also contributes to the instantaneous load, as the users act simultaneously. The instantaneous demand affects the convergence rate significantly in delay-free systems, especially under PUA.

Another important parameter in the system is the price, k. The impact of the price on the system is investigated in the next simulation. Figure 2(b) shows the

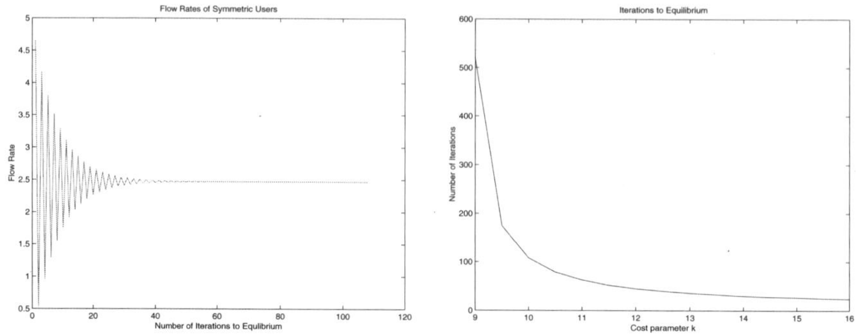

(a) (b)

Figure 2: (a) Flow rates vs. iterations to equilibrium in case of symmetric users and PUA, (b) Convergence rate of PUA for different values of k.

effect of varying the pricing parameter k under PUA. Again, there are $M = 20$ users. It can be observed that as the price increases, the convergence rate drops. An intuitive explanation for this phenomenon is based on the effect of price on the demand of users. An increase in price results in a decrease in demand and system load, leading to faster convergence. Even though the simulation here is for a delay-free linear-utility system, varying the price leads to similar results under all update schemes for both linear and nonlinear reaction functions. Theoretical calculations based on linear utility, in the previous section, show that the minimum value of k satisfying the stability criterion is 9.2 for this specific case. This bound on k is only a sufficient condition for stability, which is verified in this simulation by observing the convergence of system for $k = 9$. The large number of iterations required, on the other hand, indicates the tightness of the bound.

Next set of simulations investigate the two basic parameters of RUA: M, number of active users, and, p, the update probability. The simulation results in Figure 3(a) verify the theoretical analysis of the previous section for linear utility cost. It is observed that with the increasing number of users the optimal update probability gets closer to the value $2/3$. For completeness, the same simulation is repeated for the logarithmic utility cost. Interestingly, we obtain similar results, as shown in Figure 3(b). Due to the structure of logarithmic utility function, the demand of users is less than in the linear utility case, and hence the system is not loaded as much as in the linear case. For the same number of users, we observe that the optimal update probability shifts to higher values. In conclusion, Figure 3(b) can be considered as a stretched version of Figure 3(a), due to the change in load.

Similar to RUA, a simulation based on the relaxation parameter s is done for linear cost under GUA. The result confirms the theoretical result (38) for symmetric initial conditions. Other initial conditions, however, lead to different optimal values for s, in most cases between 0.6 and 0.8. The result can be interpreted as the variation in the amount of instantaneous demand for bandwidth. In the symmetric initial condition, all users act the same way, leading to higher simultaneous demand,

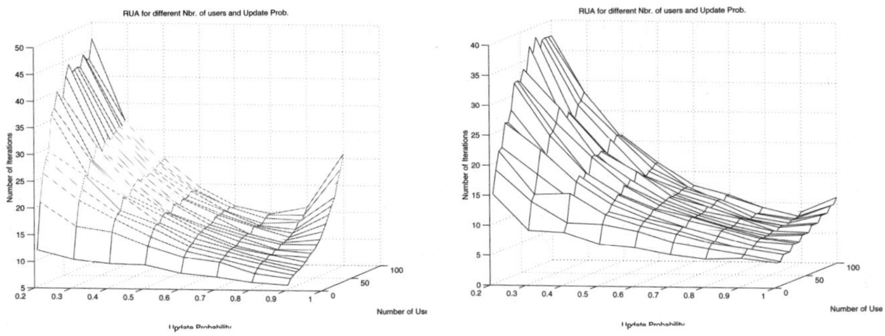

(a) (b)

Figure 3: (a) Convergence rate of RUA as M gets larger, for different update probabilities $0 < p < 1$, and linear utility, (b) Convergence rate of RUA for nonlinear utility.

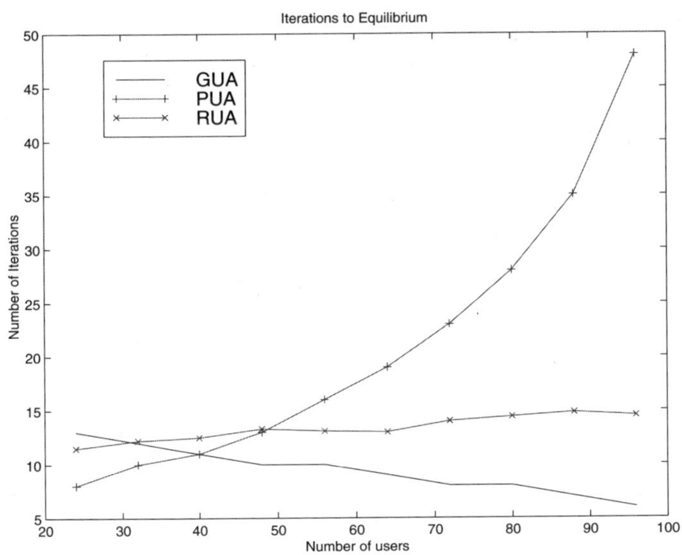

Figure 4: Comparison of convergence rates of PUA, RUA and GUA for increasing number of users, linear utility, delay-free system.

where 'being cautious' or decreasing s is advantageous. For other initial values, the instantaneous demand decreases, where increasing s affects the convergence rate positively. We conclude that GUA is only advantageous in situations with high instantaneous and total demand, which will further be verified in delayed simulations.

Finally, we conclude the simulations without delay by a comparison of the convergence rate of all three algorithms for different numbers of users. The results for the linear reaction function are depicted in Figure 4. We observe clearly that both

GUA and RUA are superior to PUA. Another important and promising observation is that the rates of convergence for GUA and PUA are almost independent of the number of users. The simulation is repeated for nonlinear utility cost and for highly and lightly loaded systems. To change the amount of load on the system, the capacity parameter m is varied this time, instead of price k. They affect, as expected, the convergence rate in opposite ways. Obviously, the smaller the capacity, the heavier the load. Similar to the linear utility case, GUA converges faster with increasing number of users. It performs, however, poorer under light load. The same trend is also observed for RUA. One interesting phenomenon is the high performance of PUA under light load. It can be interpreted as a result of the low instantaneous demand due to the variation of utility for different flow rates.

5.2 Simulations with Delay

In order to make the simulations more realistic, we next introduce the delay factor into the system in the following way: users are divided into d equal groups, where each group has an increasing number of units of delay. For example in a four-group system, the first group has no delay, the second one has one unit delay, the third group two units of delay etc. When the simulations are repeated with uniformly distributed delay as described, the results obtained are quite different from the previous ones. PUA, for example, performs better than in the linear utility case without delay. This is possibly caused by the decrease of instantaneous demand, due to the delay factor. This result strengthens the argument made on PUA in the previous section.

In RUA, however, the optimal update probability disappears in contrast to the delay-free case, as can be seen in Figure 5. Again, the underlying cause is the effect of delay factor on instantaneous demand. Another important result is the similarity of the results for linear (a) and nonlinear (b) cases in this simulation.

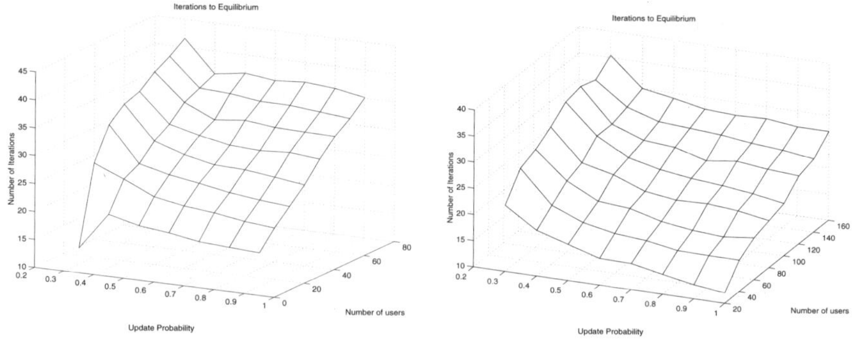

(a) (b)

Figure 5: (a) Convergence rate of RUA as M gets larger and for different update probabilities $0 < p < 1$ in a delayed system with $d = 5$; (b) same simulation with the nonlinear system.

Regarding PUA, we can conclude that it performs better when both instantaneous and total demand are low and system resources are abundant. Such conditions exist for delayed systems with users having logarithmic utility. RUA, on the other hand, performs worse with decreasing update probability in such a case.

Next, GUA is investigated under a delay incorporated system for an optimal relaxation parameter. Using the results of several simulations, we conclude that the optimal value of s decreases in the linear utility case, as the delay factor increases. Interestingly, this trend disappears for the nonlinear case. Similar to RUA, GUA also loses its advantage with nonlinear reaction functions in a delayed system under normal load. Comparison of all three algorithms in the delayed nonlinear system for high and low load can be seen in Figure 6(a). In the graph below, prices are halved, while the capacity is tripled with respect to the one above. As expected, PUA performs better than RUA for any load. Under light load, PUA is superior to GUA, with the aid of delay factor and low instantaneous delay due to logarithmic utility of users. As the load in the system increases, GUA performs comparable to PUA, verifying the observation in Figure 6(a).

In another set of simulations, we investigate the robustness of the algorithms under disturbances. Disturbance is added to the system by varying the number of users at each iteration by about 10% of the total number of users on the average. The arrival of users is modeled as a Poisson random process, and connection durations are chosen to be exponentially distributed. Hence, the number of users in the system constitute a Markov chain. Figure 6(b) shows the stability results under different update schemes in terms of the percentage distance to the ideal equilibrium for an example time window. The lower right graph is the result of the simulation with nonlinear reaction function under PUA. We observe that the average distances to the equilibrium vary between 0.5% and 1.5%, which indicate that the system is very robust under all schemes and costs.

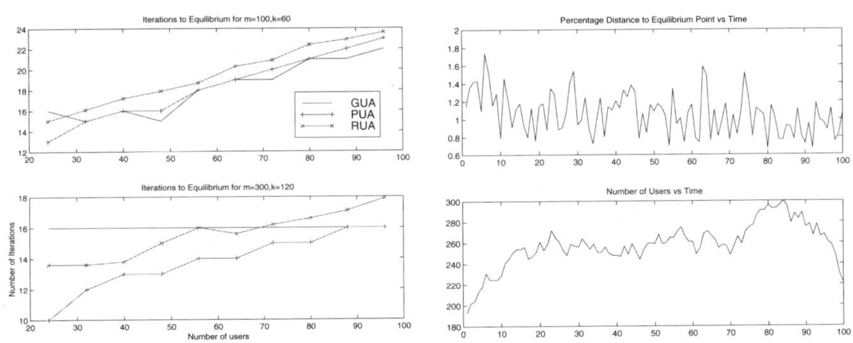

(a) (b)

Figure 6: (a) Comparison of convergence rates of PUA, RUA and GUA for increasing number of users, nonlinear cost. (b) Robustness analysis for PUA in case of the nonlinear utility. Percentage distance and number of users vs. time.

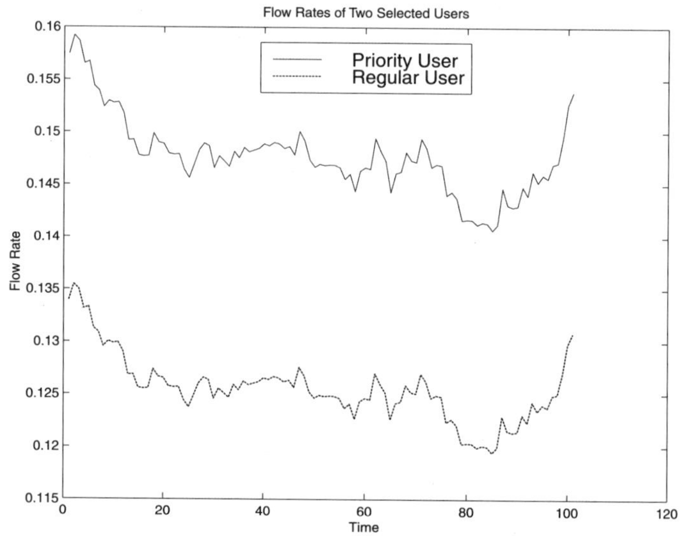

Figure 7: Flow rates of a priority user with $k = 200$ and a regular user with $k = 240$ vs. time.

Finally, a pricing scheme with two classes, a group of priority users and a group of regular users is studied. Priority users are charged less – in terms of network credits – than the regular users by setting $k = 200$ vs. $k = 240$ for the excess flow rate, x. A similar disturbance structure to the one above is used by varying the number of users in order to create a more realistic setting. Again the simulation is done with users having a nonlinear reaction function under PUA. The flow rates of two sample users from each class are shown in Figure 7. We observe that the pricing scheme is successful in differentiating the priority user from the regular one. Moreover the robustness of the model is preserved.

6 Conclusions

We have introduced a mathematical model which can be used as a basis for implementation of real time traffic on the Internet. The combination of admission control and end-to-end distributed flow control results in a very flexible framework, which captures all traffic types from low to medium elasticity. A market-based approach enables the model to address two major issues, pricing and resource allocation, simultaneously.

A unique Nash equilibrium is obtained, and convergence properties of relevant asynchronous update schemes are investigated both theoretically and numerically. Conditions for the stability of the equilibrium point under three update algorithms are obtained and analyzed in the cases of linear and nonlinear reaction functions. The simulation results suggest the use of GUA or RUA in heavy loaded systems with less delay and high demand for bandwidth. In delayed systems, however, PUA performs better than the other two. The linear analysis not only provide a local approximation to the nonlinear cost, but also establish convergence and stability results, helping to solve the general nonlinear cases.

Acknowledgements

Research reported here was supported in part by the NSF grants ANI 98-13710 and CCR 00-85917 ITR, and the MURI grant AF DC 5-36128.

REFERENCES

[1] Altman E., Başar T., Jimenez T. and Shimkin N., "Competitive routing in networks with polynomial costs," *IEEE Transactions on Automatic Control*, vol. 47(1), pp. 92–96, January 2002.

[2] Altman E. and Başar T., "Multi-user rate-based flow control," *IEEE Transactions on Communications*, vol. 46(7), pp. 940–949, July 1998.

[3] Orda A., Rom R. and Shimkin N., "Competitive routing in multiuser communication networks," *IEEE/ACM Transactions on Networking*, vol. 1, pp. 510–521, October 1993.

[4] Douligeris C. and Mazumdar R., "A game theoretic perspective to flow control in telecommunication networks," *Journal of the Franklin Institute*, vol. 329, pp. 383–402, March 1992.

[5] Başar T. and Olsder G. J., *Dynamic Noncooperative Game Theory*. 2nd edn. Philadelphia, PA: SIAM, 1999.

[6] Korilis Y. A. and Lazar A., "On the existence of equilibria in noncooperative optimal flow control," *Journal of the ACM*, vol. 42, pp. 584–613, May 1995.

[7] Hsiao M. T. and Lazar A., "Optimal decentralized flow control of markovian queueing networks with multiple controllers," *Performance Evaluation*, vol. 13, pp. 181–204, 1991.

[8] Bertsekas D. and Gallager R., *Data Networks*. 2nd edn. Upper Saddle River, NJ: Prentice Hall, 1992.

[9] Alpcan T. and Başar T., "A game-theoretic framework for congestion control in general topology networks," in *Proc. of the 41st IEEE Conference on Decision and Control*, Las Vegas, NV, December 2002, pp. 1218–1224.

[10] Alpcan T. and Başar T., "A utility-based congestion control scheme for internet-style networks with delay," in *Proc. IEEE Infocom*, San Francisco, CA, April 2003.

[11] Alpcan T., Başar T. and Tempo R., "Randomized algorithms for stability and robustness analysis of high speed communication networks," in *Proc. of IEEE Conference on Control Applications (CCA)*, Istanbul, Turkey, June 2003, pp. 397–403.

[12] Alpcan T. and Başar T., "Global stability analysis of an end-to-end congestion control scheme for general topology networks with delay," in *Proc. of the 42nd IEEE Conference on Decision and Control*, Maui, HI, December 2003, pp. 1092–1097.

[13] Shenker S., "Fundamental design issues for the future internet," *IEEE Journal on Selected Areas in Communications*, vol. 13, no. 7, pp. 1176–1188, September 1995.

[14] Bertsekas D., *Nonlinear Programming*, 2nd edn. Belmont, MA: Athena Scientific, 1999.

[15] Maheswaran R. T. and Başar T., "Multi-user flow control as a Nash game: Performance of various algorithms," in *Proc. of the 37th IEEE Conference on Decision and Control*, December 1998, pp. 1090–1095.

A Taylor Series Expansion for H^∞ Control of Perturbed Markov Jump Linear Systems

Rachid El Azouzi
University of Avignon, LIA,
339, chemin des Meinajaries, Agroparc B.P. 1228,
84911 Avignon Cedex 9, France

Eitan Altman
INRIA B.P. 93, 2004 route des Lucioles,
06902 Sophia-Antipolis Cedex, France
Eitan.Altman@sophia.inria.fr

Mohammed Abbad
Départament de Mathematiques et Informatique
Faculté des Sciences B.P. 1014
Université Mohammed V, 10000 Rabat, Morocco
Rachid.Elazouzi@sophia.inria.fr

Abstract

In a recent paper, Pan and Başar [19] have studied the H^∞ control of large scale Jump Linear systems in which the transitions of the jump Markov chain can be separated into sets having strong and weak interactions. They obtained an approximating reduced-order aggregated problem which is the limit as the rate of transitions of the faster time scale (which is a multiple of some parameter $1/\epsilon$) goes to infinity. In this paper we further investigate the solution of that problem as a function of the parameter ϵ. We show that the related optimal feedback policy and the value admit a Taylor series in terms of ϵ, and we compute its coefficients.

Key words. Singular perturbation, H^∞ control, stochastic systems, continuous-time Markov chain, averaging, aggregation, Taylor series.

1 Introduction

Several papers have studied in recent years the control of piecewise-deterministic systems in the presence of unknown (continuous) disturbances. The traditional H^∞ control setting in which the system matrices A, B and D are fixed, is extended to allow these matrices to depend on a Markov jump process with finite-state space. The corresponding dynamic programming equation leads to the analysis of a set of Riccati equations involving the generator of the underlying Markov chain. In [18,22] the case of jump linear systems was studied, and in [1] the case of nonlinear systems.

In many applications, the formulation of the problem in terms of these models leads to very high dimensional states for the Markov chain, which makes it computationally infeasible or extremely sensitive to small inaccuracies, and it is difficult to obtain solutions to the Riccati equations. To overcome this difficulty, we use singular perturbation techniques in the modeling, control design, and optimization. The resulting systems naturally display certain two-time-scale behavior, a fast time scale and a slowly varying one. Presence of such a phenomenon is best expressed mathematically by introducing a small parameter $\epsilon > 0$ and modeling the underlying system as one involving a singularly perturbed Markov chain. This motivated Pan and Başar [19] to study systems in which the transitions of the jump Markov chain can be separated into sets having strong and weak interactions. They obtained an approximating reduced-order aggregated problem which is the limit as the rate of transitions of the faster time scale (which is a multiple of $1/\epsilon$) goes to infinity. A number of examples of Markovian models with weak and strong interaction are given in Yin and Zhang [26].

In this paper we further investigate the solution of that problem as a function of the parameter ϵ. We show that the related optimal feedback policy and the value admit a Taylor series in terms of ϵ, and we compute its coefficients. We use here the methodology from [7,8] who considered the problem without disturbances.

Research on control systems with Markovian switching structure was initiated more than thirty years ago by Krassovskii and Liskii [13,15], and Florentin [11], with follow-up work in the late sixties and early seventies addressing the stochastic maximum principle [14,20], dynamic programming [20], and linear-quadratic control [24,23], in this context. The late eighties and early nineties have witnessed renewed interest in the topic, with concentrated research on theoretical issues like controllability and stabilizability [3,9,12,22,25] in linear-quadratic systems in continuous time, see also [4,17,21] for discrete-time. Perhaps the first theoretical development in differential games context was reported in [2], where a general model was adopted that allows in a multiple player environment the Markov jump process (also controlled by the players) to determine the mode of the game, in addition to affecting the system and cost structure. Results in [16] also apply, as a special case, to zero-sum differential games with Markov jump parameters and state feedback information for both players. Some selected publications on this topic are [1,22,25].

The paper is organized as follows: section 2 introduces the general model. In section 3, we present the motivation for singularly perturbed problems. Section 4 provides the Taylor expansion of the optimal control zero-sum differential game and the solution to the coupled algebraic Riccati equations. We study the computational algorithm in section 5. The paper ends with concluding remarks.

2 General Model

The class of jump linear system under consideration is described by:

$$\frac{dx}{dt} = A(\theta(t))x(t) + B(\theta(t))u(t) + D(\theta(t))w(t), \qquad x(0) = x_0, \tag{1}$$

where x is the p-dimensional system state, u is an r-dimensional control input; w is a q-dimensional disturbance, and $\theta(t)$ is a finite state Markov chain defined on the state space S, of cardinality s, with the infinitesimal generator matrix

$$\Lambda = (\lambda_{ij}), \qquad i, j \in S.$$

and an initial distribution $\pi_0 := [\pi_{01}, ..., \pi_{0s}]$.

The control input u is generated by a control policy of the form

$$u(t) = \varphi(t, x_{[0,t]}, \theta_{[0,t]}), \qquad t \in [0, +\infty), \tag{2}$$

where $\varphi : [0, +\infty] \times \mathcal{H}_x \times \Omega \to \mathcal{H}_u$ is piecewise continuous in its first argument, and piecewise Lipschitz continuous in its second argument and measurable in θ, further satisfying the given causality condition. Let us denote the class of all admissible controllers by $\mathcal{M}$.

Associated with this system is the infinite-horizon quadratic performance index:

$$L = \int_t^{+\infty} (|x(s)|^2_{Q(\theta(t))} + |u(s)|^2) ds,$$

where $Q(.) \geq 0$, and $|x|_Q$ denotes the Euclidean semi-norm.

The system state x, inputs u and w, each belong to appropriate ($\mathcal{L}^2$) Hilbert spaces $\mathcal{H}_x, \mathcal{H}_u, \mathcal{H}_w$ respectively, defined on the time interval $[0, +\infty)$. The underlying probability space is the triple (Ω, F, P). Let E denote the expectation with respect to the probability measure P.

We introduce the disturbance attenuation problem where the initial state of the system is completely unknown. The initial condition x_0 and the input w are generated by a strategy $\delta := (\delta_0, \delta_1)$, according to

$$\begin{aligned} x_0 &= \delta_0(\theta(t_0)), \\ w(t) &= \delta_1(t, x_{[0,t]}, \theta_{[0,t]}), \end{aligned} \tag{3}$$

where $\delta_0 : S \to \mathbb{R}^p$, and $\delta_1 : [0, +\infty) \times \mathcal{H}_x \times \Omega \to \mathcal{H}_w$ is piecewise continuous in its first argument, and piecewise Lipschitz continuous in its second argument and measurable in θ, further satisfying the given causality condition. Let us denote the class of all admissible policies δ by $\mathcal{D}$.

The H^∞ optimal is to find the optimal performance level γ^* due to the following $\mathcal{L}_2$ type gain:

$$\inf_{\varphi \in \mathcal{M}} \sup_{\delta \in \mathcal{D}} \frac{(EL(\varphi, \delta))^{\frac{1}{2}}}{\left(E\left\{\int_0^{+\infty} |w(t)|^2 dt + |x_0|^2_{Q_0(\theta(0))}\right\}\right)^{\frac{1}{2}}} = \gamma^*, \quad Q_0 > 0.$$

(we added (φ, δ) to the notation L to stress its dependence on the policy of both players). Moreover, we wish to find a policy f^* that guarantees this level to player 1, i.e. achieving

$$\sup_{\delta\in\mathcal{D}} \frac{(EL(f^*, \delta))^{\frac{1}{2}}}{\left(E\left\{\int_0^{+\infty} |w(t)|^2 dt + |x_0|^2_{Q_0(\theta(0))}\right\}\right)^{\frac{1}{2}}} = \gamma^*. \tag{4}$$

This H^∞ optimal control problem is known to be closely related to a class of zero-sum differential games for the jump linear system (1), with the following γ-parameterized cost function:

$$J_\gamma(\varphi, \delta) = E\left\{\int_0^\infty (|x(t)|^2_{Q(\theta(t))} + |u(t)|^2 - \gamma^2|w(t)|^2)dt - \gamma^2|x_0|^2_{Q_0(\theta(0))}\right\}, \tag{5}$$

where player 1 (the minimizer) chooses this strategy $\varphi \in \mathcal{M}$ so as to minimize the expected cost function (5), and the second player chooses the strategy $\delta \in \mathcal{D}$ so as to maximize the same expected cost function:

$$\hat{J}_\gamma = \inf_{\varphi\in\mathcal{M}} \sup_{\delta\in\mathcal{D}} J_\gamma(\varphi, \delta). \tag{6}$$

It should be noted that the upper value of the game, $\hat{J}_\gamma$, is also bounded below by 0, which can be ensured for player 2 by choosing x_0 and $w(t)$ to be zero. Because of the linear-quadratic nature of the problem, the upper value of the game will be infinite if $\hat{J}_\gamma > 0$. As in standard H^∞-control, we will study a parametrization of the solution to this problem in term of γ. The threshold γ^* is the smallest value of $\gamma > 0$ such that the above game admits a finite upper value (which is equal to zero) as well as a saddle point.

If there exists a policy $f^* \in \mathcal{M}$ that guarantees the disturbance attenuation γ^* in (4) then it also guarantees (to player one) the value of the game (6), i.e.

$$\sup_{\delta\in\mathcal{D}} J_\gamma(f^*, \delta) = \inf_{\varphi\in\mathcal{M}} \sup_{\delta\in\mathcal{D}} J_\gamma(\varphi, \delta).$$

Instead of obtaining f^* defined above, we will in fact solve a parameterized class of controllers, $\{f^\gamma, \gamma > \gamma^*\}$, where f^γ is obtained from $\sup_{\delta\in\mathcal{D}} J_\gamma(f^\gamma, \delta) = \inf_{\varphi\in\mathcal{M}} \sup_{\delta\in\mathcal{D}} J_\gamma(\varphi, \delta)$. The controller f^γ will clearly have the property that it ensures a performance level γ^2 for the index adopted for

$$\sup_{\delta\in\mathcal{D}} \frac{(EL(f^\gamma, \delta))^{\frac{1}{2}}}{\left(E\left\{\int_0^{+\infty} |w(t)|^2 dt + |x_0|^2_{Q_0(\theta(0))}\right\}\right)^{\frac{1}{2}}} = \gamma.$$

We now make the following basic assumptions for the problem formulation above.

Assumption 2.1. The initial probability distribution of the form process $\pi_{0i} > 0$ for all $i \in S$.

Assumption 2.2. The pair $(A(i), Q(i))$ are observable for each $i \in S$.

Assumption 2.3. The pair $(A(\theta(t)), B(\theta(t)))$ is stochastically stabilizable.

3 The Perturbed Solution Process

In many applications, because of the various sources of uncertainties, the Markov chain involved is inevitably large dimensional. This size brings about many numerical difficulties in the solution. Moreover, these systems may be sensitive to small perturbations of the parameter value. Often, when the Markov chain is large it is natural to think of the large number of states to be grouped into different collections of states, based on whether the interaction between any two states is weak or strong. Presence of such a phenomenon may be expressed mathematically by taking the probability transition rate matrix $\Gamma^\epsilon = (\lambda^\epsilon_{ij})$ in an appropriate singularly perturbed form, as already discussed in [5] and [6]:

$$\lambda^\epsilon_{ij} = \nu_{ij} + \frac{1}{\epsilon}\mu_{ij}, \tag{7}$$

where $(\nu_{ij})_{s\times s}$ and $(\mu_{ij})_{s\times s}$ are probability transition rate matrices corresponding to respectively weak and strong interactions within the form process. The scalar ϵ is a small positive number.

For fixed $\epsilon > 0$, for any $\gamma > \gamma^*(\epsilon)$, the associated zero-sum differential game has a zero upper value [19]. A control policy that guarantees this upper value, which is then a control policy that guarantees an H^∞ performance level of γ, is given by:

$$\varphi^*_\gamma(t, x(t), \theta(t)) = -B^T(\theta(t))P^\gamma_\epsilon(\theta(t))x(t), \tag{8}$$

where $P^\gamma_\epsilon(i)$, $i \in S$ are unique minimal positive-definite solution to the following set of coupled generalized algebraic Riccati equation: $P^\gamma(i)$ $i \in S$ to:

$$0 = A^T(i)P^\gamma_\epsilon(i) + P^\gamma_\epsilon(i)A(i) + Q(i)\sum_{i\in S}\lambda^\epsilon_{ij}P^\gamma_\epsilon(j)$$
$$- P^\gamma_\epsilon(i)\left(B(i)B^T(i) - \frac{1}{\gamma^2}D(i)D^T(i)\right)P^\gamma_\epsilon(i), \quad i \in S. \tag{9}$$

Furthermore, these solutions satisfy the condition:

$$\gamma^2 Q_0(i) - P^\gamma_\epsilon(i) > 0, \qquad i \in S, \tag{10}$$

For any fixed ϵ, one can find the optimal value and an optimal solution by solving the Riccati equation. The advantage of presenting a Taylor expansion in ϵ is to avoid having to solve a Riccati equation for each small ϵ. This expansion will be shown to have the useful feature that the number of coupled Riccati equations involved in determining the coefficients is smaller than in the original one. Thus, the optimal controller is determined for small value of $\epsilon > 0$ by solving well-behaved ϵ-independent smaller problems.

4 The Taylor Expansion

The objective addressed in this section is to find for any $\gamma > \gamma^*(\epsilon)$ a solution of (9) which satisfies (10) for any ϵ small enough. Precisely we determine explicitly the solution of Riccati equation according to the small parameter ϵ.

We consider the Markov chain associated with the transition probability rates $(\mu_{ij})_{i,j\in S}$. There exists a partition of S into a family of m recurrent classes and a transient class T, $\bar{\xi}$ denotes the class of the recurrent state,

$$S = (\cup_{n=1}^{m}\bar{\xi}_n) \cup T \quad with \quad \bar{\xi}_n \cap \bar{\xi}_{n'} = \emptyset \quad if \quad n \neq n'.$$

Hence

$$\mu_{ij} = 0 \quad if \quad i \in \bar{\xi}_n \quad and \quad j \in \bar{\xi}_{n'}, \quad n \neq n'.$$

To each class $\bar{\xi}$ is associated the invariant measure (row vector) $m_{\bar{\xi}}$ of the recurrent sub-chain defined on the class $\bar{\xi} \in \bar{S}$ where $\bar{S} = \{\bar{\xi}_1, \bar{\xi}_2, .., \bar{\xi}_m\}$. Let:

$$C = (C_{ij})_{1\leq i,j\leq s} \quad where \quad C_{ij} = \mu_{ij},$$

We shall denote by $q_{\bar{\xi}}(i)$ the probability of ending in the class $\bar{\xi}$ starting from i.

The m functions (column vectors) $q_{\bar{\xi}}(.)$ are the solutions to $Cq_{\bar{\xi}} = 0$, and form a basis of the m-dimensional subspace Ker(C).

Remark 4.1. If v is a solution to $\sum_{j\in S}\mu_{ij}v(j) = 0$, then

$$v(\bar{\xi}) := v(i), \qquad \forall i \in \bar{\xi},$$

$$v(i) = \sum_{\bar{\xi}\in\bar{S}} q_{\bar{\xi}}(i)v(\bar{\xi}), \qquad i \in T,$$

where $(v(\bar{\xi}))_{\bar{\xi}}$ are some real numbers.

We introduce the following notations:

$$\bar{A}(\bar{\xi}) = \sum_{i\in\bar{\xi}} m_{\bar{\xi}}(i)A(i), \qquad \bar{Q}(\bar{\xi}) = \sum_{i\in\bar{\xi}} m_{\bar{\xi}}(i)Q(i),$$

$$\bar{B}(\bar{\xi}) = \left(\sum_{i\in\bar{\xi}} m_{\bar{\xi}}(i)B(i)B^T(i)\right)^{1/2},$$

$$\bar{D}(\bar{\xi}) = \left(\sum_{i \in \bar{\xi}} m_{\bar{\xi}}(i) D(i) D^T(i) \right)^{1/2},$$

and

$$\nu_{\bar{\xi}\bar{\xi}'} = \sum_{i \in \bar{\xi}} \left(\sum_{j \in \bar{\xi}'} \nu_{ij} + \sum_{j \in T} q_{\bar{\xi}'}(j) \nu_{ij} \right).$$

Note that $\bar{\theta}(t)$ defines a aggregated Markov chain defined on the state space $\bar{S}$ with the infinitesimal generator matrix $\bar{\Gamma} = (\nu_{\bar{\xi}\bar{\xi}'})_{m \times m}$. For this aggregate Markov chain, we introduce the following assumption.

Assumption 4.1. The pair $(\bar{A}(\bar{\theta}), \bar{B}(\bar{\theta}))$ is stochastically stabilizable and $(\bar{A}(\bar{\xi}), \bar{Q}(\bar{\xi}))$ is observable for each $\bar{\xi} \in \bar{S}$.

In the following theorem, we show that the value and solution of the control problem have a fractional expansion in ϵ.

Theorem 4.1. *Let assumptions 2.1–4.1 hold. Then the solution of equation (9), has the following form:*

$$P_{\epsilon}^{\gamma}(i) = \sum_{n=0}^{+\infty} P_n^{\gamma}(i) \epsilon^{\frac{n}{M}}, \quad i \in S, \tag{11}$$

where $P_n^{\gamma}(i)$ is a $p \times p$ symmetric matrix for each $i \in S$ and M is positive integer.

Proof. From Puiseux's Theorem [10], the solution of equation (9) is Puiseux's series, i.e, there exists a positive integer M such that

$$P_{\epsilon}^{\gamma}(i) = \sum_{n=-K}^{+\infty} P_n^{\gamma}(i) \epsilon^{\frac{n}{M}}, \quad i \in S$$

where K is a nonnegative integer, and from Theorem 1 in [19], we have

$$P_{\epsilon}^{\gamma}(i) = \bar{P}^{\gamma}(i) + O(\epsilon),$$

then we conclude that the terms of negative power of ϵ are equal to zero. This completes the proof of theorem 4.1.

Next we shall show that in fact all coefficients corresponding to non-integer powers of ϵ vanish, and that we obtain a Taylor series.

From Theorem 4.1 and (9), it follows that

$$A^T(i)\sum_{n=0}^{+\infty} P_n^\gamma(i)\epsilon^{n/M} + \sum_{n=-M}^{+\infty} P_n^\gamma(i)\epsilon^{n/M}A(i) + Q(i)$$

$$-\sum_{n=0}^{+\infty}\sum_{k=0}^{n}\left(P_k^\gamma(i)\left(B(i)B^T(i) - \frac{1}{\gamma^2}D(i)D^T(i)\right)P_{n-k}^\gamma(i)\right)\epsilon^{n/M}$$

$$+\sum_{n=0}^{+\infty}\sum_{j\in S}\nu_{ij}P_n^\gamma(j)\epsilon^{n/M} + \sum_{n=-M}^{+\infty}\sum_{j\in S}\mu_{ij}P_{n+M}^\gamma(j)\epsilon^{n/M} = 0, \quad i \in S,$$

then we obtain the following set of equations:

If $-M \le n < 0$, then

$$\sum_{j\in S}\mu_{ij}P_{n+M}^\gamma(j) = 0 \qquad i \in S. \tag{12}$$

If $n = 0$, then

$$0 = A^T(i)P_0^\gamma(i) + P_0^\gamma(i)A(i) + Q(i) + \sum_{j\in S}\nu_{ij}P_0^\gamma(j) + \sum_{j\in S}\mu_{ij}P_M^\gamma(j)$$

$$-P_0^\gamma(i)\left(B(i)B^T(i) - \frac{1}{\gamma^2}D(i)D^T(i)\right)P_0^\gamma(i) \quad i \in S. \tag{13}$$

If $n > 0$, then

$$0 = A^T(i)P_n^\gamma(i) + P_n^\gamma(i)A(i) + \sum_{j\in S}\nu_{ij}P_n^\gamma(j) + \sum_{j\in S}\mu_{ij}P_{n+M}^\gamma(j)$$

$$-\sum_{k=0}^{n}P_k^\gamma(i)\left(B(i)B^T(i) - \frac{1}{\gamma^2}D(i)D^T(i)\right)P_{n-k}^\gamma(i), \quad i \in S.$$

Our objective in the sequel, is to show that the equations for $n \neq kM$, where $k \in \mathbb{N}$ in the above set of equations, admit solution zero and the other equation has a unique solution.

If $n = -M$, we have

$$\sum_{j\in S}\mu_{ij}P_0^\gamma(j) = 0.$$

Using the Remark 4.1, we obtain

$$\begin{cases} \bar{P}_0^\gamma(\bar{\xi}) := P_0^\gamma(i), & i \in \bar{\xi}, \\ P_0^\gamma(i) = \sum_{\bar{\xi}\in\bar{S}} q_{\bar{\xi}}(i)\bar{P}_0^\gamma(\bar{\xi}), & i \in T. \end{cases} \tag{14}$$

Substituting the first expression of (14) in (13) we get:

$$0 = A^T(i)\bar{P}_0^\gamma(\bar{\xi}) + \bar{P}_0^\gamma(\bar{\xi})A(i) + \sum_{j\in S}\mu_{ij}P_M^\gamma(j) + \sum_{j\in S}\nu_{ij}\bar{P}_0^\gamma(j)$$
$$- \bar{P}_0^\gamma(\bar{\xi})\left(B(i)B^T(i) - \frac{1}{\gamma^2}D(i)D^T(i)\right)\bar{P}_0^\gamma(\bar{\xi}), \quad i \in \bar{\xi}.$$

Multiplying (15) by $m_{\bar{\xi}}(i)$, $i \in \bar{\xi}$ and summing up over $\bar{\xi}$:

$$0 = \sum_{i\in\bar{\xi}} Q(i)m_{\bar{\xi}}(i) + \sum_{i\in\bar{\xi}} m_{\bar{\xi}}(i)A^T(i)\bar{P}_0^\gamma(\bar{\xi}) + \bar{P}_0^\gamma(\bar{\xi})\sum_{i\in\bar{\xi}} m_{\bar{\xi}}(i)A(i)$$
$$- \bar{P}_0^\gamma(\bar{\xi})\sum_{i\in\bar{\xi}}\left[m_{\bar{\xi}}(i)\left(B(i)B^T(i) - \frac{1}{\gamma^2}D(i)D^T(i)\right)\right]\bar{P}_0^\gamma(\bar{\xi})$$
$$+ \sum_{j\in S}\left(\underbrace{\sum_{i\in\bar{\xi}} m_{\bar{\xi}}(i)\mu_{ij}}_{=0}\right)P_M^\gamma(j) + \sum_{\bar{\xi}'\in\bar{S}}\left(\sum_{\substack{i\in\bar{\xi}\\ j\in\bar{\xi}'}} m_{\bar{\xi}}(i)\nu_{ij}\right)\bar{P}_0^\gamma(\bar{\xi}')$$
$$+ \sum_{j\in T}\sum_{i\in\bar{\xi}}\sum_{\bar{\xi}'\in\bar{S}} m_{\bar{\xi}}(i)\nu_{ij}q_{\bar{\xi}'}(j)\bar{P}_0^\gamma(\bar{\xi}') = 0. \tag{15}$$

Thus

$$0 = \bar{A}^T(\bar{\xi})\bar{P}_0^\gamma(\bar{\xi}) + \bar{P}_0^\gamma(\bar{\xi})\bar{A}(\bar{\xi}) + \bar{Q}(\bar{\xi}) + \sum_{\bar{\xi}'\in\bar{S}}\nu_{\bar{\xi}\bar{\xi}'}\bar{P}_0^\gamma(\bar{\xi}')$$
$$- \bar{P}_0^\gamma(\bar{\xi})\left(\bar{B}(\bar{\xi})(\bar{B}^T(\bar{\xi}) - \frac{1}{\gamma^2}\bar{D}(\bar{\xi})\bar{D}^T(\bar{\xi})\right)\bar{P}_0^\gamma(\bar{\xi}), \;\; \bar{\xi} \in \bar{S}. \tag{16}$$

Let us introduce a set of spectral radius conditions:

$$\bar{P}_0^\gamma(\bar{\xi}) = P_0^\gamma(i) < \gamma^2 Q_0(i), \quad \forall i \in \bar{\xi}, \quad \bar{\xi} \in \bar{S},$$
$$P_0^\gamma(i) = \sum_{\bar{\xi}\in\bar{S}} q_{\bar{\xi}}(i)\bar{P}_0^\gamma(\bar{\xi}) < \gamma^2 Q_0(i), \quad i \in T. \tag{17}$$

Define the quantity $\bar{\gamma} > 0$:

$$\bar{\gamma} := \inf\Big\{\, \gamma > 0 : \text{There exists a set of positive definite solutions } \bar{P}^\gamma(\bar{\xi})\ \bar{\xi} \in \bar{S} \text{ to the set of (16) coupled such that they satisfy the spectral radius conditions (17)}\Big\}.$$

By the result [18], this threshold level is finite under working assumption 4.1, since for any $\gamma > \gamma^*_\infty$, the equation (16) admits a positive solution and from theorem 4.1, we have

$$\sum_{n=0}^{\infty} P_n^\gamma(i)\epsilon^{\frac{n}{M}} < \gamma^2 Q_0(i),$$

hence, for ϵ small enough, we obtain

$$\bar{P}_0^\gamma(\bar{\xi}) = P_0^\gamma(i) < \gamma^2 Q_0(i), \quad \forall i \in \bar{\xi}, \bar{\xi} \in \bar{S},$$
$$P_0^\gamma(i) = \sum_{\bar{\xi}\in\bar{S}} q_{\bar{\xi}}(i)\bar{P}_0^\gamma(\bar{\xi}) < \gamma^2 Q_0(i), \qquad i \in T.$$

Thus, there exist $\epsilon_\gamma > 0$ such that for any $\epsilon \in (0, \epsilon_\gamma]$ we have $\gamma^*(\epsilon) \geq \bar{\gamma}$, furthermore

$$\liminf_{\epsilon\to 0} \gamma^*(\epsilon) \geq \bar{\gamma}, \tag{18}$$

and for any $\gamma > \gamma^*(\epsilon)$ where $\epsilon \in (0, \epsilon_\gamma]$, the following jump linear system is mean-square stable:

$$\dot{x}(t) = F(\bar{\theta}(t))x(t), \tag{19}$$

where $F(\bar{\xi}) = \bar{A}(\bar{\xi}) - \left(\bar{B}(\bar{\xi})\bar{B}^T(\bar{\xi}) - (1/\gamma^2)\bar{D}(\bar{\xi})\bar{D}^T(\bar{\xi})\right)\bar{P}_0^\gamma(\bar{\xi})$.

If $n = 1$, we have

$$\begin{aligned}0 = A^T(i)P_1^\gamma(i) + P_1^\gamma(i)A(i) - P_0^\gamma(i)\left(B(i)B^T(i) - \frac{1}{\gamma^2}D(i)D^T(i)\right)P_1^\gamma(i)\\ - P_1^\gamma(i)\left(B(i)B^T(i) - \frac{1}{\gamma^2}D(i)D^T(i)\right)P_0^\gamma(i) + \sum_{j\in S}\nu_{ij}P_1^\gamma(j)\\ + \sum_{j\in S}\mu_{ij}P_{M+1}^\gamma(j) = 0, \quad i \in S\end{aligned} \tag{20}$$

From (12), we have:

$$\sum_{j\in S}\mu_{ij}P_1^\gamma(j) = 0. \tag{21}$$

By using (12) and Remark 4.1, we obtain:

$$\begin{cases} \bar{P}_1^\gamma(\bar{\xi}) = P_1^\gamma(i), & i \in \bar{\xi}, \\ P_1^\gamma(i) = \sum_{\bar{\xi}\in\bar{S}} q_{\bar{\xi}}(i)\bar{P}_1^\gamma(\bar{\xi}), & i \in T. \end{cases} \tag{22}$$

Substituting the first expression of (22) in (20) to get

$$0 = A^T(i)\bar{P}_1^{\gamma}(\bar{\xi}) + \bar{P}_1^{\gamma}(\bar{\xi})A(i) - \bar{P}_0^{\gamma}(\bar{\xi}))\left(B(i)B^T(i) - \frac{1}{\gamma^2}D(i)D^T(i)\right)\bar{P}_1^{\gamma}(\bar{\xi})$$

$$- \bar{P}_1^{\gamma}(\bar{\xi}))\left(B(i)B^T(i) - \frac{1}{\gamma^2}D(i)D^T(i)\right)\bar{P}_0^{\gamma}(\bar{\xi}) + \sum_{\bar{\xi}'\in\bar{S}}\sum_{j\in\bar{\xi}'}\nu_{ij}\bar{P}_1^{\gamma}(\bar{\xi}')$$

$$+ \sum_{j\in S}\mu_{ij}P_{M+1}^{\gamma}(j) = 0. \tag{23}$$

Multiplying (23) by $m_{\bar{\xi}}(i)$, $i \in \bar{\xi}$ and summing over $\bar{\xi}$, we obtain:

$$F^T(\bar{\xi})\bar{P}_1^{\gamma}(\bar{\xi}) + \bar{P}_1^{\gamma}(\bar{\xi})F(\bar{\xi}) + \sum_{\bar{\xi}'\in\bar{S}}\nu_{\bar{\xi}\bar{\xi}'}\bar{P}_1^{\gamma}(\bar{\xi}') = 0. \tag{24}$$

Since the model (19) is mean-square stable, one applies Proposition 2 in [3] to show that equation (24) has a unique solution. Since 0 is a solution to (24), then 0 is the unique solution to (24). Thus, it follows from $P_1^{\gamma}(i) = 0$, $i \in S$ and (20) becomes

$$\sum_{j\in S}\mu_{ij}P_{M+1}^{\gamma}(j) = 0,\ i \in S. \tag{25}$$

If $M \neq 1$, then for $0 < n < M$, we have

$$0 = A^T(i)P_n^{\gamma}(i) + P_n^{\gamma}(i)A(i) - P_0^{\gamma}(i)\left(B(i)B^T(i) - \frac{1}{\gamma^2}D(i)D^T(i)\right)P_n^{\gamma}(i)$$

$$- P_n^{\gamma}(i)\left(B(i)B^T(i) - \frac{1}{\gamma^2}D(i)D^T(i)\right)P_0^{\gamma}(i) + \sum_{j\in S}\mu_{ij}P_{n+M}^{\gamma}(j)$$

$$+ \sum_{j\in S}\nu_{ij}P_n^{\gamma}(j) = 0.$$

From (12), it follows that $\sum_{j\in S}\mu_{ij}P_n^{\gamma}(j) = 0$. Then we can derive $P_1^{\gamma}(i) = \ldots = P_{M-1}^{\gamma} = 0$, and

$$\sum_{j\in S}\mu_{ij}P_{n+M}^{\gamma}(j) = 0 \qquad n = 0, 1, \ldots, M-1. \tag{26}$$

Similarly, we have that:

$$P_0^{\gamma}(i) > 0, \quad i \in S,$$
$$P_n^{\gamma}(i) = 0, \quad n \neq kM,\ \ k \in IN^*,\ \ i \in S.$$

Now, we can state the following result. □

Theorem 4.2. *Let Assumptions 2.1–4.1 hold. Then the coupled algebraic Riccati equation (9) admits a unique positive solution, which admits a Taylor series.*

It follows from Theorem 1 in [19] that the jump linear system (1) is stochastically stabilizable for sufficiently small $\epsilon > 0$, if the aggregate jump linear system (19) is stochastically stabilizable, and with assumptions 2.1-2.2 and 4.1, the solutions of the perturbed coupled algebraic Riccati equations (9) are the unique minimal positive-definite solution. A by product of Theorem 1 in [19], Theorem 4.2 and its proof, is the result given in the following corollary:

Corollary 4.1. *Let assumptions 2.1–2.2 and 4.1 hold. Then there exists ϵ_0 such that the perturbed coupled algebraic Riccati equations (9) admits a unique positive-definite solution which can be expanded as a Taylor series, for any $\epsilon \in (0, \epsilon_0]$.*

5 Computation Algorithms

The next theorem is about the computation of the terms in the expansion of P_ϵ^γ.

Notation Let Ker(C) denote the Kernel and Im(C) the range of the operator C, i.e.

$$\mathrm{Ker}(C) = \{y \in \mathbb{R}^s / Cy = 0\} \qquad \mathrm{Im}(C) = \{y \in \mathbb{R}^s / \exists x \in \mathbb{R}^s,\, y = Cx\}$$

Theorem 5.1. *Let assumptions 2.1, 2.2 and 4.1 hold. Then the solution of the algebraic Riccati equation (9) $P_\epsilon^\gamma(i)$ $i \in S$ has the expansion $\sum_{n=0}^{+\infty} P_n^\gamma(i)\epsilon^n$, where $(P_n^\gamma)_{kl} = (\tilde{P^\gamma}_n)_{kl} + (\bar{P}_n^\gamma)_{kl}$ and $(\tilde{P^\gamma}_n)_{kl} := ((\tilde{P^\gamma}_n(i))_{kl})_{i\in S} \in Im(C)$ and $(\bar{P}_n^\gamma)_{kl} = ((\bar{P}_n^\gamma(i))_{kl})_{i\in S} \in \mathrm{Ker}(C)$. The sequence $((\tilde{P^\gamma}_n), (\bar{P}_n^\gamma))$ is uniquely determined by the following iterative algorithm:*

a. P_0^γ *is given by:*

$$\begin{cases} P_0^\gamma(i) = \bar{P}_0^\gamma(\bar{\xi}), & i \in \bar{\xi}, \qquad \forall \bar{\xi} \in \bar{S}. \\ P_0^\gamma(i) = \sum_{\bar{\xi}\in\bar{S}} q_{\bar{\xi}}(i)\bar{P}_0^\gamma(\bar{\xi}), & i \in T. \end{cases}$$

where $\bar{P}_0^\gamma(\bar{\xi}), \bar{\xi} \in \bar{S}$ is a unique solution of the Coupled Algebraic Riccati Equations:

$$0 = \bar{A}^T(\bar{\xi})\bar{P}_0^\gamma(\bar{\xi}) + \bar{P}_0^\gamma(\bar{\xi})\bar{A}(\bar{\xi}) + \bar{Q}(\bar{\xi}) + \sum_{\bar{\xi}'\in\bar{S}} \nu_{\bar{\xi}\bar{\xi}'}\bar{P}_0^\gamma(\bar{\xi}')$$

$$- \bar{P}_0^\gamma(\bar{\xi})\left(\bar{B}(\bar{\xi})(\bar{B}^T(\bar{\xi}) - \frac{1}{\gamma^2}\bar{D}(\bar{\xi})\bar{D}^T(\bar{\xi})\right)\bar{P}_0^\gamma(\bar{\xi}), \quad \bar{\xi} \in \bar{S}.$$

b. $\tilde{P}_n^\gamma$, $n > 0$ *is the unique solution to a linear system:*

$$C(\tilde{P}_n^\gamma)_{kl} = (\alpha_{n-1})_{kl}, \tag{27}$$

where

$$\alpha_0(i) = -Q(i) - A^T(i)P_0^\gamma(i) - P_0^\gamma(i)A(i) - \sum_{j\in S} \nu_{if_ij} P_0^\gamma(j)$$
$$+ \left(P_0^\gamma(i)B(i)B^T(i) - \frac{1}{\gamma^2}D(i)D^T(i)\right)P_0^\gamma(i).$$

and

$$\alpha_n(i) = -A^T(i)P_n^\gamma(i) - P_n^\gamma(i)A(i) - \sum_{j\in S} \nu_{if_ij} P_n^\gamma(j) + f_n(i)$$
$$+ \left(P_0^\gamma(i)B(i)B^T(i) - \frac{1}{\gamma^2}D(i)D^T(i)\right)P_n^\gamma(i)$$
$$P_n^\gamma(i)\left(B(i)(i)B^T(i) - \frac{1}{\gamma^2}D(i)D^T(i)\right)P_0^\gamma(i), \quad n > 0,$$

c. $\bar{P}_n^\gamma$ $n > 0$ *can be computed by:*

$$F^T(\bar{\xi})\bar{P}_n^\gamma(\bar{\xi}) + \bar{P}_n^\gamma(\bar{\xi})F(\bar{\xi}) + \sum_{\bar{\xi}'\in\bar{S}} \nu_{\bar{\xi}\bar{\xi}'}\bar{P}_n^\gamma(\bar{\xi}') + T_n(\bar{\xi}) = 0.$$

where

$$T_n(\bar{\xi}) = \sum_{i\in\bar{\xi}} m_{\bar{\xi}}(i)\Bigg[A^T(i)\tilde{P}_n(i) + \tilde{P}_n(i)A(i)$$
$$- P_0(i)\left(B(i)B^T(i) - \frac{1}{\gamma^2}D(i)D^T(i)\right)\tilde{P}_n(i)$$
$$- \tilde{P}_n(i)\left(B(i)B^T(i) - \frac{1}{\gamma^2}D(i)D^T(i)\right)P_0(i)$$
$$\times \sum_{j\in S} \nu_{ij}\tilde{P}_n(j) - f_n(i)\Bigg],$$

with $f_n(i) = \sum_{k=1}^{n-1} P_k^\gamma(i)\left(B(i)B^T(i) - (1/\gamma^2)D(i)D^T(i)\right)P_{n-1-k}^\gamma(i)$.

Proof. The proof of Theorem 5.1 is similar to that of Theorem 4.1 in [8]. □

6 Conclusion

In this paper, we have presented a study of the H^∞ optimal control of perturbed Markov jump linear systems. Under perfect state measurements for infinite horizon, we have shown that the related optimal feedback policy of zero-sum differential game and the solutions of the coupled algebraic Riccati equations admit a Taylor series in terms of ϵ. This expansion has been shown to have the useful feature that the number of coupled Riccati equations involved in determining the coefficients is smaller than in the original one. A computation method for the Taylor series expansion was presented.

The starting point for showing the existence of a Taylor series expansion was the Puiseux Theorem which was applied to the Riccati equations directly. It can similarly be applied to other situations of singular perturbations, such as several (even more than two) time scales in the continuous dynamics (rather than or in addition to that of the jump parameters). An interesting open problem is to show that such a Taylor series expansion exists in the case of finite horizon problems, in which the Puiseux theorem cannot be used anymore.

REFERENCES

[1] Başar T., "Minimax control of switching systems under sampling", *System Control Letters*, 25(5): 315–325, 1995.

[2] Başar T. and Haurie A., Feedback equilibria in differential games with structural and modal uncertainties, vol. 1 of *Advances in Large Scale Systems*, pp. 163–201. Connecticut: JAI Press Inc., May 1984. (J. B. Cruz, Jr., ed.).

[3] Costa O. L. V., Do Val J. B. and Geromel J. C., "Continuous-time state-feedback H_2-control of Markovian jump linear system via convex analysis," *Automatica* 35, pp. 259–268, 1999.

[4] Czornick A., "On discrete-time linear quadratic control" *System & Control Letters*, pp. 101–107, 1999.

[5] Delebecque F., "A reduction process for perturbed Markov chains," *SIAM J. Appl. Math.*, vol. 43, pp. 325–350, 1983.

[6] Delebecque F. and Quadrat J., "Optimal control of Markov chains admitting strong and weak interactions," *Automatica*, vol. 17, pp. 281–296, 1981.

[7] El Azouzi R., Abbad M. and Altman E., "Perturbation of linear quadratic systems with jump parameters and hybrid controls", *ZOR Mathematical Methods of Operations Research*, 51, No. 3, 2000.

[8] El Azouzi R., Abbad M. and Altman E., "Perturbation of Multivariable Linear Quadratic Systems with Jump Parameters and Hybrid Controls", *IEEE Transaction Automatic Control*, vol. 46, No. 10, October, 2001.

[9] Feng X., Loparo K. A. and Chizeck H. J., " Stochastic stability properties of jump linear systems" *IEEE Trans. Automat. Control*, vol. 37 No. 1. January 1992.

[10] Filar J. and Vrieze K., *Competitive Markov Decision Processes*, Springer-Verlag, 1996.

[11] Floretin J. J., "Optimal control of continuous-time Markov, stochastic system," *J. Electron. Control*, vol. 10, 1961.

[12] Ji Y. and Chizeck H. J., "Controllability, stabilizability, and continuous-time Markovian jump linear quadratic control," *IEEE Transaction on Automat. Control*, vol. AC-35, pp. 777–778, July 1990.

[13] Krassovskii N. N. and Lidskii E. A., "Analytical design of controllers in system with random attributes I, II, III," *Automation Remote Contr.*, vol. 22, pp. 1021–1025, 1141–1146, 1289–1294, 1961.

[14] Kushner H. J., "On the stochastic maximum principle: Fixed time of control," *J. Math. Appl.*, vol. 11, pp. 78–92, 1965.

[15] Lidskii E. A., "Optimal control of systems with random properties," *Appl. Math. Mech.*, vol. 27, pp. 33–45, 1961.

[16] Mariton M., "On controllability of linear systems with stochastic jump parameter," *IEEE Trans. Automat. Control*, vol. AC-31, pp. 680–683, 1986.

[17] Moore J. B., Zhou X. Y. and Lim A. E. B., "Discrete time LQG controls with control dependent noise" *System & Control Letters* 36, pp. 199–206, 1999.

[18] Pan Z. and Başar T., "H^{∞}-Control of Markovian jump systems and solution to associated piecewise-deterministic differential game," in G.J. Olsder, editor, *Annals of Dynamic Games*, vol. 3, pp. 61–94, Birkhauser, 1995.

[19] Pan Z. and BaşarT., "H-infinity control of large scale jump linear systems via averaging and aggregation", *International J. of Control*, 72(10) pp. 866–881, 1999.

[20] Rishel R., "Dynamic programming and minimum principle for systems with jump Markov disturbances," *SIAM Journal on Control and Optimization*, vol. 13, pp. 338–371, February 1975.

[21] Shi P., Agrawal R. K. and Boukas E., "Robust stabilization of discrete time-delay Markovian jump systems" *14th World Congress of IFAC*, 1999.

[22] de Souza C. E. and Fragoso M., "H^∞ control of linear systems with Markovian jumping parameters", *Control Theory and Technology*, Vol. 9 No. 2, pp. 457–466, 1993.

[23] Wohnman W. M., "Random differential equation in control theory," in *Probabilistic Methods in Applied Mathematics*, vol. 2, A. T. Bharucha-Ried, Ed. New York: Academic, 1971.

[24] Swoder D. D., "Feedback control of a class of linear system with jump parameters," *IEEE Trans. Automat. Control*, vol. 14, pp. 9–14, 1969.

[25] Yan Y. and Lam J., " Stochastic stabilizable and H^∞ control for Markovian jump time-delay systems" *Proceeding of the 14th IFAC*, 1999.

[26] Yan G. and Zhang Q., *Continuous-time Markov chain and applications: a singular perturbed approach*, Springer-Verlag, New-York, 1998.

Advances in Parallel Algorithms for the Isaacs Equation*

Maurizio Falcone
Dipartimento di Mathematica
Universitá di Roma "La Sapienza"
Roma, Italy
falcone@mat.uniroma1.it

Paolo Stefani
CASPUR
Roma, Italy
stefanip@caspur.it

Abstract

In this paper we develop two new parallel algorithms for differential games based on the principle of "data replication". This technique is efficient on distributed memory architectures such as IBM/PS2 or Digital Alpha and is coupled with a domain decomposition techniques to construct an approximation scheme for the Isaacs equation in $\mathbb{R}^n$. The algorithms are presented for a 2-domain decomposition and some hints are given for the case of d subdomains having crossing points. The above parallel algorithms have the same fixed point as the serial algorithm so that convergence to the viscosity solution of the Isaacs equation is guaranteed by previous results. The efficiency of the above algorithms is discussed analyzing some numerical tests which include the homicidal chauffeur game.

Key words. Parallel algorithms, domain decomposition, differential games

AMS Subject Classifications. Primary 65M12; Secondary 49N25, 49L20.

1 Introduction

In this paper we report on some recent developments in the construction of parallel algorithms for the solution of pursuit-evasion games via the Dynamic Programming approach. As in the first paper [8] we will focus our attention on the solution of the Isaacs equation which characterizes the value function. It is worth noting that the value function of a pursuit-evasion game set in $\mathbb{R}^m$ depends on $n = 2m$ variables, i.e. the Isaacs equation is set in $\mathbb{R}^n$. The need for parallel algorithms is mainly

*This work has been supported by funds from M.U.R.S.T. (project "Metodologie Numeriche Avanzate per il Calcolo Scientifico") and from the TMR Network "Viscosity solutions and applications" (contract FMRX-CT98-0234). We gratefully acknowledge the technical support given by CASPUR to the development of this research.

motivated by the huge number of nodes which are necessary to solve the Isaacs equation accurately. Following [8] we deal with a class of parallel algorithms on a *fixed grid* based on the domain decomposition technique. The basic idea of the domain decomposition technique is that one can divide the computation assigning subsets of nodes to each processor of a parallel machine. The most challenging architecture is a MIMD (Multiple Instructions Multiple Data) machine where each processor can perform different tasks and has its own local memory (see [9] for an introduction to parallel architectures and algorithms). The processors are linked by a network or by a fast switch, this is the case, e.g., for the IBM/SP2 or for a cluster of workstations.

The domain decomposition usually consists of a simple geometrical splitting of the original domain Ω into d sub-domains Ω_r, $r = 1, \ldots, d$. The sub-domains are chosen in such a way that their boundaries (or interfaces) are as simple as possible, e.g. a rectangle is divided into rectangles. In order to obtain a correct global solution in Ω by a computation distributed on d processors, one has to introduce transmission boundary conditions on the interfaces and pass the information from one processor to the others. This will require fast connections between the processors and message passing. The crucial point in a domain decomposition (DD) parallel algorithm is to determine which conditions at the interfaces will guarantee the convergence to the correct global solution. A practical evaluation of the algorithm is also needed to check that the message passing overhead does not destroy all the advantages of distributing the computation load on several processors. As we said, we start from the analysis of the simple domain decomposition algorithm presented in [8] and we modify it in order to overcome some limitations reported in our numerical tests. The main modification refers to the structure of memory allocations and leads to two new algorithms: the Data Replication (DR) algorithm and the Memory Superposition (MS) algorithm. Due to these changes, the two new parallel algorithms can overcome the limitations reported in [8] as we will see in Section 5. In fact their efficiency is still very high when we increase the number of processors.

The second important change with respect to the experiments presented in [8] is that here we use the classical set-up in "relative" coordinates in order to obtain a problem with "transparent" boundary conditions. This change is essential to reduce the (negative) effect of the ficticious Dirichlet boundary conditions in the approximation of the value function and, even more important, in the approximation of optimal feedbacks, controls and trajectories. We must also mention that by this change of variable we can reduce the number of variables from $2m$ to m (see the test problems in Section 5).

Although the contribution of this paper is essentially at the algorithmic level it is important to give some background on the theoretical results which are behind these new developments. The interested reader can find there the major results and additional references. The serial algorithm which is the basis for all the above parallel versions was introduced in [4]. A series of convergence results for continuous as well as discontinuous value functions have been developed in the framework of viscosity solutions in the papers [1], [4], [2] and [14] (see also the recent survey paper [5] for a more complete list of references on the numerical approximation of pursuit–evasion games). Other approaches which lead to numerical algorithms are presented in [6] and [13]. The first numerical experiments with the DR and MS algorithms were presented in [15].

The outline of the paper is as follows. Section 2 is devoted to set up the problem and to recall some basic facts about the serial algorithm. Section 3 starts with the DD decomposition algorithm and its characterization and turns quickly to the new algorithms, the DR and the MS algorithm are introduced there. In Section 4 we present our test problems whereas Section 5 is devoted to the evaluation of the algorithms performances. We compute the speed-up and the efficiency of the three domain decomposition algorithms comparing them to the serial algorithm for an increasing number of processors and of grid points.

2 The Background for Domain Decomposition Algorithms

This section is devoted to the description of the serial algorithm based on dynamic programming because this will be the basis for the parallel algorithms of the next section. We just sketch the essential features of this approach, further details on the continuous as well as on the discrete version of differential games can be found in [4], [5].

Let us consider the dynamical system controlled by two players

$$\begin{cases} \dot{y}(t) = f(y(t), a(t), b(t)), & t > 0, \\ y(0) = x \end{cases} \tag{1}$$

where $\dot{y}(t) \in \mathbb{R}^n$ is the state, and the functions a and b are the controls respectively for the first and second players. We assume

$$\begin{array}{ll} f : \mathbb{R}^n \times A \times B \longrightarrow \mathbb{R}^n & \text{is continuous,} \\ A, B & \text{are compact metric spaces,} \end{array} \tag{2}$$

and, for simplicity, for some constant L_f

$$|f(x, a, b) - f(y, a, b)| \le L_f |x - y|, \quad \forall x, y \in \mathbb{R}^n, \quad a \in A, \quad b \in B. \tag{3}$$

Our admissible controls are $a \in \mathcal{A}$, $b \in \mathcal{B}$ where

$$\mathcal{A} := \{ a : [0, +\infty[\longrightarrow A, \text{ measurable} \}, \tag{4}$$

$$\mathcal{B} := \{ b : [0, +\infty[\longrightarrow B, \text{ measurable} \}. \tag{5}$$

A closed set $\mathcal{T} \subseteq \mathbb{R}^n$ is also given and the first time of arrival of the trajectory on $\mathcal{T}$ is defined as

$$t_x(a, b) := \begin{cases} \min\{ t : y_x(t; a, b) \in \mathcal{T} \} & \\ +\infty & \text{if } y_x(t; a, b) \notin \mathcal{T} \text{ for all } t. \end{cases} \tag{6}$$

The game is the following: the first player "a" wants to minimize the arrival time and the second player "b" wants to maximize the same cost. Let us define

$$t_x^* := \min_{a \in A} \max_{b \in B} t_x(a, b). \tag{7}$$

For computational purposes, it is convenient to rescale the time variable by the nonlinear monotone transformation

$$\Phi(r) := \begin{cases} 1 - e^{-\mu r} & \text{if } r < +\infty, \\ 1 & \text{if } r = +\infty, \end{cases} \tag{8}$$

and define $v(x) = \Phi(t_x^*)$. Note that after the rescaling the new time variable v belongs to the interval $[0, 1]$.

In our approach, the target $\mathcal{T}$ is usually defined as

$$\mathcal{T} = \{ (y_1, y_2) \in \mathbb{R}^n : |y_1 - y_2| \leq \varepsilon \}$$

for some $\varepsilon \geq 0$.

It is well known (see e.g. [3] Ch. VIII and the references therein) that, provided some local capturability assumptions are satisfied, v is Lipschitz continuous and is the unique viscosity solution of the Isaacs equation

$$\mu w + \min_{b \in B} \max_{a \in A} \{ -f(x, a, b) \cdot Dw \} - 1 = 0 \,, \quad \text{in} \quad \Omega := \mathbb{R}^n \setminus \mathcal{T}, \tag{9}$$

coupled with the homogeneous Dirichlet boundary condition on the target

$$w = 0, \quad \text{in} \quad \partial \mathcal{T}. \tag{10}$$

In the sequel we fix $\mu = 1$. The *serial algorithm* to compute the solution can be obtained via a time discretization (with time step $h := \Delta t$) of the dynamics and of the pay-off coupled with a projection on a grid with a fixed number of nodes (see [4] and [5] for details). To simplify, assume that there exists a set Ω which is invariant with respect to the dynamics (f.e. assume supp$(f) \in \Omega$) so we can actually restrict the computation to Ω and construct a finite triangulation.

Following [4], we say that a polyhedron $\Omega \subset \mathbb{R}^n$ is *"discretized with step k"* if we are given a finite family of simplices $\{S_l\}$ such that $\Omega = \cup_l S_l$, int $(S_i) \cap$ int $(S_j) = \emptyset$ for $i \neq j$, $k := \max_l \text{diam}(S_l)$. In practice, Ω will be a hypercube in dimension $2n$. Let us denote by x_i the vertices of the simplices of the triangulation and by $\mathcal{G} := \{x_i\}_{i \in I}$ the family of the vertices (the grid). Let us denote by N the number of nodes in $\mathcal{G}$. Any point $x \in \Omega$ belongs to at least one simplex S_l and it can be written as a convex combination of the vertices of S_l, that is

$$x = \sum_{m \in I} \lambda_m x_m \text{ where } \lambda_m \geq 0, \quad \sum_{m \in I} \lambda_m = 1, \quad \lambda_m = 0 \text{ if } x_m \notin S_l. \tag{11}$$

Note that, for simplicity, we are summing over the whole set of indices I although the vertices really involved in the computations are only those referring to the simplex containing the point x (i.e. only $n+1$ coefficients are really needed in $\mathbb{R}^n$).

We define the map $F : \mathbb{R}^N \to \mathbb{R}^N$ componentwise as follows:

$$F_i(V) := \begin{cases} \gamma \max_b \min_a P_i(a, b, V) + 1 - \gamma & \text{if } x_i \in \mathcal{G} \setminus \mathcal{T}, \\ 0 & \text{if } x_i \in \mathcal{T} \cap \mathcal{G}, \end{cases} \tag{12}$$

where $\gamma := e^{-h}$, $P_i(a, b, V) := \sum_m \lambda_{im}(a, b) V_m$ and the coefficients λ_{im} are such that $z_i(a, b) := x_i + hf(x_i, a, b)$,

$$\sum_m \lambda_{im}(a, b) x_m = z_i(a, b)\ , \quad \lambda_{im} \in [0, 1], \quad \sum_m \lambda_{im} = 1. \tag{13}$$

It can be proved that F has a unique fixed point V^* in $[0, 1]^N$, and that the fixed point is our approximate solution at the nodes of the grid. By linear interpolation, we can also define $w : \mathcal{Q} \to [0, 1]$ which is the local reconstruction of the fixed point V^*. Then, the function w will satisfy

$$\begin{cases} w(x) = \sum_m \lambda_m w(x_m) & \text{if } x = \sum_m \lambda_m x_m, \\ w(x_i) = \gamma \max_b \min_a w(x_i + hf(x_i, a, b)) + 1 - \gamma & \text{if } x_i \in \mathcal{G} \backslash \mathcal{T}, \\ w(x_i) = 0 & \text{if } x_i \in \mathcal{T} \cap \mathcal{G}. \end{cases} \tag{14}$$

Theorem 2.1. *Let (2) and (3) be verified. Moreover, assume that*

$$|f(x, a, b)| \leq M_f \quad \textit{for all} \quad x \in \partial\mathcal{T},\ a \in A \quad \textit{and} \quad b \in B,$$

and that $\mathcal{T}$ is the closure of an open set with Lipschitz boundary. Let h_n, k_n, and k_n/h_n converge to 0 and assume that there is a bounded continuous viscosity solution v of (9)–(10). Then w_n converges uniformly to v as $n \to \infty$ in $\overline{\Omega}$.

A rate of convergence for the above scheme has been established in [14].

Since the minmax operator has to be computed by comparing a number of different values for every node and every fixed point iteration, even a small number of controls for every player can result in a huge number of computations and comparisons. For example, taking a discretization of A and B which consists in 36 controls for each player makes $1.3 \cdot 10^3$ comparisons at every node. On a cubic grid in $\mathbb{R}^3$ with 100 nodes on each side this amounts to $1.3 \cdot 10^9$ comparisons for each fixed point iteration. Naturally, one can try to overcome this difficulty by storing in the RAM the coefficients (they are constant with respect to the fixed point iteration). This solution is often unfeasible due to memory limitations (in the above example of a cubic grid this will require a RAM of about 30 Gb).

3 Three Domain Decomposition Algorithms

The complexity of the above problem and its dimension require the development of parallel algorithms which will distribute the computing load of the fixed point algorithm onto several processors. One way is to split the problem into sub-problems dividing the nodes into subsets, this technique is called domain decomposition.

3.1 The Standard DD Algorithm

Following [8], let us take $\Omega \in \mathbb{R}^2$ in order to simplify the presentation. First we construct a domain decomposition splitting Ω into d sub-domains Ω_r, $r = 1, \dots, d$ by a number of piecewise regular curves Γ_j, $j = 1, \dots, J$. Note that in this way the domains of the decomposition cross only at the interfaces (i.e. they have an empty interior intersection).

It is easy to build from there a decomposition with overlapping between the domains; just consider the domains

$$\widehat{\Omega}_r := \Omega_r \cup B(0, \delta). \tag{15}$$

Then, the overlapping regions are neighborhoods centered at the interfaces,

$$\Gamma_j^\delta := \Gamma_j + B(0, \delta), \qquad \delta > 0. \tag{16}$$

The only restriction on δ is that δ cannot be too large, since we want to keep the overlapping regions as small as possible. In particular, we can choose δ such that

$$\overline{\Omega}_i \cap \overline{\Omega}_j = \emptyset \text{ implies } \widehat{\Omega}_i \cap \widehat{\Omega}_j = \emptyset, \forall i, j = 1, \dots, d. \tag{17}$$

The above condition means that we cannot have new neighboring domains if we enlarge the domains Ω_r, $r = 1, \dots, d$, i.e. the overlapping regions cannot include completely the neighboring sub-domains. Let us divide the nodes of $\mathcal{G}$ taking into account the sub-domains $\widehat{\Omega}_r$ and the location of the points

$$z_i(a, b) = x_i + hf(x_i, a, b).$$

For every $r = 1, \dots, d$ and $b \in B$, we define the following sets:

$$A_r(x_i, b) := \{a \in A : z_i(a, b) \in \widehat{\Omega}_r\}, \tag{18}$$

$$\mathcal{G}_r := \{x_i \in \mathcal{G} \cap \widehat{\Omega}_r : \forall b \in B\ \exists a \in A \text{ such that } z_i(a, b) \in \widehat{\Omega}_r\}, \tag{19}$$

$$\mathcal{G}_r^{in} := \{x_i \in \mathcal{G}_r : x_i \notin \mathcal{T}\} \text{and}\, \mathcal{G}_r^{out} := \{x_i \in \mathcal{G} \cap \widehat{\Omega}_r : x_i \notin (\mathcal{G}_r^{in} \cup \mathcal{T})\}. \tag{20}$$

The set $\mathcal{G}_r^{in}$ is the set of nodes in $\widehat{\Omega}_r$ such that for any choice of the second player (player "b") it is always possible for the player "a" to make a choice that keeps $z_i(a, b)$ in $\widehat{\Omega}_r$, i.e. on those nodes we are always able to compute a value just using the information on the nodes of $\widehat{\Omega}_r$. Naturally, at the nodes in $\mathcal{T}$ there is nothing to compute since we just set their value to 0. The nodes belonging to G_r^{out} would require information coming from other sub-domains no matter which control the player "a" chooses. By the above remarks, we can define a local operator that will act just on the nodes belonging to $\widehat{\Omega}_r$,

$$S_r(x_i, U, b) := \begin{cases} \min\limits_{a \in A_r(x_i, b)} \gamma P_i(a, b, U) + 1 - \gamma, & \text{for } \quad x_i \in \mathcal{G}_r^{in}, \\ 1, & \text{for } \quad x_i \in \mathcal{G}_r^{out}, \\ 0, & \text{for } \quad x_i \in \mathcal{T} \cap \widehat{\Omega}_r. \end{cases} \tag{21}$$

Then, we can define a global operator on $\mathcal{G}$ based on the family of local operators $S_r, r = 1, \dots, d$. First define $\widehat{S} : \mathcal{G} \times \mathbb{R}^N \times B \to \mathbb{R}^N$

$$\widehat{S}(x_i, U, b) := \begin{cases} S_r(x_i, U, b), & \text{for } \quad x_i \in \widehat{\Omega}_r, \\ \min\limits_{j \in J}[S_j(x_i, U, b)], & \text{for } \quad x_i \in \bigcap\limits_{j \in J} \widehat{\Omega}_j, \end{cases} \tag{22}$$

where $J := \{j \in 1, \dots, r : x_i \in \widehat{\Omega}_j\}$. Finally, we define the i-th component of the fully discrete splitting operator as

$$S(x_i, U) := \max_{b \in B} \{\widehat{S}(x_i, U, b)\}, \quad \text{for every} \quad x_i \in \mathcal{G}\,. \tag{23}$$

It should be noted that the above definition is used here to simplify the notations but in the real algorithm the max operator in (23) is computed locally in every sub-domain (i.e. separately in every processor) and does not need message passing. The only coupling between sub-domains that requires message passing appears in the definition (22).

In order to simplify the presentation, let us consider a domain decomposition based on two sub-domains with overlapping, $\widehat{\Omega}_1$ and $\widehat{\Omega}_2$ and denote by $\widehat{\Omega}_0$ their intersection.

Let us introduce the following assumptions:

A1. $\widehat{\Omega}_0 := \widehat{\Omega}_1 \cap \widehat{\Omega}_2 \neq \emptyset$;

A2. the time step h satisfies the bounds

$$0 < h < \frac{1}{M_f} \inf_{\substack{x \in \widehat{\Omega}_1 \setminus \widehat{\Omega}_0 \cap \mathcal{G} \\ y \in \widehat{\Omega}_2 \setminus \widehat{\Omega}_0 \cap \mathcal{G}}} \operatorname{dist}(x, y), \tag{24}$$

A3. $A_r(x_i, b) \neq \emptyset$, for $x_i \in \widehat{\Omega}_r$, $r = 1, 2$ and for any $b \in B$;

A4. the triangulation of Ω is such that each simplex does not cross the interface between $\widehat{\Omega}_1 \setminus \widehat{\Omega}_0$ and $\widehat{\Omega}_0$ and the interface between $\widehat{\Omega}_2 \setminus \widehat{\Omega}_0$ and $\widehat{\Omega}_0$.

The second assumption simply guarantees that the discrete dynamics cannot cross the overlapping region passing from a node $x_i \in \widehat{\Omega}_1 \setminus \widehat{\Omega}_0$ to a point in

$\widehat{\Omega}_2 \setminus \widehat{\Omega}_0$. That assumption is necessary to reduce the memory storage of the splitting algorithm since the processor computing the solution in $\widehat{\Omega}_j$, $j = 1, 2$ (i.e. applying the local operator S_r) does not require the values on the nodes belonging to the sub-domains $\widehat{\Omega}_r$, $r \neq j$. The third assumption is a compatibility condition between the domain decomposition and the vector field (we will see later how it can be removed). It implies that the regions of the domain decomposition have to be large enough. Finally, the last assumption simply means that the interfaces can be seen in the space discretization since they are formed by the sides of the triangles of the triangulation. Let us divide the nodes $x_i \in \mathcal{G}$ into three subsets depending on the regions which contain them. Let us introduce the sets of indices

$$I_0 := \{i : x_i \in \Omega_0 \setminus \mathcal{T}\}, \qquad I_{\mathcal{T}} := \{i : x_i \in \mathcal{T} \cap \mathcal{G}\}, \tag{25}$$

$$I_r := \{i \colon x_i \in \widehat{\Omega}_r \text{ and } i \in I_0 \cup I_{\mathcal{T}}\}, \quad I_r^{in} := \{i \colon x_i \in \mathcal{G}_r^{in} \setminus \Omega_0\}, \text{ for } r = 1, 2. \tag{26}$$

Let N_r, $r = 1, 2$ be the number of nodes in $\widehat{\Omega}_r$. We define the discrete restriction operators $R_r : \mathbb{R}^{\mathrm{N}} \to \mathbb{R}^{\mathrm{N_r}}$ which selects among the N components of a vector (representing the solution on the grid) those corresponding to the nodes belonging to the sub-domain $\widehat{\Omega}_r \setminus \mathcal{T}$,

$$R_r(U) = \{U_i\}_{i \in I_r \cup I_0}, \qquad \text{for} \quad r = 1, 2. \tag{27}$$

Given the vectors V^0 and W^0 in $\mathbb{R}^{\mathrm{N}}$ we define by recursion the two sequences

$$V^{n+1} = S(V^n), \qquad W^{n+1} = F(W^n), \tag{28}$$

where F is the the global operator introduced in Section 2 and S in the splitting operator defined in (22), (23). The following theorem holds true (see [8] for the proof):

Theorem 3.1. *Let the assumptions (A1)–(A4) be satisfied. Moreover, let S and F be defined as in (22)–(23) and (12) and $V^0 = W^0$. Then $V^n = W^n$ for any $n \in \mathbb{N}$ and the operators F and S have the same fixed point.*

It should be noted that if $(A3)$ is not satisfied one can obtain the same result by a slight change of the definition of S in order to take into account the possibility $A_1(x_i, b)$ or $A_2(x_i, b) = \emptyset$. Say, for instance, that $A_1(x_i, b) = \emptyset$ and $A_2(x_i, b) = A$ then we define

$$S(x_i, U, b) := \min(1, \min_{a \in A_2(x_i, b)} \gamma \Lambda_i(a, b) \cdot R_2(U) + 1 - \gamma), \tag{29}$$

i.e. we assign the value 1 to the minimum over the empty set $A_1(x_i, b)$. However, in all our test problems assumption $(A3)$ is satisfied.

Let us give a step-by-step description of the *DD algorithm* corresponding to the definition of V^n in (28) and starting at

$$V^0 := \begin{cases} 1, & \text{for} \quad x_i \in \mathcal{G} \setminus \mathcal{T}, \\ 0, & \text{for} \quad x_i \in \mathcal{G} \cap \mathcal{T}. \end{cases} \tag{30}$$

Step 0. Define $V^{1,0} = R_1(V^0) \in \mathbb{R}^{N_1}$ and $V^{2,0} = R_2(V^0) \in \mathbb{R}^{N_2}$. Set $n = 0$.
Step 1. Compute for $r = 1, 2$, $b \in B$,

$$V_i^{r,n+1/2} = \min_{a \in A_h^r(x_i)} \gamma \sum_{j \in I_r \cup I_0} \lambda_{ij}^{(r)}(a,b) V_j^{r,n} + 1 - \gamma, \qquad i \in I_r \cup I_0.$$

Step 2. Compute for $r = 1, 2$, $b \in B$,

$$V_i^{r,n+1} = \begin{cases} V_i^{r,n+1/2}, & i \in I_r, \\ \min\{V_i^{1,n+1/2},\ V_i^{2,n+1/2}\}, & i \in I_0. \end{cases}$$

Step 3. Compute

$$V_i^{n+1} = \max_{b \in B} \{V_i^{r,n+1}\}_{r=1,2}, \quad \text{for} \quad i = I \setminus I_{\mathcal{T}},$$

$$V_i^{n+1} = 0, \qquad \text{for} \qquad i \in I_{\mathcal{T}}.$$

Step 4. Check a stopping criterion.
IF it is satisfied THEN STOP
ELSE

Increase n by 1 and GO TO *Step 1*.

Note that the definition of S guarantees $V_i^{1,n+1} = V_i^{2,n+1}$ for each $i \in I_0$. The above algorithm first splits the computation in each sub-domain (Step 1) making a link at the end of each iteration (Step 2 and Step 3). Its speed of convergence to the fixed point can be quite slow since the contraction mapping coefficient is $\gamma = e^{-h}$.

The above scheme has low efficiency because it does not "scale" with the number of processors. Table 1 shows the "profile" of the DD algorithm. One can observe that the major part of the CPU time is used for the computation of the coefficients $\lambda_{i,j}$ needed in the local reconstruction (Step 1).

Table 1: Profile of the DD algorithm.

Operation	CPU Time
Computing $\lambda_{i,j}$	58.0%
Fixed point iterations	30.0%
Computing the vectorfield	10.7%
Other operations	1.3%

Note that the large amount of CPU time devoted to the computation of the coefficients $\lambda_{i,j}$ is due to the fact that the DD algorithm recomputes them in every sub-domain $\widehat{\Omega}_r$. In fact, as we noticed, a crucial point in the algorithm is the exchange of information between processors. In a domain decomposition with overlapping, the coefficients $\lambda_{i,j}$ are computed two or more times in the overlapping regions depending on the number of sub-domains which overlap. This excessive computational load makes the efficiency of the DD algorithm decrease as far as the number of processors and sub-domains increases since the area of the overlapping regions also increases.

3.2 The Data Replication Algorithm

The Data Replication (DR) algorithm allows us to avoid the cumbersome computations of the DD algorithm provided the memory space is large enough to store in every processor the values of v at every node in Ω. This allow every processor to compute on all the nodes belonging to Ω_r, but this time the computation is done only once for every node.

Let us consider a domain decomposition without overlapping (see Figure 1). Ω is divided by a number of piecewise regular curves Γ_j into d sub-domains Ω_r such that

$$\Omega = \bigcup_{r=1}^{d} \Omega_r, \qquad \text{int}\,(\Omega_i) \cap \text{int}\,(\Omega_j) \neq \emptyset \quad \forall i \neq j.$$

Moreover, let as assume that

(A5) $x_i \notin \Gamma_j$ for any $x_i \in \mathcal{G}$, $j = 1, \dots, J$,

This means that we do not allow nodes sitting on the interfaces Γ_j (Figure 3). Note that (A5) is opposite to the assumption (A4) which we have made for the DD algorithm.

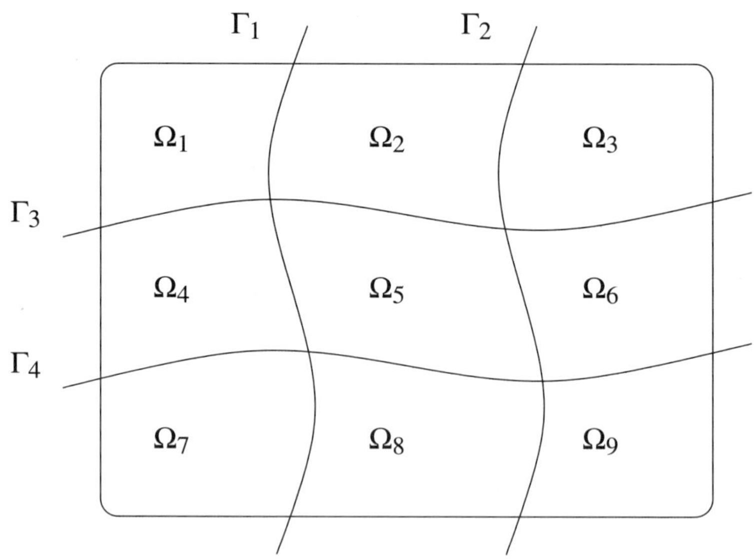

Figure 1: Ω decomposition without overlapping.

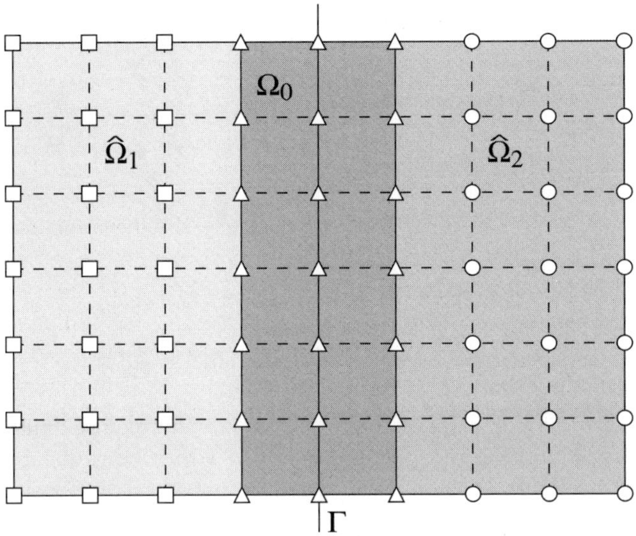

Figure 2: The domains Ω, $\widehat{\Omega}_1$, $\widehat{\Omega}_2$ and Ω_0 in the DD algorithm.

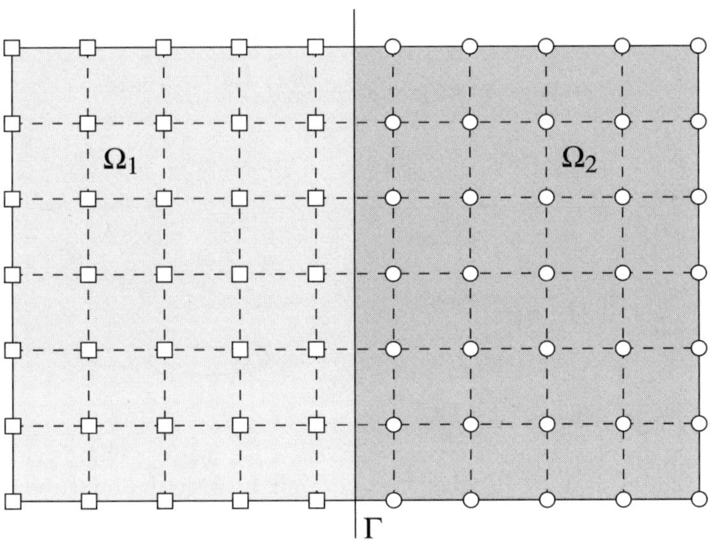

Figure 3: The domain Ω in the DR Algorithm.

Let us introduce the following subset of grid nodes:

$$\mathcal{G}_r := \{x_i \in \mathcal{G} : x_i \in \Omega_r\}, \qquad r = 1 \dots d.$$

By the definition of Ω_r we have

$$\mathcal{G}_r \cap \mathcal{G}_s = \emptyset, \qquad \forall r \neq s. \tag{31}$$

We define local operator T_r which computes on $x_i \in \mathcal{G}_r$ as,

$$T_r(x_i) := \gamma \max_b \min_a \left\{ w\big(z_i(a,b)\big) \right\} + 1 - \gamma, \tag{32}$$

and the global operator T as

$$T(x_i) := T_r(x_i), \quad \text{for } r \text{ such that } x_i \in \Omega_r. \tag{33}$$

The global operator T is uniquely defined and does not require double computations by the assumption (A5) on the grid. It is also worth noting that, by definitions (32) and (33), the operator T coincides with the operator F of the serial algorithm (defined in (12)). The convergence to the viscosity solution of the Isaacs equation is guaranteed.

The step-by-step description of the *DR algorithm* is the following:

Step 0. Set $n = 0$ and choose V^0 as in (30).

Step 1. Compute for $r = 1, 2,$

$$V_i^{r,n+1} = T_r(x_i), \text{ for any } x_i \in \mathcal{G}_r;$$

Step 2. For any $r' \neq r$ send to the processor r' the values

$$\{V_i^{r,n+1/2}\}_{x_i \in \mathcal{G}_r};$$

Step 3. Check a stopping criterion.
IF it is satisfied THEN STOP
ELSE

Increase n by 1 and GO TO *Step 1.*

The experimental results (see Section 5) show the increase of efficiency of the DR algorithm with respect to the DD algorithm. Its intrinsic limitation is due to the fact that memory allocation of every processor does not decrease with their number. Naturally there is an higher communication time with respect to the DD algorithm but this is compensated by the convergence speed of the fixed point iteration. In fact, the great advantage of the DR algorithm is that it is very flexible and can deal with problems with large vector field (f.e. the homicidal chauffeur game) where the DD algorithm would require the choice of a small time step h. Note that choosing a small h makes the number of fixed point iterations drastically increase, destroying the advantage of the parallelization.

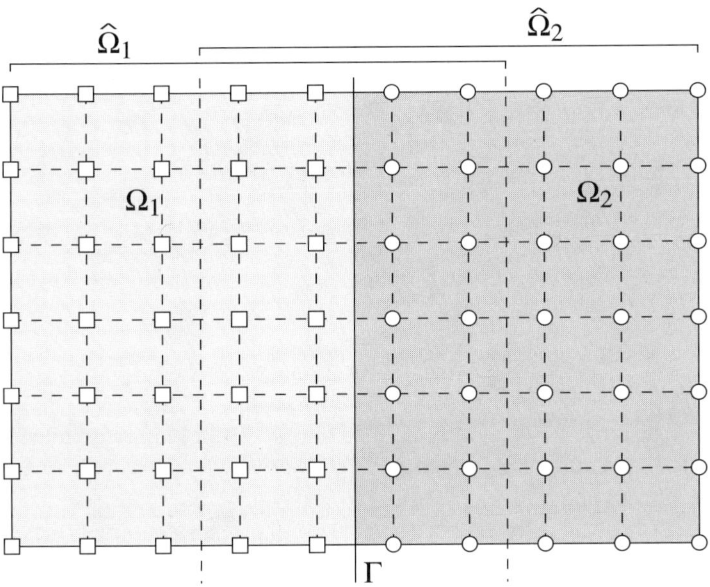

Figure 4: The domain Ω in the SR algorithm.

3.3 The Memory Superposition Algorithm

Let us introduce a method which combines the advantages of the DD and DR algorithms and produces good efficiency in the performances as well as good scalability in terms of memory allocations. Let us consider a domain decomposition of Ω with overlapping between the sub-domains (Figure 4).

Let us assume that (A2) and (A5) hold true and let $\mathcal{G}_r$ be defined as in (31). By the assumption (A2) we have,

$$x_i \in \mathcal{G}_r \text{ implies that } z_i(a,b) \in \widehat{\Omega}_r, \qquad \forall a \in A, b \in B.$$

We can redefine the local and the global operators, T_r and T as:

$$T_r(x_i) = \gamma \max_b \min_a \left\{ w\big(z_i(a,b)\big) \right\} + 1 - \gamma,$$

for $x_i \in \mathcal{G}_r$ and

$$T(x_i) := T_r(x_i), \quad \text{for every } r \text{ such that } x_i \in \Omega_r.$$

The only difference with respect to the DR algorithm is that, by assumption (A2), the local operator T_r needs only the values on the nodes belonging to $\widehat{\Omega}_r$. Then, we can reduce the memory allocation in the r processor to the values V^n corresponding to $\widehat{\mathcal{G}}_r := \{x_i : x_i \in \widehat{\Omega}_r\}$. The MS algorithm is equivalent to the serial algorithm so convergence to the viscosity solution is guaranteed.

The step-by-step description of the *MS algorithm* is the following:

Step 0. Set $n = 0$ and choose V^0 as in (30). Set $V^{r,0} := V^0|_{\widehat{\mathcal{G}}_r}$;
Step 1. Compute for $r = 1, 2, ;$

$$V_i^{r,n+1} = T_r(x_i) \text{ for any } x_i \in \widehat{\mathcal{G}}_r;$$

Step 2. For any r' such that $\widehat{\Omega}_{r'} \cap \widehat{\Omega}_r \neq \emptyset$ send to the processor r' the values

$$\{V_i^{r,n+1}\}_{x_i \in \widehat{\Omega}_{r'} \cap \Omega_r};$$

Step 3. Check a stopping criterion.
IF it is satisfied THEN STOP
ELSE

Increase n by 1 and GO TO *Step 1.*

Assumption (A2) implies, as in the DD algorithm, a limitation from above of the time step h. In practice, if the choice of the overlapping is made considering a band width Ch, where

$$C = \sup_{\substack{x \in \Omega_r \\ (a,b) \in A \times B}} \|f(x, a, b)\|, \quad .$$

(A2) is satisfied without any restriction on h.

4 Test Problems

Before giving the description of our test problems, let us give some information about the general set-up for computation of the value function and of the approximate optimal trajectories.

We said in the introduction that when each player has m state variables the value function depends on $2m$ variables. However, when the dynamics depends on the relative positions of the players, the problem can be written introducing new variables (the relative coordinates) so that the Isaacs equation can be set in $\mathbb{R}^m$. This is the case in our test problems. Usually the new system of coordinates has its origin at the pursuer position and the y axis points in the direction of the evader (this naturally means that the new system can rotate during the game). That change of coordinates is quite old and can be found for instance in the book [10]. Besides the drastic reduction of dimension, it gives also another major advantage, that of reducing the infinite target $\mathcal{T}$ to a ball of radius ε (see e.g. the tag-chase problem below in this section). Setting the problem in the new variables makes it easier to compute the value function and to deal with boundary conditions. In fact, the new target is a bounded set strictly contained in the domain of computation Ω. On the boundary $\partial\Omega$ we just impose the Dirichlet boundary condition $v = 1$.

This choice corresponds to a "transparent" boundary condition in the sense that it does not affect the solution inside Ω if the pursuer can catch the evader *inside* Ω. In general, the approximate solution will be greater or equal to the exact solution. It will be greater only when there is a possibility of capture outside the domain Ω.

Starting from the value function v, we can compute a numerical approximation of the optimal trajectory of the game as in [7]. Given $x \in \Omega$, we define $R : \Omega \to A \times B$ as

$$R(x) = \text{argmaxmin}\{w(z_i(x, a, b))\}.$$

We call a^* and b^* the optimal feedback controls for the two players. Given then the position $z(t) = \big(x(t), y(t)\big)$ of the two players at time t, their position at time $t + \Delta t$ is

$$z(t + \Delta t) = z(t) + \Delta t f(z, a^*, b^*).$$

Iterating this procedure until the evader is captured or until one of the two players exits from Ω we obtain the approximate optimal trajectories. It is important to note that the use of "transparent" boundary conditions has greatly improved accuracy in reconstruction of feedback controls. This emerges from the comparison between the trajectories illustrated in Section 5 and their analogues in [2].

4.1 The Tag-Chase Game

Let us consider two players (P and E) and the following dynamics:

$$\begin{cases} \dot{x}_P = v_P \sin\theta_P, \\ \dot{y}_P = v_P \cos\theta_P, \\ \dot{x}_E = v_E \cos\theta_E, \\ \dot{y}_E = v_E \sin\theta_E, \end{cases} \tag{34}$$

where (x_P, y_P) is the pursuer's position and (x_E, y_E) is the evader's position, and v_P and v_E are respectively their constant velocities. The controls are, respectively, $\theta_P \in [a_1, a_2] = [-\pi, \pi]$ and $\theta_E \in [b_1, b_2] = [-\pi, \pi]$. As the game just depends on the relative position of the two players, we may introduce the following change of coordinates in the space variables

$$\begin{cases} x = x_E - x_P, \\ y = y_E - y_P, \end{cases} \tag{35}$$

i.e. we choose a new dynamics with respect to a moving set of coordinates (see [10]). This new system has its origin at the pursuer's position, and is oriented in the direction of the evader. The unbounded target T in the new coordinate system becomes $\tilde{\mathcal{T}} := B(0, \varepsilon)$.

The exact solution of the Tag-chase game can be found in [2], p. 287.

4.2 The Tag-Chase Game with Control Constraints

This game has the dynamics (34). The only difference with respect to the tag-chase game is that now the pursuer P has a constraint on his displacement directions. He can choose his control in the set $\theta_P \in [a_1, a_2] \subseteq [-3/4\pi, 3/4\pi]$. The evader can still choose his control as $\theta_E \in [b_1, b_2] = [-\pi, \pi]$.

4.3 The Homicidal Chauffeur

Let us consider two players (P and E) and the following dynamics:

$$\begin{cases} \dot{x}_P = v_P \sin\theta, \\ \dot{y}_P = v_P \cos\theta, \\ \dot{x}_E = v_E \sin b, \\ \dot{y}_E = v_E \cos b, \\ \dot{\theta} = \frac{R}{v_P} a, \end{cases} \tag{36}$$

with $a \in [-1, 1]$ and $b \in [-\pi, \pi]$ being the two player's controls. Now the pursuer P is not free in is movements, he is constrained by a minimum curvature radius R. The target is defined as in the previous examples.

Also in this case one can reduce this game to a problem in R^2, this time using the following change of coordinates

$$\begin{cases} \tilde{x} = (x_E - x_P)\cos\theta - (y_E - y_P)\sin\theta, \\ \tilde{y} = (x_E - x_P)\sin\theta + (y_E - y_P)\cos\theta. \end{cases} \tag{37}$$

This means that we are now considering a new system of coordinates centered at the pursuer position and oriented along his movement direction.

A slight change in this game can be obtained moving the target $\mathcal{T} = B(0, \epsilon)$. For example, one can center the target at the points (0.2, 0.3) and (0, –0.45) as was done in [13]. The first choice produces the non-symmetric value function as we will see in the next section.

5 Numerical Tests and Algorithm Performances

We present the numerical results obtained on three tests in $\mathbb{R}^2$, comparing the numerical algorithm presented in this article with those previously available. The performance of the parallel programs are usually measured in terms of the *speed-up* A and the *efficiency* E. For readers' convenience let us define them. Let T_{ser} and T_{par} be respectively the CPU times corresponding to the execution of the serial and parallel algorithm (over N_P processors) for the solution of the same problem. We define,

$$A := \frac{T_{ser}}{T_{par}}, \qquad E := \frac{A}{N_P}. \tag{38}$$

Note that an ideal parallel algorithm without message passing loads would have $A = N_P$ and $E = 1$ so that a parallel algorithm is considered to be efficient and have good performances as far as its values for A and E are close to the above ideal values. The interested readers could find useful information on building efficient parallel codes in [9].

The platform on which the algorithm has been implemented is an IBM/SP2, a system composed of 16 nodes interconnected by a High Performance Switch (HPS); each of these nodes has one POWER2 processor running at 66.7 MHz and 128 Mbyte of local memory. The programs have been developed using the Fortran90 language and the Message Passing Interface (MPI) libraries [12]. This choice has allowed to easily port them to other parallel architectures (e.g. Digital Alpha). Figure 5 sketches the two different platforms.

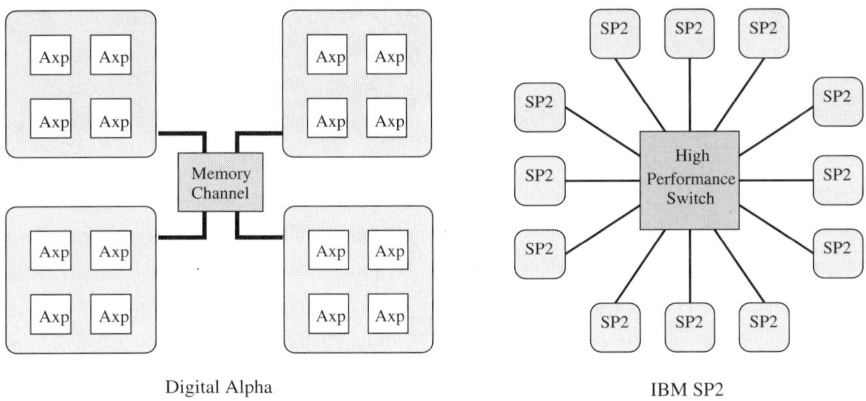

Figure 5: Parallel platforms

In the tables, we will indicate the number of processors used in every single run. In order to evaluate the performances, it is also interesting to indicate the "geometry" of the sub-domains since each sub-domain is assigned to a single processor. The following table shows this correspondence.

# Processors	Geometry
2	1×2
4	2×2
6	2×3
9	3×3
16	4×4

5.1 The Tag-Chase Game

We have considered the tag-chase game in $\Omega = [0, 1]^2$, for $v_P = 2$, $v_E = 1$. The numerical experiments for the trajectories have been obtained on a grid of 23×23

nodes for $h = 0.05$ and $\epsilon = 0.20$ and the control sets have been discretized with 41 points each.

Figure 6 shows the optimal trajectories for the problem starting in $P = (0.3, 0.3)$, $E = (0.6, -0.3)$. They are close to the exact optimal trajectories which consist of two segments of straight lines connecting the two starting points. It is interesting to note that they are close to the optimal trajectories even near the boundaries of Ω and this is due to the reformulation of the problem in relative coordinates. This shows that the boundary condition $v = 1$ is perfectly natural in the new settings.

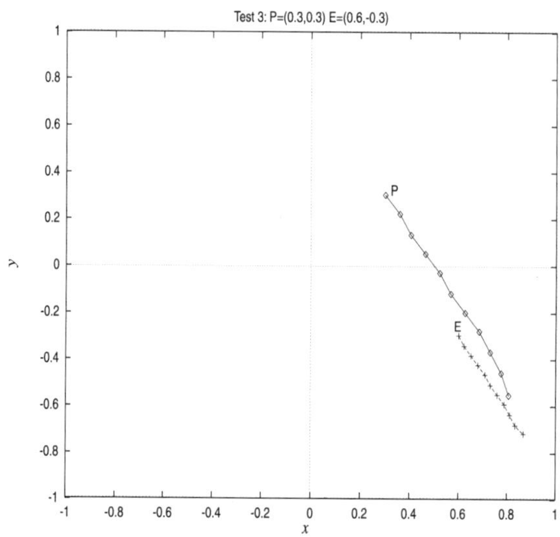

Figure 6: Tag-chase game, optimal trajectories.

In Figures 8 and 9 we compare the speed-up and the efficiency of the three domain decomposition methods (the DD, DR and MS). The comparison in made on a fixed 180×180 grid increasing the number of processors (i.e., of sub-domains) involved in the decomposition. One can observe that the speed-up of the DD algorithm (diamonds) is about 6 for 16 processors. The speed-up of the DR algorithm (+) is close to 12 for the same number of processors and the MS (squares) algorithm has even better performances. As a consequence, the efficiency of the MS algorithm stays close to 0.8 whereas that of the DD algorithm is lower than 0.4.

Figures 10 and 11 represent the values of speed-up and efficiency for the MS algorithm. The four lines in Figure 10 represent the speed-up on four different grids: 60×60 (diamonds), 120×120 (+), 180×180 (squares) and 240×240 ($\times$). Although the difference is not big, our method is slightly more efficient when the number of nodes increases. Figure 11 shows the efficiency on the same four grids for an increasing number of processors.

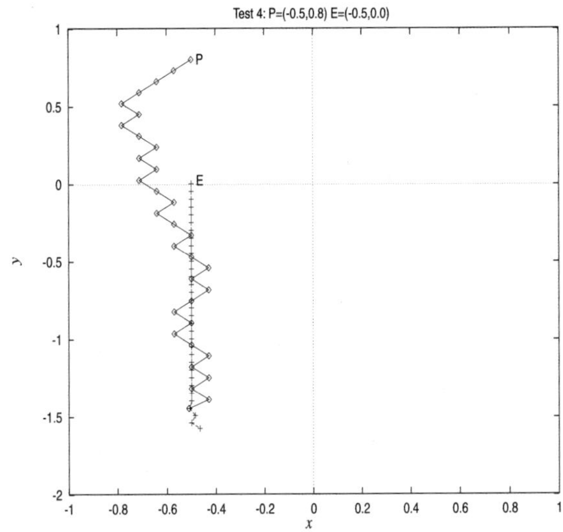

Figure 7: Tag-chase game WCC, optimal trajectories.

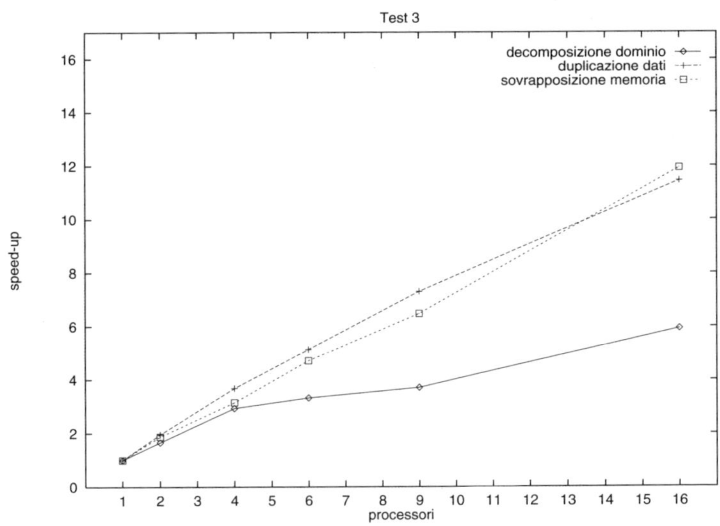

Figure 8: Tag-chase game, speed-up 180 × 180 nodes.

5.2 The Tag-Chase Game with Control Constraints

We have then considered the tag-chase game with control constraints (WCC) in $\Omega = [0, 1]^2$, for $v_P = 2$, $v_E = 1$. The numerical experiments for the trajectories have been obtained on a grid of 29×29 nodes for $h = 0.05$ and $\epsilon = 0.15$ and the control sets have been discretized with 28 points for the pursuer and 41 points for the evader.

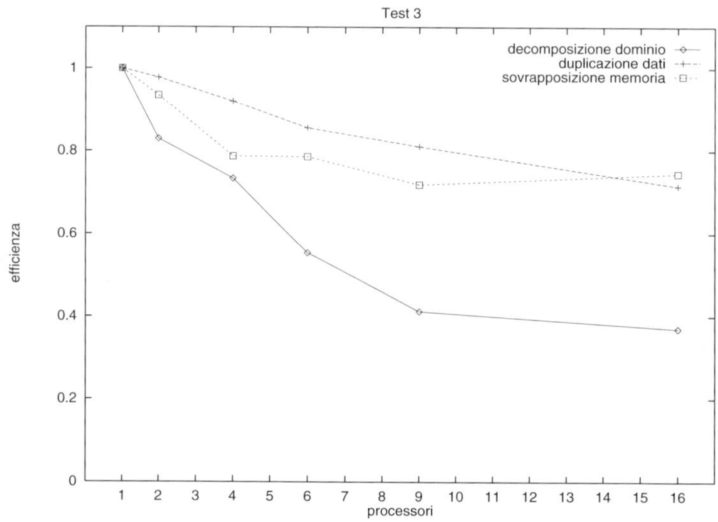

Figure 9: Tag-chase game, efficiency 180 × 180 nodes.

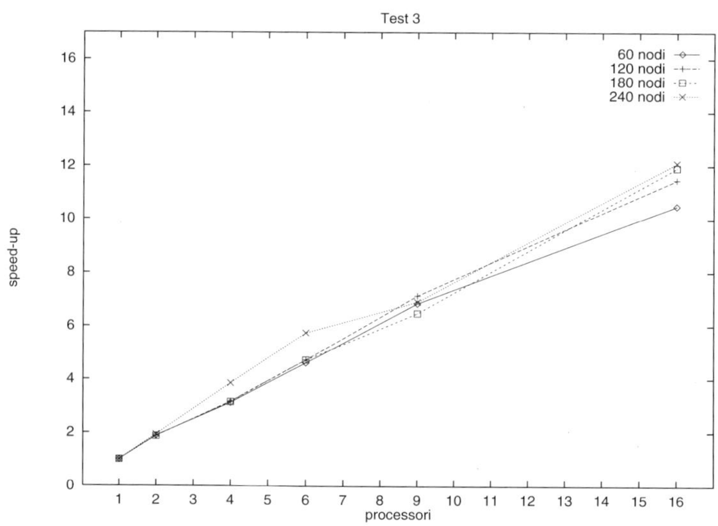

Figure 10: Tag-Chase Game, speed-up.

Note that the constraint on P's direction creates a zone behind P where the time of capture is higher than that corresponding to points in front of him. In fact, P is forced to proceed zig-zagging to catch an evader who starts behind him. That behavior can be clearly seen in Figure 7.

Figures 12 to 13 show speed-up and efficiency for the three domain decomposition algorithms on a grid of 180 × 180 nodes. On this test, the best performances are obtained by the MS algorithm which has the higher speed-up for a low number of processors also.

Figures 14 to 15 show speed-up and efficiency for the MS algorithm on the same grids as the previous test. The increase of the speed-up is clearly linear on the 240 × 240 grid.

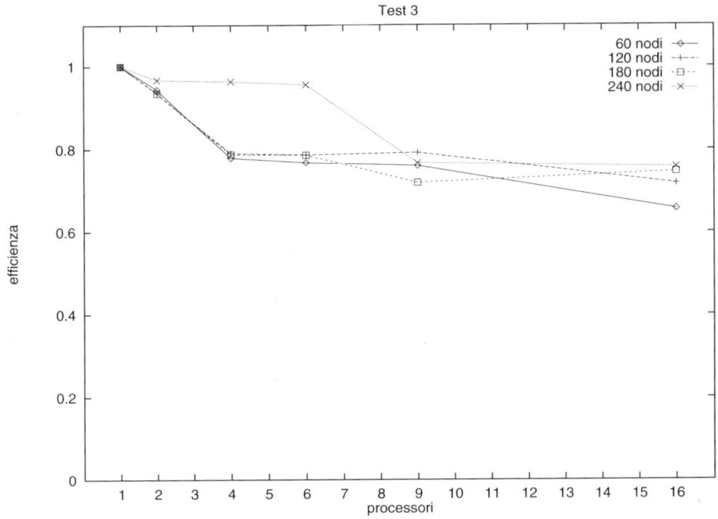

Figure 11: Tag-chase game, efficiency.

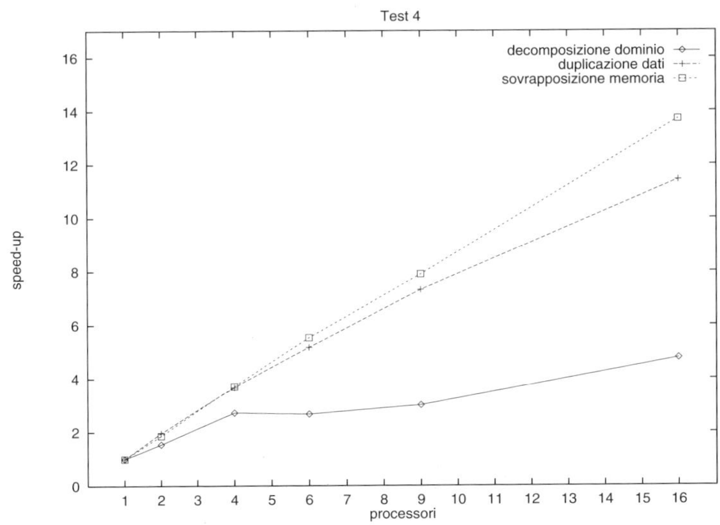

Figure 12: Tag-chase game WCC, speed-up 180 × 180 nodes.

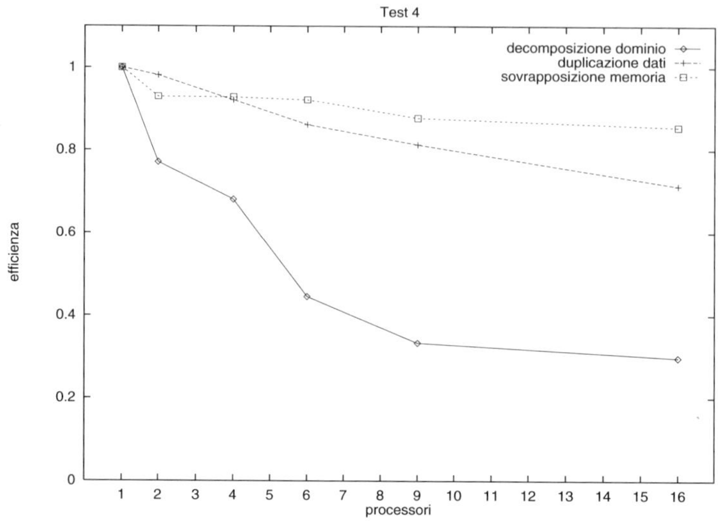

Figure 13: Tag-chase game WCC, efficiency 180×180 nodes.

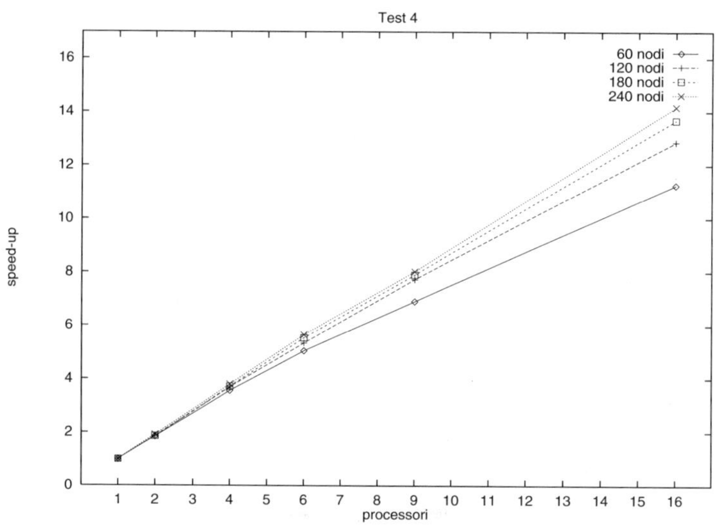

Figure 14: Tag-chase game WCC, speed-up.

5.3 The Homicidal Chauffeur

We have considered the homicidal chauffeur game where $\Omega = [-1, 1]$, $v_P = 1$, $v_E = 0.5$, $R = 0.2$. We run the algorithm on a grid made by 120×120 nodes, $h = 0.05$ and $\epsilon = 0.10$ and the control sets have been discretized with 36 points for each player to determine the optimal trajectories.

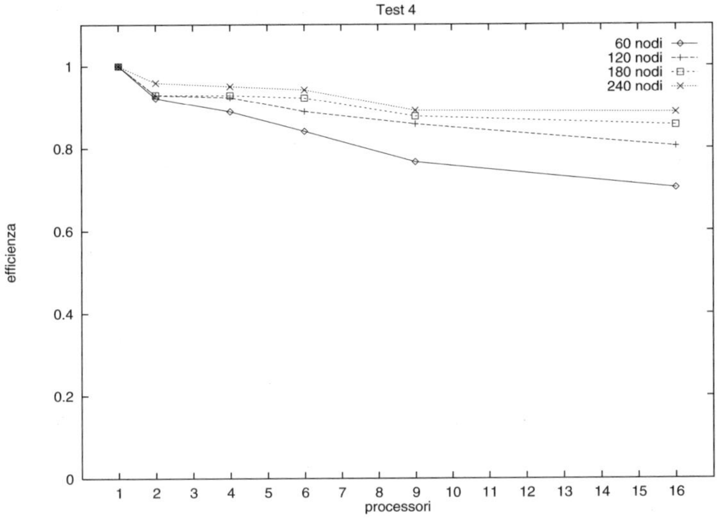

Figure 15: Tag-chase game WCC, efficiency.

Figure 16 shows the value function of the game. Note that when E is in front of P the behavior of the two players is analogous to the tag-chase game: in this case, indeed, the constraint on P's radius turn does not come into action. However, on the P sides the value function has two higher lobes. In fact, to reach the corresponding points of the domain, the pursuer must first turn around himself to be able to catch E following a straight line (see Figure 17). Finally, behind P the capture is impossible ($v = 1$) because the evader has the time to exit Ω before the pursuer can catch him. Figure 18 shows a set of optimal trajectories near a barrier in the relative coordinates

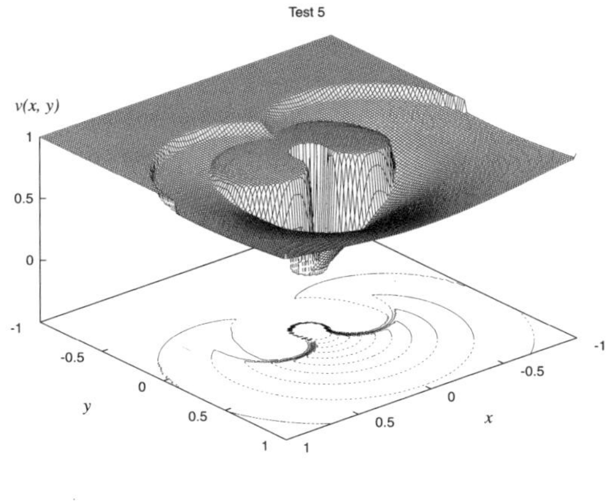

Figure 16: Homicidal chauffeur, value function.

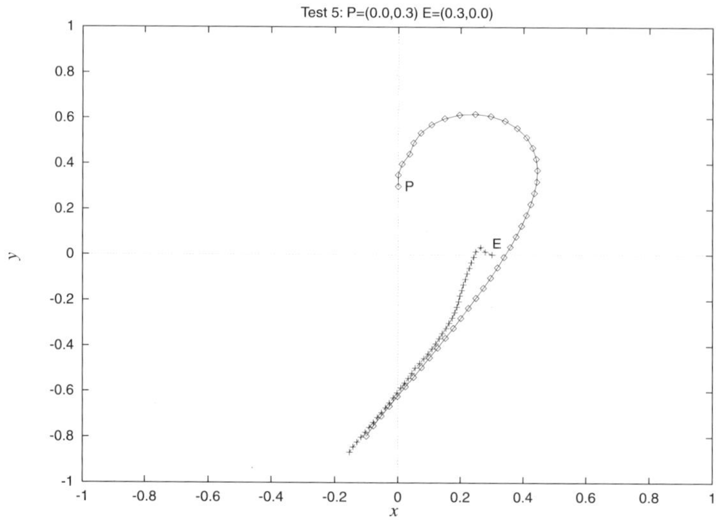

Figure 17: Homicidal chauffeur, optimal trajectories.

system (the run was made in $\Omega = [-1, 1]^2$). Figure 19 is taken from [11] and shows the optimal trajectories which have been obtained by analytical methods. One can see that our results are quite accurate since the approximate trajectories (Figure 18) look very similar to the exact solutions (Figure 19). Moreover, in the numerical approximation the barrier curve is clearly visible: that barrier cannot

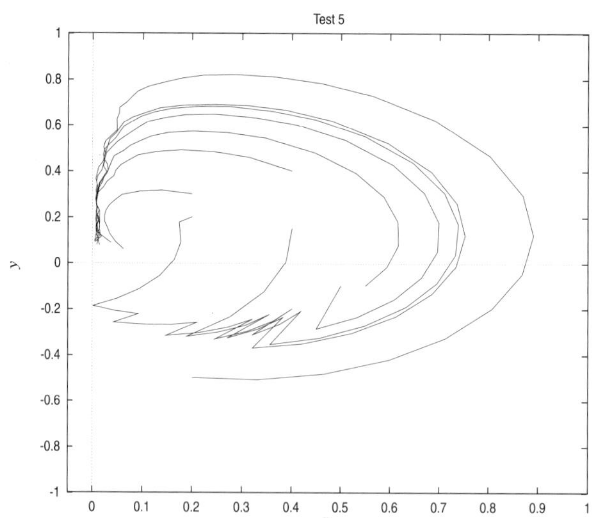

Figure 18: Homicidal chauffeur, optimal trajectories in relative coordinates

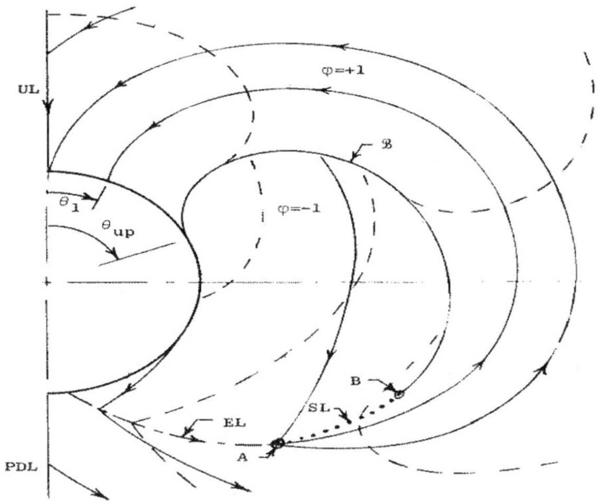

Figure 19: Homicidal chauffeur, optimal trajectories in relative coordinates [11].

be crossed if both the players behave optimally. It divides the initial positions from which the trajectories point directly to the origin from those corresponding to trajectories reaching the origin after a round trip.

Figures from 20 to 23 show speed-up and efficiency of our method in this test. From this point of view, the MS algorithm has always the best performances and the results are analogous to what we have seen in previous test problems. However, in this example the performances are lower for all the methods since the dynamics is more complicated.

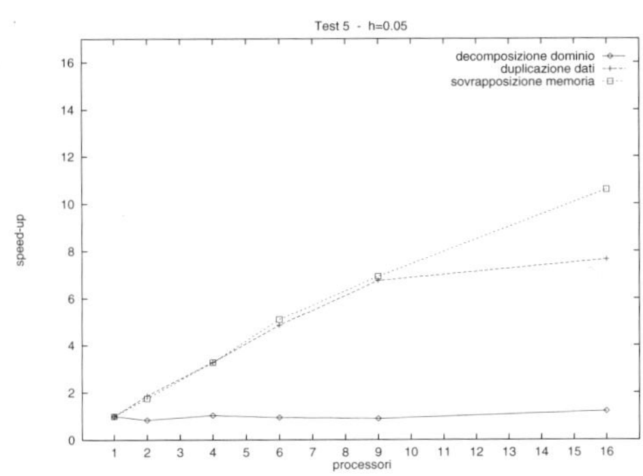

Figure 20: Homicidal chauffeur, speed-up 60 × 60 nodes.

Figure 24 shows the contours of the value functions when $v_P = 0.6$, $v_E = 0.2$, the radius turn for P is $R = 0.2$ and the target is centered at $(0.0, -0.45)$. In Figure 26 one can see the contours of the value function for the same values of the parameters and the target centered at $(0.2, 0.3)$. These pictures show that the

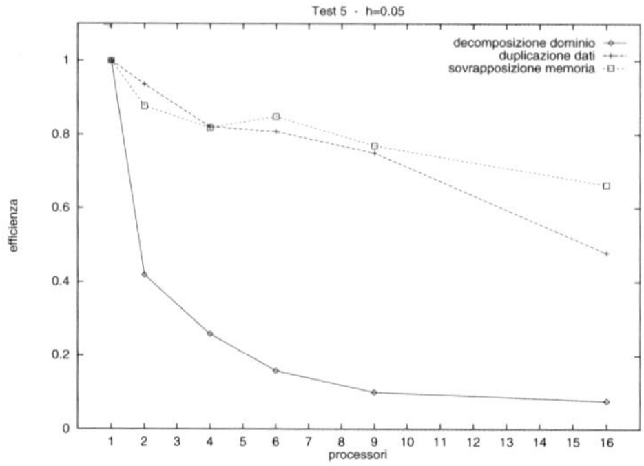

Figure 21: Homicidal chauffeur, efficiency 60×60 nodes.

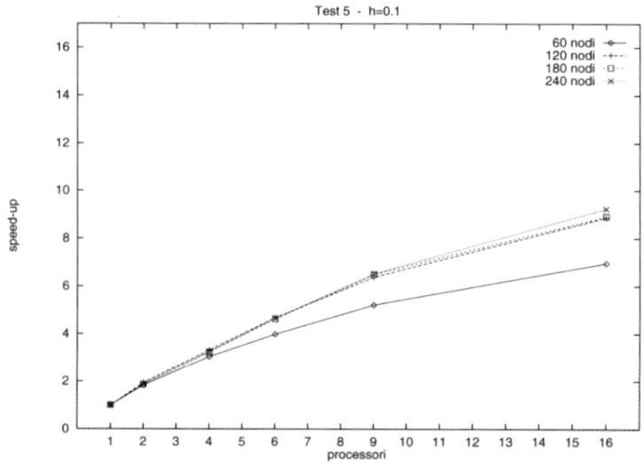

Figure 22: Homicidal chauffeur, speedup.

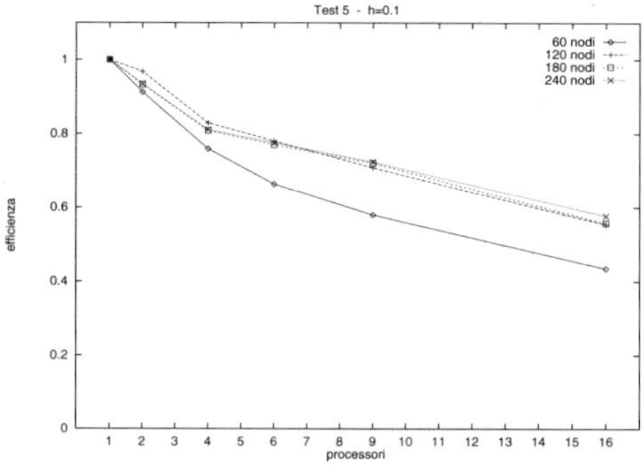

Figure 23: Homicidal chauffeur, efficiency.

results obtained by our algorithms coincide with those obtained by Patsko and Turova in [13] by the method based on the computation of "fronts" (represented in Figures 25 and 27). In these cases the value function is discontinuous but the algorithm can still compute an accurate approximate solution. The convergence of the serial approximation scheme for discontinuous value functions has been proved in [2].

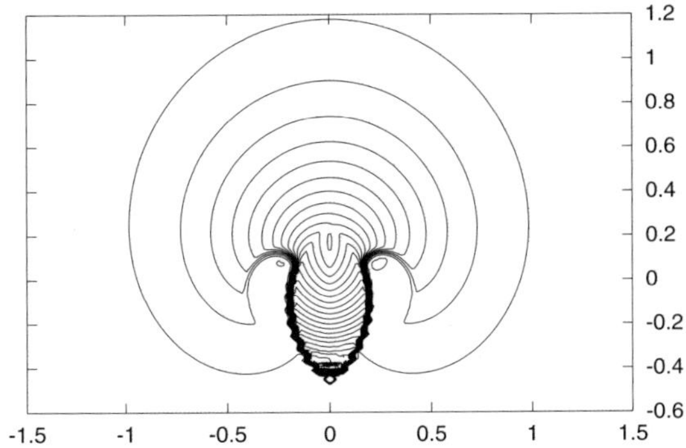

Figure 24: Homicidal chauffeur 2, level curves of the value function.

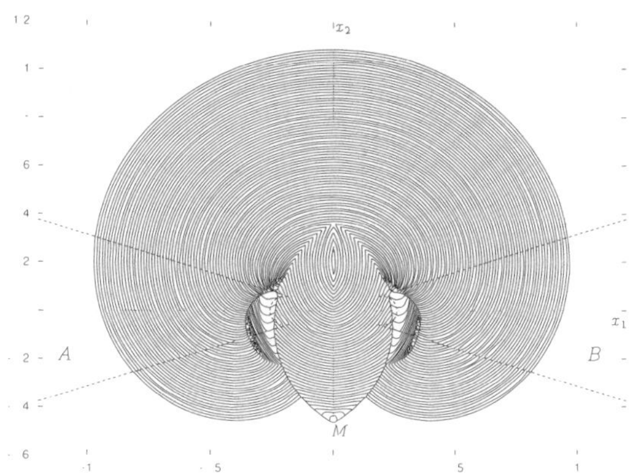

Figure 25: Homicidal chauffeur 2, level curves of the value function [13].

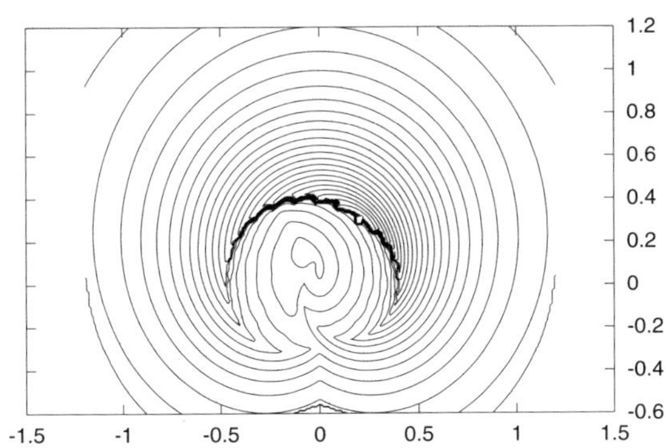

Figure 26: Homicidal chauffeur 3, level curves of the value function.

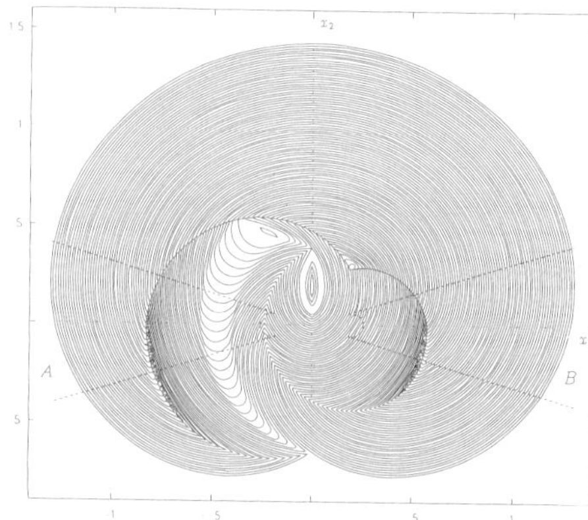

Figure 27: Homicidal chauffeur 3, level curves of the value function [13].

5.4 Conclusions

Let us make some short comments based on the analysis of the experiments and give some hints on the construction of efficient domain decomposition algorithms. In order to guarantee the convergence to the value function, it is crucial to treat carefully the internal conditions on the overlapping regions, i.e. the coupling between the local and the global operators. Moreover, for efficient parallelization we recommend the following:

a. maintain a balanced load between the processors. To this end, the target should be divided into pieces and assigned to different processors.
b. try to reduce the size of the overlapping regions. Since the width of the overlapping region strongly depends on the vector field and on the time–step h, a possibility is to reduce h. This will reduce the overlapping region to a narrow band but will also slow down the convergence of the algorithm (which is based on a fixed point iteration). A good balance depends on the problem.
c. use an efficient memory allocation to obtain a scalable algorithm. The DR and the MS algorithm are very effective whenever one can store the information they need in every processor. If this is not the case, one should use the DD algorithm and carefully verify that the global operator computes the maximum on a single node x_i in the overlapping regions, collecting all the values from the neighboring processors.

REFERENCES

[1] Alziary de Roquefort B., *Jeux différentiels et approximation numérique de fonctions valeur, 1re partie: étude théorique 2e partie: étude numérique*, RAIRO Math. Model. Numer. Anal. **25** (1991), 517–560.

[2] Bardi M., Bottacin S., Falcone M., *Convergence of discrete schemes for discontinuous value functions of pursuit-evasion games*, in G.J. Olsder (ed.), *New Trends in Dynamic Games and Applications*, Birkhäuser, Boston, (1995), 273–304.

[3] Bardi M., Capuzzo Dolcetta I., *Optimal control and viscosity solutions of Hamilton-Jacobi-Bellman equations*, Birkhäuser, 1997.

[4] Bardi M., Falcone M., Soravia P., *Fully discrete schemes for the value function of pursuit-evasion games*, Advances in dynamic games and applications, T. Basar and A. Haurie eds., Birkhäuser, (1994), 89–105.

[5] Bardi M., Falcone M., Soravia P., *Numerical Methods for Pursuit-Evasion Games via Viscosity Solutions*, in M. Bardi, T. Parthasarathy and T.E.S. Raghavan (eds.) "Stochastic and differential games: theory and numerical methods", Annals of the International Society of Differential Games, Boston: Birkhäuser, (2000), vol. 4, 289–303.

[6] Cardaliaguet P., Quincampoix M., Saint-Pierre P., *Set valued numerical analysis for optimal control and differential games*, in M. Bardi, T. Parthasarathy and T.E.S. Raghavan (eds.) "Stochastic and differential games: theory and numerical methods", Annals of the International Society of Differential Games, Boston: Birkhäuser, vol. 4, 177–247.

[7] Falcone M., *Numerical solution of Dynamic Programming equations*, Appendix A in the volume by M. Bardi and I. Capuzzo Dolcetta (eds), *Optimal control and viscosity solutions of Hamilton-Jacobi-Bellman equations*, Birkhäuser, Boston, (1997).

[8] Falcone M., Lanucara P., Marinucci M., *Parallel Algorithms for the Isaacs equation* , in O. Pourtallier and E. Altman (eds.), Annals of the International Society of Differential Games, to appear.

[9] Foster I., *Designing and building parallel programs*, Addison–Wesley, New York, (1995).

[10] Isaacs R., *Differential games*, John Wiley & Sons, (1965).

[11] Merz A. W., *The Homicidal Chauffeur - a Differential Game*, PhD Dissertation, Stanford University, (1971).

[12] Message Passing Interface Forum, *MPI: a message passing interface standard*, `http://www.mpi-forum.org/docs/mpi-11-html/mpi-report.html`, (1995).

[13] Patsko V. S., Turova V. L., *Numerical study of differential games with the homicidal chauffeur dynamics*, Scientific report of the Russian Academy of Science, Ural Branch, Ekaterinburg, (2000).

[14] Soravia P., *Estimates of convergence of fully discrete schemes for the Isaacs equation of pursuit-evasion games via maximum principle*, SIAM J. Control Optim., **36** (1998), 1–11.

[15] Stefani P., *Algoritmi paralleli per giochi di fuga-evasione*, Tesi di Laurea, Dipartimento di Matematica, Università di Roma "La Sapienza", (1998).

Numerical Algorithm for Solving Cross-Coupled Algebraic Riccati Equations of Singularly Perturbed Systems

Hiroaki Mukaidani
Graduate School of Education
Hiroshima University
Japan
mukaida@hiroshima-u.ac.jp

Hua Xu
Graduate School of Business Sciences
The University of Tsukuba
Japan

Koichi Mizukami
Graduate School of Engineering
Hiroshima Kokusai Gakuin University
Japan
mizukami@cs.hkg.ac.jp

Abstract

In this paper, we study the linear quadratic Nash games for infinite horizon singularly perturbed systems (SPS). In order to solve the generalized algebraic Lyapunov equation (GALE) corresponding to the generalized Lyapunov iterations, we propose a new algorithm which is based on the fixed point iterations. Furthermore, we also propose a new algorithm which is based on the Kleinman algorithm for solving the generalized cross-coupled algebraic Riccati equations (GCARE). It is shown that the resulting algorithm guarantees the quadratic convergence.

Key words. Singularly perturbed systems (SPS), Linear quadratic Nash games, Generalized cross-coupled algebraic Riccati equations (GCARE), Generalized Lyapunov iterations, Kleinman algorithm

1 Introduction

The linear quadratic Nash games have been studied intensively in many papers [1]–[5]. For example, Starr and Ho [1] obtained the closed-loop perfect-state linear Nash equilibrium strategies for a class of analytic differential games. In [3], a state feedback mixed H_2/H_∞ control problem has been formulated as a dynamic Nash game. In general, note that the cross-coupled algebraic Riccati equations (CARE) play an important role in problems of differential Nash Games. It is well-known that in order to obtain the Nash equilibrium strategies, we must solve the CARE. Li and Gajić [4] proposed an algorithm, called the Lyapunov iteration, to solve the linear quadratic Nash games. However, there are no results for the convergence rate of the Lyapunov iterations. It is easy to verify that the convergence speed is very slow when the simulation is carried out. In order to improve the convergence rate of the Lyapunov iterations, Mukaidani *et al.* [16] proposed an algorithm which is based on the algebraic Riccati equation (ARE) for solving the parameterized CARE related to the mixed H_2/H_∞ control problem. Freiling *et al.* [5] found the solutions to the CARE of the mixed H_2/H_∞ type by using Riccati iterations different from [16]. However, the convergence of these Riccati iterations were not proved exactly.

Linear time-invariant models of many physical systems contain slow and fast modes. Linear quadratic Nash games for such models, that is, singularly perturbed systems (SPS), have been studied by using composite controller design [6,7]. When the parameters represent small unknown perturbations, whose values are not known exactly, the composite design is very useful. However, the composite Nash equilibrium solution achieves only a performance which is $O(\varepsilon)$ close to the full-order performance. Moreover, in order to obtain the slow subsystem, the non-singularity of the fast state matrices are needed. In recent years, the recursive algorithm for solving the various control problems not only for the SPS but also for weakly coupled systems have been developed in many papers (see, e.g., [8]–[10]). It has been shown that the recursive algorithm is very effective for solving the ARE and the algebraic Lyapunov equation (ALE) when the system matrices are functions of a small perturbation parameter ε. Note that dynamic Nash games of weakly coupled systems have been studied in [8,9] by means of the recursive algorithm. Recently, the recursive algorithms for solving the CARE corresponding to both the dynamic Nash games and the mixed H_2/H_∞ control problem for the SPS have been investigated [14,15]. However, the recursive algorithm converges only to the approximation solution because the convergence solutions depend on 0-order solutions. Moreover, the recursive algorithm has the property of linear convergence. Thus, the convergence speed is very slow.

In this paper, we study the linear quadratic Nash games for infinite horizon SPS from the viewpoint of solving the CARE. After defining the generalized cross-coupled algebraic Riccati equations (GCARE), we first apply the generalized Lyapunov iterations [14]–[16]. Note that it is hard to solve the generalized algebraic Lyapunov equation (GALE) corresponding to the generalized Lyapunov iterations using the recursive algorithm because the convergence solutions depend on the 0-order solutions. Thus, we propose a new algorithm to solve the GALE. This algorithm is based on the fixed point iterations. Consequently, we reduce the computing error because we do not separate the required solution into the 0-order solution and the error term. Therefore, the solution to the Nash equilibrium can be obtained up to an arbitrary accuracy by performing iterations. Moreover, full-order Nash equilibrium solution achieves a performance which is closer to the exact performance compared with [6,7] because our new method is not based on the singular perturbation method [11]. Second, we propose a new iterative algorithm for solving the GCARE. Since the new algorithm is based on the Kleinman algorithm, it is shown that the resulting algorithm has a quadratic convergence property. The quadratic convergence of such algorithm is proved by using the Newton–Kantorovich theorem [20]. Using the new algorithm, we will improve the convergence speed compared with the previous results [5], [14]–[16]. Finally, the simulation results show that the proposed algorithm succeeds in improving the convergence rate dramatically.

Notation: The notations used in this paper are fairly standard. The superscript T denotes matrix transpose. I_n denotes the $n \times n$ identity matrix. $\|\cdot\|$ denotes its Euclidean norm for a matrix. $\det M$ denotes the determinant of M. $\mathrm{vec}M$ denotes an ordered stack of the columns of M [21]. $\otimes$ denotes Kronecker product. U_{lm} denotes a permutation matrix in Kronecker matrix sense [21] such that $U_{lm}\mathrm{vec}M = \mathrm{vec}M^T,\ (M \in \mathbf{R}^{l\times m})$.

2 Problem Formulation

Consider a linear time-invariant SPS

$$\dot{x} = A_{11}x + A_{12}z + B_{11}u_1 + B_{12}u_2,\ x(0) = x_0, \tag{1}$$

$$\varepsilon\dot{z} = A_{21}x + A_{22}z + B_{21}u_1 + B_{22}u_2,\ z(0) = z_0, \tag{2}$$

with quadratic cost functions

$$J_i(u_i,\ u_j) = \frac{1}{2}\int_0^\infty (y^T Q_i y + u_i^T R_{ii} u_i + u_j^T R_{ij} u_j)dt, i, j = 1, 2, i \neq j,$$

where

$$y = \begin{bmatrix} x \\ z \end{bmatrix},\quad Q_i = \begin{bmatrix} Q_{11i} & Q_{12i} \\ Q_{12i}^T & Q_{22i} \end{bmatrix} \geq 0,\quad R_{ii} > 0,\quad R_{ij} \geq 0,$$

and ε is a small positive parameter, $x \in \mathrm{R}^{n_1}$, $z \in \mathrm{R}^{n_2}$ and $y \in \mathbf{R}^N,\ N = n_1 + n_2$ are states, $u_i \in \mathbf{R}^{m_i},\ i = 1,\ 2$ is the control input. All matrices above are of appropriate dimensions. The system (1)–(2) is said to be in the standard form if the matrix A_{22} is nonsingular. Otherwise, it is called the nonstandard SPS [7,11].

Let us introduce the partitioned matrices

$$A_\varepsilon = \begin{bmatrix} A_{11} & A_{12} \\ \varepsilon^{-1}A_{21} & \varepsilon^{-1}A_{22} \end{bmatrix}, \ A = \begin{bmatrix} A_{11} & A_{12} \\ A_{21} & A_{22} \end{bmatrix},$$

$$B_{i\varepsilon} = \begin{bmatrix} B_{1i} \\ \varepsilon^{-1}B_{2i} \end{bmatrix}, \ B_i = \begin{bmatrix} B_{1i} \\ B_{2i} \end{bmatrix},$$

$$S_{i\varepsilon} = B_{i\varepsilon}R_{ii}^{-1}B_{i\varepsilon}^T, \ S_i = B_iR_{ii}^{-1}B_i^T = \begin{bmatrix} S_{11i} & S_{12i} \\ S_{12i}^T & S_{22i} \end{bmatrix},$$

$$G_{j\varepsilon} = B_{j\varepsilon}R_{jj}^{-1}R_{ij}R_{jj}^{-1}B_{j\varepsilon}^T, \ G_j = B_jR_{jj}^{-1}R_{ij}R_{jj}^{-1}B_j^T = \begin{bmatrix} G_{11j} & G_{12j} \\ G_{12j}^T & G_{22j} \end{bmatrix},$$

$$i, \ j = 1, 2, \ i \neq j.$$

We now consider the linear quadratic Nash games for infinite horizon SPS (1)–(2) under the following basic assumptions [4,8,9].

Assumption 2.1. There exists a small perturbation parameter $\varepsilon^* > 0$ such that the triplet $(A_\varepsilon, \ B_{1\varepsilon}, \ \sqrt{Q_1})$ and $(A_\varepsilon, \ B_{2\varepsilon}, \ \sqrt{Q_2})$ are stabilizable and detectable for all $\varepsilon \in (0, \varepsilon^*]$.

Assumption 2.2. The triplet $(A_{22}, \ B_{21}, \ \sqrt{Q_{221}})$ and $(A_{22}, \ B_{22}, \ \sqrt{Q_{222}})$ are stabilizable and detectable.

These conditions are quite natural since at least one control agent has to be able to control and observe unstable modes. The purpose is to find a linear feedback controller $(u_1^*, \ u_2^*)$ such that

$$J_i(u_i^*, \ u_j^*) \leq J_i(u_i, \ u_j^*), \ \ i, \ j = 1, 2, \ \ i \neq j. \tag{3}$$

The Nash inequality shows that u_i^* regulates the state to zero with minimum output energy. The following lemma is already known [1].

Lemma 2.1. *Under Assumptions 2.1 and 2.2, there exists an admissible controller such that (3) hold iff the following full-order CARE*

$$A_\varepsilon^T X_\varepsilon + X_\varepsilon A_\varepsilon + Q_1 - X_\varepsilon S_{1\varepsilon} X_\varepsilon - X_\varepsilon S_{2\varepsilon} Y_\varepsilon - Y_\varepsilon S_{2\varepsilon} X_\varepsilon + Y_\varepsilon G_{2\varepsilon} Y_\varepsilon = 0, \tag{4}$$

$$A_\varepsilon^T Y_\varepsilon + Y_\varepsilon A_\varepsilon + Q_2 - Y_\varepsilon S_{2\varepsilon} Y_\varepsilon - Y_\varepsilon S_{1\varepsilon} X_\varepsilon - X_\varepsilon S_{1\varepsilon} Y_\varepsilon + X_\varepsilon G_{1\varepsilon} X_\varepsilon = 0, \tag{5}$$

have stabilizing solutions $X_\varepsilon \geq 0$ and $Y_\varepsilon \geq 0$ where

$$X_\varepsilon = \begin{bmatrix} X_{11} & \varepsilon X_{21}^T \\ \varepsilon X_{21} & \varepsilon X_{22} \end{bmatrix}, \ Y_\varepsilon = \begin{bmatrix} Y_{11} & \varepsilon Y_{21}^T \\ \varepsilon Y_{21} & \varepsilon Y_{22} \end{bmatrix}.$$

Then, the closed-loop linear Nash equilibrium solutions to the full-order problem are given by

$$u_1^* = -R_{11}^{-1} B_{1\varepsilon}^T X_\varepsilon y,$$
$$u_2^* = -R_{22}^{-1} B_{2\varepsilon}^T Y_\varepsilon y.$$

Note that it is difficult to solve the CARE (4)–(5) because of the different magnitudes of their coefficients caused by the small perturbed parameter ε and high dimensions.

3 Generalized Lyapunov Iterations

To obtain the solutions of the CARE (4)–(5), we first define

$$\Pi_\varepsilon = \begin{bmatrix} I_{n_1} & 0 \\ 0 & \varepsilon I_{n_2} \end{bmatrix}.$$

Then, we introduce the following useful lemma.

Lemma 3.1. *The CARE* (4)–(5) *is equivalent to the following GCARE* (6)–(7), *respectively.*

$$(A - S_1 X - S_2 Y)^T X + X^T (A - S_1 X - S_2 Y) + Q_1 + X^T S_1 X + Y^T G_2 Y = 0, \tag{6}$$

$$(A - S_1 X - S_2 Y)^T Y + Y^T (A - S_1 X - S_2 Y) + Q_2 + Y^T S_2 Y + X^T G_1 X = 0, \tag{7}$$

where

$$X_\varepsilon = \Pi_\varepsilon X = X^T \Pi_\varepsilon, \quad Y_\varepsilon = \Pi_\varepsilon Y = Y^T \Pi_\varepsilon,$$

$$X = \begin{bmatrix} X_{11} & \varepsilon X_{21}^T \\ X_{21} & X_{22} \end{bmatrix}, \quad Y = \begin{bmatrix} Y_{11} & \varepsilon Y_{21}^T \\ Y_{21} & Y_{22} \end{bmatrix}.$$

Proof. The proof is identical to the proof of Lemma 3 in [13]. □

In [4], the Lyapunov iterations for solving CARE of the linear quadratic singularly perturbed Nash games have been presented. On the other hand, the algorithm for solving the GCARE (6)–(7) is given by (8)–(9) [14]–[16].

$$\begin{aligned}(A - S_1 X_{\text{Lya}}^{(n)} - S_2 Y_{\text{Lya}}^{(n)})^T X_{\text{Lya}}^{(n+1)} + X_{\text{Lya}}^{(n+1)T} (A - S_1 X_{\text{Lya}}^{(n)} - S_2 Y_{\text{Lya}}^{(n)}) \\ + Q_1 + X_{\text{Lya}}^{(n)T} S_1 X_{\text{Lya}}^{(n)} + Y_{\text{Lya}}^{(n)T} G_2 Y_{\text{Lya}}^{(n)} = 0,\end{aligned} \tag{8}$$

$$\begin{aligned}(A - S_1 X_{\text{Lya}}^{(n)} - S_2 Y_{\text{Lya}}^{(n)})^T Y_{\text{Lya}}^{(n+1)} + Y_{\text{Lya}}^{(n+1)T} (A - S_1 X_{\text{Lya}}^{(n)} - S_2 Y_{\text{Lya}}^{(n)}) \\ + Q_2 + X_{\text{Lya}}^{(n)T} G_1 X_{\text{Lya}}^{(n)} + Y_{\text{Lya}}^{(n)T} S_2 Y_{\text{Lya}}^{(n)} = 0,\end{aligned} \tag{9}$$

where $n = 0,\ 1,\ 2,\ \cdots$ and the initial conditions $X_{\rm Lya}^{(0)}$ and $Y_{\rm Lya}^{(0)}$ are obtained from the solutions of the following auxiliary generalized algebraic Riccati equations (GAREs) (10)–(11)

$$A^T X_{\rm Lya}^{(0)} + X_{\rm Lya}^{(0)T} A + Q_1 - X_{\rm Lya}^{(0)T} S_1 X_{\rm Lya}^{(0)} = 0, \tag{10}$$

$$\begin{aligned}(A - S_1 X_{\rm Lya}^{(0)})^T Y_{\rm Lya}^{(0)} + Y_{\rm Lya}^{(0)T} (A - S_1 X_{\rm Lya}^{(0)})\\ +Q_2 + X_{\rm Lya}^{(0)T} G_1 X_{\rm Lya}^{(0)} - Y_{\rm Lya}^{(0)T} S_2 Y_{\rm Lya}^{(0)} = 0,\end{aligned} \tag{11}$$

where

$$X_{\rm Lya}^{(n)} = \begin{bmatrix} X_{\rm Lya11}^{(n)} & \varepsilon X_{\rm Lya21}^{(n)T} \\ X_{\rm Lya21}^{(n)} & X_{\rm Lya22}^{(n)} \end{bmatrix},\quad Y_{\rm Lya}^{(n)} = \begin{bmatrix} Y_{\rm Lya11}^{(n)} & \varepsilon Y_{\rm Lya21}^{(n)T} \\ Y_{\rm Lya21}^{(n)} & Y_{\rm Lya22}^{(n)} \end{bmatrix},$$

$$X_{\rm Lya\varepsilon}^{(n)} = \Pi_\varepsilon X_{\rm Lya}^{(n)},\quad Y_{\rm Lya\varepsilon}^{(n)} = \Pi_\varepsilon Y_{\rm Lya}^{(n)}.$$

Remark 3.1. Note that the unique positive semi-definite stabilizing solution of the GARE (10)–(11) exist under Assumptions 2.1 and 2.2 [4], [8]–[10]. Moreover it is shown in [17,18] that the GARE (10)–(11) is well-conditioned.

We provide a new algorithm for solving the generalized Lyapunov iterations (8)–(9). We first introduce the following notation

$$\bar{A}^{(n)} := A - S_1 X_{\rm Lya}^{(n)} - S_2 Y_{\rm Lya}^{(n)} = \begin{bmatrix} \bar{A}_{11}^{(n)} & \bar{A}_{12}^{(n)} \\ \bar{A}_{21}^{(n)} & \bar{A}_{22}^{(n)} \end{bmatrix},$$

$$\bar{Q}^{(n)} := Q_1 + X_{\rm Lya}^{(n)T} S_1 X_{\rm Lya}^{(n)} + Y_{\rm Lya}^{(n)T} G_2 Y_{\rm Lya}^{(n)} = \begin{bmatrix} \bar{Q}_{11}^{(n)} & \bar{Q}_{12}^{(n)} \\ \bar{Q}_{12}^{(n)T} & \bar{Q}_{22}^{(n)} \end{bmatrix},$$

$$\hat{Q}^{(n)} := Q_2 + X_{\rm Lya}^{(n)T} G_1 X_{\rm Lya}^{(n)} + Y_{\rm Lya}^{(n)T} S_2 Y_{\rm Lya}^{(n)} = \begin{bmatrix} \hat{Q}_{11}^{(n)} & \hat{Q}_{12}^{(n)} \\ \hat{Q}_{12}^{(n)T} & \hat{Q}_{22}^{(n)} \end{bmatrix}.$$

Using the notation above, the generalized Lyapunov iterations (8)–(9) can be written in the following form (12)–(13).

$$\bar{A}^{(n)T} X_{\rm Lya}^{(n+1)} + X_{\rm Lya}^{(n+1)T} \bar{A}^{(n)} + \bar{Q}^{(n)} = 0, \tag{12}$$

$$\bar{A}^{(n)T} Y_{\rm Lya}^{(n+1)} + Y_{\rm Lya}^{(n+1)T} \bar{A}^{(n)} + \hat{Q}^{(n)} = 0. \tag{13}$$

We can obtain the solution of the GCARE (6)–(7) by performing generalized Lyapunov iterations (12)–(13) directly. Since the equations (12) and (13) are identical, we consider only the algorithm of the generalized Lyapunov iteration (12). We now give a perturbations analysis of the generalized Lyapunov iteration (12). Firstly, letting $\varepsilon = 0$ and using Kronecker products, the Lyapunov iteration (12) can be written as

$$\mathcal{V}\begin{bmatrix} \mathrm{vec}\bar{X}_1 \\ \mathrm{vec}\bar{X}_2 \\ \mathrm{vec}\bar{X}_3 \end{bmatrix} = \begin{bmatrix} \mathrm{vec}\bar{Q}_1 \\ \mathrm{vec}\bar{Q}_2 \\ \mathrm{vec}\bar{Q}_3 \end{bmatrix},$$

$$\mathcal{V} =$$

$$\begin{bmatrix} I_{n_1} \otimes \bar{A}_1^T + \bar{A}_1^T \otimes I_{n_1} & I_{n_1} \otimes \bar{A}_3^T + (\bar{A}_3^T \otimes I_{n_1})U_{n_2 n_1} & 0 \\ I_{n_1} \otimes \bar{A}_2^T & I_{n_1} \otimes \bar{A}_4^T & \bar{A}_3^T \otimes I_{n_2} \\ 0 & 0 & I_{n_2} \otimes \bar{A}_4^T + \bar{A}_4^T \otimes I_{n_2} \end{bmatrix},$$

$$\bar{X}_{\mathrm{Lya}}^{(n+1)}\Big|_{\varepsilon=0} = \begin{bmatrix} \bar{X}_1 & 0 \\ \bar{X}_2 & \bar{X}_3 \end{bmatrix}, \quad \bar{A}^{(n)}\Big|_{\varepsilon=0} = \begin{bmatrix} \bar{A}_1 & \bar{A}_2 \\ \bar{A}_3 & \bar{A}_4 \end{bmatrix},$$

$$\bar{Q}^{(n)}\Big|_{\varepsilon=0} = \begin{bmatrix} \bar{Q}_1 & \bar{Q}_2 \\ \bar{Q}_2^T & \bar{Q}_3 \end{bmatrix}.$$

It can be shown, after some algebra, that the determinant of $\mathcal{V}$ is expressed as

$$\det\mathcal{V} = \det[I_{n_2} \otimes \bar{A}_4^T + \bar{A}_4^T \otimes I_{n_2}] \cdot \det(I_{n_1} \otimes \bar{A}_4^T) \cdot \det[I_{n_1} \otimes \bar{A}_0^T + \bar{A}_0^T \otimes I_{n_1}].$$

where $\bar{A}_0 = \bar{A}_1 - \bar{A}_2\bar{A}_4^{-1}\bar{A}_3$.

Obviously, $\bar{A}_4$ and $\bar{A}_0$ are nonsingular matrices. Thus, there exists $\mathcal{V}^{-1}$. Therefore, the condition number [20] of $\mathcal{V}$, that is, $K(\mathcal{V}) = \|\mathcal{V}\| \cdot \|\mathcal{V}^{-1}\|$ is given by $K(\mathcal{V}) = O(1)$. Since $K(\mathcal{V})$ is not large for sufficiently small parameters ε and is independent of such parameters, the matrix $\mathcal{V} + O(\varepsilon)$ is well-conditioned for small ε.

Let us consider the method for solving the generalized Lyapunov iterations (12). We first consider the simultaneous linear equation (14) by rearranging the generalized Lyapunov iteration (12).

$$(I_n \otimes \bar{A}^{(n)T}) \cdot \mathrm{vec}X_{\mathrm{Lya}}^{(n+1)} + (\bar{A}^{(n)T} \otimes I_n) \cdot \mathrm{vec}X_{\mathrm{Lya}}^{(n+1)T} + \mathrm{vec}\bar{Q}^{(n)} = 0. \quad (14)$$

The Kronecker product method [10] on the basis of (14) is very simple and elegant. However, for large $N = n_1 + n_2$, the difficulty in solving N^2 linear equations make it impractical. Furthermore, it is difficult to solve the equation (14) because $\mathrm{vec}X_{\mathrm{Lya}}^{(n+1)}$ contains a small positive perturbation parameter ε. Thus, in order to reduce the dimension of the workspace and eliminate the influence of the small parameter ε, new algorithms for solving the generalized algebraic Lyapunov iterations are necessary.

3.1 The Fixed Point Iterations

Let us consider the generalized algebraic Lyapunov iteration (12), in a general form

$$\Lambda^T P + P^T \Lambda + \Gamma = 0, \tag{15}$$

where P is the solution of the GALE (15) and Λ and Γ are known matrices defined by

$$P = X_{\text{Lya}}^{(n+1)} = \begin{bmatrix} P_{11}(\varepsilon) & \varepsilon P_{21}(\varepsilon)^T \\ P_{21}(\varepsilon) & P_{22}(\varepsilon) \end{bmatrix}, \quad \Lambda = \bar{A}^{(n)} = \begin{bmatrix} \Lambda_{11} & \Lambda_{12} \\ \Lambda_{21} & \Lambda_{22} \end{bmatrix},$$

$$\Gamma = \bar{Q}^{(n)} = \begin{bmatrix} \Gamma_{11} & \Gamma_{12} \\ \Gamma_{12}^T & \Gamma_{22} \end{bmatrix},$$

$$P_{11} = P_{11}^T \in \mathbf{R}^{n_1 \times n_1}, \ \Gamma_{11} = \Gamma_{11}^T \in \mathbf{R}^{n_1 \times n_1},$$

$$P_{22} = P_{22}^T \in \mathbf{R}^{n_2 \times n_2}, \ \Gamma_{22} = \Gamma_{22}^T \in \mathbf{R}^{n_2 \times n_2},$$

$$\Lambda_{11} \in \mathbf{R}^{n_1 \times n_1}, \ \Lambda_{22} \in \mathbf{R}^{n_2 \times n_2}.$$

Without loss of generality, we shall make the following basic assumptions.

Assumption 3.1. The matrix Λ_{22} is nonsingular and $\Lambda_0 \equiv \Lambda_{11} - \Lambda_{12}\Lambda_{22}^{-1}\Lambda_{21}$ and Λ_{22} are Hurwitz.

The GALE (15) can be partitioned into

$$\Lambda_{11}^T P_{11} + P_{11}\Lambda_{11} + \Lambda_{21}^T P_{21} + P_{21}^T \Lambda_{21} + \Gamma_{11} = 0, \tag{16}$$

$$P_{11}\Lambda_{12} + P_{21}^T \Lambda_{22} + \Lambda_{21}^T P_{22} + \varepsilon \Lambda_{11}^T P_{21}^T + \Gamma_{12} = 0, \tag{17}$$

$$\Lambda_{22}^T P_{22} + P_{22}\Lambda_{22} + \varepsilon(\Lambda_{12}^T P_{21}^T + P_{21}\Lambda_{12}) + \Gamma_{22} = 0. \tag{18}$$

The fixed point iterations for solving the above equations (16)–(18) are given by

$$\Lambda_{22}^T P_{22}^{(i+1)} + P_{22}^{(i+1)}\Lambda_{22} + \varepsilon(\Lambda_{12}^T P_{21}^{(i)T} + P_{21}^{(i)}\Lambda_{12}) + \Gamma_{22} = 0, \tag{19}$$

$$\begin{aligned} &\Lambda_0^T P_{11}^{(i+1)} + P_{11}^{(i+1)}\Lambda_0 - \varepsilon\Lambda_{21}^T\Lambda_{22}^{-T}P_{21}^{(i)}\Lambda_0 - \varepsilon\Lambda_0^T P_{21}^{(i)T}\Lambda_{22}^{-1}\Lambda_{21} \\ &\quad + \Lambda_{21}^T\Lambda_{22}^{-T}\Gamma_{22}\Lambda_{22}^{-1}\Lambda_{21} - \Gamma_{12}\Lambda_{22}^{-1}\Lambda_{21} - \Lambda_{21}^T\Lambda_{22}^{-T}\Gamma_{12}^T + \Gamma_{11} = 0, \end{aligned} \tag{20}$$

$$P_{21}^{(i+1)} = -\Lambda_{22}^{-T}(\Lambda_{12}^T P_{11}^{(i+1)} + P_{22}^{(i+1)}\Lambda_{21} + \varepsilon P_{21}^{(i)}\Lambda_{11} + \Gamma_{12}^T), \tag{21}$$

$$P_{21}^{(0)} = 0, \quad i = 0, \ 1, \ 2, \ \cdots.$$

The following theorem indicates the convergence of the algorithms (19)–(21).

Theorem 3.1. *Under Assumption 3.1, there exists $\bar{\varepsilon} > 0$ such that for all $\varepsilon \in (0,\ \bar{\varepsilon})$, the fixed point iterations* (19)–(21) *converge to the exact solutions P_{lm} of* (16)–(18) *with the rate of convergence of $O(\varepsilon^i)$, that is*

$$\|P_{lm}^{(i)} - P_{lm}\| = O(\varepsilon^i),\ \ i = 1,\ 2,\ \cdots,\ \ lm = 11,\ 21,\ 22. \tag{22}$$

Proof. First, we need to prove the convergence property. This is done by using mathematical induction. When $i = 0$ for the equations (19)–(21), the solutions $P_{lm}^{(1)}$ are equivalent to the first-order approximations P_{lm} corresponding to the small parameters ε for equations (16)–(18). It follows from these equations that $\|P_{lm}^{(1)} - P_{lm}\| = O(\varepsilon),\ \ lm = 11,\ 21,\ 22$. When $i = \mathcal{N}$ ($\mathcal{N} \geq 1$), we assume that $\|P_{lm}^{(\mathcal{N})} - P_{lm}\| = O(\varepsilon^{\mathcal{N}})$. Subtracting (16)–(18) from (19)–(21) and using $\|P_{lm}^{(\mathcal{N})} - P_{lm}\| = O(\varepsilon^{\mathcal{N}})$, we arrive at the following equations.

$$\begin{aligned}
\Lambda_{22}^T(P_{22}^{(\mathcal{N}+1)} - P_{22}) + (P_{22}^{(\mathcal{N}+1)} - P_{22})\Lambda_{22} + O(\varepsilon^{\mathcal{N}+1}) &= 0,\\
\Lambda_0^T(P_{11}^{(\mathcal{N}+1)} - P_{11}) + (P_{11}^{(\mathcal{N}+1)} - P_{11})\Lambda_0 + O(\varepsilon^{\mathcal{N}+1}) &= 0,\\
(P_{11}^{(\mathcal{N}+1)} - P_{11})\Lambda_{12} + (P_{21}^{(\mathcal{N}+1)} - P_{21})^T\Lambda_{22}&\\
+\Lambda_{21}^T(P_{22}^{(\mathcal{N}+1)} - P_{22}) + O(\varepsilon^{\mathcal{N}+1}) &= 0.
\end{aligned}$$

Thus, using the standard properties of the ALE [19], we have $\|P_{lm}^{(\mathcal{N}+1)} - P_{lm}\| = O(\varepsilon^{\mathcal{N}+1})$. Consequently, equation (22) holds for all $i \in \mathbf{N}$. This completes the proof of the theorem concerned with the fixed point iterations.

Secondly, we show that there exists a parameter $\bar{\varepsilon}$. Subtracting (16)–(18) from (19)–(21) yields the GALE

$$\begin{aligned}
\Lambda_{22}^T(P_{22}^{(i+1)} - P_{22}) + (P_{22}^{(i+1)} - P_{22})\Lambda_{22} + \varepsilon\Lambda_{12}^T(P_{21}^{(i)} - P_{21})^T&\\
+\varepsilon(P_{21}^{(i)} - P_{21})\Lambda_{12} &= 0,
\end{aligned} \tag{23}$$

$$\begin{aligned}
\Lambda_0^T(P_{11}^{(i+1)} - P_{11}) + (P_{11}^{(i+1)} - P_{11})\Lambda_0 - \varepsilon\Lambda_{21}^T\Lambda_{22}^{-T}(P_{21}^{(i)} - P_{21})\Lambda_0&\\
-\varepsilon\Lambda_0^T(P_{21}^{(i)} - P_{21})^T\Lambda_{22}^{-1}\Lambda_{21} &= 0,
\end{aligned} \tag{24}$$

$$\begin{aligned}
P_{21}^{(i+1)} - P_{21} = -\Lambda_{22}^{-T}[\Lambda_{12}^T(P_{11}^{(i+1)} - P_{11}) + (P_{22}^{(i+1)} - P_{22})\Lambda_{21}&\\
+\varepsilon(P_{21}^{(i)} - P_{21})\Lambda_{11}].&
\end{aligned} \tag{25}$$

Thus, we have

$$\|P_{22}^{(i+1)} - P_{22}\| =$$
$$\left\|\varepsilon\int_0^{\infty} \exp[\Lambda_{22}^T t][\Lambda_{12}^T(P_{21}^{(i)} - P_{21})^T + (P_{21}^{(i)} - P_{21})\Lambda_{12}]\exp[\Lambda_{22}t]dt\right\|$$
$$\leq 2\varepsilon\|P_{21}^{(i)} - P_{21}\|\|\Lambda_{12}\|\int_0^{\infty} K_{22}\exp[-2\theta_{22}t]dt$$
$$= \frac{\varepsilon K_{22}}{\theta_{22}}\|\Lambda_{12}\|\|P_{21}^{(i)} - P_{21}\|, \tag{26}$$

$$\|P_{11}^{(i+1)} - P_{11}\| = \left\|-\varepsilon\int_0^{\infty} \exp[\Lambda_0^T t][\Lambda_{21}^T\Lambda_{22}^{-T}(P_{21}^{(i)} - P_{21})\Lambda_0\right.$$
$$\left. + \Lambda_0^T(P_{21}^{(i)} - P_{21})^T\Lambda_{22}^{-1}\Lambda_{21}]\exp[\Lambda_0 t]dt\right\|$$
$$\leq 2\varepsilon\|P_{21}^{(i)} - P_{21}\|\|\Lambda_{21}\|\|\Lambda_{22}^{-1}\|\|\Lambda_0\|\int_0^{\infty} K_0\exp[-2\theta_0 t]dt$$
$$= \frac{\varepsilon K_0}{\theta_0}\|\Lambda_{21}\|\|\Lambda_{22}^{-1}\|\|\Lambda_0\|\|P_{21}^{(i)} - P_{21}\|, \tag{27}$$

where there exist $K_{22} > 0$, $\theta_{22} > 0$ $K_0 > 0$ and $\theta_0 > 0$ depending only on Λ_{22} and Λ_0 such that

$$\|\exp[\Lambda_{22}t]\| \leq K_{22}\exp[-\theta_{22}t]dt,$$
$$\|\exp[\Lambda_0 t]\| \leq K_0\exp[-\theta_0 t]dt.$$

Using (26) and (27), we obtain from (25) that

$$\|P_{21}^{(i+1)} - P_{21}\| \leq \varepsilon\|\Lambda_{22}^{-1}\|\left[\frac{K_0}{\theta_0}\|\Lambda_{12}\|\|\Lambda_{21}\|\|\Lambda_{22}^{-1}\|\|\Lambda_0\|\right.$$
$$\left. + \frac{K_{22}}{\theta_{22}}\|\Lambda_{12}\|\|\Lambda_{21}\| + \|\Lambda_{11}\|\right]\|P_{21}^{(i)} - P_{21}\|. \tag{28}$$

It can be easily verified that if the following inequality (29) holds for ε,

$$\varepsilon\|\Lambda_{22}^{-1}\|\left[\frac{K_0}{\theta_0}\|\Lambda_{12}\|\|\Lambda_{21}\|\|\Lambda_{22}^{-1}\|\|\Lambda_0\| + \frac{K_{22}}{\theta_{22}}\|\Lambda_{12}\|\|\Lambda_{21}\| + \|\Lambda_{11}\|\right] < 1$$
$$\Leftrightarrow \varepsilon < \left(\|\Lambda_{22}^{-1}\|\left[\frac{K_0}{\theta_0}\|\Lambda_{12}\|\|\Lambda_{21}\|\|\Lambda_{22}^{-1}\|\|\Lambda_0\| + \frac{K_{22}}{\theta_{22}}\|\Lambda_{12}\|\|\Lambda_{21}\| + \|\Lambda_{11}\|\right]\right)^{-1}$$
$$\equiv \bar{\varepsilon}, \tag{29}$$

then the fixed point iterations (19)–(21) will have linear convergence. Hence, the proof is completed. □

3.2 Example

To show the value of Theorem 3.1, a scalar example is presented. Suppose $n_1 = n_2 = 1$, $\Lambda_{11} = a_1$, $\Lambda_{12} = a_2$, $\Lambda_{21} = a_3$, $\Lambda_{22} = a_4 < 0$ and $\Lambda_0 = \Lambda_{11} - \Lambda_{11}\Lambda_{22}^{-1}\Lambda_{21} = a_1 - a_2 a_4^{-1} a_3 = a_0 < 0$. The bound of (29) is given by

$$\bar{\varepsilon} = |a_4| \left(2\frac{|a_2||a_3|}{|a_4|} + |a_1| \right)^{-1}. \tag{30}$$

For this particular example, there is an easy way to check the above results. Using (23)–(25), we have

$$\begin{aligned}
p_3^{(i+1)} - p_3 &= -\varepsilon \frac{a_2}{a_4}(p_2^{(i)} - p_2),\\
p_1^{(i+1)} - p_1 &= -\varepsilon \frac{a_2 a_3^2 - a_1 a_3 a_4}{a_0 a_4^2}(p_2^{(i)} - p_2),\\
p_2^{(i+1)} - p_2 &= -\frac{a_2}{a_4}(p_1^{(i+1)} - p_1) - \frac{a_3}{a_4}(p_3^{(i+1)} - p_3) - \varepsilon \frac{a_1}{a_4}(p_2^{(i)} - p_2),
\end{aligned}$$

where

$$P = \begin{bmatrix} P_{11} & \varepsilon P_{21}^T \\ P_{21} & P_{22} \end{bmatrix} = \begin{bmatrix} p_1 & \varepsilon p_2 \\ p_2 & p_3 \end{bmatrix},$$

$$P^{(i)} = \begin{bmatrix} P_{11}^{(i)} & \varepsilon P_{21}^{(i)T} \\ P_{21}^{(i)} & P_{22}^{(i)} \end{bmatrix} = \begin{bmatrix} p_1^{(i)} & \varepsilon p_2^{(i)} \\ p_2^{(i)} & p_3^{(i)} \end{bmatrix}.$$

Hence, we arrive at the following equations.

$$p_2^{(i+1)} - p_2 = -\varepsilon \frac{a_1}{a_4}(p_2^{(i)} - p_2).$$

If $a_1 \neq 0$, the exact bound is given by

$$\tilde{\varepsilon} = \frac{|a_4|}{|a_1|}. \tag{31}$$

Therefore, the inequality (29) automatically holds because $\bar{\varepsilon} < \tilde{\varepsilon}$.

Remark 3.2. If $a_1 = 0$, we can easily get the solution of the algebraic Lyapunov equations (ALEs) (16)–(18) without any iterative technique.

In the rest of this section, we give the algorithm for solving the GCARE (6)–(7) with small positive parameter ε.

Step 1. Solve the GAREs (10)–(11) by using the algorithm proposed in [17].

Step 2. Starting with the initial matrices of $X_{\rm Lya}^{(0)}$ and $Y_{\rm Lya}^{(0)}$, compute the solutions $X_{\rm Lya}^{(n+1)}$ and $Y_{\rm Lya}^{(n+1)}$ of the generalized algebraic Lyapunov iterations (12)–(13) by using the fixed point iterations (19)–(21).

Step 3. If $\min\{\|F_1(X_{\rm Lya}^{(n)},\ Y_{\rm Lya}^{(n)})\|,\ \|F_2(X_{\rm Lya}^{(n)},\ Y_{\rm Lya}^{(n)})\|\} < O(\varepsilon^{\mathcal{M}})$ for a given integer $\mathcal{M} > 0$, go to Step 4. Otherwise, increment $n \to n+1$ and go to Step 2. Here $F_1(\cdot)$ and $F_2(\cdot)$ are defined by

$$\begin{aligned} F_1(X,\ Y) &= (A - S_1X - S_2Y)^T X + X^T(A - S_1X - S_2Y) \\ &\quad + Q_1 + X^T S_1 X + Y^T G_2 Y, \\ F_2(X,\ Y) &= (A - S_1X - S_2Y)^T Y + Y^T(A - S_1X - S_2Y) \\ &\quad + Q_2 + Y^T S_2 Y + X^T G_1 X. \end{aligned}$$

Step 4. Calculate $u_1^* = -R_{11}^{-1}B_1^T Xy$, $u_2^* = -R_{22}^{-1}B_2^T Yy$.

4 Kleinman Algorithm

In the previous section we have derived the algorithm for solving the GALE (15). However, the convergence speed of such algorithm is very slow. Moreover, so far the convergence property for the generalized Lyapunov iterations has not been investigated. In order to improve the convergence rate of the generalized Lyapunov iterations, we propose the following new algorithm which is based on the Kleinman algorithm [12].

First, we change the equations (6)–(7) to the following generalized form.

$$\begin{aligned} &\tilde{A}^T\mathcal{P} + \mathcal{P}^T\tilde{A} + \tilde{Q} - \mathcal{P}^T\tilde{S}\mathcal{P} - \mathcal{J}\mathcal{P}^T\tilde{S}\mathcal{J}\mathcal{P} - \mathcal{P}^T\mathcal{J}\tilde{S}\mathcal{P}\mathcal{J} \\ &\quad + \mathcal{J}\mathcal{P}^T\tilde{G}\mathcal{P}\mathcal{J} = 0, \end{aligned} \tag{32}$$

where

$$\mathcal{P} = \begin{bmatrix} X & 0 \\ 0 & Y \end{bmatrix},\ \tilde{A} = \begin{bmatrix} A & 0 \\ 0 & A \end{bmatrix},\ \tilde{Q} = \begin{bmatrix} Q_1 & 0 \\ 0 & Q_2 \end{bmatrix},$$

$$\tilde{S} = \begin{bmatrix} S_1 & 0 \\ 0 & S_2 \end{bmatrix},\ \tilde{G} = \begin{bmatrix} G_1 & 0 \\ 0 & G_2 \end{bmatrix},\ \mathcal{J} = \begin{bmatrix} 0 & I_N \\ I_N & 0 \end{bmatrix}.$$

Using the Kleinman algorithm [12], we give the following new iterative algorithm which consists of the generalized linear matrix equation (GLME).

$$
\begin{aligned}
&\Phi^{(n)T}\mathcal{P}^{(n+1)} + \mathcal{P}^{(n+1)T}\Phi^{(n)} - \Theta^{(n)T}\mathcal{P}^{(n+1)}\mathcal{J} - \mathcal{J}\mathcal{P}^{(n+1)T}\Theta^{(n)} \\
&\quad + \Xi^{(n)} = 0, \ \ n = 0, \ 1, \ 2, \ \cdots, \qquad (33)\\
&\Leftrightarrow \begin{cases}
\Phi_1^{(n)T}X^{(n+1)} + X^{(n+1)T}\Phi_1^{(n)} - \Theta_2^{(n)T}Y^{(n+1)} - Y^{(n+1)T}\Theta_2^{(n)} \\
\quad + \Xi_1^{(n)} = 0, \\
\Phi_2^{(n)T}Y^{(n+1)} + Y^{(n+1)T}\Phi_2^{(n)} - \Theta_1^{(n)T}X^{(n+1)} - X^{(n+1)T}\Theta_1^{(n)} \\
\quad + \Xi_2^{(n)} = 0,
\end{cases}
\end{aligned}
$$

where

$$
\Phi^{(n)} := \tilde{A} - \tilde{S}\mathcal{P}^{(n)} - \mathcal{J}\tilde{S}\mathcal{P}^{(n)}\mathcal{J} = \begin{bmatrix} \Phi_1^{(n)} & 0 \\ 0 & \Phi_2^{(n)} \end{bmatrix},
$$

$$
\Theta^{(n)} := \tilde{S}\mathcal{J}\mathcal{P}^{(n)} - \tilde{G}\mathcal{P}^{(n)}\mathcal{J} = \begin{bmatrix} 0 & \Theta_1^{(n)} \\ \Theta_2^{(n)} & 0 \end{bmatrix},
$$

$$
\begin{aligned}
\Xi^{(n)} &:= \tilde{Q} + \mathcal{P}^{(n)T}\tilde{S}\mathcal{P}^{(n)} + \mathcal{J}\mathcal{P}^{(n)T}\tilde{S}\mathcal{J}\mathcal{P}^{(n)} + \mathcal{P}^{(n)T}\mathcal{J}\tilde{S}\mathcal{P}^{(n)}\mathcal{J} \\
&\quad - \mathcal{J}\mathcal{P}^{(n)T}\tilde{G}\mathcal{P}^{(n)}\mathcal{J} \\
&= \begin{bmatrix} \Xi_1^{(n)} & 0 \\ 0 & \Xi_2^{(n)} \end{bmatrix},
\end{aligned}
$$

$$
\mathcal{P}^{(n)} = \begin{bmatrix} X^{(n)} & 0 \\ 0 & Y^{(n)} \end{bmatrix},
$$

$$
X^{(n)} = \begin{bmatrix} X_{11}^{(n)} & \varepsilon X_{21}^{(n)T} \\ X_{21}^{(n)} & X_{22}^{(n)} \end{bmatrix}, \ Y^{(n)} = \begin{bmatrix} Y_{11}^{(n)} & \varepsilon Y_{21}^{(n)T} \\ Y_{21}^{(n)} & Y_{22}^{(n)} \end{bmatrix},
$$

and the initial condition $\mathcal{P}^{(0)}$ has the following form

$$
\mathcal{P}^{(0)} = \begin{bmatrix} X^{(0)} & 0 \\ 0 & Y^{(0)} \end{bmatrix} = \begin{bmatrix} \bar{X}_{11} & \varepsilon\bar{X}_{21}^T & 0 & 0 \\ \bar{X}_{21} & \bar{X}_{22} & 0 & 0 \\ 0 & 0 & \bar{Y}_{11} & \varepsilon\bar{Y}_{21}^T \\ 0 & 0 & \bar{Y}_{21} & \bar{Y}_{22} \end{bmatrix}, \qquad (34)
$$

where

$$
\begin{aligned}
&(A - S_1\bar{X} - S_2\bar{Y})^T\bar{X} + \bar{X}^T(A - S_1\bar{X} - S_2\bar{Y}) \\
&\quad + Q_1 + \bar{X}^TS_1\bar{X} + \bar{Y}^TG_2\bar{Y} = 0, \qquad (35)
\end{aligned}
$$

$$
\begin{aligned}
&(A - S_1\bar{X} - S_2\bar{Y})^T\bar{Y} + \bar{Y}^T(A - S_1\bar{X} - S_2\bar{Y}) \\
&\quad + Q_2 + \bar{Y}^TS_2\bar{Y} + \bar{X}^TG_1\bar{X} = 0, \qquad (36)
\end{aligned}
$$

$$(A_{22} - S_{221}\bar{X}_{22} - S_{222}\bar{Y}_{22})^T\bar{X}_{22} + \bar{X}_{22}(A_{22} - S_{221}\bar{X}_{22} - S_{222}\bar{Y}_{22}) + Q_{221} + \bar{X}_{22}S_{221}\bar{X}_{22} + \bar{Y}_{22}G_{222}\bar{Y}_{22} = 0, \quad (37)$$

$$(A_{22} - S_{221}\bar{X}_{22} - S_{222}\bar{Y}_{22})^T\bar{Y}_{22} + \bar{Y}_{22}(A_{22} - S_{221}\bar{X}_{22} - S_{222}\bar{Y}_{22}) + Q_{222} + \bar{Y}_{22}S_{222}\bar{Y}_{22} + \bar{X}_{22}G_{221}\bar{X}_{22} = 0, \quad (38)$$

$$\bar{X} = \begin{bmatrix} \bar{X}_{11} & 0 \\ \bar{X}_{21} & \bar{X}_{22} \end{bmatrix}, \quad \bar{Y} = \begin{bmatrix} \bar{Y}_{11} & 0 \\ \bar{Y}_{21} & \bar{Y}_{22} \end{bmatrix}.$$

Note that the considered algorithm (34) is original. The new algorithm (34) can be constructed setting $\mathcal{P}^{(n+1)} = \mathcal{P}^{(n)} + \Delta\mathcal{P}^{(n)}$ and neglecting $O(\Delta\mathcal{P}^{(n)T}\Delta\mathcal{P}^{(n)})$ term. The Kleinman algorithm is well-known and is widely used to find a solution of the algebraic equations, and its local convergence properties are well understood. Before investigating the convergence property of the proposed algorithm, we study the asymptotic structure of the GCARE (32). The following theorem will establish the relation between the solutions X and Y and the solutions $\bar{X}_{lm}$ and $\bar{Y}_{lm}$ for the reduced-order equations (35)–(38).

Theorem 4.1. *Assume that*

$$\det\nabla\mathbf{F}(\bar{\mathbf{P}}_0) \neq 0,$$

where

$$\nabla\mathbf{F}(\bar{\mathbf{P}}_0) := \left.\frac{\partial\mathbf{F}(\mathbf{P})}{\partial\mathbf{P}^T}\right|_{\substack{\mathbf{P}=\bar{\mathbf{P}} \\ \varepsilon=0}},$$

$$\mathbf{F}(\mathbf{P}) := \begin{bmatrix} \mathrm{vec}E_{11} \\ \mathrm{vec}E_{21} \\ \mathrm{vec}E_{22} \\ \mathrm{vec}F_{11} \\ \mathrm{vec}F_{21} \\ \mathrm{vec}F_{22} \end{bmatrix}, \quad \mathbf{P} := \begin{bmatrix} \mathrm{vec}X_{11} \\ \mathrm{vec}X_{21} \\ \mathrm{vec}X_{22} \\ \mathrm{vec}Y_{11} \\ \mathrm{vec}Y_{21} \\ \mathrm{vec}Y_{22} \end{bmatrix}, \quad \bar{\mathbf{P}} := \begin{bmatrix} \mathrm{vec}\bar{X}_{11} \\ \mathrm{vec}\bar{X}_{21} \\ \mathrm{vec}\bar{X}_{22} \\ \mathrm{vec}\bar{Y}_{11} \\ \mathrm{vec}\bar{Y}_{21} \\ \mathrm{vec}\bar{Y}_{22} \end{bmatrix},$$

$$\mathcal{F}(\mathcal{P}) := \tilde{A}^T\mathcal{P} + \mathcal{P}^T\tilde{A} + \tilde{Q} - \mathcal{P}^T\tilde{S}\mathcal{P} - \mathcal{J}\mathcal{P}^T\tilde{S}\mathcal{J}\mathcal{P} - \mathcal{P}^T\mathcal{J}\tilde{S}\mathcal{P}\mathcal{J} + \mathcal{J}\mathcal{P}^T\tilde{G}\mathcal{P}\mathcal{J}$$
$$= \begin{bmatrix} E_{11} & E_{21}^T & 0 & 0 \\ E_{21} & E_{22} & 0 & 0 \\ 0 & 0 & F_{11} & F_{21}^T \\ 0 & 0 & F_{21} & F_{22} \end{bmatrix}.$$

Under Assumptions 3.1 and 3.2, the GCARE (32) admits the stabilizing solutions X and Y such that these matrices possess a power series expansion at $\varepsilon = 0$. That is,

$$X = \begin{bmatrix} \bar{X}_{11} & \varepsilon\bar{X}_{21}^T \\ \bar{X}_{21} & \bar{X}_{22} \end{bmatrix} + \sum_{k=1}^{\infty} \frac{\varepsilon^k}{k!} \begin{bmatrix} X_{11}^{(k)} & \varepsilon X_{21}^{(k)T} \\ X_{21}^{(k)} & X_{22}^{(k)} \end{bmatrix}, \tag{39}$$

$$Y = \begin{bmatrix} \bar{Y}_{11} & \varepsilon\bar{Y}_{21}^T \\ \bar{Y}_{21} & \bar{Y}_{22} \end{bmatrix} + \sum_{k=1}^{\infty} \frac{\varepsilon^k}{k!} \begin{bmatrix} Y_{11}^{(k)} & \varepsilon Y_{21}^{(k)T} \\ Y_{21}^{(k)} & Y_{22}^{(k)} \end{bmatrix}, \tag{40}$$

where

$$X_{lm}^{(k)} := \left.\frac{d^k}{d\varepsilon^k} X_{lm}\right|_{\varepsilon=0}, \quad Y_{lm}^{(k)} := \left.\frac{d^k}{d\varepsilon^k} Y_{lm}\right|_{\varepsilon=0}, \quad lm = 11,\ 21,\ 22.$$

Proof. We apply the implicit function theorem [8] to the GCARE (32). To do so, it is enough to show that the corresponding Jacobian is nonsingular at $\varepsilon = 0$. It can be shown, after some algebra, that the Jacobian of (32) in the limit as $\varepsilon \to +0$ is given by

$$J_{\bar{\mathbf{P}}} = \lim_{\varepsilon \to +0} \frac{\partial \mathbf{F}(\mathbf{P})}{\partial \mathbf{P}^T} = \nabla \mathbf{F}(\bar{\mathbf{P}}_0).$$

Therefore, $\det J_{\bar{\mathbf{P}}} \neq 0$, i.e., $J_{\bar{\mathbf{P}}}$ is nonsingular at $\varepsilon = 0$. The conclusion of Theorem 4.1 is obtained directly by using the implicit function theorem. □

We are concerned with good choices of the starting points which guarantee to find the required solution of a given GCARE (32). Our new idea is to set the initial conditions to the matrix (34). The fundamental idea is based on $\|\mathcal{P} - \mathcal{P}^{(0)}\| = O(\varepsilon)$. Consequently, we get the required solution with the rate of quadratic convergence via the Kleinman algorithm. Moreover, using the Newton–Kantorovich theorem [20], we also prove the existence of the unique solution for the GCARE (32). The main result of this section is as follows.

Theorem 4.2. *Under Assumptions 3.1 and 3.2, the new iterative algorithm* (34) *converges to the exact solution $\mathcal{P}^*$ of the GCARE* (32) *with the rate of quadratic convergence. The unique bounded solution $\mathcal{P}^{(n)}$ of the GCARE* (32) *is in the neighborhood of the exact solution $\mathcal{P}^*$. That is, the following conditions are satisfied.*

$$\|\mathcal{P}^{(n)} - \mathcal{P}^*\| \le O(\varepsilon^{2^n}),\ n = 0,\ 1,\ 2,\ \cdots, \tag{41}$$

$$\|\mathcal{P}^{(n)} - \mathcal{P}^*\| \le \frac{1}{\beta\mathcal{L}}[1 - (1 - 2\theta)^{1/2}],\ n = 0,\ 1,\ 2,\ \cdots, \tag{42}$$

where

$$\mathcal{P} = \mathcal{P}^* = \begin{bmatrix} X^* & 0 \\ 0 & Y^* \end{bmatrix},\ \mathcal{L} := 6\|\tilde{S}\| + 2\|\tilde{G}\|,\ \beta := \|[\nabla \mathbf{F}(\bar{\mathbf{P}})]^{-1}\|,$$

$$\eta := \|[\nabla \mathbf{F}(\bar{\mathbf{P}})]^{-1}\| \cdot \|\mathbf{F}(\bar{\mathbf{P}})\|,\ \theta := \beta\eta\mathcal{L},\ \nabla \mathbf{F}(\bar{\mathbf{P}}) := \left.\frac{\partial \mathbf{F}(\mathbf{P})}{\partial \mathbf{P}^T}\right|_{\mathbf{P} = \bar{\mathbf{P}}}.$$

Proof. The proof is given directly by applying the Newton–Kantorovich theorem [20] for the GCARE (32). Furthermore, using the property of the symmetrical matrix function $\mathcal{F}(\mathcal{P})$, it should be noted that the proof is done by translating the matrix into the vector.

Taking the partial derivative of the function $\mathcal{F}(\mathcal{P}) = \tilde{A}^T\mathcal{P} + \mathcal{P}^T\tilde{A} + \tilde{Q} - \mathcal{P}^T\tilde{S}\mathcal{P} - \mathcal{J}\mathcal{P}^T\tilde{S}\mathcal{J}\mathcal{P} - \mathcal{P}^T\mathcal{J}\tilde{S}\mathcal{P}\mathcal{J} + \mathcal{J}\mathcal{P}^T\tilde{G}\mathcal{P}\mathcal{J}$ with respect to $\mathcal{P}$ yields

$$\begin{aligned}\nabla\mathcal{F}(\mathcal{P}) &= \frac{\partial \mathrm{vec}\mathcal{F}(\mathcal{P})}{\partial(\mathrm{vec}\mathcal{P})^T} \\ &= (\tilde{A} - \tilde{S}\mathcal{P} - \mathcal{J}\tilde{S}\mathcal{P}\mathcal{J})^T \otimes I_{2N} + [I_{2N} \otimes (\tilde{A} - \tilde{S}\mathcal{P} - \mathcal{J}\tilde{S}\mathcal{P}\mathcal{J})^T]U_{2N2N} \\ &\quad - (\tilde{S}\mathcal{J}\mathcal{P} - \tilde{G}\mathcal{P}\mathcal{J})^T \otimes \mathcal{J} - [\mathcal{J} \otimes (\tilde{S}\mathcal{J}\mathcal{P} - \tilde{G}\mathcal{P}\mathcal{J})]^T U_{2N2N}. \quad (43)\end{aligned}$$

It is obvious that $\nabla\mathcal{F}(\mathcal{P})$ is continuous at for all $\mathcal{P}$. Thus, it is immediately obtained from the above equation (43) that

$$\begin{aligned}&\|\nabla\mathcal{F}(\mathcal{P}_1) - \nabla\mathcal{F}(\mathcal{P}_2)\| \le \mathcal{L}\|\mathcal{P}_1 - \mathcal{P}_2\| \\ \Rightarrow &\|\nabla\mathbf{F}(\mathbf{P}_1) - \nabla\mathbf{F}(\mathbf{P}_2)\| \le \mathcal{L}\|\mathbf{P}_1 - \mathbf{P}_2\|. \quad (44)\end{aligned}$$

Moreover, using the fact that

$$\nabla\mathbf{F}(\bar{\mathbf{P}}) = \nabla\mathbf{F}(\bar{\mathbf{P}}_0) + O(\varepsilon), \quad (45)$$

it follows that $\nabla\mathbf{F}(\bar{\mathbf{P}})$ is nonsingular under the condition of Theorem 4.1 for sufficiently small ε. Therefore, there exists β such that $\beta = \|[\nabla\mathbf{F}(\bar{\mathbf{P}})]^{-1}\|$. On the other hand, since $\mathbf{F}(\bar{\mathbf{P}}) = O(\varepsilon)$ from, there exists η such that $\eta = \|[\nabla\mathbf{F}(\bar{\mathbf{P}})]^{-1}\| \cdot \|\mathbf{F}(\bar{\mathbf{P}})\| = O(\varepsilon)$. Thus, there exists θ such that $\theta = \beta\eta\mathcal{L} < 2^{-1}$ because of $\eta = O(\varepsilon)$. Now, let us define

$$\begin{aligned}t^* &\equiv \frac{1}{\beta\mathcal{L}}[1 - (1 - 2\theta)^{1/2}] \\ &= \frac{1}{(6\|\tilde{S}\| + 2\|\tilde{G}\|) \cdot \|[\nabla\mathbf{F}(\bar{\mathbf{P}})]^{-1}\|}[1 - (1 - 2\theta)^{1/2}]. \quad (46)\end{aligned}$$

Using the Newton–Kantorovich theorem, we can show that $\mathcal{P}^*$ is the unique solution in the subset $\mathcal{S} \equiv \{\, \mathcal{P} \,:\, \|\mathcal{P}^{(0)} - \mathcal{P}\| \le t^* \,\}$. Moreover, using the Newton–Kantorovich theorem, the error estimate is given by

$$\|\mathcal{P}^{(n)} - \mathcal{P}^*\| \le \frac{(2\theta)^{2^n}}{2^n\beta\mathcal{L}}, \quad n = 1,\ 2,\ \cdots. \quad (47)$$

Substituting $2\theta = O(\varepsilon)$ into (47), we have (41). Furthermore, substituting $\mathcal{P}^*$ into $\mathcal{P}$ of the subset $\mathcal{S}$, we can also get (42). Therefore, (41)–(42) holds for the small ε. □

Remark 4.1. According to the reference [2], it is well-known that the solution of the GCARE (6)–(7) is not unique and several non-negative solutions exist. The Lyapunov iterations (12)–(13) guarantee that such algorithm converge to a positive semi-definite solution. In this paper, it is very important to note that if the initial conditions $\Pi_\varepsilon X^{(0)}$ and $\Pi_\varepsilon Y^{(0)}$ are the positive semi-definite solutions, the new algorithm (34) converges to the positive semi-definite solution in the same way as the Lyapunov iterations (12)–(13).

Remark 4.2. In order to obtain the initial condition (34), we have to solve the CARE and GCARE, (35)–(36) which are independent of the perturbation parameter ε. In this situation, we can also apply the Kleinman algorithm (34) to these equations, where one of the initial conditions GAREs (10)–(11) are used. In a later section, it is seen that the same algorithm can be used well to solve the GCARE (35) and (36). Moreover, it is also seen that the Kleinman algorithm (33) with the initial conditions which consist of the solutions of the AREs (10)–(11) converge to the required solution. In such cases, note that there is no guarantee of convergence for obtaining the initial condition (34).

In the rest of this section, we explain the method for solving the GLME (34) with dimension $2N$. So far, there is no argument as to the numerical method for solving the considered GLME (34). First, we convert (34) into the following form (48).

$$\begin{bmatrix} \mathcal{A}_1 & -\mathcal{B}_2 \\ -\mathcal{B}_1 & \mathcal{A}_2 \end{bmatrix} \begin{bmatrix} \mathrm{vec}X_{11}^{(n+1)} \\ \mathrm{vec}X_{21}^{(n+1)} \\ \mathrm{vec}X_{22}^{(n+1)} \\ \mathrm{vec}Y_{11}^{(n+1)} \\ \mathrm{vec}Y_{21}^{(n+1)} \\ \mathrm{vec}Y_{22}^{(n+1)} \end{bmatrix} = - \begin{bmatrix} \mathrm{vec}\,\Xi_{111}^{(n)} \\ \mathrm{vec}\,\Xi_{211}^{(n)} \\ \mathrm{vec}\,\Xi_{221}^{(n)} \\ \mathrm{vec}\,\Xi_{112}^{(n)} \\ \mathrm{vec}\,\Xi_{212}^{(n)} \\ \mathrm{vec}\,\Xi_{222}^{(n)} \end{bmatrix} \Leftrightarrow \mathcal{T}\mathcal{X} = \mathcal{Q}, \tag{48}$$

where

$$\mathcal{A}_i = \begin{bmatrix} I_{n_1} \otimes \Phi_{11i}^{(n)T} + \Phi_{11i}^{(n)T} \otimes I_{n_1} & I_{n_1} \otimes \Phi_{21i}^{(n)T} + (\Phi_{21i}^{(n)T} \otimes I_{n_1})U_{n_2 n_1} & 0 \\ I_{n_1} \otimes \Phi_{12i}^{(n)T} & I_{n_1} \otimes \Phi_{22i}^{(n)T} + \varepsilon(\Phi_{11i}^{(n)T} \otimes I_{n_2}) & \Phi_{21i}^{(n)T} \otimes I_{n_2} \\ 0 & \varepsilon[(I_{n_2} \otimes \Phi_{12i}^{(n)T})U_{n_2 n_1} + \Phi_{12i}^{(n)T} \otimes I_{n_2}] & I_{n_2} \otimes \Phi_{22i}^{(n)T} + \Phi_{22i}^{(n)T} \otimes I_{n_2} \end{bmatrix},$$

$$\mathcal{B}_i = \begin{bmatrix} I_{n_1} \otimes \Theta_{11i}^{(n)T} + \Theta_{11i}^{(n)T} \otimes I_{n_1} & I_{n_1} \otimes \Theta_{21i}^{(n)T} + (\Theta_{21i}^{(n)T} \otimes I_{n_1})U_{n_2 n_1} & 0 \\ I_{n_1} \otimes \Theta_{12i}^{(n)T} & I_{n_1} \otimes \Theta_{22i}^{(n)T} + \varepsilon(\Theta_{11i}^{(n)T} \otimes I_{n_2}) & \Theta_{21i}^{(n)T} \otimes I_{n_2} \\ 0 & \varepsilon[(I_{n_2} \otimes \Theta_{12i}^{(n)T})U_{n_2 n_1} + \Theta_{12i}^{(n)T} \otimes I_{n_2}] & I_{n_2} \otimes \Theta_{22i}^{(n)T} + \Theta_{22i}^{(n)T} \otimes I_{n_2} \end{bmatrix},$$

$$\Phi_i^{(n)} := \begin{bmatrix} \Phi_{11i}^{(n)} & \Phi_{12i}^{(n)} \\ \Phi_{21i}^{(n)} & \Phi_{22i}^{(n)} \end{bmatrix}, \ \Theta_i^{(n)} := \begin{bmatrix} \Theta_{11i}^{(n)} & \Theta_{12i}^{(n)} \\ \Theta_{21i}^{(n)} & \Theta_{22i}^{(n)} \end{bmatrix},$$

$$\Xi_i^{(n)} := \begin{bmatrix} \Xi_{11i}^{(n)} & \Xi_{12i}^{(n)} \\ \Xi_{12i}^{(n)T} & \Xi_{22i}^{(n)} \end{bmatrix}, \ i = 1, \ 2.$$

It should be noted that although the matrices $\mathcal{A}_i$ and $\mathcal{B}_i$ contain the small parameter, these matrices are well-conditioned [17,18]. Since the Newton–Kantorovich theorem guarantees the invertibility of the matrix $\mathcal{T}$, there exists the matrix $\mathcal{T}^{-1}$ for all $n, \ n = 0, \ 1, \ 2, \cdots$. Hence we have $\mathcal{X} = \mathcal{T}^{-1}\mathcal{Q}$. Note that there exists an algorithm for constructing the matrix $\mathcal{A}_i$ and $\mathcal{B}_i$. See for example [18].

5 Numerical Example

In order to demonstrate the efficiency of our proposed algorithm, we have run two simple numerical examples.

5.1 Example 1

Let us consider the following SPS,

$$\begin{bmatrix} \dot{x} \\ \varepsilon \dot{z} \end{bmatrix} = \begin{bmatrix} 1 & 2 \\ -1 & 0 \end{bmatrix} \begin{bmatrix} x \\ z \end{bmatrix} + \begin{bmatrix} 0 \\ 1 \end{bmatrix} u_1 + \begin{bmatrix} 0 \\ 2 \end{bmatrix} u_2, \tag{49}$$

with the performance index,

$$J_1 = \frac{1}{2} \int_0^\infty (x^2 + 2z^2 + u_1^2 + u_2^2) dt,$$

$$J_2 = \frac{1}{2} \int_0^\infty (x^2 + 10^{-1} z^2 + 2u_1^2 + 4u_2^2) dt.$$

The numerical results are obtained for the small parameter $\varepsilon = 10^{-2}$. Since $\det A_{22} = 0$, the system is nonstandard SPS. We give the solutions of the initial condition (34) and the GCARE (32) respectively.

$$X^{(0)} = \begin{bmatrix} 1.49367125 & 9.22242851 \times 10^{-3} \\ 9.22242851 \times 10^{-1} & 9.15631575 \times 10^{-1} \end{bmatrix},$$

$$Y^{(0)} = \begin{bmatrix} 4.63670807 \times 10^{-1} & 7.83939505 \times 10^{-3} \\ 7.83939505 \times 10^{-1} & 7.01508971 \times 10^{-1} \end{bmatrix},$$

$$X = X^{(3)} = \begin{bmatrix} 1.50568306 & 9.24880201 \times 10^{-3} \\ 9.24880201 \times 10^{-1} & 9.20906277 \times 10^{-1} \end{bmatrix},$$

$$Y = Y^{(3)} = \begin{bmatrix} 4.79012955 \times 10^{-1} & 7.90630151 \times 10^{-3} \\ 7.90630151 \times 10^{-1} & 7.14890262 \times 10^{-1} \end{bmatrix}.$$

Table 1 shows the results of the error $\|\mathcal{F}(\mathcal{P}^{(n)})\|$ per iteration. We find that the solutions of the GCARE (32) converge to the exact solution with accuracy of $\|\mathcal{F}(\mathcal{P}^{(n)})\| < 10^{-12}$ after 3 iterative iterations. Moreover, it is interesting to observe that the result of Table 1 shows that the algorithms (34) are quadratic convergence. Table 2 shows the results of comparing the number of iterations required in order to converge to the solution with the same accuracy of $\|\mathcal{F}(\mathcal{P}^{(n)})\| < 10^{-12}$ for the Lyapunov iterations [4] and the new algorithm. It can be seen that the convergence rate of the resulting algorithm is stable for all ε because the initial conditions $\mathcal{P}^{(0)}$ are quite good. On the other hand, the Lyapunov iterations converge to the exact solutions very slowly.

Table 1: $\|\mathcal{F}(\mathcal{P}^{(n)})\|$

n	$\varepsilon = 10^{-1}$	$\varepsilon = 10^{-2}$	$\varepsilon = 10^{-3}$
0	3.91×10^{-1}	3.91×10^{-2}	3.91×10^{-3}
1	3.70×10^{-2}	3.35×10^{-4}	3.31×10^{-6}
2	8.88×10^{-5}	8.66×10^{-9}	8.64×10^{-13}
3	5.31×10^{-10}	1.85×10^{-15}	–
4	1.58×10^{-15}	–	–
n	$\varepsilon = 10^{-4}$	$\varepsilon = 10^{-5}$	$\varepsilon = 10^{-6}$
0	3.91×10^{-4}	3.91×10^{-5}	3.91×10^{-6}
1	3.31×10^{-8}	3.31×10^{-10}	3.31×10^{-12}
2	2.04×10^{-15}	6.76×10^{-16}	2.02×10^{-15}
3	–	–	–
4	–	–	–

Table 2:

Number of iterations such that $\|\mathcal{F}(\mathcal{P}^{(n)})\| < 10^{-12}$.

ε	Lyapunov iterations	Kleinman algorithm
10^{-1}	46	4
10^{-2}	50	3
10^{-3}	50	2
10^{-4}	50	2
10^{-5}	50	2
10^{-6}	50	2

5.2 Example 2

Matrices A, B_1 and B_2 are chosen randomly. These matrices are given by

$$A_{11} = \begin{bmatrix} 0 & 0.4 \\ 0 & 0 \end{bmatrix}, \ A_{12} = \begin{bmatrix} 0 & 0 \\ 0.345 & 0 \end{bmatrix},$$

$$A_{21} = \begin{bmatrix} 0 & -0.524 \\ 0 & 0 \end{bmatrix}, \ A_{22} = \begin{bmatrix} 0 & 0.262 \\ 0 & -1 \end{bmatrix},$$

$$B_{11} = \begin{bmatrix} 0 \\ 0 \end{bmatrix}, \ B_{21} = \begin{bmatrix} 0 \\ 1 \end{bmatrix}, \ B_{12} = \begin{bmatrix} 0 \\ 0 \end{bmatrix}, \ B_{22} = \begin{bmatrix} 0.2 \\ 1 \end{bmatrix},$$

and a quadratic cost function

$$J_1 = \frac{1}{2}\int_0^\infty [y^T Q_1 y + u_1^2 + 2u_2^2]dt,$$

$$J_2 = \frac{1}{2}\int_0^\infty [y^T Q_2 y + 2u_1^2 + u_2^2]dt,$$

where

$$Q_1 = \begin{bmatrix} 1 & 0 & 0 & 0 \\ 0 & 0 & 0 & 0 \\ 0 & 0 & 1 & 0 \\ 0 & 0 & 0 & 0 \end{bmatrix},$$

$$Q_2 = \begin{bmatrix} 0 & 0 & 0 & 0 \\ 0 & 0 & 0 & 0 \\ 0 & 0 & 1 & 0 \\ 0 & 0 & 0 & 1 \end{bmatrix}.$$

Since $\det A_{22} = 0$, the obtained system is a nonstandard SPS. Thus, it is obvious that the existing method [6] to find the suboptimal solution is not valid for this example. On the other hand, it is solvable by using the method of this paper. Moreover, we can get the full-order Nash equilibrium solution which is closer to the exact performance compared to existing methods [7].

First, we give the solutions of the GCARE (6)–(7) by means of the Lyapunov iterations. We show the results obtained for the small parameter $\varepsilon = 10^{-4}$. We give the solutions of the GAREs (10)–(11) and the exact solutions of the GCARE (6)–(7).

$$X_{\rm Lya}^{(0)} = \begin{bmatrix} 6.28449354 & 2.89877181 & 4.71190892 \times 10^{-4} & 1.00000000 \times 10^{-4} \\ 2.89877181 & 7.28636981 & 2.21089145 \times 10^{-4} & 4.47571425 \times 10^{-6} \\ 4.71190892 & 2.21089145 & 4.71226241 & 1.00007627 \\ 1.00000000 & 4.47571425 \times 10^{-2} & 1.00007627 & 2.34520136 \times 10^{-1} \end{bmatrix},$$

$$Y_{\rm Lya}^{(0)} = \begin{bmatrix} 5.15115294 & 1.30681105 & 3.84208001 \times 10^{-4} & 3.61568037 \times 10^{-5} \\ 1.30681105 & 7.76794009 & 9.84286496 \times 10^{-5} & -4.78420574 \times 10^{-6} \\ 3.84208001 & 9.84286496 \times 10^{-1} & 4.92716725 & 4.58164313 \times 10^{-1} \\ 3.61568037 \times 10^{-1} & -4.78420574 \times 10^{-2} & 4.58164313 \times 10^{-1} & 4.34631680 \times 10^{-1} \end{bmatrix},$$

$$X = X_{\rm Lya}^{(45)} = X^{(2)} = \begin{bmatrix} 5.3735083498 & 3.6423573798 & 3.3845648024 \times 10^{-4} & 5.2601715885 \times 10^{-5} \\ 3.6423573798 & 7.0025729735 & 2.7060454965 \times 10^{-4} & 4.8753202671 \times 10^{-6} \\ 3.3845648024 & 2.7060454965 & 3.2957866169 & 7.5313947531 \times 10^{-1} \\ 5.2601715885 \times 10^{-1} & 4.8753202671 \times 10^{-2} & 7.5313947531 \times 10^{-1} & 2.3376743602 \times 10^{-1} \end{bmatrix},$$

$$
Y = Y_{\mathrm{Lya}}^{(45)} = Y^{(2)} = \begin{bmatrix}
2.5279245581 & 8.5751736777 \times 10^{-1} & 1.9861004245 \times 10^{-4} & 1.9388912118 \times 10^{-5} \\
8.5751736777 \times 10^{-1} & 6.3314203609 & 6.5949921391 \times 10^{-5} & -5.8790738357 \times 10^{-6} \\
1.9861004245 & 6.5949921391 \times 10^{-1} & 4.1834728413 & 4.0064281626 \times 10^{-1} \\
1.9388912118 \times 10^{-1} & -5.8790738357 \times 10^{-2} & 4.0064281626 \times 10^{-1} & 4.2941503846 \times 10^{-1}
\end{bmatrix}.
$$

It can be seen that the solutions of the generalized Lyapunov iterations (12)–(13) converge to the solutions with accuracy of $O(10^{-12})$ after 45 iterative iterations. In order to verify the exactitude of the solutions, we calculate the remainder by substituting $X_{\mathrm{Lya}}^{(45)}$ and $Y_{\mathrm{Lya}}^{(45)}$ into the GCARE (6)–(7).

$$
\|F_1(X_{\mathrm{Lya}}^{(45)},\ Y_{\mathrm{Lya}}^{(45)})\| = 1.28 \times 10^{-13},\ \|F_2(X_{\mathrm{Lya}}^{(45)},\ Y_{\mathrm{Lya}}^{(45)})\| = 4.59 \times 10^{-13}.
$$

Therefore, the numerical example illustrates the effectiveness of our proposed algorithm since the solutions $X_{\varepsilon}^{(n)} = \Pi_{\varepsilon} X^{(n)}$ and $Y_{\varepsilon}^{(n)} = \Pi_{\varepsilon} Y^{(n)}$ converge to the exact solutions $X_{\varepsilon} = \Pi_{\varepsilon} X$ and $Y_{\varepsilon} = \Pi_{\varepsilon} Y$ which are defined by (4)–(5). Indeed, we can obtain the solutions of the CARE (4)–(5) even if A_{22} is singular.

Second, we give the solutions of the GCARE (6)–(7) by means of the Kleinman algorithm. The initial condition (34) is given by $X^{(0)}$ and $Y^{(0)}$.

$$
X^{(0)} = \begin{bmatrix}
5.37337442 & 3.64213108 & 3.38452999 \times 10^{-4} & 5.26020763 \times 10^{-5} \\
3.64213108 & 7.00226035 & 2.70588292 \times 10^{-4} & 4.87371670 \times 10^{-6} \\
3.38452999 & 2.70588292 & 3.29550620 & 7.53105428 \times 10^{-1} \\
5.26020763 \times 10^{-1} & 4.87371670 \times 10^{-2} & 7.53105428 \times 10^{-1} & 2.33762950 \times 10^{-1}
\end{bmatrix},
$$

$$Y^{(0)} = \begin{bmatrix} 2.52797465 & 8.57518605 \times 10^{-1} & 1.98611690 \times 10^{-4} & 1.93890146 \times 10^{-5} \\ 8.57518605 \times 10^{-1} & 6.33147688 & 6.59495837 \times 10^{-5} & -5.87923154 \times 10^{-6} \\ 1.98611690 & 6.59495837 \times 10^{-1} & 4.18332895 & 4.00630362 \times 10^{-1} \\ 1.93890146 \times 10^{-1} & -5.87923154 \times 10^{-2} & 4.00630362 \times 10^{-1} & 4.29413797 \times 10^{-1} \end{bmatrix},$$

It should be noted that the proposed algorithm converges to the same solutions $X_{\mathrm{Lya}}^{(45)}$ and $Y_{\mathrm{Lya}}^{(45)}$ with accuracy of $\|\mathcal{F}(\mathcal{P}^{(n)})\| < 10^{-12}$ after 2 iterative iterations. Table 3 shows the results of the number of iterations in order to converge to the solution with the same accuracy of $\|\mathcal{F}(\mathcal{P}^{(n)})\| < 10^{-12}$ for the Lyapunov iterations [4] versus the Kleinman algorithm. It can be also seen that the convergence speed of the resulting algorithm is very fast for all ε.

Table 3:

Number of iterations such that $\|\mathcal{F}(\mathcal{P}^{(n)})\| < 10^{-12}$.

ε	Lyapunov iterations	Kleinman algorithm
10^{-1}	45	3
10^{-2}	45	3
10^{-3}	45	2
10^{-4}	45	2
10^{-5}	45	2
10^{-6}	45	1

Finally, Table 4 shows the results of the error $\|\mathcal{F}(\mathcal{P}^{(n)})\|$ per iterations via the Kleinman algorithm with the initial conditions $X_{\mathrm{Lya}}^{(0)}$ and $Y_{\mathrm{Lya}}^{(0)}$. We see that the solutions of the GCARE (32) converge to the exact solution after a few iterations. Consequently, since we do not need the initial conditions $X^{(0)}$ and $Y^{(0)}$ that are represented as the solutions of the reduced-order equations (34), we can reduce the computing time for obtaining the solutions of the GCARE (32).

6 Conclusions

The linear quadratic Nash games for infinite horizon SPS have been studied. We have proposed a new algorithm to solve the GALE. It is very important to note that the resulting algorithm is quite different from the existing method [8]. Consequently, since the convergence solutions do not depend on the 0-order solutions,

Table 4: $\|\mathcal{F}(\mathcal{P}^{(n)})\|$

n	$\varepsilon = 10^{-1}$	$\varepsilon = 10^{-2}$	$\varepsilon = 10^{-3}$
0	4.00	3.62	3.59
1	5.27×10^{-1}	4.79×10^{-1}	4.75×10^{-1}
2	1.91×10^{-1}	1.70×10^{-2}	1.68×10^{-2}
3	6.26×10^{-5}	5.77×10^{-5}	5.72×10^{-5}
4	1.27×10^{-9}	1.07×10^{-9}	1.05×10^{-9}
5	6.31×10^{-15}	1.21×10^{-14}	1.23×10^{-14}
n	$\varepsilon = 10^{-4}$	$\varepsilon = 10^{-5}$	$\varepsilon = 10^{-6}$
0	3.58	3.58	3.58
1	7.47×10^{-1}	4.74×10^{-1}	4.74×10^{-1}
2	1.67×10^{-2}	1.67×10^{-2}	1.67×10^{-2}
3	5.71×10^{-5}	5.71×10^{-5}	5.71×10^{-5}
4	1.05×10^{-9}	1.05×10^{-9}	1.05×10^{-9}
5	1.41×10^{-14}	1.22×10^{-14}	1.55×10^{-14}

we can get the exact solutions up to an arbitrary accuracy. Although the Lyapunov iterations take computation time, the resulting algorithm is quite reliable for large dimensions because the linear equations corresponding to the considered algorithms consist of the same dimension of slow and fast sub-systems. On the other hand, we have also proposed a new iterations method which is based on the Kleinman algorithm. The proposed algorithm has the property of quadratic convergence. It has been shown that the Kleinman algorithm can be used well to solve the GCARE under the appropriate initial conditions. When the dimension of the SPS is quite large, the algorithm appearing in Theorem 3.1 seems to be formidable. However, this is, in fact, quite numerically tractable for small dimensions of the SPS. Compared to Lyapunov iterations [4], even if the singular perturbation parameter is extremely small, we have succeeded in improving the convergence rate dramatically.

It is expected that the Kleinman algorithm for solving the GCARE is applicable to the wider class of control law synthesis involving the solution of the CARE with indefinite sign quadratic terms such as the mixed H_2/H_∞ control problem [3]. This problem will be addressed in future investigations.

REFERENCES

[1] Starr A. W. and Ho Y. C., Nonzero-sum differential games, J. Optimization Theory and Application, vol.3, pp.184–206 (1969)

[2] Jank G. and Kun G., Solutions of generalized Riccati differential equations and their approximation, Computational Methods and Function Theory

(CMFT'97), N. Papamichael et al. eds., pp.1–18, World Scientific Publishing Co., (1998)

[3] Limebeer D. J. N., Anderson B. D. O. and Hendel B., A Nash Ggme approach to mixed H_2/H_∞ control, IEEE Trans. Autom. Contr., vol.39, pp.69–82 (1994)

[4] Li T. and Gajić Z., Lyapunov iterations for solving coupled algebraic Lyapunov equations of Nash differential games and algebraic Riccati equations of Zero-sum games, New Trends in Dynamic Games and Applications, Birkhauser, 333–351 (1994)

[5] Freiling G., Jank G. and Abou–Kandil H., On global existence of solutions to coupled matrix Riccati equations in closed-loop Nash games, IEEE Trans. Autom. Contr., vol.41, pp.264–269 (1996)

[6] Khalil H. K. and Kokotovic P. V., Feedback and Well–Posedness of Singularly Perturbed Nash Games, IEEE Trans. Autom. Contr., vol.24, pp,699–708 (1979)

[7] Xu H., Mukaidani H. and Mizukami K., An order reduction procedure to composite Nash solution of singularly perturbed systems, IFAC World Congress, vol.F, pp.355-360, (1999)

[8] Gajić Z., Petkovski D. and Shen X., Singularly Perturbed and Weakly Coupled Linear System-a Recursive Approach, Lecture Notes in Control and Information Sciences, vol.140, Springer–Verlag (1990)

[9] Gajić Z. and Shen X., Parallel Algorithms for Optimal Control of Large Scale Linear Systems, Springer–Verlag (1993)

[10] Gajić Z. and Qureshi M. T. J., Lyapunov Matrix Equation in System Stability and Control, Mathematics in Science and Engineering, vol.195, Academic Press (1995)

[11] Kokotovic P. V., Khalil H. K. and O'Reilly J., Singular Perturbation Methods in Control, Analysis and Design, Academic Press (1986)

[12] Kleinman D. L., On the iterative technique for Riccati equation computations, IEEE Trans. Autom. Contr., vol.13, pp.114–115 (1968)

[13] Mukaidani H., Xu H. and Mizukami K., Recursive approach of H_∞ control problems for singularly perturbed systems under perfect and imperfect state measurements, Int. J. Systems Sciences, vol.30, pp.467–477 (1999)

[14] Mukaidani H., Xu H. and Mizukami K., A new algorithm for solving cross-coupled algebraic Riccati equations of singularly perturbed Nash games, Proc. 39th IEEE Conf. Decision and Control, Sydney, Australia, pp.3648–3653 (2000)

[15] Mukaidani H., Xu H. and Mizukami K., Recursive algorithm for mixed H_2/H_∞ control problem of singularly perturbed systems, Int. J. Systems Sciences, vol.31, pp.1299–1312 (2000)

[16] Mukaidani H., Xu H. and Mizukami K., A new algorithm for solving cross-coupled algebraic Riccati equations of singularly perturbed systems for mixed H_2/H_∞ control problem, Proc. 9th Int. Symp. Dynamic Games and Applications, Adelaide, Australia, pp.365–374 (2000)

[17] Mukaidani H., Xu H. and Mizukami K., New iterative algorithm for algebraic Riccati equation related to H_∞ control problem of singularly perturbed systems, IEEE Trans. Autom. Contr., vol.46, pp.1659–1666 (2001)

[18] Mukaidani H., Xu H. and Mizukami K., A revised Kleinman algorithm to solve algebraic Riccati equation of singularly perturbed systems, Automatica, vol.38, pp.553–558 (2002)

[19] Zhou K., Essentials of Robust Control, Prentice–Hall, (1998)

[20] Ortega J. M., Numerical analysis, A second course, SIAM, (1990)

[21] Magnus J. R. and Neudecker H., Matrix differential calculus with applications in statistics and econometrics, John Wiley and Sons, (1999)

Equilibrium Selection via Adaptation: Using Genetic Programming to Model Learning in a Coordination Game*

Shu-Heng Chen
AI-ECON Research Center
Department of Economics
National Chengchi University
Taipei11623, Taiwan
chchen@nccu.edu.tw

John Duffy
Department of Economics
University of Pittsburgh
Pittsburgh, PA 15260 USA
jduffy@pitt.edu

Chia-Hsuan Yeh
Department of Information Management
Yuan Ze University
Chungli, Taoyuan 320, Taiwan
imcyeh@saturn.yzu.edu.tw

Abstract

This paper models adaptive learning behavior in a simple coordination game that Van Huyck, Cook and Battalio (1994) have investigated in a controlled laboratory setting with human subjects. We consider how populations of artificially intelligent players behave when playing the same game. We use the genetic programming paradigm, as developed by Koza (1992, 1994), to model how a population of players might learn over time. In genetic programming one seeks to breed and evolve highly fit computer programs that are capable of solving a given problem. In our application, each computer program in the population can be viewed as an individual agent's forecast rule. The various forecast rules (programs) then repeatedly take part in the coordination game evolving and adapting over time according to principles of natural selection

*This project was initiated while Duffy was visiting National Chengchi University. A preliminary version of this paper, Chen, Duffy and Yeh (1996), was presented at the 1996 Evolutionary Programming Conference.

and population genetics. We argue that the genetic programming paradigm that we use has certain advantages over other models of adaptive learning behavior in the context of the coordination game that we consider. We find that the pattern of behavior generated by our population of artificially intelligent players is remarkably similar to that followed by the human subjects who played the same game. In particular, we find that a steady state that is theoretically unstable under a myopic, best-response learning dynamic turns out to be stable under our genetic-programming-based learning system, in accordance with Van Huyck et al.'s (1994) finding using human subjects. We conclude that genetic programming techniques may serve as a plausible mechanism for modeling human behavior, and may also serve as a useful selection criterion in environments with multiple equilibria.

AMS Subject Classifications. C63, D83.

1 Introduction

The empirical usefulness of static equilibrium analysis is compromised when economic models have multiple equilibria. Consequently, extensive efforts have been made to identify ways of reducing the set of equilibria that are focal in models with multiple equilibria. There seems to be some consensus emerging that a sensible selection criterion for choosing among multiple equilibria is to determine which of the candidate equilibria are stable with respect to some kind of disequilibrium, "learning" dynamic.[1] A number of such learning dynamics have been proposed and used to reduce or eliminate multiple equilibria as empirically relevant candidates. However, the notion that these learning dynamics accurately reflect the behavior of individual economic agents or groups of agents has only very recently begun to be examined through a number of controlled laboratory experiments with human subjects.[2]

This paper focuses on results obtained from one such experiment conducted by Van Huyck, Cook and Battalio (1994) that tested the predictions of a class of selection dynamics in a generic coordination game against the behavior of human subjects who played the same coordination game. Van Huyck et al. postulated that one of two candidate learning processes could describe the behavior of human subjects playing the coordination game. The first learning process is a Cournot-type, myopic best-response dynamic, and the second is an inertial learning algorithm that allows for slowing changing beliefs.[3] Both learning models are special versions of a large class of relaxation algorithms that have frequently

[1]For references, see, e.g. the surveys by Kreps (1990), Sargent (1993), and Marimon (1997).
[2]For a survey of some of these experiments, see, e.g., Kagel and Roth (1995).
[3]Van Huyck et al. (1994) refer to this inertial learning dynamic as the "L–map" which is a reference to Lucas' (1986) use of this type of learning dynamic.

appeared in the learning literature.[4] Under certain parameterizations, these two learning processes yield different predictions for the stability of one of the game's two Nash equilibria. Van Huyck et al.'s (1994) experimental results suggest that in those parameterizations where the two learning algorithms yield different predictions, the inertial learning algorithm provides a better characterization of the behavior of human subjects in the coordination game than the myopic best response dynamic.

In this paper, we adopt a computational approach, using Koza's (1992, 1994) genetic programming techniques to model the behavior of artificial economic agents playing the same simple coordination game that was studied by Van Huyck et al. (1994). The computational approach that we take to model agent behavior allows for a considerably more flexible experimental design than is possible with experiments involving human subjects. Moreover, unlike most rule-based models of adaptive learning behavior, the artificial players in our genetic programming implementation of the coordination game are explicitly endowed with the ability to "think" nonlinearly, and are given all the "building blocks" necessary to construct a vast array of both linear and nonlinear forecasting rules including the myopic best response and the inertial learning algorithms considered by Van Huyck et al. (1994). Thus we know, at the outset, that our artificial players are capable of both choosing and coordinating upon linear or nonlinear forecasting rules that may result in stationary, periodic or aperiodic trajectories. We find that our more general computational approach to modeling learning behavior in the coordination game results in behavior that is qualitatively similar to that of the subjects in Van Huyck et al.'s (1994) coordination game experiment. Indeed, we think of our genetic programming implementation of learning in the coordination game as a kind of robustness check on the experimental results reported for the same game. Finally, we argue that the genetic programming techniques we illustrate in this application have certain advantages over other artificial intelligence techniques that have been applied to economic models, namely, genetic algorithms.

The coordination game found in Van Huyck et al. (1994) differs from previous coordination games that have been studied experimentally (e.g. Cooper, DeJong, Forsythe and Ross (1990) and Van Huyck, Battalio and Beil (1990, 1991)) in that 1) the set of agent actions is considerably larger (indeed, there can be a continuum of possible actions), and 2) the stability of one of the game's two equilibria cannot be ascertained *a priori*. The first difference makes it difficult to formulate and enumerate strategies that are based upon all possible actions as is often done in adaptive learning models. The second difference arises because Van Huyck et al. (1994) entertain the notion that agents might adopt nonlinear rules to choose actions. Because of these differences, learning models that have been used to

[4]The class of relaxation algorithms includes, for example, the past averaging algorithm of Bray (1982) and Lucas (1986), and the least squares learning algorithm of Marcet and Sargent (1989).

explain behavior in the early coordination game experiments, for example, the linear learning models of Crawford (1991, 1995) and the genetic algorithm approach of Arifovic (1997) are not as well suited to the coordination game environment studied by Van Huyck et al. (1994). By contrast, we argue that the genetic-programming approach that we take to modeling learning behavior is particularly well suited to the coordination game environment of Van Huyck et al. (1994). We now turn to a description of this coordination game.

2 The Coordination Game

Consider the generic coordination game $\Gamma(\omega)$, studied by Van Huyck et al. (1994). There are n players, each of whom chooses some action $e^i \in [0, 1], i = 1, 2, \ldots n$. The individual player i's payoff function in every round of play, t, is described by:

$$\pi_{i,t}(e^i, e^{-i}) = c_1 - c_2|e^i - \omega M_t(e)[1 - M_t(e)]|, \tag{1}$$

where c_1 and c_2 are constants, $M_t(e)$ denotes the median of all n players' actions in round t, $\omega \in (1, 4]$ is a given parameter and e^{-i} denotes the vector of actions taken by the other $n - 1$ players in the same round. Both the payoff function and the set of feasible actions are assumed to be common knowledge.

It is clear from the structure of the payoff function that the individual player in this game should seek to minimize the expression that lies between the absolute value signs. That is, for a given value of the median, M, the individual player's best response function $b(M)$, is:

$$b(M) = \omega M(1 - M).$$

This best response function gives rise to two Nash equilibria: a corner equilibrium, where $e^i = 0$ for all i, and an interior equilibrium where $e^i = 1 - (1/\omega)$ for all i. The best response function $b(M)$ is easily recognized to be a member of the family of quadratic maps, where the degree of curvature is determined by the tuning parameter ω.

3 Selection Dynamics

Van Huyck et al. (1994) suggested that a certain class of "relaxation algorithms" that are frequently encountered in the learning literature could be used to characterize the evolution of play of this coordination game over time. This class of relaxation algorithms is described by the simple dynamical system:

$$M_t = b(\widehat{M}_t),$$

$$\widehat{M}_t = \widehat{M}_{t-1} + \alpha_t(M_{t-1} - \widehat{M}_{t-1}),$$

where $\widehat{M}_t$ is the representative agent's *expected* value for the median at time $t > 1$, and $\alpha_t \in [0, 1]$ is a given forgetting factor. Van Huyck et al. (1994) consider two specific versions of this relaxation algorithm: 1) a "myopic" best response algorithm where $\alpha_t = 1$ for all $t > 1$, and 2) an "inertial" algorithm, where $\alpha_t = 1/t$ for all $t > 1$.

The myopic best response version of the dynamical system gives rise to a simple first-order difference equation that characterizes the evolution of the median over time:

$$M_t = \omega M_{t-1}(1 - M_{t-1})$$

It is easily shown that for $1 < \omega < 3$, the interior equilibrium, $1-(1/\omega)$ is attracting (locally stable) while the corner equilibrium, 0, is repelling (locally unstable) under the myopic best response dynamics. However for $\omega > 3$, the dynamics of the myopic best response algorithm become increasingly more complicated, resulting in a dense set of periodic trajectories for the median that follows the Sarkovskii order. When $\omega = 3.839$, the myopic best response algorithm gives rise to a stable cycle of period 3, which according to the famous theorem of Li and Yorke implies that there are cycles of all periods and an uncountable set of non-periodic (chaotic) trajectories.[5] Thus, for $\omega > 3$, the interior equilibrium is no longer stable under the myopic best response dynamics.

The inertial version of the relaxation algorithm gives rise to the dynamical system:

$$\begin{aligned} M_t &= b(\widehat{M}_t) \\ \widehat{M}_t &= \frac{t-1}{t}\widehat{M}_{t-1} + \frac{1}{t}M_{t-1} \end{aligned}$$

Note that the inertial learning algorithm differs from the myopic best response algorithm in that the inertial algorithm gives most weight to the previous *expected* value of the median whereas the myopic best response algorithm gives all weight to the previous *realized* value of the median. It is easily shown that for all feasible values for ω ($\omega \in (1, 4]$), the interior equilibrium $1 - (1/\omega)$ is a global attractor under the dynamics of the inertial learning algorithm.

Thus if $1 < \omega \leq 3$, both the myopic best response and the inertial learning algorithms predict that the interior equilibrium $1 - (1/\omega)$ will be the equilibrium that agents eventually coordinate upon. However, for $3 < \omega \leq 4$, the myopic best response algorithm predicts that the interior equilibrium will be unstable, whereas the inertial learning algorithm predicts that the interior equilibrium will continue to be stable.

[5]For a detailed analysis of the first-order difference equation, $M_t = \omega M_{t-1}(1 - M_{t-1})$, that characterizes the myopic best response dynamic see, e.g. Devaney (1989).

4 Experimental Results and Experimental Design

Van Huyck et al. (1994) considered two experimental versions of the coordination game, $\Gamma(\omega)$, described above. In one version of the game, $\Gamma(2.4722)$, the interior equilibrium is predicted to be stable under both the myopic best response and inertial learning dynamics based on the choice of $\omega < 3$. In a second version of game, $\Gamma(3.86957)$, the interior equilibrium is predicted to be *unstable* under the myopic best response dynamics; starting from any initial condition, the myopic best response algorithm results in a chaotic trajectory for the median. By contrast, under the inertial learning dynamics, the interior equilibrium in the game, $\Gamma(3.86957)$, is predicted to remain stable since $\omega \leq 4$. Thus, the second game, $\Gamma(3.86957)$, is the more interesting one, as the stability predictions of the myopic best response and inertial learning dynamics differ for this particular game.

Van Huyck et al. (1994) report results from 2 experimental sessions of $\Gamma(2.4722)$ and 6 experimental sessions of $\Gamma(3.86957)$ using 5 subjects in each session. In all eight sessions they found that almost all subjects quickly coordinated on the interior equilibrium; that is, the interior equilibrium is judged to be stable in all treatments. The authors thus conclude that the inertial learning algorithm is a better selection device in the coordination game than the myopic best response algorithm, since the prediction of the inertial learning algorithm regarding the stability of the interior equilibrium is always consistent with the experimental findings.

Van Huyck et al.'s conclusion that the inertial learning algorithm serves as a reasonable learning model/selection criterion is subject to some criticism. First, while it is true that the inertial learning algorithm converges to the interior equilibrium in the game, $\Gamma(3.86957)$ (whereas myopic best response does not), the convergence trajectory taken by this algorithm is much too smooth when compared with the evolution of the median in the human subject experiments (see the experimental data reported in Appendix B of Van Huyck et al. (1994)). A second, related criticism is that it is apparent from the experimental data that the players in the coordination game do not all use the same learning scheme. If they all did use the same scheme, then for the same sequence of values for the median, we should expect to observe the same actions being taken. However, we observe players taking many different actions, especially in the early stages of the experiment, which suggests that they do not hold identical expectations. For this reason, it seems necessary to look beyond the predictions of representative agent learning models and to consider instead the performance of heterogeneous, multi-agent learning models. Our genetic-programming-based learning model is an example of this kind of multi-agent approach.

We note also that in implementing the coordination game experimentally, Van Huyck et al. (1994) made the simplifying assumption that the action set, e^i consists of only a *finite* set of *discrete* choices; players were asked to choose an action e^i from the set of integers $\{1, 2, \ldots, 90\}$. Each subject's action was then mapped into the unit interval using the function $f(e^i) = (90 - e^i)/89$. The discreteness of the action set, however, leads to some rather dramatic changes in the analysis of the myopic best response dynamics for the interesting case where $\omega > 3$. First, the discreteness of the action set rules out the possibility of chaotic trajectories which require the continuum of the unit interval. Indeed, the restriction that the median take on one of 90 values implies that the median must repeat itself at least once every 91 periods. Second, the discreteness of the action set also leads to the possibility that the interior equilibrium of the game, $\Gamma(3.86957)$, is locally *stable* under the myopic best response dynamics. In particular, the stability of the interior equilibrium of the discrete choice coordination game, $\Gamma(3.86957)$, now depends crucially on the initial condition, i.e. the first median value M_1. For almost all feasible values for $M_1 \in \{1, 2, \ldots, 90\}$, the myopic best response dynamics for the game, $\Gamma(3.86957)$, converge to a stable *seven* cycle, implying that the interior equilibrium is unstable. However, for some initial values the interior and the corner equilibria can also be locally *stable* under the discrete choice, myopic best response dynamics.[6]

In the genetic programming implementation of the coordination game that we explore in this paper, we do not have to discretize the action set. The computer programs that we evolve are all capable of choosing actions on the continuum of the unit interval. Therefore, unlike the experimental implementation of the coordination game we do not rule out the possibility of chaotic trajectories. Moreover, by allowing a continuum for the set of feasible actions, the coordination problem faced by our artificial agents is much more complicated than that faced by the experimental subjects whose actions were limited to a finite, discrete choice set. Finally, we consider a much larger size population of players than is practically feasible in an experiment with human subjects. This larger population size should, again, make the coordination problem more difficult. Thus, our genetic programming implementation of learning in the coordination game can be viewed as a check of whether the experimental results are robust to a continuous action set with a large number of players that would be difficult to implement in an experiment involving human subjects.

5 Genetic Programming

Before describing how we model agent behavior in the coordination game using genetic programming techniques, we first provide a brief overview of genetic programming.

[6]In particular, for $M_1 \in \{24, 67\}$, the interior equilibrium is locally stable under the discrete choice, myopic best response dynamics, and for $M_1 \in \{1, 90\}$, the corner equilibrium is locally stable under these same dynamics. It is interesting to note that in one of Van Huyck et al.'s 6 treatments of the game $\Gamma(3.86957)$ - session 7 - the initial median was 24. With this value for M_1, the discrete choice, myopic best response dynamic would predict that the system would stay at 24, the interior equilibrium forever, and indeed, this is roughly what occurred. See figure 13 of Van Huyck et al. (1994). Thus, one cannot dismiss altogether the possibility that discrete choice, myopic best response dynamic might also characterize the behavior of the experimental subjects in the game, $\Gamma(3.86957)$.

5.1 An Overview

Genetic programming (GP) represents a new field in the artificial intelligence literature that was developed only recently by Koza (1992, 1994) and others.[7] GP belongs to a class of evolutionary computing techniques based on principles of population genetics. These techniques combine Darwin's notion of natural selection through survival of the fittest with naturally occurring genetic operations of recombination (crossover) and mutation. Genetic programming techniques have already been widely applied to engineering type optimization problems (both theoretical and commercial), but have seen comparatively little application to economic problems, which are often similar in nature. The few economic applications of GP thus far include Allen and Karjalainen (1999), Chen and Yeh (1997a,b), Dworman et al. (1996) and Neely et al. (1997).

While GP techniques are often viewed as an offshoot of Holland's (1975) genetic algorithm (GA), they are perhaps more accurately viewed as a *generalization* of the genetic algorithm. The standard genetic algorithm operates on a population of structures, usually strings of bits. Each of the members of this population, the individual *bitstrings*, represents different candidate solutions to a well-defined optimization problem. The genetic algorithm evaluates the fitness of these various candidate solutions using the given objective function of the optimization problem and retains solutions that have, on average, higher fitness values. Operations of crossover (recombination) and mutation are then applied to some of these more fit solutions as a means of creating a new "generation" of candidate solutions. The whole process is repeated over many generations, in order to evolve solutions that are as close to optimal as possible. In analyzing the evolution of solutions over time, it is typical to report the solution in each generation that has the highest fitness value – this solution is designated the "best-of-generation" solution. The algorithm is ended when this best-of-generation solution satisfies a certain criterion (e.g. some tolerance) or after some maximum number of iterations has been reached.

Theoretical analyses of genetic algorithms suggest that they are capable of quickly locating regions of large and complex search spaces that yield highly fit solutions to optimization problems. That is because the genetic operators of the GA work together to optimize on the trade-off between discovering new solutions and using solutions that have worked well in the past (Holland (1975)).[8]

Koza's idea in developing genetic programming techniques was to take the genetic algorithm a step further and ask whether the same genetic operators used in GAs could be applied to a population of *computer programs* so as to evolve highly fit computer programs. There are several advantages to using computer

[7] See also Kinnear (1994).

[8] For an introduction to the theory of genetic algorithms, see, e.g. Goldberg (1989) or Mitchell (1996). Economic applications of genetic algorithms can be found in the work of Arifovic (1994–97), Arthur et al. (1997), Bullard and Duffy (1998a,b), Dawid (1996), Miller (1996) and Tesfatsion (1997) among others and are also discussed in Sargent (1993) and Birchenhall (1995).

programs rather than bitstrings as the structures to be evolved. First, the computer programs of GP have an *explicit, dynamic structure* that can be easily represented in a *decision tree format*. By contrast, the bitstrings of GAs typically encode passive yes/no type decisions or parameter values for pre-specified, often static functional forms. The dynamic nature of the computer programs of GP makes them capable of much more sophisticated and nonlinear decision-making than in generally possible using the bitstrings of GAs. Second, the computer programs of GP are *immediately implementable structures*; as such, they can be readily interpreted as the forecast rules used by a heterogeneous population of agents. For example, in GP, a computer program used by player i in round t, $gp_{i,t}$, might take the form:

$$gp_{i,t} = 0.31 + M_{t-1}(M_{t-1} - M_{t-2}).$$

Here, M_{t-j} represents the value of the median j periods in the past. Given these lagged median values, this program can be immediately executed and delivers a forecast of the median in period t, equal to the value of $0.31 + M_{t-1}(M_{t-1} - M_{t-2})$. This forecast then becomes the action taken by player i in round t. Note that this program is readily interpreted as the agent's *forecast function*. By contrast, the bitstrings used in GAs are not immediately implementable and their interpretation is less clear; these bitstrings first have to be decoded and then the decoded values must be applied to some prespecified functional form before the solution the bitstrings represent can be implemented. Finally, while the length of the bitstrings used in GAs is fixed, the length of the computer programs used in GP is free to vary (up to some limit, of course) providing for a much richer range of structures.[9]

Koza (1992, 1994) chose to develop GP techniques using the Lisp programming language because the syntax of Lisp allows computer programs to be easily manipulated like bitstrings, so that the same genetic operations used on bitstrings in GAs can also be applied to the computer programs that serve as the evolutionary structures in GP. Moreover, the new computer programs that result from application of these genetic operations are immediately executable programs.

Lisp has a single syntactic form, the symbolic expression (S-expression), that consists of a number of *atoms*. These atoms are either members of a *terminal* set, that comprise the *inputs* (e.g. data) to be used in the computer programs, or they are members of a *function* set that consists of a number of pre-specified functions or operators that are capable of processing any data value from the terminal set and any data value that results from the application of any function or operator in the function set. Each Lisp S-expression has the property that it is immediately executable as a computer program, and can be readily depicted as a rooted, point-labeled tree. Moreover, the S-expressions are easily manipulated like data; cutting a tree at any point and recombining the cut portion with another tree (S-expression) results in a new S-expression that is immediately executable.

[9]While in principle it is possible to represent dynamic, variable length expressions using the bitstrings of genetic algorithms, this has not been the practice. See Angeline (1994) for a further discussion.

As Koza and others have noted, the use of Lisp is not necessary for genetic programming; what is important for genetic programming is the implementation of a *Lisp-like environment*, where individual expressions can be manipulated like data, and are immediately executable. For the results reported in this paper, we have chosen to implement the Lisp environment using Pascal 4.0.[10]

5.2 Using Genetic Programming to Model Learning in the Coordination Game

In this section, we explain how we use genetic programming to model population learning in the coordination game. The version of genetic programming used here is the *simple genetic programming* that is described in detail in Koza (1992).

Let GP_t, denote a population of trees (S-expressions), representing a collection of players' forecasting functions. A player i, $i = 1, \dots, n$, makes a decision about his action for time t using a *parse tree*, $gp_{i,t} \in GP_t$, written over the *function* and *terminal* sets that are given in Table 1.

Table 1: Tableau for the GP–Based Learning Algorithm

Population size	500
The number of initial trees generated by the full method	250
The number of initial trees generated by the grow method	250
The maximum depth of a tree	17
Function set	$\{+, -, \times, \%, Exp, Rlog, Sin, Cos\}$
Terminal set	$\{\Re, M_{t-1}, M_{t-2}, M_{t-3}, M_{t-4}, M_{t-5}\}$
The maximum number in the domain of *Exp*	1,700
The number of trees created by reproduction	50
The number of trees created by crossover	350
The number of trees created by mutation	100
The probability of mutation	0.0033
The probability of leaf selection under crossover	0.5
The maximum number of generations	1,000
Fitness criterion	Payoff function: $\pi_{i,t}$

[10]Other programming languages, e.g. C, C++, and Mathematica have also been used to implement Lisp environments.

As Table 1 indicates, the *function set*, includes the standard mathematical operations of addition (+), subtraction (−), multiplication (×) and protected division (%), and also includes the exponential function (Exp), a protected natural logarithm function ($Rlog$) and the sin and cosine functions (Sin and Cos).[11] This set of operators and functions is the one that the artificial agents in our experiments are "endowed" with.

The *terminal set* includes the set of constants and variables that the artificial agents may use in combination with the operators and functions from the function set to build forecast rules. As indicated in Table 1, the terminal set includes the random floating-point constant $\mathbb{R}$ which is restricted to range over the interval $[-9.99, 9.99]$, as well as the population *mean* choice of action lagged up to h periods, i.e., $M_{t-1}, \dots, M_{t-h}$. Note that in our version of the coordination game, M refers to the *mean* rather than the *median* choice of action as in Van Huyck et al. (1994).[12] The choice of the lag length, h, determines players' ability to recall the past. We set h equal to 5, so that agents may consider as many as 5 past lagged values of the mean in their forecast functions.

The forecasting functions that players may construct and use are linear and nonlinear functions of $M_{t-1}, \dots, M_{t-h}, \Re$, and, as we shall see later, they may also be functions, in whole or in part, of past forecast rules $gp_{i,t}(M_{t-1}, \dots, M_{t-h})$. We note that the set of forecast functions that our artificial players may adopt includes the myopic best response, but not the inertial learning algorithm, as the latter requires knowledge of the previous period's forecast of the mean value, $\hat{M}_{t-1}$.

Indeed, Chen and Yeh (1997a) have shown that GP techniques can be used to uncover a variety of nonlinear data generating functions. In one demonstration, they generated a time series for the nonlinear, chaotic dynamical system $x_{t+1} = 4x_t(1-x_t)$, which is the same as our myopic best response law of motion with $\omega = 4$. They then used a GP-based search in an effort to recover this exogenously given system. Fitness was based on how close the forecast functions in the population

[11]The protected division operator protects against division by zero by returning the value 1 if its denominator argument is 0; otherwise, it returns the value from dividing its first argument (the numerator) by its second argument (the denominator). Similarly, the protected natural logarithm function avoids non-positive arguments by returning the natural logarithm of the absolute value of its argument, and returning the value 0 if its argument is 0. The exponential function, which takes the argument x and returns the value e^x, allows a maximum argument value of 1,700 as indicated in Table 1. Such function modifications and restrictions are necessary to avoid ill-defined forecasts; these types of modifications are quite standard in the GP literature. See, e.g., Koza (1992).

[12]Van Huyck et al. (1994) used the median rather than the mean in the coordination game because in experiments involving small numbers of human subjects (they only had 5 subjects in each experimental session), the mean can be easily influenced by the behavior of a single subject. By contrast, the computational coordination game experiments that we perform involve hundreds of artificial agents, so that the use of the mean rather than the median is no longer a concern.

came to matching the given time series behavior, and the GP function and terminal sets were nearly identical to those used in this paper. Chen and Yeh (1997a) report that the GP-based search was able to uncover the data generating process in no more than 19 generations. In this paper, by contrast, the data generating process for the mean is *endogenously* determined by the actions chosen by all of the individual players. Nevertheless, it is nice to know that a GP-based search algorithm can deduce a nonlinear data generating function such as the myopic best response law of motion.

The decoding of a parse tree $gp_{i,t}$ gives the forecasting function used by player i at time period t, i.e., $gp_{i,t}(\Omega_{t-1})$ where Ω_{t-1} is the information set containing past mean values through time $t-1$. Evaluating $gp_{i,t}(\Omega_{t-1})$ at the realization of Ω_{t-1} gives the mean action predicted by player i in round t, i.e., $gp_{i,t}$. Without any further restriction, the range of $gp_{i,t}$ is $(-\infty, \infty)$. However, since the action space for each player is restricted to $[0, 1]$, we must restrict $gp_{i,t}$ so that it also lies in $[0, 1]$. We chose to implement this restriction in two different ways. Our first approach was to use the *symmetric sigmoidal activation function* to map $(-\infty, \infty)$ to $[0, 1]$ so as to obtain a valid mean forecast, $\widehat{M}_{i,t}$, for player i in round t, i.e.

$$\widehat{M}_{i,t} = \frac{1}{1+e^{-gp_{i,t}}}.$$

A second approach that we also considered was a simple *truncated linear transformation* where player i's round t forecast was determined as follows:

$$\widehat{M}_{i,t} = \begin{cases} gp_{i,t} & \text{if } 0 \le gp_{i,t} \le 1, \\ 1 & \text{if } gp_{i,t} > 1, \\ 0 & \text{if } gp_{i,t} < 0. \end{cases}$$

Using either of these two approaches ensures that player i's mean choice of action lies in the feasible $[0, 1]$ interval.

Once we have all n players' mean action choices (equivalent to their mean forecasts), it is possible to determine the actual value of the mean in round t, $M_t = \frac{1}{n}\sum_{i=1}^{n} \widehat{M}_{i,t}$. Given this mean value, we can calculate each player's *fitness value* in round t. The *raw fitness* of a parse tree $gp_{i,t}$ is determined by the value of the player's payoffs earned in round t as determined by the payoff function $\pi_{i,t}$, given in equation (1). To avoid negative fitness values, each raw fitness value is adjusted to produce an *adjusted fitness* measure $\mu_{i,t}$ that is described as follows:

$$\mu_{i,t} = \begin{cases} \pi_{i,t} + 0.25 & \text{if } \pi_{i,t} \ge -0.25, \\ 0 & \text{if } \pi_{i,t} < -0.25. \end{cases}$$

In making this adjustment, we are effectively eliminating from the population forecast functions $gp_{i,t}$ that lose more than \$0.25, since these rules have comparatively lower adjusted fitness values (equal to 0) than rules that did not perform so poorly.

Our decision to make the above adjustment to the fitness measure was due to the following considerations. In the early rounds of a game, players have very limited experience with the environment so their expectations essentially amount to random guessing. As a consequence, many of the players will lose money. If we only considered players with forecast functions that earned positive payoffs, the selection process would quickly come to be dominated by those few players (forecast functions) that were lucky enough to earn positive payoffs in the initial stages of the game. However, we want to maintain some heterogeneity in the population and avoid the possibility of *premature convergence*, a problem that can occur in populations lacking sufficient heterogeneity. For this reason, we allow some players to earn negative payoffs, but we restrict such losses so that they do not exceed \$0.25. After a few generations, when most of the players have begun to earn positive payoffs, this protection no longer plays any effective role. We have experimented with adjustment values other than 0.25. While small adjustment values do not significantly alter our simulation results, very large adjustment values do affect the results because these large values effectively nullify the adjusted fitness measure as an indicator of the relative success of a forecast function. Later in the paper, we examine what happens when we replace the adjustment value of 0.25 in the adjustment scheme described above with the much larger value of 200.00. However, unless otherwise indicated, all of the simulation results we report below involve an adjustment value equal to 0.25.

Once all of the adjusted fitness values are determined, each adjusted fitness value $\mu_{i,t}$ is then normalized. The *normalized fitness* value $p_{i,t}$ is given by:

$$p_{i,t} = \frac{\mu_{i,t}}{\sum_{i=1}^{n} \mu_{i,t}}.$$

It is clear that the normalized fitness value is a *probability measure*. Moreover, $p_{i,t}$ varies directly with the performance of the parse tree $gp_{i,t}$; the better the parse tree performs (in terms of its payoff), the higher is its normalized fitness value. The normalized fitness values $p_{i,t}$ are used to determine the next generation of agents (parse trees) GP_{t+1} from the current generation GP_t through application of the three primary genetic operators, i.e., *reproduction, crossover*, and *mutation*. We now describe these three genetic operators.

a. **Reproduction:**
The reproduction operator makes copies of individual parse trees from generation GP_t and places them in the next generation GP_{t+1}. The criterion used for copying is the normalized fitness value $p_{i,t}$. If $gp_{i,t}$ is an individual in the population GP_t with normalized fitness value $p_{i,t}$, then each time the reproduction operator is called, $gp_{i,t}$ is copied into the next generation with probability $p_{i,t}$. The reproduction operator does not create anything new in the population and the "offspring" generated by reproduction constitute only part of the population of the next generation of trees, GP_{t+1}. As specified in Table 1, the reproduction operator is used to create only 10% (50 out of 500)

of the next generation. The rest of the offspring are generated by the other two operators, *crossover* and *mutation*.

b. **Crossover:**
The crossover operation for the genetic programming paradigm is a sexual operation that starts with two parental parse trees that have been randomly selected from the population GP_t based upon their normalized fitness values as described above. Crossover involves exchanging different parts of these "parents" to produce two new "offspring." This exchange begins by randomly and independently selecting a single point on each parental parse tree using a uniform distribution described below. By the syntax of Lisp, each point (atom) of a parse tree could be either a *leaf* (terminal) or a *inner code* (function). Thus, the point (atom) selected for crossover could either be a terminal or a function. As specified in Table 1, the probability that the crossover point is a terminal or a function is the same, i.e., one-half. Given that a terminal or function is to be the point chosen for crossover, the probability that any terminal or function is chosen as the crossover point is uniformly distributed. For example, if the crossover point is to be a terminal, and there are three terminals in the parse tree, the probability that any one of the three terminals is chosen for the crossover point is one-third. Unlike reproduction, the crossover operation adds new individuals (new forecasts rules) to the population. As indicated in Table 1, crossover is responsible for creating 70% (350 out of 500) of the next generation of parse trees, GP_{t+1}.

c. **Mutation:**
The operation of mutation also allows for the creation of new individuals. The mutation operator begins by selecting a parse tree $gp_{i,t}$ from the population GP_t based once again upon normalized fitness values $p_{i,t}$. Each point (atom) of the selected parse tree is then subjected to mutation (alteration) with a small, fixed probability. As specified in Table 1, this fixed probability of mutation is 0.0033. To ensure that the resulting expression is a syntactically and semantically valid Lisp S-expression, terminals can only be altered to another member from the terminal set and functions can only be altered to another member from the function set possessing the same number of arguments. The altered individual forecast rule (parse tree) is then copied into the next generation of the population. As indicated in Table 1, mutation is responsible for creating 20% (100 out of 500) of the next generation of parse trees.

The three operators combined create the population GP_{t+1} by copying, recombining and mutating the parse trees that make up the population GP_t. Once the new population GP_{t+1} has been created, the decoding of each parse tree $gp_{i,t+1}$ is performed to obtain the new mean, M_{t+1}. Once the new mean is determined, the raw, adjusted and normalized fitness values for each parse tree can be determined using the payoff function (1), and the GP operators can then be applied to create the population GP_{t+2}. The algorithm continues with successive generations, up to the maximum number of generations. We set the maximum number of generations equal to 1,000 as indicated in Table 1.

The initial S-expressions were randomly generated using both the methods suggested by Koza (1992) – the *full* method and the *grow* method.[13] Together, these two initialization methods provide for a great diversity of initial programs. Table 1 indicates that each method was responsible for creating one-half (250) of the initial population of trees, GP_1.

We consider the same two coordination games studied by Van Huyck et al. (1994), although as mentioned previously, we do not restrict the action set to a finite set of discrete choices. Furthermore, we have many more (artificial) players. We refer to the game, $\Gamma(2.4722)$, studied by Van Huyck et al. (1994) as Case 1 and the other game these authors considered, $\Gamma(3.86957)$, as Case 2. The exact parameterizations of these two cases are reported in Table 2.

Table 2: Parameter Values for the Coordination Game Used in the Genetic Programming Simulations

Parameter	Case 1	Case 2
ω	2.47222	3.86957
c_1	0.5	0.5
c_2	1.0	1.0
n	500	500
e_I^*	0	0
e_{II}^*	0.59551	0.74157

e_I^*: The optimal action under the strict equilibrium $e^i = 0\ \forall i$.
e_{II}^*: The optimal action under the strict equilibrium $e^i = 1 - (1/\omega)$ $\forall i$.

6 Simulation Results

Our simulation experiments were organized as follows. For each of the two different transformation functions – the symmetric sigmoidal transformation function and the truncated linear transformation function – we conducted 10 simulations for a total of 20 simulations. Within each group of 10 simulations, 5 of the simulations were conducted under the Case 1 parameterization and 5 were conducted under the Case 2 parameterization.

We focus our attention first on the 10 simulations that we conducted using the *symmetric sigmoidal transformation function.* Means and standard deviations from these 10 simulations are reported in Table 3. In this table, simulation 1.1 refers to our first simulation of Case 1, while simulation 2.1 refers to our first simulation

[13]See Koza (1992), pp. 92–93.

of Case 2, and so on. Time series for the mean, M_t, from a single simulation of Case 1 and Case 2 are plotted in Figure 1. These time series plots are typical of the other simulations we conducted for the two cases. As these figures clearly indicate, the time series for M_t in both cases of the GP-based coordination game (Case 1 and Case 2) tend to converge to a neighborhood of the strict interior equilibrium $1-(1/\omega)$, i.e., 0.59551 for Case 1 and 0.74157 for Case 2. In addition, the transition to $1-(1/\omega)$ is remarkably brief; If one considers $(0.99-(1/\omega), 1.01-(1/\omega))$ as a neighborhood of $1-(1/\omega)$ then, for all simulations, it takes no more than 50 generations to move into this neighborhood.

A second finding is that while M_t does not converge to $1-(1/\omega)$ in a strict sense, due to the constant mutation rate, there appears to be a force that serves to stabilize the movement of M_t in a very small band around the interior equilibrium. In other words, GP-based coordination games have a self-stabilizing feature. These properties are also revealed by Table 3.

As Table 3 reveals, in almost all of our simulations, the average of the means, M_t, from generation 201 to 1,000, i.e. $\overline{M}_b$, does not deviate from the interior equilibrium value, $1-(1/\omega)$, by more than 0.5%. Note also that if we compare

Table 3: GP Simulation Results Using the Symmetric Sigmoidal Transformation

Case		Simulation 1	2	3	4	5
1	$\overline{M}_a$	0.5917	0.5897	0.5910	0.5907	0.5901
	$\delta_{M,a}$	0.0118	0.0107	0.0091	0.0094	0.0122
	$\delta_{M^*,a}$	0.0124	0.0122	0.0102	0.0105	0.0133
1	$\overline{M}_b$	0.5958	0.5925	0.5936	0.5933	0.5946
	$\delta_{M,b}$	0.0037	0.0026	0.0033	0.0028	0.0021
	$\delta_{M^*,b}$	0.0038	0.0039	0.0038	0.0035	0.0023
2	$\overline{M}_a$	0.7406	0.7411	0.7415	0.7450	0.7447
	$\delta_{M,a}$	0.0061	0.0067	0.0069	0.0064	0.0064
	$\delta_{M^*,a}$	0.0061	0.0067	0.0069	0.0073	0.0071
2	$\overline{M}_b$	0.7394	0.7403	0.7399	0.7434	0.7431
	$\delta_{M,b}$	0.0033	0.0039	0.0040	0.0034	0.0024
	$\delta_{M^*,b}$	0.0039	0.0041	0.0043	0.0039	0.0029

$\overline{M}_a$ = the average of M_t of a simulation from Generation 1 to 1,000.
$\overline{M}_b$ = the average of M_t of a simulation from Generation 201 to 1,000.
$\delta_{M,a}$ = standard deviation about the M_a of a simulation from Generation 1 to 1,000.
$\delta_{M,b}$ = standard deviation about the M_b of a simulation from Generation 201 to 1,000.
$\delta_{M^*,a}$ = standard deviation about the *strict interior equilibrium* $1-(1/\omega)$ from Generation 1 to 1,000.
$\delta_{M^*,b}$ = standard deviation about the *strict interior equilibrium* $1-(1/\omega)$ from Generation 201 to 1,000.

the standard deviations, $\delta_{M,a}$ with $\delta_{M,b}$ or $\delta_{M^*,a}$ with $\delta_{M^*,b}$ for each simulation, we see that after the first 200 periods of learning, the stability of the mean in all of the GP-based coordination games improves.

A third result is that the chaotic trajectories for $\Gamma(3.86957)$ that are predicted by the myopic best response dynamic are *not apparent* in any of our simulations of Case 2. However, by comparing $\delta_{M,b}$ or $\delta_{M^*,b}$ across Case 1 and Case 2 in Table 3, we find that the standard deviations in Case 2 are generally somewhat larger than those in Case 1. Indeed, a rank order test reveals that $\delta_{M^*,b}$ is significantly larger in Case 2 as compared with Case 1 ($p \leq .10$).[14] This difference between the two cases is also apparent from a visual comparison between Figure 1. Thus, while it appears that the aggregate outcome from the GP simulations is similar for both treatments, there appears to be some evidence that the coordination problem in Case 2 is more difficult for our artificial players than is the coordination problem in Case 1.

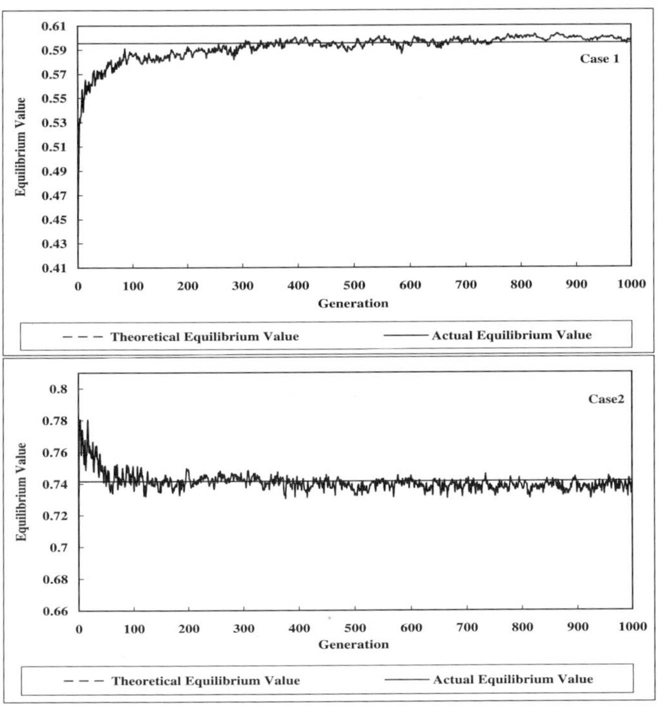

Figure 1: The Time Series of the Mean Choice of Action of the Coordination Game.

We have also considered how sensitive our results are to the use of payoff fitness as the main determinant of successive generations of forecast rules through

[14]No significant difference was found for $\delta_{M \cdot b}$ between Cases 1 and 2. See Siegel and Castellan (1988) for an explanation of the nonparametric, robust rank order test used here.

application of the reproduction, crossover and mutation operations. Recall that we made an adjustment to the raw fitness values, so as to avoid excluding rules with negative payoffs. In all our simulations, we used an adjustment factor of 0.25. We also performed a simulation exercise where we considered what happens when we used a much larger adjustment factor of 200.00. That is, we adjusted raw fitness values, $\pi_{i,t}$ as follows:

$$\mu_{i,t} = \begin{cases} \pi_{i,t} + 200.00 & \text{if } \pi_{i,t} \geq -200.00, \\ 0 & \text{if } \pi_{i,t} < -200.00. \end{cases}$$

The effect of this adjustment is to nullify the usefulness of fitness as an indicator of the relative success of individual forecast functions. That is because the raw, unadjusted fitness values, $\pi_{i,t}$ can only take on values in the range $[-.50, .50]$. (See the payoff function (1) and the parameterizations of this function given in Table 2.) Adding 200.00 to these raw fitness values makes them essentially indistinguishable from one another, even after the adjusted fitness values have been converted into the normalized fitness values that are used to determine application of the reproduction, crossover and mutation operations.

Thus, the experiment where the adjustment value is set at 200.00 rather than at 0.25 serves as a test of whether relative fitness values are the driving force behind the results reported above. Indeed, this experiment is a test of the explanatory power of GP techniques. Figure 2 presents the time series for the mean from the single experiment involving Case 1 where we set the adjustment factor equal to

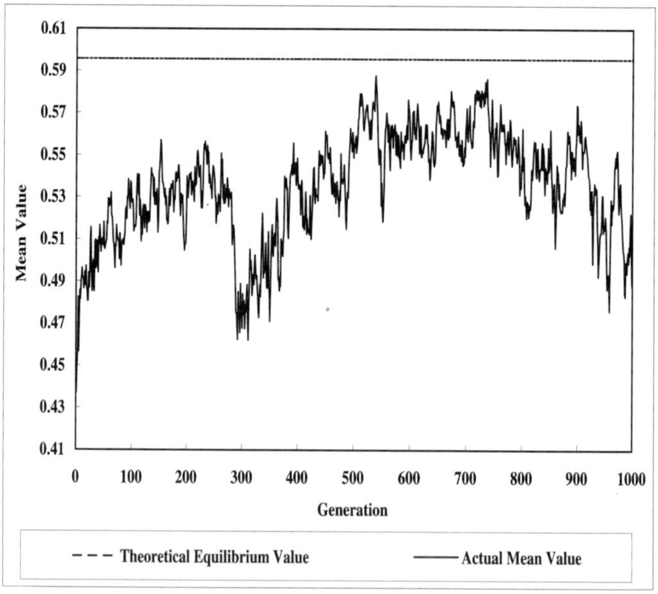

Figure 2: The Time Series of the Mean Choice of Action of their Coordination Game.

200.00 rather than 0.25. We see in this figure that the mean just wanders about randomly and has not settled down after 1,000 iterations. In particular, the mean does not approach either the corner equilibrium (0) or the interior equilibrium of Case 1 (0.59551). We may conclude from this exercise that the reliance of the genetic operators on relative fitness values is a driving force behind our simulation results, i.e. that fitness of forecast functions matters.

In addition to considering the dynamics of the mean choice of action, it is also interesting to examine the evolution of the population of forecast functions, i.e., GP_t. The *length* of the best-of-generation forecast function (Lisp S-expression) varies pretty widely. The length of a forecast function is measured by counting the number of elements (atoms) that are used in the program.[15] Programs with longer lengths are more complex than those with shorter lengths, so the length of the program serves as a measure of the complexity of the forecast rule. Initially, the length of the best-of-generation program is rather small, but over time, the length increases substantially.

Consider, Simulation 2.5 for example. (Simulation 2.5 corresponds to the fifth simulation of Case 2). The length of the shortest best-of-generation program (S-expression) in this simulation is 15 and it appears in generations 16, 23, 27, 29, 32, 35, 37, 40, 41 and 50:[16]

$$\begin{aligned}
gp_{\text{best},16} &= ((M_{t-5} + M_{t-4}) * M_{t-4}),\\
gp_{\text{best},23} &= ((M_{t-5} + M_{t-4}) * M_{t-4}),\\
gp_{\text{best},27} &= ((M_{t-5} + M_{t-3}) * M_{t-4}),\\
gp_{\text{best},29} &= (M_{t-3} * (M_{t-5} + M_{t-4})),\\
gp_{\text{best},32} &= ((M_{t-1} + M_{t-3}) * M_{t-3}),\\
gp_{\text{best},35} &= (M_{t-5} * (M_{t-5} + M_{t-4})),\\
gp_{\text{best},37} &= (M_{t-3} * (M_{t-4} + M_{t-5})),\\
gp_{\text{best},40} &= (M_{t-3} * (M_{t-4} + M_{t-5})),\\
gp_{\text{best},41} &= (M_{t-2} * (M_{t-5} + M_{t-5})),\\
gp_{\text{best},50} &= (M_{t-5} * (M_{t-4} + M_{t-5})),
\end{aligned}$$

Programs with such a small program length continue to appear frequently after generation 50 but they are no longer selected as best-of-generation programs. Instead, increasingly complicated programs with lengths over 200 are more likely to be selected as the best-of-generation; the longest best-of-generation program with a length of 459 appears in generation 554:

[15]The length of a Lisp S-expression is distinct from the *depth* of a Lisp S-expression in tree form.

[16]All the GP programs below are represented as algebraic expressions (so that they can be more easily understood) rather than in the Lisp S-expression form in which they are encoded for GP.

$$
\begin{aligned}
gp_{\text{best},554} = {} & (SinCos((CosCosCos(CosM_{t-3} * M_{t-3}) * CosSin(CosCos \\
& (Cos(CosM_{t-4}\%M_{t-4}) * Cos(M_{t-2} + SinSinM_{t-1})) + (M_{t-5} \\
& * CosM_{t-2}))) * CosCosCosSin((M_{t-3} * M_{t-3}) - CosM_{t-3})) \\
& \%CosSinCos(((Cos(CosM_{t-1} * Cos(ExpSinCosM_{t-4}*M_{t-2})) \\
& \%CosSinCosSin(M_{t-4}\%CosSinCosM_{t-1})) * CosSinRLog \\
& ((Sin(M_{t-4} * CosCosM_{t-1}) * Cos(M_{t-4}\%((CosM_{t-5} + M_{t-2}) \\
& * (CosM_{t-3} * M_{t-4})))) * Cos((Cos(M_{t-3} * CosM_{t-5}) * Cos \\
& CosM_{t-4}) * M_{t-1}))) * Cos(Sin(SinCos(CosSinCosM_{t-1} \\
& * (CosSinM_{t-2} * CosCosM_{t-4})) + ExpSinCosCosSin \\
& ((M_{t-2} * M_{t-3}) - CosM_{t-3})) * Cos((SinM_{t-3} + M_{t-5}) * \\
& ((M_{t-4} + (M_{t-3} + ((M_{t-2} * SinM_{t-2}) * M_{t-1})))\%CosM_{t-3})))))
\end{aligned}
$$

At generation 1,000 of simulation 2.5, the length of the individual programs was found to vary over the interval [3, 297]. This wide variation in program length implies that considerable heterogeneity remains in the population of forecast rules even after many generations.

Our findings that the best-of-generation programs become increasingly more complicated over time and that heterogeneity does not appear to diminish with time are perhaps attributable to our use of the symmetric sigmoidal activation function to map forecasts into the unit interval. The symmetric sigmoidal transformation function effectively "squashes" the output of forecast rules so that forecasts always lie within the unit interval. As a result, the forecasts of the various rules and their associated fitness values may not be all that distinct from one another even though the rules themselves may differ considerably. One consequence is that simple forecast rules, e.g. $gp_t = M_{t-i}, i = 1$ or 2, may be unable to effectively compete with more complicated rules (programs with longer lengths), since these more complicated rules are better able to differentiate themselves from the simpler rules after being squashed, and therefore, these more complicated rules stand a better chance of being chosen for reproduction than the simpler rules.

As an alternative to the symmetric sigmoidal transformation function, we also considered the performance of our GP–based learning algorithm when the simple truncated linear transformation (discussed above) is used in place of the symmetric sigmoidal transformation. The truncated linear transformation is essentially a *linear* mapping into the unit interval whereas the symmetric sigmoidal transformation comprises a nonlinear mapping. Thus, with the truncated linear transformation there is less "squashing" of forecasts and associated fitness values. Indeed, squashing only occurs for forecasts that exceed the bounds of the unit interval; forecasts that lie within the unit interval are unaltered, and therefore remain more distinct (in terms of fitness) than under the symmetric sigmoidal transformation.

We conducted 10 simulations using the *truncated linear transformation function* in place of the symmetric sigmoidal transformation function – 5 simulations of Case 1 and 5 simulations of Case 2.[17] Means and standard deviations from these 10 simulations are reported in Table 4. Here again, simulation 1.1 refers to our first simulation of Case 1, while simulation 2.1 refers to our first simulation of Case 2, and so on. Time series for the mean, M_t, from a single simulation of Case 1 and Case 2 are plotted in Figure 3. These time series plots are typical of the other simulations we conducted for the two cases using the truncated linear transformation.

Table 4: GP Simulation Results Using the Truncated Linear Transformation

Case		Simulation 1	2	3	4	5
1	$\overline{M}_a$	0.59281308	0.59279258	0.59328333	0.59295982	0.59249145
	$\delta_{M,a}$	0.02706819	0.02707828	0.02422021	0.02663368	0.02761726
	$\delta_{M^*,a}$	0.02720234	0.02721443	0.02432245	0.02675562	0.02778190
1	$\overline{M}_b$	0.59543389	0.59544433	0.59546691	0.59546368	0.59545463
	$\delta_{M,b}$	0.00033369	0.00033578	0.00025092	0.00028152	0.00029720
	$\delta_{M^*,b}$	0.00034227	0.00034215	0.00025460	0.00028531	0.00030232
2	$\overline{M}_a$	0.74079979	0.74081541	0.74069191	0.74081630	0.74080296
	$\delta_{M,a}$	0.01710630	0.01744866	0.01870916	0.01680324	0.01777976
	$\delta_{M^*,a}$	0.01712365	0.01746498	0.01872978	0.01682015	0.01779631
2	$\overline{M}_b$	0.74154575	0.74154649	0.74155158	0.74155041	0.74154058
	$\delta_{M,b}$	0.00043555	0.00045807	0.00044765	0.00048606	0.00050182
	$\delta_{M^*,b}$	0.00043622	0.00045867	0.00044803	0.00048645	0.00050268

$\overline{M}_a$ = the average of M_t of a simulation from Generation 1 to 1,000.
$\overline{M}_b$ = the average of M_t of a simulation from Generation 201 to 1,000.
$\delta_{M,a}$ = standard deviation about the M_a of a simulation from Generation 1 to 1,000.
$\delta_{M,b}$ = standard deviation about the M_b of a simulation from Generation 201 to 1,000.
$\delta_{M^*,a}$ = standard deviation about the *strict interior equilibrium* $1 - (1/\omega)$ from Generation 1 to 1,000.
$\delta_{M^*,b}$ = standard deviation about the *strict interior equilibrium* $1 - (1/\omega)$ from Generation 201 to 1,000.

From Table 4 and Figure 3, we see that our use of the truncated linear transformation in place of the symmetric sigmoidal transformation results in several significant differences. First, a comparison between Figures 3 and 1 and between the results in Tables 4 and 3 reveals that after 1,000 generations, the GP learning algorithm with the truncated linear transformation is generally closer to achieving the interior equilibrium than is the GP learning algorithm with the symmetric sigmoidal transformation. This quicker convergence to the interior equilibrium is

[17]Here we are using an adjustment factor of 0.25 once again to obtain adjusted fitness values.

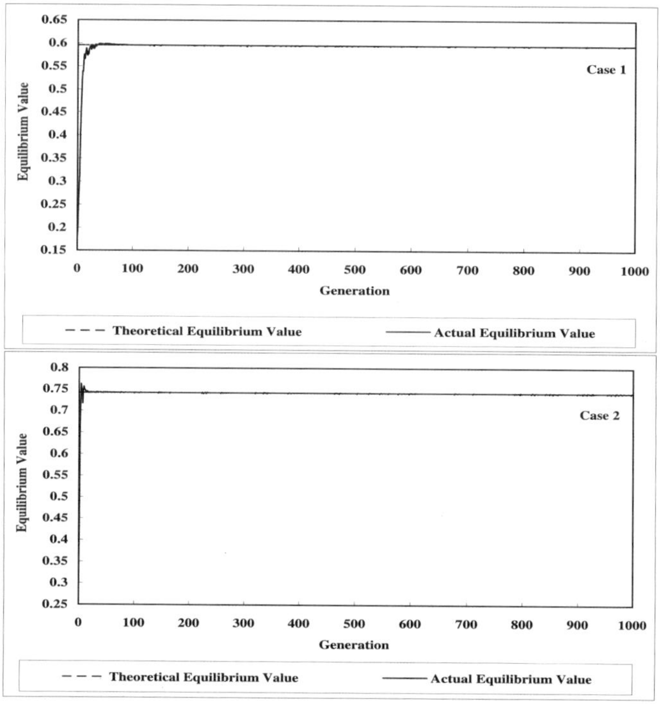

Figure 3: The Time Series of the Mean Choice of Action of the Coordination Game.

more consistent with the experimental findings of Van Huyck et al. (1994). Second, we observe that the deviation of the mean from the interior equilibrium, $1-(1/\omega)$, in both Case 1 and Case 2 is *much smaller* when we use the truncated linear transformation in place of the symmetric sigmoidal transformation; in all 10 simulations, the average of the means, M_t, from generation 201 to 1,000, i.e. $\overline{M}_b$, does not deviate from the interior equilibrium value, $1-(1/\omega)$ by more than 0.01%. This tighter distribution of forecasts around the interior equilibrium is again more consistent with the experimental findings of Van Huyck et al. (1994).[18]

As in the case of the symmetric sigmoidal transformation, we find that the chaotic trajectories for the game, $\Gamma(3.86957)$ that are predicted by the myopic best response dynamic are not apparent in any of our simulations of Case 2. We also find once again that a comparison of the standard deviations, $\delta_{M,b}$ or $\delta_{M^*,b}$ across Cases 1 and 2 in Table 4 reveals that these standard deviations are slightly larger in Case 2 than in Case 1. A rank order test confirms that both $\delta_{M,b}$ and $\delta_{M^*,b}$ are significantly larger in Case 2 than in Case 1 ($p \leq .01$ in both cases). This difference

[18]More direct comparisons between the experimental data and the simulated data from the GP-based learning algorithm are not really possible due to differences in the two experimental designs (e.g. the GP algorithm allows forecasts on the continuum of the unit interval, while the experimental subjects are limited to a finite set of discrete choices).

is also present, though difficult to see, in a visual comparison between Figure 3. We conclude that the coordination problem remains somewhat more difficult in Case 2 than in Case 1 regardless of whether we use truncated linear transformation or the symmetric sigmoidal transformation.

Finally, we note that under the truncated linear transformation, the length of the best-of-generation programs are considerably smaller than those discovered under the sigmoidal transformation. Simulation 2.2 (our second simulation of Case 2) is typical of the other simulation results we obtained using the truncated linear transformation. In this simulation, the longest best-of-generation program appeared in generation 2, and had a length of 29:

$$gp_{\text{best},2} = Exp((((M_{t-3} * M_{t-5}) - ExpM_{t-1}) * M_{t-1})$$
$$- (M_{t-2}\%Rlog(-2.66020 + M_{t-5}))).$$

Following generation 100 of simulation 2.2, no best-of-generation program had a length that exceeded 3. In fact, the best-of-generation programs after generation 100 were always of the simplest form:

$$gp_{\text{best},t>100} = M_{t-i}, \quad i = 1, 2, 3, 4 \text{ or } 5.$$

Given such simple forecast rules, it is easy to understand why the distribution of forecasts becomes more tightly concentrated around the interior equilibrium when we use the truncated linear transformation.

7 Summary and Conclusions

We have considered a simple coordination game where the actions of the individual players are modeled and updated using GP techniques. Our GP-based coordination game allows for a considerably more flexible experimental design than is possible in experiments with human subjects. In particular, we do not have to restrict the choice set to a finite set of discrete actions, and we can have large numbers of players, e.g. $n = 500$. Moreover, players in our genetic programming implementation are explicitly endowed with the ability to formulate a vast number of both linear and nonlinear forecasting rules for the mean, including the myopic best response rule. This more flexible design allows for a possibly dense set of periodic and chaotic trajectories for the mean for values of $\omega > 3$. Despite this more flexible design, the evolution of play in our GP-based coordination game remains quite similar to that observed in the experiments that Van Huyck et al. (1994) conducted with human subjects. The mean choice of action eventually settles down to a small neighborhood of the interior equilibrium, even in Case 2, where the myopic best response dynamic predicts that this interior equilibrium should be unstable. There is evidence, however, that the coordination problem that our artificial agents face in Case 2 is somewhat more difficult than the coordination problem they face in

Case 1, as indicated by the different standard deviations about the mean/interior equilibrium for these two cases.

While these results cast some doubt on the plausibility of the myopic best response dynamic as a selection criterion (or any other learning schemes that would predict the interior equilibrium to be unstable), it is not yet clear that the myopic best response dynamic should be rejected on the basis of a "bad" prediction for a single game, namely $\Gamma(3.86957)$, or that the alternative, inertial learning algorithm should be accepted as a plausible selection dynamic on the same basis. While the inertial learning dynamic predicts that the interior equilibrium is always stable, the predicted trajectory for the mean/median is much too smooth when compared with the same trajectory from the experimental data. Moreover, the notion that a single, representative-agent-type learning algorithm can be used to characterize the evolution of the mean/median is at odds with the initial heterogeneity that is apparent in the experimental subjects' actions. Finally, since our GP-based learning algorithm always "converges" to the interior equilibrium it is, by the criterion of Van Huyck et al. (1994), just as plausible a selection dynamic as the inertial learning algorithm. The initial heterogeneity of the forecasts that arise from our population-based GP algorithm makes it all the more plausible as a characterization of the experimental data.

We also note that the predictions of our GP-based learning model, especially those involving the truncated linear transformation, compare quite favorably with some new coordination game experiments that Van Huyck, Battalio and Rankin (1996) have recently conducted with human subjects. These new experiments differ from the previous experiments conducted by Van Huyck et al. (1994) in that subjects are not informed of the game's payoff function π; the only information available to subjects is their own past action/payoff history and the discrete action set that they may choose from. The purpose of this new experimental treatment is to place the human subjects in an environment that is as close as possible to that of artificial learning algorithms such as GP. In this new treatment, the human subjects learn to coordinate on the interior equilibrium *even more quickly* than in the previous treatment where subjects are informed of the payoff function π, of the game. Van Huyck et al. (1996) compare the experimental behavior in the new treatment with the behavior of a representative-agent-type, linear, stochastic reinforcement algorithm. While this algorithm eventually achieves a neighborhood of the interior equilibrium, it takes much longer to achieve this equilibrium (750 iterations) than it takes the experimental subjects. In contrast, our multi-agent GP-based learning algorithm converges much more quickly to a neighborhood of the interior equilibrium (usually within 50 iterations) so that it comes closer to mimicking the behavior of the experimental subjects.

Finally, we note that our findings for the coordination game are consistent with some other coordination experiments that have involved overlapping generations economies. Marimon, Spear and Sunder (1993) for example, report that experimental subjects are unable to coordinate on two-state sunspot equilibria, choosing instead to settle upon the steady state of an overlapping generations economy. Similarly, Bullard and Duffy (1998b) simulate behavior using a genetic algorithm–based learning model in an overlapping generations economy and find that their population of artificial agents is able to eventually coordinate on steady state and low-order cycles for inflation rates but not on the higher order periodic equilibria of their model. This paper extends these earlier findings by suggesting that it may not be possible for agents to coordinate on aperiodic, *chaotic* trajectories.

REFERENCES

[1] Allen F. and Karjalainen R., 1999. Using Genetic Algorithms to Find Technical Trading Rules. *Journal of Financial Economics*, 51(2), 245–71.

[2] Angeline P., (1994). Genetic Programming and Emergent Intelligence. Chapter 4 of Kinnear (1994).

[3] Arifovic J., (1994). Genetic Algorithm Learning and the Cobweb Model. *Journal of Economic Dynamics and Control*, 18, 3–28.

[4] Arifovic J., (1995). Genetic Algorithms and Inflationary Economies. *Journal of Monetary Economics*, 36, 219–243.

[5] Arifovic J., (1996). The Behavior of the Exchange Rate in the Genetic Algorithm and Experimental Economies. *Journal of Political Economy*, 104, 510–541.

[6] Arifovic J., (1997). Strategic Uncertainty and the Genetic Algorithm Adaptation. In H. Amman et al. (eds.), *Computational Approaches to Economic Problems*. Boston: Kluwer Academic Press, 225–36.

[7] Arthur W. B., Holland J. H., LeBaron B., Palmer R. and Tayler P., (1997). Asset Pricing Under Endogenous Expectations in an Artificial Stock Market. In W.B. Arthur et al. (eds.), *The Economy as a Evolving Complex System II*. Reading, MA: Addison-Wesley, 15–44.

[8] Birchenhall C. R., (1995). Genetic Algorithms, Classifier Systems and Genetic Programming and Their Use in Models of Adaptive Behavior and Learning. *Economic Journal*, 105, 788–795.

[9] Bray M., (1982). Learning, Estimation, and the Stability of Rational Expectations. *Journal of Economic Theory*, 26, 318–339.

[10] Bullard J. and Duffy J., (1998a). A Model of Learning and Emulation with Artificial Adaptive Agents. *Journal of Economic Dynamics and Control*, 22, 179–207.

[11] Bullard J. and Duffy J., (1998). On Learning and the Stability of Cycles. *Macroeconomic Dynamics*, 2, 22–48.

[12] Chen S. and Yeh C., (1997a). Toward a Computable Approach to the Efficient Market Hypothesis: An Application of Genetic Programming. *Journal of Economic Dynamics and Control*, 21, 1043–1063.

[13] Chen S. and Yeh C., (1997b). On the Coordination and Adaptability of the Large Economy: An Application of Genetic Programming to the Cobweb Model. In P. Angeline and K.E. Kinnear, Jr., (eds.), *Advances in Genetic Programming II*, Chapter 22. Cambridge, MA: MIT Press.

[14] Chen S., Duffy J. and Yeh C., (1996). Genetic Programming in the Coordination Game with a Chaotic Best-Response Function. In: *Proceedings of the 1996 Evolutionary Programming Conference*, San Diego, CA.

[15] Cooper R., DeJong D., Forsythe R. and Ross T., (1990). Selection Criterion in Coordination Games: Some Experimental Results. *American Economic Review*, 80, 218–233.

[16] Crawford V. P., (1991). An Evolutionary Interpretation of Van Huyck, Battalio and Beil's Experimental Results on Coordination. *Games and Economic Behavior*, 3, 25–59.

[17] Crawford V. P., (1995). Adaptive Dynamics in Coordination Games. *Econometrica*, 63, 103–143.

[18] Dawid H., (1996), *Adaptive Learning by Genetic Algorithms*, Lecture Notes in Economics and Mathematical Systems No. 441. New York: Springer.

[19] Devaney R. L., (1989). *An Introduction to Chaotic Dynamical Systems*, 2nd Ed.. Reading, MA: Addison–Wesley.

[20] Dworman G., Kimbrough S. O. and Laing J. D., (1996). On Automated Discovery of Models Using Genetic Programming: Bargaining in a Three–Agent Coalitions Game. *Journal of Management Information Systems*, 12, 97–125.

[21] Goldberg D. E., (1989). *Genetic Algorithms in Search, Optimization and Machine Learning*. Reading, MA: Addison–Wesley.

[22] Holland J. H., (1975). *Adaptation in Natural and Artificial Systems: An Introductory Analysis with Applications to Biology, Control, and Artificial Intelligence*. Ann Arbor: University of Michigan Press.

[23] Kagel J. H. and Roth A. E., (1995), eds., *Handbook of Experimental Economics*. Princeton, NJ: Princeton University Press.

[24] Kinnear K. E., Jr., (1994), (ed.) *Advances in Genetic Programming*. Cambridge, MA: MIT Press.

[25] Koza J. R., (1992). *Genetic Programming*. Cambridge, MA: MIT Press.

[26] Koza J. R., (1994). *Genetic Programming II*. Cambridge, MA: MIT Press.

[27] Kreps D. M., (1990). *Game Theory and Economic Modelling*. New York: Oxford University Press.

[28] Lucas R. E., Jr., (1986). Adaptive Behavior and Economic Theory. *Journal of Business*, 59, S401–S426.

[29] Marcet A. and Sargent T. J., (1989). Convergence of Least Squares Learning Mechanisms in Self Referential Linear Stochastic Models. *Journal of Economic Theory*, 48, 337–368.

[30] Marimon R., (1997). Learning From Learning in Economics. In D.M. Kreps and K.F. Wallis, eds., *Advances in Economics and Econometrics: Theory and Applications*, Vol. 1, Seventh World Congress, Econometric Society Monographs, No. 26, Cambridge: Cambridge University Press.

[31] Marimon R., Spear S. E. and Sunder S., (1993). Expectationally Driven Market Volatility: An Experimental Study. *Journal of Economic Theory*, 61, 74–103.

[32] Miller J. H., (1996). The Coevolution of Automata in the Repeated Prisoner's Dilemma. *Journal of Economic Behavior and Organization*, 29, 87–112.

[33] Mitchell M., (1996). *An Introduction to Genetic Algorithms*. Cambridge, MA: MIT Press.

[34] Neely C. J., Weller P. and Dittmar. R., (1997) Is Technical Analysis in the Foreign Exchange Market Profitable?: A Genetic Programming Approach. *Journal of Financial and Quantitative Analysis*, 32, 405–26.

[35] Sargent T. J., (1993), *Bounded Rationality in Macroeconomics*. New York: Oxford University Press.

[36] Siegel S. and Castellan N. J., Jr., (1988). *Nonparametric Statistics for the Behavioral Sciences*, 2nd Ed. New York: McGraw Hill.

[37] Tesfatsion L., (1997). A Trade Network Game with Endogenous Partner Selection. In H. Amman et al. (eds.), *Computational Approaches to Economic Problems*. Boston: Kluwer, 249-269.

[38] Van Huyck J. B., Battalio R. C. and Beil R., (1990). Tacit Coordination Games, Strategic Uncertainty and Coordination Failure. *American Economic Review*, 80, 234-248.

[39] Van Huyck J. B., Battalio R. C. and Beil R., (1991). Strategic Uncertainty, Equilibrium Selection Principles and Coordination Failure in Average Opinion Games. *Quarterly Journal of Economics*, 106, 885–910.

[40] Van Huyck J. B., Cook., J. P. and Battalio R .C., (1994). Selection Dynamics, Asymptotic Stability, and Adaptive Behavior. *Journal of Political Economy*, 102, 975–1005.

[41] Van Huyck J. B., Battalio., R. C. and Rankin F. W., (1996). Selection Dynamics and Adaptive Behavior Without Much Information. Working paper, Texas A&M University.

Two Issues Surrounding Parrondo's Paradox

Andre Costa
School of Applied Mathematics
University of Adelaide
Australia
acosta@maths.adelaide.edu.au

Mark Fackrell
Department of Mathematics and Statistics
University of Melbourne
Australia
mfackrel@ms.unimelb.edu.au

Peter G. Taylor
Department of Mathematics and Statistics
University of Melbourne
Australia
p.taylor@ms.unimelb.edu.au

Abstract

In the original version of Parrondo's paradox, two losing sequences of games of chance are combined to form a winning sequence. The games in the first sequence depend on a single parameter p, while those in the second depend on two parameters p_1 and p_2. The paradox is said to occur because there exist choices of p, p_1 and p_2 such that the individual sequences of games are losing but a sequence constructed by choosing randomly between the games at each step is winning.

At first sight, such behavior seems surprising. However, we contend in this paper that it should not be seen as surprising. On the contrary, we show that such behaviour is typical in situations in which we randomly create a sequence from games whose winning regions can be defined on the same parameter space.

Before we discuss this issue, we investigate in some detail the issue of when sequences of games, such as those proposed by Parrondo, should be considered to be winning, losing or fair.

1 Introduction

Different versions of Parrondo's paradox have been discussed in a number of papers [1–5,9,10]. The paradox is said to occur when two losing sequences of games of chance, say sequence A and sequence B, are combined in such a way that the resulting sequence is winning.

In the original version of Parrondo's paradox,

- sequence A consists of repetitions of a toss of a biased coin, say coin 0, which has probability p of winning, and
- in sequence B, one of two biased coins is tossed depending on the player's capital. Coin 1, with probability p_1 of winning, is tossed if the capital is a multiple of a fixed integer $M \geq 2$. Otherwise coin 2, with probability p_2 of winning, is tossed.

We can construct a third sequence, say sequence C, by choosing randomly between the coins associated with sequences A and B. Thus

- in sequence C, coin 0 is chosen with probability γ. If the capital is a multiple of M then coin 1 is chosen with probability $(1 - \gamma)$ and if the capital is not a multiple of M coin 2 is chosen with probability $(1 - \gamma)$.

It was established in [2] that there are choices of p, p_1, p_2 and γ such that sequences A and B are both losing, but sequence C is winning. This behavior, which at first sight seems surprising, has been termed *Parrondo's paradox*.

There are a number of controversial issues associated with Parrondo's paradox. One revolves around whether the activities described above should be called games at all, since they involve no strategic decisions on behalf of the player. A second, and more important, issue concerns the definition of a "winning", "losing" or "fair" sequence. In [2], it was argued that the issue of fairness should be addressed by considering the properties of a birth-and-death process associated with each sequence of games, where the current capital takes values on the set of non-negative integers and either increases or decreases by one unit each time a game is played. A sequence was defined as winning if the associated birth-and-death process is transient, fair if this birth-and-death process is null-recurrent and losing if this birth-and-death process is positive-recurrent. It has become clear to the authors, however, that this definition is not accepted by all researchers in the area. For this reason, we shall devote the first part of this paper to a discussion of this point and, in particular, to an argument as to why a generalization of the above definition is reasonable.

The second point that we wish to make about Parrondo's paradox is that it is not surprising at all. In fact we shall argue in Section 3 that whenever we have a pair of sequences of "games" for which we can define winning and losing regions in terms of the same set of parameters, it is to be expected that there exist parameter values such that a sequence constructed by randomly choosing between the original games at each step can be winning (losing) when each of the original sequences are losing (winning). This, our main result, is stated precisely in Theorem 3.1.

The paper concludes in Section 4 with an example of our reasoning involving a history-dependent version of Parrondo's game.

2 The Definition of Fairness

A classical definition of fairness was given by Doob [11]. He considered the situation in which a gambler repeatedly plays a game and where the gambler's fortune after the nth game is X_n. On page 299 of [11], he stated that it is natural to say that the game is fair if, for $n \in \mathbb{Z}_+$,

$$\mathbf{E}(X_{n+1}|X_0, X_1, \ldots X_n) = X_n.$$

In other words, assuming also the uniform integrability condition $\mathbf{E}(|X_n|) < \infty$ for all n, a game is fair if the stochastic process arising from its repeated play is a martingale. The optional stopping theorem (see Theorem 2.1 in [11]) then implies that $\mathbf{E}(X_\tau) = X_0$ for any stopping time τ. Furthermore, the optional sampling theorem (see Theorem 2.2 in [11]) implies that if we take any increasing sequence of stopping times $\tau_0, \tau_1, \tau_2, \ldots$, then the process $Y_n \equiv X_{\tau_n}$ is also a martingale. Corresponding definitions can be used to state that a game is winning if the stochastic process arising from its repeated play is sub-martingale, and losing if this stochastic process is a super-martingale. Note, however, that Doob observed that the above definition "is somewhat arbitrary, although hallowed by tradition."

Using Doob's definition, when $p_1 < 1/2$ and $p_2 > 1/2$, sequence B defined above does not correspond to a fair, winning or losing game because, when X_0 is a multiple of M, $\mathbf{E}(X_1|X_0) < X_0$ and, when X_0 is not a multiple of M, $\mathbf{E}(X_1|X_0) > X_0$. Thus the stochastic process $\{X_n\}$ is not a martingale, a super-martingale or a sub-martingale.

Given this, we might be tempted to say that it does not make sense to talk of sequences such as sequence B as being fair, winning or losing. However, this is likely to be unsatisfactory in many situations. Consider Figure 1 which appeared in [1]. The "irregular" graphs give the distribution of capital which occurred when 10,000 repetitions of sequence B played for one hundred steps in succession were simulated with parameters $p_1 = 1/10 - \epsilon$ and $p_2 = 3/4 - \epsilon$. The authors think it is reasonable to say that these results would lead most people to consider that sequence B is "fair", at least in some sense, when $\epsilon = 0$, losing when $\epsilon = 0.1$ and winning when $\epsilon = -0.1$. Our challenge is to come up with a mathematical definition that reflects this.

In [2], the authors proposed that a sequence E of games whose winning probabilities depend just on the current capital, is fair or losing if an associated birth-and-death process on $\mathbb{Z}_+$ is recurrent, and winning if it is transient. The state of this birth-and-death process is the player's current capital and the transition probabilities for $j \geq 1$ are given by

$$p_{j,j+1} = P(win|\text{the current capital is } j), \tag{1}$$

$$p_{j,j-1} = P(lose|\text{the current capital is } j), \tag{2}$$

with $p_{0,1}$ taken to be greater than zero and $p_{j,k} = 0$ otherwise.

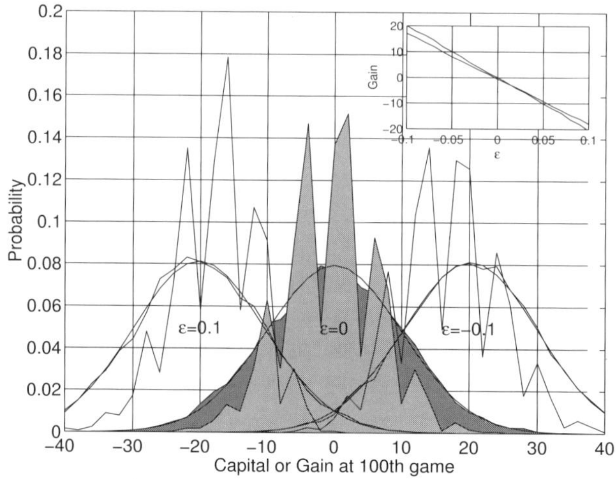

Figure 1: Results of 10,000 simulations of 100 successive games.

By applying the same definition to the sequence E' seen from the point of view of the opposition player, in which the winning and losing probabilities for each state are interchanged, we see that sequence E is fair if the birth-and-death process defined in (1) and (2) is null-recurrent, losing if it is positive-recurrent and winning if it is transient.

For sequences of games, such as sequence B above, which have a periodic structure to their transition probabilities, this can be equivalently stated in terms of the doubly-infinite birth-and-death process which is defined by extending the state space to the set $\mathbb{Z}$ of all integers. Specifically, the sequence is fair if the birth-and-death process with transition probabilities given by

$$p_{j,j+1} = P(win|\text{the current capital is} j), \tag{3}$$

$$p_{j,j-1} = P(lose|\text{the current capital is} j), \tag{4}$$

for all $j \in \mathbb{Z}$ is null-recurrent (note that, with non-zero win and loss probabilities, such a process cannot be positive-recurrent), winning if it is transient with drift towards ∞ and losing if it is transient with drift towards $-\infty$.

We believe that this definition is a natural one: over many repetitions of the game, the capital in a winning game will "drift" off to infinity with probability one and in a losing game it will "drift" off to negative infinity with probability one. In contrast, over many repetitions of a fair game, the capital will return to its starting place with probability one.

In support of this definition, we can also observe that sequence B defined above has an embedded martingale when the corresponding doubly-infinite birth-and-death process is null-recurrent, an embedded super-martingale when the birth-and-death process is transient with drift to $-\infty$ and an embedded submartingale when the birth-and-death process is transient with drift to ∞. To see this, observe first that methods similar to those in [2] can be used to show that the doubly-infinite birth-and-death process corresponding to sequence B is null-recurrent, transient with drift to ∞ and transient with drift to $-\infty$ according as

$$\frac{p_1 p_2^{M-1}}{(1-p_1)(1-p_2)^{M-1}} \tag{5}$$

is equal to, greater than or less than one.

Now consider the stochastic process constructed from the repeated play of game B by observing the value of the capital X_n only at the random times $\tau_0, \tau_1, \tau_2, \ldots$ when it is a multiple of the fixed integer M.

We can find the transition probabilities for this embedded process by considering the values of the capital in the restricted set of integers $\mathcal{N} = \{M(k-1), \ldots, M(k+1)\}$ with $M(k-1)$, Mk and $M(k+1)$ considered as absorbing states. Letting f_j be the probability that the process is absorbed in state $M(k+1)$ given that the capital starts in state j, some elementary calculation gives us that $f_{Mk} = Kp_1 p_2^{M-1}$ where K is a constant. Similarly, denoting by g_j the probability of being absorbed in state $M(k-1)$ given that the capital starts in state j we derive $g_{Mk} = K(1-p_1)(1-p_2)^{M-1}$.

It follows that

$$\begin{aligned}&\mathbf{E}(X_{\tau_{j+1}} | X_{\tau_j} = Mk)\\ &= K(p_1 p_2^{M-1} M(k+1) + (1-p_1)(1-p_2)^{M-1} M(k-1)) + LMk,\end{aligned} \tag{6}$$

where $L = 1 - Kp_1 p_2^{M-1} - K(1-p_1)(1-p_2)^{M-1}$. The right hand side of (6) is equal to, greater than or less than Mk according as (5) is equal to, greater than or less than one, which proves our assertion.

It is clear that $\mathbf{E}(|X_{\tau_j}|) < \infty$, for all j. Thus we have shown that, even though the repeated play of the games in sequence B with (5) equal to one does not correspond to a martingale, there is an optional sampling scheme under which the sampled process is a martingale. We believe that this supports the assertion that sequence B with these parameters can be considered to be fair. Similarly, it can be observed that if (5) is greater than one then there is an optional sampling scheme which is a sub-martingale and if (5) is less than one then there is an optional sampling scheme which is a super-martingale, supporting the assertions that the sequence should be considered winning and losing in these situations respectively.

Some sequences of interest do not have transition probabilities which depend only on the values of the current capital. An example is the sequence that we consider in Section 4 below in which the winning probabilities depend on the outcomes of the previous two plays. A further example has recently appeared in [8]. These sequences cannot be modeled by a simple birth-and-death process. They can, however, be modeled by a quasi-birth-and-death process, which we can think of as a Markov chain on a two-dimensional state space, one dimension of which is the non-negative integer $\mathbb{Z}_+$ representing the current capital and the second dimension of which contains the auxiliary information upon which the transition probabilities depend.

In these cases we believe that it is reasonable to extend the above criterion so that a sequence is considered winning if the corresponding quasi-birth-and-death process is transient, losing if the corresponding quasi-birth-and-death process is positive-recurrent and fair if the corresponding quasi-birth-and-death process is null-recurrent. We use this criterion in our analysis of the sequence in Section 4.

3 The Ubiquity of Parrondo's Paradox

A key observation about Parrondo's sequences A and B is that sequence A is a version of sequence B with $p_1 = p_2 = p$. Therefore, when we randomly choose between games at each step of sequences A and B, we are really choosing between different parametrizations of the game in sequence B. Furthermore, as was noted in [2], we can think of sequence C as another parametrization of sequence B with parameters $q_1 = \gamma p + (1-\gamma)p_1$ and $q_2 = \gamma p + (1-\gamma)p_2$. Thus, all three sequences are versions of sequence B, but with different values of the parameters.

Because of this, we can study the situation by examining the regions in parameter space where sequence B is winning, losing or fair, calculated according to equation (5). For the case $M = 3$, these regions are depicted in Figure 2. The line parametrized by $(x_1, x_2) = (p, p)$, for which the two winning probabilities are the same, corresponds to sequence A. For any point (p_1, p_2), the point $(\gamma p + (1-\gamma)p_1, \gamma p + (1-\gamma)p_2)$ corresponds to sequence C, whose parameters are a convex combination of those of sequence A with parameter p and sequence B with parameters p_1 and p_2.

Thus, to demonstrate the existence of Parrondo's paradox, all we have to do is find points (p, p) and (p_1, p_2) which are in the losing region of sequence B with the line between them crossing into the winning region. Due to the local concavity of the losing region, the two points (p, p) and (p_1, p_2) depicted in Figure 2 can be seen to have this property. Note also that it is not necessary for one of the points to be of the form (p, p): we can easily demonstrate a Parrondo's paradox between different parametrizations (p_1, p_2) and $(\hat{p}_1, \hat{p}_2)$ of sequence B, neither of which correspond to a parametrization of sequence A.

The above observations suggest a generalization. Consider a discrete-time stochastic process D whose transition probabilities depend on the values of a number of real parameters $\{x_1, \ldots, x_s\}$. Let $\mathcal{X}$ denote the subset of $\mathbb{R}^s$ consisting of those values of $\{x_1, \ldots, x_s\}$ which are consistent with the rules of the sequence and suppose that $\mathcal{X}$ can be partitioned by an $s-1$ dimensional manifold $\mathcal{F}$ into winning and losing regions $\mathcal{W}$ and $\mathcal{L}$ respectively. Points on $\mathcal{F}$ correspond to parameter values where the sequence is fair.

Let $\{p_1, \ldots, p_s\}$ and $\{q_1, \ldots, q_s\}$ be two parametrizations of the process D. Construct a new stochastic process as follows: at each time point n choose the parametrization $\{p_1, \ldots, p_s\}$ with probability γ and the parametrization

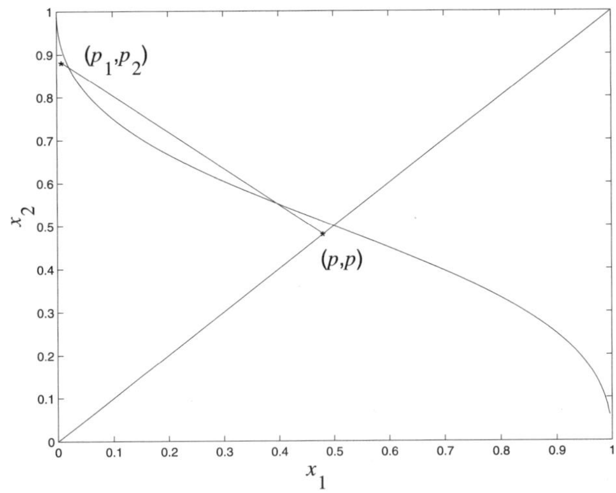

Figure 2: Plot of the curves $x_2 = x_1$ and $F(x_1, x_2) = x_1 x_2^2/(1 - x_1)(1 - x_2)^2 - 1 = 0$. Points (x_1, x_2) lying on the curve F correspond to parameters which give a fair instance of sequence B, whereas points (x_1, x_2) above and below the curve correspond to parameters which give winning and losing instances of sequence B respectively.

$\{q_1, \ldots, q_s\}$ with probability $1 - \gamma$ and then select the next state according to the transition probabilities of the process using the chosen parametrization. The state of this *mixed process* will evolve according to the process D with parametrization $\{\gamma p_1 + (1 - \gamma)q_1, \ldots, \gamma p_s + (1 - \gamma)q_s\}$. This corresponds to a point on the line between $\{p_1, \ldots, p_s\}$ and $\{q_1, \ldots, q_s\}$ in $\mathbb{R}^s$.

It follows that if we can find two sets of parameters $\{p_1, \ldots, p_s\}$ and $\{q_1, \ldots, q_s\}$ in $\mathcal{L}$ such that the line between them crosses into $\mathcal{W}$, then we shall have constructed a winning mixed parametrization from two losing parametrizations: in short we shall have demonstrated an instance of Parrondo's paradox. It is always possible to find two such points if there is a portion of $\mathcal{L}$ which is concave. Similarly, we can construct the reverse version of Parrondo's paradox in which a mixture of two winning parametrizations is losing whenever there is a portion of $\mathcal{W}$ which is concave. Since the only situations where neither $\mathcal{L}$ nor $\mathcal{W}$ has regions of concavity occurs when one of them is empty or when $\mathcal{F}$ is a hyperplane, this is the only situation in which a version of Parrondo's paradox *cannot* occur.

We thus have the following theorem.

Theorem 3.1. *Consider a process which has the form of the stochastic process D described above. Then, provided the regions $\mathcal{L}$ and $\mathcal{W}$ are non-empty and the region $\mathcal{F}$ is not a hyperplane, it is possible to demonstrate a Parrondo's paradox.*

Since only the simplest sequences of games generate a process for which the region $\mathcal{F}$ is a hyperplane, we believe that this theorem demonstrates that the existence of Parrondo's paradox should be regarded as the norm rather than an exception.

4 An Example

Using Figure 2, we have already demonstrated how the original version of Parrondo's paradox fits into the framework defined above. In this section we shall give another example which involves a process D in which the winning probabilities depend not on whether the current capital is a multiple of M or not, but rather on whether the results of the last two plays have been wins or losses. Specifically we shall assume that the player will win a single play of sequence D with probability x_1 if their previous two plays have been wins, and will win with probability x_2 otherwise. By taking $x_1 < x_2$ in this game, we can model the type of dependency that many gamblers erroneously assume is present in sequences of independent plays; that a player is less likely to win after a sequence of wins.

The easiest way in which to analyze this sequence D, is to investigate the properties of the associated quasi-birth-and-death process (see Neuts [7]) which arises when the game is played repeatedly.

A quasi-birth-and-death process, or QBD, is a discrete-time Markov chain on the state space $S = \{(k, i), k \geq 0, 1 \leq i \leq m\}$. The two dimensions of the state space are called the *level*, indexed by k, and the *phase*, indexed by i. The Markov process associated with a QBD has a transition matrix of the form

$$Q = \begin{bmatrix} B_0 & A_0 & & & & \\ A_2 & A_1 & A_0 & & & \\ & A_2 & A_1 & A_0 & & \\ & & A_2 & A_1 & A_0 & \\ & & & A_2 & A_1 & \ddots \\ & & & & \ddots & \ddots \end{bmatrix}. \tag{7}$$

The rows and columns of Q are indexed by the positive integers $\{0, 1, 2, \ldots\}$ and correspond to the level. The sub-matrices A_ℓ and B_ℓ within Q have row and column indices which correspond to the different phases.

The matrix Q defines a process for which transitions occur only between adjacent levels of the state space, hence it has a block tri-diagonal structure. The matrix A_0 governs transitions leading to an increase in the level, whereas A_2 governs transitions leading to a decrease in the level and A_1 governs state transitions for which there is no change in level. The matrix B_0 is a special case of A_1, which allows for different transition probabilities at the boundary level $k = 0$.

Let $A = A_0 + A_1 + A_2$. Then A is a stochastic matrix with stationary probability vector $\boldsymbol{\alpha}$ given by the solution to

$$\boldsymbol{\alpha} A = \boldsymbol{\alpha}, \tag{8}$$

with $\boldsymbol{\alpha e} = 1$. For m finite, it is a consequence of Latouche and Ramaswami [6] (Theorem 7.2.3) that, a QBD defined by matrices A_0, A_1 and A_2 is positive-recurrent, null-recurrent or transient according as

$$f = \boldsymbol{\alpha} A_0 \boldsymbol{e} - \boldsymbol{\alpha} A_2 \boldsymbol{e}, \tag{9}$$

is less than 0, equal to 0 or greater than 0.

We can model the process D with a QBD by letting the level correspond to the current capital and the phase to the outcomes of the previous two plays, labelled (l, l), (l, w), (w, l) and (w, w) with l and w representing a loss and win respectively. The current state of a player is thus denoted by (k, i), where k is the current capital and i represents an element of the set $\{(l, l), (l, w), (w, l), (w, w)\}$.

With this representation of state, the matrix $A_1 = 0$ reflecting the fact that the player's capital cannot stay the same after a single play of the game and the matrices A_0 and A_2 are given by

$$A_0 = \begin{bmatrix} 0 & x_2 & 0 & 0 \\ 0 & 0 & 0 & x_2 \\ 0 & x_2 & 0 & 0 \\ 0 & 0 & 0 & x_1 \end{bmatrix}$$

and

$$A_2 = \begin{bmatrix} 1-x_2 & 0 & 0 & 0 \\ 0 & 0 & 1-x_2 & 0 \\ 1-x_2 & 0 & 0 & 0 \\ 0 & 0 & 1-x_1 & 0 \end{bmatrix}.$$

The vector $\boldsymbol{\alpha}$ is found by solving the equation

$$\boldsymbol{\alpha}(A_0 + A_2) = \boldsymbol{\alpha}. \tag{10}$$

and the function f is

$$\begin{aligned} f(x_1, x_2) &= \boldsymbol{\alpha} A_0 \boldsymbol{e} - \boldsymbol{\alpha} A_2 \boldsymbol{e} \\ &= \frac{x_2^2 - x_1 x_2 + x_2 + x_1 - 1}{x_2^2 - x_1 x_2 + x_2 - x_1 + 1}. \end{aligned}$$

The sign of $f(x_1, x_2)$ tells us whether the point (x_1, x_2) is in the winning region $\mathcal{W}$ ($f > 0$), the losing region $\mathcal{L}$ ($f < 0$) or the fair region $\mathcal{F}$ ($f = 0$). For $(x_1, x_2) \in \mathcal{X} = [0, 1]^2$, the regions $\mathcal{W}$, $\mathcal{L}$ and $\mathcal{F}$ are shown in Figure 3.

It is easy to see that the region $\mathcal{F}$ is not a hyperplane. In fact, the region $\mathcal{L}$ is convex. Thus, by the arguments leading to Theorem 3.1, it is possible to choose points $(x_1, x_2) = (p, p)$, $(x_1, x_2) = (p_1, p_2)$ and a mixing parameter γ to produce a Parrondo's paradox for process D in which two winning sequences are mixed to produce a losing sequence. Examples of parameter sets where this occurs are $p = 51/101$, $p_1 = 1/100$, $p_2 = 5/8$ and $\gamma = 1/2$, and $p = 51/101$, $p_1 = 99/100$, $p_2 = 1/4$ and $\gamma = 1/2$. The fact that $\mathcal{L}$ is convex shows also that, for process D, it is not possible to produce a Parrondo's paradox in which two losing sequences are mixed to produce a winning sequence.

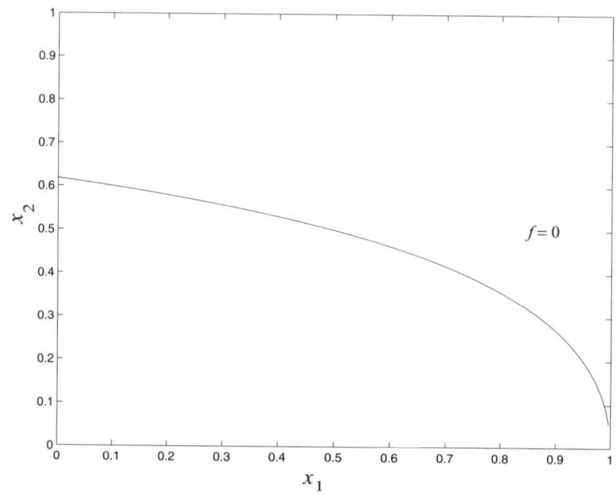

Figure 3: The regions $\mathcal{W}$, $\mathcal{L}$ and $\mathcal{F}$ for process D. The region $\mathcal{F}$ is given by the curve $f = 0$. The regions $\mathcal{W}$ and $\mathcal{L}$ lie above and below the curve $f = 0$, respectively.

5 Conclusion

In this paper we have discussed two issues of relevance to the study of Parrondo's paradox. The first of these concerns the definition of a fair sequence of games. We have argued that a reasonable definition is that the sequence is winning, losing or fair according as the associated birth-and-death process, or quasi-birth-and-death process, denoting the player's capital is transient, positive recurrent or null recurrent respectively.

Second, we have observed that it is possible to construct a Parrondo's paradox for a process which depends on s parameters whenever the $(s - 1)$ dimensional manifold $\mathcal{F}$ which separates the winning and losing regions is not a hyperplane. Since $\mathcal{F}$ is a hyperplane for only very simple processes, this means that we should expect Parrondo's paradox to be ubiquitous.

Acknowledgement

The authors would like to thank two anonymous referees whose helpful suggestions greatly improved the paper.

REFERENCES

[1] Harmer G. P. and Abbott D., (1999) Parrondo's paradox. *Statistical Science* **14,** 206–213.

[2] Harmer G. P., Abbott D. and Taylor P. G., (2000) The paradox of Parrondo's games. *Proceedings of the Royal Society of London: Series A*, **456,** 247–259.

[3] Harmer G. P., Abbott D., Taylor P. G. and Parrondo J. M. R., (2000) Parrondo's paradoxical games and the discrete Brownian ratchet. *Proceedings of the Second International Conference on Unsolved Problems of Noise and Fluctuations, UPoN'99*, University of Adelaide, July 12-16, 1999 (eds. D. Abbott & L.B. Kish), American Institute of Physics, Melville, NY, USA, **511** 2000, 189–200.

[4] Harmer G. P., Abbott D., Taylor P. G., Pearce C. E. M. and Parrondo J. M. R., (2000) Information Entropy and Parrondo's Discrete-Time Ratchet, In *Stochastic and Chaotic Dynamics in the Lakes (Stochaos)* (eds. D.S. Broomhead, E.A. Luchinskaya, P.V.E. McClintock and T. Mullin), American Institute of Physics, Melville, NY, USA, **502** 2000, 544–549.

[5] Key E. S., Kosek M. M. and Abbott D., (2000) On Parrondo's paradox: how to construct unfair games by composing fair games. Preprint. math.PR/0206151.

[6] Latouche G. and Ramaswami V., (1999) *Introduction to Matrix Analytic Methods*. ASA-SIAM.

[7] Neuts M. F., (1981) *Matrix-Geometric Solutions in Stochastic Models*. Dover, New York.

[8] Parrondo, Juan M.R., Harmer G.P. and Abbott, Derek (2000) New paradoxical games based on Brownian ratchets. *Physical Review Letters*, **85**, 5526–5529.

[9] Pearce C. E. M., (2000) Entropy, Markov information sources and Parrondo's games. *Proceedings of the Second International Conference on Unsolved Problems of Noise and Fluctuations, UPoN'99*, University of Adelaide, July 12-16, 1999 (eds. D.A. Abbott & L.B. Kish), American Institute of Physics, Melville, NY, USA, **511** 2000, 207–212.

[10] Pearce C. E. M., (2000) On Parrondo's paradoxical games. *Proceedings of the Second International Conference on Unsolved Problems of Noise and Fluctuations, UPoN'99*, University of Adelaide, July 12-16, 1999 (eds. D. Abbott & L.B. Kish), American Institute of Physics, Melville, NY, USA, **511** 2000, 201–206.

[11] Doob J. L., (1953) *Stochastic Processes*. John Wiley and Sons, Inc.

PART VI
Parrondo's Games and Related Topics

State-Space Visualization and Fractal Properties of Parrondo's Games

Andrew Allison
Centre for Biomedical Engineering (CBME)
Department of Electrical and Electronic Engineering
The University of Adelaide
Australia
aallison@eleceng.adelaide.edu.au

Derek Abbott
Centre for Biomedical Engineering (CBME)
Department of Electrical and Electronic Engineering
The University of Adelaide
Australia
dabbott@eleceng.adelaide.edu.au

Charles Pearce
Department of Applied Mathematics
University of Adelaide
Australia
cpearce@maths.adelaide.edu.au

Abstract

Parrondo's games are essentially Markov games. They belong to the same class as Snakes and Ladders. The important distinguishing feature of Parrondo's games is that the transition probabilities may vary in time. It is as though "snakes," "ladders" and "dice" were being added and removed while the game was still in progress. Parrondo's games are not homogeneous in time and do not necessarily settle down to an equilibrium. They model non-equilibrium processes in physics.

We formulate Parrondo's games as an inhomogeneous sequence of Markov transition operators, with rewards. Parrondo's "paradox" is shown to be equivalent to saying that the expected value of the reward, from the whole process, is not a linear function of the Markov operators. When we say that a *game* is "winning" or "losing" then we must be careful to include the whole process in our definition of the word "'game." Specifically, we must include the time varying probability vector in our calculations. We give practical rules for calculating the expected value of the return from sequences of Parrondo's games. We include a worked example and a comparison between the theory and a simulation.

We apply visualization techniques, from physics and engineering, to an inhomogeneous Markov process and show that the limiting set or "attractor" of this process has fractal geometry. This is in contrast to the relevant theory for *homogeneous* Markov processes where the stable, equilibrium limiting set is a single point in the state space. We show histograms of simulations and describe methods for calculating the capacity dimension and the moments of the fractal attractors. We indicate how to construct optimal forms of Parrondo's games and describe a symmetrical family of games which includes the optimal form, as a limiting case. We investigate the fractal geometry of the attractors for this symmetrical family of games. The resulting geometry is very interesting, even beautiful.

1 Introduction

In Parrondo's games, the apparently paradoxical situation occurs where individually losing games combine to win [1,2]. The basic formulation and definitions of Parrondo's games are described in Harmer *et al.* [3–7]. These games have recently gained considerable attention as they are physically motivated and have been related to physical systems such as the Brownian ratchet [4], lattice gas automata [8] and spin systems [9]. Various authors have pointed out interest in these games for areas as diverse as biogenesis [10], political models [9], small-world networks [11], economics [9] and population genetics [12].

In this chapter, we will first introduce the relevant properties of Markov transition operators and then introduce some terminology and visualization techniques from the theory of dynamical systems. We will then use these tools, later in the chapter, to define and investigate some interesting properties of Parrondo's games.

We must first discuss and introduce the mathematical machinery, terms and notation that we will use. The key concepts are:

state: This contains all the information that we need to uniquely specify what is happening in the system at any given time. In Parrondo's original games, the state can be represented by a single integer.

time-varying probability vector: This is a time-varying probability distribution which specifies the probabilities that the system will be in certain states at any given time.

state space: For many physical systems, the state variables satisfy all the transformations required for a vector [13] and form a vector space [14] which is referred to as "state space" [15]. In this paper, we regard the time-varying probability vector as a state-vector within a state-space.

transition matrix: This is a Markov operator which determines the way in which the time varying probability vector will evolve over time.

These concepts are defined and discussed at length in many of the standard text books on stochastic processes [16–19].

Time-homogeneous sequences of regular Markov transition operators have unique stable limiting state-probabilities. The state-space representations of the associated time-varying probability vectors converge to unique points. If the sequence of Markov transition operators is not homogeneous in time, then the sequence of time-varying probability vectors generated by the products of these different operators need not converge to a single point, in the original state space. We construct quite simple examples to show that this is the case.

If the sequences are periodic, then it is possible to incorporate the finite memory of these systems into a new definition of "state." The new systems can be re-defined as strictly homogeneous Markov processes. These new Markov processes, with new states, will generally have unique limiting probability vectors.

If we allow the sequence to become indefinitely long, then the amount of memory required grows without bound. In principle, it is still possible to define these indefinitely long periodic sequences as homogeneous Markov processes although the definition, and encoding, of the states would require great care. We can consider any one indefinite sequence of operators as being one of many possible indefinite sequences of operators, in which case most of the possible sequences will appear to be "random." We can learn something about the general case, or arbitrary long sequences, by studying indefinitely long random sequences.

If the sequence of operators is chosen at random, then the time-varying probability vector, as defined in the original state-space, does not generally converge to a single unique value. Simulations show that the time-varying probability vector assumes a distribution in the original state-space which is self-similar, or "fractal," in appearance. We establish the existence of fractal geometry with rigor, for some particular Markov games. We also establish a transcendental equation which allows the calculation of the capacity dimensions of these fractal objects.

If state-transitions of the time-inhomogeneous Markov chains are associated with rewards then it is possible to show that even simple, "two-state," Markov chains can generate a Parrondo effect, as long as we are free to choose the reward matrix. Homogeneous sequences of the individual games generate a net loss over time. Inhomogeneous mixtures of two games can generate a net gain.

We show that the expected rates of return, or moments of the reward process, for the time-inhomogeneous games are identical to the expected rates of return from a homogeneous sequence of a time-averaged game. This is a logical consequence of the Law of Total Probability and the definition of expected value.

Two different views of the time-inhomogeneous process emerge, depending on the viewpoint that one takes:

- If you have access to the history of the time-varying probability vector and you have the memory to store this information and you choose to represent this data in a state-space, then you will see distributions with fractal geometry. This is more or less the view that a large casino might have if they were to visualize the average states of their many customers, all playing the one randomized sequence of Parrondo's games.
- If you do not have access to the time-varying probability vector or you have no memory in which to store this information, then all that you can see is a sequence of rewards from a stochastic process. The internal details of this

process are hidden from you. You have no way of knowing precisely how this process was constructed from an inhomogeneous sequence of Markov operators. There is no experiment that you can perform to distinguish between the time-inhomogeneous process and the time-averaged process. The time-averaged process is a homogeneous sequence of a single operator. We can calculate a single unique limiting value for the probability vector. This is more or less the view that a single, mathematically inclined, casino patron might have if they were playing against an elaborate poker machine, based on Parrondo's games.The internal workings of the machine would be hidden from the customer but it would still be possible to analyze the outcomes and estimate the parameters for the time-averaged model.

We show that the time inhomogeneous process is consistent in the sense that the "casino" and the "customer" will always agree on the expected winnings or losses of the customer. In more technical terms, the time-average, which the customer sees, is the same as the ensemble-average over state-space, which the casino can calculate.

2 Time-Homogeneous Markov Chains and Notation

Finite discrete-time Markov chains can be represented in terms of matrices of conditional transition probabilities. These matrices are called Markov transition operators. We denote these by capital letters in brackets, eg: $[A]$ where $A_{i,j} = \Pr\{K_{t+1} = j | K_t = i\}$ and $K \in \mathcal{Z}$ is a measure of displacement or the "state" of the system. The Markov property requires that $A_{i,j}$ cannot be a function of K but it can be a function of time, t. In Parrondo's original games, K, represents the amount of capital that a player has. There is a one-to-one mapping between Markov games and the Markov transition operators for these games. We will refer to the games and the transition operators interchangeably.

The probability that the system will be in any one state at a given instant of time can be represented by a distribution called the time-varying probability vector. We represent this probability mass function, at time t, using a row vector, $\mathbf{V_t}$. We can represent the evolution of the Markov chain in time using a simple Matrix equation,

$$\mathbf{V_{t+1}} = \mathbf{V_t} \cdot [A]\,. \tag{1}$$

This can be viewed as a multi-dimensional finite difference equation. The initial value problem can be solved using generating function, or Z transform, methods. Sequences of identical Markov transition operators, where $[A]$ does not vary, are said to be time-homogeneous. A Markov transition operator is said to be regular if some positive power of that operator has all positive elements. Time-homogeneous sequences of regular Markov transition operators always have stable limiting probability vectors, $\lim_{t\to\infty}(\mathbf{V_t}) = \Pi$. The time varying probability vector reliably converges to a point [16–19].

We can think of the space which contains the time-varying probability vectors, and the stable limiting probability vector, as a vector space which has a strong analogy to the state-space which is used in the theory of control. We shall refer to this space as "state-space," $[0, 1]^N$, and we will refer to the time-varying probability vector as a "state-vector," within a "state-space." This terminology is used in the engineering literature [19]. We emphasize that the "*state-vector*," $\mathbf{V_t} \in \Re^N$ is distinct from the "*state*"' of the system, $K \in \mathcal{Z}$, which we defined earlier. As a simple example, we can consider the regular Markov transition operator

$$[A] = \begin{bmatrix} \frac{13}{16} & \frac{3}{16} \\ \frac{1}{16} & \frac{15}{16} \end{bmatrix}, \tag{2}$$

using the initial condition

$$\mathbf{V_t} = [V_0, V_1] \; = \left[\frac{3}{4}, \frac{1}{4}\right] \text{ when } t = 0. \tag{3}$$

The components of $\mathbf{V_t}$ are V_0 and V_1 and these can be considered to be the dimensions of a Cartesian space. This "state-space" has clear analogies with the state-space used in the theory of control [15], with the phase-space of Poincare [20], with the "γ" or gaseous phase-space of Gibbs [21] and with the configuration-space used in Lagrangian dynamics [22]. We can freely borrow some of the visualization techniques from these other disciplines although we must be careful not to press these analogies too far since the state-spaces of physics and of Markov chains use different transition operators and obey different conservation laws.

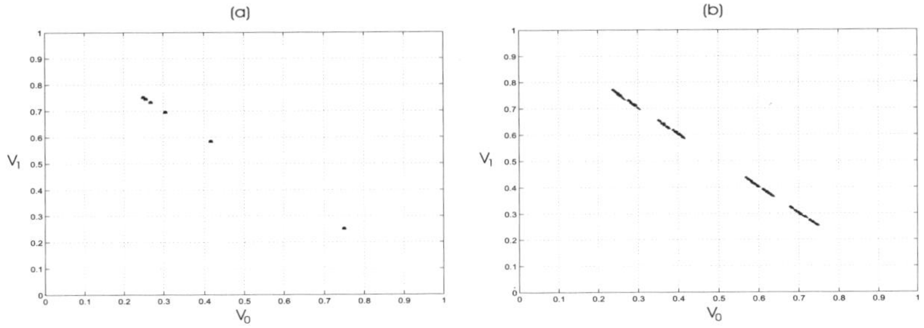

Figure 1: State-space trajectories: In part (a) we see the response to a homogeneous sequence of games "T," from (8). The time varying probability vector converges geometrically to a single fixed point. There is only one point of accumulation at $\left[\frac{1}{4}, \frac{3}{4}\right]$. In part (b) we see the response to an inhomogeneous randomized sequence of games "S" and "T," from (7) and (8). The time varying probability vector does not converge geometrically to any fixed point. There are an infinite number of points of accumulation along the line segment between points [1/4, 3/4] and [3/4, 1/4]. The points of accumulation form an attractor that corresponds to Cantor's fractal.

A fundamental question in the study of dynamical systems is to classify how they behave as $t \to \infty$ and all initial transient effects have decayed. The evolution of the state-vector of a discrete-time Markov chain generally traces out a sequence of points or "trajectory" in the state-space. The natural technique would be to draw a graph of this trajectory. As an example of this, we can consider the trajectory of a time homogeneous Markov chain shown in Figure 1(a). The state-vector, $\mathbf{V_t}$, always satisfies the constraint, $V_0 + V_1 = 1$, which follows from the law of total probability. The dynamics of the system all occur within a sub-space on the entire state-space. This is clearly visible in Figure 1(a). We can think of the set

$$M = \{[V_0, V_1] \mid (0 \leq V_0 \leq 1) \wedge (0 \leq V_1 \leq 1) \wedge (V_0 + V_1 = 1)\}, \qquad (4)$$

as a state manifold for the dynamical system defined by (2) and (3). The dynamics, within the state manifold, always converge to a single stable fixed point, as long as the Markov transition operators are regular and time-homogeneous [16–19]. The convergent point is the correct state-space representation of the stable limiting probability for the Markov chain.

3 Time-Inhomogeneous Markov Chains

The existence, uniqueness and dynamical stability of the fixed point are important parts of the theory of Markov chains but we must be careful not to apply these theorems to systems where the basic premises are not satisfied. If the Markov transition operators are not homogeneous in time then there may no longer a single fixed point in state-space. The state-vector can perpetually move through two or more points without ever converging to any single stable value. To demonstrate this important point, we present a simple example, using two regular Markov transition operators:

$$[S] = \begin{bmatrix} \frac{3}{4} & \frac{1}{4} \\ \frac{3}{4} & \frac{1}{4} \end{bmatrix} \qquad (5)$$

and

$$[T] = \begin{bmatrix} \frac{1}{4} & \frac{3}{4} \\ \frac{1}{4} & \frac{3}{4} \end{bmatrix}. \qquad (6)$$

The rows of these matrices are all identical. This indicates that the outcome of each game is completely independent of the initial state. The limiting stable probabilities for these regular Markov transition operators are $\Pi_\mathbf{S} = \left[\frac{3}{4}, \frac{1}{4}\right]$ and $\Pi_\mathbf{T} = \left[\frac{1}{4}, \frac{3}{4}\right]$ respectively. The time-varying probability vector immediately moves to the stable limiting value after even a single play of each game:

$[Q] \cdot [S] = [S]$ and $[Q] \cdot [T] = [T]$ for any conformable stochastic matrix $[Q]$. This leads to some interesting corollaries: $[T] \cdot [S] = [S]$ and $[S] \cdot [T] = [T]$. If we play an indefinite alternating sequence of these games, $\{STST \cdots\}$, then there are two simple ways in which we can associatively group the terms:

$$\begin{aligned}\mathbf{V_{2N}} &= \mathbf{V_0} ([S][T]) ([S][T]) \cdots ([S][T]) \\ &= \mathbf{V_0} [T] \\ &\Rightarrow \Pi = \Pi_{\mathbf{T}}\end{aligned}$$

and

$$\begin{aligned}\mathbf{V_{2N+1}} &= (\mathbf{V_0} [S]) ([T][S]) ([T][S]) \cdots ([T][S]) \\ &= \mathbf{V_0} [S] \\ &\Rightarrow \Pi = \Pi_{\mathbf{S}}.\end{aligned}$$

If we *assume* that there is a unique probability limit then we must conclude that $\Pi_{\mathbf{S}} = \Pi_{\mathbf{T}}$ and hence $\frac{1}{4} = \frac{3}{4}$ which is a contradiction. We can invoke the principle of excluded middle *(reductio ad absurdum)* to conclude that the assumption of a single limiting stable value for $\lim_{t \to \infty} (\mathbf{V_t})$ is false. In the limit as $t \to \infty$, the state-vector alternately assumes one of the *two* values $\Pi_{\mathbf{S}}$ or $\Pi_{\mathbf{T}}$. We refer to the set of all recurring state-vectors of this type, $\{\Pi_{\mathbf{S}}, \Pi_{\mathbf{T}}\}$, as the *attractor* of the system. In more general terms an attractor is a set of points in the state-space which is invariant and stable under the dynamics of the system as $t \to \infty$.

3.1 Reduction of the Periodic Case to a Time-Homogeneous Markov Chain

In the last section, we considered a short sequence of length 2. This can be generalized to an arbitrary length, $N \in \mathcal{Z}$. It is possible to associatively group the operators into sub-sequences of length N. As with the sequences of length two, the choice of time origin is not unique. We are free to make an arbitrary choice of time origin with the initial condition at $t = 0$. We can think of the operators as having an offset of $n \in \mathcal{Z}$, where $0 \leq n \leq N - 1$ within the sub-sequence. We can also calculate a new equivalent operator to represent the entire sequence, $\left[A_{\mathrm{eq}}\right] = \prod_{n=0}^{N-1} [A_n]$. We can then calculate the steady-state probabilities associated with this operator, $\Pi_{\mathrm{eq}} = \Pi_{\mathrm{eq}} \cdot \left[A_{\mathrm{eq}}\right]$ in the limit as $t \to \infty$. We can refer the asymptotic trajectory of the time-varying probability vector to this fixed point, $\mathbf{V}_{(\mathbf{t} \pmod{\mathbf{N}})} = \Pi_{\mathrm{eq}} \cdot \Pi_{n=0}^{(t \pmod{N})-1} [A_n]$. In the periodic case, there is generally not a single fixed point in the original state-space but the time-varying probability vector settles into a stable limit cycle of length N. If we aggregate time, modulo N, then we can re-define what we mean by "state" and we can define a new state-space in which the time-varying vector does converge to a single point.

If we allow the length of the period, N, to become indefinitely long $N \to \infty$ then our new definition of "state" becomes infinitely complicated. We would have to contemplate indefinitely large offsets, $n \to \infty$, within the infinitely long cycle. If we wish to avoid the many paradoxes that infinity can conceal, then we really should consider the case with "infinite" period as being qualitatively different from the case with finite period, N.

4 Random Selection of Markov Transition Operators

4.1 Two Simple Markov Games that Generate a Simple Fractal in State-Space

We proceed to construct a simple system in which operators are selected at random and we use the standard theories regarding probability and expected values to derive some useful results. If we modify the system specified by (5) and (6):

$$[S] = \begin{bmatrix} \frac{5}{6} & \frac{1}{6} \\ \frac{1}{2} & \frac{1}{2} \end{bmatrix} \tag{7}$$

and

$$[T] = \begin{bmatrix} \frac{1}{2} & \frac{1}{2} \\ \frac{1}{6} & \frac{5}{6} \end{bmatrix} \tag{8}$$

and select the sequence of transition operators at random then the attractor becomes an infinite set. If we were to play a homogeneous sequence of either of these games, then they would have the same stable limiting probabilities as before, Π_{S} and Π_{T}, and the dynamics would be similar to those shown in Figure 1(a). In contrast, if we play an indefinite *random* sequence of the new games S and T, $\{STSSTSTTSTT\cdots\}$, then there are no longer any stable limiting probabilities and the attractor has a fractal or "self-similar" appearance which is shown in Figure 1(b).

4.2 The Cantor Middle-Third Fractal

These games have been constructed in such a way that they generate the Cantor middle-third fractal.

It should be noted that the Cantor Middle-Third fractal is an uncountable set and so a countably infinite random sequence of operators will never generate enough points to cover the entire set. The solution to this problem is to consider the uncountably infinite set generated by all possible infinite, random sequences of operators. We could construct a probability measure on the resulting set and then we could calculate probabilities and expected values. It is also reasonable to talk about the probability density function of the time-varying probability vector in the state-space.

In order to stimulate intuition, we can simulate the process and generate a histogram, showing the distribution of the time-varying probability vector. The result is shown in Figure 2. For the x axis in this figure, we *could* have chosen the first element of the time-varying probability vector, V_0, but this would not have been the easiest way to analyze the dynamics. We choose another parameterization which reveals the simplicity of the underlying process. If we examine the eigenvectors of the matrices in (7) and (8), then we find that a better re-parameterization is: $x = (V_0 - V_1)$ and $y = (V_0 + V_1)$. Of course, we always have $y = 1$ and x is a new variable in the range $-1/2 \leq x \leq +1/2$. The Cantor fractal lies in the unit interval $-1/2 \leq x \leq 1/2$ which is the x interval shown in Figure 2. The transformation for matrix $[S]$, in (7) reduces to:

$$\left(+\frac{1}{2} - x_{t+1}\right) = \frac{1}{3} \cdot \left(+\frac{1}{2} - x_t\right) \tag{9}$$

and the transformation for matrix $[T]$, in (8) reduces to:

$$\left(-\frac{1}{2} - x_{t+1}\right) = \frac{1}{3} \cdot \left(-\frac{1}{2} - x_t\right). \tag{10}$$

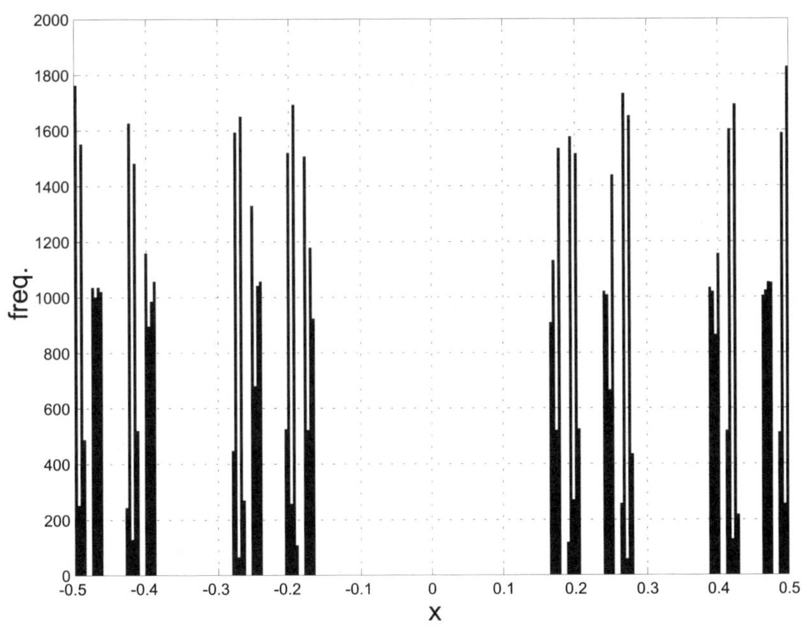

Figure 2: A histogram of the distribution of $\mathbf{V_t}$ in state-space. This is a finite approximation of the Cantor Fractal.

The transformation S has a fixed point at $x = +\frac{1}{2}$ and the transformation T has a fixed point at $x = -\frac{1}{2}$. If we choose these transformations at random then the recurrent values of x lie in the interval between the fixed points, $-\frac{1}{2} \leq x \leq \frac{1}{2}$. This is precisely the iterated function system for the Cantor Middle-Third Fractal. These are described in Barnsley [23].

The most elementary analysis that we can perform is to calculate the dimension of this set. If we assume conservation of measure, then every time we perform a transformation, we reduce the diameter by a factor of $\frac{1}{3}$ but the transformed object is geometrically half of the original object so we can write $\frac{1}{2} = \left(\frac{1}{3}\right)^D$ where D is the fractional dimension. This is the law of conservation of measure for this particular system. We can solve this equation for D to get $D = \log(2)/\log(3) \approx 0.630929\cdots$.

We can invert the rules described in (9) and (10) giving: $x_t = 3x_{t+1} - 1$ and $x_t = 3x_{t+1} + 1$. If we consider these equations, together with the law of conservation of total probability, then we get a self-similarity rule for the PDF (or Probability Density Function), $p(x)$, of the time-varying probability vector, $\mathbf{V_t}$:

$$\frac{3}{2} p\left(3x - 1\right) + \frac{3}{2} p\left(3x + 1\right) = p\left(x\right).$$

This PDF, $p(x)$, is the density function towards which the histogram in Figure 2 would converge if we could collect enough samples. The self-similarity rule for the PDF gives rise to a recursion rule for the moment generating function, $\Phi\left(\Omega\right) = E\left(e^{j\Omega x}\right)$:

$$\Phi\left(\Omega\right) = \Phi\left(\frac{\Omega}{3}\right) \cdot \cos\left(\frac{\Omega}{3}\right). \tag{11}$$

We can evaluate the derivatives at $\Omega = 0$ and calculate as many of the moments as we wish. We can calculate the mean, μ, and the variance, σ^2, giving:

$$\mu = 0 \tag{12}$$

$$\sigma^2 = \frac{1}{8}. \tag{13}$$

These algebraic results are consistent with results from numerical simulations.

4.3 Iterated Function Systems

The cause of the fractal geometry is best understood if we realize that Markov transition operators perform affine transformations on the state-space. An indefinite sequence of different Markov transition operators is equivalent to an indefinite sequence of different affine transformations which is called an "Iterated Function System." We refer the reader to the work of Michael Barnsley [23] and the theory of Iterated Function Systems to show that fractal geometry is quite a general property of a system of randomly selected affine transformations.

5 An Equivalent Representation of the Random Selection of Markov Transition Operators

Consider two mutually exclusive events, $A \cap B = \emptyset$, embedded within a probability space $(\Omega, \mathcal{F}, P)$. Consider any third event $C \subseteq A \cup B$. These events are represented in Figure 3. The law of total probability asserts that

$$\Pr(C) = \Pr(C|A) \cdot \Pr(A) + \Pr(C|B) \cdot \Pr(B). \tag{14}$$

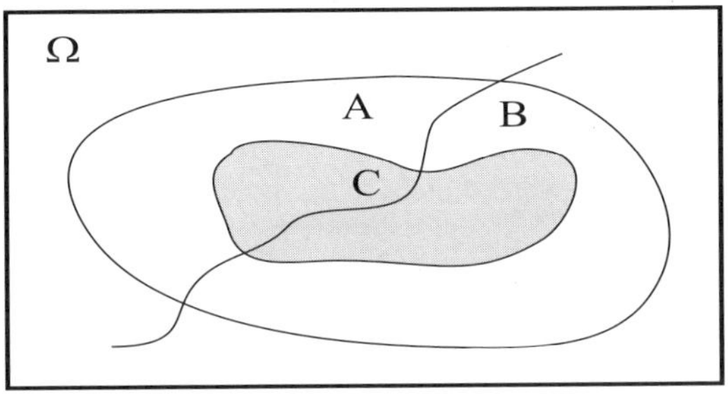

Figure 3: Set relationships and change of probability.

We can now make the following particular identifications:

$$C \equiv \{X \in \Omega \mid K_{t+1} = i \ \wedge \ K_t = j\},$$
$$A \equiv \{\text{played game } A\},$$
$$B \equiv \{\text{played game } B\}.$$

If we select games A and B at random with probabilities of γ and $(1 - \gamma)$ respectively, then we can write $\Pr(A) = \gamma$ and $\Pr(B) = (1 - \gamma)$. By definition, the Markov matrices for games A and B contain conditional probabilities for state transitions:

$$A_{i,j} = \Pr\{(K_{t+1} = j \mid K_t = i) \ \wedge \ \text{played game } A\},$$
$$B_{i,j} = \Pr\{(K_{t+1} = j \mid K_t = i) \ \wedge \ \text{played game } B\}.$$

Note that in this case $C = A \cup B$. We can define a new operator corresponding to the events $C_{i,j}$:

$$C_{i,j} = \Pr\{K_{t+1} = j \mid K_t = i\},$$

and (14) reduces to

$$C_{i,j} = A_{i,j} \cdot \gamma + B_{i,j} \cdot (1 - \gamma). \tag{15}$$

The conditional probabilities of state transitions of the inhomogeneous Markov process generated by games A and B are the same as the conditional probabilities of a new equivalent game called "Game C." The transition matrix for Game C is a linear convex combination of the matrices for the original basis games, A and B. Even if we have complete access to the state of the system, there is no function that we can perform on the state, or state transitions, which could allow us to distinguish between a homogeneous sequence of games C and an inhomogeneous *random* sequence of games A and B. We refer to game C as the time-average model. This is analogous to the state-space averaged model found in the theory of control [24].

6 The Phenomenon of Parrondo's Games

6.1 Markov Chains with Rewards

Suppose that we apply a reward matrix to the process:

$$R_{i,j} = \text{reward if } (K_{t+1} = j) \mid (K_t = i).$$

There is a specific reward associated with each specific state transition. We can think of $R_{i,j}$ as the reward that we earn when a transition occurs from state i to state j. The state transitions, rewards and probabilities of transition, for "Game A" are shown in Figure 4. The state transition diagrams for "Game B" and the time averaged "Game C" would have identical topology and have identical reward structure, although the probabilities of transition between states would be different. Systems of this type have been analyzed by Howard [25] although we use different matrix notation to perform the necessary multiplications and summations.

The expected contribution to the reward from each transition of the time-averaged homogeneous process is:

$$Y_{i,j} = E\left[R_{i,j} \circ C_{i,j}\right], \tag{16}$$

where "$\circ$" represents the Hadamard, or element by element, product.

If we wish to calculate the mean expected reward, then we must sum over all recurrent states in proportion to their probability of occurrence. This will be a function of the transition matrix, C, and the relevant steady state probability vector, $\Pi_{\mathbf{C}}$:

$$Y(C) = \Pi_{\mathbf{C}} \cdot ([R] \circ [C]) \cdot \mathbf{U}^T, \tag{17}$$

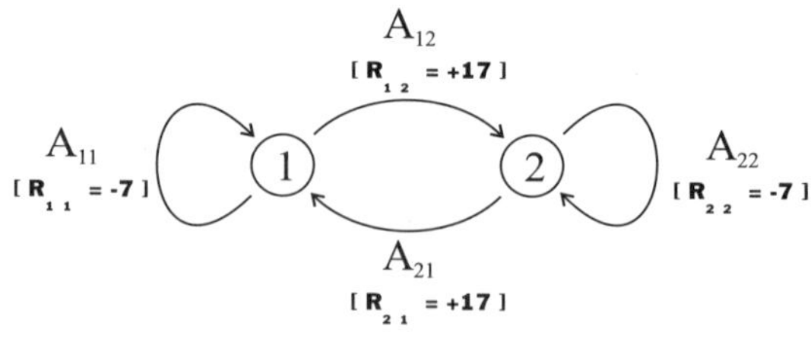

Figure 4: State transition diagram for "Game A" with rewards.

where $\mathbf{U}^T$ is a unit column vector of dimension N. Post-multiplication by $\mathbf{U}^T$ has the effect of performing the necessary summation. We recall that $\Pi_{\mathbf{C}}$ represents the steady state probability vector for matrix C. The function $Y(C)$ represents the expected asymptotic return, in units of "reward," per unit time when the games are played.

If we include the definition of C from (15) in (16) then we can write:

$$Y_{i,j} = E\left[R_{i,j} \cdot \left(\gamma A_{i,j} + (1-\gamma) B_{i,j}\right)\right] \tag{18}$$

$$= \gamma E\left[R_{i,j} \cdot A_{i,j}\right] + (1-\gamma) E\left[R_{i,j} \cdot B_{i,j}\right]. \tag{19}$$

We can also define:

$$Y(A) = \Pi_{\mathbf{A}} \left([R] \circ [A]\right) \mathbf{U}^T \tag{20}$$

and

$$Y(B) = \Pi_{\mathbf{B}} \left([R] \circ [B]\right) \mathbf{U}^T, \tag{21}$$

and we might *falsely* conclude that

$$Y(C) = \gamma Y(A) + (1-\gamma) Y(B). \tag{22}$$

This would be equivalent to saying that:

$$Y(C) = \gamma \left(\Pi_{\mathbf{A}} \left([R] \circ [A]\right) \mathbf{U}^T\right) + (1-\gamma) \left(\Pi_{\mathbf{B}} \left([R] \circ [B]\right) \mathbf{U}^T\right), \tag{23}$$

but these equations (22) and (23) are in *error* because (19) must be summed over all of the recurrent states of the *mixed* inhomogeneous games but in the *false* equation, (23), the first term is summed with respect to the recurrent states of Game "A" and the second term is summed with respect to the recurrent states of game "B" which is a mistake! The dependency on state makes the reward process nonlinear in C. The correct expression for $Y(C)$ would be:

$$Y(C) = \gamma \left(\Pi_{\mathbf{C}} \left([R] \circ [A]\right) \mathbf{U}^T\right) + (1-\gamma) \left(\Pi_{\mathbf{C}} \left([R] \circ [B]\right) \mathbf{U}^T\right). \tag{24}$$

The difference between the intuitively appealing but *false* equations (22) and (23) and the correct equation (24) is the cause of "Parrondo's paradox."

6.2 Parrondo's Paradox Defined

The essence of the problem is that when we say that "Game A is losing" or "Game B is losing" we perform summation with respect to the steady state probability vectors for games "A" and "B" respectively. When we say that "a random sequence of games A and B is winning," we perform the summation with respect to the steady state probability vector for the time-averaged game, Game "C."

We can say that the "paradox" exists whenever we can find two games A and B and a reward matrix R such that:

$$Y\left(\gamma A + (1-\gamma)\, B\right) \neq \gamma Y(A) + (1-\gamma)\, Y(B). \tag{25}$$

The "paradox" is equivalent to saying that the reward process is not a linear function of the Markov transition operators.

6.3 A Simple "Two-State" Example of Parrondo's Games

We can show that Parrondo's paradox does exist by constructing a simple example. We can define

$$[A] = \begin{bmatrix} \frac{5}{6} & \frac{1}{6} \\ \frac{1}{2} & \frac{1}{2} \end{bmatrix} \tag{26}$$

and

$$[B] = \begin{bmatrix} \frac{1}{2} & \frac{1}{2} \\ \frac{1}{6} & \frac{5}{6} \end{bmatrix}. \tag{27}$$

The steady state probability vectors are: $\Pi_{\mathbf{A}} = \left[\frac{3}{4}, \frac{1}{4}\right]$ and $\Pi_{\mathbf{B}} = \left[\frac{1}{4}, \frac{3}{4}\right]$. These games are the same as games "S" and "T" defined earlier but we analyse them using the theory of Markov chains with rewards. We can define a reward matrix

$$[R] = \begin{bmatrix} -7 & +17 \\ +17 & -7 \end{bmatrix}, \tag{28}$$

and we can apply (20), (21) and (24) to get:

$$Y(A) = \begin{bmatrix} \frac{3}{4} & \frac{1}{4} \end{bmatrix} \left(\begin{bmatrix} -7 & +17 \\ +17 & -7 \end{bmatrix} \circ \begin{bmatrix} \frac{5}{6} & \frac{1}{6} \\ \frac{1}{2} & \frac{1}{2} \end{bmatrix} \right) \begin{bmatrix} 1 \\ 1 \end{bmatrix} = -1 \tag{29}$$

and

$$Y(B) = \begin{bmatrix} \frac{1}{4} & \frac{3}{4} \end{bmatrix} \left(\begin{bmatrix} -7 & +17 \\ +17 & -7 \end{bmatrix} \circ \begin{bmatrix} \frac{1}{2} & \frac{1}{2} \\ \frac{1}{6} & \frac{5}{6} \end{bmatrix} \right) \begin{bmatrix} 1 \\ 1 \end{bmatrix} = -1, \tag{30}$$

and for the time-average we get:

$$Y(C) = \begin{bmatrix} \frac{1}{2} & \frac{1}{2} \end{bmatrix} \left(\begin{bmatrix} -7 & +17 \\ +17 & -7 \end{bmatrix} \circ \begin{bmatrix} \frac{2}{3} & \frac{1}{3} \\ \frac{1}{3} & \frac{2}{3} \end{bmatrix} \right) \begin{bmatrix} 1 \\ 1 \end{bmatrix} = +1. \tag{31}$$

Games "A" and "B" are losing and the mixed time-average game, $C = \frac{1}{2}(A+B)$, is winning and (25) is satisfied and so we have Parrondo's "paradox" for the two-state games "A" and "B" as defined in (26) and (27). We can simulate the dynamics of this two-state version of Parrondo's games. Some typical sample paths are shown in Figure 5. The results from the simulations are consistent with the algebraic results.

Figure 5: Simulation of a Two-State version of Parrondo's games: Note that the homogeneous sequences of games "A" or "B" are losing but the inhomogeneous mixture of games "A" and "B" is winning. This is an example of Parrondo's paradox.

If we refer back to Figure 4 then an intuitive explanation for this phenomenon is possible. The negative or "punishing" rewards are associated with transitions that do not change state. The good positive rewards are associated with the changes of state. If we play a homogeneous sequence of Games "A" or "B", then there are relatively few changes of state and the resulting weighted sum of all the rewards is negative. If we play the mixed game, then the rewarding changes of state are much more frequent and the resulting weighted sum of rewards is positive.

7 Consistency between State-Space and Time Averages

In order for the "fractal view" of the process, in state-space, to be consistent with the time average view of the process we require:

$$E\left[\mathbf{V_t}\right] = \Pi_{\mathbf{C}} \tag{32}$$

The value of $E[\mathbf{V_t}]$ follows from the argument in Section 4.2. We can use the mean, evaluated in (12), and construct the relation

$$E\left[\mathbf{V_t}\right] = \left[\frac{1}{2} + \frac{1}{2}E\left[x\right], \frac{1}{2} - \frac{1}{2}E\left[x\right]\right] \tag{33}$$

$$= \left[\frac{1}{2} + \frac{1}{2}\mu, \frac{1}{2} - \frac{1}{2}\mu\right] \tag{34}$$

$$= \left[\frac{1}{2}, \frac{1}{2}\right]. \tag{35}$$

The value of $\Pi_{\mathbf{C}}$ follows from the arguments in Section 6.3. Specifically we require $\Pi_{\mathbf{C}} = \Pi_{\mathbf{C}} \cdot C$ which gives:

$$\Pi_{\mathbf{C}} = \left[\frac{1}{2}, \frac{1}{2}\right], \tag{36}$$

which is consistent with (35). This proves this special case. To prove the more general case we need to have some notation for an entire fractal set, like the one shown in Figure 1(b). We use $\{F\}$ to denote the attractor generated by two operators A and B. We can write:

$$E\left[\{F\}\right] = \gamma E\left[\{F\}\right] A + (1-\gamma)\, E\left[\{F\}\right] B. \tag{37}$$

This follows from conservation of measure under the affine transformations A and B. We note that *everything* in these equations is linear and so we can write

$$\begin{aligned} E\left[\{F\}\right] &= E\left[\{F\}\right](\gamma A + (1-\gamma)\, B) \\ &= E\left[\{F\}\right] \cdot C, \end{aligned}$$

which is the defining property of $\Pi_{\mathbf{C}}$ which implies that

$$E\left[\{F\}\right] = \Pi_{\mathbf{C}}. \tag{38}$$

The two ways of viewing the situation are consistent which means that we can use the time-averaged game to calculate expected values of returns from Parrondo's games.

8 Parrondo's Original Games

8.1 Original Definition of Parrondo's Games

In their original form, Parrondo's games spanned infinite domains, of all integers or all non-negative integers [3]. If our interest is to examine the asymptotic behaviour of the games as $t \to \infty$ and to study asymptotic rates of return or moments, then it is possible to reduce these games by aggregating states of the Markov chain modulo three. We can do this without losing any information about the rate of return from the games. After reduction, the Markov transition operators take the form:

$$[A] = \begin{bmatrix} 0 & a_0 & (1-a_0) \\ (1-a_1) & 0 & a_1 \\ a_2 & (1-a_2) & 0 \end{bmatrix}, \tag{39}$$

where a_0, a_1 and a_2 are the conditional probabilities of winning, given the current state modulo three. This form of the games has been described by Pearce [6].

8.2 Optimised form of Parrondo's Games

Simulations reveal that *periodic* inhomogeneous sequences of Parrondo's games have the strongest effect. Further investigation by the authors, using Genetic Algorithms, suggest that the most powerful form of the games is a set of three games that are played in a strict periodic sequence $\{G_0, G_1, G_2, G_0, G_1, G_2, \cdots\}$. The transition probabilities are as follows:

Game G_0**:** $[a_0, a_1, a_2] = [\mu, (1-\mu), (1-\mu)]$,
Game G_1**:** $[a_0, a_1, a_2] = [(1-\mu), \mu, (1-\mu)]$,
Game G_2**:** $[a_0, a_1, a_2] = [(1-\mu), (1-\mu), \mu]$,

where μ is a small probability, $0 < \mu < 1$. We can think of μ as being a very small, ideally "microscopic", positive number. The rate of return from any *pure sequence* of these games is approximately $Y \approx \frac{1}{2} \cdot \mu$ which is close to zero and yet the return from the *cyclic combination* of these games is approximately $Y \approx 1 - 3 \cdot \mu$ which is close to a certain win. We can engineer a situation where we can deliver an almost certain win every time using games that, on their own, would deliver almost no benefit at all! These games clearly work better as a team than on their own. Just as team players may pass the ball in a game of soccer, the games $\{G_0, G_1, G_2\}$ carefully pass the state-vector from one trial to the next as this sequence of Parrondo's games unfolds.

8.3 An Exquisite Fractal Object

We can de-rate these games by increasing μ. In the limit as $\mu \to \frac{1}{2}$ the Parrondo effect vanishes and the attractor collapses to a single point in state-space. Just before this limit the attractor takes the form of the very small and exquisite fractal shown in Figure 6. This fractal is embedded in a two-dimensional sub-space of the three-dimensional state-space of the games $\{G_0, G_1, G_2\}$. The two-dimensional sub-space has been projected onto the page in order to make it easier to view. The projection preserves dot product, length and angle measure. The coordinates "x" and "y" are linear combinations of the components of the original state-vector, $\mathbf{V_t} = [V_0, V_1, V_2]$. The orientation of the image is such that the original "V_2" axis is projected onto the new "y" axis. (The direction of "up" is preserved.) The negative numbers on the axes represent negative offsets rather than negative probabilities. This is the same concept that is used when we write down a probability $(1 - p)$. If p is a valid probability, then so is $(1 - p)$. The number $-p$ is an offset that just happens to be negative.

The dimension of this fractal is $D \approx \log(9)/\log(4) \approx 1.585$. We define the amount of Parrondo effect, Δp, as the difference in rate of return, Y, between the mixed sequence of games $\{G_0, G_1, G_2\}$ and the best performance from any pure sequence of a single game. For this limiting case, $\Delta p \approx 0$. There are some interesting qualitative relationships between the capacity dimension and the amount of Parrondo effect that deserve further investigation to see if it is possible to state a general quantitative law.

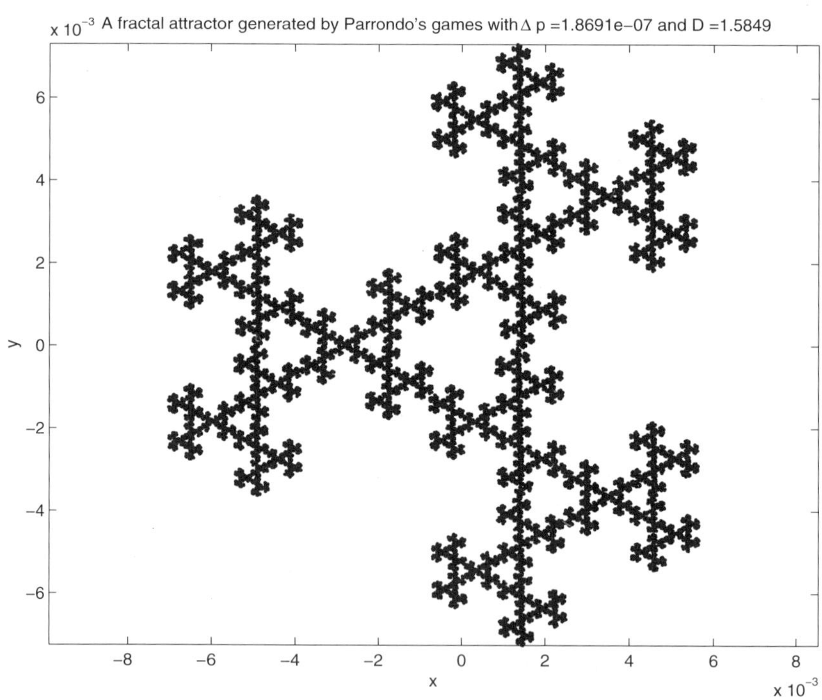

Figure 6: A 2D projection of a fractal attractor generated by the **"last gasp"** of Parrondo's games. Note the similarity of this fractal to the fractals of Sierpinski and of Koch.

9 Summary

In this paper we have analyzed Parrondo's games in terms of the theory of Markov chains with rewards. We have illustrated the concepts constructively, using a very simple two-state version of Parrondo's games and we have shown how this gives rise to fractal geometry in the state-space. We have arrived at a simple method for calculating the expected value of the asymptotic rate of reward from these games and we have shown that this can be calculated in terms of an equivalent time-averaged game. We have used graphic representations of trajectories and attractors in state-space to motivate the arguments.

The use of state-space concepts opens up new lines of inquiry. Simulation and visualization encourage intuition and help us to grasp the essential features of a new system. This would be much more difficult if we were to use a purely formal algebraic approach at the start. We do not propose visualization as a *replacement* for rigorous analysis. We see it as a guide to help us to decide which problems are worthy of more detailed attention and which problems might later yield to a more formal approach. We believe that state-space visualization will be as useful for the study of the dynamics of Markov chains as it has already been for the study of other dynamical systems.

Finally, we conclude that Parrondo's games are not really "paradoxical" in the true sense. The anomaly arises because the reward process is a nonlinear function of the Markov transition operators and our "common sense" tells us the reward process "ought" to be linear. When we combine the games by selecting them at random, we perform a linear convex combination of the operators but the expected asymptotic value of the rewards from this combined process is not a linear combination of the rewards from the original games.

REFERENCES

[1] Harmer G. P. and Abbott D., Parrondo's paradox: losing strategies cooperate to win. *Nature* (London), **402**, 864 (1999).

[2] Harmer G. P., Abbott D. and Taylor P. G., The paradox of Parrondo's games *Proc. Royal Soc., Series A, (Math. Phys. and Eng. Science)*, **456(99)**, 247–259 (2000).

[3] Harmer G. P., Abbott D., Taylor P. G. and Parrondo J. M. R., "Parrondo's Paradoxical Games and the Discrete Brownian Ratchet," in *Proc. 2nd Int. Conf. Unsolved Problems of Noise and Fluctuations (UPoN'99)* D. Abbott and L. B. Kish editors, vol. 511, American Inst. Phys., 2000, pp. 189–200.

[4] Harmer G. P. and Abbott D., Parrondo's paradox. *Statistical Science*, **14(2)**, 206–213, 1999.

[5] Pearce C. E. M., "Parrondo's paradoxical games," in *Proc.* $2^{\underline{nd}}$ *Int. Conf. Unsolved Problems of Noise and Fluctuations (UPoN'99)* D. Abbott and L. B. Kish editors, vol. 511, American Inst. Phys., 2000, pp. 201–206.

[6] Pearce C. E. M., "Entropy, Markov Information Sources and Parrondo's Games," in *Proc.* $2^{\underline{nd}}$ *Int. Conf. Unsolved Problems of Noise and Fluctuations (UPoN'99)* D. Abbott and L. B. Kish editors, vol. 511, American Inst. Phys., 2000, pp. 207–212.

[7] Harmer G. P. and Abbott D., A review of Parrondo's paradox, *Fluctuation and Noise Letters*, **2**, R71–R107 (2002).

[8] Meyer D. A. and Blumer H., Parrondo's games as lattice gas automata, *J. Stat. Phys.*, **107**, 225–239 (2002).

[9] Moraal H., Counterintuitive behaviour in games based on spin models, *J. Phys. A, Math. Gen.*, **33**, L203–L206 (2000).

[10] Davies P. C. W., "Physics and Life, The Abdus Salam Memorial Lecture," in *Sixth Trieste Conference on Chemical Evolution*, Trieste, Italy, eds.: J. Chela-Flores and T. Tobias and F. Raulin., Kluwer Academic Publishers (2001), 13–20.

[11] Toral R., Cooperative Parrondo's games, *Fluctuation and Noise Letters*, **1**, L7–L12 (2001).

[12] McClintock P. V. E., Unsolved problems of noise, *Nature (London)*, **401**, 23 (1999).

[13] Borisenko A. I., and Tarapov I. E., *Vector and Tensor Analysis, with Applications*, Dover Publications, Inc., 1968.

[14] McCoy N. H., *Fundamentals of Abstract Algebra*, Allyn and Bacon, Inc., 1972.

[15] DeRusso P. M., Roy R. J., and Close C. M., *State Variables for Engineers*, John Wiley and Sons, Inc., 1965.

[16] Karlin S., and Taylor, H. M., *A First Course in Stochastic Processes*, Academic Press, 1975.

[17] Karlin S., and Taylor, H. M., *An Introduction to Stochastic Modeling*, Academic Press, 1998.

[18] Norris J. R., *Markov chains*, Cambridge University Press, 1997.

[19] Yates R. D., and Goodman, D. J., *Probability and Stochastic Processes*, John Wiley and Sons Inc., 1999.

[20] Diacu F., and Holmes P., *Celestial Encounters*, Princeton University Press, 1996, chap. 1.

[21] Baierlein R., *Thermal Physics*, Cambridge University Press, 1999, chap. 13.

[22] Lanczos C., *The Variational Principles of Mechanics*, Dover Publications Inc., 1949, chap. 1.

[23] Barnsley M., *Fractals Everywhere*, Academic Press, 1988.

[24] Middlebrook R. D., and Ćuk, S. A., "A general unified approach to modelling switching converter power stages," in *IEEE Power Electronics Specialist's Conf. Rec*, 1976, pp. 18–34, in Proceedings of IEEE Power Electronics Specialist's Conf. Rec.

[25] Howard R. A., *Dynamic Programming and Markov Processes*, John Wiley and Sons Inc., 1960.

Parrondo's Capital and History-Dependent Games

Gregory P. Harmer
Centre for Biomedical Engineering (CBME)
Department of Electrical and Electronic Engineering
The University of Adelaide
Australia
gpharmer@eleceng.adelaide.edu.au

Derek Abbott
Centre for Biomedical Engineering (CBME)
Department of Electrical and Electronic Engineering
The University of Adelaide
Australia
dabbott@eleceng.adelaide.edu.au

Juan M. R. Parrondo
Departamento de Física Atómica, Molecular y Nuclear
Universidad Complutense de Madrid
Spain
parr@seneca.fis.ucm.es

Abstract

It has been shown that it is possible to construct two games that when played individually lose, but alternating randomly or deterministically between them can win. This apparent paradox has been dubbed "Parrondo's paradox." The original games are capital-dependent, which means that the winning and losing probabilities depend on how much capital the player currently has. Recently, new games have been devised, that are not capital-dependent, but history-dependent. We present some analytical results using discrete-time Markov-chain theory, which is accompanied by computer simulations of the games.

1 Introduction

It has recently been shown [1,2] that a discrete-time version of the flashing ratchet [3–5] can be interpreted as simple gambling games. There exist two losing games that can be combined to form a game with a winning expectation, much in the same way as a flashing ratchet can be made to move Brownian particles uphill with the use of mechanisms that individually let the particles move downhill. More information regarding this analogy can be found in [6].

However, this original incarnation of the games has probabilities that depend on the value of the current capital of the player, that is, the games are capital-dependent. Though this is useful in certain applications [7], a version of the games that does not depend on capital is more natural. This led to a construction of the games where the probabilities depend on the results of the previous two games, referred to as history-dependent games [8].

In this chapter, we analyze the games using simple discrete-time Markov chain theory and show analytical results from numerical simulations of the games. We also offer an explanation of the games in terms of their equilibrium distributions.

2 Parrondo's Capital-Dependent Games

In this section we construct capital-dependent games and explain how the concept of fairness applies to these games. Certain results of playing the games are also shown. The results have been found analytically, that is, they are what would be expected if we averaged then over almost an infinite number of games.

2.1 Construction of the Games

Game A is straightforward and can be thought of as tossing a weighted coin that has probability p of winning. Game B is a little more complex and can be generally described by the following statement. If the present capital is a multiple of M, then the chance of winning is p_1, if it is not a multiple of M, the chance of winning is p_2. Thus, the respective losing probabilities are $1 - p_1$ and $1 - p_2$.

The two games can be represented diagrammatically using branching elements, shown in Fig. 1. The notation (x, y) at the top of the branch gives the probability or condition for taking left and right branch respectively.

Figure 1: Construction of the capital-dependent games. The games could be formed using three biased coins.

If we require to control the three probabilities p, p_1 and p_2 via a single variable, a biasing parameter ϵ can be used to represent a subset of the probability space with the transformation

$$p = 1/2 - \epsilon,$$

$$p_1 = 1/10 - \epsilon \quad \text{and}$$
$$p_2 = 3/4 - \epsilon. \tag{1}$$

This parameterization along with $M = 3$ gives Parrondo's original numbers for the games [1].

2.1.1 The Randomised Game

Dealing with the randomized game is not as difficult as it first appears. Let us define a mixing parameter γ that gives the probability of playing game A, which is assumed to be a $1/2$ unless otherwise stated. When the capital is a multiple of M, the probability of winning is

$$q_1 = gp + (1 - g)p_1. \tag{2}$$

This is the chance of playing game A multiplied by the chance of winning it and correspondingly the chance of playing game B multiplied by the chance of winning. Alternatively, when the capital is not a multiple of M, the probability of winning is

$$q_2 = gp + (1 - g)p_2. \tag{3}$$

The respective losing probabilities are $1 - q_1$ and $1 - q_2$. Using these probabilities we can treat the randomized game exactly the same as game B, except replace each p_i with a q_i.

2.1.2 Fairness

An issue that needs to be clarified is the question of how to define whether the games are losing, fair or winning. To classify a game as either winning or losing is trivial, but when it comes to deciding if it is fair, the issue can become controversial. The reason is that the behavior of game B differs from that of game A as we are likely to win or lose a small amount depending on the value of the capital that we start with. If the starting capital is a multiple of M, then it is likely we will lose a little, if not, it is likely we will gain a little.

A brief discussion of fairness follows. A more detailed mathematical formulation of fairness relating to Parrondo's games is given by [9]. Consider a gambler repeatedly playing a game and after the nth game has capital $X(n)$, or X_n for short. Classically, as defined by [10], a fair game is one where given all the past results, the expectation of the next result is the same as the present result for any given game. That is, the game has to be a martingale where the expected value of capital after playing a game is the same as the present value.

The difficulty with game B is when X_0 is a multiple of M, $E[X_1|X_0] < X_0$ and correspondingly when X_0 is not a multiple of M, $E[X_1|X_0] > X_0$. This makes it troublesome to classify game B as either winning, losing or fair [9]. Suffice to say it is argued in [9] that fairness can be defined in terms of drift rates. Thus, if the capital tends to drift toward infinity then it classifies as winning ($\epsilon < 0$) or if it drifts towards negative infinity it is losing ($\epsilon > 0$). If there is no drift, then the game is fair ($\epsilon = 0$).

Therefore, using the above criterion, both games A and B are fair when ϵ set to zero in (1). This is true of game A because the probabilities of moving up and down in capital are equal for all n. It is also true of game B even though the value of starting capital influences the probability of going up and down for small values of n because as $n \to \infty$, there is no change in capital. The transient response actually decays to almost nothing very quickly, after about 20 games. The drift rates that determine fairness can be easily verified by considering a detained balance [11] of the corresponding system.

Although there is some concern over whether game B is technically fair, it is not that important in the context of the apparently paradoxical nature of the games as they definitely lose when $\epsilon > 0$. This is satisfactory since the only prerequisite we have in later sections are that games A and B lose when $\epsilon > 0$.

2.2 Playing the Games Analytically

As has been implied in the introduction, the mode of analysis for the games is via discrete-time Markov chains (DTMCs). Each value of capital is represented by a state, and the transition probabilities are determined by the rules of the games. Since in every game we must either incrementally win or lose, i.e. go up or down the chain by one state, the DTMC is referred to as *skip-free*.

The transition probabilities p_{ij} form the entries of the transition matrix $\mathbb{P}$, which defines the DTMC. Since the matrix represents a skip-free DTMC, $\mathbb{P}$ is tridiagonal with the main diagonal all zeros and all the columns summing to unity. Since the DTMC that represents the games is doubly-infinite, the dimensions of $\mathbb{P}$ also extend to $\pm\infty$. However, in practice the dimensions only need to extend to twice the number of games that are being played.

The transition matrix modeling game B is given by

$$\mathbb{P}_B = \begin{bmatrix} 0 & 1-p_2 & & & & \\ p_1 & 0 & \ddots & & & \\ & p_2 & \ddots & 1-p_1 & & \\ & & \ddots & 0 & 1-p_2 & \\ & & & p_1 & 0 & \ddots \\ & & & & p_2 & \ddots \\ & & & & & \ddots \end{bmatrix}. \tag{4}$$

This matrix shows the state dependency that is exhibited with the probabilities p_1 and $1 - p_1$ leaving the state that are divisible by M.

Since game A is a specific case of game B where $p_1 = p_2 = p$, $\mathbb{P}_A$ can be easily found from $\mathbb{P}_B$. Recalling from (2) and (3), anything derived for game B equally holds true for the randomized game, thus $\mathbb{P}_R$ can be determined. This is sufficient for all the analysis since the combination of two DTMCs simply forms another DTMC that obeys Markov chain theory.

From the transition matrices representing the games, the equilibrium probabilities (or stationary distribution) $\boldsymbol{\pi} = [\dots, \pi_{-1}, \pi_0, \pi_1, \dots]^T$ can be found. This contains the probabilities of finding the capital in each of the states. The expected outcome when playing a game can then be found by applying $\mathbb{P}$ to $\boldsymbol{\pi}$. Hence, the posterior distribution after playing n games is given by

$$\boldsymbol{\pi}_n = \mathbb{P}^n \boldsymbol{\pi}_0,$$

where the $\boldsymbol{\pi}_0$ is the starting capital. As $n \to \infty$ this gives the stationary distribution. To initially start (i.e. $n = 0$) with zero capital we would have $\boldsymbol{\pi}_0 = [0, \dots, 0, 1, 0, \dots, 0]^T$. By using the appropriate transition matrix, the individual or randomly mixed games can be played.

To play a deterministic mix of games, the appropriate $\mathbb{P}$ must be substituted. Thus, we could have

$$\boldsymbol{\pi}_n^{[a,b]} = \mathbb{P}_X^n \boldsymbol{\pi}_0,$$

where the notation $[a, b]$ represents playing game A a times, game B b times and so on, thus

$$\mathbb{P}_X = \begin{cases} \mathbb{P}_A & \text{if } n \bmod (a+b) < a, \\ \mathbb{P}_B & \text{otherwise.} \end{cases}$$

The deterministically mixed games can be implemented using a single transition matrix by grouping the periodic sequence. For example, $\mathbb{P}_{2,2} = \mathbb{P}_B^2 \mathbb{P}_A^2$ represents the equivalent transition matrix of playing $AABB$. Applying $\mathbb{P}_{2,2}$ is then equivalent to playing four consecutive games. Due to the multiple paths the capital can take within those four games, the algebra becomes tedious – a symbolic programming language is most advantageous.

Using the stationary distribution we can determine some statistical properties of the games, namely the mean μ, and standard deviation σ. We define a capital vector $\mathbf{x} = [-n, \dots, -1, 0, 1, \dots, n]$ so that the values correspond to the stationary probabilities in $\boldsymbol{\pi}$, thus the 0 in $\mathbf{x}$ should be aligned with the 1 in $\boldsymbol{\pi}_0$. The mean is then given by

$$\mu_n \equiv E[X_n] = \mathbf{x}\boldsymbol{\pi}_n \tag{5}$$

and the standard deviation by

$$\sigma_n = ((\mathbf{x} - \mu_n)^2 \boldsymbol{\pi}_n)^{1/2}, \tag{6}$$

where the squared vector term is an element-wise operation.

Several characteristics of the games are plotted in Fig. 2. The probability density functions (PDF) $p(x, n)$ of the games, which are equivalent to the stationary probabilities $\boldsymbol{\pi}$ are shown in Fig. 2a. However, since the capital must increase or decrease after each game, it leaves every second state with a zero stationary probability. To correct for this misleading characteristic a centered mean is taken, denoted by a hat,

$$\hat{p}(\mathbf{x}, n) = \frac{p(\mathbf{x}, n-1) + 2p(\mathbf{x}, n) + p(\mathbf{x}, n+1)}{4}, \tag{7}$$

which is the quantity plotted in Fig. 2a.

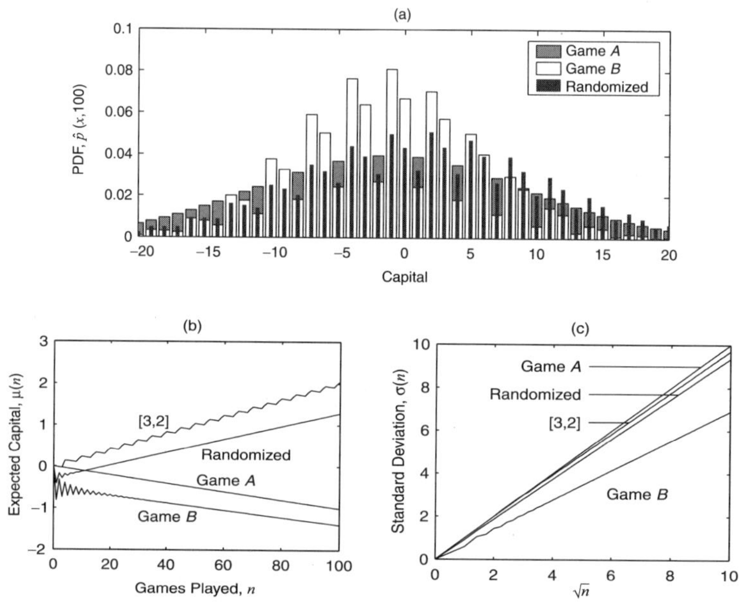

Figure 2: Characteristics for the capital-dependent games using (1). (a) The probability density function of the games using the centered mean of (7) with $\epsilon = 0$. (b) The expected outcome when playing the games individually and mixing with $\epsilon = 0.05$. The notation [3, 2] for example, refers to playing the sequence $AAABB\ldots$. (c) The standard deviations of the games, which are proportional to $\sqrt{n}$.

To better observe the ratchet potential that is exhibited by game B, a higher value of M is preferable, $M = 7$ with $p_1 = 0.075$ and $p_2 = 0.6032$ for example. This clearly shows the Brownian ratchet mechanism that the games are based on from [1].

In Fig. 2b the expected outcome of the games using (5) is plotted against the number of games played. This shows clearly the paradoxical result of the games – two losing games can combine to form a game with a winning expectation. One should note, however, that this is an apparent paradox, even though it has a counter-intuitive result that even experienced mathematicians find surprising; a proof is available that explains the situation.

Fig. 2c plots the standard deviations using (6) against $\sqrt{n}$ for the same games in Fig. 2b. This shows the behavior of the games does not diverge rapidly, but in fact the standard deviations of the games are all proportional to $\sqrt{n}$ and less than that of game A's.

2.3 Analysis using Equilibrium Distributions

When analyzing the games, it is sufficient to only consider whether the capital $X(n)$ is in a state relative to the modulus rule. Thus we can define a cyclic DTMC by

$$Y(n) \equiv X(n) \bmod M, \tag{8}$$

where $Y(t)$ has the states $\{0, \ldots, M-1\}$. If we win at the highest state $M-1$ we go back to state 0, and vice versa from state 0 to $M-1$. Thus, given an initial distribution of capital among the states and as $n \to \infty$ the probability of the capital being in any of the states reaches an equilibrium, $\boldsymbol{\pi}_n \to \boldsymbol{\pi} = [\pi_0, \ldots, \pi_{M-1}]$. From this equilibrium distribution, many properties of the games can be found analytically. The transition matrix associated with $Y(t)$ is

$$\mathbb{P}_B = \begin{bmatrix} 0 & 1-p_2 & & & p_2 \\ p_1 & 0 & \ddots & & \\ & p_2 & \ddots & 1-p_2 & \\ & & \ddots & 0 & 1-p_2 \\ 1-p_1 & & & p_2 & 0 \end{bmatrix}, \tag{9}$$

which is used to represent game B (or the randomised game by replacing each p with a q). This is restricted to $M \times M$ in size and the two extra entries (c.f. (4)) provide the cyclic nature of the chain.

From the transition matrix, there are many ways to find the stationary distribution, see [12,13] for example. Using $M = 3$ to simplify the algebra, the stationary distribution is

$$\boldsymbol{\pi}^B = \frac{1}{D}\begin{bmatrix} 1-p_2+p_2^2 \\ 1-p_2+p_1p_2 \\ 1-p_1+p_1p_2 \end{bmatrix}, \tag{10}$$

where $D = 3 - p_1 - 2p_2 + 2p_1p_2 + p_2^2$ is the normalization constant. If we let $p_1 = p_2 = p$ represent game A, then the stationary distribution simplifies to

$\boldsymbol{\pi}^A = (1/3)[1, 1, 1]^T$ as expected for a three-state chain. Using the probabilities of (1) with $\epsilon = 0$, the stationary distribution for game B turns out to be

$$\boldsymbol{\pi}^B = (1/13)[5, 2, 6]^T. \tag{11}$$

2.3.1 Capital-Dependent Games Constraints

It would be desirable, given a set of parameters, if constraints could be found to determine if Parrondo's paradox exists. An intuitive approach is finding the probability of winning using the stationary distribution, which is given by

$$p_{\text{win}} = \sum_{i=0}^{M-1} \pi_i p_i, \tag{12}$$

where p_i is the winning probability in state π_i. The games are winning, losing or fair when p_{win} is greater than, less than or equal to a half, which implies that $\langle X(n) \rangle$ is a decreasing, increasing or constant function with respect to n respectively.

For game A to lose, from (12) we get $p < 1/2$, or alternatively

$$\frac{1-p}{p} > 1. \tag{13}$$

The probability of winning game B by expanding (12) is

$$p_{\text{win}} = \pi_0 p_1 + (1 - \pi_0) p_2, \tag{14}$$

recalling that $\sum \pi_i = 1$. Subjecting $p_{\text{win}} < 1/2$ and using the stationary probabilities $\boldsymbol{\pi}^{B'}$ of (10) yields

$$\frac{(1-p_1)(1-p_2)^2}{p_1 p_2^2} > 1, \tag{15}$$

for $M = 3$. This is the condition that needs to be satisfied for game B to be losing.

For the randomized game we use the expression for game B, except replacing each p_i with a q_i, and conditioning the game to win by setting $p_{\text{win}} > 1/2$ leads to

$$\frac{(1-q_1)(1-q_2)^2}{q_1 q_2^2} < 1. \tag{16}$$

This is the condition for the randomized game to win. Therefore, in order for Parrondo's paradox to be exhibited we require probabilities and parameters to satisfy (13), (15) (i.e. to make games A and B lose) and (16) (i.e. make the randomized game win). This happens to be the case for $p = 5/11$, $p_1 = 1/121$, $p_2 = 10/11$ and $\gamma = 1/2$.

This type of analysis becomes tedious as M becomes larger due to the necessity of finding the equilibrium distribution. An alternative analysis, which can be solved for the general modulo M game, considers the conditions for recurrence of the corresponding DTMC and is given in [6]. The conditions that need to be satisfied for the generalized games are

$$\frac{1-p}{p} > 1, \tag{17}$$

$$\frac{(1-p_1)(1-p_2)^{M-1}}{p_1 p_2^{M-1}} > 1 \quad \text{and} \tag{18}$$

$$\frac{(1-q_1)(1-q_2)^{M-1}}{q_1 q_2^{M-1}} < 1. \tag{19}$$

Using this type of analysis it is possible to find other properties such as rate of return, range of ϵ where the paradox occurs and the probability space for example.

2.4 Explanation in Terms of Distributions

When investigating game B *prima facie*, it can be mistakenly interpreted as a winning game, thus invalidating the paradoxical result. This is due to taking the wrong line of analysis by considering the games statistically. This approach assumes the capital spends an equal amount of time in all states. When $M = 3$ it would be mistakenly assumed the capital is in each of the three states a third of the time. Then using the probabilities in (1) with $\epsilon = 0$ so the games are fair, the winning probability is calculated as

$$p_{\text{win}} = \frac{1}{3} \cdot \frac{1}{10} + \frac{1}{3} \cdot \frac{3}{4} + \frac{1}{3} \cdot \frac{3}{4} = \frac{16}{30},$$

which is greater than a half. This implies that the game B is winning, which is incorrect – it is actually fair.

As we have seen, the correct analysis is via DTMCs. Using the correct distribution probabilities from (11) the probability of winning is

$$p_{\text{win}} = \frac{5}{13} \cdot \frac{1}{10} + \frac{2}{13} \cdot \frac{3}{4} + \frac{6}{13} \cdot \frac{3}{4} = \frac{1}{2},$$

which correctly dictates that the game is fair. Subtracting a small amount ϵ from each of the probabilities makes $p_{\text{win}} < 1/2$ and the game is losing.

We notice that the construction of the game keeps the stationary distribution π^B locked at these values and manages to weight the probabilities so game B is losing. We can think of game B as consisting of two coins, a bad (C_1) and a good (C_2) coin biased to win according to p_1 and p_2 respectively. Then we use coin C_1, $5/13$ of the time and C_2 for the remaining time. If we can somehow 'flatten' the distribution of the game it can be made to win. This is achieved by mixing game B with something completely random like game A. This has the effect of playing the better coin C_2 more often than C_1, and hence producing a winning game.

This can be related to several observations in Fig. 2. The distributions of game B have a very definite shape whereas that of game A is smooth. It is this well-defined shape of game B that allows it to lose using both good and bad coins.

When mixing the two games evenly together the new PDF loses some of its shape. This is enough to allow the new game to be more evenly distributed, as seen from the skinny bars in Fig. 2a, to produce a winning game. It is this breaking up of the PDF of game B that leads to the paradoxical result. Note, in Fig. 2a the PDFs for games A and B have drifted to the left and that for the randomized game to the right.

The consequences of breaking the distribution are seen in the standard deviations in Fig. 2c. Since game A essentially represents free diffusion it has the largest standard deviation, whereas the standard deviation of game B is the smallest due to the capital being caught by the rules of the game. It should be of no then surprise then to find the standard deviation of the games formed by mixing A and B to lie between the standard deviation of the individual games.

3 The History Dependent Games

It has been shown that two losing capital dependent games can win, but are there any other types of games that have this characteristic? Although state dependent games are applicable in certain areas (see [7] for examples), it may be desirable to have a version of the games independent of capital. The answer to the aforementioned question is in the affirmative, in the form of history dependent games. These were also devised by Parrondo [8], although other implementations are possible [9].

3.1 Construction and Results

Game A is the same as before and we introduce game B', the modified version of the original game B. The probabilities that we use for the new game depend on the results of the two previous games, hence there are four options. Game B' is shown by a branching process in Fig. 3 and could be played using four biased coins.

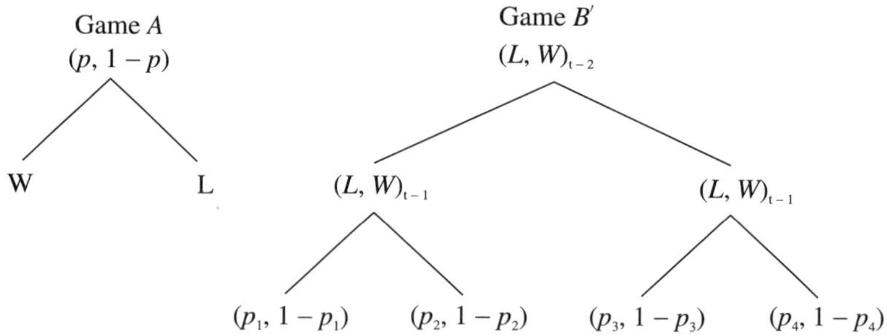

Figure 3: Construction of the history dependent games, game B' has four possible options {LL,LW,WL,WW}.

We can also parameterize the history-dependent games as

$$\begin{aligned} p &= 1/2 - \epsilon, \\ p_1 &= 9/10 - \epsilon, \\ p_2 &= p_3 = 1/4 - \epsilon \quad \text{and} \\ p_4 &= 7/10 - \epsilon. \end{aligned} \tag{20}$$

This parameterization gives Parrondo's original probabilities for the history-dependent games [8], which behave in much the same way as the parameterization of the capital-dependent games in (1). That is, games are fair when $\epsilon = 0$, losing when $\epsilon > 0$ and winning when $\epsilon < 0$. The method of analysis closely follows that of the capital-dependent games.

The same counter-intuitive result occurs when playing games A and B', that is, when playing the games individually they are losing, but switching between them creates a winning expectation. The switching can be either stochastic or deterministic as shown by various games that are plotted in Fig. 4. Similarly, there are initial stating transients, the magnitude and shape depending on the initial conditions used, i.e. LL, LW, WL or WW. The sequences shown in Fig. 4 are averaged from each of the four starting conditions, thus eliminating much of the transient behavior.

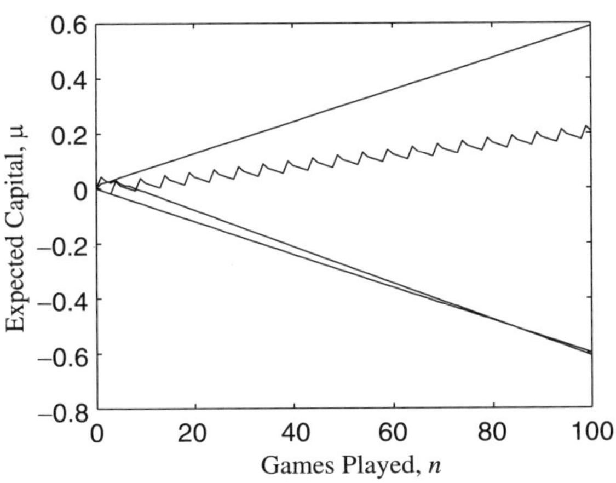

Figure 4: Games were played using the probabilities in (20) with $\epsilon = 0.003$, the results were averaged over each of the four starting conditions.

3.2 Analysis using DTMCs

When analyzing the chain that is associated with game B' we notice the capital $X(n)$ is not a Markovian process [8]. However, there are two ways to overcome this limitation; to model the game as a quasi-birth-and-death (QBD) process or define a state space $Y'(t)$ similar in nature to $Y(n)$ for the capital-dependent games.

With either method we require to record the past two events to determine what probability to use for the current game. Using a QBD process this is achieved by the use of phases, the second index in the state space E [14]. Details of the QBD formulation and the transition matrix directly representing game B' can be found in [9].

If we consider $Y(n)$ used to analyse game B, it only records where the capital is in each period in the periodic structure, not the absolute value of the capital. Similarly, we can define $Y'(n)$ as

$$Y'(n) = [X(n-1) - X(n-2), X(n) - X(n-1)], \tag{21}$$

which records the past events of the game. This gives four states as $[-1,-1]$, $[-1,+1]$, $[+1,-1]$ and $[+1,+1]$, where $+1$ represents winning and -1 losing. Using this representation we can perform the same types of analysis as for the capital-dependent games. The corresponding transition matrix to $Y'(n)$ is

$$\mathbb{P}_{B'} = \begin{bmatrix} 1-p_1 & 0 & 1-p_3 & 0 \\ p_1 & 0 & p_3 & 0 \\ 0 & 1-p_2 & 0 & 1-p_4 \\ 0 & p_2 & 0 & p_4 \end{bmatrix}, \tag{22}$$

with the rows and columns representing the four states LL, LW, WL and WW labelling from the top left corner. This matrix is always 4×4 as only the results of the previous two games are recorded. The stationary probabilities can be calculated as

$$\boldsymbol{\pi}^{B'} = \frac{1}{D'} \begin{bmatrix} (1-p_3)(1-p_4) \\ (1-p_4)p_1 \\ (1-p_4)p_1 \\ p_1 p_2 \end{bmatrix}, \tag{23}$$

where the normalization constant $D' = p_1 p_2 + (1+2p_1-p_3)(1-p_4)$. Using the probabilities of (20) with $\epsilon = 0$ gives $\boldsymbol{\pi}^{B'} = (1/22)[5, 6, 6, 5]^T$.

When randomly mixing the games, the probabilities can be given by

$$q_i = \gamma p + (1-\gamma) p_i, \tag{24}$$

for $i = 1, \ldots, 4$ and γ is the mixing parameter.

Thus, we can simply use the probability of winning to find constraints for the games paradox to exist. Using

$$p_{\text{win}} = \sum_{i=1}^{4} \pi_i p_i, \tag{25}$$

with the stationary probabilities of game B' in (23) yields

$$p_{\text{win}}^{B} = \frac{p_1(1+p_2-p_4)}{p_1 p_2 + (1-p_4)(1+2p_1-p_3)}. \tag{26}$$

Subjecting this to the constraint $p_{\text{win}} > 1/2$ for a winning game or $p_{\text{win}} < 1/2$ for a losing game, we have the following conditions,

$$\frac{1-p}{p} > 1, \tag{27}$$

$$\frac{(1-p_4)(1-p_3)}{p_1 p_2} > 1 \quad \text{and} \tag{28}$$

$$\frac{(1-q_4)(1-q_3)}{q_1 q_2} < 1 \tag{29}$$

for games A and B' to lose and the randomized game to win.

The explanation of the games in terms of the equilibrium distribution is the same as that for the capital-dependent games with the only difference being that for each value of capital there are four amounts to plot, LL, LW, WL and WW. If one plots the PDFs for history-dependent games it is easy to see how the introduction of game A breaks up the equilibrium distribution of game B'.

4 Summary

We have described two versions of Parrondo's games and given simple DTMC analysis of them. The analytical results derived match closely with computer simulations. The analysis was performed via use of simple Markov-chain theory and the apparent paradox explained in terms of breaking the equilibrium distribution set by game B. Observing the similarity between the capital-and history-dependent games, one may assume that further investigation may reveal other settings where the games can be applied.

Acknowledgements

This work was supported by the Dirección General de Enseñanza Superior e Investigación Científica (Spain) Project No. PB97-0076-C02, GTECH (U.S.A.), the Sir Ross and Sir Keith Smith Fund (Australia), and the Australian Research Council (ARC).

REFERENCES

[1] Harmer G. P. and Abbott D., Parrondo's paradox. *Statistical Science*, 14(2):206–213, 1999.

[2] Harmer G. P. and Abbott D., Parrondo's paradox: losing strategies cooperate to win. *Nature (London)*, 402:864, 1999.

[3] Adjari A. and Prost J., Drift induced by a periodic potential of low symmetry: pulsed dielectrophoresis. *C. R. Academy of Science Paris, Série II*, 315:1635–1639, 1992.

[4] Astumian R. D. and Bier M., Fluctuation driven ratchets: Molecular motors. *Physical Review Letters*, 72(11):1766–1769, 1994.

[5] Doering C. R., Randomly rattled ratchets. *Nuovo Cimento*, 17D(7–8):685–697, 1995.

[6] Harmer G. P., Abbott D., Taylor P. G. and Parrondo J. M. R., Parrondo's paradoxical games and the discrete Brownian ratchet. In D. Abbott and L. B. Kish, editors, *Second International Conference on Unsolved Problems of Noise and Fluctuations*, volume 511, pages 189–200, Adelaide, Australia, American Institute of Physics, 2000.

[7] Harmer G. P., Abbott D., Taylor P. G. and Parrondo J. M. R., Parrondo's games and Brownian ratchets. *Chaos* 11(3):705–714.

[8] Parrondo J. M. R., Harmer G. P. and Abbott D., New paradoxical games based on Brownian ratchets. *Physical Review Letters*, 85(24):5226–5229, 2000.

[9] Costa A., Fackrell M. and Taylor P. G., Two issues surrounding Parrondo's paradox. *Birkhäuser Annals of Dynamic Games*, This volume, 2004.

[10] Doob J. L., *Stochastic Processes*. John Wiley & Sons, Inc., New York, 1953.

[11] Onsager L., Reciprocal relations in irreversible processes I. *Physical Review*, 37:405–426, 1931.

[12] Pearce C. E. M., Entropy, Markov information sources and Parrondo games. In D. Abbott and L. B. Kish, editors, *Second International Conference on Unsolved Problems of Noise and Fluctuations*, volume 511, pages 207–212, Adelaide, Australia, American Institute of Physics, 2000.

[13] Pyke R., On random walks related to Parrondo's games. *Preprint math.PR/0206150*, 2001.

[14] Neuts M. F., *Matrix-Geometric Solutions in Stochastic Models: An Algorithmic Approach*. The John Hopkins University Press, USA, 1981.

Introduction to Quantum Games and a Quantum Parrondo Game*

Joseph Ng
Centre for Biomedical Engineering (CBME)
Department of Electrical and Electronic Engineering
The University of Adelaide
Australia
jng@eleceng.adelaide.edu.au

Derek Abbott
Centre for Biomedical Engineering (CBME)
Department of Electrical and Electronic Engineering
The University of Adelaide
Australia
dabbott@eleceng.adelaide.edu.au

Abstract

In this paper, we provide an introduction to quantum game theory through discussion of ways of converting classical games into the quantum regime. We illustrate how a quantum-based approach can simulate all possible classical game histories in parallel, for the example of Parrondo's games.

1 Introduction

Game theory has been used to describe and model the world in a variety of ways. A large group of these models rely, at some level, on probability or stochastic modeling. Quantum mechanics is a theory based on probability. So it is natural to extend classical game theory into the quantum regime. It has been suggested that classical game theory is, in fact, a limiting subset of quantum game theory.

Meyer introduced the idea of quantum game theory for two-person zero-sum games using deterministic and probabilistic strategies [6]. He proved that, in dynamic games[1], quantum strategies are indeed always at least as good as classical ones. Eisert *et al.* later published corresponding principles for static games[2] through examining the prisoner's dilemma in the quantum regime [2]. This work

*Presented at the 9th International Symposium on Dynamic Games and Applications held in Adelaide, South Australia in December 18–21, 2000

[1]Dynamic games are games where players play sequential moves in turn, e.g. chess

[2]Static games are where players make simultaneous decisions, e.g. the game of paper-stone-scissors

was later generalized by Benjamin and Hayden [1]. Attempts have even been made to produce a quantum Monty Hall game [4]. For a summary of quantum game theory, see Marinatto [5].

Section two describes briefly the classical Parrondo's paradox [3], in which we have the counter-intuitive phenomenon where two losing games combine to result in a winning game. For simplicity we have used Parrondo's original parameters – but note, however, that the games need not be restricted to this particular parameter set. The third section provides a brief introduction to quantum mechanics and general principles of quantum game theory. These have been simplified greatly but, we believe, provide sufficient explanation for those readers with no previous quantum mechanics background.

Sections four and five describe one method of playing classical Parrondo games through a quantum computer. The results are classical, but the ability of quantum computers to simulate classical systems efficiently is demonstrated. Classical systems require 2^n bits to represent the parameter space of an n bit system, where quantum computers only require n qubits.

2 Classical Parrondo's Paradox

Parrondo's Paradox [3] is a counter-intuitive result in which two statistically "losing" games (Game A and Game B) combine to create a "winning" game. This is best demonstrated by tossing coins where the coins are biased one way or another (towards winning or losing). The original game [3] is a capital-dependent (CD) game requiring feedback loops. Parrondo *et al.* later published a capital-independent but history-dependent (HD) game requiring feed-forward loops [7].

In both CD and HD games, Game A involves tossing a single coin which is slightly biased towards losing, i.e. $p_{1,\text{win}} = 1 - p_{1,\text{lose}} = 0.5 - \epsilon$, where ϵ is a small number.

Game B is where CD and HD games differ. In the CD game, there are two coins, biased at $p_{2,\text{win}} = 0.1 - \epsilon$ and the other at $p_{3,\text{win}} = 0.75 - \epsilon$. We play either coin 2 or coin 3 depending on the amount of capital (money), C, that we have at the moment, hence a Capital-Dependent (CD) game. If C is a multiple of 3, then we play coin 2, otherwise we play coin 3. This means that on average, coin 3 will be played more often than coin 2, but because coin 2 has such a poor probability of winning, it outweighs coin 3. This makes Game B a losing game overall.

On the other hand, 3 coins are used in Game B of the HD game. Depending on whether we won or lost in the previous game history, we choose one of the 3 coins to toss (see Table 1). The probabilities are given as $p_{2,\text{win}} = 0.9 - \epsilon$, $p_{3,\text{win}} = 0.25 - \epsilon$, $p_{4,\text{win}} = 0.7 - \epsilon$. As coin 3 is played much more frequently than the other coins, this is a losing game as well. It can be shown that the starting condition does not influence subsequent games, and so it is convenient to start the game with coin 2, and then coin 2 or 3 depending on the result of the first game.

It has been shown that for both the CD [3] and HD [7] games, by combining their respective losing games A and B, the combined game has a winning expectation overall.

Table 1: The choice of the next coin to play, Game_n, depends on the results of the previous two games. This table shows which coin to play.

Game_{n-2}	Game_{n-1}	Coin Played
Loss	Loss	2
Loss	Win	3
Win	Loss	3
Win	Win	4

3 Basic Quantum Mechanics and Quantum Game Theory

The difference between quantum game theory and conventional classical game theory comes from the ability to place the bits into a superposition of states and the ability to entangle the bits. These bits are thus called qubits (quantum bits). A qubit has two distinct states. They may be arbitrarily labeled "Heads" and "Tails" if we are dealing with coin tosses, or more generally, Win/Lose. For the purpose of computation, we shall label them "1" and "0". These are orthogonal states in Hilbert space. Now, when a qubit is in a superposition, we can think of it as being both 0 and 1 at the same time. However, when we measure the qubit, the superposition will collapse into one of the two states with the probability defined by the nature of the superposition.

The standard notation for expressing these quantum states is the Dirac Bra-Ket notation. Each state is written as $|\psi\rangle$. So, the 0 state is $|0\rangle$, called the 0 ket. A ket is a complex vector in Hilbert space. Superpositions are expressed as vector sums of state kets. In the case of qubits, it is $a|0\rangle + b|1\rangle$, where a and b are, in general, complex probability amplitudes [3] of the respective kets. A measurement is a projection of this superposition onto the basis kets, with each of the magnitudes being the probability that we will find the qubit in a particular state. In other words, $|a|^2$ and $|b|^2$ are the probabilities that when we measure the qubit, we will find 0 and 1 respectively. From this, we can also conclude that $|a|^2 + |b|^2 = 1$.

One of the easiest ways to picture a qubit is by considering photon polarizations (Fig. 1). We can define vertical polarization of the photon as $|0\rangle$ and horizontal polarization as $|1\rangle$. Now imagine that we have a single photon of 45° polarization – this is a superposition of $|0\rangle$ and $|1\rangle$. What happens when this photon arrives at a

[3] Quantum probability amplitudes differ from classical probability by obeying Feynman's rules rather than the classical Bayesian rules. In fact, a complex probability amplitude multiplied by its conjugate results in classical probability.

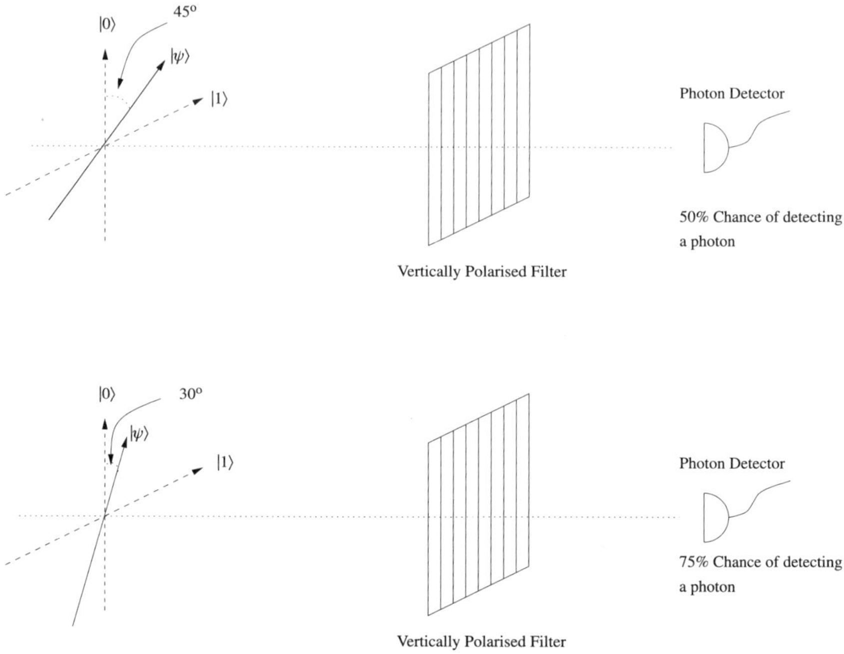

Figure 1: Using photon polarization as a qubit. For the 45° polarized photon, the photon detector behind the filter has a 50% chance of detecting that the photon has passed through the filter. For the 30° polarized photon, the chance is increased to 75%.

vertically polarized filter? This is a measurement of the photon, and thus the superposition will collapse. The photon will collapse into either a vertically polarized or a horizontally polarized state with 50/50 probability. This is an even superposition of $|0\rangle$ and $|1\rangle$. i.e. $(1/\sqrt{2})|0\rangle + (1/\sqrt{2})|1\rangle$. Obviously if the photon is vertically polarized, it will pass through, otherwise, it will not. Now if the photon is 30° polarized, then we can see that this is $(\sqrt{3}/2)|0\rangle + (1/2)|1\rangle$. This means that we have a $|(\sqrt{3}/2)|^2 = (3/4)$ chance of detecting that the photon has passed through the filter.

Quantum mechanically, we place a qubit into a superposition by rotating the state ket in Hilbert space. Suppose we start with $|\psi_i\rangle = |0\rangle$. To create $|\psi_f\rangle = (1/\sqrt{2})|0\rangle + (1/\sqrt{2})|1\rangle$, we rotate this ket by 45°. In our example, this can be achieved by rotating a vertically (or horizontally) polarized photon by 45° through a nonlinear medium or waveguide.

This can be thought of as a simple gambling game, where one person can bet on whether the detector behind the filter will register a photon or not. Since a 45° polarized photon has a 50% of passing through the filter, this is a fair gamble. A 30° photon on the other hand represents uneven odds, and so this is equivalent to, say, tossing a biased, weighted coin.

4 Quantum Parrondo's Paradox

We have chosen to simulate the HD Parrondo's paradox game. This is because the feedback loop required for the CD game will generally, but not necessarily, be irreversible. In quantum terms, irreversibility means that some information must be taken out of the system, which can be regarded as a measurement. As noted earlier, a measurement on a quantum superposition will cause the superposition to collapse into one of the eigenstates depending on the probability amplitudes and thus lose its strange quantum properties.

4.1 Simulating Game A

Suppose we have a ket, $|\psi\rangle = |T\rangle$, representing a single coin initially in the "Tails" state. Fig. 2 shows this in the two-dimensional Hilbert Space.

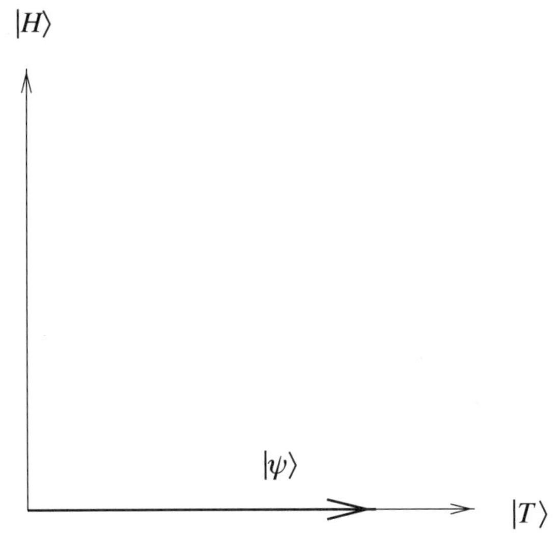

Figure 2: $|\psi\rangle = |T\rangle$ in 2-D Hilbert Space. When we measure the system, we will get $|T\rangle$ with probability 1.

If we do nothing to it, $|\psi\rangle$ will always be in the eigenstate $|T\rangle$ every time we measure it. This is analogous to leaving a coin sitting on the table (or having a vertically or horizontally polarized photon); it is either heads or tails (in this case, tails), and will stay that way until we do something to it.

A fair coin toss will be to rotate $|\psi\rangle$ by $0.5*(\pi/2) = (\pi/4)$ (Fig. 3). Fig. 4 shows this in a quantum circuit form. If we want to bias the coin, we just change the probability from 0.5 to some other probability, but we will assume an unbiased coin at the moment for simplicity. This puts $|\psi\rangle$ into the state $(1/\sqrt{2})|T\rangle + (1/\sqrt{2})|H\rangle$ and so if we measure the system now, the superposition has a probability $|(1/\sqrt{2})|^2 = 0.5$

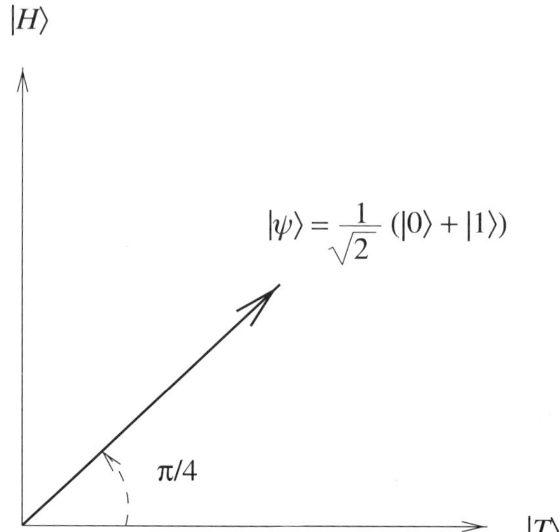

Figure 3: Rotating $|\psi\rangle$ by $\pi/4$ to create a superposition.

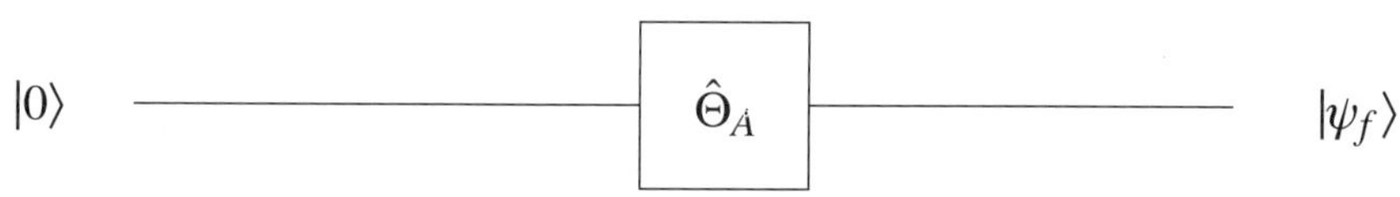

Figure 4: Quantum circuit representation of a single coin toss. For Parrondo's Game A, $\hat{\Theta}_A$ will rotate the qubit by $(0.5-\epsilon)*(\pi/2)$, assuming $\epsilon=0$, so $|\psi_f\rangle = (1/\sqrt{2})(|0\rangle + |1\rangle)$

of collapsing into the eigenstate $|T\rangle$ and a 0.5 chance of collapsing into the eigenstate $|H\rangle$. So this is a fair coin toss. A simplified analogy is if we sit the coin on its side, representing a superposition of heads and tails, and then slap a hand onto the coin and see what we are left with (measurement). Under fair circumstances, when we lift our hand, half the time, we find that we have heads, and tails the other half of the time. Also, if we manipulate the coin no further, it does not matter how many more times we slap our hand onto the coin, it will remain as heads or tails[4].

Algebraically this is a multiplication of the state ket, $|\psi\rangle = \begin{bmatrix} 1 \\ 0 \end{bmatrix} \begin{matrix} |T\rangle \\ |H\rangle \end{matrix}$ by a unitary rotation matrix $\hat{\Theta}(\theta) = \begin{bmatrix} \cos(\theta_A) & \sin(\theta_A) \\ -\sin(\theta_A) & \cos(\theta_A) \end{bmatrix}$ where the parameter θ_A gives the angle of rotation of the ket. Applying $\hat{\Theta}(\theta)$ on $|\psi\rangle$ results in:

$$\hat{\Theta}(\theta)|\psi\rangle = a|T\rangle + b|H\rangle,$$

[4]It must be stressed that the example given is only an analogy and not a true quantum superposition. A coin is a classical object. A 45° polarized photon, however, is a true quantum superposition.

$$\begin{bmatrix} \cos(\theta_A) & \sin(\theta_A) \\ -\sin(\theta_A) & \cos(\theta_A) \end{bmatrix} \begin{bmatrix} 1 \\ 0 \end{bmatrix} = \begin{bmatrix} a \\ b \end{bmatrix},$$

where $a = (1/\sqrt{2}), b = -(1/\sqrt{2})$ for $\theta_A = \pi/4$. To simulate the Game A described by Parrondo *et al.* [7], we simply choose $\theta_A = (0.5 - \epsilon) * \pi/2$. We can see that in this example, the probability amplitude b is negative. As mentioned earlier, probability amplitudes are, in general, complex quantities. This reflects that they have both magnitude and phase components. The only limiting factor, in this case, being that $|b|^2 = 1/2$.

4.2 Two or More Coin Tosses

For two tosses of the coin, we cannot just use the rotation matrix on the same qubit again. If we did, we would put $|\psi\rangle$ into the eigenstate $|H\rangle$, which is obviously not representative of tossing the coin twice (Fig. 5).

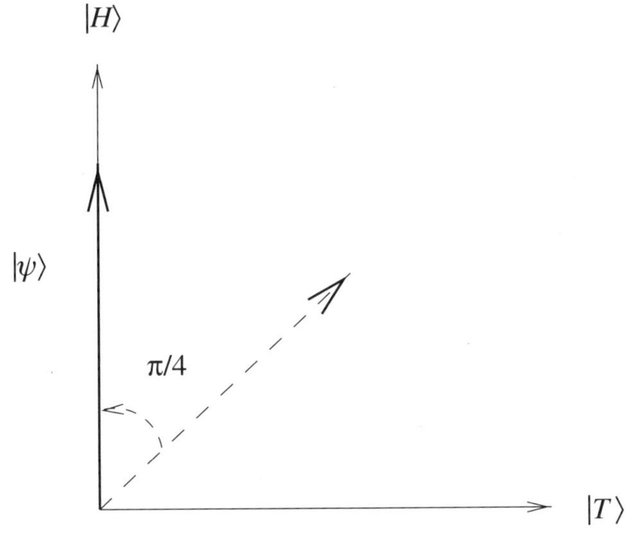

Figure 5: Rotating $|\psi\rangle$ by $\pi/4$ again will place the quantum coin into the $|H\rangle$ state. So when we now measure the system, we will get Heads with probability 1.

What we need is to use another qubit, $|\psi_2\rangle$, and rotate that one instead. Now the total state of the system can be described as $|\psi\rangle = |\psi_1\psi_2\rangle$ in the 4-dimensional Hilbert space with $|TT\rangle, |TH\rangle, |HT\rangle, |HH\rangle$ as its base kets. The rotation matrices in this case would be the tensor product of $\hat{\Theta}(\theta)$ and $\hat{I}$, i.e. for the first qubit,

$$\hat{\Theta}_1(\theta) = \hat{\Theta}(\theta) \otimes \hat{I}$$
$$= \begin{bmatrix} \cos(\theta_A) & \sin(\theta_A) \\ -\sin(\theta_A) & \cos(\theta_A) \end{bmatrix} \otimes \begin{bmatrix} 1 & 0 \\ 0 & 1 \end{bmatrix}$$

$$= \begin{bmatrix} \cos(\theta_A) & 0 & \sin(\theta_A) & 0 \\ 0 & \cos(\theta_A) & 0 & \sin(\theta_A) \\ -\sin(\theta_A) & 0 & \cos(\theta_A) & 0 \\ 0 & -\sin(\theta_A) & 0 & \cos(\theta_A) \end{bmatrix},$$

and for the second qubit,

$$\begin{aligned} \hat{\Theta}_2(\theta) &= \hat{I} \otimes \hat{\Theta}(\theta) \\ &= \begin{bmatrix} 1 & 0 \\ 0 & 1 \end{bmatrix} \otimes \begin{bmatrix} \cos(\theta_A) & \sin(\theta_A) \\ -\sin(\theta_A) & \cos(\theta_A) \end{bmatrix} \\ &= \begin{bmatrix} \cos(\theta_A) & \sin(\theta_A) & 0 & 0 \\ -\sin(\theta_A) & \cos(\theta_A) & 0 & 0 \\ 0 & 0 & \cos(\theta_A) & \sin(\theta_A) \\ 0 & 0 & -\sin(\theta_A) & \cos(\theta_A) \end{bmatrix}. \end{aligned}$$

So the total system, $|\psi\rangle$, becomes $|\psi\rangle = \hat{\Theta}_2\hat{\Theta}_1|\psi_1\psi_2\rangle$, which is a superposition of the base kets. i.e. $a|TT\rangle + b|TH\rangle + c|HT\rangle + d|HH\rangle$. As before, $|a|^2$, $|b|^2$, $|c|^2$, $|d|^2$ represent the classical probability of obtaining the respective states if we measure the system and $|a|^2 + |b|^2 + |c|^2 + |d|^2 = 1$.

So what does it all mean if we find that the system is in the state, say, $|TH\rangle$?

Now the T represents the 1st qubit and the H, the 2nd qubit. As we have defined qubit 1 as the result of the 1st toss, and qubit 2 as the 2nd, $|TH\rangle$ means that we have a tail at the 1st toss, followed by a head. It gives us the toss history of the set of games. So, if a head is considered a win, and tails, a loss, then $|b|^2$ is the probability of losing the first game, and then winning the second game[5].

5 Simulating Game B

For Game B, we employ a very similar approach to Game A. However, the difference is that we will now use a Controlled-Controlled-Rotation ($CCRot$) matrix. A $CCRot$ gate is a 3-qubit gate, where the 3rd bit is rotated when the first 2 qubits are 1 (Fig. 6). The truth table of a $CCRot$ gate is given in Table 2:

Table 2: The truth table of a $CCRot$ gate. The rotation is given by θ.

$\|q_0\rangle$	$\|q_1\rangle$	$\|q_2\rangle$
$\|0\rangle$	$\|0\rangle$	$\|0\rangle$
$\|0\rangle$	$\|1\rangle$	$\|0\rangle$
$\|1\rangle$	$\|0\rangle$	$\|0\rangle$
$\|1\rangle$	$\|1\rangle$	$\cos(\theta)\|0\rangle + \sin(\theta)\|1\rangle)$

[5]From now on, we shall represent Heads as 1, and Tails as 0, i.e. $|00\rangle \equiv |TT\rangle$, $|11\rangle \equiv |HH\rangle$ etc.

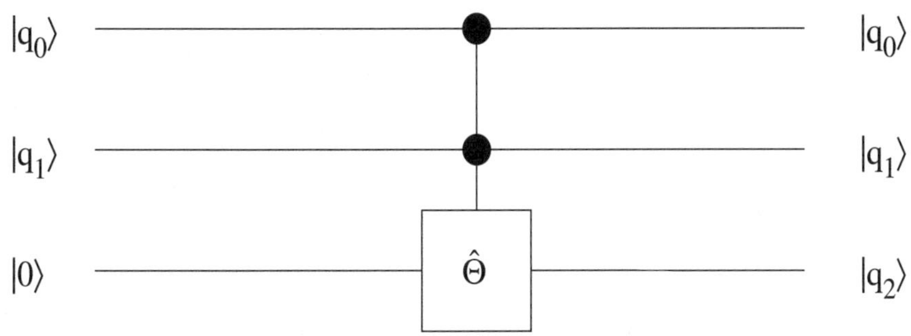

Figure 6: A Controlled-Controlled-Rotation gate/matrix. If qubit 2, $|q_2\rangle$, starts off in the $|0\rangle$ state, it will be rotated if both $|q_0\rangle$ and $|q_1\rangle$ are in the $|1\rangle$ state.

In matrix form, this is

$$\begin{matrix} |000\rangle \\ |001\rangle \\ |010\rangle \\ |011\rangle \\ |100\rangle \\ |101\rangle \\ |110\rangle \\ |111\rangle \end{matrix} \begin{bmatrix} 1 & 0 & 0 & 0 & 0 & 0 & 0 & 0 \\ 0 & 1 & 0 & 0 & 0 & 0 & 0 & 0 \\ 0 & 0 & 1 & 0 & 0 & 0 & 0 & 0 \\ 0 & 0 & 0 & 1 & 0 & 0 & 0 & 0 \\ 0 & 0 & 0 & 0 & 1 & 0 & 0 & 0 \\ 0 & 0 & 0 & 0 & 0 & 1 & 0 & 0 \\ 0 & 0 & 0 & 0 & 0 & 0 & \cos(\theta_B) & \sin(\theta_B) \\ 0 & 0 & 0 & 0 & 0 & 0 & -\sin(\theta_B) & \cos(\theta_B) \end{bmatrix}.$$

The $CCRot$ matrix is perfect for what we need to do because in Game B of the HD game, the choosing of the coin for the 3rd toss (qubit) is dependent on the previous 2 results (qubits). But Game B is a little more than the above. As mentioned earlier, the above matrix will rotate the 3rd bit if the first 2 qubits are 1. In our context, this means that the state of the system is only changed if we won the previous 2 games, i.e. this simulates choosing and tossing coin 4 in Game B. What we need is to obtain 3 other variations of the $CCRot$ gate to simulate the other coins for the different possible histories. So for coins 2, 3 and 4, their respective matrices are:

$$\hat{\Theta}_{\text{win,win}} = \begin{bmatrix} 1 & 0 & 0 & 0 & 0 & 0 & 0 & 0 \\ 0 & 1 & 0 & 0 & 0 & 0 & 0 & 0 \\ 0 & 0 & 1 & 0 & 0 & 0 & 0 & 0 \\ 0 & 0 & 0 & 1 & 0 & 0 & 0 & 0 \\ 0 & 0 & 0 & 0 & 1 & 0 & 0 & 0 \\ 0 & 0 & 0 & 0 & 0 & 1 & 0 & 0 \\ 0 & 0 & 0 & 0 & 0 & 0 & \cos(\theta_{B_4}) & \sin(\theta_{B_4}) \\ 0 & 0 & 0 & 0 & 0 & 0 & -\sin(\theta_{B_4}) & \cos(\theta_{B_4}) \end{bmatrix},$$

$$\hat{\Theta}_{\text{win,lose}} = \begin{bmatrix} 1 & 0 & 0 & 0 & 0 & 0 & 0 & 0 \\ 0 & 1 & 0 & 0 & 0 & 0 & 0 & 0 \\ 0 & 0 & 1 & 0 & 0 & 0 & 0 & 0 \\ 0 & 0 & 0 & 1 & 0 & 0 & 0 & 0 \\ 0 & 0 & 0 & 0 & \cos(\theta_{B_3}) & \sin(\theta_{B_3}) & 0 & 0 \\ 0 & 0 & 0 & 0 & -\sin(\theta_{B_3}) & \cos(\theta_{B_3}) & 0 & 0 \\ 0 & 0 & 0 & 0 & 0 & 0 & 1 & 0 \\ 0 & 0 & 0 & 0 & 0 & 0 & 0 & 1 \end{bmatrix},$$

$$\hat{\Theta}_{\text{lose,win}} = \begin{bmatrix} 1 & 0 & 0 & 0 & 0 & 0 & 0 & 0 \\ 0 & 1 & 0 & 0 & 0 & 0 & 0 & 0 \\ 0 & 0 & \cos(\theta_{B_3}) & \sin(\theta_{B_3}) & 0 & 0 & 0 & 0 \\ 0 & 0 & -\sin(\theta_{B_3}) & \cos(\theta_{B_3}) & 0 & 0 & 0 & 0 \\ 0 & 0 & 0 & 0 & 1 & 0 & 0 & 0 \\ 0 & 0 & 0 & 0 & 0 & 1 & 0 & 0 \\ 0 & 0 & 0 & 0 & 0 & 0 & 1 & 0 \\ 0 & 0 & 0 & 0 & 0 & 0 & 0 & 1 \end{bmatrix},$$

$$\hat{\Theta}_{\text{lose,lose}} = \begin{bmatrix} \cos(\theta_{B_2}) & \sin(\theta_{B_2}) & 0 & 0 & 0 & 0 & 0 & 0 \\ -\sin(\theta_{B_2}) & \cos(\theta_{B_2}) & 0 & 0 & 0 & 0 & 0 & 0 \\ 0 & 0 & 1 & 0 & 0 & 0 & 0 & 0 \\ 0 & 0 & 0 & 1 & 0 & 0 & 0 & 0 \\ 0 & 0 & 0 & 0 & 1 & 0 & 0 & 0 \\ 0 & 0 & 0 & 0 & 0 & 1 & 0 & 0 \\ 0 & 0 & 0 & 0 & 0 & 0 & 1 & 0 \\ 0 & 0 & 0 & 0 & 0 & 0 & 0 & 1 \end{bmatrix}.$$

Now Game B is obtained by putting these 4 $CCRot$ matrices one after another (Fig. 7). i.e.

$$\hat{\Theta}_B = \hat{\Theta}_0\hat{\Theta}_1\hat{\Theta}_2\hat{\Theta}_3 = \begin{bmatrix} B_2 & 0 & \dots & 0 \\ 0 & B_3 & 0 & \vdots \\ \vdots & 0 & B_3 & 0 \\ 0 & \dots & 0 & B_4 \end{bmatrix},$$

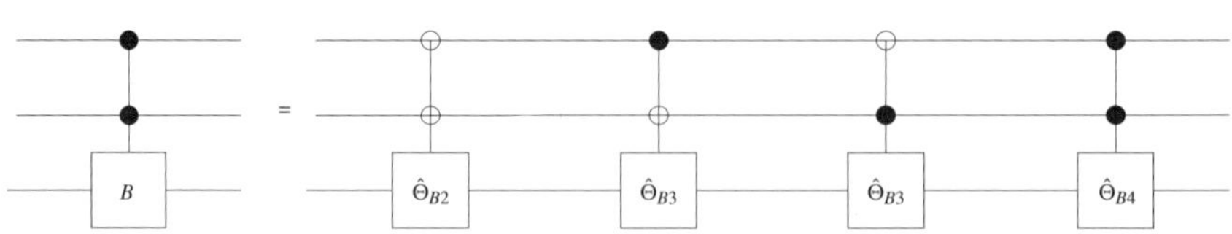

Figure 7: Combining the different rotation matrices to form the Game B quantum gate. A solid dot means that the control qubit must be in the $|1\rangle$ state for the gate to rotate the target qubit. An open circle means that the control qubit must be in the $|0\rangle$ state.

where

$$B_2 = \begin{bmatrix} \cos(\theta_{B2}) & \sin(\theta_{B2}) \\ -\sin(\theta_{B2}) & \cos(\theta_{B2}) \end{bmatrix},$$

$$B_3 = \begin{bmatrix} \cos(\theta_{B3}) & \sin(\theta_{B3}) \\ -\sin(\theta_{B3}) & \cos(\theta_{B3}) \end{bmatrix},$$

$$B_4 = \begin{bmatrix} \cos(\theta_{B4}) & \sin(\theta_{B4}) \\ -\sin(\theta_{B4}) & \cos(\theta_{B4}) \end{bmatrix}.$$

As the $\hat{\Theta}_i$ matrices commute with each other, the order is not important. All we need to do is to vary the amount of rotation for each $\hat{\Theta}_i$, this gives us the required matrix for simulating Game B which we will denote by $\hat{\Theta}_B$.

5.1 Combining Games A and B

To combine the two games, all we need to do is to decide on the number of games, create the correct rotation matrices for these games, and then apply these matrices to an initial state ket, $|\psi_0\rangle = |00\ldots0\rangle$. For example, to play two games of Game A and a game of Game B, the final state of the system is $|\psi_f\rangle = \hat{\Theta}_1\hat{\Theta}_2\hat{\Theta}_3|000\rangle$ (see Fig. 8), which will be a superposition of all possible outcomes, so $|\psi_f\rangle = a|000\rangle + b|001\rangle + c|010\rangle + \ldots + h|111\rangle$. This means that we can now plot a graph of probability vs. outcome, and thus work out the most likely histories if we play the game infinite times (Fig. 9). Fig. 10 shows the results for playing two games of A's followed by two games of B's followed by two games of A's etc. for 10 games.

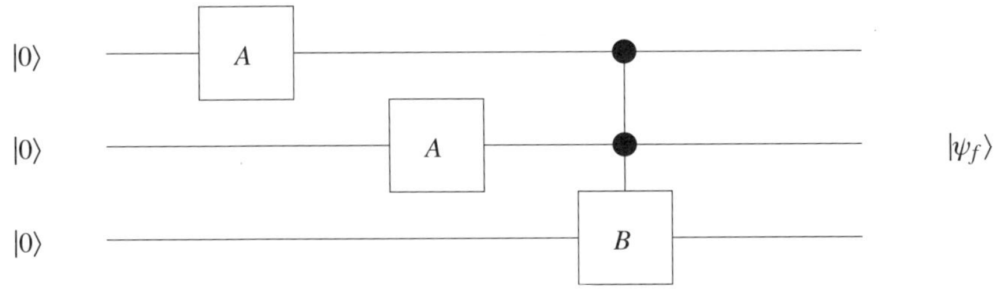

Figure 8: A 3-toss game, where we play two games of A followed by one game of B.

6 Discussion

What we have done so far is essentially the same as calculating each of the possible outcomes by multiplying the respective probabilities. So what we have here is a quantum game that produces classical results, where the two losing games combine to create a winning game (Fig. 11). It is a quantum simulation of a classical system. However, on a classical computer, to calculate every possible history for n games

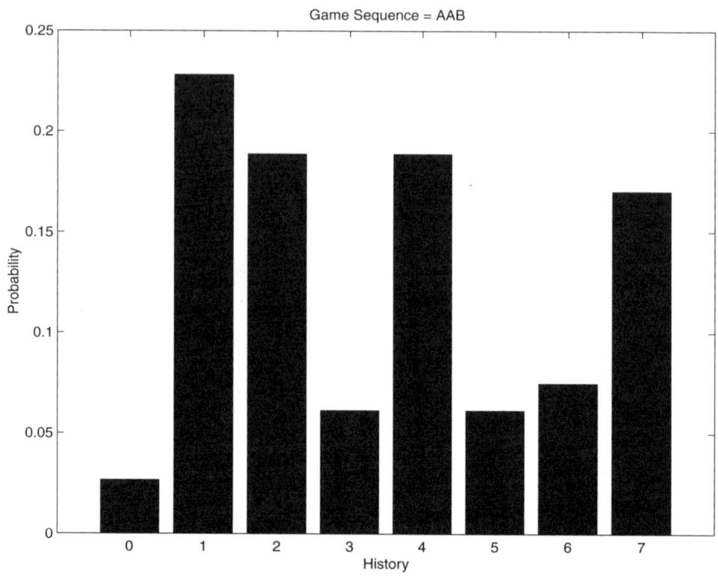

Figure 9: Classical probability vs history plot of playing two games of A's followed by one game of B. The indices are binary strings converted into decimals. For example, the bar for 2 (010) is the probability of losing the first game, winning the second, and losing the third.

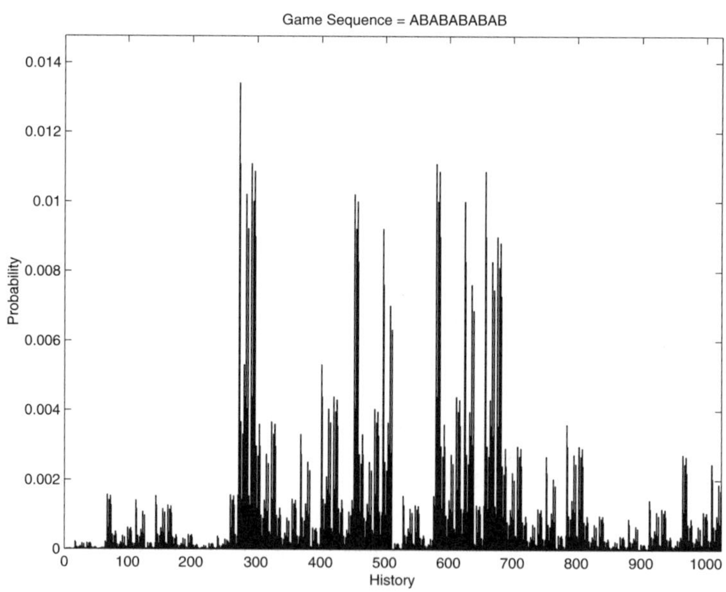

Figure 10: Probability vs history plot of playing 10 games. These games are played by playing two games of A's, followed by two games of B's and so on.

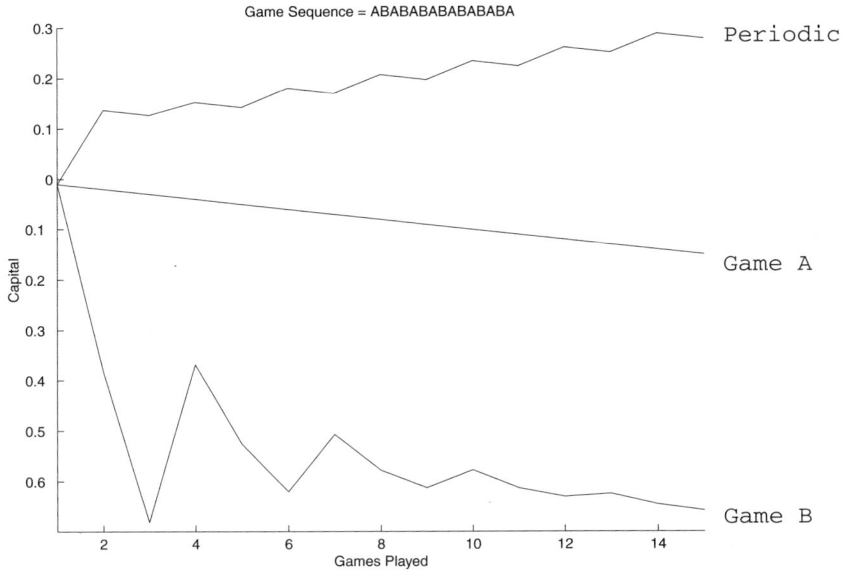

Figure 11: When we combine the two losing games (Game A and Game B) in a periodic fashion (alternate game A and B), a winning game results. This result is identical to that of classical simulations [7].

require 2^n bits. On a quantum computer, on the other hand, only n qubits are required: an improvement of $\log(n)$. The rotation matrices are $2^n \mathrm{x} 2^n$ because it is a classical way of representing, calculating and simulating quantum processes.

Looking back at Fig. 2, it is natural to ask what happens if we extend the axes to allow the coefficients of $|0\rangle$ and $|1\rangle$, a and b, to be negative and/or complex? These complex amplitudes are taken into account by a phase factor $e^{i\phi}$ which is inserted into the rotation matrices. So the basic rotation matrix, $\hat{\Theta}(\theta)$, becomes

$$\hat{\Theta}(\theta, \phi) = \begin{bmatrix} e^{i\phi}\cos(\theta) & \sin(\theta) \\ \sin(\theta) & e^{-i\phi}\cos(\theta) \end{bmatrix}.$$

Since the actual probabilities are the lengths of the state kets, these complex amplitudes will still produce classical results under normal circumstance. However, if made to interact with each other, they can produce radically different results. Two probability amplitudes of the same magnitude and different signs can cancel each other out, resulting in destructive interference. This will not happen in classical game theory because classical probabilities are always real and positive.

So how do we cause these quantum probability amplitudes to interact? Eisert's approach [2] was to employ an entangling gate, J, which calculates the payoff of the two parties. By varying the entangling parameter (which is essentially a phase parameter) in J, Eisert's result showed that the classical problem of Prisoner's dilemma is a subset of the quantum game, and there is no longer a dilemma when the game is fully explored in the quantum regime.

Parrondo's game can be seen as a competition between two players: Casino (C) and Parrondo (P). Both Parrondo's and Casino's aim is to maximize the winnings or minimize the losses. As mentioned above, despite the Casino's Game A and Game B being originally unfairly biased against Parrondo, he can construct a combined winning game by playing the games in certain sequences. However, one of the Casino's business strategists has read Parrondo's published paper, and brought the issue up at the next Casino board meeting. At that meeting, it was decided that the Casino should employ quantum mechanics to help them turn the tables back in their favor. This was done by implementing a Casino Gate (Fig. 12), $C(\phi)$, and a "de-Casino" gate $C^{\dagger}(\phi)$, where $C(\phi) = C^{\dagger}(\phi) = \begin{bmatrix} \cos(\gamma) & \sin(\gamma) \\ -\sin(\gamma) & \cos(\gamma) \end{bmatrix}$. For n games, the resultant state is $|\psi_f\rangle = C^{\dagger}(\gamma)G(n)C(\gamma)|0\rangle^{\otimes n}$ where $G(n)$ is the collection of quantum gates that describes the sequence of games played. This can be thought of as the player walking through the doors from the classical world into the quantum Casino, and later, from the Casino back into the classical world. By setting $\gamma = \pi/2$, Game A remains the same but Game B becomes a winning

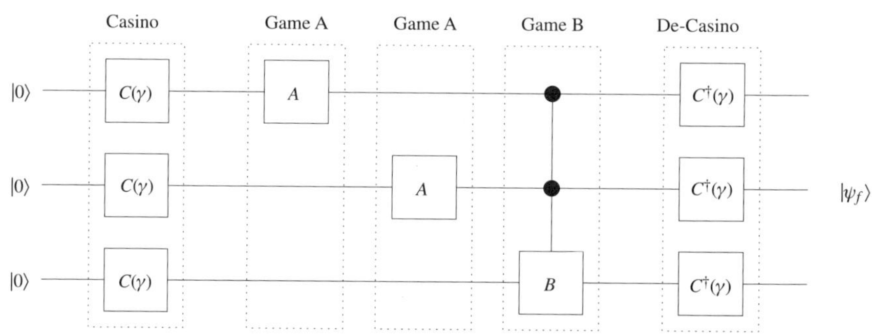

Figure 12: The Casino adopts a quantum strategy by employing entangling gates, $J(\gamma)$. The degree of entanglement is determined by γ, with $\gamma = 0$ representing no entanglement, and $\gamma = \pi/2$ representing maximum entanglement.

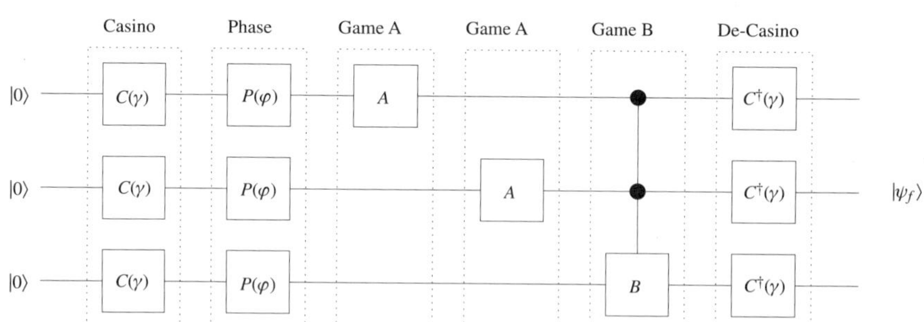

Figure 13: The player also adopts a quantum strategy by employing phase gates, $P(\phi)$.

game, yet through Parrondo's strategy, the combined game is now a losing game (Fig. 14). Interestingly, although Game B wins faster than Game A, the combined game is still losing. In fact, it loses faster than just playing Game A on its own. This is a different paradox to the original!

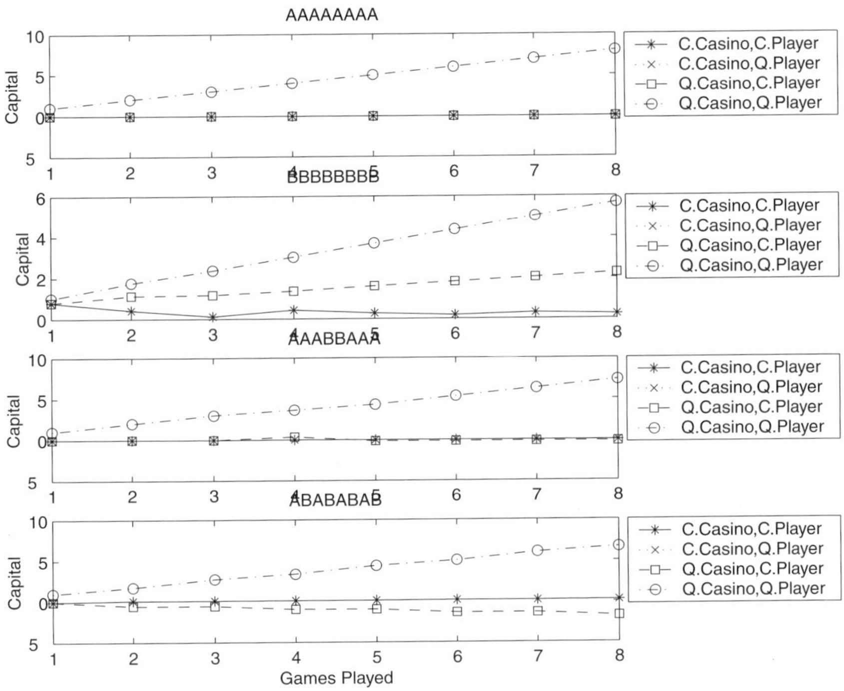

Figure 14: Plotting Capital vs Games for varying γ and ϕ and varying game sequences (as labeled). Classical Casino uses $\gamma = 0$, Quantum Casino uses $\gamma = \pi/2$. Classical Parrondo uses $\phi = 0$, Quantum Parrondo uses $\phi = \pi/2$. As can be seen, for a classical casino, the results are the same regardless of whether Parrondo uses a quantum coin or not.

However, it did not take long for Parrondo to realise that this sudden change of fortune is not simply a statistical abnormality, but rather, due to the Casino's quantum strategy. So he decides to beat the Casino at their own game again, and adopts a quantum strategy as well. This is done with a phase-shift gate (Fig. 13), $P(\phi) = \begin{bmatrix} e^{i\phi} & 0 \\ 0 & e^{-i\phi} \end{bmatrix}$. Now, the resultant state is $|\psi_f\rangle = C^\dagger(\gamma)G(n)P(\phi)C(\gamma)|0\rangle^{\otimes n}$. This causes both Game A and Game B to become winning games, and combine to create a winning overall game. So when both the Casino and Parrondo adopt quantum strategies, the original Parrondo game is no longer a paradox.

So as can be seen, Parrondo's best strategy lies in employing a quantum strategy. He is guaranteed to win regardless of the strategy used by the Casino (Fig. 15).

Unfortunately (or fortunately, depending on how one prefers to see the situation), the same cannot be said for the Casino however. If the casino adopts a quantum strategy, Parrondo can choose a classical strategy and play only the winning Game B or a quantum strategy and still win.

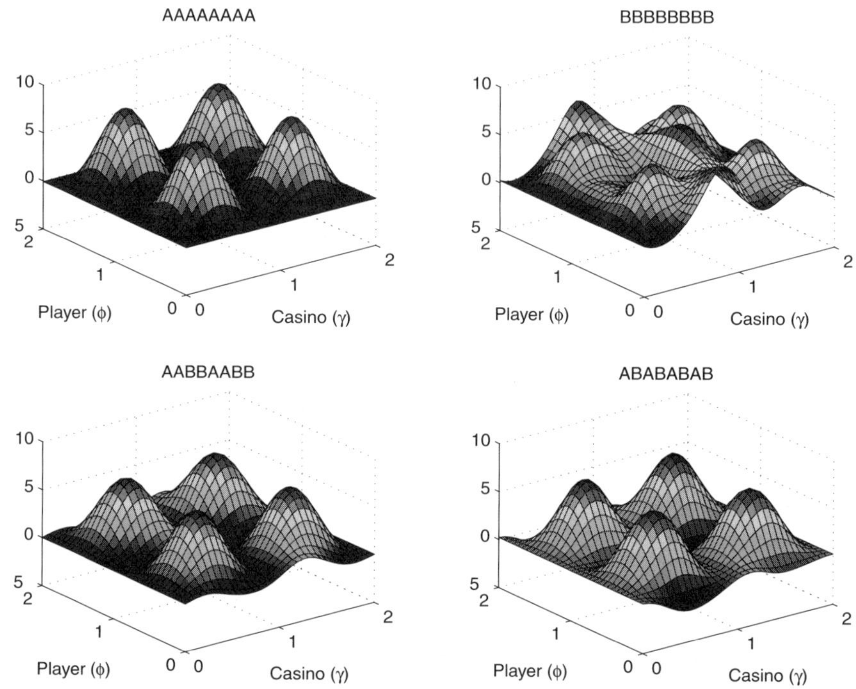

Figure 15: Plotting the capital after playing 8 games as labeled. The axes are plotted from 0 to 2π. γ is the casino's parameter, while ϕ is Parrondo's parameter. At $\gamma = \phi = 0$, we have an entirely classical game. In fact, for all $\gamma = 0$, the results are the same as classical results, so in the plots, we have a straight line at $\gamma = 0$, regardless of what ϕ is.

7 Conclusion

Parrondo's games are of general interest as they illustrate how two losing coin tossing games can win when combined either in deterministic or non-deterministic sequences. For this phenomenon to occur, there must be coupling between the games. In Section 2 we saw that the CD games couple via capital-based state-dependence and the HD games couple via history-based state-dependence. The open question is, can a quantum Parrondo game be designed such that the coupling is via quantum entanglement?

For the case of non-deterministic sequences of games A and B, Game A can be thought of as "noise" that breaks up the state-dependent rules that are biasing

Game B to lose – and this is why the combination of A and B wins ("the Boston Interpretation"). So another open question for quantum Parrondo games is, can the effect of Game A be in fact replaced by some form of decoherence such as a measurement?

Acknowledgements

We would like to thank Gerard Milburn, Bill Munro, Ben Travaglione and Michael Nielsen of the SRC for Quantum Computer Technology, University of Queensland for all the inspirational and educational discussions over the course of this work. Thanks are also due to Wanli Li, Dept. of Physics, Princeton, for a number of manuscript suggestions. Funding from GTECH and the Sir Ross and Sir Keith Smith Fund is gratefully acknowledged.

REFERENCES

[1] Benjamin S. C. and Hayden P. M., Multi-player quantum games. *Phys. Rev. A* 64:030301(R), 2001 See also LANL Preprint quant-ph/0007038.

[2] Eisert J., Wilkens M. and Lewenstein M., Quantum games and quantum strategies. *Phys. Rev. Lett.*, 83:3077, 1999. See also LANL Preprint quant-ph/9806088.

[3] Harmer G. P. and Abbott D., Losing strategies can win by Parrondo's paradox. *Nature (London)*, 402:864, 1999.

[4] Li C-F., Zhang Y-S., Huang Y-F. and Guo G-C., Quantum strategies of quantum measurement *Phys. Lett.* A 280:257, 2001 See also LANL Preprint quant-ph/0007120.

[5] Marinatto L. and Weber T., A quantum approach to static games of complete information. *Phys. Lett.* A 272:291, 2000 See also LANL Preprint quant-ph/0004081.

[6] Meyer D.A., Quantum strategies. *Phys. Rev. Lett.*, 82:1052, 1999. See also LANL Preprint quant-ph/9804010.

[7] Parrondo J. M. R., Harmer G. P. and Abbott D., New paradoxical games based on brownian ratchets. *Phys. Rev. Lett.*, 85(24):5226–5229, December 2000. See also LANL Preprint cond-mat/0003386.

A Semi-quantum Version of the Game of Life

Adrian P. Flitney
Centre for Biomedical Eng. (CBME)
Department of Electrical and Electronic Engineering
The University of Adelaide
Australia
aflitney@physics.adelaide.edu.au

Derek Abbott
Centre for Biomedical Engineering (CBME)
Department of Electrical and Electronic Engineering
The University of Adelaide
Australia
dabbott@eleceng.adelaide.edu.au

Abstract

A version of John Conway's game of Life is presented where the normal binary values of the cells are replaced by oscillators which can represent a superposition of states. The original game of Life is reproduced in the classical limit, but in general additional properties not seen in the original game are present that display some of the effects of a quantum mechanical Life. In particular, interference effects are seen.

Key words. Cellular automata, quantum games, quantum cellular automata

AMS Subject Classifications. Primary 68Q80; Secondary 37B15

1 Introduction

John Conway's game of Life [10] is a well-known two-dimensional cellular automaton where cells are arranged in a square grid and have binary values generally known as dead or alive. The status of the cells change in a discrete fashion, each "generation" depending upon the number of neighboring cells that are alive, the general idea being that a cell dies if there is either overcrowding or isolation. There are many different rules that can be applied for birth or survival of a cell and a number of these give rise to interesting properties such as still lives (stable patterns), oscillators (patterns that periodically repeat), spaceships or gliders (fixed shapes that move across the Life universe), glider guns, and so on [3,12,11]. Conway's original rules are some of the few that are balanced between survival and extinction of the Life "organisms." In this version a dead (or empty) cell

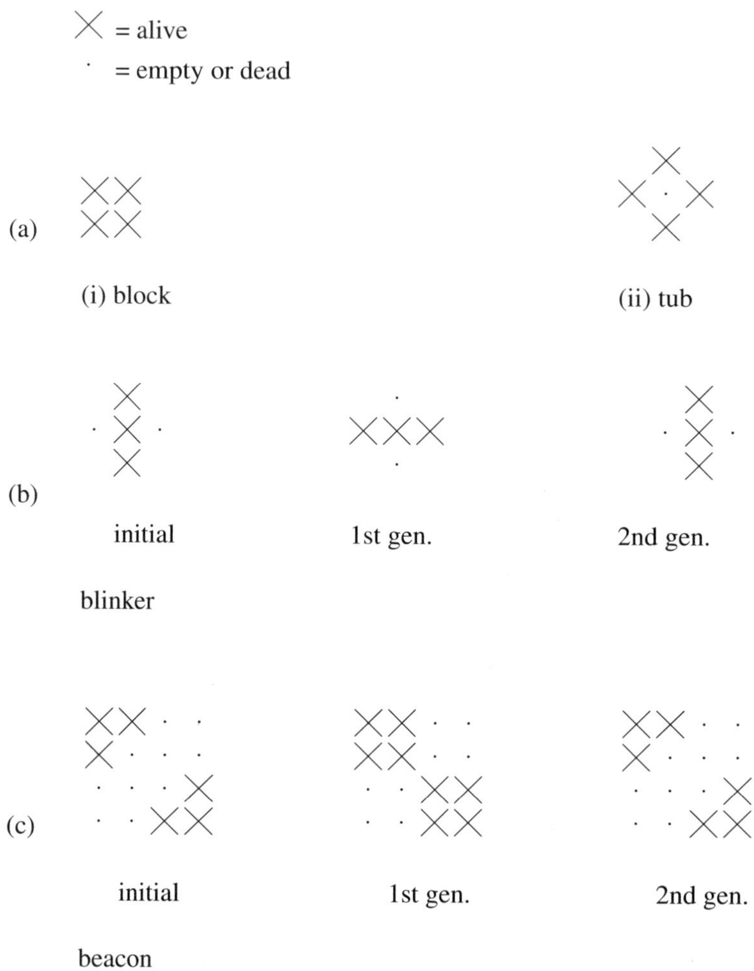

Figure 1: A small sample of the simplest structures in Conway's Life: (a) the simplest still-lives (stable patterns), the block and the tub, and the simplest oscillators (periodic patterns), (b) the blinker and (c) the beacon, both of period two. A number of blocks and blinkers will normally evolve from any moderate-sized random collection of alive and dead cells.

becomes alive if it has exactly three living neighbors, while an alive cell survives if and only if it has two or three living neighbors. Much literature on the game of Life and its implications exists. For a recent discussion on the possibilities of this and other cellular automata the interested reader is referred to reference [24]. The simplest still lives and oscillators are given in figure 1, while figure 2 shows a glider, the simplest and most common moving form. A large enough random collection of alive and dead cells will, after a period of time, usually decay into a

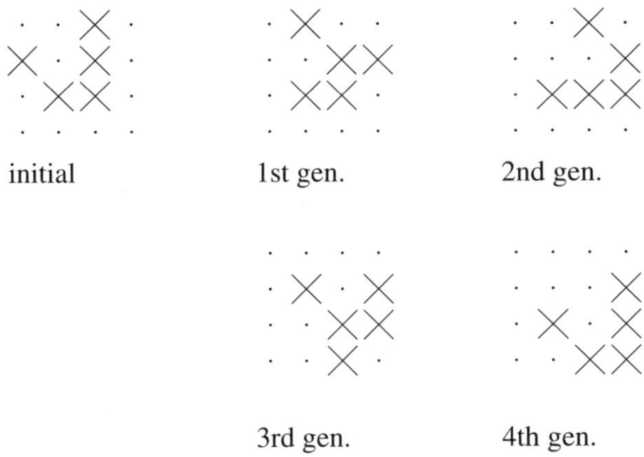

Figure 2: In Conway's Life, the simplest spaceship (a pattern that moves continuously through the Life universe), the glider. The figure shows how the glider moves one cell diagonally over a period of four generations.

collection of still lives and oscillators like those shown here while firing a number of gliders off towards the outer fringes of the Life universe.

The recent interest in quantum games [1,2,4,6,7,9,16–20] suggests the possibility of applying the idea of superposition of states in quantum mechanics to the game of Life. Unfortunately Conway's Life is irreversible while, in the absence of a measurement, quantum mechanics is reversible. In particular, operators that represent measurable quantities must be unitary. A full quantum Life would be problematic given the known difficulties of quantum cellular automata [21]. Recently, in an attempt to generalize von Neumann's universal constructor [22] to quantum mechanics, it was found that a quantum universal constructor capable of self-reproduction cannot exist with finite resource in a deterministic universe [23]. This could have important bearing in understanding life from a quantum theoretic viewpoint.

Interesting behavior can still be obtained in a semi-quantum mechanical Life by representing the cells by classical sine-wave oscillators with a period equal to one generation, an amplitude between zero and one, and a variable phase. The amplitude of the oscillation represents the coefficient of the alive state so that the square of the amplitude gives the probability of finding the cell in the alive state when a measurement of the "health" of the cell is taken. If the initial state of the system contains at least one cell that is in a superposition of eigenstates the neighboring cells will be influenced according to the coefficients of the respective eigenstates, propagating the superposition to the surrounding region.

If the coefficients of the superpositions are restricted to positive real numbers we do not expect to see qualitatively new phenomena. By allowing the coefficients to be complex, that is, by allowing phase differences between the oscillators, qualitatively new phenomena, for example interference effects, may arise. The interference effects we see are those due to an array of classical oscillators with phase shifts and are not fully quantum mechanical. Our cellular automaton should be distinguished from quantum cellular automata discussed in references [5,8,13–15].

2 A First Model

To represent the state of a cell we introduce the following notation: [1]

$$|\psi\rangle = a|\text{alive}\rangle + b|\text{dead}\rangle, \tag{1}$$

subject to the normalization condition

$$|a|^2 + |b|^2 = 1. \tag{2}$$

$|a|^2$ and $|b|^2$ represent the probabilities of measuring the cell as alive or dead respectively. If the values of a and b are restricted to non-negative real numbers we cannot get destructive interference. The model still differs from a classical probabilistic mixture since it is the amplitudes that are added and not the probabilities. In our model $|a|$ is the amplitude of the oscillator. Restricting a to non-negative real numbers corresponds to the oscillators all being in phase.

The birth, death and survival operators have the following effects

$$\begin{aligned} B|\psi\rangle &= |\text{alive}\rangle, \\ D|\psi\rangle &= |\text{dead}\rangle, \\ S|\psi\rangle &= |\psi\rangle. \end{aligned} \tag{3}$$

A cell can be represented by the vector

$$\begin{pmatrix} a \\ b \end{pmatrix}.$$

The B and D operators are not unitary. Indeed they can be represented in matrix form by

$$\begin{aligned} B &\propto \begin{pmatrix} 1 & 1 \\ 0 & 0 \end{pmatrix}, \\ D &\propto \begin{pmatrix} 0 & 0 \\ 1 & 1 \end{pmatrix}, \end{aligned} \tag{4}$$

where the proportionality constant is not relevant for our purposes. After applying B or D (or some mixture) the new state will require (re-) normalization so that the probabilities of being dead or alive still sum to unity.

[1] $|\ldots\rangle$ is the standard quantum mechanical notation to be read as "the state of . . . "

A new generation is obtained by determining the number of living neighbors each cell has and then applying the appropriate operator to that cell. The number of living neighbors in our model is the amplitude of the superposition of the oscillators representing the surrounding eight cells. This process is carried out on all cells simultaneously in effect. When the cells are permitted to take a superposition of states, the number of living neighbors need not be an integer. Thus a mixture of the B, D and S operators may need to be applied. For consistency with standard Life, the following conditions will be imposed upon the operators that produce the next generation:

- If there are an integer number of living neighbors the operator applied must be the same as that in standard Life.
- The operator that is applied to a cell must continuously change from one of the basic forms to another, as the sum of the a coefficients from the neighboring cells changes from one integer to another.
- The operators can only depend upon this sum and not on the individual coefficients.

If the sum of the a coefficients of the surrounding eight cells is

$$A = \sum_{i=1}^{8} a_i, \tag{5}$$

then the following set of operators, depending upon the value of A, is the simplest that has the required properties

$$\begin{aligned}
0 \leq A \leq 1;\ & G_0 = D, \\
1 < A \leq 2;\ & G_1 = (\sqrt{2}+1)(2-A)D \ + \ (A-1)S, \\
2 < A \leq 3;\ & G_2 = (\sqrt{2}+1)(3-A)S \ + \ (A-2)B\ , \\
3 < A < 4;\ & G_3 = (\sqrt{2}+1)(4-A)B \ + \ (A-3)D\ , \\
A \geq 4;\ & G_4 = D.
\end{aligned} \tag{6}$$

For integer values of A, the G operators are the same as the basic operators of standard Life, as required. For non-integer values in the range $(1, 4)$, the operators are a linear combination of the standard operators. The factors of $\sqrt{2}+1$ have been inserted to give more appropriate behavior in the middle of each range. For example, consider the case where $A = 3 + 1/\sqrt{2}$, a value that may represent three neighboring cells that are alive and one that has a probability of one half of being alive. The operator in this case is

$$G = \frac{1}{\sqrt{2}}B + \frac{1}{\sqrt{2}}D, \tag{7}$$

or in matrix form

$$M = \frac{1}{\sqrt{2}} \begin{pmatrix} 1 & 1 \\ 1 & 1 \end{pmatrix} . \tag{8}$$

Applying this to either a cell in the alive, $\begin{pmatrix} 1 \\ 0 \end{pmatrix}$ or dead, $\begin{pmatrix} 0 \\ 1 \end{pmatrix}$ states will produce the state,

$$|\psi\rangle = \frac{1}{\sqrt{2}} |\text{alive}\rangle + \frac{1}{\sqrt{2}} |\text{dead}\rangle, \tag{9}$$

which represents a cell with a 50% probability of being alive. That is, G is an equal combination of the birth and death operators, as we might have expected given the possibility that A represents an equal probability of three or four living neighbors. Of course the same value of A may have been obtained by other combinations of neighbors that do not lie half way between three and four living neighbors, but one of our requirements is that the operators can only depend on the sum of the a coefficients of the neighboring cells and not on how the sum was obtained.

The new state of a cell is obtained by calculating A, applying the matrix G corresponding to the appropriate operator:

$$\begin{pmatrix} a' \\ b' \end{pmatrix} = G \begin{pmatrix} a \\ b \end{pmatrix} , \tag{10}$$

and then normalizing the resulting state so that $|a'|^2 + |b'|^2 = 1$. It is this process of normalization that means that multiplying the matrix by a constant has no effect. Hence, for example, G_2 for $A = 3$ has the same effect as G_3 in the limit as $A \to 3$, despite differing by the constant factor $(\sqrt{2} + 1)$.

3 Semi-Quantum Life

To get qualitatively different behavior from classical Life we need to introduce a phase associated with the coefficients, that is, a phase difference between the oscillators. We require the following features from this version of Life:

- It must smoothly approach the classical mixture of states, as all the phases are taken to zero.
- Interference, that is the partial or complete cancellation between cells of different phases, must be possible.
- The overall phase of the Universe must not be measurable. That is, multiplying all cells by $e^{i\phi}$ for some real ϕ should have no measurable consequences.
- The symmetry between $(B, |\text{alive}\rangle)$ and $(D, |\text{dead}\rangle)$ that is a feature of the original game of Life should be retained. That is, if the state of all cells is reversed ($|\text{alive}\rangle \longleftrightarrow |\text{dead}\rangle$) and the operation of the B and D operators is reversed, the system should behave in the same manner.

In order to incorporate complex coefficients while keeping the above properties, the basic operators are modified in the following way:

$$\begin{aligned} B|\text{dead}\rangle &= e^{i\phi}|\text{alive}\rangle, \\ B|\text{alive}\rangle &= |\text{alive}\rangle, \\ D|\text{alive}\rangle &= e^{i\phi}|\text{dead}\rangle, \\ D|\text{dead}\rangle &= |\text{dead}\rangle, \\ S|\psi\rangle &= |\psi\rangle, \end{aligned} \tag{11}$$

where the superposition of the surrounding oscillators is

$$\alpha = \sum_{i=1}^{8} a_i = Ae^{i\phi}, \tag{12}$$

A and ϕ being real positive numbers. That is, the birth and death operators are modified so that the new alive or dead state has the phase of the sum of the surrounding cells. The operation of the B and D operators on the state $\begin{pmatrix} a \\ b \end{pmatrix}$ can be written as

$$\begin{aligned} B\begin{pmatrix} a \\ b \end{pmatrix} &= \begin{pmatrix} a + |b|e^{i\phi} \\ 0 \end{pmatrix}, \\ D\begin{pmatrix} a \\ b \end{pmatrix} &= \begin{pmatrix} 0 \\ |a|e^{i\phi} + b \end{pmatrix}, \end{aligned} \tag{13}$$

with S leaving the cell unchanged. The modulus of the sum of the neighboring cells, A, determines which operators apply, in the same way as before (see Eqn. (6)). The addition of the phase factors for the cells allows for interference effects since the coefficients of alive cells may not always reinforce in taking the sum, $\alpha = \sum a_i$. A cell with $a = -1$ still has a unit probability of being measured in the alive state but its effect on the sum will cancel that of a cell with $a = 1$. We are free to make the phase of the dead cell have some effect, but this does not fit the physical model presented in the introduction. Also, we wish to ensure that standard Life, in which empty cells have no effect, is a subset of our model. Hence we have chosen for the phase of the dead cells to have no effect. It is retained in order to maintain the alive $\longleftrightarrow$ dead symmetry.

A useful notation to represent semi-quantum Life is to use an arrow whose length represents the amplitude of the a coefficient and whose angle with the horizontal is a measure of the phase of a. That is, the arrow represents the phaser of the oscillator at the beginning of that generation. For example

$$\begin{aligned}
\longrightarrow &= \begin{pmatrix} 1 \\ 0 \end{pmatrix}, \\
\uparrow &= e^{i\pi/2} \begin{pmatrix} 1/2 \\ \sqrt{3}/2 \end{pmatrix} = \begin{pmatrix} i/2 \\ i\sqrt{3}/2 \end{pmatrix}, \\
\nearrow &= e^{i\pi/4} \begin{pmatrix} 1/\sqrt{2} \\ 1/\sqrt{2} \end{pmatrix} = \begin{pmatrix} (1+i)/2 \\ (1+i)/2 \end{pmatrix},
\end{aligned} \tag{14}$$

etc. Then α is the vector sum of the arrows. This notation includes no information about the b coefficient. The magnitude of this coefficient can be determined from a and the normalization condition. As noted previously, the phase of the b coefficient has no effect on the future progression of the game so it is not necessary to represent this.

4 Results and Discussion

The above rules have been implemented [2] in *Mathematica* [25]. All the structures of standard Life can be recreated by making the phase of all the alive cells equal. We are interested in whether there are new effects in our model or whether existing effects can be reproduced in simpler or more generalized structures.

The most important aspect of our model, not present in standard Life, is interference. Two live cells can work against each other as indicated in figure 3 that shows an elementary example in a block still life with one cell out of phase with its neighbors. In standard Life there are linear structures called wicks that die or "burn" at a constant rate. The simplest such structure is a diagonal line of live cells as indicated in figure 4a. In this, it is not possible to stabilize an end without introducing other effects. In our model a line of cells of alternating phase, that is of units of $\longrightarrow\longleftarrow$'s, is a generalization of this effect (figures 4b and 4c) since it can be in any orientation and the ends can be stabilized easily. A line of alternating phase live cells can be used to create other structures such as the loop in figure 5a. This is a generalization of the boat still life (figure 5b) in the standard model that is of a fixed size and shape. The stability of the line of $\longrightarrow\longleftarrow$'s results from the fact that while each cell in the line has exactly two living neighbors, the cells above or below this line have a net of zero (or one at a corner) living neighbors, due to the canceling effect of the opposite phases. No new births around the line will occur unlike the case where all the cells are in phase.

Oscillators (figure 1) and spaceships (figure 2) cannot be made simpler than the minimal examples shown for standard Life. Figure 6 shows a stable boundary that results from the appropriate adjustment of the phase differences between the cells.

[2] A version is available from the leading author.

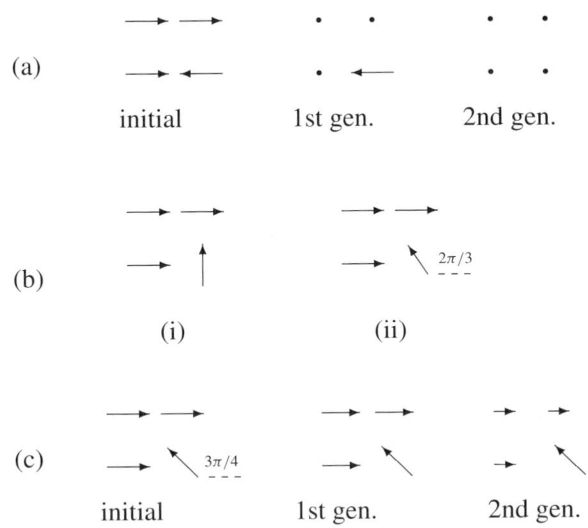

Figure 3: (a) A simple example of destructive interference in semi-quantum Life: a block with one cell out of phase by π dies in two generations. (b) Blocks where the phase difference of the fourth cell is insufficient to cause complete destructive interference; each cell maintains a net of at least two living neighbors and so the patterns are stable. In the second of these, the fourth cell is at a critical angle. Any greater phase difference causes instability resulting in eventual death as seen in (c), which dies in the fourth generation.

The angles have been chosen so that each cell in the line has between two and three living neighbors, while the empty cells above and below the line have either two or four living neighbors and so remain life-less. Such boundaries are known in standard Life but require a more complex structure.

In Conway's Life interesting effects can be obtained by colliding gliders. In our model we can obtain additional effects from colliding gliders and "anti-gliders," where all the cells have a phase difference of π with those of the original glider. For example, a head-on collision between a glider and an anti-glider as indicated in figure 7, causes annihilation, whereas the same collision between two gliders leaves a block. However, there is no consistency with this effect since other glider–anti-glider collisions produce alternative effects, sometimes being the same as those from the collision of two gliders.

5 Conclusion

John Conway's game of Life is a two-dimensional cellular automaton where the new state of a cell is determined by the sum of the neighboring states that are in one particular state generally referred to as "alive." In semi-quantum Life cells may be in a superposition of the alive and dead states with the coefficient of the alive state being represented by an oscillator. The equivalent of evaluating the number of living neighbors of a cell is to take the superposition of the oscillators of

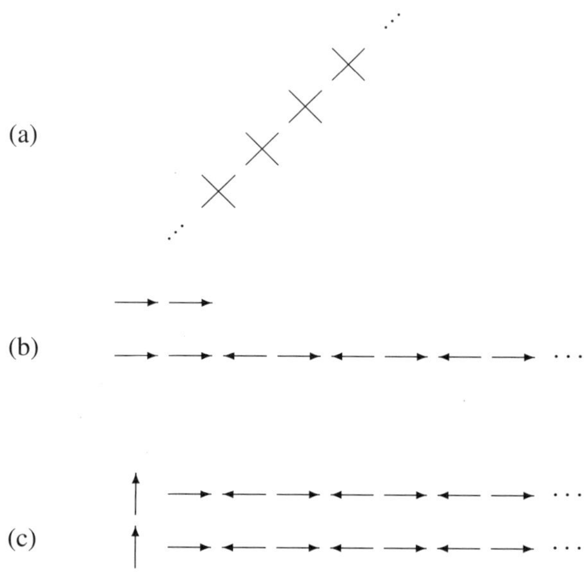

Figure 4: (a) A wick (an extended structure that dies, or "burns", at a constant rate) in standard Life that burns at the speed of light (one cell per generation), in this case from both ends. It is impossible to stabilize one end without giving rise to other effects. (b) In semi-quantum Life an analogous wick can be in any orientation. The block on the left-hand end stabilizes that end; a block on both ends would give a stable line; the absence of the block would give a wick that burns from both ends. (c) Another example of a light-speed wick in semi-quantum Life showing one method of stabilizing the left-hand end.

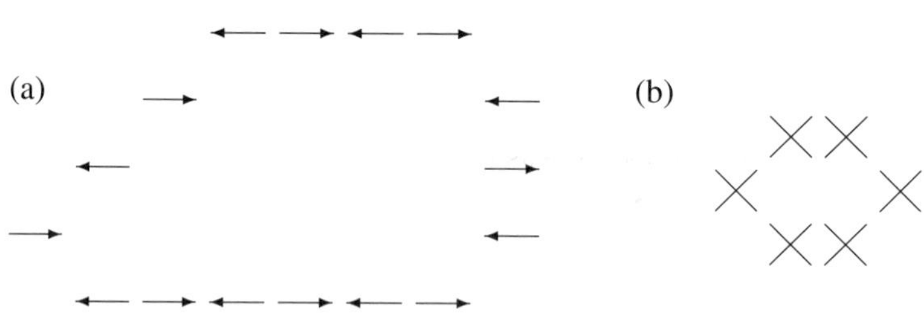

Figure 5: (a) An example of a stable loop made from cells of alternating phase. Above a certain minimum, such structures can be made of arbitrary size and shape. Compare this with (b), the boat still life in Conway's scheme, that cannot be extended without added complexity.

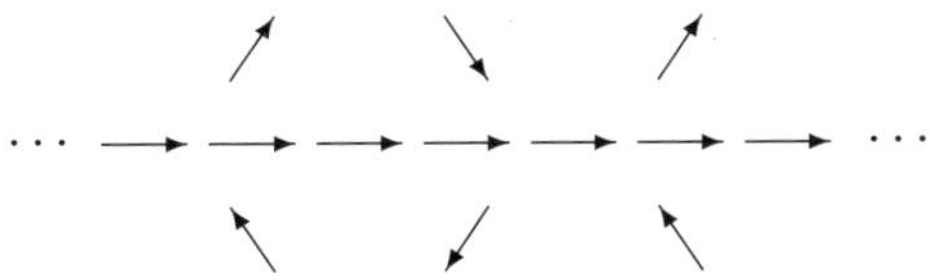

Figure 6: A boundary utilizing appropriate phase differences to produce stability. The upper cells are out of phase by $\pm\pi/3$ and the lower by $\pm 2\pi/3$ with the central line.

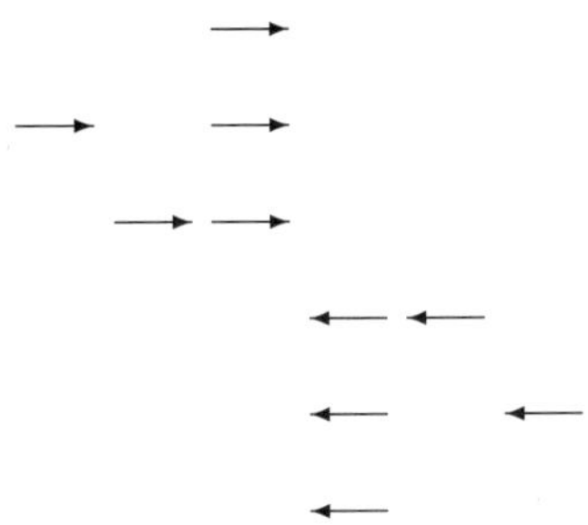

Figure 7: A head-on collision between a glider and its phase-reversed counter-part, an anti-glider, produces annihilation in six generations.

the surrounding states. The amplitude of this superposition will determine which operator(s) to apply to the central cell to determine its new state, while the phase gives the phase of any new state produced. Such a system is able to reproduce some of the aspects of quantum mechanics such as interference.

Obviously this paper just touches on some of the results that can be obtained with this new scheme but it can be seen that some new effects and structures occur and that some of the known effects in Conway's Life can occur in a simpler manner.

Acknowledgement

Arun K. Pati of the Institute of Physics, Orissa, India is gratefully acknowledged for useful suggestions. Funding was provided by GTECH Corporation Australia with the assistance of the SA Lotteries Commission (Australia).

REFERENCES

[1] Benjamin S. C. and Hayden P. M., Comment on "A quantum approach to static games of complete information," *Phys. Rev. Lett.* **87**, 069801 (2001).

[2] Benjamin S. C. and Hayden P. M., Comment on "Quantum games and quantum strategies," *Phys. Rev.* A **64**, 030301(R) (2001).

[3] Berlekamp E. R., Conway J. H. and Guy R. K., *Winning Ways for your Mathematical Plays, Vol. 2* (Academic Press, London, 1982).

[4] Du J., Xu X., Li H., Zhou X. and Han R., Entanglement playing a dominating role in quantum games, *Phys. Lett.* A **289**, 9 (2001).

[5] Dürr C. and Santha M., A decision procedure for well-formed unitary linear quantum cellular automata, *SIAM J. Comp.* **31**, 1076 (2001).

[6] Eisert J., Wilkens M. and Lewenstein M., Quantum games and quantum strategies, *Phys. Rev. Lett.* **83**, 3077 (1999).

[7] Eisert J. and Wilkens M., Quantum games, *J. Mod. Opt.* **47**, 2543 (2000).

[8] Fitzpatrick M., Smith K., Belousek D. W., Delgado A., Roos K. R. and Kenny J. P., The quantum cellular automata as a Markov process, *Chaos Soliton Fract.* **10**, 1375 (1999).

[9] Flitney A. P. and Abbott D., Quantum version of the Monty Hall problem, *Phys. Rev.* A **65**, 062318 (2002).

[10] Gardiner M., Mathematical games: The fantastic combinations of John Conway's new solitaire game 'Life,' *Sci. Am.* **223**, Oct. 120 (1970).

[11] Gardiner M., Mathematical games: On cellular automata, self-reproduction, the Garden of Eden and the game of 'Life,' *Sci. Am.* **224** Feb. 116 (1971).

[12] Gardner M., *Wheels, Life and Other Mathematical Amusements* (W.H. Freeman, New York, 1983).

[13] Grossing G. and Zeilinger A., Structures in quantum cellular automata, *Physica* B **151**, 366 (1988).

[14] 't Hooft G., Isler K. and Kalitzin S., Quantum field theoretic behavior of a deterministic cellular automata, *Nucl. Phys.* B **386**, 495 (1992).

[15] Wu Hua and Sprung D. W. L., Three-dimensional simulation of quantum cellular automata and the zero-dimensional approximation, *J. Appl. Phys.* **84**, 4000 (1998).

[16] Iqbal A. and Toor A. H., Evolutionary stable strategies in quantum games, *Phys. Lett.* A **280**, 249 (2001).

[17] Iqbal A. and Toor A. H., Quantum mechanics gives stability to Nash equilibrium, *Phys. Rev.* A **65**, 022036 (2002).

[18] Johnson N. F., Playing a quantum game with a corrupted source, *Phys. Rev.* A **63** 020302(R) (2001).

[19] Marinatto L. and Weber T., A quantum approach to static games of complete information, *Phys. Lett.* A **272**, 291 (2000).

[20] Meyer D. A., Quantum strategies, *Phys. Rev. Lett.* **82**, 1052 (1999).

[21] Meyer D. A., From quantum cellular automata to quantum lattice gases, *J. Stat. Phys.* **85**, 551 (1996).

[22] von Neumann J., *The Theory of Self-Replicating Automata* (Univ. of Illinois Press, Urbana, IL, 1966).

[23] Pati A. K. and Braunstein S. L., Quantum mechanical universal constructor, preprint xxx.lanl.gov/quant-ph/0303124.

[24] Wolfram S., *A New Kind of Science* (Wolfram Media Inc., Champaign, IL, USA, 2002).

[25] Wolfram S., *Mathematica: A System for Doing Mathematics by Computer* (Addison-Wesley, Redwood City, California, 1988, 2000).